Progress in Molecular and Subcellular Biology

Volume 56

This series gives an insight into the most current, cutting edge topics in molecular biology, including applications in biotechnology and molecular medicine. In the recent years, the progress of research in the frontier area of molecular and cell biology has resulted in an overwhelming amount of data on the structural components and molecular machineries of the cell and its organelles and the complexity of intra- and intercellular communication. The molecular basis of hereditary and acquired diseases is beginning to be unravelled, and profound new insights into development and evolutionary biology, as well as the genetically driven formation of 3D biological architectures, have been gained from molecular approaches. Topical volumes, written and edited by acknowledged experts in the field, present the most recent findings and their implications for future research. This series is indexed in PubMed.

More information about this series at http://www.springer.com/series/388

Ben E. Black
Editor

Centromeres and Kinetochores

Discovering the Molecular Mechanisms Underlying Chromosome Inheritance

Springer

Editor
Ben E. Black
Department of Biochemistry and Biophysics
Perelman School of Medicine
at the University of Pennsylvania
Philadelphia, PA
USA

ISSN 0079-6484 ISSN 2197-8484 (electronic)
Progress in Molecular and Subcellular Biology
ISBN 978-3-319-58591-8 ISBN 978-3-319-58592-5 (eBook)
DOI 10.1007/978-3-319-58592-5

Library of Congress Control Number: 2017939904

Printed on acid-free paper

This Springer imprint is published by Springer Nature
The registered company is Springer International Publishing AG
The registered company address is: Gewerbestrasse 11, 6330 Cham, Switzerland

Preface

The inheritance of our chromosomes through the germline has fascinated biologists for over a century. Likewise, the faithful transmission of chromosomes every time somatic cells divide is a process that has attracted the attention of top scientists from many areas of the biosciences—*many of whom contributed chapters to this book.* As we get deeper into the twenty-first century, there has been a wonderful convergence of science coming from experts in very diverse areas: computational genomics, genetics of model organisms, evolutionary biology, proteomics, biochemistry, structural biology, cell biology, and others. In focusing on the chromosomal locus—*the centromere*—that is key to chromosome segregation and the mammoth meiotic and mitotic machine—*the kinetochore*—that connects each chromosome via the centromere to the microtubule-based spindle, one aim of this book is to span a large swath of the research being done in this area. The book highlights historical and recent progress, but it is also intended as a reference for those involved in chromosome segregation research and as guide for those outside the field. It aims to allow the reader to gain access to the history and future questions that will require creative and innovative approaches to unlock remaining mysteries of the processes that drive faithful chromosome inheritance.

Part I covers the approaches and model systems that have led to the identification, organization, and regulation of centromere and kinetochore components. In "Use of Mass Spectrometry to Study the Centromere and Kinetochore", Samejima, Platani, and Earnshaw explore the history of the identification of the proteins of the centromere and kinetochore, highlighting the role of proteomic methodologies in not just building the "parts list" but also giving us a sophisticated view on the organization of centromeres and kinetochores. In "Critical Foundation of the Kinetochore: The Constitutive Centromere-Associated Network (CCAN)", Hara and Fukagawa provide an up-to-date view of the current understanding of the organization and function of members of the constitutive protein components of the centromere, a complex called the "constitutive centromere associated network". In "The Power of *Xenopus* Egg Extract for Reconstitution of Centromere and Kinetochore Function", French and Straight highlight the biochemical powerhouse model system in *Xenopus* egg extracts and specifically how it has aided in the mechanistic understanding of centromeres and kinetochores. In “Centrochromatin of Fungi”, Friedman and Freitag describe the great diversity and intriguing biology that is represented within fungi, from the first centromere cloned in budding yeast to species that are important human pathogens. In “Evolutionary Lessons from Species with Unique Kinetochores”, Drinnenberg and Akioshi supply a compendium of the species where wild deviation from the major lessons learned from traditional model organisms promise to unlock new mechanisms for specifying centromeres and segregating chromosomes during cell division. In “Quantitative Microscopy Reveals Centromeric Chromatin Stability, Size and Cell Cycle Mechanisms to Maintain Centromere Homeostasis”, Stankovic and Jansen highlight how innovations and applications of quantitative microscopy have transformed our understanding of many critical aspects of the formation, maintenance, and function of centromeric chromatin.

Part II focuses on how centromere location is defined in most eukaryotes, not by DNA sequence, but rather by epigenetic information where the histone H3 variant, CENP-A, plays in the leading role. In “Orchestrating the Specific Assembly of Centromeric Nucleosomes”, Zasadzińska and Foltz cover the current view of how CENP-A nucleosomes are assembled at centromeres and how this process is carefully regulated to ensure that centromere identity is not compromised. In “Artificial Chromosomes and Strategies to Initiate Epigenetic Centromere Establishment”, Barrey and Heun discuss the lessons learned about centromere formation and the epigenetic processes underlying centromere identity through innovations and applications of artificial chromosomes and other systems where *de novo* centromere formation can be monitored. In “Post-translational Modifications of Centromeric Chromatin”, García del Arco and Erhardt tackle the emerging view from studies of post-translational modification of CENP-A and canonical histones in the chromatin of functional centromeres in diverse model systems. In “Centromere Silencing Mechanisms”, McNulty and Sullivan discuss what is known about how a functional centromere can be ‘turned off’ and what has been learned about the epigenetic processes at centromeres by studying what happens when a centromere is silenced. In “Centromere Transcription: Means and Motive”, Duda, Trusiak, and O’Neill cover many documented instances of centromeric transcription, give an overview of the diverse proposals for the outcome of the centromeric transcripts, and provide ideas about the potential role(s) of transcription in centromere identity and function.

Part III explores our understanding of the sequence, genomic organization, and role of DNA sequence at centromeres; an area of centromere biology where many baffling observations pointedly demonstrate how much is left to be learned about centromeres. In “The Promises and Challenges of Genomic Studies of Human Centromeres”, Miga explains why centromeres have remained the “final frontier” of the human genome, how special tools and approaches must be used for attacking centromere sequence and organization using genomic data, the recent major progress in human centromere genomics, and the outlook for future centromere genomics. In “DNA Sequences in Centromere Formation and Function”, Dumont and Fachinetti cover experiments that support the idea that repetitive centromeric DNA sequences, while they can be bypassed for centromere identity and basic function, nonetheless play important functional roles in these processes. In "The Unique DNA Sequences Underlying Equine Centromeres", Giulotto, Raimondi, and Sullivan describe how many equine species have one or more centromere residing on non-repetitive DNA and how these species provide a potentially powerful natural milieu for exploring the relationship between epigenetic and genetic contributions to centromere identity and function.

Part IV get to the meat of how centromeres and kinetochores work in segregating our chromosomes and their role in reproduction and in healthy somatic cell divisions and their dysfunction in cancer. In "Centromere Dynamics in Male and Female Germ Cells", Dunleavy and Collins review what is known about the behavior of centromeres in the germline, the challenges that the germline presents to faithfully passing along to our offspring the location of the centromere, and recent progress on understanding the molecular processes that maintain centromere identity from one generation to the next. In "Cell Biology of Cheating—Transmission of Centromeres and Other Selfish Elements Through Asymmetric Meiosis", Chmátal, Schultz, Black, and Lampson present our current understanding of how the asymmetric process of female meiosis opens the door to selfish centromeres or other genetic elements to cheat and provide the chromosome that they lie upon an advantage in making it to the next generation (i.e., non-Mendelian ratios of chromosome inheritance). In "Biophysics of Microtubule End Coupling at the Kinetochore", Grishchuk gets right down to the essential issue of how kinetochores physically connect to spindle microtubules and the biophysical understanding of how this connection and the polymerization/depolymerization of microtubules work together to orchestrate the steps of mitotic chromosome segregation. In "Molecular Mechanisms of Spindle Assembly Checkpoint Activation and Silencing ", Corbett describes the current molecular understanding of the intricate cell signaling pathway—*the spindle assembly checkpoint*—that allows both the sensitivity to checkpoint arrest an entire cell if a single kinetochore remains unattached to the spindle and the ability to rapidly silence the checkpoint when all attachments have been made. In "A Kinase-Phosphatase Network that Regulates Kinetochore-Microtubule Attachments and the SAC", Vallardi, Cordeiro, and Saurin focus specifically on the complex network of kinases and phosphatases that directly modulate kinetochore–microtubule interactions in a manner that both promotes the ultimate formation of proper spindle connections and also integrates directly into the aforementioned spindle assembly checkpoint mechanism. In "Centromeric Cohesin: Molecular Glue and Much More", Mirkovic and Oliveira cover the role of the cohesin complex, how it is retained as the final point of cohesion between sister chromatids until anaphase onset, the functional importance of the cohesin cycle that is intertwined with mitotic progression, and how it interacts in an important way with inner centromeric signaling complexes that help ensure proper chromosome segregation. In "Centromere Structure and Function", Bloom and Costanzo discuss the role of chromatin, itself, in the physical act of chromosome segregation and also explore the role of centromeric DNA repeats in this process. In "The Role of Centromere Defects in Cancer", Beh and Kalitsis discuss the relevance to cancer genesis and proliferation of chromosome mis-segregation, the aneuploidy it causes, mutation in kinetochore proteins, and abnormal centromere structures that form through chromosome rearrangement.

Breaking down the barriers to understanding the molecular underpinnings of a biological process as beautiful, complex, and fundamentally important as chromosome segregation has proven an endeavor worthy of a global effort from some of the most talented scientists, hard-working scholars, and exceptionally brilliant minds in the biosciences. In achieving the goals of providing an up-to-date and sophisticated text on the breakthroughs and pressing challenges in the study of centromeres and kinetochores and also supplying an educational resource for a more general audience to learn about this area of investigation, the writing of this book was an endeavor worthy of a similarly global effort from many of the field's shining stars.

Philadelphia, USA Ben E. Black

Contents

Part I
Identification, Organization, and Regulation of Centromere and Kinetochore Components

Use of Mass Spectrometry to Study the Centromere and Kinetochore

Itaru Samejima, Melpomeni Platani and William C. Earnshaw

Abstract A number of paths have led to the present list of centromere proteins, which is essentially complete for constitutive structural proteins, but still may be only partial if we consider the many other proteins that briefly visit the centromere and kinetochore to fine-tune the chromatin and adjust other functions. Elegant genetics led to the description of the budding yeast point centromere in 1980. In the same year was published the serendipitous discovery of antibodies that stained centromeres of human mitotic chromosomes in antisera from CREST patients. Painstaking biochemical analyses led to the identification of the human centromere antigens several years later, with the first yeast proteins being described 6 years after that. Since those early days, the discovery and cloning of centromere and kinetochore proteins has largely been driven by improvements in technology. These began with expression cloning methods, which allowed antibodies to lead to cDNA clones. Next, functional screens for kinetochore proteins were made possible by the isolation of yeast centromeric DNAs. Ultimately, the completion of genome sequences for humans and model organisms permitted the coupling of biochemical fractionation with protein identification by mass spectrometry. Subsequent improvements in mass spectrometry have led to the current state where virtually all structural components of the kinetochore are known and where a high-resolution map of the entire structure will likely emerge within the next several years.

1 Discovery of Centromere Proteins Using Anti-centromere Autoantibodies

Using newly discovered anti-centromere autoantibodies (here called ACA) (Moroi et al. 1980), the Tan and Brinkley labs showed that the antigen was present at centromeres throughout the cell cycle (Brenner et al. 1981; Moroi et al. 1981).

I. Samejima · M. Platani · W.C. Earnshaw (✉)
Wellcome Trust Centre for Cell Biology,
University of Edinburgh, Edinburgh, Scotland, UK
e-mail: bill.earnshaw@ed.ac.uk

B.E. Black (ed.), *Centromeres and Kinetochores*, Progress in Molecular and Subcellular Biology 56, DOI 10.1007/978-3-319-58592-5_1

However, its molecular characterisation was initially elusive. Eventually, immunoblotting using purified human chromosomes revealed that many ACA recognised three antigens, namely CENP-A, CENP-B and CENP-C (Guldner et al. 1984; Earnshaw and Rothfield 1985). Certain sera recognised a fourth antigen, CENP-D, which turned out to be RCC1 (Bischoff et al. 1990), the guanine nucleotide exchange factor (GEF) for the small GTPase Ran (Bischoff and Ponstingl 1991a, b). The role of CENP-D at centromeres is not known.

The next steps in characterising centromere proteins came with the development of expression vector cloning in lambda phage—a technique in which DNA molecules encoding protein epitopes could be cloned using specific antibodies (Young and Davis 1983). Use of ACA led to cloning of cDNAs encoding CENP-B, CENP-C, CENP-F and HP1alpha (Earnshaw et al. 1987; Saitoh et al. 1992; Saunders et al. 1993; Liao et al. 1995). Experimental antibodies raised against mitotic chromosome scaffolds (the protein-rich residue remaining when most proteins and nucleic acids are extracted from isolated mitotic chromosomes—(Adolph et al. 1977)) were used to clone INCENP and CENP-E cDNAs (Cooke et al. 1987; Yen et al. 1991).

Several biochemical studies had suggested that CENP-A was likely to be a core histone (Palmer and Margolis 1985; Palmer et al. 1991). When the CENP-A cDNA was eventually cloned by degenerate PCR based on peptide sequences from chymotryptic fragments of CENP-A protein (Palmer et al. 1991), and the protein was confirmed to be a centromere-specific variant of histone H3 (Sullivan et al. 1994).

2 Identification of Centromere Proteins Using Yeast Genetic Screens

The budding yeast point centromere is a DNA sequence that is necessary and sufficient to confer centromere function in *Saccharomyces cerevisiae* (Clarke and Carbon 1980; Hieter et al. 1985; Clarke and Baum 1990). Availability of cloned yeast centromeres revealed a 125 bp core sequence with three conserved DNA elements: CDE I, CDE II, CDE III (Fitzgerald-Hayes et al. 1982). CDE III was essential, and point mutations could abolish centromere activity. A heroic effort subsequently led to the biochemical isolation of the CBF3 complex, which binds to CDE III (Lechner and Carbon 1991). This complex turned out to be specific to budding yeasts with a point centromere. Other organisms with regional centromeres (Pluta et al. 1995) do not use proteins related to CBF3, but instead use proteins related to the human CENP proteins. [An exception to this are the Trypanosomatids, which assemble an inner centromere from a completely divergent group of at least 20 proteins—(Akiyoshi and Gull 2014; Nerusheva and Akiyoshi 2016). Their outer kinetochore, however, contains proteins distantly related to members of the Ndc80 complex (see below), so their kinetochore is not entirely divergent from that of other eukaryotes (D'Archivio and Wickstead 2017).]

The availability of cloned centromeres and assays for centromere function led to several screens for centromere proteins in budding yeast. Several complementation groups were isolated in a genetic screen for mutants defective for minichromosome maintenance (MCM) at 35 °C (Maine et al. 1984; Roy et al. 1997). Of these, mcm2–mcm7 mutants were defective in the initiation of DNA replication and are now known to make up part of the essential CMG helicase (Moyer et al. 2006). Others had defects in chromosome segregation into daughter cells. MCM16, MCM17 (CHL4), MCM18 (CTF19), MCM19 (IML3), MCM21, MCM22 are components of the yeast kinetochore.

A screen for genes that reduced the fidelity of chromosome transmission at high gene dosage yielded MIF2, which showed genetic interactions with CDE II of centromere DNA (Meeks-Wagner et al. 1986). The Mif2 amino acid sequence had limited homology with CENP-C, and indeed Mif2 is budding yeast CENP-C (Meluh and Koshland 1995). ctf19, mcm21 and okp1 of the COMA complex (De Wulf et al. 2003) were isolated from a one hybrid system screen that was designed to identify proteins that localise to CEN DNA (Ortiz et al. 1999). Ctf19 had been identified in a previous large screen for chromosome transmission fidelity mutants (Spencer et al. 1990; Doheny et al. 1993; Hyland et al. 1999). The budding yeast homologue of CENP-A, Cse4, was identified in 1996 in a genetic screen for mutants affecting chromosome segregation efficiency and interacting genetically with the point centromere (Stoler et al. 1995).

The Ndc80 complex is found in the outer kinetochore, where it has a key role in microtubule attachment. This complex was discovered when the budding yeast spindle pole body (SPB—equivalent to the metazoan centrosome) was purified biochemically and its components identified by mass spectrometry (Rout and Kilmartin 1990; Wigge et al. 1998). In contrast to genuine SPB components, Ndc80, Spc24 and Spc25 associated with only a subset of nuclear microtubules and it was subsequently realised that they localised to the outer kinetochore rather than the SPB (Wigge and Kilmartin 2001).

Schizosaccharomyces pombe regional centromeres are ≥ 30 times larger than their budding yeast counterparts (Fishel et al. 1988), rich in repetitive DNAs (Nakaseko et al. 1986; Fishel et al. 1988; Chikashige et al. 1989), and require an epigenetic component to acquire centromere activity (Steiner and Clarke 1994). Functional studies turned out to be challenging, and it was some time before it was possible to design artificial chromosomes in this organism (Hahnenberger et al. 1989). Like *S. pombe* centromeres, higher eukaryote regional centromeres also require an epigenetic component for centromere activity (Earnshaw and Migeon 1985; Karpen and Allshire 1997; Vafa and Sullivan 1997; Warburton et al. 1997). Thus, proteins essential for *S. pombe* centromere function have turned out to be essential for vertebrate centromeres as well.

S. pombe temperature-sensitive mutants were screened visually by fluorescence microscopy to look for chromosome segregation defects (Takahashi et al. 1994). Genes identified through characterisation of mutants isolated in this screen included Mis12, Mis13(Dsn1), Mis14(Nsl1) of the MIS12 complex, a key structural component of the outer kinetochore—(Goshima et al. 1999); Mis16 and Mis18, key factors in loading the critical centromeric histone CENP-A—(Hayashi et al. 2004); and Mis6, a component of the inner centromere known as CENP-I in vertebrates (Goshima et al. 1999).

By 2003, most yeast kinetochore proteins had been identified (Westermann et al. 2007; Biggins 2013). At that time, only a handful of vertebrate kinetochore proteins were known. Unfortunately, it was not easy to use sequence homologies to identify metazoan proteins corresponding to the yeast genes (or vice versa). In fact, CENP-I and Mis12 were the only vertebrate centromere proteins identified via sequence homologies with fission yeast counterparts (Goshima et al. 1999; Nishihashi et al. 2002).

3 Post-genomics Approaches to Discover Centromere Proteins: RNAi Screening in *C. elegans*

More comprehensive approaches that led to discovery of centromere proteins emerged following the completion of whole genome sequencing for the major model organisms. One such RNAi screen in *C. elegans* identified ~250 genes whose depletion resulted in chromosome segregation defects (Gonczy et al. 2000; Sonnichsen et al. 2005). A kinetochore null (KNL) phenotype was observed with RNAi of five genes. This phenotype is a set of defects consistent with complete loss of kinetochore function: chromosome segregation failure, defective chromosome alignment, precocious spindle pole separation, failure to assemble stable mitotic spindles and no recruitment of other kinetochore components. The KNL phenotype was first observed with RNAi of CENP-A and CENP-C (Oegema et al. 2001).

The large-scale screen identified CENP-A and CENP-C plus three novel KNL genes (Desai et al. 2003; Cheeseman et al. 2004; Maddox et al. 2007). Affinity-purification of KNL-1 and KNL-3 followed by mass spectrometry revealed physical association of 10 members of what was called the KMN network (Cheeseman et al. 2004). Their depletion phenotype defined three subcomplexes that assemble into the outer kinetochore: the Ndc80 complex, the Mis12 complex and KNL-1.

4 Identification of Centromere/Kinetochore Proteins by Affinity-Purification Mass Spectrometry (AP-MS)

Mass spectrometry (MS) is a technique that allows the unbiased identification and quantitation of a wide range of molecules in biochemical samples. The ability to identify unexpected or unknown proteins in such samples is a great advantage of MS. Advances in MS were key to determining the protein composition of centromeres. Initial studies focused on CENP-A nucleosomes and their associated proteins. Later studies turned to shotgun characterisation of the proteome of whole chromosomes.

During the time frame covering these studies, the types of mass spectrometers underwent a significant evolution. The main method of mass spectrometry used in the early days was MALDI-TOF (Matrix assisted laser desorption/inonization coupled with time of flight mass analyser) (Yates 1998). MALDI-TOF, can successfully handle relatively simple samples such as a band or a spot cut out from a polyacrylamide gel that contains one or a few polypeptides (Jensen et al. 1997). Mass spectrometry on complex samples became possible only after the development of LC-MS/MS (Liquid chromatography coupled tandem mass spectrometry) systems (Washburn et al. 2001; Aebersold and Mann 2003). Peptide identification by direct sequencing using tandem MS was a great improvement from peptide mass fingerprinting previously available with the single TOF mass analyser (Jensen et al. 1999). Furthermore, fractionation of samples by liquid chromatography before ionisation reduced the complexity of the sample prior to injection into the mass spectrometer. Thus, more complex samples such as whole chromosomes could be analysed. Development of instruments such as Orbitrap mass spectrometers provided higher signal resolution and better mass accuracy, allowing the separation of many more peptides in protein mixtures (Adachi et al. 2006; Macek et al. 2006; Han et al. 2008). This led to the identification of many more proteins in complex samples.

Affinity pull-down followed by mass spectrometry (AP-MS) allowed the identification of kinetochore proteins that bridge CENP-A nucleosomes and the outer kinetochore. Most structural kinetochore proteins were identified by this approach.

The first breakthrough study pulled down interphase chromatin from HeLa cells using a CENP-A-specific monoclonal antibody (Obuse et al. 2004). The authors used micrococcal nuclease digestion to solubilise centromere chromatin and obtain complexes derived from the interphase centromere (ICEN). This analysis recovered all human centromere proteins known at the time (CENP-A, CENP-B, CENP-C, CENP-H, CENP-I and Mis12) plus >30 other proteins that had not previously been linked with centromeres. The ICEN contained several Polycomb group proteins as well as other proteins that function as chromatin remodelers. The authors speculated that these proteins might help establish or maintain heterochromatin in or around centromeres.

A follow-up study published 2 years later described functional analysis of seven ICEN components of unknown function (ICEN22, 24, 32, 33, 36, 37 and 39) (Izuta et al. 2006). All seven localised to centromeres when tagged with GFP. RNAi studies revealed abnormal mitotic phenotypes, with defects in chromosome alignment or segregation when the proteins were depleted. CENP-H and CENP-I were depleted from centromeres in these knock-downs (with the exception of ICEN24 and ICEN36), but CENP-A and CENP-C were not affected. These results were in good agreement with later functional studies of the CCAN (constitutive centromere-associated network—(Cheeseman and Desai 2008)) and these ICEN proteins turned out to be components of the CCAN as follows (ICEN22/CENP-T, ICEN24/CENP-U, ICEN32/CENP-N, ICEN33/CENP-L, ICEN36/CENP-O, ICEN37/CENP-K and ICEN39/CENP-M). These pioneering studies ended with the retirement of Kinya Yoda.

In 2006, two studies using AP-MS to identify proteins associated with CENP-A nucleosomes proved to be definitive for identification of structural components of the mammalian centromere. In the first, the authors labelled CENP-A with a tandem affinity-purification LAP tag (GFP–TEV protease cleavage site—S peptide fusion) and pulled down CENP-A nucleosomes, identifying the CENP-A nucleosome associated complex (NAC) (Foltz et al. 2006). In addition to CENP-A, the NAC contained the known centromere proteins CENP-C, CENP-H and CENP-U(50) plus three novel proteins, designated CENP-M, CENP-N and CENP-T. The latter three were required for assembly of this complex of CENP-A-proximal proteins.

When CENP-M, CENP-N and CENP-T were each LAP-tagged and pull downs were performed, this yielded an additional set of centromere-associated proteins, known as CENP-A-nucleosome distal (CAD). These proteins were designated CENP-K, CENP-L, CENP-O, CENP-P, CENP-Q, CENP-R and CENP-S. They were not directly associated with the CENP-A nucleosome, and were assumed to localise further out in the kinetochore. Depletion of NAC components caused profound problems with kinetochore function and chromosome segregation. Depletion of CAD component and CENP-U did not affect NAC assembly and cells traversed mitosis, albeit with many mitotic abnormalities.

In a second paper published in 2006 CENP-H and CENP-I were tagged by knock-in of FLAG and GFP tags to the endogenous genes in chicken DT40 cells (Okada et al. 2006). Pulldowns with the corresponding antibodies yielded five associated proteins: CENP-K, CENP-L, CENP-M, CENP-O and CENP-P. LAP-tagged CENP-O and CENP-P localised to HeLa centromeres across the cell cycle. A pulldown in HeLa cells with LAP-tagged CENP-O yielded two more novel centromere proteins—CENP-Q and CENP-R. Overall, the CENP-H/CENP-I complex in chicken and HeLa cells was found to contain 11 proteins. Analysis of the distributions of proteins in various knockout mutants together with the resulting cellular phenotypes led to the suggestion that CENP-H/CENP-I/CENP-K and CENP-L form one centromeric subcomplex, while CENP-O/CENP-P/CENP-Q and CENP-U(50) form another. The first of these complexes was required for efficient assembly of the outer, but not the inner, kinetochore. The second appeared to be non-essential for life in vertebrates (though the corresponding COMA complex is essential for life in budding yeast—(Biggins 2013; Yamagishi et al. 2014)).

At the end of 2006, most of the major structural components of what came to be called the constitutive centromere-associated network (Cheeseman and Desai 2008) had been identified. Just one important complex remained to be described.

Fractionation of cell extracts by gel filtration revealed that CENP-T migrated differently from CENP-H and was probably in an independent complex (Hori et al. 2008). The endogenous CENP-T gene was tagged and pulled down under conditions more stringent than those used in previous studies. In these pulldowns, only extremely low levels of CENP-H and CENP-O were observed, suggesting that these proteins formed distinct complexes. Instead, CENP-T was associated with a small protein previously known as CUG2, renamed here CENP-W. Depletion of CENP-W caused severe chromosome segregation defects, and electron microscopy revealed that CENP-W is required for normal formation of the kinetochore outer plate.

In other experiments, gel filtration fractionation of cell extracts revealed that CENP-S also migrated differently from CENP-O (Amano et al. 2009). This suggested that CENP-S might also form an independent CCAN subcomplex. CENP-S turned out to be associated with a novel protein, named CENP-X. Both proteins were bona fide components of the CCAN, but were not essential for the life of DT40 cells, although their knockdown in HeLa cells resulted in more severe mitotic phenotypes. In DT40, CENP-S knockouts were found to have smaller kinetochore plates by thin-section electron microscopy, and to recruit reduced numbers of outer kinetochore proteins.

It has since emerged that CENP-S and CENP-X are in at least two functional complexes in vertebrate cells. One complex of CENP-T/CENP-W/CENP-S/CENP-X functions primarily at kinetochores (Nishino et al. 2012, 2013). All of these proteins have histone folds, and the complex has been reported to associate with DNA directly and to introduce super-coils into it (Takeuchi et al. 2013). CENP-S/MHF1 and CENP-X/MHF2 have also been found to associate with FANCM and to promote the association of the FANCM complex with chromatin during resolution of interstrand DNA crosslinks and sister chromatid exchanges by the Fanconi Anemia-mediated pathway (Singh et al. 2010; Yan et al. 2010).

As described above, mass spectrometry was critical to identification of most structural components of the kinetochore. A recent attempt to assemble a functional kinetochore capable of interacting in vitro with microtubules assembled a 21 subunit complex containing a CENP-A nucleosome plus CENP-C/CENP-H/CENP-I/CENP-K/CENP-L/CENP-M/CENP-N and the KMN network of the KNL1 complex (KNL1, Zwint), the Mis12 complex (Dsn1, Mis12, Nsl1, Nnf1) and the Ndc80 complex (Ndc80, Nuf2, Spc24, Spc25) (Weir et al. 2016). This can be thought of as a minimal constitutive kinetochore, and is not much less complex than the isolated budding yeast kinetochore (see next section). But these minimal kinetochores lack many of the finer regulatory aspects and they are not associated with native centromeric chromatin. In order to characterise the entire kinetochore in its native context, proteomic analysis of kinetochores associated with mitotic chromosomes is required.

5 Shotgun Proteomics of Isolated Yeast Kinetochores

A goal for many years was the isolation of kinetochores from mitotic chromosomes, however, despite the availability of specific antibodies, this goal remained elusive, largely as a result of the inability to cleanly excise the centromeric DNA from metaphase chromosomes of metazoans. This goal was eventually reached in two studies performed in the budding yeast.

In the first, whole chromosome with only 2 kb of DNA including a lac operator array were isolated by pulling down with antibodies to FLAG-tagged lac repressor protein (Akiyoshi et al. 2009). Centromere-containing and control minichromosomes containing mutations that blocked centromere assembly were pulled-down and only the centromeric minichromosomes had associated CENP-A/Cse4 and Ndc80 proteins. A total of 329 proteins were identified on the centromeric minichromosomes including 35 of 38 known constitutive centromere proteins. One novel centromere protein was found: a protein phosphatase 1-targeting subunit called Fin1 that associates with kinetochores in a PP1-dependent manner and is required for normal function of the spindle assembly checkpoint.

In a second study, FLAG-tagged Mis12 complex subunit Dsn1 was used to pull down a soluble protein complex from yeast lysates under physiological salt conditions following shearing of the DNA (Akiyoshi et al. 2010). This protein complex contained the 39 core kinetochore proteins plus other regulatory proteins, including spindle checkpoint components. In this study, the purpose of MS analysis was essentially for quality control, and the main focus of the paper was on functional analysis of the purified kinetochores. The isolated complexes appeared to correspond to functional kinetochores, as they could bind to taxol-stabilised microtubules and make load-bearing attachments to disassembling microtubules that were stabilised by tension in the physiological range.

6 Shotgun Proteomics of Whole Isolated Mitotic Chromosomes

Early attempts to catalogue non-histone protein components of chromosomes by brute force proteomic approaches missed a significant fraction of known mitotic chromosomal proteins, including kinetochore components. This is because kinetochore proteins are among the least abundant proteins on chromosomes. Mass spectrometers have a finite rate with which they can fragment peptides and identify sequences. Thus if, for example one protein is hundreds or thousands of times more abundant than another, simple probability will mean that few or no peptides from the less abundant protein will be identified. Hence, the relatively low abundance of kinetochore proteins in such a complex sample environment made it a serious challenge to detect them. Detection required improvements in instruments plus better fractionation of samples prior to their introduction into the machine.

Uchiyama et al. (2005) made a pioneering effort to comprehensively identify mitotic chromosome proteins (Uchiyama et al. 2005). They obtained a list of proteins from purified mitotic chromosomes analysed by 1- and 2-dimensional SDS-PAGE followed by peptide mass fingerprinting of each band/spot by MALDI-TOF. Comparison of two chromosome isolation methods allowed the investigators to focus on 107 proteins resolved by SDS-PAGE that showed a high affinity for mitotic chromosomes. This list contained 10 mitochondrial proteins, but many of the other proteins were well-known chromosomal proteins.

The authors classified chromosomal proteins into four "layers" according to their localisation: chromosome-coating proteins (mostly mitochondrial and cytoplasmic proteins), chromosome peripheral proteins (mostly nucleolar proteins), chromosome structural proteins (including histones and other proteins known to be involved in chromosome structure, such as condensin), and chromosome fibrous proteins (cytoskeletal proteins including actin, tubulin and vimentin). Although many novel chromosome-associated proteins were identified, no known kinetochore proteins were found.

In a follow-up study (Takata et al. 2007) chromosomes were isolated from two human cell lines and proteins resolved by 1D and 2D gel electrophoresis as before. This time, 189 proteins were identified. These data were analysed in terms of the four layer model described above and similar results were obtained. This analysis identified several centromere-associated proteins, including three components of the chromosomal passenger complex (CPC), HP1 and CENP-B. No novel centromere proteins were described.

Gassmann et al. took advantage of protocols for biochemical purification of mitotic chromosomes developed in the Laemmli lab (Lewis and Laemmli 1982; Gassmann et al. 2004, 2005). Their study analysed the chromosome scaffold fraction instead of whole chromosomes (Lewis and Laemmli 1982; Gassmann et al. 2004, 2005). Chromosome scaffolds are essentially isolated metaphase chromosomes depleted of histones (Adolph et al. 1977; Earnshaw and Laemmli 1983).

Gassmann et al. identified 79 proteins, 30 of which were as-yet uncharacterised. Several of the latter were tested for their localisation on mitotic chromosomes and two were further studied. These were Borealin (Gassmann et al. 2004) and CENP-V (Tadeu et al. 2008). Neither is a component of the CCAN. Borealin is a component of the CPC. CENP-V encodes a GFA—an enzyme that catabolizes and detoxifies formaldehyde, which is a byproduct of (histone) demethylation. CENP-V was present on the prometaphase chromosome axis but was particularly concentrated at heterochromatic regions of the inner centromere. The protein was required for centromeric localisation of the CPC and Sgo1 and for normal compaction of heterochromatin and formation of the primary constriction.

Gassmann et al. failed to observe CENP-C in their proteomic screen despite previous reports that CENP-C and other kinetochore proteins were present in the chromosome scaffold fraction (Earnshaw et al. 1984; Gassmann et al. 2005). Nor did they find other centromere proteins in their search. Identification of these non-abundant proteins in complex samples required further instrument development and improvements in software and sample preparation protocols.

In a comprehensive chromosome proteomics study, Ohta et al. identified >4000 proteins including essentially all chromosomal proteins previously described (Ohta et al. 2010a). These fell into 28 ontological categories, including 562 uncharacterised proteins. The problem that confronted these experimenters was how to determine which of those proteins were bona fide chromosomal proteins and which were hitchhikers. The latter were defined as cytoplasmic proteins that associate with chromosomes in vivo after nuclear envelope breakdown but are not functionally relevant to mitotic chromosomes.

This study depended on quantitative proteomics: an application of mass spectrometry to detect differences in abundance of each protein in at least two samples (Ranish et al. 2003). For example, the relative abundance of each protein can be measured by comparison of two samples labelled with different isotopes. Stable isotope labelling of amino acids in cell culture (SILAC) is one such technique for in vivo labelling of cellular proteins (Ong et al. 2003; Ong and Mann 2006). In SILAC, samples are prepared from cells grown with heavy and light amino acids (e.g. using ^{13}C and ^{15}N). The relative amounts of corresponding heavy and light peptides measured in the mass spectrometer correspond to the relative amounts of the proteins in the two cell cultures (Ong et al. 2003) (Fig. 1). The SILAC method was used in several studies described below.

The approach taken by Ohta and co-workers was termed multi-classifier combinatorial proteomics (MCCP) (Ohta et al. 2010a). In this case SILAC-based quantitative proteomics methods were used to determine the protein composition of chromosomes isolated under a number of different conditions. Equal numbers of isolated chromosomes rather than equal numbers of starting cells (as in traditional SILAC) were used in order to correct for the effect that different conditions might have on the efficiency of chromosome isolation.

Each list of proteins quantitated under different conditions was termed a classifier. The classifiers used in the initial study included the copy number of individual proteins in chromosomes; the relative abundance of each protein in cytosol versus isolated chromosomes; the tendency of individual proteins to exchange between chromosomes and cytosol when the two were incubated together; the relative abundance of proteins found in chromosomes isolated from cells depleted of condensin and the relative abundance of proteins in chromosomes isolated from cells depleted of the kinetochore component Ska3 (Fig. 2). The problem that immediately confronted the investigators was that when these lists were plotted out, it was obvious that although some chromosomal proteins were enriched towards the top of each list, others were distributed throughout the entire list. Thus, no unique cutoff could be distinguished between chromosomal proteins and hitchhikers.

In an initial attempt to solve this problem, different classifiers were plotted against one another in two-dimensional plots. Regions of these plots could be identified that were enriched in chromosomal proteins (Fig. 3). However, this analysis was not quantitative and it was not evident how multi-dimensional comparisons could be conducted.

A breakthrough came when it was realised that the multiple classifiers could formally be regarded as analogous to data from multiple microarray experiments that compare quantitation of RNA levels across a range of experimental conditions (corresponding to classifiers). A powerful machine-learning method, random forest analysis, had been developed to compare results of such microarray experiments. Random forest analysis has the advantage that it can make comparisons between lists of numbers and can deal with lists having missing values—a plague of proteomics experiments where not every protein is identified in every experiment.

To perform random forest analysis, two training sets were defined, corresponding to known chromosomal proteins and known cytoplasmic proteins (e.g. for the latter, mitochondrial and membrane proteins plus other proteins of known cytoplasmic localisation). The random forest designer then set up a series of decision trees where each branch point is a classifier (chosen at random from the

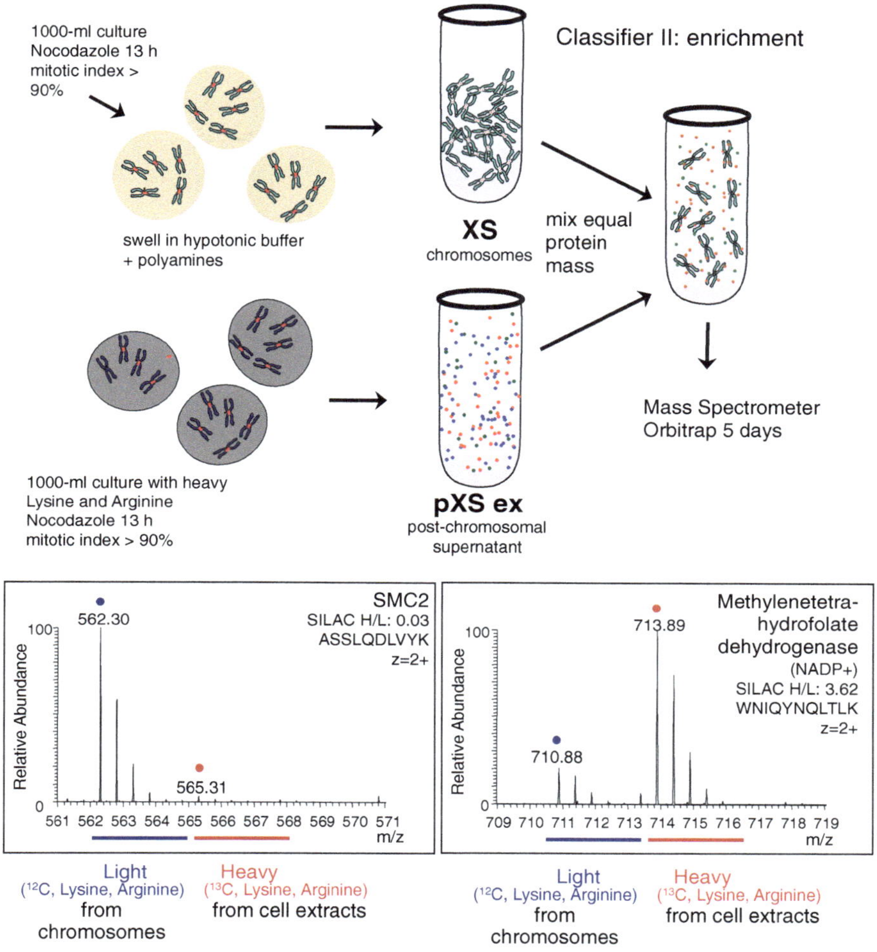

Fig. 1 Use of quantitative proteomics to create a classifier for analysis of chromosomal proteins. Mitotic chromosomes isolated from cells grown in normal medium were mixed with an equal mass of post-chromosomal supernatant (cytosol) from a culture grown in the presence of ^{13}C-Lysine and ^{13}C-Arginine. Heavy peptides come from the cytosol and light peptides from the chromosomes. Sample spectra are shown for chromosomal protein SMC2 and cytosolic protein methylenetetrahydrofolate dehydrogenase. The ratio between the areas under the peaks (SILAC ratio) is a measure of the enrichment of each protein in chromosomes. Unpublished data used courtesy of Shinya Ohta and Juri Rappsilber

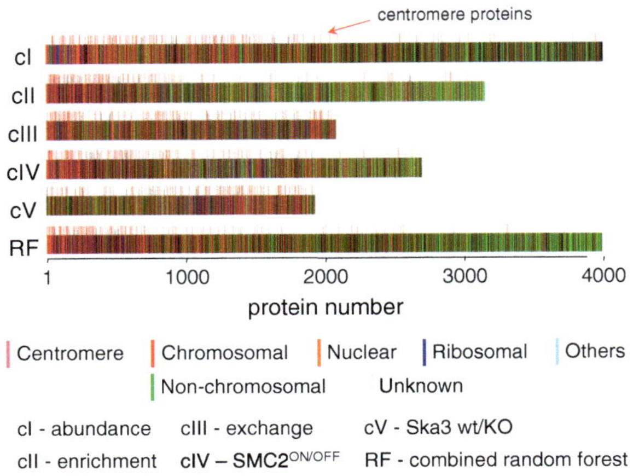

Fig. 2 The five classifiers used in reference (Ohta et al. 2010a). Centromere proteins (*pink bars*) are distributed all along the classifier lists (ranked by SILAC ratio and arranged so that highest ratios are to the *left*) with no clear cut-off between chromosomal and cytoplasmic proteins. In the list of proteins ranked by random forest analysis (RF), there is a much more significant stratification of the data with chromosomal proteins to the *left*. Figure excerpted from Ohta et al. (2010a), used with permission

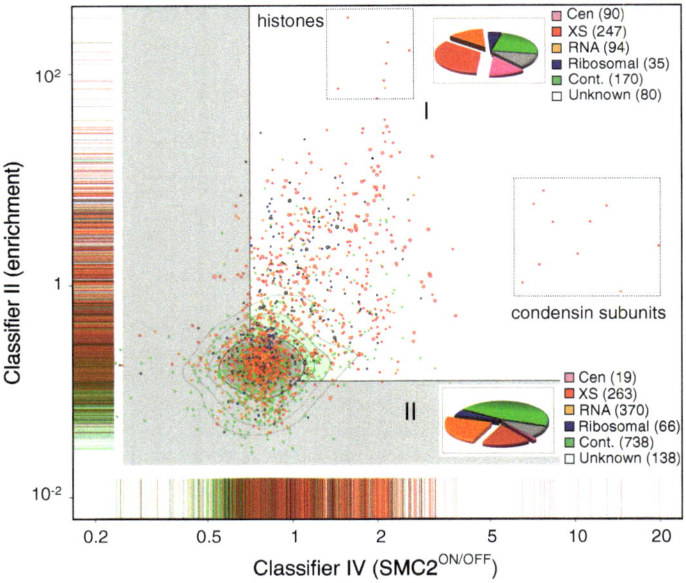

Fig. 3 Plotting two classifiers against one another identifies a zone enriched in chromosomal proteins (Zone I) and one enriched in cytoplasmic proteins (Zone II), but this analysis is not quantitative and difficult to extend to include further classifiers (dimensions). Unpublished data used courtesy of Shinya Ohta and Juri Rappsilber

list) for which a threshold value was defined from the training set. If the value for a test protein exceeds the threshold value (a "true" chromosomal protein), then the protein goes one way at the branch. If the value is less than the threshold value, then the protein goes the other way. After a given number of branches (defined by the training set), the protein ends up assigned as either chromosomal or non-chromosomal. Each protein is run through many randomly generated decision trees (the forest), and in the end, the majority "vote" of the trees decides whether the protein is classified as chromosomal or cytoplasmic.

This analysis was extremely powerful, and could separate essentially all known chromosomal proteins from all known cytoplasmic proteins. Of the training samples, only two cytoplasmic proteins were classified as chromosomal and only one chromosomal protein was classified as cytoplasmic. Overall, the analysis identified 1331 hitchhikers and contaminants and classified the unknowns into cytoplasmic and chromosomal proteins.

To test the random forest, 50 uncharacterised proteins were tagged with GFP and tested their localisation in U2OS cells. Of these 18 were found to be cytoplasmic, 12 localised generally with chromosomes, 7 localised to the chromosome periphery and 13 localised to centromeres in at least a subpopulation of the cells. Importantly, of 16 proteins predicted to be cytoplasmic by random forest analysis, 14 were indeed cytoplasmic. Of 34 proteins predicted to be chromosomal, only 4 GFP-tagged proteins were cytoplasmic. Of the remaining 511 uncharacterised proteins, this analysis predicted that 224 were likely to be chromosomal and 287 to be cytoplasmic, suggesting by analogy that overall 90 chromosomal, 46 periphery and 97 centromere-associated proteins remained to be discovered.

Importantly, with the exception of Ska3 (which had not previously been described at the time this work was done), none of the other novel centromere-associated proteins turned out to be structural components of the centromere (Ohta et al. 2010a, b). Most were associated more generally with chromatin, and 5 of the 13 new centromere localising proteins were predicted to be subunits of complexes that bind to and/or modify histones. Thus, this analysis strongly suggested that all structural components of kinetochores have been identified, but many more modifiers of centromeric chromatin remain to be described.

7 Use of Mass Spectrometry to Study Kinetochore Protein Complexes

In addition to being used for protein discovery, mass spectrometry can also be used to identify members of protein complexes and to map the interactions between them. Protein complexes are readily identified in the MCCP approach. In the different classifiers, their multiple members tend to behave in tandem (Ohta et al. 2010a, c; Kustatscher et al. 2016). For example, depletion of SMC2 causes an equal depletion of all five members of the condensin complex from mitotic chromosomes

(Ohta et al. 2010a, 2016). In a follow-up methodological study, it was shown that far from being limited to use with "big data", so-called "nano" random forests could also be used in a targeted analysis, training with small data sets to follow the behaviour of protein complexes and to identify proteins that associate with those protein complexes (Ohta et al. 2016; Montaño-Gutierrez et al. 2017).

The composition of kinetochore protein complexes was explored extensively in an analysis of whole chromosome proteomes in nine mutants of kinetochore structure proteins plus two mutants of factors required for CENP-A assembly (Samejima et al. 2015). This analysis used profile plots prepared using the software Perseus (Tyanova et al. 2016) as well as Pearson correlation analysis to look at coordinated behaviour of all known kinetochore proteins. Known complexes such as CENP-H/CENP-I/CENP-K/CENP-M and CENP-O/CENP-P/CENP-Q/CENP-R/CENP-U were readily detected and shown to be distinct by this analysis, but so was an apparent complex—yet to be described structurally—between CENP-N/CENP-L/CENP-_T/CENP-W (Figs. 4 and 5). The use of correlation analysis also allowed a more subtle analysis, suggesting that only a subpopulation of the chromosome-associated Ndc80 complex is part of the KMN network (Fig. 5). Some of this complex may interact with CENP-N/CENP-L/CENP-T/CENP-W without requiring the Mis12 complex. Furthermore, two populations of CENP-N/CENP-L/CENP-T/CENP-W complex were suggested. Ndc80 complex and RZZ complex/CENP-E/Mad1 cohort appeared to be alternative partners to extend the CENP-N/CENP-L/CENP-T/CENP-W core complex. Ndc80 complex/Ska2/CENP-N/CENP-L/CENP-T/CENP-W and RZZ/CENP-E/Mad1// CENP-N/CENP-L/CENP-T/CENP-W may represent microtubule-binding and unattached kinetochores, respectively. Such predictions can be tested by functional studies in cells (Gascoigne et al. 2011; Schleiffer et al. 2012; Nishino et al. 2013) and also by studies in which mass spectrometry is used to probe kinetochore ultrastructure (see below).

Protein complexes present in mitotic chromosomes were revealed by correlation analysis of the mass spectrometry data (Fig. 6). Protein cohorts emerged by selecting pairs of proteins with correlation coefficients above a certain threshold value. More than 30 centromere/kinetochore proteins comprised a single network. These included members of the Mis12, Ndc80, RZZ, Ska and CCAN (CENP-H/CENP-I/CENP-K/CENP-M/CENP-N/CENP-L/CENP-T) complexes. However, CENP-A, CENP-C and CENP-O/CENP-P/CENP-Q/CENP-U/CENP-R were not included in this kinetochore network. Members of the well-characterised complexes were connected within each subcomplex. However, not every protein in the entire network was directly connected to each other, suggesting a mosaic organisation of the kinetochore.

Inter-subunit interactions were more visible if correlation pairs were filtered at less stringent thresholds. As the stringency was decreased, many solitary networks became connected, culminating in two major networks of chromosome-associated proteins, plus several minor isolated networks. One major network was composed of chromosomal protein complexes. The other contained largely ribosomal proteins and those involved in RNA metabolism. CENP-O, CENP-P, CENP-Q, CENP-R

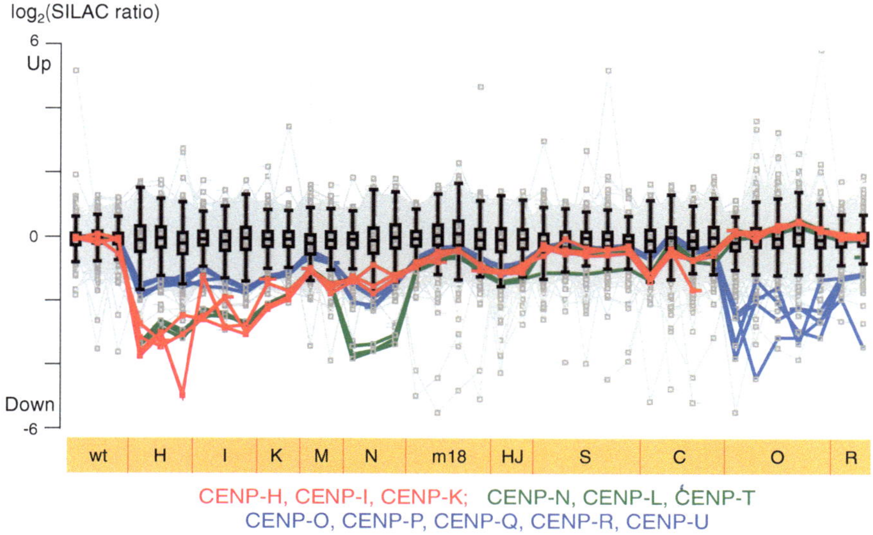

Fig. 4 Profile plot showing $\log_2$(SILAC ratios) of all proteins (grey lines) in a series of experiments analysing the total protein composition of chromosomes isolated from a series of cell lines mutant for various kinetochore proteins. The interquartile (IQR) population for all proteins detected in each experiment is contained in the box. Whiskers extend to 1.5 times IQR away from the edge of the box. A datapoint beyond a wisker is considered as a significant change, and the one within a box is a negligible change. The *y*-axis is oriented so that proteins that decrease in mutant chromosomes are plotted *downwards* and those that increase are plotted *upwards*. The behaviour of CENP-H/CENP-I/CENP-K, is shown in *red*; CENP-N/CENP-L/CENP-T (*green*); and CENP-O/CENP-P/CENP-Q/CENP-R/CENP-U (*blue*). The mutant cell lines used for isolation of chromosomes are designated in abbreviated form across the bottom of the profile plot. *wt* wild type; *m18* Mis18alpha; *HJ* HJURP

and CENP-U were not included in the kinetochore network at a stringent threshold, instead constituting a solitary network. At a lower threshold, the OPQRU complex was linked to the larger network of chromosome proteins via plk1. Plk1 phosphorylates CENP-U and CENP-Q and regulates their association with chromosomes (Kang et al. 2011).

8 Use of CLMS Mass Spectrometry to Study the Ultra-structure of Kinetochore Protein Complexes

Chemical crosslinking coupled with mass spectrometry (CLMS/XL-MS) can be used to study protein structures and organisation in macromolecular complexes (Fischer et al. 2013; Chen et al. 2016; Leitner et al. 2016). When two residues are crosslinked, the demonstration that they are in close proximity is useful for

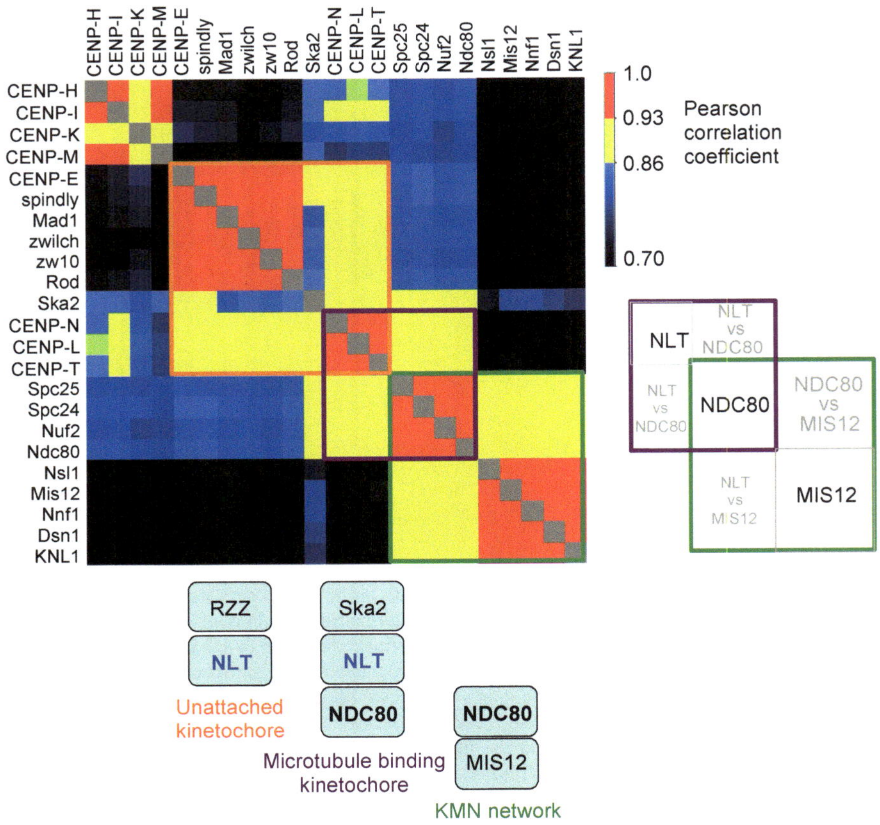

Fig. 5 Pearson correlation coefficient analysis reveals protein complexes and super-complexes in kinetochores. (*Left*) Pearson correlation coefficients between kinetochore proteins are shown as a heat map table. The heat map is colour coded as shown in the bar on *right*. Proteins in *red blocks* have the highest correlation and are likely to form protein complexes. *Green*, *purple* or *orange* frames enclosing *red* and *yellow* areas represent probable super-complexes containing two or more protein complexes. (*Right*) Key to protein–protein interactions in *red* and *yellow* areas within *purple* and *green* frames. (*Bottom*) Super-complexes that contain the NDC80 complex or CENP-N/CENP-L/CENP-T complex. The NDC80 or CENP-N/CENP-L/CENP-T complexes are shared by more than one super-complexes. The NDC80 complex and Ska2 are microtubule-binding adaptors at outer kinetochore. RZZ is released from kinetochore upon binding to microtubule

understanding protein structure or protein–protein interactions. Such residues in close physical proximity may occur at distant positions in primary sequence or even in different proteins (Maiolica et al. 2007). This can be particularly useful for protein complexes that do not yield readily to crystallographic approaches because of regions of peptide flexibility such as coiled coils (Barysz et al. 2015; Pekgoz Altunkaya et al. 2016). In a pioneering study, CLMS was used to determine the

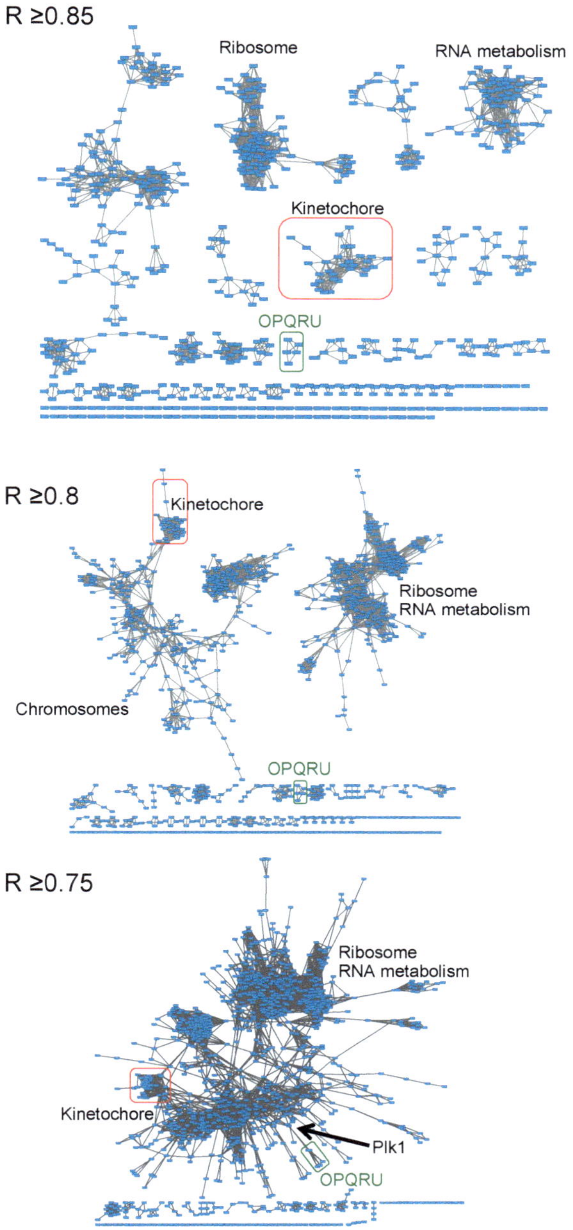

Fig. 6 Protein interaction networks present in mitotic chromosomes. Pearson correlation coefficients were filtered with different thresholds (R = 0.85, 0.80 or 0.75) prior to plotting using Cytoscape (Shannon et al. 2003) (www.cytoscape.org). The kinetochore network and CENP-O/CENP-P/CENP-Q/CENP-R/CENP-U are indicated by the *red* and *green* boxes, respectively

pattern of protein contacts within the Ndc80 complex (Maiolica et al. 2007; Ciferri et al. 2008). Pairs of residues in proximity were identified by crosslinked peptides and the arrays of the cross links revealed the orientation of coiled coils. This allowed the design of a shortened "bonsai" complex that could link kinetochores to microtubules, but was amenable to crystallographic analysis.

Subsequent studies have looked at protein interactions within the CCAN and between kinetochores and microtubules. Mapping of proximity relationships in the CENP-H/CENP-I/CENP-K/CENP-M subcomplex revealed close proximity between CENP-H and CENP-K but no crosslinks were observed with CENP-I, which was found instead to crosslink with CENP-M (Basilico et al. 2014). A second study revealed that CENP-H and CENP-K contacted CENP-C (Klare et al. 2015). Indeed, these and other studies have suggested that CENP-C makes extensive contacts with a number of inner and outer kinetochore proteins (Weir et al. 2016).

In the budding yeast kinetochore, CENP-T^{Cnn1} was crosslinked to CENP-H^{Mcm16}, CENP-I^{Ctf3} and CENP-K^{Mcm22} (Pekgoz Altunkaya et al. 2016). This led to the proposal that CENP-I^{Ctf3} embraces and stabilises the Ndc80 binding domain of CENP-T^{Cnn1}. Indeed, extensive crosslinks were also observed between CENP-T^{Cnn1} and Ndc80. In other studies, CLMS was used to examine protein interactions between the inner and outer kinetochore. These studies suggested that CENP-U^{Ame1} may link the Mis12 complex to the inner kinetochore (Hornung et al. 2014). This is quite different from the situation in metazoans, where CENP-C forms the link between the inner kinetochore and the Mis12 complex (Przewloka et al. 2011; Screpanti et al. 2011).

CLMS technology is now beginning to be much more widely applied. For example, it was used to examine the interaction between proteins of the Ska complex and microtubules (Abad et al. 2014). This analysis revealed that the Ska and Ndc80 complexes contact different faces of the tubulin dimer. Another study looked at the interactions of the SKAP:astrin heterodimer with spindle microtubules. CLMS was used to identify and map a microtubule-binding domain in SKAP, which was found to interact with both alpha- and beta tubulins (Friese et al. 2016).

9 Future Prospects

We now probably know the identity of all major structural proteins of the kinetochore, but it is clear that many chromatin modifiers and other auxiliary factors remain to be described. Furthermore, the recognition that kinetochores are transcribed during mitosis (Chan et al. 2012), and the realisation that this transcription may prevent pericentric heterochromatin from invading the kinetochore chromatin (Molina et al. 2016) point to an increasing complexity of kinetochore models in years to come. Together with increased understanding of the biology, improvements in the instrumentation and associated software mean that shotgun approaches in the future may allow truly comprehensive and quantitative lists of kinetochore components to be made. Challenges for the future include analysis of changes in bulk kinetochore protein composition at different phases of mitosis, comprehensive identification of the posttranslational modifications of all kinetochore proteins and determination of their functional consequences, and continued development of CLMS to allow the preparation of near-atomic maps of kinetochore protein structures and interactions.

Acknowledgements We thank Shinya Ohta and Juri Rappsilber for allowing us to reproduce their unpublished data. Work in the Earnshaw lab is funded by the Wellcome Trust, of which W.C. E. is a Principal Research Fellow (grant number 073915). The Wellcome Trust Centre for Cell Biology is supported by core grant numbers 077707 and 092076.

References

Abad MA, Medina B, Santamaria A, Zou J, Plasberg-Hill C, Madhumalar A, Jayachandran U, Redli PM, Rappsilber J, Nigg EA et al (2014) Structural basis for microtubule recognition by the human kinetochore Ska complex. Nat Commun 5:2964

Adachi J, Kumar C, Zhang Y, Olsen JV, Mann M (2006) The human urinary proteome contains more than 1500 proteins, including a large proportion of membrane proteins. Genome Biol 7: R80

Adolph KW, Cheng SM, Paulson JR, Laemmli UK (1977) Isolation of a protein scaffold from mitotic HeLa cell chromosomes. Proc Natl Acad Sci (USA) 11:4937–4941

Aebersold R, Mann M (2003) Mass spectrometry-based proteomics. Nature 422:198–207

Akiyoshi B, Gull K (2014) Discovery of unconventional kinetochores in kinetoplastids. Cell 156:1247–1258

Akiyoshi B, Nelson CR, Ranish JA, Biggins S (2009) Quantitative proteomic analysis of purified yeast kinetochores identifies a PP1 regulatory subunit. Genes Dev 23:2887–2899

Akiyoshi B, Sarangapani KK, Powers AF, Nelson CR, Reichow SL, Arellano-Santoyo H, Gonen T, Ranish JA, Asbury CL, Biggins S (2010) Tension directly stabilizes reconstituted kinetochore-microtubule attachments. Nature 468:576–579

Amano M, Suzuki A, Hori T, Backer C, Okawa K, Cheeseman IM, Fukagawa T (2009) The CENP-S complex is essential for the stable assembly of outer kinetochore structure. J Cell Biol 186:173–182

Barysz H, Kim JH, Chen ZA, Hudson DF, Rappsilber J, Gerloff DL, Earnshaw WC (2015) Three-dimensional topology of the SMC2/SMC4 subcomplex from chicken condensin I revealed by cross-linking and molecular modelling. Open Biol 5

Basilico F, Maffini S, Weir JR, Prumbaum D, Rojas AM, Zimniak T, De Antoni A, Jeganathan S, Voss B, van Gerwen S et al (2014) The pseudo GTPase CENP-M drives human kinetochore assembly. Elife 3:e02978

Biggins S (2013) The composition, functions, and regulation of the budding yeast kinetochore. Genetics 194:817–846

Bischoff FR, Maier G, Tilz G, Ponstingl H (1990) A 47-kDa human nuclear protein recognized by antikinetochore autoimmune sera is homologous with the protein encoded by RCC1, a gene implicated in onset of chromosome condensation. Proc Nat Acad Sci (USA) 87:8617–8621

Bischoff FR, Ponstingl H (1991a) Catalysis of guanine nucleotide exchange on Ran by the mitotic regulator RCC1. Nature 354:80–82

Bischoff FR, Ponstingl H (1991b) Mitotic regulator protein RCC1 is complexed with a nuclear ras-related polypeptide. Proc Natl Acad Sci (USA) 88:10830–10834

Brenner S, Pepper D, Berns MW, Tan E, Brinkley BR (1981) Kinetochore structure, duplication and distribution in mammalian cells: analysis by human autoantibodies from scleroderma patients. J Cell Biol 91:95–102

Chan FL, Marshall OJ, Saffery R, Kim BW, Earle E, Choo KH, Wong LH (2012) Active transcription and essential role of RNA polymerase II at the centromere during mitosis. Proc Natl Acad Sci (USA) 109:1979–1984

Cheeseman IM, Desai A (2008) Molecular architecture of the kinetochore-microtubule interface. Nat Rev Mol Cell Biol 9:33–46

Cheeseman IM, Niessen S, Anderson S, Hyndman F, Yates JR, Oegema K, Desai A (2004) A conserved protein network controls assembly of the outer kinetochore and its ability to sustain tension. Genes Dev 18:2255–2268

Chen ZA, Fischer L, Cox J, Rappsilber J (2016) Quantitative cross-linking/mass spectrometry using isotope-labeled cross-linkers and MaxQuant. Mol Cell Proteomics 15:2769–2778

Chikashige Y, Kinoshita N, Nakaseko Y, Matsumoto T, Murikami S, Niwa O, Yanagida M (1989) Composite motifs and repeat symmetry in S. pombe centromeres: direct analysis by integration of Notl restriction sites. Cell 57:739–751

Ciferri C, Pasqualato S, Screpanti E, Varetti G, Santaguida S, Dos Reis G, Maiolica A, Polka J, De Luca JG, De Wulf P et al (2008) Implications for kinetochore-microtubule attachment from the structure of an engineered Ndc80 complex. Cell 133:427–439

Clarke L, Baum MP (1990) Functional analysis of a centromere from fission yeast: a role for centromere-specific repeated DNA sequences. Mol Cell Biol 10:1863–1872

Clarke L, Carbon J (1980) Isolation of a yeast centromere and construction of functional small circular chromosomes. Nature 287:504–509

Cooke CA, Heck MM, Earnshaw WC (1987) The inner centromere protein (INCENP) antigens: movement from inner centromere to midbody during mitosis. J Cell Biol 105:2053–2067

D'Archivio S, Wickstead B (2017) Trypanosome outer kinetochore proteins suggest conservation of chromosome segregation machinery across eukaryotes. J Cell Biol 216:379–391

De Wulf P, McAinsh AD, Sorger PK (2003) Hierarchical assembly of the budding yeast kinetochore from multiple subcomplexes. Genes Dev 17:2902–2921

Desai A, Rybina S, Muller-Reichert T, Shevchenko A, Hyman A, Oegema K (2003) KNL-1 directs assembly of the microtubule-binding interface of the kinetochore in *C. elegans*. Genes Dev 17:2421–2435

Doheny KF, Sorger PK, Hyman AA, Tugendreich S, Spencer F, Hieter P (1993) Identification of essential components of the *S. cerevisiae* kinetochore. Cell 73:761–774

Earnshaw WC, Halligan N, Cooke C, Rothfield N (1984) The kinetochore is part of the chromosome scaffold. J Cell Biol 98:352–357

Earnshaw WC, Laemmli UK (1983) Architecture of metaphase chromosomes and chromosome scaffolds. JCell Biol 96:84–93

Earnshaw WC, Migeon B (1985) A family of centromere proteins is absent from the latent centromere of a stable isodicentric chromosome. Chromosoma (Berl) 92:290–296

Earnshaw WC, Rothfield N (1985) Identification of a family of human centromere proteins using autoimmune sera from patients with scleroderma. Chromosoma (Berl) 91:313–321

Earnshaw WC, Sullivan KF, Machlin PS, Cooke CA, Kaiser DA, Pollard TD, Rothfield NF, Cleveland DW (1987) Molecular cloning of cDNA for CENP-B, the major human centromere autoantigen. J Cell Biol 104:817–829

Fischer L, Chen ZA, Rappsilber J (2013) Quantitative cross-linking/mass spectrometry using isotope-labelled cross-linkers. J Proteomics 88:120–128

Fishel B, Amstitz H, Baum M, Carbon J, Clarke L (1988) Structural organization and functional analysis of centromeric DNA in the fission yeast *Schizosaccharomyces pombe*. Mol Cell Biol 8:754–763

Fitzgerald-Hayes M, Clarke L, Carbon J (1982) Nucleotide sequence comparisons and functional analysis of yeast centromere DNAs. Cell 29:235–244

Foltz DR, Jansen LE, Black BE, Bailey AO, Yates JR, Cleveland DW (2006) The human CENP-A centromeric nucleosome-associated complex. Nat Cell Biol 8:458–469

Friese A, Faesen AC, Huis in't Veld PJ, Fischbock J, Prumbaum D, Petrovic A, Raunser S, Herzog F, Musacchio A (2016) Molecular requirements for the inter-subunit interaction and kinetochore recruitment of SKAP and Astrin. Nat Commun 7:11407

Gascoigne KE, Takeuchi K, Suzuki A, Hori T, Fukagawa T, Cheeseman IM (2011) Induced ectopic kinetochore assembly bypasses the requirement for CENP-A nucleosomes. Cell 145:410–422

Gassmann R, Carvalho A, Henzing AJ, Ruchaud S, Hudson DF, Honda R, Nigg EA, Gerloff DL, Earnshaw WC (2004) Borealin: a novel chromosomal passenger required for stability of the bipolar mitotic spindle. J Cell Biol 166:179–191

Gassmann R, Henzing AJ, Earnshaw WC (2005) Novel components of human mitotic chromosomes identified by proteomic analysis of the chromosome scaffold fraction. Chromosoma 113:385–397

Gonczy P, Echeverri C, Oegema K, Coulson A, Jones SJ, Copley RR, Duperon J, Oegema J, Brehm M, Cassin E et al (2000) Functional genomic analysis of cell division in *C. elegans* using RNAi of genes on chromosome III. Nature 408:331–336

Goshima G, Saitoh S, Yanagida M (1999) Proper metaphase spindle length is determined by centromere proteins Mis12 and Mis6 required for faithful chromosome segregation. Genes Dev 13:1664–1677

Guldner HH, Lakomek H-J, Bautz FA (1984) Human anti-centromere sera recognise a 19.5 kD non-histone chromosomal protein from HeLa cells. Clin Exp Immunol 58:13–20

Hahnenberger KM, Baum MP, Polizzi CM, Carbon J, Clarke L (1989) Construction of functional artificial minichromosomes in the fission yeast *Schizosaccharomyces pombe*. Proc Nat Acad Sci (USA) 86:577–581

Han X, Aslanian A, Yates JR 3rd (2008) Mass spectrometry for proteomics. Curr Opin Chem Biol 12:483–490

Hayashi T, Fujita Y, Iwasaki O, Adachi Y, Takahashi K, Yanagida M (2004) Mis16 and Mis18 are required for CENP-A loading and histone deacetylation at centromeres. Cell 118:715–729

Hieter P, Pridmore D, Hegemann JH, Thomas M, Davis RW, Philippsen P (1985) Functional selection and analysis of yeast centromeric DNA. Cell 42:913–921

Hori T, Amano M, Suzuki A, Backer CB, Welburn JP, Dong Y, McEwen BF, Shang WH, Suzuki E, Okawa K et al (2008) CCAN makes multiple contacts with centromeric DNA to provide distinct pathways to the outer kinetochore. Cell 135:1039–1052

Hornung P, Troc P, Malvezzi F, Maier M, Demianova Z, Zimniak T, Litos G, Lampert F, Schleiffer A, Brunner M et al (2014) A cooperative mechanism drives budding yeast kinetochore assembly downstream of CENP-A. J Cell Biol 206:509–524

Hyland KM, Kingsbury J, Koshland D, Hieter P (1999) Ctf19p: a novel kinetochore protein in *Saccharomyces cerevisiae* and a potential link between the kinetochore and mitotic spindle. J Cell Biol 145:15–28

Izuta H, Ikeno M, Suzuki N, Tomonaga T, Nozaki N, Obuse C, Kisu Y, Goshima N, Nomura F, Nomura N et al (2006) Comprehensive analysis of the ICEN (interphase centromere complex) components enriched in the CENP-A chromatin of human cells. Genes Cells 11:673–684

Jensen ON, Podtelejnikov AV, Mann M (1997) Identification of the components of simple protein mixtures by high-accuracy peptide mass mapping and database searching. Anal Chem 69:4741–4750

Jensen ON, Wilm M, Shevchenko A, Mann M (1999) Peptide sequencing of 2-DE gel-isolated proteins by nanoelectrospray tandem mass spectrometry. Methods Mol Biol 112:571–588

Kang YH, Park CH, Kim TS, Soung NK, Bang JK, Kim BY, Park JE, Lee KS (2011) Mammalian polo-like kinase 1-dependent regulation of the PBIP1-CENP-Q complex at kinetochores. J Biol Chem 286:19744–19757

Karpen GH, Allshire RC (1997) The case for epigenetic effects on centromere identity and function. Trends Genet 13:489–496

Klare K, Weir JR, Basilico F, Zimniak T, Massimiliano L, Ludwigs N, Herzog F, Musacchio A (2015) CENP-C is a blueprint for constitutive centromere-associated network assembly within human kinetochores. J Cell Biol 210:11–22

Kustatscher G, Grabowski P, Rappsilber J (2016) Multiclassifier combinatorial proteomics of organelle shadows at the example of mitochondria in chromatin data. Proteomics 16:393–401

Lechner J, Carbon J (1991) A 240 kd multisubunit protein complex, CBF3, is a major component of the budding yeast centromere. Cell 64:717–725

Leitner A, Faini M, Stengel F, Aebersold R (2016) Crosslinking and mass spectrometry: an integrated technology to understand the structure and function of molecular machines. Trends Biochem Sci 41:20–32

Lewis CD, Laemmli UK (1982) Higher order metaphase chromosome structure: evidence for metalloprotein interactions. Cell 29:171–181

Liao H, Winkfein RJ, Mack G, Rattner JB, Yen TJ (1995) CENP-F is a protein of the nuclear matrix that assembles onto kinetochores at late G2 and is rapidly degraded after mitosis. J Cell Biol 130:507–518

Macek B, Waanders LF, Olsen JV, Mann M (2006) Top-down protein sequencing and MS3 on a hybrid linear quadrupole ion trap-orbitrap mass spectrometer. Mol Cell Proteomics 5:949–958

Maddox PS, Hyndman F, Monen J, Oegema K, Desai A (2007) Functional genomics identifies a Myb domain-containing protein family required for assembly of CENP-A chromatin. J Cell Biol 176:757–763

Maine GT, Sinha P, Tye BK (1984) Mutants of *S. cerevisiae* defective in the maintenance of minichromosomes. Genetics 106:365–385

Maiolica A, Cittaro D, Borsotti D, Sennels L, Ciferri C, Tarricone C, Musacchio A, Rappsilber J (2007) Structural analysis of multiprotein complexes by cross-linking, mass spectrometry, and database searching. Mol Cell Proteomics 6:2200–2211

Meeks-Wagner D, Wood JS, Garvik B, Hartwell LH (1986) Isolation of two genes that affect mitotic chromosome transmission in *S. cerevisiae*. Cell 44:53–63

Meluh PB, Koshland D (1995) Evidence that the MIF2 gene of *Saccharomyces cerevisiae* encodes a centromere protein with homology to the mammalian centromere protein CENP-C. Mol Biol Cell 6:793–807

Molina O, Vargiu G, Abad MA, Zhiteneva A, Jeyaprakash AA, Masumoto H, Kouprina N, Larionov V, Earnshaw WC (2016) Epigenetic engineering reveals a balance between histone modifications and transcription in kinetochore maintenance. Nat Commun 7:13334

Montaño-Gutierrez LF, Ohta S, Kustatscher G, Earnshaw WC, Rappsilber J (2017) Nano random forests to mine protein complexes and their relationships in quantitative proteomics data. Mol Biol Cell 28:673–680

Moroi Y, Hartman AL, Nakane PK, Tan EM (1981) Distribution of kinetochore (centromere) antigen in mammalian cell nuclei. J Cell Biol 90:254–259

Moroi Y, Peebles C, Fritzler MJ, Steigerwald J, Tan EM (1980) Autoantibody to centromere (kinetochore) in scleroderma sera. Proc Nat Acad Sci (USA) 77:1627–1631

Moyer SE, Lewis PW, Botchan MR (2006) Isolation of the Cdc45/Mcm2-7/GINS (CMG) complex, a candidate for the eukaryotic DNA replication fork helicase. Proc Natl Acad Sci U S A 103:10236–10241

Nakaseko Y, Adachi Y, Funahashi S, Niwa O, Yanagida M (1986) Chromosome walking shows a highly homologous repetitive sequence present in all the centromere regions of fission yeast. The EMBO J 5:1011–1021

Nerusheva OO, Akiyoshi B (2016) Divergent polo box domains underpin the unique kinetoplastid kinetochore. Open Biol 6

Nishihashi A, Haraguchi T, Hiraoka Y, Ikemura T, Regnier V, Dodson H, Earnshaw WC, Fukagawa T (2002) CENP-I is essential for centromere function in vertebrate cells. Dev Cell 2:463–476

Nishino T, Rago F, Hori T, Tomii K, Cheeseman IM, Fukagawa T (2013) CENP-T provides a structural platform for outer kinetochore assembly. EMBO J 32:424–436

Nishino T, Takeuchi K, Gascoigne KE, Suzuki A, Hori T, Oyama T, Morikawa K, Cheeseman IM, Fukagawa T (2012) CENP-T-W-S-X forms a unique centromeric chromatin structure with a histone-like fold. Cell 148:487–501

Obuse C, Yang H, Nozaki N, Goto S, Okazaki T, Yoda K (2004) Proteomics analysis of the centromere complex from HeLa interphase cells: UV-damaged DNA binding protein 1 (DDB-1) is a component of the CEN-complex, while BMI-1 is transiently co-localized with the centromeric region in interphase. Genes Cells 9:105–120

Oegema K, Desai A, Rybina S, Kirkham M, Hyman AA (2001) Functional analysis of kinetochore assembly in *Caenorhabditis elegans*. J Cell Biol 153:1209–1226

Ohta S, Bukowski-Wills JC, Sanchez-Pulido L, Alves Fde L, Wood L, Chen ZA, Platani M, Fischer L, Hudson DF, Ponting CP et al (2010a) The protein composition of mitotic chromosomes determined using multiclassifier combinatorial proteomics. Cell 142:810–821

Ohta S, Bukowski-Wills JC, Wood L, de Lima Alves F, Chen Z, Rappsilber J, Earnshaw WC (2010b) Proteomics of isolated mitotic chromosomes identifies the kinetochore protein Ska3/Rama1. Cold Spring Harb Symp Quant Biol 75:433–438

Ohta S, Montano-Gutierrez LF, de Lima Alves F, Ogawa H, Toramoto I, Sato N, Morrison CG, Takeda S, Hudson DF, Rappsilber J et al (2016) Proteomics analysis with a nano random forest approach reveals novel functional interactions regulated by SMC complexes on mitotic chromosomes. Mol Cell Proteomics 15:2802–2818

Ohta S, Wood L, Bukowski-Wills JC, Rappsilber J, Earnshaw WC (2010c) Building mitotic chromosomes. Curr Opin Cell Biol 23:114–121

Okada M, Cheeseman IM, Hori T, Okawa K, McLeod IX, Yates JR, Desai A, Fukagawa T (2006) The CENP-H-I complex is required for the efficient incorporation of newly synthesized CENP-A into centromeres. Nat Cell Biol 8:446–457

Ong SE, Foster LJ, Mann M (2003) Mass spectrometric-based approaches in quantitative proteomics. Methods 29:124–130

Ong SE, Mann M (2006) A practical recipe for stable isotope labeling by amino acids in cell culture (SILAC). Nat Protoc 1:2650–2660

Ortiz J, Stemmann O, Rank S, Lechner J (1999) A putative protein complex consisting of Ctf19, Mcm21, and Okp1 represents a missing link in the budding yeast kinetochore. Genes Dev 13:1140–1155

Palmer DK, Margolis RL (1985) Kinetochore components recognized by human autoantibodies are present on mononucleosomes. Mol Cell Biol 5:173–186

Palmer DK, O'Day K, Le Trong H, Charbonneau H, Margolis RL (1991) Purification of the centromeric protein CENP-A and demonstration that it is a centromere specific histone. Proc Nat Acad Sci (USA) 88:3734–3738

Pekgoz Altunkaya G, Malvezzi F, Demianova Z, Zimniak T, Litos G, Weissmann F, Mechtler K, Herzog F, Westermann S (2016) CCAN assembly configures composite binding interfaces to promote cross-linking of Ndc80 complexes at the Kinetochore. Curr Biol 26:2370–2378

Pluta AF, Mackay AM, Ainsztein AM, Goldberg IG, Earnshaw WC (1995) The centromere: hub of chromosomal activities. Science 270:1591–1594

Przewloka MR, Venkei Z, Bolanos-Garcia VM, Debski J, Dadlez M, Glover DM (2011) CENP-C is a structural platform for kinetochore assembly. Curr Biol 21:399–405

Ranish JA, Yi EC, Leslie DM, Purvine SO, Goodlett DR, Eng J, Aebersold R (2003) The study of macromolecular complexes by quantitative proteomics. Nat Genet 33:349–355

Rout MP, Kilmartin JV (1990) Components of the yeast spindle and spindle pole body. J Cell Biol 111:1913–1927

Roy N, Poddar A, Lohia A, Sinha P (1997) The mcm17 mutation of yeast shows a size-dependent segregational defect of a mini-chromosome. Curr Genet 32:182–189

Saitoh H, Tomkiel JE, Cooke CA, Ratrie HR, Maurer M, Rothfield NF, Earnshaw WC (1992) CENP-C, an autoantigen in scleroderma, is a component of the human inner kinetochore plate. Cell 70:115–125

Samejima I, Spanos C, Alves Fde L, Hori T, Perpelescu M, Zou J, Rappsilber J, Fukagawa T, Earnshaw WC (2015) Whole-proteome genetic analysis of dependencies in assembly of a vertebrate kinetochore. J Cell Biol 211:1141–1156

Saunders WS, Chue C, Goebl M, Craig C, Clark RF, Powers JA, Eissenberg JC, Elgin SC, Rothfield NF, Earnshaw WC (1993) Molecular cloning of a human homologue of Drosophila heterochromatin protein HP1 using anti-centromere autoantibodies with anti-chromo specificity. J Cell Sci 104:573–582

Schleiffer A, Maier M, Litos G, Lampert F, Hornung P, Mechtler K, Westermann S (2012) CENP-T proteins are conserved centromere receptors of the Ndc80 complex. Nat Cell Biol 14:604–613

Screpanti E, De Antoni A, Alushin GM, Petrovic A, Melis T, Nogales E, Musacchio A (2011) Direct binding of Cenp-C to the Mis12 complex joins the inner and outer kinetochore. Curr Biol 21:391–398

Shannon P, Markiel A, Ozier O, Baliga NS, Wang JT, Ramage D, Amin N, Schwikowski B, Ideker T (2003) Cytoscape: a software environment for integrated models of biomolecular interaction networks. Genome Res 13:2498–2504

Singh TR, Saro D, Ali AM, Zheng XF, Du CH, Killen MW, Sachpatzidis A, Wahengbam K, Pierce AJ, Xiong Y et al (2010) MHF1-MHF2, a histone-fold-containing protein complex, participates in the Fanconi anemia pathway via FANCM. Mol Cell 37:879–886

Sonnichsen B, Koski LB, Walsh A, Marschall P, Neumann B, Brehm M, Alleaume AM, Artelt J, Bettencourt P, Cassin E et al (2005) Full-genome RNAi profiling of early embryogenesis in Caenorhabditis elegans. Nature 434:462–469

Spencer F, Gerring SL, Connelly C, Hieter P (1990) Mitotic chromosome transmission fidelity mutants in *Saccharomyces cerevisiae*. Genetics 124:237–249

Steiner N, Clarke L (1994) A novel epigenetic effect can alter centromere function in fission yeast. Cell 79:865–874

Stoler S, Keith KC, Curnick KE, Fitzgerald-Hayes M (1995) A mutation in CSE4, an essential gene encoding a novel chromatin-associated protein in yeast, causes chromosome nondisjunction and cell cycle arrest at mitosis. Genes Dev 9:573–586

Sullivan KF, Hechenberger M, Masri K (1994) Human CENP-A contains a histone H3 related histone fold domain that is required for targeting to the centromere. J Cell Biol 127:581–592

Tadeu AM, Ribeiro S, Johnston J, Goldberg I, Gerloff D, Earnshaw WC (2008) CENP-V is required for centromere organization, chromosome alignment and cytokinesis. EMBO J 27:2510–2522

Takahashi K, Yamada H, Yanagida M (1994) Fission yeast minichromosome loss mutants mis cause lethal aneuploidy and replication abnormality. Mol Biol Cell 5:1145–1158

Takata H, Uchiyama S, Nakamura N, Nakashima S, Kobayashi S, Sone T, Kimura S, Lahmers S, Granzier H, Labeit S et al (2007) A comparative proteome analysis of human metaphase chromosomes isolated from two different cell lines reveals a set of conserved chromosome-associated proteins. Genes Cells 12:269–284

Takeuchi K, Nishino T, Mayanagi K, Horikoshi N, Osakabe A, Tachiwana H, Hori T, Kurumizaka H, Fukagawa T (2013) The centromeric nucleosome-like CENP-T-W-S-X complex induces positive supercoils into DNA. Nucleic Acids Res 42:1644–1655

Tyanova S, Temu T, Sinitcyn P, Carlson A, Hein MY, Geiger T, Mann M, Cox J (2016) The perseus computational platform for comprehensive analysis of (prote)omics data. Nat Methods 13:731–740

Uchiyama S, Kobayashi S, Takata H, Ishihara T, Hori N, Higashi T, Hayashihara K, Sone T, Higo D, Nirasawa T et al (2005) Proteome analysis of human metaphase chromosomes. J Biol Chem 280:16994–17004

Vafa O, Sullivan KF (1997) Chromatin containing CENP-A and a-satellite DNA is a major component of the inner kinetochore plate. Curr Biol 7:897–900

Warburton PE, Cooke C, Bourassa S, Vafa O, Sullivan BA, Stetten G, Gimelli G, Warburton D, Tyler-Smith C, Sullivan KF et al (1997) Immunolocalization of CENP-A suggests a distinct nucleosome structure at the inner kinetochore plate of active centromeres. Curr Biol 7:901–904

Washburn MP, Wolters D, Yates JR 3rd (2001) Large-scale analysis of the yeast proteome by multidimensional protein identification technology. Nat Biotechnol 19:242–247

Weir JR, Faesen AC, Klare K, Petrovic A, Basilico F, Fischbock J, Pentakota S, Keller J, Pesenti ME, Pan D et al (2016) Insights from biochemical reconstitution into the architecture of human kinetochores. Nature 537:249–253

Westermann S, Drubin DG, Barnes G (2007) Structures and functions of yeast kinetochore complexes. Annu Rev Biochem 76:563–591

Wigge PA, Jensen ON, Holmes S, Soues S, Mann M, Kilmartin JV (1998) Analysis of the Saccharomyces spindle pole by matrix-assisted laser desorption/ionization (MALDI) mass spectrometry. J Cell Biol 141:967–977

Wigge PA, Kilmartin JV (2001) The Ndc80p complex *from Saccharomyces cerevisiae* contains conserved centromere components and has a function in chromosome segregation. J Cell Biol 152:349–360

Yamagishi Y, Sakuno T, Goto Y, Watanabe Y (2014) Kinetochore composition and its function: lessons from yeasts. FEMS Microbiol Rev 38:185–200

Yan Z, Delannoy M, Ling C, Daee D, Osman F, Muniandy PA, Shen X, Oostra AB, Du H, Steltenpool J et al (2010) A histone-fold complex and FANCM form a conserved DNA-remodeling complex to maintain genome stability. Mol Cell 37:865–878

Yates JR 3rd (1998) Mass spectrometry and the age of the proteome. J Mass Spectrom 33:1–19

Yen TJ, Compton DA, Wise D, Zinkowski RP, Brinkley BR, Earnshaw WC, Cleveland DW (1991) CENP-E, a novel human centromere-associated protein required for progression from metaphase to anaphase. EMBO J 10:1245–1254

Young RA, Davis RB (1983) Yeast polymerase II genes: isolation with antibody probes. Science 222:778–782

Critical Foundation of the Kinetochore: The Constitutive Centromere-Associated Network (CCAN)

Masatoshi Hara and Tatsuo Fukagawa

Abstract The kinetochore is a large protein complex, which is assembled at the centromere of a chromosome to ensure faithful chromosome segregation during M-phase. The centromere in most eukaryotes is epigenetically specified by DNA sequence-independent mechanisms. The constitutive centromere-associated network (CCAN) is a subcomplex in the kinetochore that localizes to the centromere throughout the cell cycle. The CCAN has interfaces bound to the centromeric chromatin and the spindle microtubule-binding complex; therefore, it functions as a foundation of kinetochore formation. Here, we summarize recent progress in our understanding of the structure and organization of the CCAN. We also discuss an additional role of the CCAN in the maintenance of centromere position and dynamic reorganization of the CCAN.

1 Introduction

One of the distinguishing features of living organisms is self-replication. To maintain the continuity of life, genetic materials have to be faithfully inherited by successive generations. In the early 1900s, Walter Sutton and Theodor Boveri independently suggested that chromosomes carry genetic materials, using insect germ cells and echinoderm embryos, respectively (Sutton 1902, 1903; Boveri 1904). Their "chromosome theory" explained the mechanism underlying Mendel's laws that were rediscovered at the same time (de Vries 1900; Tschermak 1900; Correns 1900; Birchler 2015). Thereafter, the chromosome theory was experimentally verified by Thomas Morgan Hunt based on fly genetics (Morgan 1915). These findings led to the next important question: how were the chromosomes correctly segregated into daughter cells? A clue to answering this question was described by Walter Flemming decades before the theory was established

M. Hara · T. Fukagawa (✉)
Graduate School of Frontier Biosciences,
Osaka University, Suita, Osaka 565-0871, Japan
e-mail: tfukagawa@fbs.osaka-u.ac.jp

B.E. Black (ed.), *Centromeres and Kinetochores*, Progress in Molecular and Subcellular Biology 56, DOI 10.1007/978-3-319-58592-5_2

(Flemming 1882). He found primary constrictions on chromosomes, where the mitotic spindle attached to deliver these chromosomes into daughter cells (Flemming 1882).

In 1936, Cyril Darlington defined the term "centromere" (Darlington 1936). Centromeres are now known as regions on chromosomes where a macro proteinaceous complex, called the kinetochore, is assembled to connect centromeres with spindle microtubules during mitosis (Cheeseman and Desai 2008; McKinley and Cheeseman 2016; Pesenti et al. 2016; Nagpal and Fukagawa 2016). The kinetochore also contributes to correct chromosome segregation by producing a signal upon incorrect attachment of kinetochores with microtubules (Foley and Kapoor 2013; London and Biggins 2014; Stukenberg and Burke 2015; Musacchio 2015).

The kinetochore in vertebrates is made-up of more than 100 proteins (Tipton et al. 2012). These proteins are divided into sub-protein complexes. The constitutive centromere-associated network (CCAN) is one of the major subcomplexes in the kinetochore (Cheeseman and Desai 2008; Perpelescu and Fukagawa 2011; Takeuchi and Fukagawa 2012). The CCAN proteins constitutively localize to the centromeres throughout the cell cycle and form a foundation for kinetochore assembly. Another major subcomplex is the KMN (the KNL1, the Ndc80 and the Mis12 complexes) network, which is recruited to the CCAN during M-phase (Cheeseman et al. 2006; Cheeseman and Desai 2008; Varma and Salmon 2012; Foley and Kapoor 2013; Nagpal et al. 2015). The CCAN is associated with centromere chromatin and the KMN network binds directly to the spindle microtubule; therefore, the kinetochore effectively mediates the interaction between the chromosomes and the microtubules.

The centromere is specified at a particular position on a chromosome in many species and the kinetochore is formed within the centromeric region. In this chapter, we introduce centromere specification and then describe and discuss the CCAN structure and function and its dynamic regulations for kinetochore assembly.

2 Centromere

2.1 Centromere Organization

The centromere is a genome region, where the kinetochore is assembled (Fig. 1a). Although the centromere is crucial for faithful chromosome segregation, genome organization of centromeres is diverse among various species. The budding yeast, *Saccharomyces cerevisiae*, has simple and small centromeres, which are defined by a 125-basepair-specific DNA sequence in each chromosome (Fig. 1b) (Hegemann and Fleig 1993; Pluta et al. 1995; Clarke 1998). The short budding yeast centromere DNA is sufficient to assemble the kinetochore for efficient chromosome segregation (Hegemann and Fleig 1993). The sequence-dependent centromere in the budding yeast is known as a point centromere, which is bound to a single microtubule through the kinetochore (Pluta et al. 1995).

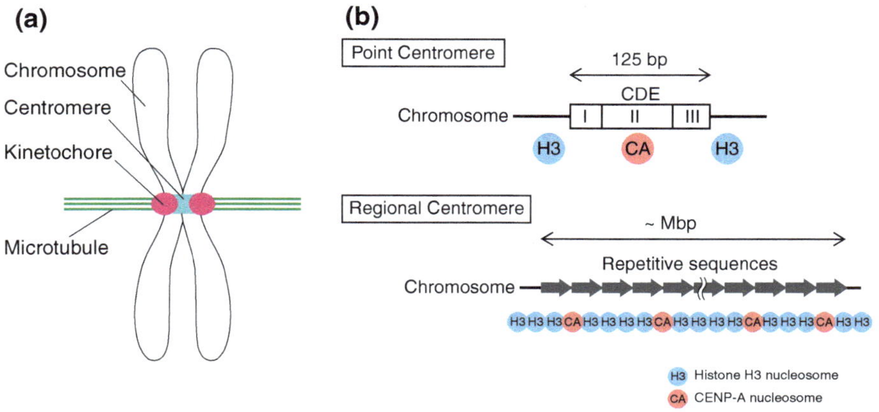

Fig. 1 Centromere structure and organization. **a** A schematic representation of vertebrate chromosome. The centromere is a specific genomic region where the kinetochore is assembled to establish a microtubule-binding interface for faithful chromosome segregation. The centromeres are found on sites where constriction is formed in chromosomes during mitosis. **b** The point centromere in budding yeast, *Saccharomyces cerevisiae* (*top*). It is specified by a 125-bp sequence, which contains centromere DNA elements (CDE) I, II, and III. The short DNA motif is occupied with a nucleosome, which contains a centromere-specific histone H3 variant, CENP-A (Ces4 in *Saccharomyces cerevisiae*). *Bottom* shows the regional centromere. It stretches over large regions that comprise repetitive sequences in most species (e.g., alpha-satellite repeats DNA in human). Although the repetitive sequences facilitate centromere formation, the position of a regional centromere is specified by the CENP-A nucleosome, which is an epigenetic marker of the centromere

In contrast to the budding yeast, the majority of other organisms, with the exception of *Caenorhabditis elegans* and some insects that have holocentromeres, have a regional centromere, which spans a much larger chromosomal region (several kilobasepairs to megabasepairs) (Fukagawa and Earnshaw 2014a; Kursel and Malik 2016). The regional centromeres typically have repetitive DNA sequences and are bound to multiple microtubules (Fukagawa and Earnshaw 2014a; Kursel and Malik 2016). Human centromeres span hundreds of kilobasepairs to several megabasepairs and consist of arrays of alpha-satellite DNA repeats (Fig. 1b) (Aldrup-Macdonald and Sullivan 2014). The biological significance of the repetitive DNA for centromere function is controversial. Human artificial chromosomes (HACs), which have the alpha-satellite DNA repeats from human centromeres, are efficiently generated and stably maintained in human cells (Harrington et al. 1997; Ikeno et al. 1998). Although the alpha-satellite repeats promote the functional kinetochore assembly in the experimental condition of the HAC formation (Ohzeki et al. 2002), observations in several species suggest that the repetitive sequences are not always necessary to specify centromere regions on chromosomes. Species in which this was observed include horse (Wade et al. 2009; Piras et al. 2010), chicken (Shang et al. 2010), and orangutan (Locke et al. 2011).

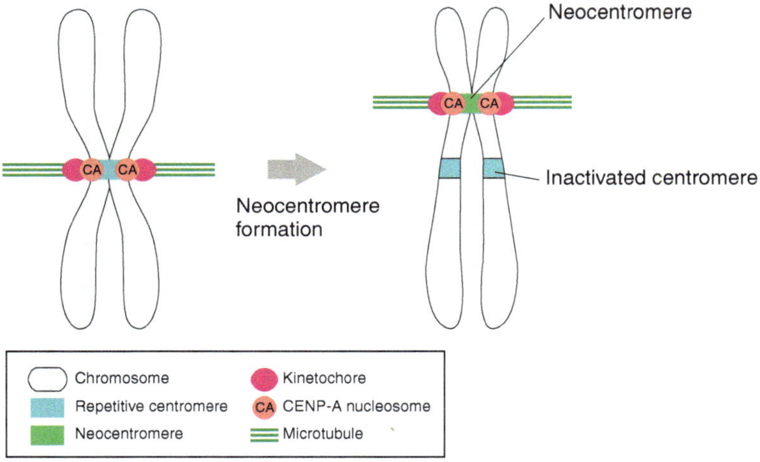

Fig. 2 CENP-A exists in active centromeres. A neocentromere is a newly formed centromere at non-centromeric region, when the native centromere is inactivated or disrupted. The repetitive sequence is not necessary for the neocentromere formation, indicating that the regional centromere is specified by sequence-independent epigenetic mechanisms. CENP-A is found in active neocentromeres, but not in inactive centromeres, which contain the repetitive DNA sequences, in dicentric chromosomes

A neocentromere is a newly formed centromere within a non-centromeric locus on a chromosome (Fig. 2) (Marshall et al. 2008; Fukagawa and Earnshaw 2014b). The first human neocentromere, which lacks the alpha-satellite repeats, was discovered in 1993 (Voullaire et al. 1993). Since then, over 90 cases of human neocentromeres have been reported (Marshall et al. 2008). Discovery of these neocentromeres supports the idea that repetitive sequences are not essential for centromere specification. Neocentromeres can be formed naturally on various DNA sequences on chromosomes upon disruption or inactivation of a native centromere. The neocentromere formation process was experimentally reproduced using genetic engineering to remove native centromeres in fungi and chicken DT40 cells (Fig. 3a) (Ishii et al. 2008; Ketel et al. 2009; Shang et al. 2013). Based on these observations, the locus for the regional centromeres does not seem to be specified with genetic marks, such as particular DNA sequences. This suggests that epigenetic marks play a key role in centromere specification.

2.2 *CENP-A Is a Critical Epigenetic Mark for Centromere Specification*

The insight of centromere specification in the regional centromere was derived from the discovery of centromere proteins. Centromere protein (CENP)-A was originally

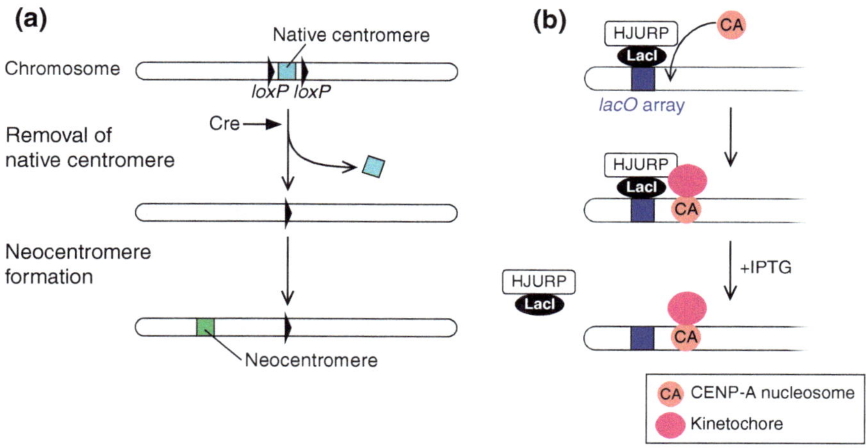

Fig. 3 The experimental systems to generate neocentromeres artificially. Because efficiency of natural neocentromere formation is very low, the experimental systems, which generate neocentromeres artificially, help to understand mechanisms of neocentromere formation. **a** The inducible centromere removal system by which the native centromere flanked with *loxP* sequences are excised by Cre recombinase and neocentromeres are formed on non-centromere loci. The system is applied to fungi and chicken DT40 cells. **b** Neocentromeres are also induced by ectopic CENP-A nucleosome deposition on non-centromeric region. Artificial tethering of CENP-A chaperones (HJURP in vertebrate and CAL1 in *Drosophila*) using *lacO*/LacI system deposits the CENP-A nucleosomes on a non-centromere locus. Kinetochore proteins are recruited to the ectopic site to form a functional kinetochore. The artificial kinetochore is functional, because the artificial kinetochore is replaceable in the native centromere (see also Fig. 5a)

identified as an antigen for anti-centromere autoimmune sera from patients with CREST syndrome (Earnshaw and Rothfield 1985). CENP-A is a centromere-specific histone H3 variant that forms a nucleosome, replacing the canonical histone H3 (Fig. 1b) (Palmer et al. 1987; Yoda et al. 2000). CENP-A orthologs are found in most eukaryotes, including the budding yeast, which have the point centromere (Stoler et al. 1995; Buchwitz et al. 1999; Takahashi et al. 2000; Blower and Karpen 2001; Talbert et al. 2002). CENP-A null mice exhibit early embryonic lethality (Howman et al. 2000) and inactivation or depletion of CENP-A in most organisms causes chromosome mis-segregation during M-phase (Stoler et al. 1995; Howman et al. 2000; Takahashi et al. 2000; Oegema et al. 2001; Blower and Karpen 2001; Goshima et al. 2003; Regnier et al. 2005), indicating that CENP-A is an essential gene for faithful chromosome segregation.

CENP-A is only localized onto the active centromere in human dicentric chromosomes. Dicentric chromosomes have two centromeres on a chromosome—one is an active centromere and the other is inactive (Fig. 2) (Earnshaw and Rothfield 1985). CENP-A never localizes on the inactive centromere, regardless of the presence of the repetitive alpha-satellite DNA (Fig. 2) (Warburton et al. 1997). These observations lead to the idea that CENP-A is an epigenetic mark for specification of active centromeres.

The idea that centromeres are specified epigenetically but not genetically is further supported by artificial kinetochore assembly on chromosome arms using a *lacO*/LacI system in which LacI-fused centromere proteins are tethered on a *lacO* array on non-centrometric region in a chromosome arm (Fig. 3b). Targeting the CENP-A-specific chaperones, HJURP and CAL1 in vertebrates and *Drosophila melanogaster* respectively, to their non-centromeric region induces CENP-A deposition and kinetochore assembly on the targeting sites (Fig. 3b) (Barnhart et al. 2011; Hori et al. 2013; Chen et al. 2014). Once the artificial CENP-A chromatin is established, the CENP-A and kinetochore proteins are maintained without the LacI-fused CENP-A chaperone (Fig. 3b) (Barnhart et al. 2011; Hori et al. 2013; Chen et al. 2014). The artificially CENP-A-deposited chromatin forms a functional active centromere, because the induced artificial kinetochore can be replaced with an endogenous centromere in chicken DT40 cells (Hori et al. 2013). These tethering experiments (Fig. 3b), combined with the centromere removal experiments (Fig. 3a) provide evidence that CENP-A is an epigenetic mark for centromere specification independent of DNA sequence. Similarly, ectopic *Drosophila* CENP-A, CID, localization to non-centromeric region via overexpression or using the *lacO*/LacI system induces centromeres with the CENP-A nucleosomes (Heun et al. 2006; Mendiburo et al. 2011).

In human cells, LacI-fused CENP-A array on the *lacO* repeat is likely to induce nucleosomes with the LacI-fused CENP-A into the tethered sites. However, it is inefficient, possibly as a result of an indirect consequence of recruitment of HJURP or other kinetochore proteins onto the high-density LacI-CENP-A array. In fact, overexpression of CENP-A results in CENP-A misincorporation into non-centromere chromosome loci, but does not cause ectopic centromere formation in human cells (Van Hooser et al. 2001; Gascoigne et al. 2011). This suggests that additional regulation may be involved in the formation of active centromeres in CENP-A-incorporated chromatin. Indeed, we recently demonstrated that histone H4 Lys20 is mono-methylated specifically in the CENP-A nucleosome in the centromeric region and that this methylation is essential for kinetochore assembly (Hori et al. 2014). It is possible that additional modifications occur in CENP-A-containing chromatin (Blower et al. 2002; Sullivan and Karpen 2004; Bergmann et al. 2011; Bailey et al. 2016; Shang et al. 2016), and that the combination of such modifications would function as additional epigenetic marks for centromere specification and kinetochore assembly.

3 CCAN Organization

Once a centromere is specified by CENP-A, additional proteins are assembled on a centromere region to form a functional kinetochore. Extensive genetic and biochemical approaches have identified kinetochore proteins in vertebrate cells. Among them, CCAN proteins are constitutively localized to centromeres throughout the cell cycle (Cheeseman and Desai 2008; Perpelescu and Fukagawa

2011; Nagpal and Fukagawa 2016). The CCAN is composed of at least 16 proteins in vertebrates (CENP-C, -H, -I, -K, -L, -M, -N, -O, -P, -Q, -R, -S, -T, -U, -W, -X). These proteins make subfunctional complexes, which interact together to form the entire CCAN assembly (Nagpal and Fukagawa 2016). Since the CCAN is bound to the CENP-A chromatin and the KMN network is bound to the spindle microtubules (Pesenti et al. 2016; Nagpal and Fukagawa 2016), the CCAN forms a base of the kinetochore to link between the centromere and microtubules.

3.1 CCAN Subcomplexes

3.1.1 CENP-C

Like CENP-A, CENP-C was originally identified as an antigen for the autoimmune sera from patients with CREST syndrome (Earnshaw and Rothfield 1985). Using electron microscopy, CENP-C was the first protein found to localize in the inner kinetochore (Saitoh et al. 1992). CENP-C homologs are found in most organisms, including the budding yeast, and their depletion causes chromosome misalignment (Saitoh et al. 1992; Brown et al. 1993; Tomkiel et al. 1994; Brown 1995; Meluh and Koshland 1995; Fukagawa and Brown 1997; Fukagawa et al. 1999; Moore and Roth 2001; Holland et al. 2005; Heeger et al. 2005). Although entire sequence homology between yeast and vertebrate CENP-Cs is not high, several domains, which are predicted to be functionally important domains, are highly conserved (Brown 1995; Meluh and Koshland 1995; Fukagawa and Brown 1997; Moore and Roth 2001; Holland et al. 2005; Heeger et al. 2005; Milks et al. 2009; Carroll et al. 2010; Przewloka et al. 2011; Screpanti et al. 2011; Kato et al. 2013; Klare et al. 2015; Nagpal et al. 2015). The N-terminal region of CENP-C is one such conserved domain. It binds to the Mis12 complex, which is a member of the KMN network (see Fig. 5) (Przewloka et al. 2011; Screpanti et al. 2011). Ectopic targeting of the CENP-C N-terminus, including the Mis12 complex-binding domain into a non-centromere locus, induces formation of the functional kinetochore, which does not contain other CCAN proteins (see Fig. 5b, c) (Hori et al. 2013). The KMN network directly binds to microtubules (Cheeseman and Desai 2008); therefore, CENP-C connects with microtubules through association with the KMN network. The central domain and the C-terminal region of CENP-C, which are also conserved, have been shown to associate with the CENP-A nucleosomes (see Fig. 5) (Carroll et al. 2010; Kato et al. 2013; Falk et al. 2015, 2016). This interaction reshapes the CENP-A nucleosome structure and results in stabilization within the centromere (Falk et al. 2015, 2016). These conserved domains make CENP-C an important bridge molecule between the centromeric chromatin and the microtubules. Another functional domain, which is also conserved among vertebrates, is found in the middle region (also known as PEST rich domain) of CENP-C (Klare et al. 2015; Nagpal et al. 2015). This domain is critical for interaction with other CCAN proteins such as CENP-H and CENP-L/N (Klare et al. 2015;

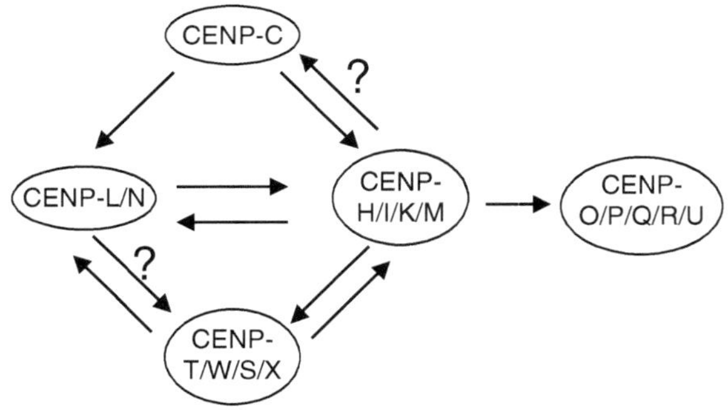

Fig. 4 An interaction map among the CCAN subcomplexes in vertebrates. The *arrows* indicate dependency for centromere localization among the CCAN subcomplexes in vertebrate cells

Nagpal et al. 2015; McKinley et al. 2015). CENP-C is required for localization of other CCAN members into the centromere in human cells (Klare et al. 2015; McKinley et al. 2015). This leads to the idea that CENP-C is a blueprint for kinetochore assembly (Klare et al. 2015). However, CENP-C does not simply stand at the hierarchical top for kinetochore assembly in chicken DT40 cells, because CENP-C depletion does not result in the complete loss of other CCAN proteins (Fukagawa et al. 2001; Kwon et al. 2007; Hori et al. 2008a). In addition, our data suggest that interaction between CENP-C and CENP-H and the CENP-L/N is not a simple linear pathway (Fukagawa et al. 2001; Kwon et al. 2007; Nagpal et al. 2015); rather, this appears to be a complex and dynamic process (see Sect. 4). This idea is also supported by a recent study in human cells (McKinley et al. 2015). Therefore, although CENP-C plays a central role in CCAN organization, we believe that CCAN is formed through a complex process (Fig. 4).

3.1.2 CENP-H/I/K/M

CENP-H has been identified as a coiled-coil protein, which has been shown to constitutively localize to the centromere in mouse cells (Sugata et al. 1999) and thereafter human and chicken homologs were identified (Sugata et al. 2000; Fukagawa et al. 2001). CENP-I was cloned from chicken cells using a homologous region of fission yeast centromere protein Mis6 (Saitoh et al. 1997), and the interaction between CENP-H and CENP-I was shown by a two-hybrid assay (Nishihashi et al. 2002).

We performed immunoprecipitation experiments with CENP-H and CENP-I to identify additional centromere-associated proteins and found CENP-K and -M in chicken and human cells (Okada et al. 2006). CENP-H, -I, -K, and -M were also identified as CENP-A chromatin-associated proteins in human cells (Obuse et al. 2004; Foltz et al. 2006). CENP-M is proposed to be a pseudo GTPase, based on crystal structure analysis (Basilico et al. 2014), but its detailed function in centromeres remains unknown.

Biochemical and functional experiments suggest that CENP-H, -I, -K and -M form a complex, which is essential for chromosomal alignment and segregation, as well as cell viability (Nishihashi et al. 2002; Foltz et al. 2006; Okada et al. 2006; Izuta et al. 2006; Basilico et al. 2014). Depletion of each factor disrupts the centromere localization of other factors in the complex, suggesting that their localization is interdependent (Nishihashi et al. 2002; Foltz et al. 2006; Okada et al. 2006; Izuta et al. 2006; Basilico et al. 2014; McKinley et al. 2015). Consistent with this observation, these proteins interact together to form a complex in vivo, which can be reconstituted with recombinant proteins (Nishihashi et al. 2002; Foltz et al. 2006; Okada et al. 2006; Izuta et al. 2006; Basilico et al. 2014; McKinley et al. 2015). The CENP-H/I/K/M complex interacts with other subcomplexes of the CCAN (Basilico et al. 2014; McKinley et al. 2015; Weir et al. 2016); therefore, one of its functions could be maintenance of CCAN integrity (Fig. 4).

3.1.3 CENP-L/N

CENP-L and -N were also identified as interacting proteins of CENP-H and -I in chicken cells (Okada et al. 2006) and as CENP-A chromatin-associated proteins in human cells (Obuse et al. 2004; Foltz et al. 2006). Depletion of these proteins results in chromosome misalignment (Foltz et al. 2006; Okada et al. 2006; Izuta et al. 2006; McClelland et al. 2007). The CENP-N N-terminus directly binds to the CENP-A nucleosome (Carroll et al. 2009; McKinley et al. 2015). CENP-L forms a heterodimer with CENP-N through association with the C-terminus of CENP-N (Carroll et al. 2009; Nagpal et al. 2015) and the recombinant CENP-L/N dimer interacts with the CENP-H/I/K/M subcomplex in vitro (McKinley et al. 2015; Weir et al. 2016). The CENP-N needs to interact with other CCAN members to be localized to the centromere in vivo, despite its direct interaction with the CENP-A nucleosomes in vitro (Carroll et al. 2009).

3.1.4 CENP-O/P/Q/R/U

Extensive proteomic analysis identified additional CCAN members. The members of the CENP-O/P/Q/R/U subcomplex were co-purified with CENP-H and -I or the CENP-A chromatin (Obuse et al. 2004; Foltz et al. 2006; Okada et al. 2006). CENP-U was originally identified as a constitutive centromere protein called CENP-50 (Minoshima et al. 2005). In contrast to other CCAN subcomplexes, the members of the CENP-O/P/Q/R/U are not essential for cell viability in chicken DT40 cells, although disruption of the members (except for CENP-R) causes cell cycle delay and deficiency in recovery from spindle damages (Hori et al. 2008b). CENP-R appears to be downstream of other members, because centromere localization of those proteins is independent of CENP-R (Hori et al. 2008b). Nonetheless, in vitro biochemical experiments clearly show that these five proteins form a complex (Hori et al. 2008b; McKinley et al. 2015). The members of this subcomplex (except for CENP-R) are conserved as the COMA (Ctf19-Okp1-Mcm21-Ame1) complex in the budding yeast (De Wulf et al. 2003). In contrast to the vertebrate cells, the COMA complex is essential in the budding yeast (De Wulf et al. 2003).

Depletion of the members of the CENP-O/P/Q/R/U does not dramatically change the centromere localization of other CCAN proteins in vertebrate cells, as those cells are viable; however, some proteins are subtly reduced in cells with depletion of the CENP-O/P/Q/R/U complex (Hori et al. 2008b). In contrast, centromere localization of CENP-O/P/Q/R/U is completely abolished in CENP-H- or CENP-I-knockout cells, indicating that centromere recruitment of the CENP-O/P/Q/R/U depends on CENP-H and -I (Okada et al. 2006; Izuta et al. 2006; Hori et al. 2008b). Although it has been shown that CENP-Q and -U directly interact with microtubules (Amaro et al. 2010; Hua et al. 2011), the significance of this interaction and the functional role of the subcomplex for kinetochore assembly is largely unknown.

3.1.5 CENP-T/S/W/X

CENP-T was originally identified as a protein that interacts with CENP-A chromatin, whereas CENP-S was found as a CENP-M- and -U- associated protein (Foltz et al. 2006; Izuta et al. 2006). Subsequent immunopurification of CENP-T and -S identified CENP-W and -X as their binding proteins, respectively (Hori et al. 2008a; Amano et al. 2009). Although CENP-S and -X constitutively localize to centromeres throughout the cell cycle (Amano et al. 2009), they also interact with Fanconi Anemia M, which is involved in the response to and repair of DNA damage (Singh et al. 2010; Yan et al. 2010). CENP-T, -W, -S and -X have a histone fold domain (HFD) (Hori et al. 2008a; Nishino et al. 2012), which distinguishes them from other CCAN members. Biochemical and structural analysis revealed that CENP-T and -W form a heterodimer by binding with their HFD (Nishino et al. 2012). The CENP-T/W heterodimer has DNA-binding activity, which requires their HFDs in vitro (Nishino et al. 2012; Takeuchi et al. 2014). This DNA-binding ability is essential for the CENP-T/W to localize to the centromere (Nishino et al. 2012; Takeuchi et al. 2014).

CENP-S and -X form a heterotetramer (Nishino et al. 2012). Strikingly, when CENP-T/W heterodimer and CENP-S/X heterotetramer are mixed, a dimer part of CENP-S/X is replaced with the CENP-T/W heterodimer to form the CENP-T/W/S/X heterotetramer, which has a nucleosome-like structure (Nishino et al. 2012). The nucleosome-like complex binds to 80–100 bp of DNA and introduces positive supercoils into DNA in vitro, whereas canonical histone nucleosome induces negative supercoils (Takeuchi et al. 2014). These data suggest that the DNA-binding of the CENP-T/W/S/X complex might contribute to a distinct feature of centromeric chromatin. Centromere localization of the CENP-T/W depends on the CENP-H/I/K/M subcomplex proteins in human cells (Basilico et al. 2014; McKinley et al. 2015), although CENP-T/W is localized in CENP-H-depleted chicken DT40 cells despite a reduction in their levels (Hori et al. 2008a).

These findings suggest that the CENP-H/I/K/M is likely to support centromere localization of the CENP-T/W/S/X subcomplex in addition to its direct DNA-binding activity.

Although the CENP-T/W and the CENP-S/X form the complex, the CENP-T/W is essential for cell viability, whereas the CENP-S/X is not (Hori et al. 2008a; Amano et al. 2009; Nishino et al. 2012). CENP-S- or -X-deficient cells show mild mitotic defects as observed in knockout cells for CENP-O complex proteins (Amano et al. 2009). This suggests that, although the CENP-T/W can be recruited onto the centromere without the CENP-S/X to form a scaffold in the kinetochore, the CENP-S/X is required for proper kinetochore formation and to support proper mitotic progression (Amano et al. 2009).

CENP-W, -S, and -X are small proteins (~ 100 aa) and entire regions of these proteins make up the HFD. In contrast, CENP-T has the HFD in its C-terminus, but contains a long N-terminal region (~ 500 aa). Interestingly, the CENP-T N-terminus directly binds to the Ndc80 complex, which is a key microtubule-binding complex, through phosphorylation of CENP-T (see Fig. 5) (Gascoigne et al. 2011; Nishino et al. 2013; Rago et al. 2015). Since CENP-T associates with centromeric DNA and the Ndc80 complex, which binds to the microtubules, in its C- and N-termini, respectively, CENP-T would function as a bridge between chromatin and microtubules (Gascoigne et al. 2011; Nishino et al. 2012, 2013; Rago et al. 2015). Recent studies showed that the budding yeast CENP-T homolog Cnn1 has similar functions (Bock et al. 2012; Schleiffer et al. 2012).

As discussed below, ectopic localization of the CENP-T N-terminus into a non-centromere locus recruits the Ndc80 complex and induces a functional artificial kinetochore on the targeting site in chicken DT40 cells (Hori et al. 2013). Interestingly, other CCAN proteins, including CENP-C, are not detected in the CENP-T-derived artificial kinetochore (Hori et al. 2013). Considering that the CENP-C N-terminus, which recruits the Ndc80 complex through interaction with the Mis12 complex, also establishes a functional kinetochore without other CCAN proteins on it (Hori et al. 2013), CENP-T would make a pathway to recruit the Ndc80 complex independently of CENP-C in the CCAN. This implies the existence of two independent parallel pathways to recruit the Ndc80 complex to kinetochores (Hori et al. 2013; Nishino et al. 2013; Kim and Yu 2015; Rago et al. 2015).

3.2 CCAN Organization and Functions

3.2.1 CCAN Interaction

The 16 CCAN proteins are assembled on the centromere, where they interact together in both inter and intra subcomplexes (Fig. 4). The complicated features of their interaction network and their interdependency for centromere localization was characterized via extensive knockdown and conditional knockout studies

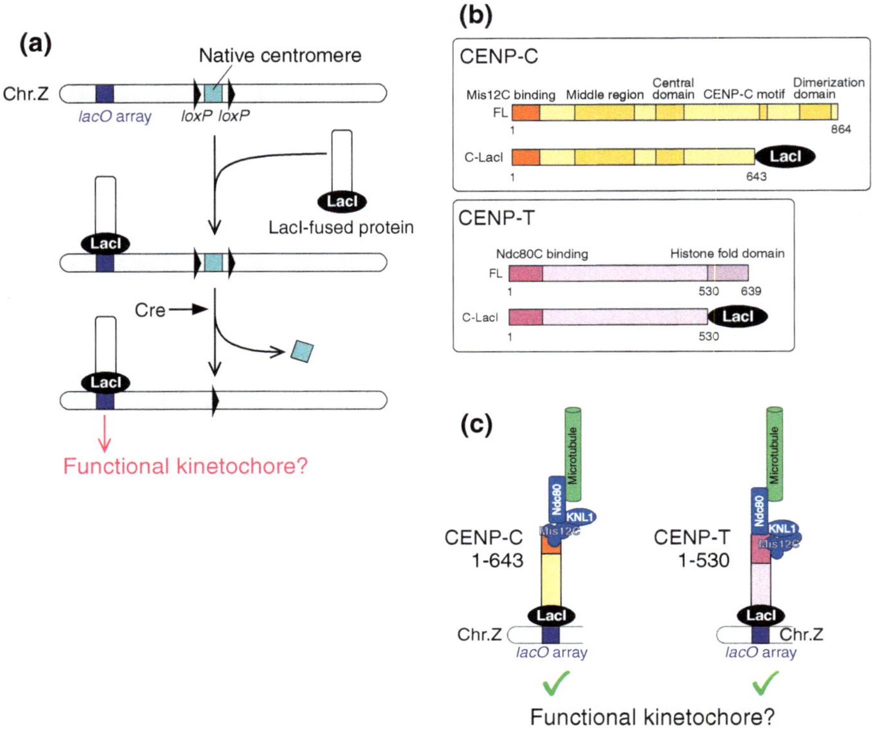

Fig. 5 An experimental design for generation of a functional artificial kinetochore. **a** A schematic representation of the experimental procedure in chicken DT40 cells. A protein of interest fused with LacI is tethered onto the *lacO* array inserted in a non-centromeric region of a Z chromosome. The native centromere flanked by *loxP* sites is removed by Cre recombinase. Functionality of the artificial kinetochore can be tested by cell viability. **b** Schematic representation of the domain organization of CENP-C and -T. **c** CENP-C N-terminal fragment including the Mis12 complex (Mis12C)-binding domain and the conserved middle region recruits the KMN (KNL1, Mis12, and Ndc80 complexes) network and induces functional kinetochore formation on a non-centromeric region (*left*). CENP-T N-terminus including Ndc80 complex (Ndc80C)-binding domain but not histone fold domain also induces functional kinetochore formation (*right*). Given the fragments can establish the functional kinetochores without other CCAN members, each N-terminus of CENP-C and -T is sufficient to form a functional kinetochore. This suggests that the CCAN could set two independent-pathways to recruit the KMN network, which binds to microtubules, in the kinetochore

(Fukagawa et al. 2001; Foltz et al. 2006; Okada et al. 2006; Izuta et al. 2006; Kwon et al. 2007; Hori et al. 2008a, b; Amano et al. 2009; Basilico et al. 2014; Klare et al. 2015; McKinley et al. 2015). In addition, biochemical reconstitution of the CCAN with recombinant proteins complements our understanding of the molecular interaction in the CCAN (Nishino et al. 2012, 2013; Basilico et al. 2014; Klare et al. 2015; Nagpal et al. 2015; McKinley et al. 2015; Weir et al. 2016). As described above, CENP-C is a key factor for the CCAN localization to the centromere, because CENP-C has multiple functional domains (see Fig. 5b) (Milks et al. 2009; Carroll et al. 2010; Przewloka et al. 2011; Screpanti et al. 2011; Hori et al. 2013; Kato et al. 2013; Klare et al. 2015; Nagpal et al. 2015), which include the KMN network-binding, the DNA-binding, and the CENP-A nucleosome binding domains, as well as domains to interact with other CCAN subunits.

Centromere localization of most CCAN proteins depends on CENP-C in human cells (Klare et al. 2015; McKinley et al. 2015). CENP-C binds to the CENP-H/I/K/M and the CENP-L/N through its middle region (Klare et al. 2015; McKinley et al. 2015; Weir et al. 2016). Although CENP-C is required for centromeric localization of the CENP-T/W/S/X in human cells (Klare et al. 2015; McKinley et al. 2015), the interaction is indirect through the CENP-H/I/K/M sub-complex, because the CENP-T/W/S/X does not directly bind to CENP-C in vitro (McKinley et al. 2015) and the CENP-T/W requires the CENP-H/I/K/M for its centromeric localization (Basilico et al. 2014; McKinley et al. 2015). In addition to playing an important role in CCAN organization, CENP- C binds both the KMN network and the CENP-A nucleosome; therefore, CENP-C is proposed to be a central component of the kinetochore assembly (Klare et al. 2015; Weir et al. 2016).

However, when CENP-C was conditionally knocked-out in chicken DT40 cells, the CENP-H/I/K/M and -T/W/S/X remained on the kinetochores despite a slight reduction of their levels (Hori et al. 2008a). In addition, dependency of CENP-C localization on the CENP-H/I/K/M varies between interphase and M-phase in both human cells and chicken DT40 cells (Fukagawa et al. 2001; Kwon et al. 2007; Nagpal et al. 2015; McKinley et al. 2015) (see in Sect. 4). This suggests that the CCAN assembly is not mediated by a simple linear pathway, but rather a complicated meshwork among the subcomplexes with multiple binding interfaces (Fig. 4).

3.2.2 The CCAN as a Bridge Between Centromere and Microtubule

The kinetochore is assembled on the centromeric chromatin. One of the key functions of CCAN is generation of a basement to build the kinetochore on the CENP-A-containing chromatin. CENP-C and -N have been shown to directly interact with the CENP-A nucleosome (Carroll et al. 2009, 2010; Kato et al. 2013; Nagpal et al. 2015; McKinley et al. 2015). These bindings to CENP-A nucleosomes might trigger the formation of CCAN on the centromeric chromatin (Fig. 6). Although centromeric localization of the CENP-T/W/S/X depends on other CCAN subunits (Hori et al. 2008a; Basilico et al. 2014; McKinley et al. 2015), it has its own DNA-binding activity through their HFD, which is essential for its centromere localization (Hori et al. 2008a; Nishino et al. 2012). Thus, the CENP-T/W/S/X might recognize a specific structure of centromeric chromatin attributed to the CENP-A nucleosomes possibly with CENP-C and -N binding (Fig. 6). Once the CENP-T/W/S/X is targeted into centromeric chromatin, it could contribute to formation of its specific features. Understanding how the CENP-T/W/S/X complex recognizes the centromere to localize there specifically is an important issue.

While a recent study suggested that FACT, a histone chaperone, might bring the CENP–T/W/S/X to the centromere in human cells (Prendergast et al. 2016), further studies are needed to clarify this issue.

The CCAN also functions as a platform to recruit the KMN network, which is directly bound to microtubules, onto the kinetochore (Fig. 6) (Cheeseman and Desai 2008; Przewloka and Glover 2009; Varma and Salmon 2012; Foley and Kapoor 2013). CENP-C is one of the scaffolds in the CCAN, because it recruits the Mis12 complex, a subcomplex of the KMN network, onto the CCAN via the conserved N-terminus of CENP-C (Fig. 6) (Screpanti et al. 2011; Hori et al. 2013; Kim and Yu 2015; Rago et al. 2015). Direct interaction of the Mis12 complex with the conserved N-terminus has been demonstrated with biochemical reconstitution and structural analysis (Przewloka et al. 2011; Screpanti et al. 2011; Petrovic et al. 2016).

CENP-T also provides another scaffold for microtubule-binding (Fig. 6). The disordered N-terminus of CENP-T directly binds the Ndc80 complex (Gascoigne et al. 2011; Hori et al. 2013; Nishino et al. 2013; Rago et al. 2015). Structural studies revealed that phosphorylation on the CENP-T N-terminus by Cdk1 stabilized its binding to the Ndc80 complex (Nishino et al. 2013). When the Ndc80 complex-binding domain in CENP-T is disrupted, Ndc80 recruitment is prevented,

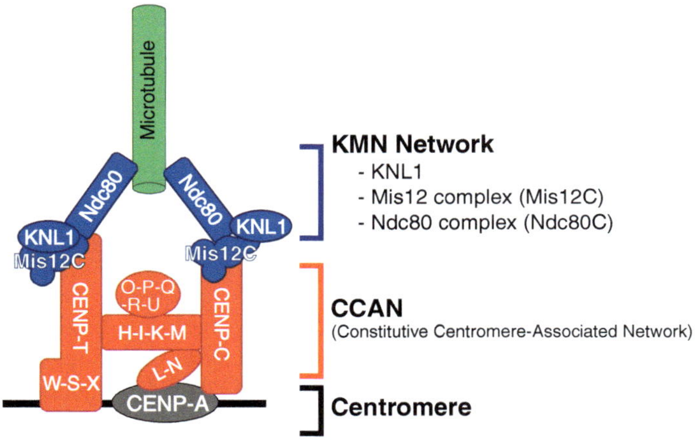

Fig. 6 An organization of the vertebrate kinetochore. The CCAN constitutively localizes to the centromere throughout the cell cycle and functions as a foundation to create a linkage between the centromere and microtubules in the kinetochore. CENP-C and the CENP-L/N subcomplex make direct interactions with the CENP-A nucleosomes. The CENP-T/W/S/X subcomplex has DNA-binding activity via its histone fold domains that form a nucleosome-like structure; therefore, the CCAN would interact with the centromere through three interfaces, CENP-C, the CENP-L/N and the CENP-T/W/S/X. In M-phase, the KMN network, which binds to microtubules, is recruited onto the CCAN. CENP-C and -T recruit the KMN network via their N-termini, independently. This implies existence of two pathways from the centromere to the microtubules, because both CENP-C and -T would be bound to the centromere through direct binding to the CENP-A nucleosome and DNA, respectively

resulting in cell death (Nishino et al. 2013). Together with the CENP-A nucleosomes and DNA-binding ability of CENP-C and -T, respectively, these results indicate that CENP-T together with CENP-C create the Ndc80 complex-binding platforms to bridge the centromere and the microtubules (Fig. 6).

When the CENP-C N-terminus and the CENP-T N-terminus are tethered onto a non-centromeric region using the *lacO*/LacI system, both fragments recruit the KMN network and induce the functional artificial kinetochores (Gascoigne et al. 2011; Hori et al. 2013). Importantly, both the CENP-C- and the CENP-T-derived kinetochores contain none of the other CCAN members, nor CENP-A (Hori et al. 2013). This indicates that CENP-C and -T have the ability to make a functional kinetochore independently. Presumably, the two pathways could form an independent parallel pathway to recruit the Ndc80 complex on the CCAN in the native kinetochore (Hori et al. 2013; Nishino et al. 2013; Kim and Yu 2015; Rago et al. 2015). If so, other challenging questions arise: why do the two pathways to recruit the Ndc80 complex exist in the native kinetochore? How do the two pathways coordinate with each other in the kinetochore assembly?

3.2.3 In Vitro Reconstitution of the CCAN

Recently, all subcomplexes of the vertebrate CCAN and the KMN network have been reconstituted with recombinant proteins (Hori et al. 2008b; Screpanti et al. 2011; Nishino et al. 2012, 2013; Basilico et al. 2014; Nagpal et al. 2015; McKinley et al. 2015; Weir et al. 2016). Using these as building blocks, interaction of the CCAN and the KMN subcomplexes was assembled in vitro (Weir et al. 2016). Combined with cross link mass spectrometry analysis (Basilico et al. 2014; Weir et al. 2016), the complicated meshwork of the CCAN and the KMN network that has multiple binding interfaces among the subunits have been revealed.

The most recent efforts have successfully assembled the large part of the CCAN in vitro; this includes CENP-C and the CENP-H/I/KM and the CENP-L/N subcomplexes (Weir et al. 2016). This reconstituted CCAN is bound to the CENP-A nucleosome through CENP-C and -N (Weir et al. 2016), as previously shown by genetic assay and in vitro binding studies (Carroll et al. 2009, 2010; Kato et al. 2013; Nagpal et al. 2015; McKinley et al. 2015). Since CENP-C has the KMN-binding activity in its N-terminus, the reconstituted CCAN binds the KMN network and established microtubule-binding via the KMN network (Weir et al. 2016). The in vitro assembly of linkage from the CENP-A nucleosome to microtubules through the CCAN and the KMN network is a seminal work for reconstitution of the functional vertebrate kinetochores in the future.

The current reconstituted kinetochore misses another essential subcomplex, the CENP-T/W/S/X. The structural analyses and ectopic tethering experiments in various systems showed that the CENP-T/W/S/X is also directly bound to the Ndc80 complex to establish microtubule-binding on the kinetochore as well as DNA-binding activity through the HFD (Hori et al. 2008a, 2013; Gascoigne et al. 2011; Nishino et al. 2012, 2013); therefore, to understand the entire kinetochore structure, it is interesting to include the CENP-T/W/S/X in the reconstituted complexes. The reconstituted whole kinetochore would allow us to understand how the CCAN utilizes the two pathways from the centromere to the microtubules. In fact, a functional kinetochore, which is assembled onto the point kinetochore, has been purified from *S. cerevisiae* (Akiyoshi et al. 2010). The purified kinetochore includes the CCAN and the KMN network as well as other kinetochore factors (Akiyoshi et al. 2010). The biophysics studies with the purified yeast kinetochores provided a great deal of mechanistic insight into the kinetochore regulation (Akiyoshi et al. 2010; Miller et al. 2016). Future studies with the reconstituted vertebrate kinetochore will unveil new molecular insight into functions of the kinetochore assembled on the regional centromere.

3.2.4 The CCAN-Dependent Stabilization of Centromere Position

It has been shown that the centromeric CENP-A chromatin can move its location in horse fibroblasts (Purgato et al. 2015). Recently, we demonstrated a new role of CCAN in the stabilization of centromere position through centromere movement suppression (Fig. 7) (Hori et al. 2017). One of unique features of chicken DT40 cells is that they have non-repetitive centromeres in chromosomes 5, 27, and Z (Shang et al. 2010). This attribute allows us to examine precise centromere position and size in the chromosomes by chromatin immunoprecipitation using anti-CENP-A antibody combined with deep-sequencing (ChIP-Seq). The extensive CENP-A ChIP-Seq from DT40 cells found evidence that centromere position on the chromosomes could move during many passages (Hori et al. 2017). Centromere size of the Z chromosome from a laboratory stock was about 50 kbp (Hori et al. 2017). In contrast, isolated clones from the laboratory stock as a parental line had smaller centromeres of about 30 kbp on the chromosome Z. This suggests that the

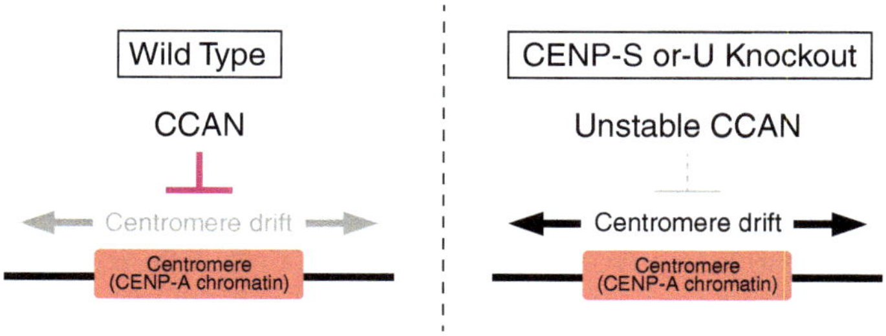

Fig. 7 A model for stabilization of centromere position by the complete CCAN structure. ChIP-Seq analyses with anti-CENP-A antibody using DT40 cells have demonstrated that the centromeres (chromatin region associated with the CENP-A nucleosomes) are mobile, although the centromeres scarcely move in the wild type cells. In contrast, conditional knockout of CENP-S and -U, which are the nonessential CCAN subunits in DT40 cells, increases frequency of the centromere drift due to incomplete structure of CCAN in these knockout cells. The complete CCAN organization stabilizes the centromere position

50 kbp centromere size in the parental cell line is made from a mixture of cells with various centromere positions (Hori et al. 2017). Interestingly, although the repetitive DNA in the centromere is thought to contribute to centromere position specification, the centromere drift was also found in a repetitive centromere on chromosome 1 (Hori et al. 2017).

However, as seen with the smaller sized centromere in the freshly isolated lines, the centromere barely moves during the relative short-term culture, which is at least 2–3 weeks (Hori et al. 2017). Strikingly, the frequent centromere drift was found in knockout of the nonessential CCAN proteins, CENP-U or -S (Hori et al. 2008b; Amano et al. 2009), even during the short-term culture, suggesting that depletion of those CENPs increased mobility of the centromere (Fig. 7) (Hori et al. 2017). Considering that depletion of CENP-U or -S destabilized interaction among other CCAN proteins in the kinetochore, the intact CCAN organization may be required for stable centromere positioning (Fig. 7) (Hori et al. 2017). In other words, when cells exhibit stress or damage in their CCAN, centromere position moves during the cell cycle, presumably resulting in deleterious effects. This may occur because the centromere and kinetochore formation could be affected by the chromatin environment, such as histone modifications and transcription levels (Bergmann et al. 2012), and transcription is suppressed in a neocentromere in chicken DT40 cells (Shang et al. 2013).

4 Dynamic Rearrangement of CCAN Organization

4.1 Reorganization of the CCAN During the Cell Cycle

Given that the CCAN is constitutively localized to the centromere throughout the cell cycle and that the recombinant CCAN subcomplexes are successfully reconstituted, one might think the CCAN could form stable interaction networks on the centromere. However, this is not the case. The CCAN interaction appears to be dynamically reorganized along the cell cycle progression (Fukagawa et al. 2001; Kwon et al. 2007; Nagpal et al. 2015; McKinley et al. 2015).

It has been demonstrated that CENP-C changes dependency on CENP-H/K for its kinetochore localization during the cell cycle in chicken DT40 cells (Fukagawa et al. 2001; Kwon et al. 2007). Although recombinant CENP-C interacts with the CENP-H/I/K/M subcomplex in vitro (McKinley et al. 2015; Weir et al. 2016), CENP-C remains bound to the mitotic centromere in CENP-H- or -K-deficient chicken DT40 cells (Kwon et al. 2007; Hori et al. 2008a), suggesting that CENP-C binds to the centromere independently of the CENP-H/I/K/M in M-phase. However, when CENP-H or -K is depleted, CENP-C is dissociated from the centromere in interphase cells (Fukagawa et al. 2001; Kwon et al. 2007). In contrast, CENP-H localization is not completely abolished in CENP-C-deficient DT40 cells (Fukagawa et al. 2001; Hori et al. 2008a). Similar CENP-C regulation is also found in human cells (McKinley et al. 2015) using the auxin-inducible degron system by which a target protein can be rapidly degraded after auxin treatment (Nishimura et al. 2009). CENP-C localization relies on CENP-I, a member of the CENP-H/I/K/M as well as CENP-L in interphase cells (McKinley et al. 2015). However, when cells enter into M-phase, CENP-C binds to the kinetochore independently of CENP-I and -L (McKinley et al. 2015). It is worth mentioning that CENP-C is required for centromeric localization of all other subcomplexes of the CCAN in human mitotic cells (McKinley et al. 2015).

The changes in localization dependency of CENP-C on the CENP-H/I/K/M and CENP-L/N during cell cycle progression suggest that CENP-C could alter its binding partners in the CCAN (Fig. 8). Indeed, when the CENP-C C-terminus, which includes the conserved CENP-C motif and the dimerization domain (Fig. 5b), was expressed in CENP-C-deficient DT40 cells, the fragment was targeted onto the kinetochore in M-phase but not in interphase cells (Nagpal et al. 2015). In contrast, a CENP-C N-terminal fragment, which contained the conserved middle region (Fig. 5b), was localized to the kinetochore restrictively in interphase but not in M-phase (Nagpal et al. 2015). Biochemical studies showed that the middle region and the C-terminal fragment of CENP-C are bound to the CENP-H/I/K/M and -L/N subcomplexes and the CENP-A nucleosomes, respectively (Kato et al. 2013; Klare et al. 2015; Nagpal et al. 2015; McKinley et al. 2015). These observations suggest that CENP-C would change its major interacting domains with the kinetochore between interphase and M-phase (Fig. 8). Since CENP-C has multiple functional domains (Brown 1995; Meluh and Koshland 1995; Fukagawa and Brown 1997; Moore and Roth 2001; Holland et al. 2005; Heeger et al. 2005; Milks et al. 2009; Carroll et al. 2010; Przewloka et al. 2011; Screpanti et al. 2011; Kato et al. 2013; Klare et al. 2015; Nagpal et al. 2015), another domain might also contribute to CENP-C-targeting onto the kinetochore. Nevertheless, CENP-C could dynamically alter its kinetochore-binding mode via cell cycle-dependent regulations (Fig. 8) (Nagpal et al. 2015; Nagpal and Fukagawa 2016).

Although the significance of cell cycle-dependent CCAN reorganization has not been elucidated yet, it might be related to fine localization of the CCAN subunits in kinetochores. In fact, electron microscopy observation suggests that the CCAN subunits change their distribution in the kinetochore when tension is applied to the kinetochore from microtubules during M-phase (Suzuki et al. 2011). Consistent with this observation, a super-resolution microscopy analysis also shows that distance between CENP-A and the KMN network is stretched when the tension is applied during M-phase (Wan et al. 2009), suggesting that in order to resist the tension generated from microtubules, the CCAN might form an elastic and strong interaction meshwork (Suzuki et al. 2014). The structural rearrangement of the CCAN could be important for proper M-phase progression, because the rearrangement in the kinetochore is proposed to play a role in silencing of the spindle assembly checkpoint (Maresca and Salmon 2009; Uchida et al. 2009). Following M-phase completion, newly synthesized CENP-A is deposited onto the centromere during early G1 phase (Jansen et al. 2007; Dunleavy et al. 2009; Foltz et al. 2009).

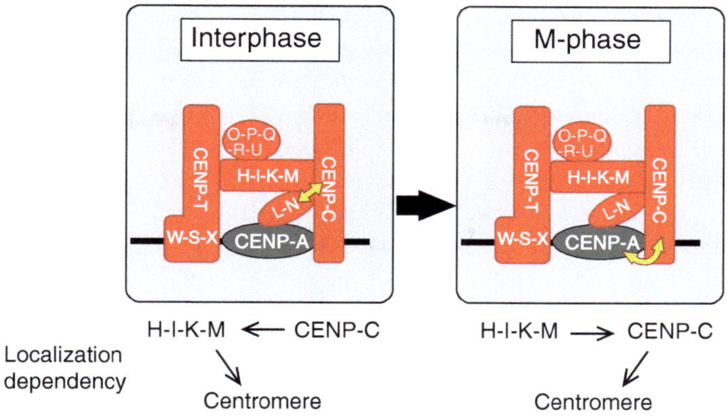

Fig. 8 A model for reorganization of the CCAN during the cell cycle progression. The CCAN constitutively localizes to the centromere during the cell cycle. However, interaction network among its subunits appears to be dynamically reorganized along cell cycle progression. In interphase, centromeric localization of CENP-C depends on the CENP-L/N and -H/I/K/M subcomplexes. In contrast, when cells enter into M-phase, CENP-C is localized to the centromere independently of the CENP-L/N and -H/I/K/M subcomplexes probably through its direct binding to the CENP-A nucleosome. This suggests that CENP-C switches its major interaction interfaces during the cell cycle

It is also suggested that new CCAN proteins are likely to be recruited during interphase (Hemmerich et al. 2008; Prendergast et al. 2011). Studies using fluorescence recovery after photobleaching assays suggest that the CCAN subunits are dynamically incorporated into the centromere during interphase but not during M-phase (Hemmerich et al. 2008; Prendergast et al. 2011). The CCAN rearrangement in interphase might convert the CCAN meshwork to a more open structure to facilitate the targeting of the new centromere protein to the kinetochore.

4.2 *CCAN Organization During Development*

The CCAN displays different regulation and function when viewed from a developmental point of view. The depletion of subunits of the CENP-O/P/Q/R/U sub-complex did not affect cell viability in chicken DT40 cells despite a slight mitotic defect (Hori et al. 2008b). In contrast, CENP-U null mice died during early embryo development (Kagawa et al. 2014). The CENP-$U^{+/-}$ intercross never gave birth to homozygous CENP-$U^{-/-}$ mice, rather the CENP-U null mice died after late gastrulation stage (E7.5) (Kagawa et al. 2014). However, mouse embryonic fibroblast cells isolated from CENP-U null mice were viable with mild mitotic defects as seen in the CENP-U-deficient DT40 cells (Kagawa et al. 2014). Interestingly, when CENP-U was depleted in mouse embryonic stem (mES) cells, the mES cells died after showing chromosome segregation errors (Kagawa et al. 2014).

Despite the cell-proliferation defects in the CENP-U null mES, CENP-H localization into the kinetochore was not affected as seen in the CENP-U-deficient DT40 cells (Kagawa et al. 2014). This suggests that although the CCAN composition in mES cells is similar to that in DT 40 cells, the CCAN without CENP-U is unfunctional in the undifferentiated mES cells or early embryonic cells (Hori et al. 2008b; Kagawa et al. 2014).

CENP-U depletion appeared to have an effect on the CCAN meshwork formation and stability of the centromere position in DT40 cells (Hori et al. 2008b, 2017). Spindle assembly checkpoint may be weaker in mES cells than in somatic cells; therefore, mES cells could fail chromosome segregation more often with the impaired CENP-U-deficient CCAN (Kagawa et al. 2014). An alternate possibility is that the difference in pericentromeric heterochromatin (PCH) between differentiated and undifferentiated cells is responsible. It has been shown that a transcription factor, NANOG (Chambers et al. 2003; Mitsui et al. 2003), which is a key pluripotency factor, establishes and maintains open PCH structure in mES cells (Novo et al. 2016). Because PCH is thought to be involved in faithful chromosome segregation and centromere stability (Peters et al. 2001; Yamagishi et al. 2008), open and active forms of PCH combined with the impaired CCAN might cause cell death in undifferentiated cells with depletion of CENP-U.

5 Conclusion

Since establishment of the chromosome theory from studies with various systems including echinoderm embryos, insect germ lines, and plant cells (Benson 2001; Satzinger 2008), our understanding of the molecular mechanisms responsible for the correct segregation of chromosomes to the next generations has been accelerated. These large efforts revealed the basic structure of the CCAN organization. Although reconstitution of the CCAN assembly in vitro is a key step in building a functional recombinant kinetochore for further biochemical and biophysics studies, this structure might show only one aspect of the CCAN structure as a snapshot. Indeed, accumulating data suggest that dynamic remodeling occurs in the CCAN organization during the cell cycle. How the CCAN organization is remodeled and what is the significance of the remodeling are future questions to be addressed combining various experimental approaches including genetics, cell biology, molecular biology, genome science, biochemistry, structural biology, and biophysics.

Most of the CCAN members are found in both yeasts and vertebrates; therefore, the CCAN looks to be a conserved structure among various species (Przewloka and Glover 2009). However, the CCAN organization has been dynamically rewired during evolution. For example, CENP-C is the sole CCAN component in *D. melanogaster* (Drinnenberg et al. 2014, 2016). CENP-C could make a platform of kinetochore assembly in the fly. On the other hand, the silkworm, *Bombyx mori*, lacks in CENP-C (Drinnenberg et al. 2016). Since most of the KMN network subunits exist in *B. mori*, the silkworm should have another scaffold to the KMN other than CENP-C. Moreover, *B. mori* has holocentromeres (Murakami and Imai 1974), a type of centromere, which are extended through entire chromosomes (Dernburg 2001). In fact, the holocentromere is found in many insects or plants as well as *C. elegans* (Albertson and Thomson 1982; Drinnenberg et al. 2014). The CCAN could adapt to be assembled on the holocentric chromatin in those species.

Centromere structures and CCAN organizations are highly diverse among species. Studies on molecular functions and regulatory mechanisms of the kinetochore in various species in addition to the model systems should lead us to a comprehensive understanding of the chromosome segregation, similar to when the chromosome theory was established.

Acknowledgements The authors thank Tetsuya Hori and other members in the Fukagawa Lab for helpful discussions. This work is supported by JSPS KAKENHI Grant Numbers P25221106, JP16H06279, and JP15H05972 to TF and JP16K18491 to MH.

References

Akiyoshi B, Sarangapani KK, Powers AF, Nelson CR, Reichow SL, Arellano-Santoyo H, Gonen T, Ranish JA, Asbury CL, Biggins S (2010) Tension directly stabilizes reconstituted kinetochore-microtubule attachments. Nature 468:576–579. doi:10.1038/nature09594

Albertson DG, Thomson JN (1982) The kinetochores of *Caenorhabditis elegans*. Chromosoma. doi:10.1007/BF00292267

Aldrup-Macdonald ME, Sullivan BA (2014) The past, present, and future of human centromere genomics. Genes (Basel) 5:33–50. doi:10.3390/genes5010033

Amano M, Suzuki A, Hori T, Backer C, Okawa K, Cheeseman IM, Fukagawa T (2009) The CENP-S complex is essential for the stable assembly of outer kinetochore structure. J Cell Biol 186:173–182. doi:10.1083/jcb.200903100

Amaro AC, Samora CP, Holtackers R, Wang E, Kingston IJ, Alonso M, Lampson M, McAinsh AD, Meraldi P (2010) Molecular control of kinetochore-microtubule dynamics and chromosome oscillations. Nat Cell Biol 12:319–329. doi:10.1038/ncb2033

Bailey AO, Panchenko T, Shabanowitz J, Lehman SM, Bai DL, Hunt DF, Black BE, Foltz DR (2016) Identification of the post-translational modifications present in centromeric chromatin. Mol Cell Proteomics 15:918–931. doi:10.1074/mcp.M115.053710

Barnhart MC, Kuich PHJL, Stellfox ME, Ward JA, Bassett EA, Black BE, Foltz DR (2011) HJURP is a CENP-A chromatin assembly factor sufficient to form a functional de novo kinetochore. J Cell Biol 194:229–243. doi:10.1083/jcb.201012017

Basilico F, Maffini S, Weir JR, Prumbaum D, Rojas AM, Zimniak T, De Antoni A, Jeganathan S, Voss B, van Gerwen S, Krenn V, Massimiliano L, Valencia A, Vetter IR, Herzog F, Raunser S, Pasqualato S, Musacchio A (2014) The pseudo GTPase CENP-M drives human kinetochore assembly. Elife 3:e02978. doi:10.7554/eLife.02978

Benson KR (2001) T.H. Morgan's resistance to the chromosome theory. Nat Rev Genet 2:469–474. doi:10.1038/35076532

Bergmann JH, Martins NMC, Larionov V, Masumoto H, Earnshaw WC (2012) HACking the centromere chromatin code: insights from human artificial chromosomes. Chromosome Res 20:505–519. doi:10.1007/s10577-012-9293-0

Bergmann JH, Rodríguez MG, Martins NMC, Kimura H, Kelly DA, Masumoto H, Larionov V, Jansen LET, Earnshaw WC (2011) Epigenetic engineering shows H3K4me2 is required for HJURP targeting and CENP-A assembly on a synthetic human kinetochore. EMBO J 30:328–340. doi:10.1038/emboj.2010.329

Birchler JA (2015) Mendel, mechanism, models, marketing, and more. Cell 163:9–11. doi:10.1016/j.cell.2015.09.008

Blower MD, Karpen GH (2001) The role of Drosophila CID in kinetochore formation, cell-cycle progression and heterochromatin interactions. Nat Cell Biol 3:730–739. doi:10.1038/35087045

Blower MD, Sullivan BA, Karpen GH (2002) Conserved organization of centromeric chromatin in flies and humans. Dev Cell 2:319–330

Bock LJ, Pagliuca C, Kobayashi N, Grove RA, Oku Y, Shrestha K, Alfieri C, Golfieri C, Oldani A, Dal Maschio M, Bermejo R, Hazbun TR, Tanaka TU, De Wulf P (2012) Cnn1 inhibits the interactions between the KMN complexes of the yeast kinetochore. Nat Cell Biol 14:614–624. doi:10.1038/ncb2495

Boveri T (1904) Ergebnisse über die Konstitution der chromatischen Substanz des Zellkerns. Verlag von Gustav Fischer, Jena

Brown MT (1995) Sequence similarities between the yeast chromosome segregation protein Mif2 and the mammalian centromere protein CENP-C. Gene 160:111–116. doi:10.1016/0378-1119(95)00163-Z

Brown MT, Goetsch L, Hartwell LH (1993) MIF2 is required for mitotic spindle integrity during anaphase spindle elongation in *Saccharomyces cerevisiae*. J Cell Biol 123:387–403

Buchwitz BJ, Ahmad K, Moore LL, Roth MB, Henikoff S (1999) A histone-H3-like protein in *C. elegans*. Nature 401:547–548. doi:10.1038/44062

Carroll CW, Milks KJ, Straight AF (2010) Dual recognition of CENP-A nucleosomes is required for centromere assembly. J Cell Biol 189:1143–1155. doi:10.1083/jcb.201001013

Carroll CW, Silva MCC, Godek KM, Jansen LET, Straight AF (2009) Centromere assembly requires the direct recognition of CENP-A nucleosomes by CENP-N. Nat Cell Biol 11:896–902. doi:10.1038/ncb1899

Chambers I, Colby D, Robertson M, Nichols J, Lee S, Tweedie S, Smith A (2003) Functional expression cloning of Nanog, a pluripotency sustaining factor in embryonic stem cells. Cell 113:643–655. doi:10.1016/S0092-8674(03)00392-1

Cheeseman IM, Chappie JS, Wilson-Kubalek EM, Desai A (2006) The conserved KMN network constitutes the core microtubule-binding site of the kinetochore. Cell 127:983–997. doi:10.1016/j.cell.2006.09.039

Cheeseman IM, Desai A (2008) Molecular architecture of the kinetochore-microtubule interface. Nat Rev Mol Cell Biol 9:33–46. doi:10.1038/nrm2310

Chen C-C, Dechassa ML, Bettini E, Ledoux MB, Belisario C, Heun P, Luger K, Mellone BG (2014) CAL1 is the Drosophila CENP-A assembly factor. J Cell Biol 204:313–329. doi:10.1083/jcb.201305036

Clarke L (1998) Centromeres: proteins, protein complexes, and repeated domains at centromeres of simple eukaryotes. Curr Opin Genet Dev 8:212–218. doi:10.1016/S0959-437X(98)80143-3

Correns CE (1900) G. Mendel's Regel über das Verhalten der Nachkommenschaft der Rassenbastarde. Berichte der Deutsche Botanischen 18:158–168. doi:10.1111/j.1438-8677.1900.tb04893.x

Darlington CD (1936) The external mechanics of the chromosomes. Proc R Soc Lond 121:264–319. doi:10.2307/82059

de Vries H (1900) Sur la loi de disjonction des hybrides. Comptes rendus de l'Académie des Sciences 130:845–847

De Wulf P, McAinsh AD, Sorger PK (2003) Hierarchical assembly of the budding yeast kinetochore from multiple subcomplexes. Genes Dev 17:2902–2921. doi:10.1101/gad.1144403

Dernburg AF (2001) Here, there, and everywhere. J Cell Biol 153:F33–F38. doi:10.1083/jcb.153.6.F33

Drinnenberg IA, deYoung D, Henikoff S, Malik HS (2014) Recurrent loss of CenH3 is associated with independent transitions to holocentricity in insects. Elife 3:2104–2119. doi:10.7554/eLife.03676

Drinnenberg IA, Henikoff S, Malik HS (2016) Evolutionary turnover of kinetochore proteins: a Ship of Theseus? Trends Cell Biol 26:498–510. doi:10.1016/j.tcb.2016.01.005

Dunleavy EM, Roche D, Tagami H, Lacoste N, Ray-Gallet D, Nakamura Y, Daigo Y, Nakatani Y, Almouzni-Pettinotti G (2009) HJURP is a cell-cycle-dependent maintenance and deposition factor of CENP-A at centromeres. Cell 137:485–497. doi:10.1016/j.cell.2009.02.040

Earnshaw WC, Rothfield N (1985) Identification of a family of human centromere proteins using autoimmune sera from patients with scleroderma. Chromosoma 91:313–321

Falk SJ, Guo LY, Sekulic N, Smoak EM, Mani T, Logsdon GA, Gupta K, Jansen LET, Van Duyne GD, Vinogradov SA, Lampson MA, Black BE (2015) Chromosomes. CENP-C reshapes and stabilizes CENP-A nucleosomes at the centromere. Science 348:699–703. doi:10.1126/science.1259308

Falk SJ, Lee J, Sekulic N, Sennett MA, Lee T-H, Black BE (2016) CENP-C directs a structural transition of CENP-A nucleosomes mainly through sliding of DNA gyres. Nat Struct Mol Biol 23:204–208. doi:10.1038/nsmb.3175

Flemming W (1882) Zellsubstanz, kern und zelltheilung. F.C.W. Vogel, Leipzig

Foley EA, Kapoor TM (2013) Microtubule attachment and spindle assembly checkpoint signalling at the kinetochore. Nat Rev Mol Cell Biol 14:25–37. doi:10.1038/nrm3494

Foltz DR, Jansen LET, Bailey AO, Yates JR, Bassett EA, Wood S, Black BE, Cleveland DW (2009) Centromere-specific assembly of CENP-a nucleosomes is mediated by HJURP. Cell 137:472–484. doi:10.1016/j.cell.2009.02.039

Foltz DR, Jansen LET, Black BE, Bailey AO, Yates JR, Cleveland DW (2006) The human CENP-A centromeric nucleosome-associated complex. Nat Cell Biol 8:458–469. doi:10.1038/ncb1397

Fukagawa T, Brown W (1997) Efficient conditional mutation of the vertebrate CENP-C gene. Hum Mol Genet. doi:10.1093/hmg/6.13.2301

Fukagawa T, Earnshaw WC (2014a) The centromere: chromatin foundation for the kinetochore machinery. Dev Cell 30:496–508. doi:10.1016/j.devcel.2014.08.016

Fukagawa T, Earnshaw WC (2014b) Neocentromeres. Curr Biol 24:R946–R947. doi:10.1016/j.cub.2014.08.032

Fukagawa T, Mikami Y, Nishihashi A, Regnier V, Haraguchi T, Hiraoka Y, Sugata N, Todokoro K, Brown W, Ikemura T (2001) CENP-H, a constitutive centromere component, is required for centromere targeting of CENP-C in vertebrate cells. EMBO J 20:4603–4617. doi:10.1093/emboj/20.16.4603

Fukagawa T, Pendon C, Morris J, Brown W (1999) CENP-C is necessary but not sufficient to induce formation of a functional centromere. EMBO J 18:4196–4209. doi:10.1093/emboj/18.15.4196

Gascoigne KE, Takeuchi K, Suzuki A, Hori T, Fukagawa T, Cheeseman IM (2011) Induced ectopic kinetochore assembly bypasses the requirement for CENP-A nucleosomes. Cell 145:410–422. doi:10.1016/j.cell.2011.03.031

Goshima G, Kiyomitsu T, Yoda K, Yanagida M (2003) Human centromere chromatin protein hMis12, essential for equal segregation, is independent of CENP-A loading pathway. J Cell Biol 160:25–39. doi:10.1083/jcb.200210005

Harrington JJ, Van Bokkelen G, Mays RW, Gustashaw K, Willard HF (1997) Formation of de novo centromeres and construction of first-generation human artificial microchromosomes. Nat Genet 15:345–355. doi:10.1038/ng0497-345

Heeger S, Leismann O, Schittenhelm R, Schraidt O, Heidmann S, Lehner CF (2005) Genetic interactions of separase regulatory subunits reveal the diverged Drosophila Cenp-C homolog. Genes Dev 19:2041–2053. doi:10.1101/gad.347805

Hegemann JH, Fleig UN (1993) The centromere of budding yeast. BioEssays 15:451–460. doi:10.1002/bies.950150704

Hemmerich P, Weidtkamp-Peters S, Hoischen C, Schmiedeberg L, Erliandri I, Diekmann S (2008) Dynamics of inner kinetochore assembly and maintenance in living cells. J Cell Biol 180:1101–1114. doi:10.1083/jcb.200710052

Heun P, Erhardt S, Blower MD, Weiss S, Skora AD, Karpen GH (2006) Mislocalization of the Drosophila centromere-specific histone CID promotes formation of functional ectopic kinetochores. Dev Cell 10:303–315. doi:10.1016/j.devcel.2006.01.014

Holland S, Ioannou D, Haines S, Brown WRA (2005) Comparison of Dam tagging and chromatin immunoprecipitation as tools for the identification of the binding sites for S. pombe CENP-C. Chromosome Res 13:73–83. doi:10.1007/s10577-005-7062-z

Hori T, Amano M, Suzuki A, Backer CB, Welburn JP, Dong Y, McEwen BF, Shang W-H, Suzuki E, Okawa K, Cheeseman IM, Fukagawa T (2008a) CCAN makes multiple contacts with centromeric DNA to provide distinct pathways to the outer kinetochore. Cell 135:1039–1052. doi:10.1016/j.cell.2008.10.019

Hori T, Kagawa N, Toyoda A, Fujiyama A, Misu S, Monma N, Makino F, Ikeo K, Fukagawa T (2017) Constitutive centromere-associated network controls centromere drift in vertebrate cells. J Cell Biol 216:101–113. doi:10.1083/jcb.201605001

Hori T, Okada M, Maenaka K, Fukagawa T (2008b) CENP-O class proteins form a stable complex and are required for proper kinetochore function. Mol Biol Cell 19:843–854. doi:10.1091/mbc.E07-06-0556

Hori T, Shang W-H, Takeuchi K, Fukagawa T (2013) The CCAN recruits CENP-A to the centromere and forms the structural core for kinetochore assembly. J Cell Biol 200:45–60. doi:10.1083/jcb.201210106

Hori T, Shang W-H, Toyoda A, Misu S, Monma N, Ikeo K, Molina O, Vargiu G, Fujiyama A, Kimura H, Earnshaw WC, Fukagawa T (2014) Histone H4 Lys 20 monomethylation of the CENP-A nucleosome is essential for kinetochore assembly. Dev Cell 29:740–749. doi:10.1016/j.devcel.2014.05.001

Howman EV, Fowler KJ, Newson AJ, Redward S, MacDonald AC, Kalitsis P, Choo KH (2000) Early disruption of centromeric chromatin organization in centromere protein A (Cenpa) null mice. Proc Natl Acad Sci USA 97:1148–1153

Hua S, Wang Z, Jiang K, Huang Y, Ward T, Zhao L, Dou Z, Yao X (2011) CENP-U cooperates with Hec1 to orchestrate kinetochore-microtubule attachment. J Biol Chem 286:1627–1638. doi:10.1074/jbc.M110.174946

Ikeno M, Grimes B, Okazaki T, Nakano M, Saitoh K, Hoshino H, McGill NI, Cooke H, Masumoto H (1998) Construction of YAC-based mammalian artificial chromosomes. Nat Biotechnol 16:431–439. doi:10.1038/nbt0598-431

Ishii K, Ogiyama Y, Chikashige Y, Soejima S, Masuda F, Kakuma T, Hiraoka Y, Takahashi K (2008) Heterochromatin integrity affects chromosome reorganization after centromere dysfunction. Science 321:1088–1091. doi:10.1126/science.1158699

Izuta H, Ikeno M, Suzuki N, Tomonaga T, Nozaki N, Obuse C, Kisu Y, Goshima N, Nomura F, Nomura N, Yoda K (2006) Comprehensive analysis of the ICEN (interphase centromere complex) components enriched in the CENP-A chromatin of human cells. Genes Cells 11:673–684. doi:10.1111/j.1365-2443.2006.00969.x

Jansen LET, Black BE, Foltz DR, Cleveland DW (2007) Propagation of centromeric chromatin requires exit from mitosis. J Cell Biol 176:795–805. doi:10.1083/jcb.200701066

Kagawa N, Hori T, Hoki Y, Hosoya O, Tsutsui K, Saga Y, Sado T, Fukagawa T (2014) The CENP-O complex requirement varies among different cell types. Chromosome Res 22:293–303. doi:10.1007/s10577-014-9404-1

Kato H, Jiang J, Zhou B-R, Rozendaal M, Feng H, Ghirlando R, Xiao TS, Straight AF, Bai Y (2013) A conserved mechanism for centromeric nucleosome recognition by centromere protein CENP-C. Science 340:1110–1113. doi:10.1126/science.1235532

Ketel C, Wang HSW, McClellan M, Bouchonville K, Selmecki A, Lahav T, Gerami-Nejad M, Berman J (2009) Neocentromeres form efficiently at multiple possible loci in *Candida albicans*. PLoS Genet 5:e1000400. doi:10.1371/journal.pgen.1000400

Kim S, Yu H (2015) Multiple assembly mechanisms anchor the KMN spindle checkpoint platform at human mitotic kinetochores. J Cell Biol 208:181–196. doi:10.1083/jcb.201407074

Klare K, Weir JR, Basilico F, Zimniak T, Massimiliano L, Ludwigs N, Herzog F, Musacchio A (2015) CENP-C is a blueprint for constitutive centromere-associated network assembly within human kinetochores. J Cell Biol 210:11–22. doi:10.1083/jcb.201412028

Kursel LE, Malik HS (2016) Centromeres. Curr Biol 26:R487–R490. doi:10.1016/j.cub.2016.05.031

Kwon M-S, Hori T, Okada M, Fukagawa T (2007) CENP-C is involved in chromosome segregation, mitotic checkpoint function, and kinetochore assembly. Mol Biol Cell 18:2155–2168. doi:10.1091/mbc.E07-01-0045

Locke DP, Hillier LW, Warren WC, Worley KC, Nazareth LV, Muzny DM, Yang S-P, Wang Z, Chinwalla AT, Minx P, Mitreva M, Cook L, Delehaunty KD, Fronick C, Schmidt H, Fulton LA, Fulton RS, Nelson JO, Magrini V, Pohl C, Graves TA, Markovic C, Cree A, Dinh HH, Hume J, Kovar CL, Fowler GR, Lunter G, Meader S, Heger A, Ponting CP, Marques-Bonet T, Alkan C, Chen L, Cheng Z, Kidd JM, Eichler EE, White S, Searle S, Vilella AJ, Chen Y, Flicek P, Ma J, Raney B, Suh B, Burhans R, Herrero J, Haussler D, Faria R, Fernando O, Darré F, Farré D, Gazave E, Oliva M, Navarro A, Roberto R, Capozzi O, Archidiacono N, Valle Della G, Purgato S, Rocchi M, Konkel MK, Walker JA, Ullmer B, Batzer MA, Smit AFA, Hubley R, Casola C, Schrider DR, Hahn MW, Quesada V, Puente XS, Ordoñez GR, López-Otín C, Vinar T, Brejova B, Ratan A, Harris RS, Miller W, Kosiol C, Lawson HA, Taliwal V, Martins AL, Siepel A, Roychoudhury A, Ma X, Degenhardt J, Bustamante CD, Gutenkunst RN, Mailund T, Dutheil JY, Hobolth A, Schierup MH, Ryder OA, Yoshinaga Y, de Jong PJ, Weinstock GM, Rogers J, Mardis ER, Gibbs RA, Wilson RK (2011) Comparative and demographic analysis of orang-utan genomes. Nature 469:529–533. doi:10.1038/nature09687

London N, Biggins S (2014) Signalling dynamics in the spindle checkpoint response. Nat Rev Mol Cell Biol 15:736–747. doi:10.1038/nrm3888

Maresca TJ, Salmon ED (2009) Intrakinetochore stretch is associated with changes in kinetochore phosphorylation and spindle assembly checkpoint activity. J Cell Biol 184:373–381. doi:10.1083/jcb.200808130

Marshall OJ, Chueh AC, Wong LH, Choo KHA (2008) Neocentromeres: new insights into centromere structure, disease development, and karyotype evolution. Am J Hum Genet 82:261–282. doi:10.1016/j.ajhg.2007.11.009

McClelland SE, Borusu S, Amaro AC, Winter JR, Belwal M, McAinsh AD, Meraldi P (2007) The CENP-A NAC/CAD kinetochore complex controls chromosome congression and spindle bipolarity. EMBO J 26:5033–5047. doi:10.1038/sj.emboj.7601927

McKinley KL, Cheeseman IM (2016) The molecular basis for centromere identity and function. Nat Rev Mol Cell Biol 17:16–29

McKinley KL, Sekulic N, Guo LY, Tsinman T, Black BE, Cheeseman IM (2015) The CENP-L-N complex forms a critical node in an integrated meshwork of interactions at the centromere-kinetochore interface. Mol Cell 60:886–898. doi:10.1016/j.molcel.2015.10.027

Meluh PB, Koshland D (1995) Evidence that the MIF2 gene of *Saccharomyces cerevisiae* encodes a centromere protein with homology to the Mammalian centromere protein CENP-C. Mol Biol Cell 6:793–807. doi:10.1091/mbc.6.7.793

Mendiburo MJ, Padeken J, Fülöp S, Schepers A, Heun P (2011) Drosophila CENH3 is sufficient for centromere formation. Science 334:686–690. doi:10.1126/science.1206880

Milks KJ, Moree B, Straight AF (2009) Dissection of CENP-C-directed centromere and kinetochore assembly. Mol Biol Cell 20:4246–4255. doi:10.1091/mbc.E09-05-0378

Miller MP, Asbury CL, Biggins S (2016) A TOG protein confers tension sensitivity to kinetochore-microtubule attachments. Cell 165:1428–1439. doi:10.1016/j.cell.2016.04.030

Minoshima Y, Hori T, Okada M, Kimura H, Haraguchi T, Hiraoka Y, Bao Y-C, Kawashima T, Kitamura T, Fukagawa T (2005) The constitutive centromere component CENP-50 is required for recovery from spindle damage. Mol Cell Biol 25:10315–10328. doi:10.1128/MCB.25.23.10315-10328.2005

Mitsui K, Tokuzawa Y, Itoh H, Segawa K, Murakami M, Takahashi K, Maruyama M, Maeda M, Yamanaka S (2003) The homeoprotein Nanog is required for maintenance of pluripotency in mouse epiblast and ES cells. Cell 113:631–642. doi:10.1016/S0092-8674(03)00393-3

Moore LL, Roth MB (2001) Hcp-4, a Cenp-C–Like protein in *Caenorhabditis elegans*, is required for resolution of sister centromeres. J Cell Biol 153:1199–1208. doi:10.1083/jcb.153.6.1199

Morgan TH (1915) The mechanism of Mendelian heredity. Holt

Murakami A, Imai HT (1974) Cytological evidence for holocentric chromosomes of the silkworms, *Bombyx mori* and *B. mandarina*, (Bombycidae, Lepidoptera). Chromosoma 47:167–178. doi:10.1007/BF00331804

Musacchio A (2015) The molecular biology of spindle assembly checkpoint signaling dynamics. Curr Biol 25:R1002–R1018. doi:10.1016/j.cub.2015.08.051

Nagpal H, Fukagawa T (2016) Kinetochore assembly and function through the cell cycle. Chromosoma 125:645–659. doi:10.1007/s00412-016-0608-3

Nagpal H, Hori T, Furukawa A, Sugase K, Osakabe A, Kurumizaka H, Fukagawa T (2015) Dynamic changes in CCAN organization through CENP-C during cell-cycle progression. Mol Biol Cell 26:3768–3776. doi:10.1091/mbc.E15-07-0531

Nishihashi A, Haraguchi T, Hiraoka Y, Ikemura T, Regnier V, Dodson H, Earnshaw WC, Fukagawa T (2002) CENP-I is essential for centromere function in vertebrate cells. Dev Cell 2:463–476

Nishimura K, Fukagawa T, Takisawa H, Kakimoto T, Kanemaki M (2009) An auxin-based degron system for the rapid depletion of proteins in nonplant cells. Nat Methods 6:917–922. doi:10.1038/nmeth.1401

Nishino T, Rago F, Hori T, Tomii K, Cheeseman IM, Fukagawa T (2013) CENP-T provides a structural platform for outer kinetochore assembly. EMBO J 32:424–436. doi:10.1038/emboj.2012.348

Nishino T, Takeuchi K, Gascoigne KE, Suzuki A, Hori T, Oyama T, Morikawa K, Cheeseman IM, Fukagawa T (2012) CENP-T-W-S-X forms a unique centromeric chromatin structure with a histone-like fold. Cell 148:487–501. doi:10.1016/j.cell.2011.11.061

Novo CL, Tang C, Ahmed K, Djuric U, Fussner E, Mullin NP, Morgan NP, Hayre J, Sienerth AR, Elderkin S, Nishinakamura R, Chambers I, Ellis J, Bazett-Jones DP, Rugg-Gunn PJ (2016) The pluripotency factor Nanog regulates pericentromeric heterochromatin organization in mouse embryonic stem cells. Genes Dev 30:1101–1115. doi:10.1101/gad.275685.115

Obuse C, Yang H, Nozaki N, Goto S, Okazaki T, Yoda K (2004) Proteomics analysis of the centromere complex from HeLa interphase cells: UV-damaged DNA binding protein 1 (DDB-1) is a component of the CEN-complex, while BMI-1 is transiently co-localized with the centromeric region in interphase. Genes Cells 9:105–120

Oegema K, Desai A, Rybina S, Kirkham M, Hyman AA (2001) Functional analysis of kinetochore assembly in *Caenorhabditis elegans*. J Cell Biol 153:1209–1226

Ohzeki J-I, Nakano M, Okada T, Masumoto H (2002) CENP-B box is required for de novo centromere chromatin assembly on human alphoid DNA. J Cell Biol 159:765–775. doi:10.1083/jcb.200207112

Okada M, Cheeseman IM, Hori T, Okawa K, McLeod IX, Yates JR, Desai A, Fukagawa T (2006) The CENP-H–I complex is required for the efficient incorporation of newly synthesized CENP-A into centromeres. Nat Cell Biol 8:446–457. doi:10.1038/ncb1396

Palmer DK, O'Day K, Wener MH, Andrews BS, Margolis RL (1987) A 17-kD centromere protein (CENP-A) copurifies with nucleosome core particles and with histones. J Cell Biol 104:805–815

Perpelescu M, Fukagawa T (2011) The ABCs of CENPs. Chromosoma 120:425–446. doi:10.1007/s00412-011-0330-0

Pesenti ME, Weir JR, Musacchio A (2016) Progress in the structural and functional characterization of kinetochores. Curr Opin Struct Biol 37:152–163. doi:10.1016/j.sbi.2016.03.003

Peters AHFM, O'Carroll D, Scherthan H, Mechtler K, Sauer S, Schöfer C, Weipoltshammer K, Pagani M, Lachner M, Kohlmaier A, Opravil S, Doyle M, Sibilia M, Jenuwein T (2001) Loss of the Suv39h histone methyltransferases impairs mammalian heterochromatin and genome stability. Cell 107:323–337. doi:10.1016/S0092-8674(01)00542-6

Petrovic A, Keller J, Liu Y, Overlack K, John J, Dimitrova YN, Jenni S, van Gerwen S, Stege P, Wohlgemuth S, Rombaut P, Herzog F, Harrison SC, Vetter IR, Musacchio A (2016) Structure of the MIS12 complex and molecular basis of its interaction with CENP-C at human kinetochores. Cell 167(1028–1040):e15. doi:10.1016/j.cell.2016.10.005

Piras FM, Nergadze SG, Magnani E, Bertoni L, Attolini C, Khoriauli L, Raimondi E, Giulotto E (2010) Uncoupling of satellite DNA and centromeric function in the genus Equus. PLoS Genet 6:e1000845. doi:10.1371/journal.pgen.1000845

Pluta AF, Mackay AM, Ainsztein AM, Goldberg IG, Earnshaw WC (1995) The centromere: hub of chromosomal activities. Science 270:1591–1594

Prendergast L, Müller S, Liu Y, Huang H, Dingli F, Loew D, Vassias I, Patel DJ, Sullivan KF, Almouzni G (2016) The CENP-T/-W complex is a binding partner of the histone chaperone FACT. Genes Dev 30:1313–1326. doi:10.1101/gad.275073.115

Prendergast L, van Vuuren C, Kaczmarczyk A, Doering V, Hellwig D, Quinn N, Hoischen C, Diekmann S, Sullivan KF (2011) Premitotic assembly of human CENPs -T and -W switches centromeric chromatin to a mitotic state. PLoS Biol 9:e1001082. doi:10.1371/journal.pbio.1001082

Przewloka MR, Glover DM (2009) The kinetochore and the centromere: a working long distance relationship. Annu Rev Genet 43:439–465. doi:10.1146/annurev-genet-102108-134310

Przewloka MR, Venkei Z, Bolanos-Garcia VM, Debski J, Dadlez M, Glover DM (2011) CENP-C is a structural platform for kinetochore assembly. Curr Biol 21:399–405. doi:10.1016/j.cub.2011.02.005

Purgato S, Belloni E, Piras FM, Zoli M, Badiale C, Cerutti F, Mazzagatti A, Perini G, Valle Della G, Nergadze SG, Sullivan KF, Raimondi E, Rocchi M, Giulotto E (2015) Centromere sliding on a mammalian chromosome. Chromosoma 124:277–287. doi:10.1007/s00412-014-0493-6

Rago F, Gascoigne KE, Cheeseman IM (2015) Distinct organization and regulation of the outer kinetochore KMN network downstream of CENP-C and CENP-T. Curr Biol 25:671–677. doi:10.1016/j.cub.2015.01.059

Regnier V, Vagnarelli P, Fukagawa T, Zerjal T, Burns E, Trouche D, Earnshaw W, Brown W (2005) CENP-A is required for accurate chromosome segregation and sustained kinetochore association of BubR1. Mol Cell Biol 25:3967–3981. doi:10.1128/MCB.25.10.3967-3981.2005

Saitoh H, Tomkiel J, Cooke CA, Ratrie H, Maurer M, Rothfield NF, Earnshaw WC (1992) CENP-C, an autoantigen in scleroderma, is a component of the human inner kinetochore plate. Cell 70:115–125

Saitoh S, Takahashi K, Yanagida M (1997) Mis6, a fission yeast inner centromere protein, acts during G1/S and forms specialized chromatin required for equal segregation. Cell 90:131–143. doi:10.1016/S0092-8674(00)80320-7

Satzinger H (2008) Theodor and Marcella Boveri: chromosomes and cytoplasm in heredity and development. Nat Rev Genet 9(3):231–238

Schleiffer A, Maier M, Litos G, Lampert F, Hornung P, Mechtler K, Westermann S (2012) CENP-T proteins are conserved centromere receptors of the Ndc80 complex. Nat Cell Biol 14:604–613. doi:10.1038/ncb2493

Screpanti E, De Antoni A, Alushin GM, Petrovic A, Melis T, Nogales E, Musacchio A (2011) Direct binding of Cenp-C to the Mis12 complex joins the inner and outer kinetochore. Curr Biol 21:391–398. doi:10.1016/j.cub.2010.12.039

Shang W-H, Hori T, Martins NMC, Toyoda A, Misu S, Monma N, Hiratani I, Maeshima K, Ikeo K, Fujiyama A, Kimura H, Earnshaw WC, Fukagawa T (2013) Chromosome engineering allows the efficient isolation of vertebrate neocentromeres. Dev Cell 24:635–648. doi:10.1016/j.devcel.2013.02.009

Shang W-H, Hori T, Toyoda A, Kato J, Popendorf K, Sakakibara Y, Fujiyama A, Fukagawa T (2010) Chickens possess centromeres with both extended tandem repeats and short non-tandem-repetitive sequences. Genome Res 20:1219–1228. doi:10.1101/gr.106245.110

Shang W-H, Hori T, Westhorpe FG, Godek KM, Toyoda A, Misu S, Monma N, Ikeo K, Carroll CW, Takami Y, Fujiyama A, Kimura H, Straight AF, Fukagawa T (2016) Acetylation of histone H4 lysine 5 and 12 is required for CENP-A deposition into centromeres. Nat Commun 7:13465. doi:10.1038/ncomms13465

Singh TR, Saro D, Ali AM, Zheng X-F, Du C-H, Killen MW, Sachpatzidis A, Wahengbam K, Pierce AJ, Xiong Y, Sung P, Meetei AR (2010) MHF1-MHF2, a histone-fold-containing protein complex, participates in the Fanconi anemia pathway via FANCM. Mol Cell 37:879–886. doi:10.1016/j.molcel.2010.01.036

Stoler S, Keith KC, Curnick KE, Fitzgerald-Hayes M (1995) A mutation in CSE4, an essential gene encoding a novel chromatin-associated protein in yeast, causes chromosome nondisjunction and cell cycle arrest at mitosis. Genes Dev 9:573–586

Stukenberg PT, Burke DJ (2015) Connecting the microtubule attachment status of each kinetochore to cell cycle arrest through the spindle assembly checkpoint. Chromosoma 124:463–480. doi:10.1007/s00412-015-0515-z

Sugata N, Li S, Earnshaw WC, Yen TJ, Yoda K, Masumoto H, Munekata E, Warburton PE, Todokoro K (2000) Human CENP-H multimers colocalize with CENP-A and CENP-C at active centromere–kinetochore complexes. Hum Mol Genet 9:2919–2926

Sugata N, Munekata E, Todokoro K (1999) Characterization of a novel kinetochore protein, CENP-H. J Biol Chem 274:27343–27346

Sullivan BA, Karpen GH (2004) Centromeric chromatin exhibits a histone modification pattern that is distinct from both euchromatin and heterochromatin. Nat Struct Mol Biol 11:1076–1083. doi:10.1038/nsmb845

Sutton WS (1902) On the morphology of the chromosome group in *Brachystola magna*. Biol Bull 4:24–39

Sutton WS (1903) The chromosomes in heredity. Biol Bull 4:231–250

Suzuki A, Badger BL, Wan X, DeLuca JG, Salmon ED (2014) The architecture of CCAN proteins creates a structural integrity to resist spindle forces and achieve proper intrakinetochore stretch. Dev Cell 30:717–730. doi:10.1016/j.devcel.2014.08.003

Suzuki A, Hori T, Nishino T, Usukura J, Miyagi A, Morikawa K, Fukagawa T (2011) Spindle microtubules generate tension-dependent changes in the distribution of inner kinetochore proteins. J Cell Biol 193:125–140. doi:10.1083/jcb.201012050

Takahashi K, Chen ES, Yanagida M (2000) Requirement of Mis6 centromere connector for localizing a CENP-A-like protein in fission yeast. Science 288:2215–2219

Takeuchi K, Fukagawa T (2012) Molecular architecture of vertebrate kinetochores. Exp Cell Res 318:1367–1374. doi:10.1016/j.yexcr.2012.02.016

Takeuchi K, Nishino T, Mayanagi K, Horikoshi N, Osakabe A, Tachiwana H, Hori T, Kurumizaka H, Fukagawa T (2014) The centromeric nucleosome-like CENP-T-W-S-X complex induces positive supercoils into DNA. Nucleic Acids Res 42:1644–1655. doi:10.1093/nar/gkt1124

Talbert PB, Masuelli R, Tyagi AP, Comai L, Henikoff S (2002) Centromeric localization and adaptive evolution of an Arabidopsis histone H3 variant. Plant Cell 14:1053–1066. doi:10.1105/tpc.010425

Tipton AR, Wang K, Oladimeji P, Sufi S, Gu Z, Liu S-T (2012) Identification of novel mitosis regulators through data mining with human centromere/kinetochore proteins as group queries. BMC Cell Biol 13(1):15. doi:10.1186/1471-2121-13-15

Tomkiel J, Cooke CA, Saitoh H, Bernat RL, Earnshaw WC (1994) CENP-C is required for maintaining proper kinetochore size and for a timely transition to anaphase. J Cell Biol 125:531–545

Tschermak E (1900) Über künstliche Kreuzung bei Pisum sativum. Berichte der Deutsche Botanischen Gessellschaft 28:232–239

Uchida KSK, Takagaki K, Kumada K, Hirayama Y, Noda T, Hirota T (2009) Kinetochore stretching inactivates the spindle assembly checkpoint. J Cell Biol 184:383–390. doi:10.1083/jcb.200811028

Van Hooser AA, Ouspenski II, Gregson HC, Starr DA, Yen TJ, Goldberg ML, Yokomori K, Earnshaw WC, Sullivan KF, Brinkley BR (2001) Specification of kinetochore-forming chromatin by the histone H3 variant CENP-A. J Cell Sci 114:3529–3542. doi:10.1007/BF00332792

Varma D, Salmon ED (2012) The KMN protein network—chief conductors of the kinetochore orchestra. J Cell Sci 125:5927–5936. doi:10.1242/jcs.093724

Voullaire LE, Slater HR, Petrovic V, Choo KH (1993) A functional marker centromere with no detectable alpha-satellite, satellite III, or CENP-B protein: activation of a latent centromere? Am J Hum Genet 52:1153–1163

Wade CM, Giulotto E, Sigurdsson S, Zoli M, Gnerre S, Imsland F, Lear TL, Adelson DL, Bailey E, Bellone RR, Blöcker H, Distl O, Edgar RC, Garber M, Leeb T, Mauceli E, MacLeod JN, Penedo MCT, Raison JM, Sharpe T, Vogel J, Andersson L, Antczak DF, Biagi T, Binns MM, Chowdhary BP, Coleman SJ, Valle Della G, Fryc S, Guérin G, Hasegawa T, Hill EW, Jurka J, Kiialainen A, Lindgren G, Liu J, Magnani E, Mickelson JR, Murray J, Nergadze SG, Onofrio R, Pedroni S, Piras MF, Raudsepp T, Rocchi M, Røed KH, Ryder OA, Searle S, Skow L, Swinburne JE, Syvänen AC, Tozaki T, Valberg SJ, Vaudin M, White JR, Zody MC, Broad institute genome sequencing platform, broad institute whole genome assembly team, Lander ES, Lindblad-Toh K (2009) Genome sequence, comparative analysis, and population genetics of the domestic horse. Science 326:865–867. doi:10.1126/science.1178158

Wan X, O'Quinn RP, Pierce HL, Joglekar AP, Gall WE, DeLuca JG, Carroll CW, Liu S-T, Yen TJ, McEwen BF, Stukenberg PT, Desai A, Salmon ED (2009) Protein architecture of the human kinetochore microtubule attachment site. Cell 137:672–684. doi:10.1016/j.cell.2009.03.035

Warburton PE, Cooke CA, Bourassa S, Vafa O, Sullivan BA, Stetten G, Gimelli G, Warburton D, Tyler-Smith C, Sullivan KF, Poirier GG, Earnshaw WC (1997) Immunolocalization of CENP-A suggests a distinct nucleosome structure at the inner kinetochore plate of active centromeres. Curr Biol 7:901–904

Weir JR, Faesen AC, Klare K, Petrovic A, Basilico F, Fischböck J, Pentakota S, Keller J, Pesenti ME, Pan D, Vogt D, Wohlgemuth S, Herzog F, Musacchio A (2016) Insights from biochemical reconstitution into the architecture of human kinetochores. Nature 537:249–253. doi:10.1038/nature19333

Yamagishi Y, Sakuno T, Shimura M, Watanabe Y (2008) Heterochromatin links to centromeric protection by recruiting shugoshin. Nature 455:251–255. doi:10.1038/nature07217

Yan Z, Delannoy M, Ling C, Daee D, Osman F, Muniandy PA, Shen X, Oostra AB, Du H, Steltenpool J, Lin T, Schuster B, Décaillet C, Stasiak A, Stasiak AZ, Stone S, Hoatlin ME, Schindler D, Woodcock CL, Joenje H, Sen R, de Winter JP, Li L, Seidman MM, Whitby MC, Myung K, Constantinou A, Wang W (2010) A histone-fold complex and FANCM form a conserved DNA-remodeling complex to maintain genome stability. Mol Cell 37:865–878. doi:10.1016/j.molcel.2010.01.039

Yoda K, Ando S, Morishita S, Houmura K, Hashimoto K, Takeyasu K, Okazaki T (2000) Human centromere protein A (CENP-A) can replace histone H3 in nucleosome reconstitution in vitro. PNAS 97:7266–7271. doi:10.1073/pnas.130189697

The Power of *Xenopus* Egg Extract for Reconstitution of Centromere and Kinetochore Function

Bradley T. French and Aaron F. Straight

Abstract Faithful transmission of genetic information during cell division requires attachment of chromosomes to the mitotic spindle via the kinetochore. In vitro reconstitution studies are beginning to uncover how the kinetochore is assembled upon the underlying centromere, how the kinetochore couples chromosome movement to microtubule dynamics, and how cells ensure the site of kinetochore assembly is maintained from one generation to the next. Here we give special emphasis to advances made in *Xenopus* egg extract, which provides a unique, biochemically tractable in vitro system that affords the complexity of cytoplasm and nucleoplasm to permit reconstitution of the dynamic, cell cycle-regulated functions of the centromere and kinetochore.

1 Introduction

All organisms must accurately segregate their genomes during cell division. In eukaryotes, duplicated sister chromosomes condense as cells enter mitosis, attach to the microtubules of the bipolar mitotic spindle, and then use microtubule-dependent forces to move to opposite spindle poles before separation into two new cells during cytokinesis. These events were first described nearly 150 years ago (Flemming 1880), yet we still lack a detailed molecular understanding of the processes that orchestrate chromosome segregation.

Central to the process of chromosome segregation is the multiprotein complex that links chromosomes to microtubules—the kinetochore. Kinetochores provide the primary interaction sites between microtubules and chromosomes for chromosome movement and monitor proper chromosome alignment through the activities of the spindle assembly checkpoint (Cheeseman 2014; Foley and Kapoor 2013). The process of kinetochore formation occurs in mitosis as numerous proteins and

B.T. French · A.F. Straight (✉)
Department of Biochemistry, Stanford University, 279 Campus Drive, Beckman 409, Stanford, CA 94305, USA
e-mail: astraigh@stanford.edu

B.E. Black (ed.), *Centromeres and Kinetochores*, Progress in Molecular and Subcellular Biology 56, DOI 10.1007/978-3-319-58592-5_3

protein complexes localize to a specialized domain of the chromosome termed the centromere (McKinley and Cheeseman 2016; Westhorpe and Straight 2015). Centromeres are defined by a specialized chromatin domain in which the histone H3 variant CENP-A replaces histone H3 in nucleosomes. CENP-A nucleosomes recruit a complex of $\sim$17 centromere-specific proteins, the constitutive centromere-associated network (CCAN), which associates with the chromosome throughout the cell cycle and provides the essential substrate for mitotic kinetochore assembly (Fig. 1). The centromere and kinetochore thus play structural, mechanical, and signaling roles that are essential for proper chromosome segregation.

Over the last thirty years, efforts from numerous groups have identified many of the core components of centromeres and kinetochores. The first centromere proteins were discovered as autoantigens recognized by sera from scleroderma patients (Earnshaw and Rothfield 1985; Earnshaw et al. 1986, 1987). Subsequently, genetic studies of chromosome missegregation and biochemical purification (Foltz et al. 2006; Obuse et al. 2004; Izuta et al. 2006; Okada et al. 2006) have provided a relatively complete list of centromere and kinetochore proteins: by current estimates, more than 80 different proteins have been identified (Westhorpe and Straight 2015; McKinley and Cheeseman 2016).

As we now have many of the proteins of the centromere and kinetochore in hand, a major challenge going forward is to understand how those components come together to give rise to the activities that segregate chromosomes. In vitro reconstitution studies are beginning to uncover the biochemical functions of these components, how they are connected, and how their activities are regulated and coordinated to achieve proper chromosome segregation. As comprehensive coverage of the functions of the centromere and kinetochore is provided in other chapters of this book, the goal of this chapter is to highlight the specific contributions of reconstitution studies. Special emphasis is given to *Xenopus* egg extract, which provides a biochemically tractable in vitro system that affords the complexity of cytoplasm and nucleoplasm, uniquely permitting reconstitution of the dynamic, cell cycle-regulated functions of the centromere and kinetochore.

2 Reconstituting Centromere and Kinetochore Functions In Vitro

A key challenge in studying centromeres and kinetochores in vivo is that their activities are essential for cell viability. Thus deletion or mutation of protein components often results in cell death and give rise to phenotypes that are often difficult to interpret. Cell free or reconstituted systems bypass this problem and thus enable the study of centromere and kinetochore protein function without the limitations of cell viability.

Early studies of kinetochore activities in cell free systems showed that the kinetochores of purified mitotic chromosomes, when incubated with tubulin,

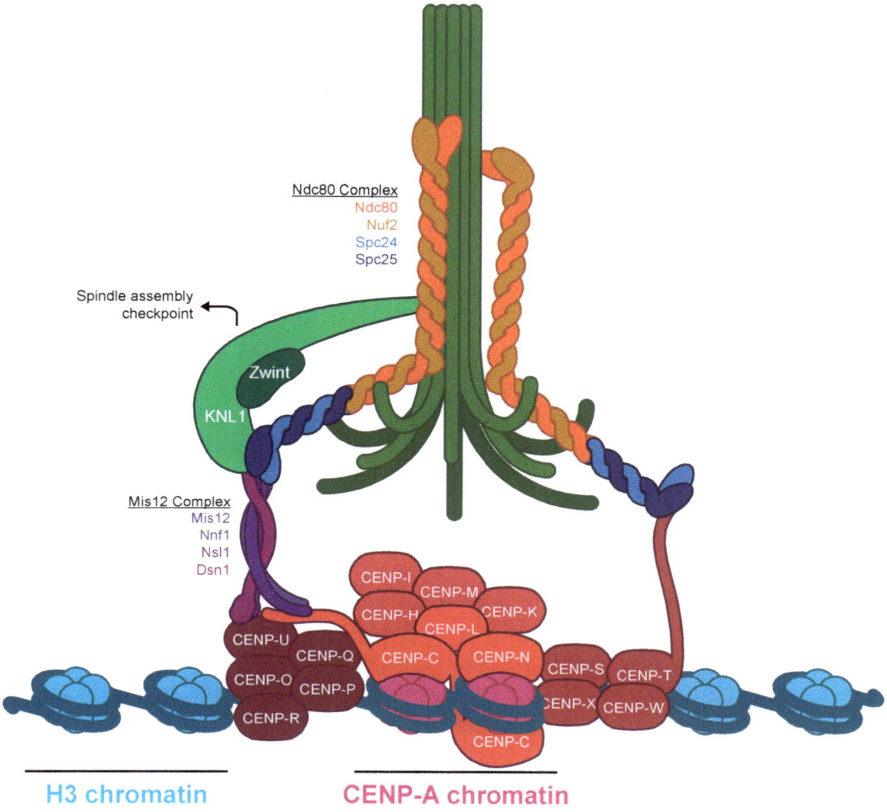

Fig. 1 General schematic of a vertebrate kinetochore. The point of attachment between a chromosome and the microtubules of the mitotic spindle is a region called the centromere. During mitosis, the centromere recruits a multiprotein complex called the kinetochore that physically links chromosomes to microtubules, monitors its own attachment to microtubules, and delays anaphase onset until chromosomes are properly attached to opposite poles of the spindle (*SAC* the spindle assembly checkpoint). At the heart of the kinetochore is the KMN network: KNL1, which can bind microtubules and serves as the hub for SAC signaling; the Mis12 complex (Mis12, Nnf1, Nsl1, Dsn1) which anchors the kinetochore to the centromere; and the Ndc80 complex (Ndc80, Nuf2, Spc24, Spc25) which provides the core microtubule-binding activity of the kinetochore. The location of the centromere is specified by chromatin containing nucleosomes in which histone H3 has been replaced with the centromere-specific histone H3 variant CENP-A. CENP-A nucleosomes are bound directly by the constitutive centromere-associated network (CCAN), a complex of ~17 "CENP" proteins that associate with the centromere throughout the cell cycle and provide the platform for kinetochore assembly in mitosis. Of these, only CENP-C and CENP-N are known to directly interact with the CENP-A nucleosome. The CCAN is further divided into several subcomplexes: CENP-C, CENP-LN, CENP-HIKM, CENP-OPQRU, CENP-TWSX

bind microtubules, regulate microtubule dynamic instability, and couple chromosome movement to microtubule depolymerization (Mitchison and Kirschner 1985a, b; Koshland et al. 1988; Hyman and Mitchison 1990). In budding yeast, where a

125 bp DNA sequence is sufficient to define a centromere, incubation of centromere DNA containing plasmids in yeast extract can recapitulate the assembly of multiple centromere proteins on DNA thereby providing a biochemically tractable system that benefits from the wealth of mutants generated in genetic studies of chromosome segregation (Hyman et al. 1992; Kingsbury and Koshland 1991; Sorger et al. 1994; Sandall et al. 2006). These studies eventually led to the purification of intact kinetochore particles from yeast which made possible elegant structural, biophysical, and proteomic studies (Sorger et al. 1994; Akiyoshi et al. 2009, 2010; Gonen et al. 2012).

Although purification of kinetochores from cells has provided a powerful approach to studying the functions of kinetochore proteins, mutant analysis using this approach is still limited by host viability. Full reconstitution of kinetochores using purified components offers one potential solution to this problem. But despite major advances toward this goal (Pesenti et al. 2016), many centromere and kinetochore proteins have resisted expression and purification and it has been difficult to recapitulate post-translational regulation and cell cycle dependent assembly of centromeres and kinetochores in vitro.

Cell free approaches to centromere and kinetochore reconstitution that can capture the dynamic properties of centromeres and kinetochores and their cell cycle dependent regulation stand to provide significant new insight into their function. One of the most powerful approaches in this class is the use of extracts from *Xenopus* eggs, in which functional centromeres can be reconstituted de novo on defined chromatin templates, and which support spindle assembly, checkpoint control, and anaphase chromosome segregation. Below we discuss first the general advantages of egg extract for in vitro reconstitution studies, and then how its strengths have been specifically leveraged for the study of centromeres and kinetochores.

2.1 Xenopus *Egg Extract as a Versatile System for In Vitro Reconstitution*

Xenopus egg extracts were originally developed as a starting material for biochemical fractionation following the demonstration that cytoplasmic components were responsible for controlling cell cycle progress and for regulating the ensuing changes in chromosomes during DNA replication and mitosis (Gurdon and Woodland 1968; Gurdon 1968; Masui and Markert 1971; Benbow and Ford 1975; Benbow et al. 1975). *Xenopus* eggs are an ideal starting material for biochemical studies. First, because each egg contains ~ 1 μL of cytoplasm and a single frog will lay 1000–2000 eggs, a few frogs will provide several milliliters of cytoplasm. Second, because new RNA synthesis does not start until the twelfth division after fertilization each egg is stockpiled with maternal proteins and RNAs to support chromatin assembly, nuclear assembly, and chromosome segregation (Newport and

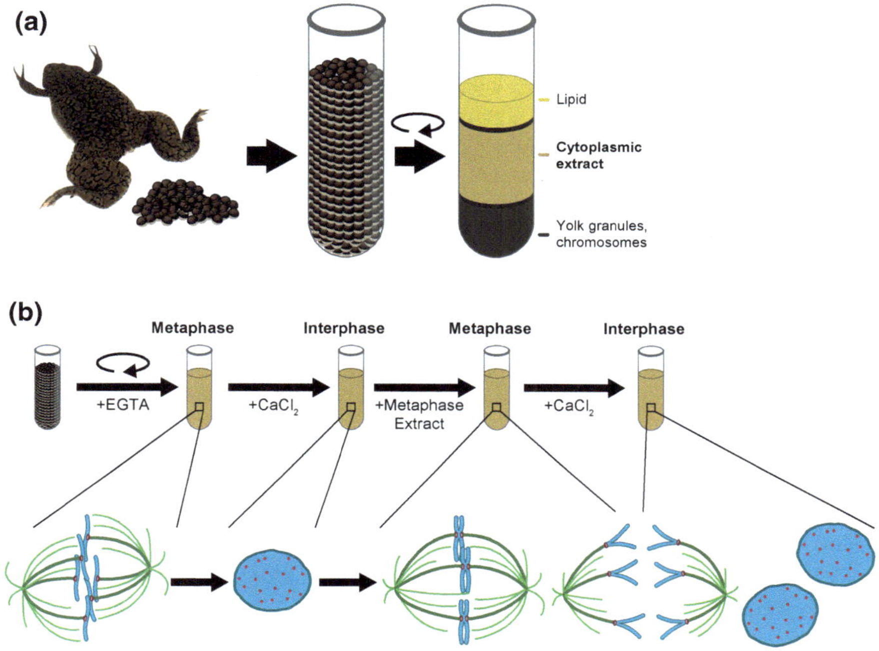

Fig. 2 Extract and cell cycle manipulation. Preparation and manipulation of extracts from *Xenopus laevis* eggs. **a** Preparation of egg extract. *Xenopus laevis* females are injected with human chorionic gonadotropin (hCG) to stimulate ovulation. Eggs laid overnight are collected, treated with cysteine to remove their sticky jelly coat, and subjected to a low-speed packing spin to remove interstitial buffer. Centrifugation at higher speed gently ruptures the cell membrane and causes the maternal chromosomes to pellet with the pigment granules, resulting in chromatin-free cytoplasmic extract that contains soluble cytoplasmic components including ribosomes as well as the membranes required to support nuclear envelope assembly. Egg extract is isolated following centrifugation by side puncture using a syringe. **b** Cell cycle manipulation of extract. Crushing unfertilized eggs in the presence of the calcium-chelator EGTA results in extract that preserves the metaphase II meiotic arrest of the intact eggs. This metaphase extract assembles bipolar spindles around unreplicated chromosomes following addition of demembranated sperm chromatin. Addition of calcium mimics fertilization and causes release of the metaphase II arrest and entry into interphase. Chromosomes decondense, form interphase nuclei, and undergo DNA replication. Addition of fresh metaphase extract to interphase extract drives it forward into the next mitosis. Replicated chromosomes condense and form bioriented spindles in this cycled mitotic extract. Addition of calcium again drives exit from mitosis, with replicated chromosomes undergoing microtubule-direct anaphase segregation

Kirschner 1982a, b). Therefore, extract from a batch of *Xenopus* eggs (Fig. 2a) which maintains RNA and protein concentrations at or near normal cytoplasmic levels provides abundant material for biochemical studies.

In addition to their biochemical advantages, egg extracts recapitulate complex cellular processes in vitro (Lohka and Masui 1983, 1984; Iwao and Katagiri 1984;

Miake-Lye and Kirschner 1985). Lohka and Maller first demonstrated that addition of demembranated sperm chromatin to *Xenopus* egg extracts reconstituted pronuclear assembly, mitotic chromosome condensation, and spindle assembly (Lohka and Maller 1985). Over the following decades, reconstitution in *Xenopus* egg extracts has informed studies of sperm decondensation and chromatin assembly (Earnshaw et al. 1980; Almouzni and Mechali 1988a, b; Ohsumi and Katagiri 1991), nuclear envelope assembly (Sheehan et al. 1988), nucleocytoplasmic transport (Saitoh et al. 1996; Moore and Blobel 1992; Newmeyer and Forbes 1988), DNA replication (Blow and Laskey 1986), mitotic chromosome condensation (Shintomi et al. 2015), mitotic spindle assembly (Lohka and Maller 1985), spindle assembly checkpoint signaling (Minshull et al. 1994; Chen 2008), and anaphase chromosome segregation (Murray et al. 1996). This is in addition to other activities recapitulated by extract including the DNA damage response (Kornbluth et al. 1992; Matsumoto and Bogenhagen 1989; Garner and Costanzo 2009; Lupardus et al. 2007), developmental signaling (Salic et al. 2000), apoptosis (Newmeyer et al. 1994; Kornbluth and Evans 2001), and organelle scaling (Levy and Heald 2015).

A significant advantage of working in egg extracts for dissecting such complex phenomena is that the cytoplasmic composition can be easily manipulated. Proteins can be removed by immunodepletion and complemented with in vitro translated or purified proteins at varying concentrations. Furthermore, complementation with mutants to assess phenotypes, fluorescently labeled proteins to image protein dynamics, inhibitors (small molecules, peptides, antibodies, etc.) to functionally manipulate the extract or radiolabeled metabolites to follow the flux of molecules through the extract are all straightforward.

A major advantage of the egg extract system for studying centromere and kinetochore function is the ability to manipulate the cell cycle state (Murray 1991; Desai et al. 1998) (Fig. 2b). Xenopus eggs are arrested at metaphase of meiosis II and initiate division once fertilized. Egg extracts prepared in the absence of calcium are naturally arrested in metaphase and can be cycled into interphase simply by addition of calcium to mimic fertilization. Sperm chromatin added to calcium-released extracts undergoes chromatid decondensation and subsequent DNA replication (Lohka and Maller 1985). Furthermore, interphase extracts can be driven back into metaphase by the addition of metaphase arrested extracts, thereby causing microtubule assembly around sperm chromatin to generate metaphase spindles with bioriented sister kinetochores that attach to opposite spindle poles and support spindle assembly checkpoint signaling (Sawin and Mitchison 1991; Minshull et al. 1994). Addition of calcium to these spindle-containing extracts drives anaphase chromosome segregation (Murray et al. 1996; Shamu and Murray 1992). Because of the short duration of embryonic cell cycles (30–90 min) (Newport and Kirschner 1982a), the events of multiple successive cell cycles can be tracked easily in egg extract (Chang and Ferrell 2013; Murray and Kirschner 1989).

The properties described above make *Xenopus* egg extract the ideal system for detailed dissection of complex cell cycle-regulated processes. In the next section, we discuss how the unique advantages *Xenopus* egg extract have and continue to contribute to the elucidation of various aspects of centromere and kinetochore function; including maintenance of centromere identity, recruitment of CCAN proteins onto chromatin, assembly of kinetochores during mitosis, and regulation of the spindle assembly checkpoint.

3 Application of Frog Egg Extracts to Study Centromere and Kinetochore Function

Xenopus egg extract is the only cell free system that assembles functional centromeres and kinetochores de novo (Desai et al. 1997). During spermatogenesis in frogs, CENP-A is retained on sperm chromatin while the rest of the centromere is disassembled (Milks et al. 2009; Palmer et al. 1990). Thus when demembranated sperm chromatin is added into egg extract it acts as a naïve template for centromere assembly. Because centromere proteins must be assembled from the egg cytoplasm, depleting proteins from the extract makes it possible to assay depletion phenotypes within a single cell cycle without the confounding secondary effects often associated with RNAi or gene deletion experiments in living cells. These unique advantages of egg extract have enabled insights into several important aspects of centromere and kinetochore function.

3.1 Epigenetic Maintenance of Centromere Identity

CENP-A nucleosomes appear to provide a true epigenetic mark that dictates the position of the centromere (McKinley and Cheeseman 2016). DNA sequence is neither necessary nor sufficient for centromere identity in vertebrates, whereas loss of CENP-A prevents centromere formation (McKinley and Cheeseman 2016). Moreover, targeting CENP-A to a noncentromeric locus promotes assembly of a neocentromere that is epigenetically maintained through subsequent generations (Mendiburo et al. 2011; McKinley and Cheeseman 2016). To understand the persistence of centromere identity, it is necessary to understand how CENP-A chromatin directs new CENP-A assembly. Perturbing the self-renewing mechanisms that maintain CENP-A chromatin is challenging in vivo because centromere inheritance is required for viability. *Xenopus* egg extract is the only in vitro system that recapitulates epigenetic CENP-A inheritance. Short cell cycles and facile control of cell cycle transitions have facilitated temporal dissection of the stages of CENP-A assembly. In addition, these studies benefit from the ability to manipulate the chromatin-associated and the soluble pools of CENP-A separately to understand the distinct mechanisms involved in recognizing template chromatin and in recognizing prenucleosomal CENP-A. Together, these features have permitted detailed studies of the mechanisms by which CENP-A chromatin is propagated, including what chaperones are involved, what proteins recognize the pre-existing centromere, and what features of CENP-A chromatin are recognized.

Extract-based CENP-A assembly assays recapitulate interphase- and centromere-specific CENP-A deposition on demembranated sperm chromatin that requires the CENP-A chaperone HJURP (Bernad et al. 2011; Moree et al. 2011; Foltz et al. 2009; Dunleavy et al. 2009) (Fig. 3a). Biochemical studies demonstrate that HJURP binds a region on CENP-A called the CATD (CENP-A Targeting Domain; Fig. 4b) first identified because transplanting this domain into histone H3 caused the chimeric molecule to assemble at centromeres (Black et al. 2004). Artificially tethering HJURP to an arbitrary chromosomal locus is sufficient to promote local CENP-A assembly (Barnhart et al. 2011). However, other H3 chaperones including RbAp46/48 have also been implicated in CENP-A assembly by studies in human cells and *Drosophila* (Dunleavy et al. 2009; Furuyama et al. 2006), though their roles have been difficult to study in vivo due to their importance in H3 chromatin assembly. In egg extract, inviability due to defective H3 chromatin assembly is not a concern. Depletion of RbAp48 from egg extract has no effect on CENP-A assembly, and depletion of CAF1p150 or HIRA—components of the H3.1 and H3.3 assembly complexes, respectively—have only mild effects (Bernad et al. 2011). Thus, HJURP appears to be the predominant CENP-A chaperone.

To understand how CENP-A serves as an epigenetic mark for centromere identity, it is necessary to understand how HJURP and the proteins that assemble new CENP-A recognize the pre-existing centromere. In egg extract, immunodepletion of CENP-C, which directly binds nucleosomal CENP-A, causes defective HJURP localization and decreased CENP-A assembly (Moree et al. 2011). The facile cell cycle control available in extract enabled identification of two roles for CENP-C in promoting new CENP-A assembly: recruitment of HJURP in interphase, and recruitment of another essential CENP-A assembly factor, the Mis18 complex, during mitosis (Moree et al. 2011) (Fig. 3b). The Mis18 complex, comprised of Mis18α, Mis18β, and M18BP1 (Maddox et al. 2007; Fujita et al. 2007) is also required for HJURP localization, as immunodepletion of M18BP1 causes a loss of both centromeric HJURP and CENP-A loading (Moree et al. 2011). Its mitotic function, however, remains mysterious. The Mis18 complex has been proposed to "prime" centromeric chromatin for assembly by modulating post-translational modification of chromatin (Fujita et al. 2007; Hayashi et al. 2004; Ohzeki et al. 2016; Kim et al. 2012) which could contribute to a proposed licensing mechanism for CENP-A assembly. Indeed, while M18BP1 is recruited to centromeres in metaphase egg extract by a direct interaction with CENP-C, M18BP1 switches to a CENP-C-independent mechanism for interphase localization (Moree et al. 2011; Westhorpe et al. 2015), possibly analogous to cell cycle-regulated licensing of DNA replication origins (Walter and Arias 2004). Facile cell cycle manipulation of egg extract and the ability to transfer chromatin between extracts of different composition or cell cycle states were essential for identifying the mechanisms of DNA replication licensing (Blow and Laskey 2016) and will likewise be valuable in elucidating the mechanisms that may license new CENP-A assembly.

While immunodepletion analysis has identified CENP-C as a key recognition factor for centromeric chromatin and ruled out other CCAN components such as CENP-T and CENP-W (Krizaic et al. 2015; Moree et al. 2011), the fundamental

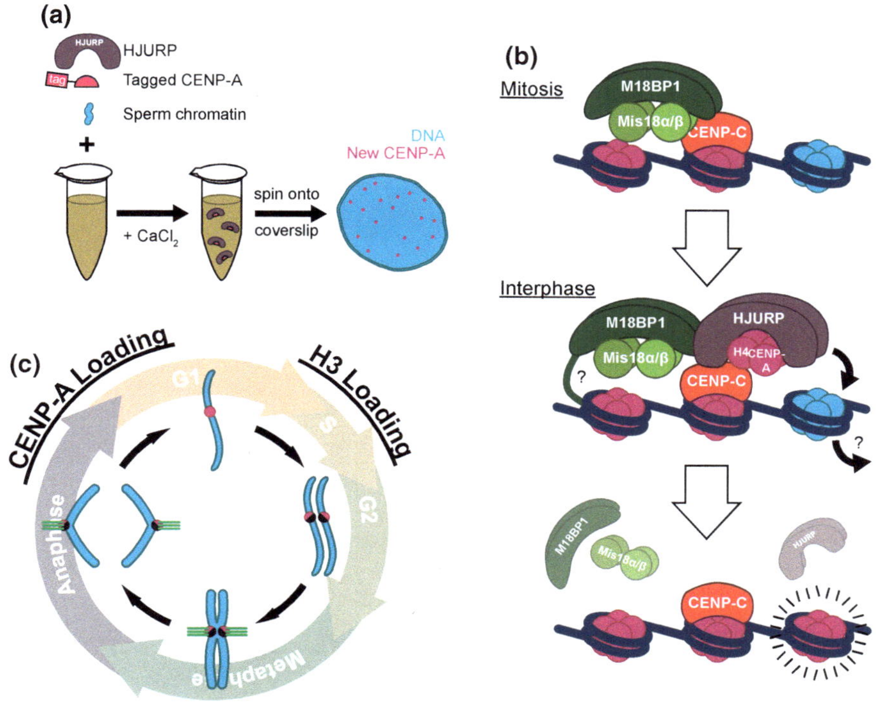

Fig. 3 CENP-A assembly. Epigenetic maintenance of CENP-A. **c** Changes in centromeric CENP-A content during the cell cycle. During S-phase, parental nucleosomes are partitioned equally between the daughter chromatids as DNA replicates. While new H3 nucleosome assembly is coupled to the replication machinery, new CENP-A nucleosome assembly does not proceed until the following G1. This results in halving of centromeric CENP-A from G1 to G2, indicated by transition of the *pink circles* to *pink half-circles* at the centromere. **b** Schematic of CENP-A loading. In *Xenopus*, initial recruitment of the CENP-A assembly machinery takes place in mitosis. The Mis18 complex (Mis18α, Mis18β, and M18BP1) localizes to centromeres via a cell cycle-regulated interaction between CENP-C and M18BP1. In interphase, CENP-C and the Mis18 complex interact with the CENP-A-specific chaperone HJURP to recruit a prenucleosomal complex of CENP-A/H4 to the centromere. The details of CENP-A nucleosome assembly following HJURP recruitment remain unclear, but may entail disassembly of H3 nucleosomes at sites where new CENP-A nucleosomes will form. **a** Schematic of CENP-A assembly assays in egg extract. Because CENP-A is retained on sperm chromatin through spermatogenesis, tagged CENP-A is added to extract to differential new and old CENP-A. Demembranated sperm chromatin or reconstituted CENP-A chromatin can serve as the template for CENP-A assembly. The assembly reaction is initiated by the addition of calcium and the CENP-A chaperone HJURP, recapitulating the in vivo requirements for assembly. Incorporation of tagged CENP-A at centromeres is assessed by immunofluorescence

question underpinning the epigenetic inheritance of CENP-A is what features of CENP-A chromatin are recognized by the assembly machinery and how. Extract affords the only system in which the composition of the template chromatin can be

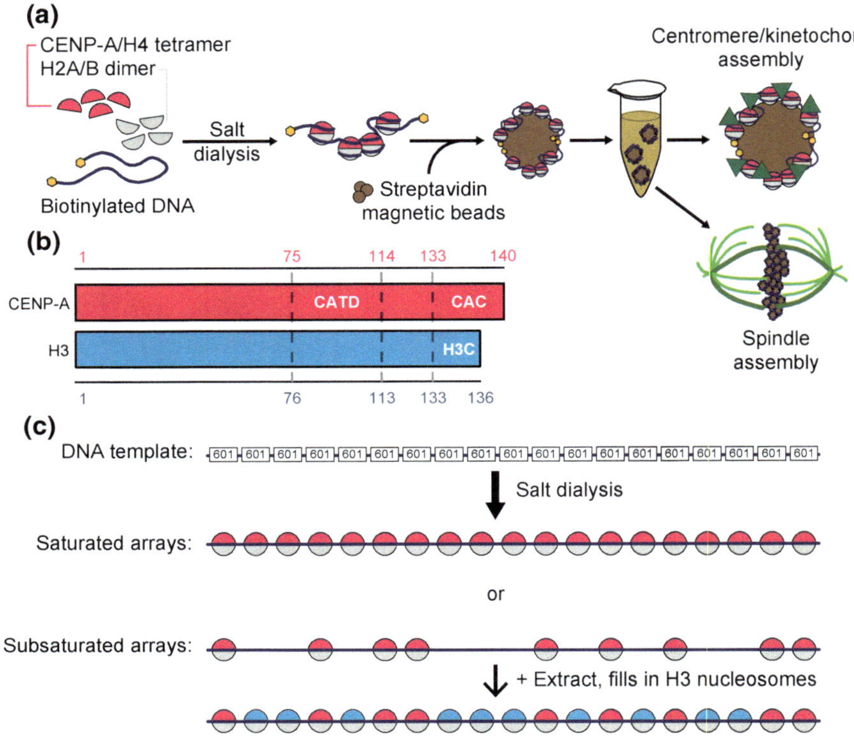

Fig. 4 Reconstitution of chromatin arrays. Reconstitution of kinetochore function with reconstituted chromatin arrays. **a** Schematic of chromatin reconstitution and kinetochore assembly assay. Purified CENP-A/H4 tetramer and H2A/B dimer are mixed in high salt buffer with biotinylated DNA containing high affinity nucleosome positioning sequences. Gradual dialysis of the assembly reaction to low salt causes spontaneous assembly of octameric nucleosomes on the DNA, which can then be coupled to streptavidin-coated magnetic beads. Following their recovery from egg extract, centromere and kinetochore protein recruitment or spindle assembly on chromatin-coated beads is assessed by immunofluorescence. **b** Schematic of CENP-A and H3 domains. Reconstitution of chromatin from chimeric histones in which domains of CENP-A have been exchanged with the corresponding domains of H3 provides a method for understanding the roles of specific CENP-A domains in centromere function. **c** The DNA substrate for chromatin reconstitution contains nineteen high affinity "601" nucleosome positioning sequences. Performing the reconstitution with higher concentrations of histone (2 μM) results in "saturated" chromatin arrays in which every positioning site is occupied by a CENP-A nucleosome. At lower concentrations (0.5 μM), only a subset of the sites are occupied, yielding "subsaturated" chromatin arrays. In egg extract, H3 nucleosomes assemble at these unoccupied sites resulting in mixed CENP-A/H3 arrays that may more accurately reflect the admixture of CENP-A and H3 nucleosomes observed at centromeres in vivo

freely manipulated to directly address this question. Chromatin reconstitution methods have been extensively developed over the past three decades, making it possible to assemble nucleosomes on arbitrary DNA sequences (Luger et al. 1999). This includes centromeric nucleosomes which have been reconstituted both as mononucleosomes and tandem nucleosome arrays (Guse et al. 2011; Yoda et al. 2000) (Fig. 4a). Using this approach, we altered both the composition and density of nucleosomes in the chromatin template underlying the centromere to probe the effects of centromeric chromatin composition on the CENP-A assembly process in egg extract.

The ability of CENP-A to promote the assembly of the centromere and kinetochore arises from key structural features of the CENP-A molecule that differ from H3 including the CATD—a loop and alpha helix within the histone fold domain (Black et al. 2004)—and a six amino acid C-terminal extension (the CENP-A C-terminus; CAC) (Carroll et al. 2010) (Fig. 4b). CENP-C and CENP-N are the two proteins known to selectively bind CENP-A nucleosomes. The CATD was shown to be the determinant in centromeric nucleosomes recognized by CENP-N (Carroll et al. 2009), and CENP-C recognizes the CAC (Carroll et al. 2010). Hydrophobic amino acids in the CAC contact hydrophobic residues in a conserved domain of CENP-C termed the "CENP-C motif" while basic amino acids in the CENP-C motif interact with negatively charged amino acids on H2A/B termed the "acidic patch"(Kato et al. 2013).

By generating chromatin arrays that contained chimeric nucleosomes with the CATD domain and/or the CAC replacing the corresponding domains of histone H3, we determined the functions of those CENP-A domains in promoting CENP-A assembly. Consistent with the essential role of CENP-C in CENP-A assembly, reconstitution of CENP-A assembly on chimeric templates requires the CAC (Westhorpe et al. 2015). However, M18BP1 localization and CENP-A assembly are only supported by chimeras containing both the CATD and the CAC (Westhorpe et al. 2015). This suggests a CATD-binding protein like CENP-N—whose role has been implied in cell-based studies (Carroll et al. 2009)—may be required, or that CENP-A assembly machinery itself recognizes CATD-containing chromatin (Sandmann et al. 2017). As nascent CENP-A requires the CATD for HJURP-mediated incorporation into chromatin (Foltz et al. 2009; Fachinetti et al. 2013; Logsdon et al. 2015), extract reconstitution provides an advantage over similar cell-based experiments that cannot distinguish such separate roles for the CATD in HJURP recognition and localization of the CENP-A assembly machinery.

Reconstituted chromatin also affords the opportunity to ask how the density of CENP-A nucleosomes contributes to CENP-A assembly. Centromeric nucleosomes are partitioned equally between sister chromatids during DNA replication in the absence of new CENP-A assembly (Jansen et al. 2007), with H3.3 proposed to fill the "gaps" in centromeric chromatin left by this redistribution (Dunleavy et al. 2011) (Fig. 3c). Performing chromatin reconstitution with lower histone concentrations results in nucleosome arrays that are "subsaturated," and egg extract assembles H3 nucleosomes in the gaps between CENP-A nucleosomes (Westhorpe et al. 2015) (Fig. 4c). While new CENP-A assembly occurs to some degree on chromatin templates that are saturated with CENP-A, it is twice as efficient on templates only 40% saturated with CENP-A (Westhorpe et al. 2015). This could reflect a role for H3 nucleosome disassembly in the CENP-A assembly process

(Dunleavy et al. 2011), and may suggest the existence of mechanisms to distinguish CENP-A from H3 nucleosomes.

Reconstitution of nucleosome arrays with different histone compositions has provided several insights into centromere assembly, but relatively untapped is the role of DNA sequence in centromere specification. Nucleosome arrays can be reconstituted on arbitrary DNA sequences, providing an ideal tool for addressing this question. In cells, human artificial chromosome formation requires CENP-B box-containing alpha satellite sequences (Ohzeki et al. 2002; Masumoto et al. 1998). Recently, mass spectrometry analysis of bacterial artificial chromosomes (BACs) containing human α-satellite DNA recovered from egg extract revealed association of several centromere proteins including CENP-A, suggesting sequence-dependent, de novo centromere assembly may be recapitulated in egg extract (Aze et al. 2016). Understanding the role of DNA sequence in centromere maintenance and neocentromere formation represents an important goal for the field.

3.2 CCAN Assembly

An essential function of CENP-A chromatin is to assemble the CCAN which serves as the platform for kinetochore assembly in mitosis (Fig. 1). Experiments in frog egg extract in which purified plasmid DNA was coupled to beads and added to metaphase extract showed that DNA alone promotes the formation of bipolar microtubule spindles. However, these DNA beads could not support the formation of a centromere or kinetochore (Heald et al. 1996). However, if CENP-A nucleosomes are preassembled on DNA before addition to egg extract then beads coated with CENP-A chromatin acquire the capacity to form centromeres and kinetochores that attach to microtubules and elicit mitotic arrest in response to microtubule depolymerization (Guse et al. 2011). This reconstituted chromatin system demonstrates that CENP-A is sufficient to build a functional centromere in frog egg extract and provides a powerful tool to study how CENP-A specifies centromere and kinetochore assembly.

Two CCAN proteins, CENP-C and CENP-N, bind directly to CENP-A nucleosomes (see above) and, through interactions with other centromere proteins, recruit the other members of the CCAN to assemble the constitutive centromere (Fig. 1). A key advantage of egg extract is that their functions in CENP-A nucleosome assembly can be uncoupled from their functions in CENP-A nucleosome assembly. Whereas the epigenetic inheritance of CENP-A requires both the CATD and the CAC (see above), chromatin reconstituted with chimeric histones containing only the CAC is sufficient for centromere and kinetochore assembly in mitotic extracts (Guse et al. 2011). These kinetochores bind microtubules and exhibit spindle assembly checkpoint activity (Guse et al. 2011). Despite direct binding of CENP-N to the CATD (Carroll et al. 2009), CAC-containing chimeras recruit more CENP-N in extract than CATD-containing chimeras, suggesting that CENP-C provides the critical link between centromeric chromatin and the CCAN (Guse et al. 2011). Multivalent interactions between CENP-C and several CCAN components including CENP-N could stabilize their localization to centromeres. Work in several labs is beginning to elucidate this complex network of interactions with purified proteins and how their collective assembly gives rise to selective centromere formation at CENP-A chromatin (McKinley et al. 2015; Hinshaw and Harrison 2013; Nishino et al. 2012; Weir et al. 2016; Basilico et al. 2014; Pesenti et al. 2016; Klare et al. 2015).

Cell-based studies have indicated that the mode of CENP-C recruitment changes from mitosis to interphase (Nagpal et al. 2015; Kwon et al. 2007), suggesting that CENP-C localization may be influenced either by cell cycle-regulated post-translational modification or by changes in the configuration of chromatin during chromosome condensation. In addition, the presence of H3 nucleosomes interspersed with CENP-A nucleosomes in centromeric chromatin may influence centromere assembly. Direct manipulation of chromatin configurations or of the amount of H3 at centromeres is not possible in vivo. But by reconstituting subsaturated chromatin templates (Fig. 4c), we have shown that the mechanism of CENP-C localization can change with the density of CENP-A nucleosomes at the centromere. Reconstituted chromatin saturated with CAC-containing chimeric nucleosomes recruits CENP-C similar to wild-type CENP-A chromatin (Westhorpe et al. 2015; Guse et al. 2011). In subsaturated arrays where chimeric CENP-A nucleosomes are interspersed with H3 nucleosomes; however, the CAC is not sufficient for full CENP-C recruitment (Westhorpe et al. 2015). In this case, the CATD is also required to recruit the same amount of CENP-C as CENP-A chromatin. CENP-C contains two nucleosome binding motifs (Kato et al. 2013) in addition to a dimerization motif (Sugimoto et al. 1997) and therefore may be able to engage adjacent CENP-A nucleosomes and localize exclusively by the CAC where the CENP-A density is high. At lower CENP-A densities, it may require interactions with other CCAN components such as CENP-N for robust localization. Additional studies are required to understand this regulation, and more detailed studies of in vivo chromatin organization will shed light on the biological relevance of different saturation states. However, alteration of centromere protein recruitment in response to chromatin organization poses an exciting possibility.

3.3 Kinetochore Assembly

The primary function of the centromere and kinetochore in mitosis is to couple chromosomes to the mitotic spindle and translate the force generated by depolymerizing microtubules into chromosome segregation (Cheeseman 2014). For many years, it was unclear how chromosomes were coupled to depolymerizing microtubules (Gudimchuk et al. 2013; Lombillo et al. 1995). Eventually genetic studies in budding yeast demonstrated that the Ndc80 complex (composed of Ndc80, Nuf2, Spc24, and Spc25) was required for microtubule attachment during anaphase (He et al. 2001; Wigge and Kilmartin 2001) (Fig. 1). But while vertebrate homologues of Ndc80 and Nuf2 could be identified by homology, the vertebrate homologues of Spc24/25 remained elusive. Unlike Ndc80 and Nuf2 mutants, Spc24/25 mutants in yeast not only failed to bind microtubules but also failed to arrest at the spindle assembly checkpoint in response to lack of microtubule attachment (Janke et al. 2001), highlighting the importance of characterizing the entire vertebrate Ndc80 complex. And although many parts of the centromere and kinetochore are conserved from yeast to vertebrates, it was not clear the microtubule-binding mechanism would be the same between yeast point centromeres, in which a single CENP-A nucleosome form the microtubule-binding site, and the larger, regional centromeres of vertebrates.

Egg extract provides concentrated starting material for biochemical purification of intact protein complexes, facilitating the identification of functionally related proteins. Using antibodies raised against *Xenopus* Ndc80 and Nuf2, McCleland and colleagues purified the 190-kD Ndc80 complex from *Xenopus* egg extract and demonstrated it was composed of four species: Ndc80, Nuf2, and highly divergent homologues of Spc24/25 (McCleland et al. 2004). Purification and cloning of the *Xenopus* proteins facilitated the cloning of the human genes and dissection of their functions in chromosome microtubule attachment and metaphase alignment in vivo (Bharadwaj et al. 2004; Ciferri et al. 2008).

The Ndc80 complex is the major microtubule-binding component of the KMN network, comprised of three subcomplexes: KNL1, which also binds microtubules and serves as the platform for spindle assembly checkpoint signaling (see below); Mis12, which connects the KMN network to the underlying centromere; and Ndc80 (Fig. 1). Purification and reconstitution of the Ndc80 complex demonstrated that cooperative arrays of Ndc80 maintain attachment to individual microtubules (Alushin et al. 2010; Ciferri et al. 2008). Reconstitution of the entire KMN network, however, synergistically enhances in vitro binding to purified microtubules (Cheeseman et al. 2006). An outstanding problem is understanding the basis for this synergy, as well as to understand how this cooperation plays out during chromosome segregation and gives rise to the mechanical properties of the kinetochore under load, a problem extract is well suited to address (see below). Overall, these studies support the model that microtubule binding by the kinetochore is largely conserved between yeast and vertebrates and the proposal that regional centromeres may effectively function as an array of point centromeres (Zinkowski et al. 1991).

Essential to proper chromosome segregation is that kinetochore assembly is restricted to the single centromeric locus (Westhorpe and Straight 2015). De novo centromere assembly in conjunction with immunodepletion studies have made extract an ideal system for dissecting how centromere proteins assemble the kinetochore. This has been particularly useful to address different models regarding the relative contributions of CENP-C and another essential CCAN component, CENP-T, to kinetochore recruitment (Basilico et al. 2014; Nishino et al. 2013; Rago et al. 2015) (Fig. 1). Centromeres on demembranated sperm rapidly assemble CENP-C upon incubation in metaphase extract (Milks et al. 2009). CENP-T does not localize to these centromeres until interphase, but then persists at centromeres after extract cycles into the next mitosis (Krizaic et al. 2015). This feature provides a convenient system for separating the respective contributions of CENP-C and CENP-T to kinetochore assembly.

In the absence of CENP-T, immunodepletion of CENP-C results in complete loss of Mis12 complex and spindle assembly checkpoint components in metaphase extract (Milks et al. 2009). By interacting with the CENP-C N-terminus, the Mis12 complex bridges the CCAN and the KMN network (Milks et al. 2009; Petrovic et al. 2010, 2014, 2016; Przewloka et al. 2011; Screpanti et al. 2011). In cycled mitotic extract in which both CENP-C and CENP-T localize to centromeres, CENP-C depletion only partially reduces Mis12 and Ndc80 localization (Krizaic et al. 2015). In turn, depletion of CENP-T also results in decreased Mis12 and Ndc80 localization (Krizaic et al. 2015), complementing in vitro studies where CENP-T binds both these components (Huis In 't Veld et al. 2016; Gascoigne et al. 2011; Nishino et al. 2013). While this supports a model in which CENP-C and CENP-T recruit Mis12 and Ndc80 by separate mechanisms, CENP-T localization is largely lost upon CENP-C depletion (Krizaic et al. 2015). This supports an emerging model that CENP-C plays a predominant role in organizing both the Mis12-Ndc80 and CENP-T-Ndc80 branches of kinetochore assembly (Klare et al. 2015), and suggests that additional factors contribute to kinetochore recruitment.

Our understanding of how the CCAN promotes kinetochore assembly and microtubule binding is still emerging. Efforts over the last decade to reconstitute kinetochore complexes with purified proteins have culminated in reconstitution of kinetochore particles comprising the KMN network bound to CENP-CHIKMLN bound to CENP-A mononucleosomes (Weir et al. 2016). This work demonstrated that the presence of the CCAN enhances in vitro microtubule binding by the KMN network, raising new questions about how these components integrate to form stable, load-bearing microtubule attachments. Moving forward, egg extract offers a powerful and unique system in which such detailed biochemical analyses of microtubule binding and kinetochore structure can be directly coupled to phenotypic assays of spindle formation and chromosome segregation by supplementation of extract with purified proteins.

3.4 The Spindle Assembly Checkpoint

Another essential function of the kinetochore is its ability to monitor and regulate microtubule attachment to ensure biorientation, or attachment to opposite spindle poles (Cheeseman 2014). Inter-kinetochore tension generated at bioriented chromosomes by opposing microtubule-pulling forces has been proposed to regulate microtubule dynamics and attachment which, in turn, regulates the metaphase/anaphase transition via the spindle assembly checkpoint (SAC). The SAC delays exit from mitosis in response to unattached kinetochores by inhibiting the E3 ubiquitin ligase activity of the anaphase promoting complex (APC) to permit establishment of biorientation (Lara-Gonzalez et al. 2012).

Disruption of the SAC in cells frequently results in chromosome missegregation and death.

Reconstituting spindle assembly checkpoint function in vitro generally poses a particular challenge because it requires reconstitution of both the kinetochore-microtubule interface and cell cycle regulation. These features, however, are easy to recapitulate in egg extract. Assembly of checkpoint protein complexes at kinetochores can be measured by immunofluorescence or biochemically. Furthermore, exit from mitosis in extract is easily monitored using radioactive kinase assays to measure the decline of Cdk activity where SAC activation presents as a delay in calcium-induced loss of Cdk phosphorylation.

Although the SAC is inactive in the early embryo and in extracts containing lower amounts of sperm chromatin, egg extract recapitulates the mitotic checkpoint when sperm is increased to 9000 sperm/μL (Minshull et al. 1994). SAC activation upon increased sperm concentration reflects a proportionality between the amount of checkpoint signal generated and the number of unattached kinetochores in the cytoplasm (Collin et al. 2013; Heinrich et al. 2013). While several genes required for spindle assembly checkpoint function had been identified in budding yeast (Hoyt et al. 1991; Li and Murray 1991), the molecular nature of the SAC signal and the mechanism by which the SAC monitored biorientation were unknown at that time (Minshull et al. 1994).

Xenopus Mad2 was among the first SAC components identified in vertebrates (Chen et al. 1996) (Fig. 5). Crucial to demonstrating the role of Mad2 in checkpoint arrest, immunodepletion of Mad2 from egg extract results in failure to delay mitotic exit upon nocodazole treatment, whereas supplementation with purified Mad2 stimulates mitotic arrest (Li et al. 1997; Fang et al. 1998). Unattached kinetochores recruit *Xenopus* Mad2 (Chen et al. 1996) and catalyze the formation of a soluble inhibitory complex: the mitotic checkpoint complex (MCC; Fig. 5), comprised of Mad2, Bub3, BubR1, and the APC co-activator subunit Cdc20 (Li et al. 1997; Fang et al. 1998; Rieder et al. 1995). Co-purification from extract and structural analyses have shown the MCC binds the APC as a pseudo-substrate, blocking cyclin degradation while simultaneously sequestering Cdc20 which is required to initiate anaphase (Fang et al. 1998; Izawa and Pines 2015). Association of the MCC with the APC ultimately leads to ubiquitination and degradation of Cdc20 followed by recycling of MCC components, resulting in robust but rapidly reversible inhibition of mitotic exit.

The ability to directly supplement egg extract with purified proteins afforded the first opportunity to investigate the role of Mad2 conformational changes in checkpoint control. Assembly of Mad2 into the MCC requires its conversion at unattached kinetochores from an "open" conformation to a "closed" conformation (Lara-Gonzalez et al. 2012; Luo et al. 2004) (Fig. 5). Purification of recombinant Mad2 from bacteria yields two forms: a monomeric, inactive form and an active, oligomeric form (Fang et al. 1998). Addition of monomeric Mad2 ("open" Mad2) to egg extract fails to stimulate checkpoint arrest; however, the conformation of oligomeric Mad2 recapitulates that of "closed" Mad2 and inhibits APC activity (Fang et al. 1998). The ease with which kinetochores can be manipulated in extract

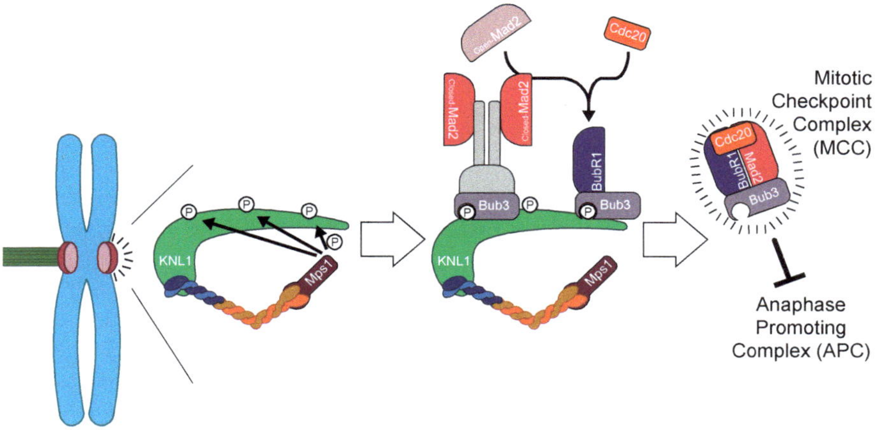

Fig. 5 Spindle assembly checkpoint regulation at kinetochores. Spindle assembly checkpoint (SAC) signaling prevents anaphase onset in response to unattached kinetochores. KNL1 serves as the platform for assembling SAC proteins. In the absence of microtubules, the checkpoint kinase Mps1 binds to Ndc80 and phosphorylates KNL1. Phosphorylated KNL1 recruits Bub3:BubR1, Mad2, and the APC coactivator Cdc20 to unattached kinetochores. Mad2 exists in both an "open" inactive conformation, and a "closed" active conformation. Open Mad2 must be converted to closed Mad2 at unattached kinetochores in order to bind Cdc20 and promote SAC arrest. Bub3, BubR1, Mad2, and Cdc20 assemble into the mitotic checkpoint complex (MCC) which inhibits the anaphase promoting complex by sequestering Cdc20, promoting Cdc20 degradation, and preventing cyclin degradation

and chromatin isolated has facilitated dissection of the catalytic cycle occurring on kinetochores by which Mad2 coverts between "open" to "closed" forms, a cycle which largely governs the responsiveness of the SAC to microtubule attachment.

Arguably the most important question regarding the spindle assembly checkpoint is how the kinetochore senses the status of its attachment to microtubules. This appears to be mediated in part by a kinase, Mps1, which binds competitively to the microtubule-binding surface of Ndc80 to identify unattached kinetochores (Hiruma et al. 2015; Ji et al. 2015) (Fig. 5). Mps1 phosphorylation of KNL1 promotes assembly of SAC proteins including Mad2 (Abrieu et al. 2001; Chen 2002; Zhao and Chen 2006; Hewitt et al. 2010). In addition, the absence of tension between kinetochores that are not bioriented promotes disassembly of improper microtubule attachments, contributing indirectly to SAC function. Because tension and microtubule attachment are intimately linked in vivo, however, it has been challenging to dissect their separate contributions to regulation of microtubule attachment and to SAC function. Indeed, elongation of Ndc80 experiencing tension along the kinetochore axis may directly affect SAC function by separating Mps1 from KNL1 and preventing its phosphorylation (Wan et al. 2009; Aravamudhan et al. 2015). While the role of tension in stabilizing kinetochore microtubules and regulating anaphase onset was first described fifty years ago in micromanipulation studies of grasshopper spermatocyte chromosomes (Nicklas and Koch 1969), in the absence of straightforward methods to simultaneously manipulate tension applied to kinetochores and perturb centromere/kinetochore composition, tension-dependent changes in kinetochore function have been challenging to dissect.

Purification and micromanipulation of yeast kinetochores have begun to provide insight into novel mechanisms by which tension stabilizes kinetochore microtubules (Akiyoshi et al. 2009, 2010; Miller et al. 2016). However, tolerance of the host to mutation of kinetochore components is still a limitation of this method. This limitation does not apply in egg extract, where tension can be applied directly to sperm or reconstituted kinetochores (Almagro and Dimitrov 2005; Yan et al. 2007). Extract also affords the unique opportunity to understand how cytoplasmic factors or those that do not copurify with kinetochores may contribute to their response to tension. In egg extract, Aurora substrate phosphorylation or SAC activation can be monitored directly, permitting careful measurement of force-response curves that would provide a quantitative, molecular description of how microtubule dynamics/attachment and cell cycle arrest are regulated by tension, measurements that have been challenging to make in vivo. Biophysical studies of kinetochores in egg extract will also provide a unique method for comparing the mechanical properties of regional vertebrate centromeres to the point centromeres of budding yeast.

4 Conclusion

Reconstitution studies have given us tremendous insight into the many functions of the centromere and kinetochore. Egg extract in particular has provided a versatile system for probing these questions biochemically because of the ease with which the composition of centromere, the composition of the cytoplasm, and the cell cycle state can be controlled. Extract is unique in its ability to capture the dynamically regulated aspects of this structure in vitro.

Historically, extract has been valued for the opportunity to validate observations by comparing processes reconstituted in vitro with the corresponding processes in *Xenopus* tissue culture cells or in intact embryos. Because of the difficulties associated with genetic manipulations in *Xenopus*, however, most in vivo work has shifted to human tissue culture cells or other model systems. But with the recent publication of the *Xenopus laevis* genome (Session et al. 2016) and rapid innovations in gene editing technologies, a new dawn for in vivo studies of chromosome segregation in frogs could lie ahead. The ability to genetically encode tagged centromere proteins would also circumvent the often laborious process of antibody generation that precedes most work in extract.

Our ability to reconstitute functional kinetochores from purified components has progressed from a pipe dream of the 1960s and 70s to reality. Although many questions remain regarding centromere recognition, centromere propagation, and kinetochore assembly and function, integrating reconstitution of specific structures from purified proteins with reconstitution of dynamic and cell cycle-regulated aspects of chromosome segregation in extract offers a wide array of possibilities for probing their mechanisms.

References

Abrieu A, Magnaghi-Jaulin L, Kahana JA, Peter M, Castro A, Vigneron S, Lorca T, Cleveland DW, Labbe J (2001) Mps1 is a kinetochore-associated kinase essential for the vertebrate mitotic checkpoint. Cell 106:83–93

Akiyoshi B, Nelson CR, Ranish JA, Biggins S (2009) Quantitative proteomic analysis of purified yeast kinetochores identifies a PP1 regulatory subunit. Genes Dev 23(24):2887–2899. doi:10.1101/gad.1865909

Akiyoshi B, Sarangapani KK, Powers AF, Nelson CR, Reichow SL, Arellano-Santoyo H, Gonen T, Ranish JA, Asbury CL, Biggins S (2010) Tension directly stabilizes reconstituted kinetochore-microtubule attachments. Nature 468(7323):576–579. doi:10.1038/nature09594

Almagro S, Dimitrov S (2005) Assembly and micromanipulation of Xenopus in vitro-assembled mitotic chromosomes. Curr Protoc Cell Biol. doi:10.1002/0471143030.cb2209s26 (Chapter 22:Unit 22 29)

Almouzni G, Mechali M (1988a) Assembly of spaced chromatin involvement of ATP and DNA topoisomerase activity. EMBO J 7(13):4355–4365

Almouzni G, Mechali M (1988b) Assembly of spaced chromatin promoted by DNA synthesis in extracts from Xenopus eggs. EMBO J 7(3):665–672

Alushin GM, Ramey VH, Pasqualato S, Ball DA, Grigorieff N, Musacchio A, Nogales E (2010) The Ndc80 kinetochore complex forms oligomeric arrays along microtubules. Nature 467(7317):805–810. doi:10.1038/nature09423

Aravamudhan P, Goldfarb AA, Joglekar AP (2015) The kinetochore encodes a mechanical switch to disrupt spindle assembly checkpoint signalling. Nat Cell Biol 17(7):868–879. doi:10.1038/ncb3179

Aze A, Sannino V, Soffientini P, Bachi A, Costanzo V (2016) Centromeric DNA replication reconstitution reveals DNA loops and ATR checkpoint suppression. Nat Cell Biol 18(6):684–691. doi:10.1038/ncb3344

Barnhart MC, Kuich PH, Stellfox ME, Ward JA, Bassett EA, Black BE, Foltz DR (2011) HJURP is a CENP-A chromatin assembly factor sufficient to form a functional de novo kinetochore. J Cell Biol 194(2):229–243. doi:10.1083/jcb.201012017

Basilico F, Maffini S, Weir JR, Prumbaum D, Rojas AM, Zimniak T, De Antoni A, Jeganathan S, Voss B, van Gerwen S, Krenn V, Massimiliano L, Valencia A, Vetter IR, Herzog F, Raunser S, Pasqualato S, Musacchio A (2014) The pseudo GTPase CENP-M drives human kinetochore assembly. Elife 3:e02978. doi:10.7554/eLife.02978

Benbow RM, Ford CC (1975) Cytoplasmic control of nuclear DNA synthesis during early development of Xenopus laevis: a cell-free assay. Proc Natl Acad Sci USA 72(6):2437–2441

Benbow RM, Pestell RQW, Ford CC (1975) Appearance of DNA-polymerase activities during early development of Xenopus-Laevis. Dev Biol 43(1):159–174. doi:10.1016/0012-1606(75)90138-4

Bernad R, Sanchez P, Rivera T, Rodriguez-Corsino M, Boyarchuk E, Vassias I, Ray-Gallet D, Arnaoutov A, Dasso M, Almouzni G, Losada A (2011) Xenopus HJURP and condensin II are required for CENP-A assembly. J Cell Biol 192(4):569–582. doi:10.1083/jcb.201005136

Bharadwaj R, Qi W, Yu H (2004) Identification of two novel components of the human NDC80 kinetochore complex. J Biol Chem 279(13):13076–13085. doi:10.1074/jbc.M310224200

Black BE, Foltz DR, Chakravarthy S, Luger K, Woods VL Jr, Cleveland DW (2004) Structural determinants for generating centromeric chromatin. Nature 430(6999):578–582. doi:10.1038/nature02766

Blow JJ, Laskey RA (1986) Initiation of DNA replication in nuclei and purified DNA by a cell-free extract of Xenopus eggs. Cell 47(4):577–587

Blow JJ, Laskey RA (2016) Xenopus cell-free extracts and their contribution to the study of DNA replication and other complex biological processes. Int J Dev Biol 60(7-8-9):201–207. doi:10.1387/ijdb.160142jb

Carroll CW, Silva MC, Godek KM, Jansen LE, Straight AF (2009) Centromere assembly requires the direct recognition of CENP-A nucleosomes by CENP-N. Nat Cell Biol 11(7):896–902. doi:10.1038/ncb1899

Carroll CW, Milks KJ, Straight AF (2010) Dual recognition of CENP-A nucleosomes is required for centromere assembly. J Cell Biol 189(7):1143–1155. doi:10.1083/jcb.201001013

Chang JB, Ferrell JE Jr (2013) Mitotic trigger waves and the spatial coordination of the Xenopus cell cycle. Nature 500(7464):603–607. doi:10.1038/nature12321

Cheeseman IM (2014) The kinetochore. Cold Spring Harb Perspect Biol 6(7):a015826. doi:10.1101/cshperspect.a015826

Cheeseman IM, Chappie JS, Wilson-Kubalek EM, Desai A (2006) The conserved KMN network constitutes the core microtubule-binding site of the kinetochore. Cell 127(5):983–997. doi:10.1016/j.cell.2006.09.039

Chen RH (2002) BubR1 is essential for kinetochore localization of other spindle checkpoint proteins and its phosphorylation requires Mad1. J Cell Biol 158(3):487–496. doi:10.1083/jcb.200204048

Chen RH (2008) The spindle checkpoint in Xenopus laevis. Front Biosci 13:2231–2237

Chen RH, Waters JC, Salmon ED, Murray AW (1996) Association of spindle assembly checkpoint component XMAD2 with unattached kinetochores. Science 274(5285):242–246

Ciferri C, Pasqualato S, Screpanti E, Varetti G, Santaguida S, Dos Reis G, Maiolica A, Polka J, De Luca JG, De Wulf P, Salek M, Rappsilber J, Moores CA, Salmon ED, Musacchio A (2008) Implications for kinetochore-microtubule attachment from the structure of an engineered Ndc80 complex. Cell 133(3):427–439. doi:10.1016/j.cell.2008.03.020

Collin P, Nashchekina O, Walker R, Pines J (2013) The spindle assembly checkpoint works like a rheostat rather than a toggle switch. Nat Cell Biol 15(11):1378–1385. doi:10.1038/ncb2855

Desai A, Deacon HW, Walczak CE, Mitchison TJ (1997) A method that allows the assembly of kinetochore components onto chromosomes condensed in clarified Xenopus egg extracts. Proc Natl Acad Sci USA 94:12378–12383

Desai A, Murray A, Mitchison TJ, Walczak CE (1998) Chapter 20 The use of xenopus egg extracts to study mitotic spindle assembly and function in vitro. Methods Cell Biol 61:385–412. doi:10.1016/s0091-679x(08)61991-3

Dunleavy EM, Roche D, Tagami H, Lacoste N, Ray-Gallet D, Nakamura Y, Daigo Y, Nakatani Y, Almouzni-Pettinotti G (2009) HJURP is a cell-cycle-dependent maintenance and deposition factor of CENP-A at centromeres. Cell 137(3):485–497. doi:10.1016/j.cell.2009.02.040

Dunleavy EM, Almouzni G, Karpen GH (2011) H3.3 is deposited at centromeres in S phase as a placeholder for newly assembled CENP-A in G(1) phase. Nucleus 2(2):146–157. doi:10.4161/nucl.2.2.15211

Earnshaw WC, Rothfield N (1985) Identification of a family of human centromere proteins using autoimmune sera from patients with scleroderma. Chromosoma 91(3–4):313–321. doi:10.1007/Bf00328227

Earnshaw WC, Honda BM, Laskey RA, Thomas JO (1980) Assembly of nucleosomes: the reaction involving *X. Laevis* nucleoplasmin. Cell 21(2):373–383

Earnshaw W, Bordwell B, Marino C, Rothfield N (1986) Three human chromosomal autoantigens are recognized by sera from patients with anti-centromere antibodies. J Clin Invest 77(2):426–430. doi:10.1172/JCI112320

Earnshaw WC, Sullivan KF, Machlin PS, Cooke CA, Kaiser DA, Pollard TD, Rothfield NF, Cleveland DW (1987) Molecular cloning of cDNA for CENP-B, the major human centromere autoantigen. J Cell Biol 104(4):817–829

Fachinetti D, Diego Folco H, Nechemia-Arbely Y, Valente LP, Nguyen K, Wong AJ, Zhu Q, Holland AJ, Desai A, Jansen LE, Cleveland DW (2013) A two-step mechanism for epigenetic specification of centromere identity and function. Nat Cell Biol 15(9):1056–1066. doi:10.1038/ncb2805

Fang G, Yu H, Kirschner MW (1998) The checkpoint protein MAD2 and the mitotic regulator CDC20 form a ternary complex with the anaphase-promoting complex to control anaphase initiation. Genes Dev 12(12):1871–1883

Flemming W (1880) Beitrage zur Kenntniss der Zelle und ihrer Lebenserscheinungen, Theil II. Archivfiir Mikroskopische Anatomie 18:151–259

Foley EA, Kapoor TM (2013) Microtubule attachment and spindle assembly checkpoint signalling at the kinetochore. Nat Rev Mol Cell Biol 14(1):25–37. doi:10.1038/nrm3494

Foltz DR, Jansen LE, Black BE, Bailey AO, Yates JR 3rd, Cleveland DW (2006) The human CENP-A centromeric nucleosome-associated complex. Nat Cell Biol 8(5):458–469. doi:10.1038/ncb1397

Foltz DR, Jansen LE, Bailey AO, Yates JR 3rd, Bassett EA, Wood S, Black BE, Cleveland DW (2009) Centromere-specific assembly of CENP-A nucleosomes is mediated by HJURP. Cell 137(3):472–484. doi:10.1016/j.cell.2009.02.039

Fujita Y, Hayashi T, Kiyomitsu T, Toyoda Y, Kokubu A, Obuse C, Yanagida M (2007) Priming of centromere for CENP-A recruitment by human hMis18alpha, hMis18beta, and M18BP1. Dev Cell 12(1):17–30. doi:10.1016/j.devcel.2006.11.002

Furuyama T, Dalal Y, Henikoff S (2006) Chaperone-mediated assembly of centromeric chromatin in vitro. Proc Natl Acad Sci USA 103(16):6172–6177. doi:10.1073/pnas.0601686103

Garner E, Costanzo V (2009) Studying the DNA damage response using in vitro model systems. DNA Repair (Amst) 8(9):1025–1037. doi:10.1016/j.dnarep.2009.04.015

Gascoigne KE, Takeuchi K, Suzuki A, Hori T, Fukagawa T, Cheeseman IM (2011) Induced ectopic kinetochore assembly bypasses the requirement for CENP-A nucleosomes. Cell 145(3):410–422

Gonen S, Akiyoshi B, Iadanza MG, Shi D, Duggan N, Biggins S, Gonen T (2012) The structure of purified kinetochores reveals multiple microtubule-attachment sites. Nat Struct Mol Biol 19(9):925–929. doi:10.1038/nsmb.2358

Gudimchuk N, Vitre B, Kim Y, Kiyatkin A, Cleveland DW, Ataullakhanov FI, Grishchuk EL (2013) Kinetochore kinesin CENP-E is a processive bi-directional tracker of dynamic microtubule tips. Nat Cell Biol 15(9):1079–1088. doi:10.1038/ncb2831

Gurdon JB (1968) Changes in somatic cell nuclei inserted into growing and maturing amphibian oocytes. J Embryol Exp Morphol 20(3):401–414

Gurdon JB, Woodland HR (1968) The cytoplasmic control of nuclear activity in animal development. Biol Rev Camb Philos Soc 43(2):233–267

Guse A, Carroll CW, Moree B, Fuller CJ, Straight AF (2011) In vitro centromere and kinetochore assembly on defined chromatin templates. Nature 477(7364):354–358. doi:10.1038/nature10379

Hayashi T, Fujita Y, Iwasaki O, Adachi Y, Takahashi K, Yanagida M (2004) Mis16 and Mis18 are required for CENP-A loading and histone deacetylation at centromeres. Cell 118(6):715–729. doi:10.1016/j.cell.2004.09.002

He X, Rines DR, Espelin CW, Sorger PK (2001) Molecular analysis of kinetochore-microtubule attachment in budding yeast. Cell 106(2):195–206

Heald R, Tournebize R, Blank T, Sandaltzopoulos R, Becker P, Hyman A, Karsenti E (1996) Self-organization of microtubules into bipolar spindles around artificial chromosomes in Xenopus egg extracts. Nature 382(6590):420–425. doi:10.1038/382420a0

Heinrich S, Geissen EM, Kamenz J, Trautmann S, Widmer C, Drewe P, Knop M, Radde N, Hasenauer J, Hauf S (2013) Determinants of robustness in spindle assembly checkpoint signalling. Nat Cell Biol 15(11):1328–1339. doi:10.1038/ncb2864

Hewitt L, Tighe A, Santaguida S, White AM, Jones CD, Musacchio A, Green S, Taylor SS (2010) Sustained Mps1 activity is required in mitosis to recruit O-Mad2 to the Mad1-C-Mad2 core complex. J Cell Biol 190(1):25–34. doi:10.1083/jcb.201002133
Hinshaw SM, Harrison SC (2013) An Iml3-Chl4 heterodimer links the core centromere to factors required for accurate chromosome segregation. Cell Rep. doi:10.1016/j.celrep.2013.08.036
Hiruma Y, Sacristan C, Pachis ST, Adamopoulos A, Kuijt T, Ubbink M, von Castelmur E, Perrakis A, Kops GJPL (2015) Competition between MPS1 and microtubules at kinetochores regulates spindle checkpoint signaling. Science 348(6240):1264–1267. doi:10.1126/science.aaa4055
Hoyt MA, Totis L, Roberts BT (1991) *S. cerevisiae* genes required for cell cycle arrest in response to loss of microtubule function. Cell 66(3):507–517
Huis In 't Veld PJ, Jeganathan S, Petrovic A, Singh P, John J, Krenn V, Weissmann F, Bange T, Musacchio A (2016) Molecular basis of outer kinetochore assembly on CENP-T. Elife 5: e21007. doi:10.7554/eLife.21007
Hyman AA, Mitchison TJ (1990) Modulation of microtubule stability by kinetochores in vitro. J Cell Biol 110(5):1607–1616
Hyman AA, Middleton K, Centola M, Mitchison TJ, Carbon J (1992) Microtubule-motor activity of a yeast centromere-binding protein complex. Nature 359(6395):533–536. doi:10.1038/359533a0
Iwao Y, Katagiri C (1984) In vitro induction of sperm nucleus decondensation by cytosol from mature toad eggs. J Exp Zool 230(1):115–124. doi:10.1002/jez.1402300115
Izawa D, Pines J (2015) The mitotic checkpoint complex binds a second CDC20 to inhibit active APC/C. Nature 517(7536):631–634. doi:10.1038/nature13911
Izuta H, Ikeno M, Suzuki N, Tomonaga T, Nozaki N, Obuse C, Kisu Y, Goshima N, Nomura F, Nomura N, Yoda K (2006) Comprehensive analysis of the ICEN (Interphase Centromere Complex) components enriched in the CENP-A chromatin of human cells. Genes Cells 11 (6):673–684. doi:10.1111/j.1365-2443.2006.00969.x
Janke C, Ortiz J, Lechner J, Shevchenko A, Shevchenko A, Magiera MM, Schramm C, Schiebel E (2001) The budding yeast proteins Spc24p and Spc25p interact with Ndc80p and Nuf2p at the kinetochore and are important for kinetochore clustering and checkpoint control. EMBO J 20 (4):777–791. doi:10.1093/emboj/20.4.777
Jansen LE, Black BE, Foltz DR, Cleveland DW (2007) Propagation of centromeric chromatin requires exit from mitosis. J Cell Biol 176(6):795–805
Ji Z, Gao H, Yu H (2015) Kinetochore attachment sensed by competitive Mps1 and microtubule binding to Ndc80C. Science 348(6240):1260–1264. doi:10.1126/science.aaa4029
Kato H, Jiang J, Zhou BR, Rozendaal M, Feng H, Ghirlando R, Xiao TS, Straight AF, Bai Y (2013) A conserved mechanism for centromeric nucleosome recognition by centromere protein CENP-C. Science 340(6136):1110–1113. doi:10.1126/science.1235532
Kim IS, Lee M, Park KC, Jeon Y, Park JH, Hwang EJ, Jeon TI, Ko S, Lee H, Baek SH, Kim KI (2012) Roles of Mis18alpha in epigenetic regulation of centromeric chromatin and CENP-A loading. Mol Cell 46(3):260–273. doi:10.1016/j.molcel.2012.03.021
Kingsbury J, Koshland D (1991) Centromere-dependent binding of yeast minichromosomes to microtubules in vitro. Cell 66(3):483–495
Klare K, Weir JR, Basilico F, Zimniak T, Massimiliano L, Ludwigs N, Herzog F, Musacchio A (2015) CENP-C is a blueprint for constitutive centromere-associated network assembly within human kinetochores. J Cell Biol 210(1):11–22. doi:10.1083/jcb.201412028
Kornbluth S, Evans EK (2001) Analysis of apoptosis using Xenopus egg extracts. Curr Protoc Cell Biol. doi:10.1002/0471143030.cb1112s09 (Chapter 11:Unit 11 12)
Kornbluth S, Smythe C, Newport J (1992) In vitro cell cycle arrest induced by using artificial DNA templates. Mol Cell Biol 12(7):3216–3223
Koshland DE, Mitchison TJ, Kirschner MW (1988) Polewards chromosome movement driven by microtubule depolymerization in vitro. Nature 331(6156):499–504. doi:10.1038/331499a0

Krizaic I, Williams SJ, Sanchez P, Rodriguez-Corsino M, Stukenberg PT, Losada A (2015) The distinct functions of CENP-C and CENP-T/W in centromere propagation and function in Xenopus egg extracts. Nucleus 6(2):133–143. doi:10.1080/19491034.2014.1003509

Kwon MS, Hori T, Okada M, Fukagawa T (2007) CENP-C is involved in chromosome segregation, mitotic checkpoint function, and kinetochore assembly. Mol Biol Cell 18 (6):2155–2168. doi:10.1091/mbc.E07-01-0045

Lara-Gonzalez P, Westhorpe FG, Taylor SS (2012) The spindle assembly checkpoint. Curr Biol 22 (22):R966–R980. doi:10.1016/j.cub.2012.10.006

Levy DL, Heald R (2015) Biological scaling problems and solutions in amphibians. Cold Spring Harb Perspect Biol 8(1):a019166. doi:10.1101/cshperspect.a019166

Li R, Murray AW (1991) Feedback control of mitosis in budding yeast. Cell 66(3):519–531

Li Y, Gorbea C, Mahaffey D, Rechsteiner M, Benezra R (1997) MAD2 associates with the cyclosome/anaphase-promoting complex and inhibits its activity. Proc Natl Acad Sci USA 94 (23):12431–12436

Logsdon GA, Barrey EJ, Bassett EA, DeNizio JE, Guo LY, Panchenko T, Dawicki-McKenna JM, Heun P, Black BE (2015) Both tails and the centromere targeting domain of CENP-A are required for centromere establishment. J Cell Biol 208(5):521–531. doi:10.1083/jcb.201412011

Lohka MJ, Masui Y (1983) Formation in vitro of sperm pronuclei and mitotic chromosomes induced by amphibian ooplasmic components. Science 220(4598):719–721

Lohka MJ, Masui Y (1984) Effects of $Ca2^+$ ions on the formation of metaphase chromosomes and sperm pronuclei in cell-free preparations from unactivated Rana pipiens eggs. Dev Biol 103 (2):434–442

Lohka MJ, Maller JL (1985) Induction of nuclear envelope breakdown, chromosome condensation, and spindle formation in cell-free extracts. J Cell Biol 101(2):518–523

Lombillo VA, Stewart RJ, McIntosh JR (1995) Minus-end-directed motion of kinesin-coated microspheres driven by microtubule depolymerization. Nature 373(6510):161–164. doi:10.1038/373161a0

Luger K, Rechsteiner TJ, Richmond TJ (1999) Expression and purification of recombinant histones and nucleosome reconstitution. Methods Mol Biol 119:1–16. doi:10.1385/1-59259-681-9:1

Luo X, Tang Z, Xia G, Wassmann K, Matsumoto T, Rizo J, Yu H (2004) The Mad2 spindle checkpoint protein has two distinct natively folded states. Nat Struct Mol Biol 11(4):338–345. doi:10.1038/nsmb748

Lupardus PJ, Van C, Cimprich KA (2007) Analyzing the ATR-mediated checkpoint using Xenopus egg extracts. Methods 41(2):222–231. doi:10.1016/j.ymeth.2006.07.024

Maddox PS, Hyndman F, Monen J, Oegema K, Desai A (2007) Functional genomics identifies a Myb domain-containing protein family required for assembly of CENP-A chromatin. J Cell Biol 176(6):757–763. doi:10.1083/jcb.200701065

Masui Y, Markert CL (1971) Cytoplasmic control of nuclear behavior during meiotic maturation of frog oocytes. J Exp Zool 177(2):129–145. doi:10.1002/jez.1401770202

Matsumoto Y, Bogenhagen DF (1989) Repair of a synthetic abasic site in DNA in a Xenopus laevis oocyte extract. Mol Cell Biol 9(9):3750–3757

Masumoto H, Ikeno M, Nakano M, Okazaki T, Grimes B, Cooke H, Suzuki N (1998) Assay of centromere function using a human artificial chromosome. Chromosoma 107(6–7):406–416. doi:10.1007/s004120050324

McCleland ML, Kallio MJ, Barrett-Wilt GA, Kestner CA, Shabanowitz J, Hunt DF, Gorbsky GJ, Stukenberg PT (2004) The Vertebrate Ndc80 complex contains Spc24 and Spc25 homologs, which are required to establish and maintain kinetochore-microtubule attachment. Curr Biol 14 (2):131–137. doi:10.1016/j.cub.2003.12.058

McKinley KL, Cheeseman IM (2016) The molecular basis for centromere identity and function. Nat Rev Mol Cell Biol 17(1):16–29. doi:10.1038/nrm.2015.5

McKinley KL, Sekulic N, Guo LY, Tsinman T, Black BE, Cheeseman IM (2015) The CENP-L-N complex forms a critical node in an integrated meshwork of interactions at the centromere-kinetochore interface. Mol Cell 60(6):886–898. doi:10.1016/j.molcel.2015.10.027

Mendiburo MJ, Padeken J, Fulop S, Schepers A, Heun P (2011) Drosophila CENH3 is sufficient for centromere formation. Science 334(6056):686–690. doi:10.1126/science.1206880
Miake-Lye R, Kirschner MW (1985) Induction of early mitotic events in a cell-free system. Cell 41(1):165–175
Milks KJ, Moree B, Straight AF (2009) Dissection of CENP-C-directed centromere and kinetochore assembly. Mol Biol Cell 20(19):4246–4255. doi:10.1091/mbc.E09-05-0378
Miller MP, Asbury CL, Biggins S (2016) A TOG protein confers tension sensitivity to kinetochore-microtubule attachments. Cell 165(6):1428–1439. doi:10.1016/j.cell.2016.04.030
Minshull J, Sun H, Tonks NK, Murray AW (1994) A MAP kinase-dependent spindle assembly checkpoint in Xenopus egg extracts. Cell 79(3):475–486
Mitchison TJ, Kirschner MW (1985a) Properties of the kinetochore in vitro. I. Microtubule nucleation and tubulin binding. J Cell Biol 101(3):755–765
Mitchison TJ, Kirschner MW (1985b) Properties of the kinetochore in vitro. II. Microtubule capture and ATP-dependent translocation. J Cell Biol 101(3):766–777
Moore MS, Blobel G (1992) The two steps of nuclear import, targeting to the nuclear envelope and translocation through the nuclear pore, require different cytosolic factors. Cell 69(6):939–950
Moree B, Meyer CB, Fuller CJ, Straight AF (2011) CENP-C recruits M18BP1 to centromeres to promote CENP-A chromatin assembly. J Cell Biol 194(6):855–871. doi:10.1083/jcb.201106079
Murray AW (1991) Chapter 30 cell cycle extracts. 36:581–605. doi:10.1016/s0091-679x(08)60298-8
Murray AW, Kirschner MW (1989) Cyclin synthesis drives the early embryonic-cell cycle. Nature 339(6222):275–280. doi:10.1038/339275a0
Murray AW, Desai AB, Salmon ED (1996) Real time observation of anaphase in vitro. Proc Natl Acad Sci USA 93(22):12327–12332
Nagpal H, Hori T, Furukawa A, Sugase K, Osakabe A, Kurumizaka H, Fukagawa T (2015) Dynamic changes in the CCAN organization through CENP-C during cell-cycle progression. Mol Biol Cell 26(21):3768–3776
Newmeyer DD, Forbes DJ (1988) Nuclear import can be separated into distinct steps in vitro, nuclear pore binding and translocation. Cell 52(5):641–653
Newmeyer DD, Farschon DM, Reed JC (1994) Cell-free apoptosis in Xenopus egg extracts: inhibition by Bcl-2 and requirement for an organelle fraction enriched in mitochondria. Cell 79(2):353–364
Newport J, Kirschner M (1982a) A major developmental transition in early xenopus embryos: I. characterization and timing of cellular changes at the midblastula stage. Cell 30(3):675–686
Newport J, Kirschner M (1982b) A major developmental transition in early Xenopus embryos: II. Control of the onset of transcription. Cell 30(3):687–696
Nicklas RB, Koch CA (1969) Chromosome micromanipulation. 3. Spindle fiber tension and the reorientation of mal-oriented chromosomes. J Cell Biol 43(1):40–50
Nishino T, Takeuchi K, Gascoigne KE, Suzuki A, Hori T, Oyama T, Morikawa K, Cheeseman IM, Fukagawa T (2012) CENP-T-W-S-X forms a unique centromeric chromatin structure with a histone-like fold. Cell 148(3):487–501. doi:10.1016/j.cell.2011.11.061
Nishino T, Rago F, Hori T, Tomii K, Cheeseman IM, Fukagawa T (2013) CENP-T provides a structural platform for outer kinetochore assembly. EMBO J 32(3):424–436. doi:10.1038/emboj.2012.348
Obuse C, Yang H, Nozaki N, Goto S, Okazaki T, Yoda K (2004) Proteomics analysis of the centromere complex from HeLa interphase cells: UV-damaged DNA binding protein 1 (DDB-1) is a component of the CEN-complex, while BMI-1 is transiently co-localized with the centromeric region in interphase. Genes Cells 9(2):105–120
Ohsumi K, Katagiri C (1991) Characterization of the ooplasmic factor inducing decondensation of and protamine removal from toad sperm nuclei—involvement of nucleoplasmin. Dev Biol 148(1):295–305. doi:10.1016/0012-1606(91)90338-4

Ohzeki J, Nakano M, Okada T, Masumoto H (2002) CENP-B box is required for de novo centromere chromatin assembly on human alphoid DNA. J Cell Biol 159(5):765–775. doi:10.1083/jcb.200207112

Ohzeki J, Shono N, Otake K, Martins NM, Kugou K, Kimura H, Nagase T, Larionov V, Earnshaw WC, Masumoto H (2016) KAT7/HBO1/MYST2 regulates CENP-A chromatin assembly by antagonizing Suv39h1-mediated centromere inactivation. Dev Cell 37(5):413–427. doi:10.1016/j.devcel.2016.05.006

Okada M, Cheeseman IM, Hori T, Okawa K, McLeod IX, Yates JR 3rd, Desai A, Fukagawa T (2006) The CENP-H-I complex is required for the efficient incorporation of newly synthesized CENP-A into centromeres. Nat Cell Biol 8(5):446–457. doi:10.1038/ncb1396

Palmer DK, O'Day K, Margolis RL (1990) The centromere specific histone CENP-A is selectively retained in discrete foci in mammalian sperm nuclei. Chromosoma 100(1):32–36. doi:10.1007/Bf00337600

Pesenti ME, Weir JR, Musacchio A (2016) Progress in the structural and functional characterization of kinetochores. Curr Opin Struct Biol 37:152–163. doi:10.1016/j.sbi.2016.03.003

Petrovic A, Pasqualato S, Dube P, Krenn V, Santaguida S, Cittaro D, Monzani S, Massimiliano L, Keller J, Tarricone A, Maiolica A, Stark H, Musacchio A (2010) The MIS12 complex is a protein interaction hub for outer kinetochore assembly. J Cell Biol 190(5):835–852. doi:10.1083/jcb.201002070

Petrovic A, Mosalaganti S, Keller J, Mattiuzzo M, Overlack K, Krenn V, De Antoni A, Wohlgemuth S, Cecatiello V, Pasqualato S, Raunser S, Musacchio A (2014) Modular assembly of RWD domains on the Mis12 complex underlies outer kinetochore organization. Mol Cell 53 (4):591–605. doi:10.1016/j.molcel.2014.01.019

Petrovic A, Keller J, Liu Y, Overlack K, John J, Dimitrova YN, Jenni S, van Gerwen S, Stege P, Wohlgemuth S, Rombaut P, Herzog F, Harrison SC, Vetter IR, Musacchio A (2016) Structure of the MIS12 complex and molecular basis of its interaction with CENP-C at human kinetochores. Cell 167 (4):1028–1040 e1015. doi:10.1016/j.cell.2016.10.005

Przewloka MR, Venkei Z, Bolanos-Garcia VM, Debski J, Dadlez M, Glover DM (2011) CENP-C is a structural platform for kinetochore assembly. Curr Biol 21(5):399–405. doi:10.1016/j.cub.2011.02.005

Rago F, Gascoigne KE, Cheeseman IM (2015) Distinct organization and regulation of the outer kinetochore KMN network downstream of CENP-C and CENP-T. Curr Biol 25(5):671–677. doi:10.1016/j.cub.2015.01.059

Rieder CL, Cole RW, Khodjakov A, Sluder G (1995) The checkpoint delaying anaphase in response to chromosome monoorientation is mediated by an inhibitory signal produced by unattached kinetochores. J Cell Biol 130(4):941–948. doi:10.1083/jcb.130.4.941

Saitoh H, Cooke CA, Burgess WH, Earnshaw WC, Dasso M (1996) Direct and indirect association of the small GTPase ran with nuclear pore proteins and soluble transport factors: studies in Xenopus laevis egg extracts. Mol Biol Cell 7(9):1319–1334

Salic A, Lee E, Mayer L, Kirschner MW (2000) Control of beta-catenin stability: reconstitution of the cytoplasmic steps of the wnt pathway in Xenopus egg extracts. Mol Cell 5(3):523–532

Sandall S, Severin F, McLeod IX, Yates JR 3rd, Oegema K, Hyman A, Desai A (2006) A Bir1-Sli15 complex connects centromeres to microtubules and is required to sense kinetochore tension. Cell 127(6):1179–1191. doi:10.1016/j.cell.2006.09.049

Sandmann M, Talbert P, D Demidov, Kuhlmann M, Rutten T, Conrad U, Lermontova I (2017) Targeting of A. thaliana KNL2 to centromeres depends on the conserved CENPC-k motif in its C-terminus. Plant Cell. doi:10.1105/tpc.16.00720

Sawin KE, Mitchison TJ (1991) Mitotic spindle assembly by two different pathways in vitro. J Cell Biol 112(5):925–940

Screpanti E, De Antoni A, Alushin GM, Petrovic A, Melis T, Nogales E, Musacchio A (2011) Direct binding of Cenp-C to the Mis12 complex joins the inner and outer kinetochore. Curr Biol 21(5):391–398. doi:10.1016/j.cub.2010.12.039

Session AM, Uno Y, Kwon T, Chapman JA, Toyoda A, Takahashi S, Fukui A, Hikosaka A, Suzuki A, Kondo M, van Heeringen SJ, Quigley I, Heinz S, Ogino H, Ochi H, Hellsten U, Lyons JB, Simakov O, Putnam N, Stites J, Kuroki Y, Tanaka T, Michiue T, Watanabe M, Bogdanovic O, Lister R, Georgiou G, Paranjpe SS, van Kruijsbergen I, Shu S, Carlson J, Kinoshita T, Ohta Y, Mawaribuchi S, Jenkins J, Grimwood J, Schmutz J, Mitros T, Mozaffari SV, Suzuki Y, Haramoto Y, Yamamoto TS, Takagi C, Heald R, Miller K, Haudenschild C, Kitzman J, Nakayama T, Izutsu Y, Robert J, Fortriede J, Burns K, Lotay V, Karimi K, Yasuoka Y, Dichmann DS, Flajnik MF, Houston DW, Shendure J, DuPasquier L, Vize PD, Zorn AM, Ito M, Marcotte EM, Wallingford JB, Ito Y, Asashima M, Ueno N, Matsuda Y, Veenstra GJ, Fujiyama A, Harland RM, Taira M, Rokhsar DS (2016) Genome evolution in the allotetraploid frog Xenopus laevis. Nature 538(7625):336–343. doi:10.1038/nature19840

Shamu CE, Murray AW (1992) Sister chromatid separation in frog egg extracts requires DNA topoisomerase II activity during anaphase. J Cell Biol 117(5):921–934

Sheehan MA, Mills AD, Sleeman AM, Laskey RA, Blow JJ (1988) Steps in the assembly of replication-competent nuclei in a cell-free system from Xenopus eggs. J Cell Biol 106(1):1–12

Shintomi K, Takahashi TS, Hirano T (2015) Reconstitution of mitotic chromatids with a minimum set of purified factors. Nat Cell Biol 17(8):1014–1023. doi:10.1038/ncb3187

Sorger PK, Severin FF, Hyman AA (1994) Factors required for the binding of reassembled yeast kinetochores to microtubules in vitro. J Cell Biol 127(4):995–1008

Sugimoto K, Kuriyama K, Shibata A, Himeno M (1997) Characterization of internal DNA-binding and C-terminal dimerization domains of human centromere/kinetochore autoantigen CENP-C in vitro: role of DNA-binding and self-associating activities in kinetochore organization. Chromosome Res 5(2):132–141

Walter JC, Arias EE (2004) Initiation of DNA replication in Xenopus egg extracts. Front Biosci 9:3029–3045

Wan X, O'Quinn RP, Pierce HL, Joglekar AP, Gall WE, DeLuca JG, Carroll CW, Liu ST, Yen TJ, McEwen BF, Stukenberg PT, Desai A, Salmon ED (2009) Protein architecture of the human kinetochore microtubule attachment site. Cell 137(4):672–684. doi:10.1016/j.cell.2009.03.035

Weir JR, Faesen AC, Klare K, Petrovic A, Basilico F, Fischbock J, Pentakota S, Keller J, Pesenti ME, Pan D, Vogt D, Wohlgemuth S, Herzog F, Musacchio A (2016) Insights from biochemical reconstitution into the architecture of human kinetochores. Nature 537(7619):249–253. doi:10.1038/nature19333

Westhorpe FG, Straight AF (2015) The centromere: epigenetic control of chromosome segregation during mitosis. Cold Spring Harb Perspect Biol 7(1):a015818. doi:10.1101/cshperspect.a015818

Westhorpe FG, Fuller CJ, Straight AF (2015) A cell-free CENP-A assembly system defines the chromatin requirements for centromere maintenance. J Cell Biol 209(6):789–801. doi:10.1083/jcb.201503132

Wigge PA, Kilmartin JV (2001) The Ndc80p complex from Saccharomyces cerevisiae contains conserved centromere components and has a function in chromosome segregation. J Cell Biol 152(2):349–360. doi:10.1083/jcb.152.2.349

Yan J, Maresca TJ, Skoko D, Adams CD, Xiao B, Christensen MO, Heald R, Marko JF (2007) Micromanipulation studies of chromatin fibers in Xenopus egg extracts reveal ATP-dependent chromatin assembly dynamics. Mol Biol Cell 18(2):464–474. doi:10.1091/mbc.E06-09-0800

Yoda K, Ando S, Morishita S, Houmura K, Hashimoto K, Takeyasu K, Okazaki T (2000) Human centromere protein A (CENP-A) can replace histone H3 in nucleosome reconstitution in vitro. P Natl Acad Sci USA 97(13):7266–7271. doi:10.1073/pnas.130189697

Zhao Y, Chen RH (2006) Mps1 phosphorylation by MAP kinase is required for kinetochore localization of spindle-checkpoint proteins. Curr Biol 16(17):1764–1769. doi:10.1016/j.cub.2006.07.058

Zinkowski RP, Meyne J, Brinkley BR (1991) The centromere-kinetochore complex: a repeat subunit model. J Cell Biol 113(5):1091–1110

Centrochromatin of Fungi

Steven Friedman and Michael Freitag

Abstract The centromere is an essential chromosomal locus that dictates the nucleation point for assembly of the kinetochore and subsequent attachment of spindle microtubules during chromosome segregation. Research over the last decades demonstrated that centromeres are defined by a combination of genetic and epigenetic factors. Recent work showed that centromeres are quite diverse and flexible and that many types of centromere sequences and centromeric chromatin ("centrochromatin") have evolved. The kingdom of the fungi serves as an outstanding example of centromere plasticity, including organisms with centromeres as diverse as 0.15–300 kb in length, and with different types of chromatin states for most species examined thus far. Some of the species in the less familiar taxa provide excellent opportunities to help us better understand centromere biology in all eukaryotes, which may improve treatment options against fungal infection, and biotechnologies based on fungi. This review summarizes the current knowledge of fungal centromeres and centrochromatin, including an outlook for future research.

1 Introduction

The location of centromeres is functionally defined by the presence of a specialized histone H3 variant, CENP-A, or the presence of kinetochore complex components (Cleveland et al. 2003; Ohzeki et al. 2015). This hypothesis has been confirmed in the overwhelming majority of eukaryotes with canonical chromatin, i.e., DNA wrapped around nucleosome cores formed by histones. Much recent work has focused on the recruitment and maintenance of CENP-A, the constitutive centromere-associated network (CCAN), which forms the inner kinetochore, and the KNL1-MIS12-NDC80 complexes (KMN), which form the outer kinetochore and serve as the attachment point of chromatin to the microtubule

S. Friedman · M. Freitag (✉)
Department of Biochemistry and Biophysics, Oregon State University,
2011 ALS Bldg, Corvallis, OR 97331, USA
e-mail: freitagm@cgrb.oregonstate.edu

B.E. Black (ed.), *Centromeres and Kinetochores*, Progress in Molecular and Subcellular Biology 56, DOI 10.1007/978-3-319-58592-5_4

spindles. In combination, the CCAN and KMN form the "kinetochore interaction network" (KIN); the fungal KIN has been reviewed recently (Freitag 2016). The "centromere" has come to denote DNA and DNA-binding chromatin proteins, whereas "kinetochore" has come to mean proteins that do not directly contact DNA (Fukagawa and Earnshaw 2014), though this is an arbitrary division.

Based on the functional definition, "regional" centromeres are epigenetically determined, which holds true in most organisms studied. Notable exceptions are yeasts in the Saccharomycetaceae, in which "point" centromeres with specific DNA sequence features are the rule. This apparent conflict in how centromeric DNA sequences are recognized during the deposition of centromere and kinetochore proteins raises many questions about the role of centromere sequence: What are the defining features of the underlying centromere sequences? How has the centromeric DNA sequence affected the evolution of centromere proteins, their epigenetics, and the structure of kinetochore proteins that interact with this locus? What mechanisms keep centromeres at the same locus on a chromosome across many generations? We will address some of these questions in this chapter on fungal centromeric DNA and centromeric chromatin, or "centrochromatin" (Sullivan and Karpen 2004). As high-throughput sequencing has improved, many novel genomes from diverse taxa are being assembled, and the extent of centromere sequences examined in detail is increasing rapidly. New technologies also enable us to turn species with few traditional genetic or molecular resources into model organisms, which allows us to address specific questions at a mechanistic level. The combination of new sequencing, proteomics, and cytological techniques are thus providing a more complete picture of what is required to faithfully segregate chromosomes, and the fungi prove to be a very diverse kingdom to study. By the end of this decade, well over 1000 different taxa will have complete or almost complete genome sequences available (see the "1000 Fungal Genomes Project"; http://1000.fungalgenomes.org/home/), perhaps the best sampling of an entire kingdom.

The Mycota, or fungi, span a wide range of lifestyles and serve important roles in biotechnology and biomedical research. Fungi also include some of the most devastating plant diseases that threaten food supplies on a regular basis, in addition to animal and human pathogens that are especially dangerous to immunocompromised patients. *Saccharomyces cerevisiae*, the budding or bakers yeast, is the best studied fungus because of its widespread use in the food, beverage, industrial, and biofuel industries, as well as its history as a model organism for genetics and cell biology. The fission yeast, *Schizosaccharomyces* pombe, is only distantly related to budding yeast and has become a prominent research organism for the study of mitosis and meiosis, DNA repair, chromatin, and chromosome structure. Two filamentous fungi, *Neurospora crassa* and *Aspergillus nidulans,* are classic genetic model organism instrumental for the development of biochemical genetics and eukaryotic molecular biology. *Aspergillus niger* is an industrial workhorse that produces many enzymes and most of the citric acid used in the world. In addition, there are lignocellulose degrading fungi, like the ascomycete *Trichoderma* and the basidiomycete *Phlebia*, that are used for the production of biofuels and enzymes.

An increasing number of fungi are studied as pathogens to humans, animals, plants, and even other fungi. Human fungal pathogens, like *Cryptoccocus neoformans*, *Candida albicans*, and *Aspergillus fumigatus*, are known for infecting mostly immunosuppressed patients. There is a shortage of effective antifungal treatments (Kilani and Fillinger 2016), which is especially critical when treating infections with basal taxa, like the "zygomycete" *Mucor circinelloides* and other members of this large and diverse group (Riley et al. 2016). Plant pathogenic fungi, such as *Zymoseptoria tritici*, *Magnaporthe oryzae*, and members of the diverse genus *Fusarium,* pose recurring problems to agriculture, causing devastating and costly diseases of all major industrial crops. By their diversity of lifestyles and uses fungi have far reaching impacts on humanity through their roles in biotechnology and biomedicine.

Centromere and kinetochore research in fungi has been focused intensely on the classical genetically and molecularly accessible workhorses, *S. cerevisiae* and *S. pombe.* More recently, work on the human pathogens, *C. albicans* and *C. neoformans*, the best-studied model for filamentous fungi, *N. crassa*, as well as the plant pathogens *Fusarium* and *Z. tritici* has revealed highly diverse centromere arrangements and sets of kinetochore proteins within the kingdom. As more is understood about centromere dynamics in human, animal, and plant populations, and particularly in human cancer cells, a diverse set of fungi may provide key insights as genetically, cytologically, and biochemically tractable systems. Future research on the centromeres of more basal lineages, fungi previously called "zygomycetes" (see http://zygolife.org/home/) and "chytrids" that have now been split into several large groups, will provide an unprecedented look into the diversity of functional centromeres and kinetochores across a whole kingdom. Preliminary results indicate that fungal pathogens have sufficiently divergent kinetochores, allowing for some components of this critical complex to be utilized as drug targets in human and agricultural pathogens (Sanyal 2012). Improved understanding of centromere biology in industrially relevant fungi promises new methods of genome engineering with stable minichromosomes and multi-organism shuttle vectors. Here we highlight what is known—and what remains unknown—about centromeric DNA sequences and centrochromatin of fungi by examining representatives of major taxa to compare and contrast the genetic and epigenetic requirements for proper chromosome segregation.

2 Centromeric DNA Sequence: Many Ways to Shape a Remarkable Genetic Locus

Diversity is a hallmark of centromeres. Centromere sequences range from discrete sequence motifs in *S. cerevisiae,* and related fungi, to the megabase-long arrays of simple repeats in human, many animal, and plant centromeres. In depth studies of model organisms have revealed features of centromeric DNA sequences, including

AT-richness and a propensity for repetitive elements, which seem to be conserved in many—but not all—eukaryotes. Centromeres are split into two broad categories, "point" and "regional" centromeres, based on the role of DNA sequences in centromere function and on the number of spindle microtubules recruited to the centromeric region of each chromosome.

2.1 *The Dark Matter of Fungal Centromeres Lies in the Basal Lineages*

The 1000 Fungal Genomes Project allows for an unprecedented look into the extent of centromere diversity in any kingdom by providing a resource to examine the variability of centromere sequences and conservation of kinetochore proteins. In addition to the traditional model organisms discussed below, we now have access to the genomes of many pathogens of interest for plant, animal, and human health studies. While this work is still underway, preliminary results support the idea that the centromere is an extremely divergent chromosomal locus. The last common ancestor of fungi split from the animals around ~ 1.5 billion years ago, and the most basal fungal lineages, the Microsporidia and Cryptomycota, separated from the precursors of the "zygomycetes", Ascomycota and Basidiomycota, ~ 1.3 billion years ago (Fig. 1) (Hibbett et al. 2007). Studies with the two most widely used ascomycete yeast species were—and still are—trail blazing, yet the largest diversity of the fungi is found in the other Ascomycota, Basidiomycota, and the more basal and poorly studied "zygomycetes", whose classification has recently been updated (Spatafora et al. 2016). No molecular studies, e.g., by ChIP with CENP-A homologs, have investigated the position of centromeres in the basal taxa (Fig. 1). Widespread absence of CENP-A or CENP-C, like in some insects and trypanosomes (Drinnenberg 2014; D'Archivio and Wickstead 2017), has not been observed in fungi (S. Friedman and M. Freitag, unpublished results), suggesting kinetochores that are similar to those in the Ascomycota and animals, though not all of the constituents of the most centrochromatin-proximal KIN subcomplexes can be identified by homology alone (Freitag 2016). Microsporidia are obligate commensals or parasites found in many animals; they have reduced genome size and mitochondrial function, and are overall poorly studied. Typical centromeric DNA elements similar to those found in the ascomycetes have not been found in the taxa for which high-quality genome drafts are available, e.g. the human parasite *Enzephalitozoon cuniculi* (Meraldi et al. 2006) or *Nosema spp.*, which infect bees, silkmoth, and other insects (Pombert et al. 2013). Many taxa in the Mucormycotina (e.g., *Mucor circinelloides*), Entomophthoromycotina (e.g., *Conidiobolus coronatus*, *Basidiobolus ranarum*) are opportunistic human pathogens that are difficult to control once an infection has been established (Mendoza 2014; Riley et al. 2016). Fungi in the Chytridiomycota (e.g., *Batrachochytrium dendrobatidis*) are well-known pathogens of frogs (James et al. 2015; Rosenblum et al. 2013). For many of these species genomes are available, but these organisms are often difficult to culture, have poorly developed molecular methods or genetics, and thus the wealth of information on these diverse taxa has been largely inaccessible for centromere and kinetochore studies. To understand the diversity of fungal centromeres and uncover underlying uniting principles more efforts should be directed toward understanding the biology of basal fungi.

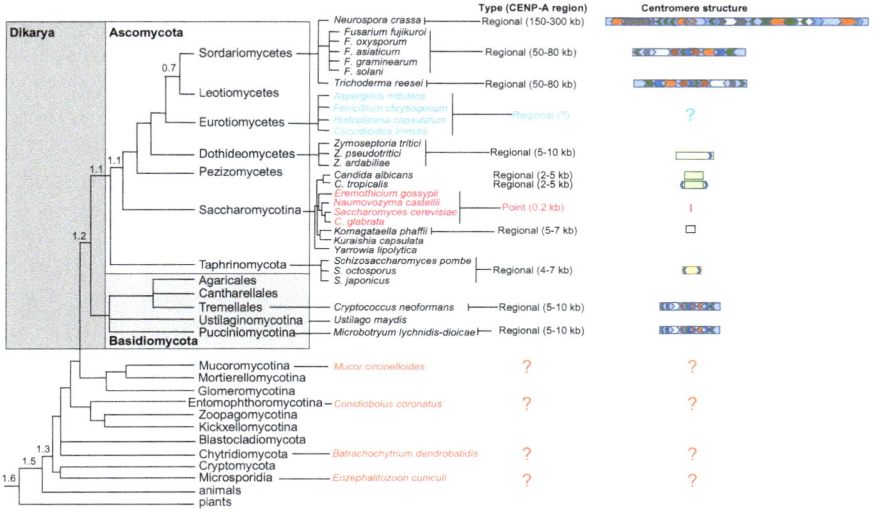

Fig. 1 Phylogenetic relationships between major clades of fungi, extent of CENP-A-enriched regions, and centromere repeat structure of taxa that have been studied. Basal taxa, including the microsporidia, chytrids and "zygomycetes" have not been examined at all (*orange question marks*); species named are mentioned in the text and important animal or human pathogens. In the Dikarya, Basidiomycota are not well studied; only *Cryptococcus* has been examined in detail and presumed centromeric DNA has been identified by Hi-C in *M. lychnidis-dioicae*. Ascomycota include several model organisms for genetics and molecular biology, e.g. budding yeast (*S. cerevisiae*), fission yeast (*S. pombe*), the dimorphic opportunistic pathogen *C. albicans*, and the molds *A. nidulans* and *N. crassa.* Evolution of point centromeres appears to be monophyletic in the Saccharomycetaceae, as all other taxa have some form of regional centromere. In fission yeast a conserved core region (*yellow*) is bordered by inner and outer repeats (*blue chevrons*), and a similar arrangement has been found in *C. tropicalis* (*green* non-conserved regions flanked by repeats). Repeats, either inverted or direct, are mostly absent from centromeric regions of other *Candida* species and the Dothideomycete genus *Zymoseptoria.* There is a large knowledge gap regarding centromeres in the industrially and medically important Eurotiomycetes—centromeric regions have not yet been identified by ChIP with CENP-A in any of these species (*blue question mark*). The Sordariomycetes that have been examined all have medium to large regional centromeres consisting of retrotransposons or relicts of retrotransposons from various families (colored chevrons). Based on sequence analyses of the most likely centromeric regions, several Dothideomycete genera (*Parastagonospora*, *Pyrenophora*) and several Erotiomycetes (*Aspergillus*, *Talaromyces*, *Penicillium*) seem to have centromeric regions that are similar to those in *Fusarium*; the arrangement in *Zymoseptoria* appears to be atypical. *Numbers* indicate time to last common ancestor in billion years

2.2 Fission Yeast, Budding Yeast, and the Dimorphic *Candida* as Models for Centromere Research

Earlier diverging species of ascomycetes fall into the Taphrinomycota. The fission yeast, *Schizosaccharomyces pombe*, is one of the best studied model organisms, whose centromeric core and pericentric regions range from 30 to 100 kb in size. Well-defined pericentric flanking regions include the *outer repeat* (*otr*) and *inner most repeat* (*imr*) surround the *central core* (*cc* or *cnt*; 4–7 kb) (Steiner, Hahnenberger, and Clarke 1993), which contains the majority of nucleosomes with CENP-A^{Cnp1} (Thakur et al. 2015). The discovery of *S. pombe* centromeres showed early on that fungi can have regional centromeres that do not depend on conserved recognition sequences for kinetochore complexes, and that they can be excellent genetic models for animal centromeres (Clarke 1998). The ill-conserved *cc* sequences alone are not sufficient to allow de novo CENP-A^{Cnp1} deposition and to assemble a functional kinetochore (Folco et al. 2008). The *otr* and *imr* repeats are tied to an siRNA-based pathway for de novo heterochromatin assembly that is also involved in proper recruitment of cohesins for binding sister chromatids (Bernard et al. 2001; Volpe et al. 2002; Hall et al. 2002; Allshire and Ekwall 2015). A role for centromere sequence to generate *cis*-acting short or long noncoding RNA is a novel concept that seems to be shared by other taxa as well (Choi et al. 2011; Koo et al. 2016; Scott 2013; Du et al. 2010; Rosic and Erhardt 2016; Rosic et al. 2014).

Other species in the genus *Schizosaccharomyces* have been investigated as some offer advantages for research on cell division or cell polarity. The genomes of *S. octosporus*, *S. japonicas,* and *S. cryophilus* have been sequenced and comparisons of centromere structure showed that *S. japonicus*, *S. octosporus,* and *S. cryophilus* all have repeat elements in their centromeric regions but that the *S. pombe cc* regions are not conserved and all repeat structures and arrangements are diversified between these species (Rhind et al. 2011). Some repeats in the centromeric regions of these species are longer than short reads obtained by high-throughput sequencing, thus there still remains some uncertainty about the exact placement and number of some repeat elements. Gene order and overall synteny around the centromeric and pericentric regions between *S. pombe* and *S. octosporus* are conserved (Rhind et al. 2011).

Almost 35 years ago, searches for centromere consensus sequences revealed that the budding yeast, *S. cerevisiae*, has a genetically defined centromere with three conserved Centromere Determining Elements (CDEI, CDEII, and CDEIII) (Clarke and Carbon 1980; Clarke 1985). CDEI, an 8-bp palindromic sequence, is bound by Cbf1 and CDEIII, a conserved 26-bp motif, is bound by the CBF3 complex, interrupted by 75–86 bp of AT-rich CDEII sequence; Cbf1 and CBF3 are conserved only in the Saccharomycotina. CDEI is not essential to form functional centromeres and some mutations within CDEII are tolerated. CDEIII, however, is essential to recruit centromere foundation and kinetochore proteins for proper spindle microtubule attachment. Discoveries with budding yeast facilitated trail blazing research into the requirements for kinetochore formation and “portable” centromere signals (Biggins 2013). *Candida glabrata* has recognizable CDE motifs, though swapping its CDE motifs with those of *S. cerevisiae* does not allow for normal centromere function (Kitada et al. 1997), and the same is true for *Kluyveromyces lactis* (Heus et al. 1994), though in both yeasts the cloned centromere elements stabilized plasmids with autonomously replicating sequences (ARS), like in *S. cerevisiae*. *Naumovozyma castellii*, a related budding yeast, contains CDE-like regions that have diverged even more from the consensus sequence of CDEI-III of other budding yeasts (e.g., *Ashbya gossypii* (Dietrich et al. 2004), *Vanderwaltozyma* (*Kluyveromyces*) *polyspora,* and *S. bayanus* (Gordon et al. 2011), but *Naumovozyma* species still utilize a CBF3 complex to recruit kinetochore components (Kobayashi et al. 2015). This implies that the point centromeres of Saccharomycetaceae can undergo accelerated evolution or may have evolved more than once (Fig. 1).

In addition to the presence of CDEs, kinetochores of *S. cerevisiae* bind a single microtubule per chromosome (Winey et al. 1995; Gonen et al. 2012); together, these two characteristics define what is a point centromere. The emergence of point centromeres is considered to have occurred before the whole genome duplication event that occurred in the ancestors of *Saccharomyces*, *Candida glabrata*, *N. castellii*, and *V. polyspora* but after divergence of the "true" *Candida* species and *Komagataella phaffii* (*Pichia pastoris*), which have short regional centromeres (Gordon et al. 2011). The simplest model for regional centromeres is thus a "repeat subunit model," in which several point centromeres with attached kinetochores attract binding of several spindle microtubules to a larger centromeric region (Zinkowski et al. 1991; Joglekar et al. 2008). In summary, although the point centromeres of budding yeast were discovered first, largely because of the genetic tractability of the system, based on the combination of molecular and phylogenetic evidence point centromeres of many Saccharomycotina represent a more recently evolved state (Malik and Henikoff 2009).

Genome-wide analyses of many yeast genomes from the Saccharomycotina suggested that searching for GC-poor troughs may indicate positions of centromeres. In *Yarrowia lipolytica* they coincide with the five experimentally identified centromeres (Vernis et al. 1999), in *Pichia stipitis* and *Debaryomyces hansenii* the troughs contain clusters of the retrotransposon Tps5 (Lynch et al. 2010), and in *Kuraishia capsulata*, a nitrate-assimilating yeast more closely related to *K. phaffii* (*P. pastoris*), 2–6 kb regions with a 200-bp conserved motif were identified. These putative centromeres were confirmed as interacting regions by chromosome conformation capture (3C) analysis (Morales et al. 2013) but ChIP experiments with the CENP-A homolog were not carried out. In the methylotrophic yeast *Hansenula polymorpha* AT-rich regions on the chromosomes were identified but functional studies to validate centromeric regions are lacking (Ravin et al. 2013). In *K. phaffii*, however, *mid* regions are flanked by 2-kb inverted repeats on all four chromosomes and have the highest occupancy of CENP-A^{Cse4}; some CENP-A^{Cse4} is also found within the flanking repeats. Because of the presence of inverted repeats, the authors suggested the existence of a third type of centromere, the regional "IR centromere" in addition to the classical point and the epigenetically determined regional centromeres (Coughlan et al. 2016). *Yarrowia lipolytica*, an important species for biofuels research (Ledesma-Amaro, Dulermo, and Nicaud 2015), contains small (0.2–1.2 kb), AT-rich centromeres but conserved CDE1 and CDEIII elements are lacking, suggesting that this may be the smallest regional centromere found thus far (Fournier et al. 1993; Vernis et al. 2001).

The genus *Candida* includes some of the best studied human pathogens within the fungi. *Candida albicans* and *S. cerevisiae* shared a common ancestor >145 million years ago, while *C. albicans* and *C. dubliniensis* diverged ~20 million years ago (Mishra et al. 2007). Recent work has thoroughly examined the centromeres of *C. albicans*, *C. dubliniensis*, *C. lusitaniae,* and *C. tropicalis*. *Candida albicans* centromeres were identified by immunoprecipitation of DNA with antibodies against CENP-A^{Cse4}, followed by cloning and sequencing (Sanyal et al. 2004). This direct identification of centromeric DNA showed that *C. albicans* has small (~3–5 kb) centromeres composed of non-repetitive sequence elements that are not conserved, even on chromosomes of the same strain or species (Fig. 1). Syntenic centromeric regions have different sequences in *C. dubliniensis* and *C. tropicalis* (Chatterjee et al. 2016; Padmanabhan et al. 2008). Centromere cores of *C. tropicalis* are flanked by inverted repeats, similar to the organization found in *S. pombe* and *K. phaffii* (Chatterjee et al. 2016). The seven centromeres of *C. tropicalis* are more similar than those of the other species, suggesting ongoing homogenization by gene conversion (Chatterjee et al. 2016). In all *Candida* species studied thus far, the core regions have the highest CENP-A^{Cse4} occupancy, but when (inverted) repeats are present they are often enriched with CENP-A^{Cse4}. At the same time, *Candida* centromeres very likely have—like *Saccharomyces* and other Saccharomycotina—a single kinetochore-microtubule spindle attachment per chromosome (Joglekar et al. 2008; Burrack e al. 2011), thus blurring the line between point and regional centromeres.

2.3 Filamentous Fungi Contain Diverse Types of Centromeres

The filamentous fungi comprise most of the taxa in the basal lineages or Dikarya (Fig. 1). In the fungi that have been used as model organisms for genetics, biochemistry, and molecular biology (e.g., *N. crassa*, *A. nidulans,* and *Coprinopsis cinerea*), centromere location on chromosomes had been determined by genetic mapping. In *N. crassa*, a 16 kb region that mapped to the centromeric region of LG VII was identified in a YAC library and sequenced (Cambareri, Aisner, and Carbon 1998; Centola and Carbon 1994). Some of the inactive retrotransposons appeared centromere-specific but after the genome of the most widely used lab strain of *N. crassa* was sequenced (Galagan et al. 2003) it became clear that *Tcen*, *Tgl1*, and *Tgl2* regions also occur in subtelomeric regions and dispersed heterochromatin (Selker et al. 2003). The approximate sizes of all centromeres were determined by genome sequencing as most of the centromeric and pericentric regions were captured and correctly assembled (Borkovich et al. 2004). This was aided by the presence of Repeat-Induced Point mutation (RIP), a premeiotic mutator system that affects duplicated sequences as short as 150 bp (Selker 1990; Gladyshev and Kleckner 2016). After RIP, both copies show transitions from C:G to T:A, and a DNA methyltransferase homologue, called RID, is involved (Freitag et al. 2002), though the process may not involve cytosine DNA methylation. RIP is active in successive premeiotic cycles and continues as long as duplications show more than ~85% identity. While RIP need not be complete (i.e., some repeats may "escape" one or several rounds of mutagenesis in individual strains), the effect on the population is heterogenization of previously identical repeated retrotransposons. This heterogeneity has made assembly of *N. crassa* centromeres tractable since the extensive repeat structures confounding assembly in other long regional centromeres do not exist.

Neurospora has large, 170–300 kb, regional centromeres with high AT content that are enriched with inactivated retroelements, much like the centromeres of plants and animals, but unlike in plants and some animals satellite repeats are absent (Smith et al. 2011). Based on ChIP-seq with CENP-A^{CenH3} the centromeric core regions are surrounded by short (2–45 kb) regions of pericentric heterochromatin. Limited examination of the Mauriceville strain showed large differences in the centromere sequences (Pomraning et al. 2011). Further comparative studies on DNA sequence variation within centromeres of different strains of *N. crassa* revealed several different centromere sequence types that shaped the centromeres within the pedigree of the most commonly used lab strain (Friedman and Freitag, in preparation). SNP mapping shows that centromeres are passed on as intact recombination blocks, lending further support to the long-standing hypothesis of recombination suppression within and near centromeres (Beadle 1932; Choo 1998). Gene conversion, however, does occur within centromeres of strains that are part of a pedigree of laboratory strains, or even within centromeres during single crosses (S. Friedman and M. Freitag, in preparation), as has also been observed in plants and insects (Shi et al. 2010; Miller et al. 2016). Centromere sequence variation has also been found in humans (Aldrup-MacDonald et al. 2016), and studies on natural populations will add greatly to our understanding of centromere–kinetochore evolution. Fungi seem the best-suited model organisms to use for these investigations.

Taxa most closely related to *Neurospora* all have shorter regional centromeres, with active or incapacitated retrotransposons, similar to those found in the putative centromeric regions of *Magnaporthe oryzae* (Thon et al. 2006) and *Verticillium* (Faino 2015; Seidl et al. 2015). For many of these species CENP-A^{CenH3} has not yet been mapped by ChIP-seq, though several species in the genus *Fusarium* (*F. graminearum*, *F. asiaticum*, *F. oxysporum*, *F. solani*, *F. fujikuroi*) have been examined more closely and revealed centromeric DNA of 50–80 kb. All have centromeres that are enriched with retrotransposons, some of which appear to be active. The centromeric regions also are more homogeneous within individual strains and between species than those of *Neurospora*, showing numerous short direct and inverted repeats (L. Connolly, S. Shahi, M. Rep, L. Fokkens, S.-H. Yun and M. Freitag, unpublished results). Like in *Candida*, centromeres of the sister species are embedded in regions with high synteny, even though centromeric sequence has diverged.

Genomes of several strains of the industrially important species *Trichoderma reesei* and its sister species had been assembled on scaffolds (Martinez et al. 2008; Schmoll et al. 2016; Mukherjee et al. 2013). Recent chromosomes conformation capture (3C) followed by high-throughput sequencing ("Hi-C") in combination with information from the existing 77 contigs allowed mapping to seven super-scaffolds by application of the GRAAL algorithm (Marie-Nelly et al. 2014). As this algorithm makes no a priori assumptions about centromere clustering (which is often seen in fungi, see below), the application of Hi-C techniques in the near future will allow the precise mapping and assembly of centromeres of many fungi. This technique will be faster and far more accessible for organisms for which tagging of CENP-A homologues or production of antibodies for ChIP-seq is too expensive or cumbersome.

Centromeres of Eurotiomycetes, a medically and industrially important class of ascomcyetes, are ill-described. Centromeric sequences from important industrial genera such as *Penicillium* or the human pathogens *Histoplasma* and *Coccidioides* remain unknown. Centromere location for species of the genus *Aspergillus* is conserved and largely syntenic (Fedorova et al. 2008), though centromere sequence has not been assembled yet. ChIP-seq with a tagged CENP-A^{CenH3} protein followed by assembly or mapping found the predicted edges of the centromeres but was insufficient to assemble the complete regions, suggesting the presence of near-identical long repeats (L. Connolly, J. Larsen, K. Smith, S. Osmani and M. Freitag, unpublished results). These had been predicted based on analyses of repeat elements, e.g., the *Dane1* and *Dane2* LTR elements (Aleksenko et al. 2001). Based on comparative genomics, centromeric regions are thought to be between 8 and 80 kb long (Fedorova et al. 2008).

The Dothideomycetes harbor some destructive pathogens, mostly of cereal crops. In some species the locations of centromeres had been predicted based on the longest AT-rich regions on each chromosome. Surprisingly, such predictions turned out to be wrong for the genus *Zymoseptoria*, where most centromeres are not associated with AT-rich DNA (Schotanus et al. 2015). Instead, ChIP-seq with *Z. tritici* CENP-A^{CenH3} revealed short (5–10 kb) CENP-A enriched regions without distinct sequence patterns; some centromeres contain expressed genes, while others harbor active or silent retroelements. Studies with two sister species, *Z. pseudotritici* and *Z. ardabiliae*, yielded similar results and also revealed the presence of at least two dicentric chromosomes (K. Schotanus, E. Stukenbrock and M. Freitag, in preparation). The surprisingly short and variable centromeres of this clade of true filamentous fungi will require future functional studies. This genus presents a molecular model for co-evolution of a pathogen (*Z. tritici*) with a domesticated host (wheat), while wild grass pathogens within the genus *Zymoseptoria* can serve as comparison.

Among the Basidiomycota only *Cryptococcus neoformans* var. *grubii* has been examined in any detail for centromere sequences, both by sequence comparisons and ChIP-seq with CENP-C (Loftus et al. 2005; Janbon et al. 2014). These regions are between 20 and 65 kb long and enriched for active or disabled Tcn1-Tcn6 retrotransposons. Flanking regions between *C. neoformans* var. *grubii* and *C. neoformans* var. *neoformans* are mostly syntenic though the chromosome numbering is changed between the two subspecies, and a similar arrangement was observed with the more distantly related *C. gattii* (Janbon et al. 2014). These centromeric regions are not heavily, if at all, transcribed. Mapping of CENP-C showed that this centromere foundation protein binds to a core region within the centromeric and pericentric region, covering ~5 kb on *CEN14* (Janbon et al. 2014). *Microbotryum lychnidis-dioicae* (Pucciniomycotina) has been sequenced by PacBio long sequencing read methods and the almost complete sequence predicts repeat-rich centromeres of ~100 kb (Badouin et al. 2015). As in many other genomes, the large contigs of several chromosomes show breaks within the centromeres. The well-studied *Ustilago maydis* seems to have short centromeres that are associated with ARS sequences (Meksem et al. 2005; Kamper et al. 2006) but, again, no detailed studies have been undertaken.

The classic genetic model organisms are by far the best studied organisms as far as fungal centromere sequences are concerned, and *N. crassa* is still one of the few systems to allow study of the role of DNA sequences in fully assembled large and repetitive centromeres. New genomics efforts, e.g., as part of the 1000 Fungal Genomes Project, usually involve Illumina short-read sequencing. To build complete assemblies this approach will need to be complemented with PacBio or Oxford Nanopore methods (Thomma et al. 2016), though Hi-C methods may also yield almost complete de novo genomes with centromeres identified as regions with the strongest interchromosomal interactions (Marie-Nelly et al. 2014; Galazka et al. 2016). While these methods provide important information, the extent of centromeres should always be confirmed by examining the localization and cell cycle-dependent behavior of the defining epigenetic marker, CENP-A, or other conserved KIN proteins.

What emerges as a common theme from the investigation of short point or regional centromeres in the Saccharomycotina and the longer regional centromeres in *Fusarium* and *Aspergillus* is that genes surrounding the centromeric core or repeat regions of related species are syntenic. This suggests that centromeric sequences undergo mutation without repair at a higher frequency than the surrounding sequences. What still remains to be determined is whether these observations imply positive selection toward functional sequence signals for deposition of CENP-A (and the kinetochore) and thus specific adaptation or even speciation (Malik and Henikoff 2002; Henikoff et al. 2001), or rather increased drift because the location of CENP-A deposition is determined epigenetically by pre-existing CENP-A nucleosomes, making DNA sequences immaterial. Investigations on neocentromere formation and function of centrochromatin, outlined below, attempt to shed light on this process.

3 Centrochromatin

That centromeric chromatin is different from bulk chromatin had been suspected since the realization that CENP-A is a variant of histone H3 (Earnshaw and Rothfield 1985; Palmer et al. 1987, 1991). Because of the sparsity of genes, enrichment of repeat elements and active or disabled retrotransposons, transcriptional repression, and compaction during interphase, centrochromatin has long been considered constitutive heterochromatin. This view was challenged when a mixture of histone modifications that are associated with both active and silent states was found in fission yeast, *Drosophila*, and human cells (Cam et al. 2005; Lam et al. 2006; Sullivan and Karpen 2004). Thus, together with euchromatin and heterochromatin, "centrochromatin" seems to constitute a distinct third form of chromatin.

The fundamental role of CENP-A in defining centromeres made this protein the focus of studies in every organism in which CENP-A-containing centromeres have been studied. Unlike the highly conserved canonical histone H3, CENP-A is rapidly evolving, especially the N-terminal tail, Loop 1 of the histone fold domain (HFD), and the CENP-A targeting domain (CATD), which are all involved in CENP-A localization and kinetochore interactions; in fungi this has been studied in the Taphrinomycotina (Folco et al. 2015), Saccharomycetacae (Baker and Rogers 2006), and Sordariomycetes (P. Phatale, S. Friedman and M. Freitag, unpublished results). Studies with yeast and human cells have exposed the C-terminal tail as important for recruitment of essential CCAN components, CENP-C and CENP-N (Westhorpe et al. 2015; Carroll et al. 2010; Fachinetti et al. 2013; Fang et al. 2015). The controversy about the shape and size of centromeric nucleosomes across the cell cycle has been reviewed extensively (Biggins 2013), and is discussed elsewhere. The balance of CENP-A nucleosomes and post-translationally modified H3 nucleosomes is a topic of ongoing investigations; earlier studies uncovered roles for histone modifications in de novo establishment of stable fission yeast centromeres (Folco et al. 2008). How histone modifications may aid in centromere maintenance, however, remains to be uncovered in most organisms. Because the complement of histone genes is very simple [in most fungi there are single genes for H2A, H2B and H3, and only two genes encoding identical H4 proteins; (Hays et al. 2002)], and because most fungi are genetically and molecularly tractable organisms this may be the area where they can contribute the most to advances in the near future.

While there are some similarities among eukaryotes, most organisms have evolved their own flavor of centrochromatin (Fig. 2). Thus, centrochromatin ranges from a single nucleosome in the budding yeasts to dozens or hundreds of kilobases of constitutive heterochromatin in the few filamentous fungi that have been examined. Even though some CENP-A^{Cse4} may incorporate into nucleosomes near the single nucleosome that is well positioned over the CDEII element and, along with the CBF3 complex, recruits the kinetochore, *S. cerevisiae* centrochromatin

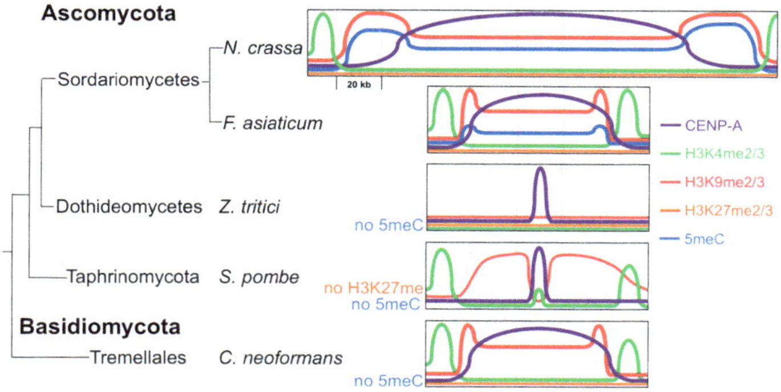

Fig. 2 Types of centrochromatin in reference fungi with regional centromeres. Representative species are arranged according to phylogeny (see Fig. 1). Species in the Sordariomycetes show CENP-A-enriched regions (*purple*) that stretch across 30–50 kb; *Neurospora* seems to be an exception with CENP-A-enriched regions that can be as long as 300 kb. In these species, H3K4me2/3 (*green*) and H3K27me2/3 (*orange*) are absent from the CENP-A regions, while H3K9me3 (*red*) nucleosomes and cytosine DNA methylation (5meC, *blue*) are interspersed with CENP-A nucleosomes. The pericentric regions, defined by presence of 5meC and H3K9me3 but absence of CENP-A, is usually short (5–20 kb). It is still unclear whether there is widespread 5meC in *Fusarium* species. Among the Dothideomycetes only *Zymoseptoria* species have been examined, and their CENP-A-enriched regions are short (5–10 kb); none of the histone modifications tested thus far are correlated with CENP-A enrichment; the relative enrichment of histone modifications across all centromeric regions is shown by the arrangement of the *colored lines* (i.e., H3K9me3 > H3K27me3 > H3K4me2). The two yeasts shown here, *S. pombe* and *C. neoformans*, have similar distributions of CENP-A and H3K9me2 or -me3 enrichments. Differences between the two species are the slight enrichment of H3K4me2 in the CENP-A-enriched region and the absence of all H3K27me2/3 in *S. pombe*. Not all species are capable to catalyze all chromatin modifications shown here (indicated to the *left of the panels*)

encompasses a single nucleosome (Biggins 2013). Functionally, however, the region involved in chromosome segregation is much larger (Lawrimore et al. 2016; Bloom 2014) and involves rapid exchange of canonical pericentric histones by chromatin remodeling (Verdaasdonk et al. 2012).

Candida species contain sequence-independent, regional centromeres. Centromere inactivation or deletion experiments in *C. albicans* demonstrated the utility of this system for the study of "neocentromere" formation and inheritance (Burrack and Berman 2012). Neocentromeres are previously naïve chromatin regions that give rise to functional kinetochores competent for chromosome segregation. Upon inactivation or deletion of the original centromeres, *Candida* neocentromeres form almost anywhere on the chromosome, though there is some preference for pericentric and subtelomeric regions (Burrack et al. 2016; Koren et al. 2010; Ketel et al. 2009; Thakur and Sanyal 2013). Like in *S. cerevisiae*, centromeres and neocentromeres in *Candida* interfere with transcription of nearby genes (Burrack et al. 2016), suggesting heterochromatin characteristics. The role of histone modifications or other classical markers for heterochromatin in *Candida* is unresolved; like *S. cerevisiae, Candida* lacks the conserved heterochromatin pathways that rely on methylation of H3K9 and H3K27.

Heterochromatin, defined by the presence of H3K9 di- or trimethylation (H3K9me2/3) in most and cytosine DNA methylation in many organisms, is a shared characteristic of most regional pericentric or centromeric regions. In *S. pombe*, the RNAi machinery is essential to recruit H3K9me2 to the pericentric repeat and help to incorporate CENP-A^{Cnp1} on naïve plasmid sequences (Folco et al. 2008), though the significance of the RNAi pathway in centromere function during the normal cell cycle is still uncertain. Once assembled, the heterochromatin machinery (the histone methyltransferase Clr4^{SUV39} and the H3K9me2 adapter Swi6^{HP1}) is not required for CENP-A^{Cnp1} inheritance on plasmid DNA. The RNAi requirement for de novo CENP-A^{Cnp1} deposition can be circumvented by tethering Clr4^{SUV39} to specific regions (Kagansky et al. 2009). Nevertheless, heterochromatin seems necessary for proper chromosome segregation and chromosome structure, likely by the recruitment of cohesins (Pidoux and Allshire 2005; Bernard et al. 2001; Mizuguchi 2014; Nonaka et al. 2002). Early genome-wide ChIP studies showed that the *imr* and *cc* (*cnt*) regions were enriched for a euchromatic mark, H3K4me2, and interspersed with CENP-A^{Cnp1} nucleosomes, which further motivated studies on the role of ncRNA in centromere function (Choi et al. 2011; Djupedal et al. 2009) that are still ongoing. Conversely, recent experiments suggest that H3 nucleosomes are depleted from the centromeric core and that CENP-A^{Cnp1} nucleosomes and CENP-T complex dominate (Thakur et al. 2015).

Studies on centrochromatin of *N. crassa* revealed that centromeric regions of *N. crassa* contain blocks of canonical nucleosomes that are enriched with H3K9me3 interspersed with blocks of CENP-A^{CenH3} nucleosomes (Smith et al. 2011, 2012). In mutants lacking the H3K9me3 methyltransferase DIM-5^{SUV39} and the H3K9me3-binding protein HP1 the regions enriched for CENP-A^{CenH3} were smaller. H3K4me2/3 did not associate with the formerly H3K9me3-enriched nucleosomes though overall nucleosome occupancy seemed unaltered (Smith et al. 2011). Further studies revealed re-localization of H3K27me2/3 to regions usually occupied by H3K9me3 in DIM-5^{SUV39} and HP1 mutants, suggesting that HP1 prohibits H3K27 di- and trimethylation in H3K9me3-regions (Basenko et al. 2015; Jamieson et al. 2016). Single mutants lacking DIM-5^{SUV39} and HP1 show growth and chromosome segregation phenotypes and are homozygously sterile or result in aberrant progeny (Freitag et al. 2004; Tamaru and Selker 2001), while SET-7^{EZH2} mutants show no overt defects (Jamieson et al. 2013). In DIM-5^{SUV39} SET-7^{EZH2} or HP1 SET-7^{EZH2} double mutants these phenotypes are largely suppressed (Basenko et al. 2015; Jamieson et al. 2016), suggesting that it is the presence of H3K27me2/3 at centromeres or its absence from normal facultative heterochromatin that results in the chromosome segregation defects.

Hi-C and cytological studies with wild type and mutants defective in constitutive (DIM-5^{SUV39}, HP1) or facultative (SET-7^{EZH2}) heterochromatin showed that, like in budding (Duan et al. 2010) and fission yeast (Tanizawa et al. 2010; Mizuguchi 2014), all centromeres are co-localized within the nucleus, revealing Rabl orientation (Klocko 2016; Galazka et al. 2016). Surprisingly, DIM-5^{SUV39}, HP1, and

SET-7^{EZH2} mutants showed relatively minor changes by Hi-C; the overall organization of centromeres was maintained and no gross chromosomal defects were observed. A novel role for importin alpha (*N. crassa* DIM-3) in chromosome organization was found (Galazka et al. 2016); importin alpha may act by its role in targeting of the DIM-5-containing DCDC complex (Klocko et al. 2015), though mechanisms remain to be uncovered. Cytology showed altered position and increased numbers of centromere foci in the SET-7^{EZH2} but not the DIM-5^{SUV39} and HP1 single or SET-7^{EZH2} DIM-5^{SUV39} double mutants (Klocko 2016). These data suggest a role for H3K27me2/3 in the control of centromere maintenance.

A link between the RNAi and meiotic silencing pathways and heterochromatin establishment in *N. crassa* has not been found (Freitag et al. 2004), unlike in fission yeast, and no long ncRNA transcripts or H3K4me2/3 have so far been detected (P. Phatale, K. Smith, M. Freitag, unpublished results). This suggests that RNAi or long ncRNA may not play an essential role in the formation or maintenance of centrochromatin in all fungi.

Fusarium species that have been studied show depletion of H3K4me2/3 and little transcription in their centromeric regions. In *F. fujikuroi* centromeric regions are 50–80 kb long and associated with H3K9me3 (Wiemann et al. 2013), similar to *F. oxysporum*, *F. asiaticum*, *F. solani,* and *F. graminearum* (L. Connolly, S. Shahi, M. Rep, L. Fokkens, S.-H. Yun and M. Freitag, unpublished results). *Fusarium* species use H3K27me3 to silence ~30% of their genomes, and most of these regions are in large subtelomeric blocks. No H3K27me3 has been detected in centromeric regions (Studt 2016; Connolly et al. 2013) and there is very little H3K27me2 (L. Connolly, R. Gonçalves, M. Freitag, unpublished results). Re-localization of H3K27me2/3, which has been observed in *Neurospora* DIM-5^{SUV39} and HP1 mutants as well as *C. neoformans ccc1* mutants, does not occur in *F. graminearum* (L. Connolly and M. Freitag, unpublished results). Overt phenotypes are also drastically different from those observed in *Neurospora* (Jamieson et al. 2013), as H3K9me3-defective strains have no discernable phenotypes under standard growth conditions while single mutants lacking H3K27me3 or double mutants lacking H3K9me3 and H3K27me3 show numerous developmental and other defects (Connolly et al. 2013).

In *Z. tritici*, centrochromatin cannot easily be defined because CENP-A-enriched regions do not show any obvious DNA sequence or chromatin pattern (Schotanus et al. 2015). So far only H3K4, H3K9 and H3K27 methylation have been tested but none of these marks overlap reliably with CENP-A localization. In contrast to *Neurospora* or *Fusarium*, many H3K9me3 regions also are enriched for H3K27me3; the control for deposition of both histone marks in this species remains to be deciphered.

Currently, the only basidiomycete that has been studied in regard to centrochromatin is *C. neoformans.* H3K9me2 is found almost exclusively in the centromeric, pericentric, and subtelomeric regions. While no H3K27me3 was found in centromeric regions, it overlaps with H3K9me2 at the subtelomeric loci (Dumesic et al. 2015), similar to what had been found with *Neurospora* (Basenko et al. 2015; Jamieson et al. 2016) and *Zymoseptoria* (Schotanus et al. 2015). As in

Table 1 Characteristics of fungal centromeric DNA and centrochromatin

Organism	*CEN* (kb)	Sequence content	A + T (%)	Histone mod.	DNA mod.
S. pombe	4–7	Unique core, *imr*	70	None specific	None known
S. cerevisiae	0.2	CDEI-III	65–95	None specific	None known
C. albicans	2–5	Unique seq.	65	None specific	Uncertain
C. tropicalis	2–5	Core flanked by repeats	65	None known	None known
Z. tritici	5–10	Unique seq.	53	None specific	Uncertain
F. asiaticum	50–80	TE, SSR	75	H3K9me3	None known
N. crassa	150–300	TE, SSR	7	H3K9me3	Cytosine methylation
C. neoformans	60–110	TE, SSR	52	H3K9me3	None known

Species are arranged by accepted phylogeny (see Fig. 1). While centromeric regions in most fungi are AT rich, *C. tropicalis* and *Z. tritici* have CENP-A-enriched regions (*CEN*) that have A-T% below genome average. Status of DNA methylation in *C. albicans* is uncertain; there is a single report of cytosine methylation. In *Z. tritici*, the gene for the cytosine DNA methyltransferase homologue of *N. crassa* DIM-2 has been duplicated and underwent RIP, thus abolishing DNA methylation. No specific histone modifications have been associated with the CENP-A nucleosome or surrounding canonical nucleosomes in the CENP-A-enriched regions in *S. cerevisiae*, *Candida* and *Zymoseptoria* species, though this has not been exhaustively tested. In *Neurospora*, *Fusarium*, and *Cryptococcus* the CENP-A-enriched regions are also enriched for H3K9me3 and depleted of H3K4me2 and H3K27me2/3, though other histone modifications have not been thoroughly tested yet. References for the data shown are mentioned in the text

Neurospora, re-localization of H3K27me3 has been found under certain conditions. The *Cryptococcus* Ezh2^{EZH2} complex contains three conserved proteins (Ezh2, Eed and Msl1) and two proteins unique to *Cryptococcus* (Bnd1, Ccc1), but it lacks a recognizable SUZ12 homologue that is usually associated with E(z) (Dumesic et al. 2015). While there are no clear homologues of Bnd1 in ascomycetes, the best Ccc1 homologues in *Neurospora* and *Fusarium* are not involved in H3K27 methylation (Jamieson et al. 2016) (L. Connolly and M. Freitag, unpublished results). Ccc1 is involved in H3K27me3 recognition and binding. Disruption of *ccc1* results in redistribution of H3K27me3 into H3K9me2 domains, especially the centromeric regions. If H3K9me2 deposition is abolished by deletion of the Clr4^{SUV39} homologue, H3K27me3 is also lost from these regions in the *ccc1* background (Dumesic et al. 2015). These results suggest that normal binding of the Ezh2 complex to its product, H3K27me3, via Ccc1 suppresses an inherent activity toward H3K9me2-modified centromeric chromatin regions (Dumesic et al. 2015).

4 Summary

Many fungi are genetically tractable and accessible to modern molecular techniques. This allows mechanistic studies of the balance of chromatin marks, transcription, and CENP-A deposition at stable centromere loci; such studies will continue to provide deep insights into centromere regulation. Fungi represent an opportunity to test centrochromatin plasticity that is still difficult to carry out in many other model organisms (Table 1). Species with large regional centromeres can be leveraged to improve understanding of the centromere defects that may play a role in tumorigenesis and other genomic instabilities that affect plants and humans. In addition, many fungi play a significant role in biotechnology applications, including the textiles, food, biofuels, and drug industries. Due to the role of centromeres in chromosome stability, centromere research should provide opportunities to genetically engineer strains with improved functionality for these industries. Research on the centromeres of fungi is just gaining speed and improved techniques in molecular biology and genomics are allowing insights into the functional diversity of this critical chromosomal locus. Future research on the varying sizes and types of centromeres in the filamentous fungi will provide a clearer picture of how these factors determine the number and recruitment of spindle microtubules. Increasing microtubule attachments may provide a remedy for chromosome segregation defects, while eliminating aberrant microtubule attachments may attenuate meiotic drive defects that occur in cancer cells.

Centrochromatin seems to be as diverse as all other aspects of centromere biology studied to this point. Though several types of centrochromatin have been found within the fungi some common themes have emerged. All known variants of centrochromatin in fungi are more similar to heterochromatin than euchromatin, whether due to the presence of silencing marks (e.g., H3K9me2/3) or the involvement of the RNAi machinery in *S. pombe*. Hi-C studies suggest that there is a higher order organization to centrochromatin, though the precise role of heterochromatin and condensin or other scaffolding proteins is still uncertain. Future studies will need to define functional roles of the different centrochromatin states in model fungi from major understudied clades. Examination of diverse fungi will help to reveal the full extent of centrochromatin states in nature, which should provide insights into possible links between centrochromatin and lifestyle, genomic architecture, and other aspects of biology. While it is clear that chromatin modification enzymes are important for centromere establishment more work must be done to determine what, if any, negative effects on chromosome segregation occur in chromatin mutants. Hi-C studies carried out across the cell cycle in synchronized cells will provide deeper understanding of the relationships between chromatin state and centromere function from a chromosome organization perspective.

References

Aldrup-MacDonald ME, Kuo ME, Sullivan LL, Chew K, Sullivan BA (2016) Genomic variation within alpha satellite DNA influences centromere location on human chromosomes with metastable epialleles. Genome Res 26:1301–1311

Aleksenko A, Nielsen ML, Clutterbuck AJ (2001) Genetic and physical mapping of two centromere-proximal regions of chromosome IV in Aspergillus nidulans. Fungal Genet Biol 32:45–54

Allshire RC, Ekwall K (2015) Epigenetic regulation of chromatin states in schizosaccharomyces pombe. Cold Spring Harb Perspect Biol 7:a018770

Badouin H, Hood ME, Gouzy J, Aguileta G, Siguenza S, Perlin MH, Cuomo CA, Fairhead C, Branca A, Giraud T (2015) Chaos of rearrangements in the mating-type chromosomes of the anther-smut fungus microbotryum lychnidis-dioicae. Genetics 200:1275–1284

Baker RE, Rogers K (2006) Phylogenetic analysis of fungal centromere H3 proteins. Genetics 174:1481–1492

Basenko EY, Sasaki T, Ji L, Prybol CJ, Burckhardt RM, Schmitz RJ, Lewis ZA (2015) Genome-wide redistribution of H3K27me3 is linked to genotoxic stress and defective growth. Proc Natl Acad Sci U S A 112:E6339–E6348

Beadle GW (1932) A possible influence of the spindle fibre on crossing-over in drosophila. Proc Natl Acad Sci U S A 18:160–165

Bernard P, Maure JF, Partridge JF, Genier S, Javerzat JP, Allshire RC (2001) Requirement of heterochromatin for cohesion at centromeres. Science 294:2539–2542

Biggins S (2013) The composition, functions, and regulation of the budding yeast kinetochore. Genetics 194:817–846

Bloom KS (2014) Centromeric heterochromatin: the primordial segregation machine. Annu Rev Genet 48:457–484

Borkovich KA, Alex LA, Yarden O, Freitag M, Turner GE, Read ND, Seiler S, Bell-Pedersen D, Paietta J, Plesofsky N, Plamann M, Goodrich-Tanrikulu M, Schulte U, Mannhaupt G, Nargang FE, Radford A, Selitrennikoff C, Galagan JE, Dunlap JC, Loros JJ, Catcheside D, Inoue H, Aramayo R, Polymenis M, Selker EU, Sachs MS, Marzluf GA, Paulsen I, Davis R, Ebbole DJ, Zelter A, Kalkman ER, ORourke R, Bowring F, Yeadon J, Ishii C, Suzuki K, Sakai W, Pratt R (2004) Lessons from the genome sequence of *Neurospora crassa*: tracing the path from genomic blueprint to multicellular organism. Microbiol Mol Biol Rev 68:1–108

Burrack LS, Applen SE, Berman J (2011) The requirement for the dam1 complex is dependent upon the number of kinetochore proteins and microtubules. Curr Biol 21:889–896

Burrack LS, Berman J (2012) Neocentromeres and epigenetically inherited features of centromeres. Chromosome Res 20:607–619

Burrack LS, Hutton HF, Matter KJ, Clancey SA, Liachko I, Plemmons AE, Saha A, Power EA, Turman B, Thevandavakkam MA, Ay F, Dunham MJ, Berman J (2016) Neocentromeres provide chromosome segregation accuracy and centromere clustering to multiple loci along a candida albicans chromosome. PLoS Genet 12:e1006317

Cam HP, Sugiyama T, Chen ES, Chen X, FitzGerald PC, Grewal SI (2005) Comprehensive analysis of heterochromatin- and RNAi-mediated epigenetic control of the fission yeast genome. Nat Genet 37:809–819

Cambareri EB, Aisner R, Carbon J (1998) Structure of the chromosome VII centromere region in *Neurospora crassa*: degenerate transposons and simple repeats. Mol Cell Biol 18:5465–5477

Carroll CW, Milks KJ, Straight AF (2010) Dual recognition of CENP-A nucleosomes is required for centromere assembly. J Cell Biol 189:1143–1155

Centola M, Carbon J (1994) Cloning and characterization of centromeric DNA from *Neurospora crassa*. Mol Cell Biol 14:1510–1519

Chatterjee G, Sankaranarayanan SR, Guin K, Thattikota Y, Padmanabhan S, Siddharthan R, Sanyal K (2016) Repeat-associated fission yeast-like regional centromeres in the ascomycetous budding yeast candida tropicalis. PLoS Genet 12:e1005839

Choi ES, Stralfors A, Castillo AG, Durand-Dubief M, Ekwall K, Allshire RC (2011) Identification of noncoding transcripts from within CENP-A chromatin at fission yeast centromeres. J Biol Chem 286:23600–23607

Choo KH (1998) Why is the centromere so cold? Genome Res 8:81–82

Clarke L (1998) Centromeres: proteins, protein complexes, and repeated domains at centromeres of simple eukaryotes. Curr Opin Genet Dev 8:212–218

Clarke L, Carbon J (1980) Isolation of a yeast centromere and construction of functional small circular chromosomes. Nature 287:504–509

Clarke L (1985) The structure and function of yeast centromeres. Ann Rev Genet 19:29–55

Cleveland DW, Mao Y, Sullivan KF (2003) Centromeres and kinetochores: from epigenetics to mitotic checkpoint signaling. Cell 112:407–421

Connolly LR, Smith KM, Freitag M (2013) The fusarium graminearum histone h3 k27 methyltransferase kmt6 regulates development and expression of secondary metabolite gene clusters. PLoS Genet 9:e1003916

Coughlan AY, Hanson SJ, Byrne KP, Wolfe KH (2016) Centromeres of the yeast komagataella phaffii (pichia pastoris) have a simple inverted-repeat structure. Genome Biol Evol 8: 2482–2492

D'Archivio S, Wickstead B (2017) Trypanosome outer kinetochore proteins suggest conservation of chromosome segregation machinery across eukaryotes. J Cell Biol 216:379–391

Dietrich FS, Voegeli S, Brachat S, Lerch A, Gates K, Steiner S, Mohr C, Pohlmann R, Luedi P, Choi S, Wing RA, Flavier A, Gaffney TD, Philippsen P (2004) The *Ashbya gossypii* genome as a tool for mapping the ancient saccharomyces cerevisiae genome. Science 304:304–307

Djupedal I, Kos-Braun IC, Mosher RA, Soderholm N, Simmer F, Hardcastle TJ, Fender A, Heidrich N, Kagansky A, Bayne E, Wagner EG, Baulcombe DC, Allshire RC, Ekwall K (2009) Analysis of small RNA in fission yeast; centromeric siRNAs are potentially generated through a structured RNA. EMBO J 28:3832–3844

Drinnenberg IA, deYoung D, Henikoff S, Malik HS (2014) Recurrent loss of CenH3 is associated with independent transitions to holocentricity in insects. Elife 3

Du Y, Topp CN, Dawe RK (2010) DNA binding of centromere protein C (CENPC) is stabilized by single-stranded RNA. PLoS Genet 6:e1000835

Duan Z, Andronescu M, Schutz K, McIlwain S, Kim YJ, Lee C, Shendure J, Fields S, Blau CA, Noble WS (2010) A three-dimensional model of the yeast genome. Nature 465:363–367

Dumesic PA, Homer CM, Moresco JJ, Pack LR, Shanle EK, Coyle SM, Strahl BD, Fujimori DG, Yates JR 3rd, Madhani HD (2015) Product binding enforces the genomic specificity of a yeast polycomb repressive complex. Cell 160:204–218

Earnshaw WC, Rothfield N (1985) Identification of a family of human centromere proteins using autoimmune sera from patients with scleroderma. Chromosoma 91:313–321

Fachinetti D, Folco HD, Nechemia-Arbely Y, Valente LP, Nguyen K, Wong AJ, Zhu Q, Holland AJ, Desai A, Jansen LE, Cleveland DW (2013) A two-step mechanism for epigenetic specification of centromere identity and function. Nat Cell Biol 15:1056–1066

Faino L, Seidl MF, Datema E, van den Berg GC, Janssen A, Wittenberg AH, Thomma BP (2015) Single-molecule real-time sequencing combined with optical mapping yields completely finished fungal genome. MBio 6

Fang J, Liu Y, Wei Y, Deng W, Yu Z, Huang L, Teng Y, Yao T, You Q, Ruan H, Chen P, Xu RM, Li G (2015) Structural transitions of centromeric chromatin regulate the cell cycle-dependent recruitment of CENP-N. Genes Dev 29:1058–1073

Fedorova ND, Khaldi N, Joardar VS, Maiti R, Amedeo P, Anderson MJ, Crabtree J, Silva JC, Badger JH, Albarraq A, Angiuoli S, Bussey H, Bowyer P, Cotty PJ, Dyer PS, Egan A, Galens K, Fraser-Liggett CM, Haas BJ, Inman JM, Kent R, Lemieux S, Malavazi I, Orvis J, Roemer T, Ronning CM, Sundaram JP, Sutton G, Turner G, Venter JC, White OR, Whitty BR, Youngman P, Wolfe KH, Goldman GH, Wortman JR, Jiang B, Denning DW, Nierman WC (2008) Genomic islands in the pathogenic filamentous fungus *Aspergillus fumigatus*. PLoS Genet 4:e1000046

Folco HD, Campbell CS, May KM, Espinoza CA, Oegema K, Hardwick KG, Grewal SI, Desai A (2015) The CENP-A N-tail confers epigenetic stability to centromeres via the CENP-T branch of the CCAN in fission yeast. Curr Biol 25:348–356

Folco HD, Pidoux AL, Urano T, Allshire RC (2008) Heterochromatin and RNAi are required to establish CENP-A chromatin at centromeres. Science 319:394–397

Fournier P, Abbas A, Chasles M, Kudla B, Ogrydziak DM, Yaver D, Xuan JW, Peito A, Ribet AM, Feynerol C et al (1993) Colocalization of centromeric and replicative functions on autonomously replicating sequences isolated from the yeast *Yarrowia lipolytica*. Proc Natl Acad Sci U S A 90:4912–4916

Freitag M (2016) The kinetochore interaction network (KIN) of ascomycetes. Mycologia 108: 485–505

Freitag M, Hickey PC, Khlafallah TK, Read ND, Selker EU (2004a) HP1 is essential for DNA methylation in Neurospora. Mol Cell 13:427–434

Freitag M, Lee DW, Kothe GO, Pratt RJ, Aramayo R, Selker EU (2004b) DNA methylation is independent of RNA interference in Neurospora. Science 304:1939

Freitag M, Williams RL, Kothe GO, Selker EU (2002) A cytosine methyltransferase homologue is essential for repeat-induced point mutation in *Neurospora crassa*. Proc Natl Acad Sci U S A 99:8802–8807

Fukagawa T, Earnshaw WC (2014) The centromere: chromatin foundation for the kinetochore machinery. Dev Cell 30:496–508

Galagan JE, Calvo SE, Borkovich KA, Selker EU, Read ND, Jaffe D, FitzHugh W, Ma LJ, Smirnov S, Purcell S, Rehman B, Elkins T, Engels R, Wang S, Nielsen CB, Butler J, Endrizzi M, Qui D, Ianakiev P, Bell-Pedersen D, Nelson MA, Werner-Washburne M, Selitrennikoff CP, Kinsey JA, Braun EL, Zelter A, Schulte U, Kothe GO, Jedd G, Mewes W, Staben C, Marcotte E, Greenberg D, Roy A, Foley K, Naylor J, Stange-Thomann N, Barrett R, Gnerre S, Kamal M, Kamvysselis M, Mauceli E, Bielke C, Rudd S, Frishman D, Krystofova S, Rasmussen C, Metzenberg RL, Perkins DD, Kroken S, Cogoni C, Macino G, Catcheside D, Li W, Pratt RJ, Osmani SA, DeSouza CP, Glass L, Orbach MJ, Berglund JA, Voelker R, Yarden O, Plamann M, Seiler S, Dunlap J, Radford A, Aramayo R, Natvig DO, Alex LA, Mannhaupt G, Ebbole DJ, Freitag M, Paulsen I, Sachs MS, Lander ES, Nusbaum C, Birren B (2003) The genome sequence of the filamentous fungus *Neurospora crassa*. Nature 422: 859–868

Galazka JM, Klocko AD, Uesaka M, Honda S, Selker EU, Freitag M (2016) Neurospora chromosomes are organized by blocks of importin alpha-dependent heterochromatin that are largely independent of H3K9me3. Genome Res 26:1069–1080

Gladyshev E, Kleckner N (2016) Recombination-independent recognition of DNA homology for repeat-induced point mutation (rip) is modulated by the underlying nucleotide sequence. PLoS Genet 12:e1006015

Gonen S, Akiyoshi B, Iadanza MG, Shi D, Duggan N, Biggins S, Gonen T (2012) The structure of purified kinetochores reveals multiple microtubule-attachment sites. Nat Struct Mol Biol 19:925–929

Gordon JL, Byrne KP, Wolfe KH (2011) Mechanisms of chromosome number evolution in yeast. PLoS Genet 7:e1002190

Hall IM, Shankaranarayana GD, Noma KI, Ayoub N, Cohen A, Grewal SI (2002) Establishment and maintenance of a heterochromatin domain. Science 297:2232–2237

Hays SM, Swanson J, Selker EU (2002) Identification and characterization of the genes encoding the core histones and histone variants of *Neurospora crassa*. Genetics 160:961–973

Henikoff S, Ahmad K, Malik HS (2001) The centromere paradox: stable inheritance with rapidly evolving DNA. Science 293:1098–1102

Heus JJ, Zonneveld BJ, Steensma HY, Van den Berg JA (1994) Mutational analysis of centromeric DNA elements of *Kluyveromyces lactis* and their role in determining the species specificity of the highly homologous centromeres from *K. lactis* and *Saccharomyces cerevisiae*. Mol Gen Genet 243:325–333

Hibbett DS, Binder M, Bischoff JF, Blackwell M, Cannon PF, Eriksson OE, Huhndorf S, James T, Kirk PM, Lucking R, Thorsten Lumbsch H, Lutzoni F, Matheny PB, McLaughlin DJ, Powell MJ, Redhead S, Schoch CL, Spatafora JW, Stalpers JA, Vilgalys R, Aime MC, Aptroot A, Bauer R, Begerow D, Benny GL, Castlebury LA, Crous PW, Dai YC, Gams W, Geiser DM, Griffith GW, Gueidan C, Hawksworth DL, Hestmark G, Hosaka K, Humber RA, Hyde KD, Ironside JE, Koljalg U, Kurtzman CP, Larsson KH, Lichtwardt R, Longcore J, Miadlikowska J, Miller A, Moncalvo JM, Mozley-Standridge S, Oberwinkler F, Parmasto E, Reeb V, Rogers JD, Roux C, Ryvarden L, Sampaio JP, Schussler A, Sugiyama J, Thorn RG, Tibell L, Untereiner WA, Walker C, Wang Z, Weir A, Weiss M, White MM, Winka K,

Yao YJ, Zhang N (2007) A higher-level phylogenetic classification of the Fungi. Mycol Res 111:509–547

James TY, Toledo LF, Rodder D, da Silva Leite D, Belasen AM, Betancourt-Roman CM, Jenkinson TS, Soto-Azat C, Lambertini C, Longo AV, Ruggeri J, Collins JP, Burrowes PA, Lips KR, Zamudio KR, Longcore JE (2015) Disentangling host, pathogen, and environmental determinants of a recently emerged wildlife disease: lessons from the first 15 years of amphibian chytridiomycosis research. Ecol Evol 5:4079–4097

Jamieson K, Rountree MR, Lewis ZA, Stajich JE, Selker EU (2013) Regional control of histone H3 lysine 27 methylation in Neurospora. Proc Natl Acad Sci U S A 110:6027–6032

Jamieson K, Wiles ET, McNaught KJ, Sidoli S, Leggett N, Shao Y, Garcia BA, Selker EU (2016) Loss of HP1 causes depletion of H3K27me3 from facultative heterochromatin and gain of H3K27me2 at constitutive heterochromatin. Genome Res 26:97–107

Janbon G, Ormerod KL, Paulet D, Byrnes EJ 3rd, Yadav V, Chatterjee G, Mullapudi N, Hon CC, Billmyre RB, Brunel F, Bahn YS, Chen W, Chen Y, Chow EW, Coppee JY, Floyd-Averette A, Gaillardin C, Gerik KJ, Goldberg J, Gonzalez-Hilarion S, Gujja S, Hamlin JL, Hsueh YP, Ianiri G, Jones S, Kodira CD, Kozubowski L, Lam W, Marra M, Mesner LD, Mieczkowski PA, Moyrand F, Nielsen K, Proux C, Rossignol T, Schein JE, Sun S, Wollschlaeger C, Wood IA, Zeng Q, Neuveglise C, Newlon CS, Perfect JR, Lodge JK, Idnurm A, Stajich JE, Kronstad JW, Sanyal K, Heitman J, Fraser JA, Cuomo CA, Dietrich FS (2014) Analysis of the genome and transcriptome of Cryptococcus neoformans var. grubii reveals complex RNA expression and microevolution leading to virulence attenuation. PLoS Genet 10:e1004261

Joglekar AP, Bouck D, Finley K, Liu X, Wan Y, Berman J, He X, Salmon ED, Bloom KS (2008) Molecular architecture of the kinetochore-microtubule attachment site is conserved between point and regional centromeres. J Cell Biol 181:587–594

Kagansky A, Folco HD, Almeida R, Pidoux AL, Boukaba A, Simmer F, Urano T, Hamilton GL, Allshire RC (2009) Synthetic heterochromatin bypasses RNAi and centromeric repeats to establish functional centromeres. Science 324:1716–1719

Kamper J, Kahmann R, Bolker M, Ma LJ, Brefort T, Saville BJ, Banuett F, Kronstad JW, Gold SE, Muller O, Perlin MH, Wosten HA, de Vries R, Ruiz-Herrera J, Reynaga-Pena CG, Snetselaar K, McCann M, Perez-Martin J, Feldbrugge M, Basse CW, Steinberg G, Ibeas JI, Holloman W, Guzman P, Farman M, Stajich JE, Sentandreu R, Gonzalez-Prieto JM, Kennell JC, Molina L, Schirawski J, Mendoza-Mendoza A, Greilinger D, Munch K, Rossel N, Scherer M, Vranes M, Ladendorf O, Vincon V, Fuchs U, Sandrock B, Meng S, Ho EC, Cahill MJ, Boyce KJ, Klose J, Klosterman SJ, Deelstra HJ, Ortiz-Castellanos L, Li W, Sanchez-Alonso P, Schreier PH, Hauser-Hahn I, Vaupel M, Koopmann E, Friedrich G, Voss H, Schluter T, Margolis J, Platt D, Swimmer C, Gnirke A, Chen F, Vysotskaia V, Mannhaupt G, Guldener U, Munsterkotter M, Haase D, Oesterheld M, Mewes HW, Mauceli EW, DeCaprio D, Wade CM, Butler J, Young S, Jaffe DB, Calvo S, Nusbaum C, Galagan J, Birren BW (2006) Insights from the genome of the biotrophic fungal plant pathogen *Ustilago maydis*. Nature 444:97–101

Ketel C, Wang HS, McClellan M, Bouchonville K, Selmecki A, Lahav T, Gerami-Nejad M, Berman J (2009) Neocentromeres form efficiently at multiple possible loci in *Candida albicans*. PLoS Genet 5:e1000400

Kilani J, Fillinger S (2016) Phenylpyrroles: 30 years, two molecules and (nearly) no resistance. Front Microbiol 7:2014

Kitada K, Yamaguchi E, Hamada K, Arisawa M (1997) Structural analysis of a *Candida glabrata* centromere and its functional homology to the *Saccharomyces cerevisiae* centromere. Curr Genet 31:122–127

Klocko AD, Ormsby T, Galazka JM, Leggett NA, Uesaka M, Honda S, Freitag M, Selker EU (2016) Normal chromosome conformation depends on subtelomeric facultative heterochromatin in *Neurospora crassa*. Proc Natl Acad Sci U S A 113(52):15048–15053

Klocko AD, Rountree MR, Grisafi PL, Hays SM, Adhvaryu KK, Selker EU (2015) Neurospora importin alpha is required for normal heterochromatic formation and DNA methylation. PLoS Genet 11:e1005083

Kobayashi N, Suzuki Y, Schoenfeld LW, Muller CA, Nieduszynski C, Wolfe KH, Tanaka TU (2015) Discovery of an unconventional centromere in budding yeast redefines evolution of point centromeres. Curr Biol 25:2026–2033

Koo DH, Zhao H, Jiang J (2016) Chromatin-associated transcripts of tandemly repetitive DNA sequences revealed by RNA-FISH. Chromosome Res 24(4):467–480

Koren A, Tsai HJ, Tirosh I, Burrack LS, Barkai N, Berman J (2010) Epigenetically-inherited centromere and neocentromere DNA replicates earliest in S-phase. PLoS Genet 6:e1001068

Lam AL, Boivin CD, Bonney CF, Rudd MK, Sullivan BA (2006) Human centromeric chromatin is a dynamic chromosomal domain that can spread over noncentromeric DNA. Proc Natl Acad Sci U S A 103:4186–4191

Lawrimore J, Aicher JK, Hahn P, Fulp A, Kompa B, Vicci L, Falvo M, Taylor RM 2nd, Bloom K (2016) ChromoShake: a chromosome dynamics simulator reveals that chromatin loops stiffen centromeric chromatin. Mol Biol Cell 27:153–166

Ledesma-Amaro R, Dulermo T, Nicaud JM (2015) Engineering Yarrowia lipolytica to produce biodiesel from raw starch. Biotechnol Biofuels 8:148

Loftus BJ, Fung E, Roncaglia P, Rowley D, Amedeo P, Bruno D, Vamathevan J, Miranda M, Anderson IJ, Fraser JA, Allen JE, Bosdet IE, Brent MR, Chiu R, Doering TL, Donlin MJ, DSouza CA, Fox DS, Grinberg V, Fu J, Fukushima M, Haas BJ, Huang JC, Janbon G, Jones SJ, Koo HL, Krzywinski MI, Kwon-Chung JK, Lengeler KB, Maiti R, Marra MA, Marra RE, Mathewson CA, Mitchell TG, Pertea M, Riggs FR, Salzberg SL, Schein JE, Shvartsbeyn A, Shin H, Shumway M, Specht CA, Suh BB, Tenney A, Utterback TR, Wickes BL, Wortman JR, Wye NH, Kronstad JW, Lodge JK, Heitman J, Davis RW, Fraser CM, Hyman RW (2005) The genome of the basidiomycetous yeast and human pathogen *Cryptococcus neoformans*. Science 307:1321–1324

Lynch DB, Logue ME, Butler G, Wolfe KH (2010) Chromosomal G + C content evolution in yeasts: systematic interspecies differences, and GC-poor troughs at centromeres. Genome Biol Evol 2:572–583

Malik HS, Henikoff S (2002) Conflict begets complexity: the evolution of centromeres. Curr Opin Genet Dev 12:711–718

Malik HS, Henikoff S (2009) Major evolutionary transitions in centromere complexity. Cell 138:1067–1082

Marie-Nelly H, Marbouty M, Cournac A, Flot JF, Liti G, Parodi DP, Syan S, Guillen N, Margeot A, Zimmer C, Koszul R (2014) High-quality genome (re)assembly using chromosomal contact data. Nat Commun 5:5695

Martinez D, Berka RM, Henrissat B, Saloheimo M, Arvas M, Baker SE, Chapman J, Chertkov O, Coutinho PM, Cullen D, Danchin EG, Grigoriev IV, Harris P, Jackson M, Kubicek CP, Han CS, Ho I, Larrondo LF, de Leon AL, Magnuson JK, Merino S, Misra M, Nelson B, Putnam N, Robbertse B, Salamov AA, Schmoll M, Terry A, Thayer N, Westerholm-Parvinen A, Schoch CL, Yao J, Barabote R, Nelson MA, Detter C, Bruce D, Kuske CR, Xie G, Richardson P, Rokhsar DS, Lucas SM, Rubin EM, Dunn-Coleman N, Ward M, Brettin TS (2008) Genome sequencing and analysis of the biomass-degrading fungus *Trichoderma reesei* (syn. Hypocrea jecorina). Nat Biotechnol 26:553–560

Meksem K, Shultz J, Tebbji F, Jamai A, Henrich J, Kranz H, Arenz M, Schlueter T, Ishihara H, Jyothi LN, Zhang HB, Lightfoot DA (2005) A bacterial artificial chromosome based physical map of the *Ustilago maydis* genome. Genome 48:207–216

Mendoza L, Vilela R, Voelz K, Ibrahim AS, Voigt K, Lee SC (2014) Human fungal pathogens of mucorales and entomophthorales. Cold Spring Harb Perspect Med 5

Meraldi P, McAinsh AD, Rheinbay E, Sorger PK (2006) Phylogenetic and structural analysis of centromeric DNA and kinetochore proteins. Genome Biol 7:R23

Miller DE, Smith CB, Kazemi NY, Cockrell AJ, Arvanitakas AV, Blumenstiel JP, Jaspersen SL, Hawley RS (2016) Whole-genome analysis of individual meiotic events in drosophila

melanogaster reveals that noncrossover gene conversions are insensitive to interference and the centromere Effect. Genetics 203:159–171
Mishra PK, Baum M, Carbon J (2007) Centromere size and position in Candida albicans are evolutionarily conserved independent of DNA sequence heterogeneity. Mol Genet Genomics 278:455–465
Mizuguchi T, Fudenberg G, Mehta S, Belton JM, Taneja N, Folco HD, FitzGerald P, Dekker J, Mirny L, Barrowman J, Grewal SI (2014) Cohesin-dependent globules and heterochromatin shape 3D genome architecture in *S. pombe*. Nature 516(7531):432–435
Morales L, Noel B, Porcel B, Marcet-Houben M, Hullo MF, Sacerdot C, Tekaia F, Leh-Louis V, Despons L, Khanna V, Aury JM, Barbe V, Couloux A, Labadie K, Pelletier E, Souciet JL, Boekhout T, Gabaldon T, Wincker P, Dujon B (2013) Complete DNA sequence of Kuraishia capsulata illustrates novel genomic features among budding yeasts (Saccharomycotina). Genome Biol Evol 5:2524–2539
Mukherjee PK, Horwitz BA, Herrera-Estrella A, Schmoll M, Kenerley CM (2013) Trichoderma research in the genome era. Annu Rev Phytopathol 51:105–129
Nonaka N, Kitajima T, Yokobayashi S, Xiao G, Yamamoto M, Grewal SI, Watanabe Y (2002) Recruitment of cohesin to heterochromatic regions by Swi6/HP1 in fission yeast. Nat Cell Biol 4:89–93
Ohzeki J, Larionov V, Earnshaw WC, Masumoto H (2015) Genetic and epigenetic regulation of centromeres: a look at HAC formation. Chromosome Res 23:87–103
Padmanabhan S, Thakur J, Siddharthan R, Sanyal K (2008) Rapid evolution of Cse4p-rich centromeric DNA sequences in closely related pathogenic yeasts, *Candida albicans* and *Candida dubliniensis*. Proc Natl Acad Sci U S A 105:19797–19802
Palmer DK, ODay K, Trong HL, Charbonneau H, Margolis RL (1991) Purification of the centromere-specific protein CENP-A and demonstration that it is a distinctive histone. Proc Natl Acad Sci U S A 88:3734–3738
Palmer DK, ODay K, Wener MH, Andrews BS, Margolis RL (1987) A 17-kD centromere protein (CENP-A) copurifies with nucleosome core particles and with histones. J Cell Biol 104: 805–815
Pidoux AL, Allshire RC (2005) The role of heterochromatin in centromere function. Philos Trans R Soc Lond B Biol Sci 360:569–579
Pombert JF, Xu J, Smith DR, Heiman D, Young S, Cuomo CA, Weiss LM, Keeling PJ (2013) Complete genome sequences from three genetically distinct strains reveal high intraspecies genetic diversity in the microsporidian *Encephalitozoon cuniculi*. Eukaryot Cell 12:503–511
Pomraning KR, Smith KM, Freitag M (2011) Bulk segregant analysis followed by high-throughput sequencing reveals the Neurospora cell cycle gene, ndc-1, to be allelic with the gene for ornithine decarboxylase, spe-1. Eukaryot Cell 10:724–733
Ravin NV, Eldarov MA, Kadnikov VV, Beletsky AV, Schneider J, Mardanova ES, Smekalova EM, Zvereva MI, Dontsova OA, Mardanov AV, Skryabin KG (2013) Genome sequence and analysis of methylotrophic yeast *Hansenula polymorpha* DL1. BMC Genom 14:837
Rhind N, Chen Z, Yassour M, Thompson DA, Haas BJ, Habib N, Wapinski I, Roy S, Lin MF, Heiman DI, Young SK, Furuya K, Guo Y, Pidoux A, Chen HM, Robbertse B, Goldberg JM, Aoki K, Bayne EH, Berlin AM, Desjardins CA, Dobbs E, Dukaj L, Fan L, FitzGerald MG, French C, Gujja S, Hansen K, Keifenheim D, Levin JZ, Mosher RA, Muller CA, Pfiffner J, Priest M, Russ C, Smialowska A, Swoboda P, Sykes SM, Vaughn M, Vengrova S, Yoder R, Zeng Q, Allshire R, Baulcombe D, Birren BW, Brown W, Ekwall K, Kellis M, Leatherwood J, Levin H, Margalit H, Martienssen R, Nieduszynski CA, Spatafora JW, Friedman N, Dalgaard JZ, Baumann P, Niki H, Regev A, Nusbaum C (2011) Comparative functional genomics of the fission yeasts. Science 332:930–936
Riley TT, Muzny CA, Swiatlo E, Legendre DP (2016) Breaking the mold: a review of mucormycosis and current pharmacological treatment options. Ann Pharmacother 50:747–757
Rosenblum EB, James TY, Zamudio KR, Poorten TJ, Ilut D, Rodriguez D, Eastman JM, Richards-Hrdlicka K, Joneson S, Jenkinson TS, Longcore JE, Parra Olea G, Toledo LF,

Arellano ML, Medina EM, Restrepo S, Flechas SV, Berger L, Briggs CJ, Stajich JE (2013) Complex history of the amphibian-killing chytrid fungus revealed with genome resequencing data. Proc Natl Acad Sci U S A 110:9385–9390

Rosic S, Erhardt S (2016) No longer a nuisance: long non-coding RNAs join CENP-A in epigenetic centromere regulation. Cell Mol Life Sci 73:1387–1398

Rosic S, Kohler F, Erhardt S (2014) Repetitive centromeric satellite RNA is essential for kinetochore formation and cell division. J Cell Biol 207:335–349

Sanyal K (2012) How do microbial pathogens make CENs? PLoS Pathog 8:e1002463

Sanyal K, Baum M, Carbon J (2004) Centromeric DNA sequences in the pathogenic yeast Candida albicans are all different and unique. Proc Natl Acad Sci U S A 101:11374–11379

Schmoll M, Dattenbock C, Carreras-Villasenor N, Mendoza-Mendoza A, Tisch D, Aleman MI, Baker SE, Brown C, Cervantes-Badillo MG, Cetz-Chel J, Cristobal-Mondragon GR, Delaye L, Esquivel-Naranjo EU, Frischmann A, Gallardo-Negrete Jde J, Garcia-Esquivel M, Gomez-Rodriguez EY, Greenwood DR, Hernandez-Onate M, Kruszewska JS, Lawry R, Mora-Montes HM, Munoz-Centeno T, Nieto-Jacobo MF, Nogueira Lopez G, Olmedo-Monfil V, Osorio-Concepcion M, Pilsyk S, Pomraning KR, Rodriguez-Iglesias A, Rosales-Saavedra MT, Sanchez-Arreguin JA, Seidl-Seiboth V, Stewart A, Uresti-Rivera EE, Wang CL, Wang TF, Zeilinger S, Casas-Flores S, Herrera-Estrella A (2016) The genomes of three uneven siblings: Footprints of the lifestyles of three trichoderma species. Microbiol Mol Biol Rev 80:205–327

Schotanus K, Soyer JL, Connolly LR, Grandaubert J, Happel P, Smith KM, Freitag M, Stukenbrock EH (2015) Histone modifications rather than the novel regional centromeres of *Zymoseptoria tritici* distinguish core and accessory chromosomes. Epigenetics Chromatin 8:41

Scott KC (2013) Transcription and ncRNAs: at the cent(rome)re of kinetochore assembly and maintenance. Chromosome Res 21:643–651

Seidl MF, Faino L, Shi-Kunne X, van den Berg GC, Bolton MD, Thomma BP (2015) The genome of the saprophytic fungus verticillium tricorpus reveals a complex effector repertoire resembling that of its pathogenic relatives. Mol Plant Microbe Interact 28:362–373

Selker EU, Tountas NA, Cross SH, Margolin BS, Murphy JG, Bird AP, Freitag M (2003) The methylated component of the *Neurospora crassa* genome. Nature 422:893–897

Selker EU (1990) Premeiotic instability of repeated sequences in *Neurospora crassa*. Annu Rev Genet 24:579–613

Shi J, Wolf SE, Burke JM, Presting GG, Ross-Ibarra J, Dawe RK (2010) Widespread gene conversion in centromere cores. PLoS Biol 8:e1000327

Smith KM, Galazka JM, Phatale PA, Connolly LR, Freitag M (2012) Centromeres of filamentous fungi. Chromosome Res 20:635–656

Smith KM, Phatale PA, Sullivan CM, Pomraning KR, Freitag M (2011) Heterochromatin is required for normal distribution of *Neurospora crassa* CenH3. Mol Cell Biol 31:2528–2542

Spatafora JW, Chang Y, Benny GL, Lazarus K, Smith ME, Berbee ML, Bonito G, Corradi N, Grigoriev I, Gryganskyi A, James TY, ODonnell K, Roberson RW, Taylor TN, Uehling J, Vilgalys R, White MM, Stajich JE (2016) A phylum-level phylogenetic classification of zygomycete fungi based on genome-scale data. Mycologia 108:1028–1046

Steiner NC, Hahnenberger KM, Clarke L (1993) Centromeres of the fission yeast *Schizosaccharomyces pombe* are highly variable genetic loci. Mol Cell Biol 13:4578–4587

Studt L, Rosler SM, Burkhardt, Arndt B, Freitag M, Humpf HU, Dickschat JS, Tudzynski B (2016) Knock-down of the methyltransferase Kmt6 relieves H3K27me3 and results in induction of cryptic and otherwise silent secondary metabolite gene clusters in Fusarium fujikuroi. Environ Microbiol 18(11):4037–4054

Sullivan BA, Karpen GH (2004) Centromeric chromatin exhibits a histone modification pattern that is distinct from both euchromatin and heterochromatin. Nat Struct Mol Biol 11:1076–1083

Tamaru H, Selker EU (2001) A histone H3 methyltransferase controls DNA methylation in *Neurospora crassa*. Nature 414:277–283

Tanizawa H, Iwasaki O, Tanaka A, Capizzi JR, Wickramasinghe P, Lee M, Fu Z, Noma K (2010) Mapping of long-range associations throughout the fission yeast genome reveals global genome organization linked to transcriptional regulation. Nucleic Acids Res 38:8164–8177

Thakur J, Sanyal K (2013) Efficient neocentromere formation is suppressed by gene conversion to maintain centromere function at native physical chromosomal loci in *Candida albicans*. Genome Res 23:638–652

Thakur J, Talbert PB, Henikoff S (2015) Inner kinetochore protein interactions with regional centromeres of fission yeast. Genetics 201:543–561

Thomma BP, Seidl MF, Shi-Kunne X, Cook DE, Bolton MD, van Kan JA, Faino L (2016) Mind the gap; seven reasons to close fragmented genome assemblies. Fungal Genet Biol 90:24–30

Thon MR, Pan H, Diener S, Papalas J, Taro A, Mitchell TK, Dean RA (2006) The role of transposable element clusters in genome evolution and loss of synteny in the rice blast fungus *Magnaporthe oryzae*. Genome Biol 7:R16

Verdaasdonk JS, Gardner R, Stephens AD, Yeh E, Bloom K (2012) Tension-dependent nucleosome remodeling at the pericentromere in yeast. Mol Biol Cell 23:2560–2570

Vernis L, Chasles M, Pasero P, Lepingle A, Gaillardin C, Fournier P (1999) Short DNA fragments without sequence similarity are initiation sites for replication in the chromosome of the yeast *Yarrowia lipolytica*. Mol Biol Cell 10:757–769

Vernis L, Poljak L, Chasles M, Uchida K, Casaregola S, Kas E, Matsuoka M, Gaillardin C, Fournier P (2001) Only centromeres can supply the partition system required for ARS function in the yeast *Yarrowia lipolytica*. J Mol Biol 305:203–217

Volpe TA, Kidner C, Hall IM, Teng G, Grewal SI, Martienssen RA (2002) Regulation of heterochromatic silencing and histone h3 lysine-9 methylation by RNAi. Science 297: 1833–1837

Westhorpe FG, Fuller CJ, Straight AF (2015) A cell-free CENP-A assembly system defines the chromatin requirements for centromere maintenance. J Cell Biol 209:789–801

Wiemann P, Sieber CM, von Bargen KW, Studt L, Niehaus EM, Espino JJ, Huss K, Michielse CB, Albermann S, Wagner D, Bergner SV, Connolly LR, Fischer A, Reuter G, Kleigrewe K, Bald T, Wingfield BD, Ophir R, Freeman S, Hippler M, Smith KM, Brown DW, Proctor RH, Munsterkotter M, Freitag M, Humpf HU, Guldener U, Tudzynski B (2013) Deciphering the cryptic genome: genome-wide analyses of the rice pathogen fusarium fujikuroi reveal complex regulation of secondary metabolism and novel metabolites. PLoS Pathog 9:e1003475

Winey M, Mamay CL, OToole ET, Mastronarde DN, Giddings TH, McDonald JrKL, McIntosh JR (1995) Three-dimensional ultrastructural analysis of the Saccharomyces cerevisiae mitotic spindle. J Cell Biol 129:1601–1615

Zinkowski RP, Meyne J, Brinkley BR (1991) The centromere-kinetochore complex: a repeat subunit model. J Cell Biol 113:1091–1110

Evolutionary Lessons from Species with Unique Kinetochores

Ines A. Drinnenberg and Bungo Akiyoshi

Abstract The kinetochore is the multi-protein complex that drives chromosome segregation in eukaryotes. It assembles onto centromeric DNA and mediates attachment to spindle microtubules. Kinetochore research over the last several decades has been focused on a few animal and fungal model organisms, which revealed a detailed understanding of the composition and organization of their kinetochores. Yet, these traditional model organisms represent only a small fraction of all eukaryotes. To gain insights into the actual degree of kinetochore diversity, it is critical to extend these studies to nontraditional model organisms from evolutionarily distant lineages. In this chapter, we review the current knowledge of kinetochores across diverse eukaryotes with an emphasis on variations that arose in nontraditional model organisms. In addition, we also review the literature on species, in which the subcellular localization of kinetochores has changed from the nucleoplasm to the nuclear membrane. Finally, we speculate on the organization of the chromosome segregation machinery in an early eukaryotic ancestor to gain insights into fundamental principles of the chromosome segregation machinery, which are common to all eukaryotes.

1 Introduction

Mitosis is the process that partitions newly replicated chromosomes from the mother cell into the two emerging daughter cells (McIntosh 2016). Fundamental to this process is the kinetochore, a macromolecular protein complex that assembles onto specialized chromosomal regions called centromeres to mediate the attachment

I.A. Drinnenberg (✉)
Institut Curie, Paris Sciences et Lettres Research university, Centre de la Recherche Scientifique UMR 3664, 75005 Paris, France
e-mail: ines.drinnenberg@curie.fr

B. Akiyoshi (✉)
Department of Biochemistry, University of Oxford, Oxford, UK
e-mail: bungo.akiyoshi@bioch.ox.ac.uk

B.E. Black (ed.), *Centromeres and Kinetochores*, Progress in Molecular and Subcellular Biology 56, DOI 10.1007/978-3-319-58592-5_5

of sister chromatids to spindle microtubules (Cheeseman and Desai 2008; Santaguida and Musacchio 2009). Kinetochores also promote the recruitment of cohesin complexes around centromeres to hold duplicated sister chromatids together until anaphase (Nasmyth and Haering 2009). At their DNA-binding interface, kinetochores need to ensure stable attachment to tolerate the pulling forces exerted by kinetochore microtubules (Allshire and Karpen 2008; Fukagawa and Earnshaw 2014; Westhorpe and Straight 2015; McKinley and Cheeseman 2016). In contrast to this more static attachment, the binding to spindle microtubules must be dynamically regulated (Foley and Kapoor 2013; Cheerambathur and Desai 2014; London and Biggins 2014; Etemad and Kops 2016). Faithful chromosome segregation requires that sister kinetochores form bioriented attachments to spindle microtubules emanating from opposite poles (Nicklas 1997). Biorientation is necessary for the accurate distribution of sister chromatids into daughter cells during anaphase.

Research on kinetochores has mainly been performed on a few model organisms, such as fungi, worms, flies, and vertebrates. Their studies have been instrumental in informing us about the basic composition and organization of kinetochores among these species. However, these “traditional” model organisms only represent a small fraction of the entire eukaryotic biodiversity. In fact, both animals and fungi are members of the Opisthokonta, that is only one out of six major supergroups of eukaryotes (Fig. 1) (Walker et al. 2011; Adl et al. 2012). While extensive analyses have not been performed on kinetochores in non-opisthokonts, glances into kinetochores from additional species scattered across the eukaryotic phylogenetic tree have revealed extraordinary levels of variations in kinetochore composition and subcellular location. This stands in sharp contrast to many other cell cycle machines that are highly conserved among diverse eukaryotes (e.g., Cyclin/CDK, cohesin, condensin, the anaphase promoting complex, and proteasomes). In this chapter, we will first discuss the extent of similarity and variation in kinetochore composition among animals and fungi. We will then review kinetochores in select organisms from different supergroups, as well as unique kinetochores that evolved in kinetoplastids. Following up on that, we will highlight membrane-bound kinetochores found in some unicellular organisms. Finally, we will speculate on the organization of the chromosome segregation machinery in early eukaryotes.

2 The Kinetochore Complex in Animals and Fungi

Genetic and biochemical analyses in fungi and vertebrates have led to the identification of more than 80 proteins that are part of the kinetochore (Biggins 2013; Cheeseman 2014). The structural core of the kinetochore consists of an inner and an outer complex. The inner kinetochore complex binds centromeric chromatin. It serves as a platform for the recruitment of the outer kinetochore complex that binds spindle microtubules during mitosis and meiosis. Both complexes are characterized

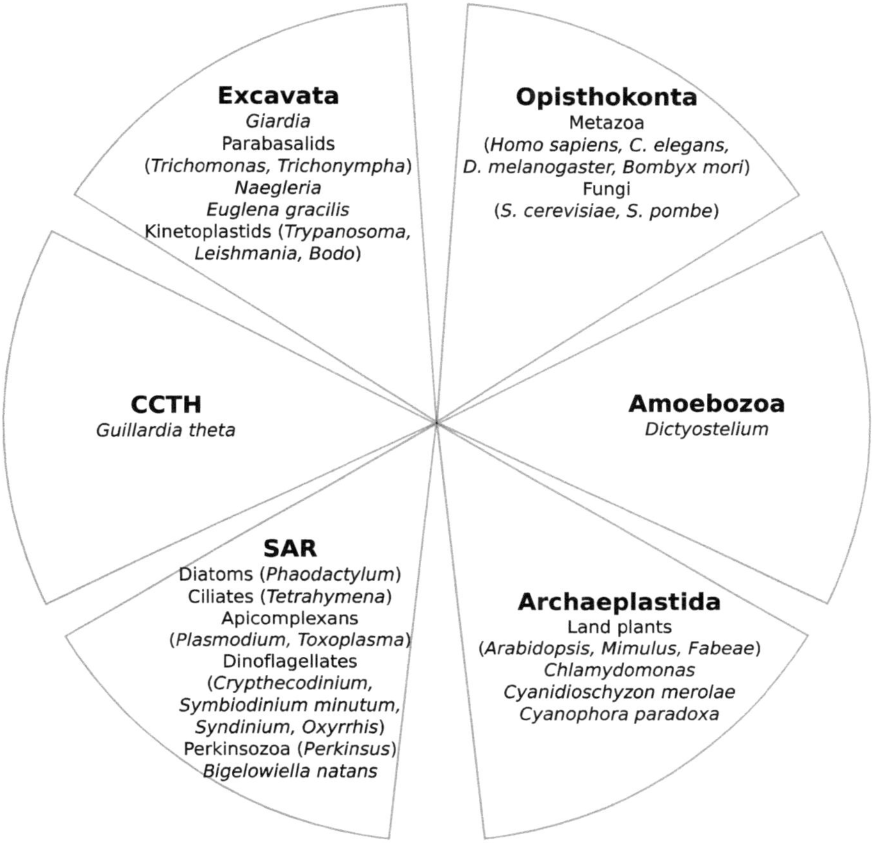

Fig. 1 Six eukaryotic supergroups. Representative organisms from each supergroup are shown as examples

by a network of several protein complexes that work in concert to regulate the proper attachment of kinetochore microtubules to centromeric DNA.

2.1 *Similarities and Variations of the Inner Kinetochore in Animals and Fungi*

In vertebrates and fungi, the inner kinetochore consists of ~16 members that are commonly referred to as the Constitutive Centromere Associated Network or CCAN (Cheeseman and Desai 2008; Westermann and Schleiffer 2013; Westhorpe and Straight 2013; Fukagawa and Earnshaw 2014) (Table 1). It is generally agreed that the recruitment of all CCAN members in these species depends on a specialized centromeric histone H3 variant, CENP-A (also known as CenH3—see the Note on nomenclature at the end of this chapter) (Black and Cleveland 2011; Müller and Almouzni 2014; Earnshaw 2015) and its direct DNA-binding partner CENP-C (Carroll et al. 2010; Basilico et al. 2014). In addition to CENP-A and CENP-C, other CCAN components also make DNA contacts, including the histone-fold proteins CENP-T and CENP-W as well as CENP-U^{Ame1} and CENP-Q^{Okp1} in budding yeast (Hori et al. 2008; Hornung et al. 2014).

Given their central role in kinetochore function, it is surprising that several inner kinetochore components undergo rapid evolution at the amino acid level, which complicates homology-based predictions even in well-sequenced species (Henikoff et al. 2001; Talbert et al. 2009; Malik and Henikoff 2009). While sequence similarity of several CCAN components between vertebrates and budding yeast was revealed early on (Meraldi et al. 2006), the identification of phylogenetic relationship for other CCAN components often required advanced bioinformatics tools due to limited sequence similarities (Schleiffer et al. 2012; Westermann and Schleiffer 2013). For example, the budding yeast CENP-T^{Cnn1} was only identified using a combination of proteomic approaches and remote homology predictions (Schleiffer et al. 2012). Thus, experimental approaches as well as advanced bioinformatics are required to obtain a comprehensive picture of kinetochores.

While most CCAN components are conserved between vertebrates and fungi, CCAN proteins appear to be absent in *Caenorhabditis elegans* and *Drosophila melanogaster* except for CENP-C (Table 1). While it is formally possible that these species have highly divergent CCAN members, the wealth of extensive genetic screens for chromosome segregation defects and biochemical purifications of kinetochore components makes this unlikely (Cheeseman et al. 2004; Goshima et al. 2007; Przewloka et al. 2007, 2011). Therefore, it appears that nematodes and Diptera have "simpler" inner kinetochore complexes that just consist of CENP-C, which connects the CENP-A-containing chromatin to outer kinetochore proteins. The reason for this potential reduction in kinetochore complexity remains unknown. In contrast to *D. melanogaster*, homologous CCAN members have been identified in other insects (see below), showing that the near-complete loss of CCAN is not common to all insects and instead occurred in a dipteran ancestor around 250 Mya (Hedges et al. 2006).

While CENP-A was thought to be essential for kinetochore assembly in all animals and fungi, recent studies showed that a number of insects have recurrently lost CENP-A (Drinnenberg et al. 2014). Intriguingly, all CENP-A-deficient insects analyzed are derived from independent transitions from monocentric chromosomes (where microtubules attach to a single chromosomal region) to holocentric chromosomes (where microtubules attach along the entire length of the chromosome) (Melters et al. 2012; Drinnenberg et al. 2014). This strong correlation between the change in centromeric architecture and the loss of CENP-A supports a causal relationship between the two events in that the transition to holocentromeres facilitated the loss of CENP-A or vice versa. While CENP-A and its binding partner CENP-C are lost, several of the CCAN components continue to be present even in CENP-A-deficient insects (e.g., the silkworm *Bombyx mori* (Table 1)). These findings suggest that the

Table 1 List of putative kinetochore proteins in select eukaryotes

Centromere/kinetochore protein	Opisthokonta					
	Homo sapiens	*Saccharomyces cerevisiae*	*Schizosaccharomyces pombe*	*Caenorhabditis elegans*	*Drosophila melanogaster*	*Bombyx mori*
CENP-A	CENP-A	Cse4	Cnp1	HCP-3	Cid	
CENP-C	CENP-C	Mif2	Cnp3	HCP-4	CENP-C	
CENP-H	CENP-H	Mcm16	Fta3			
CENP-I	CENP-I	Ctf3	Mis6			XP_004931054.2
CENP-K	CENP-K	Mcm22	Sim4			
CENP-L	CENP-L	Iml3	Fta1			NP_001266344.1
CENP-M	CENP-M					XP_004928337.1
CENP-N	CENP-N	Chl4	Mis15			XP_004922229.1
CENP-O	CENP-O	Mcm21	Mal2			
CENP-P	CENP-P	Ctf19	Fta2			
CENP-Q	CENP-Q	Okp1	Fta7			
CENP-U	CENP-U	Ame1	Mis17			
CENP-T	CENP-T	Cnn1	Cnp20			
CENP-W	CENP-W	Wip1	New1			
CENP-S	CENP-S	Mhf1	Mhf1	CCD31143		NP_001266386.1
CENP-X	CENP-X	Mhf2	Mhf2	CCD69638		NP_001268381.1
Mis12	Mis12	Mtw1	Mis12	Mis12	Mis12	fcaL18J21
Nnf1	PMF1	Nnf1	Nnf1	KBP-1	Nnf1a, Nnf1b	XP_004928275.1
Dsn1	Dsn1	Dsn1	Mis13	KNL3		XP_004928630.1
Nsl1	DC8	Nsl1	Mis14	KBP-2	Kmn1	
Knl1	KNL1	Spc105	Spc7	KNL1	Spc105R	XP_012546088.1
Ndc80	Hec1	Ndc80	Ndc80	Ndc80	Ndc80	XP_012545504.1
Nuf2	Nuf2	Nuf2	Nuf2	Nuf2	Nuf2	XP_004929441.1
Spc24	Spc24	Spc24	Spc24	KBP-4	Kmn2	XP_004925111.1
Spc25	Spc25	Spc25	Spc25	KBP-3	Spc25	NP_001040214.1

Table 1 (continued)

Centromere/kinetochore protein	Amoebozoa	Archaeplastida			
	Dictyostelium discoideum	*Arabidopsis thaliana*	*Chlamydomonas reinhardtii*	*Cyanidioschyzon merolae*	*Cyanophora paradoxa*
CENP-A	Q54F38.1	HTR12	EDP07681.1 and/or EDO99752.1	BAM79321.1	Contig11189-snap-gene-0.0
CENP-C		CENP-C	XP_001701669.1	BAM80204.1	
CENP-H					
CENP-I	XP_635116.1				
CENP-K					
CENP-L	XP_635393.1				
CENP-M					
CENP-N					
CENP-O		NP_001190285.1			
CENP-P					
CENP-Q					
CENP-U					
CENP-T					
CENP-W					
CENP-S	XP_638695.1	NP_199906.1			
CENP-X	XP_003290871.1	NP_178000.2		BAM78991.1	
Mis12	XP_638733.1	Mis12		BAM81376.1	Contig39033-abinit-gene-0.2
Nnf1	XP_641683.1	AAO39938.1	XP_001697791.1	BAM78899.1	Contig54317-snap-gene-0.0
Dsn1					Contig25943-abinit-gene-0.0
Nsl1			XP_001703375.1		
Knl1	XP_639120.1	NP_671774.4	XP_001690815.1	BAM81078.1	
Ndc80	XP_638830.1	NP_191024.1	XP_001699834.1	BAM79598.1	Contig10691-abinit-gene-0.1
Nuf2	XP_641696.1	NP_176296.2	XP_001700269.1	BAM78839.1	Contig53383-abinit-gene-0.8
Spc24	XP_638781.1	NP_187500.2	XP_001697066.1	BAM79682.1	Contig54569-abinit-gene-0.4
Spc25	XP_638784.1	NP_566900.1	XP_001697914.1	BAM79992.1	Contig26077-abinit-gene-0.7

Table 1 (continued)

Centromere/ kinetochore protein	SAR					
	Phaeodactylum tricornutum	*Tetrahymena thermophila*	*Plasmodium falciparum*	*Toxoplasma gondii*	*Symbiodinium minutum*	*Bigelowiella natans*
CENP-A	EEC43222.1	Q2N2K6.1	CAD52476.1	EPT27047.1	symbB1.v1.2.034671.t1	aug1.3_g1557
CENP-C	XP_002184047.1		CZT98473.1			fgenesh1_pg.119_#_34
CENP-H	XP_002177997.1					
CENP-I						
CENP-K						
CENP-L	XP_002185884.1					
CENP-M						
CENP-N						
CENP-O						
CENP-P						
CENP-Q						
CENP-U						
CENP-T						
CENP-W						
CENP-S	XP_002180057.1					aug1.3_g1673
CENP-X						aug1.96_g19929
Mis12	XP_002179310.1				symbB.v1.2.039506.t1	fgenesh1_pg.45_#_162
Nnf1	XP_002180232.1					
Dsn1						
Nsl1						
Knl1						fgenesh1_pg.8_#_205
Ndc80	XP_002177760.1	XP_001024017.1	XP_966148.1	XP_002365215.1	symbB.v1.2.002582.t1	estExt_fgenesh1_pg.C_190200
Nuf2	XP_002185912.1		XP_001351235.1	XP_002364205.1	symbB.v1.2.033878.t1	e_gw1.128.32.1
Spc24	XP_002182127.1					estExt_fgenesh1_pg.C_680067
Spc25	XP_002178778.1	XP_001030098.2	XP_002809072.1	XP_002368113.1	symbB.v1.2.035905.t1	aug1.38_g12030

Table 1 (continued)

Centromere/ kinetochore protein	CCTH	Excavata			
	Guillardia theta	*Giardia intestinalis*	*Trichomonas vaginalis*	*Naegleria gruberi*	*Euglena gracilis*
CENP-A	EKX34792.1 (nucleus) AAK39657.1? (nucleomorph)	EFO62248.1	EAY15370.1	EFC49963.1	dark_m.11114 and/or dark_m.83792
CENP-C	EKX50218.1		XP_001311235.1	EFC46326.1	
CENP-H					
CENP-I					
CENP-K	XP_005834533.1				
CENP-L					
CENP-M					
CENP-N					
CENP-O					
CENP-P					
CENP-Q					
CENP-U					
CENP-T					
CENP-W					
CENP-S			XP_001306373.1		
CENP-X			XP_001579665.1		
Mis12	XP_005839738.1		XP_001308419.1	EFC50450.1	
Nnf1	XP_005820862.1		XP_001307740.1		
Dsn1					
Nsl1	XP_005822661.1				
Knl1	XP_005832827.1			EFC50203.1	
Ndc80	EKX45170.1		XP_001311732.1		dark_m.19013
Nuf2	CAC27079.1	EFO61302.1			dark_m.38633
Spc24	EKX52051.1	EDO78311.1			
Spc25	EKX40188.1	EFO63967.1	XP_001323375.1	EFC47967.1	light_m.34007

The composition of the outer kinetochore, not inner kinetochore, is widely conserved. In those cases where multiple homologous proteins are identified, only one is listed unless otherwise specified. Note that kinetochore localization has yet to be established for many of the proteins listed here. CENP-S and CENP-X have independent roles in DNA repair and may not be involved in CCAN function in some organisms. KNL1 homologs were predicted in Tromer et al. (2015)

assembly of the inner kinetochore has been altered in CENP-A-deficient insects, allowing CENP-A-independent kinetochore formation. Whether or not new kinetochore components have evolved to compensate for the loss of CENP-A is an open question. It is important to note that other holocentric organisms including nematodes (e.g., *C. elegans*) have retained CENP-A$^{HCP\text{-}3}$. Thus, despite the usage of the generic term "holocentromere", the basic architecture and regulation of holocentromeres is likely to be diverse among different species.

2.2 *The Composition of the Outer Kinetochore Is Highly Conserved in Animals and Fungi*

The outer kinetochore complex is recruited to centromeres upon the onset of mitosis to connect to spindle microtubules. This interaction is accomplished by the ~10-subunit KMN network that consists of the Knl1, Mis12, and Ndc80 complexes (Cheeseman et al. 2006; Petrovic et al. 2014). In contrast to the inner kinetochore, the composition of the outer kinetochore is widely conserved across animals and fungi (Meraldi et al. 2006; Tromer et al. 2015). Even CENP-A-deficient insects encode the same repertoire of outer kinetochore components, implying similar means of attaching to microtubules while utilizing alternate inner kinetochore assembly pathways (Drinnenberg et al. 2016). A notable exception to the otherwise conserved composition of the KMN network is found in Diptera. *D. melanogaster* has lost Dsn1, a subunit of the Mis12 complex (Przewloka and Glover 2009). In addition, the Nnf1 subunit of the Mis12 complex underwent a duplication event giving rise to two paralogs, Nnf1a and Nnf1b, that are part of two distinct Mis12 complexes with similar biochemical behaviors (Przewloka et al. 2007; Schittenhelm et al. 2007; Liu et al. 2016; Richter et al. 2016; Blattner et al. 2016). The loss of Dsn1 could have been compensated by the C-terminal part of the *Drosophila* Knl1 homolog (Przewloka et al. 2009). Indeed, the overall organization of this complex appears to resemble the human and yeast counterparts (Hornung et al. 2011; Przewloka et al. 2011; Screpanti et al. 2011). Whether these changes have any functional consequences on the *Drosophila* KMN complex is currently unclear.

3 Glimpses into Kinetochore Compositions in Diverse Eukaryotes

While research on kinetochores in fungi and animals has revealed a paradigm for the basic organization of kinetochores, it remains unclear whether other eukaryotes have similar kinetochores. Comparative studies in additional eukaryotic lineages are a key to revealing the degree of conservation and divergence of kinetochores among eukaryotes. Although bioinformatic analyses have identified some homologous kinetochore proteins in diverse eukaryotes (Table 1), very few studies have characterized the function of individual kinetochore proteins. Furthermore, extensive proteomic screens have not been carried out in most organisms, leaving open the possibility of lineage-specific evolution of additional kinetochore proteins. Below we summarize the current knowledge of kinetochores in select eukaryotes from different supergroups to highlight their peculiarities.

3.1 Supergroup Amoebozoa

The only kinetochore protein that has been characterized in the supergroup Amoebozoa is the centromere-specific histone H3 variant in *Dictyostelium discoideum* (Dubin et al. 2010). In contrast to nearly all other characterized CENP-A proteins that have at least one extra amino acid in the loop 1 region within the histone fold compared to histone H3 (Malik and Henikoff 2003), *D. discoideum* CENP-A^{CenH3} does not have a longer loop 1. While alterations or shortening of residues in loop 1 in other species can impair centromere targeting (Vermaak et al. 2002), cytological studies of *D. discoideum* CENP-A^{CenH3} revealed incorporation into centromeric DNA (Dubin et al. 2010). Therefore, the insertion of extra amino acids in loop 1 is not an obligatory feature of CENP-A.

3.2 Supergroup Archaeplastida

Several kinetochore proteins have been characterized in land plants (e.g., *Arabidopsis*, maize, and barley) (Dawe et al. 1999; ten Hoopen et al. 2000; Sato et al. 2005). For example, homologous kinetochore proteins (such as CENP-C and Mis12) identified by bioinformatics searches were analyzed by means of cytological and mutational studies, confirming their importance for chromosome segregation in mitosis and meiosis. Although most eukaryotes have a single CENP-A protein, multiple CENP-A^{CENH3} variants are found in *Arabidopsis halleri*, *A. lyrata* (Kawabe et al. 2006), *Brassica* sp. (Wang et al. 2011), *Mimulus* monkeyflowers (Finseth et al. 2015), barley (Ishii et al. 2015), and *Fabeae* sp. (Neumann et al. 2012; Neumann et al. 2015). While it is currently unclear whether the individual CENP-A^{CENH3} variants are functionally distinct, it has been hypothesized that CENP-A^{CENH3} duplications occurred to counteract the evolutionary force from centromere drive (Finseth et al. 2015) (centromere drive is discussed in the chapter "Cell Biology of Cheating—Transmission of Centromeres and Other Selfish Elements Through Asymmetric Meiosis" by Chmátal et al.).

Compared to land plants, much less is known about kinetochores in other Archaeplastida species. *Cyanidioschyzon merolae* is a thermoacidiphilic red alga that is thought to be one of the most primitive photosynthetic eukaryotes. Its simple cellular architecture and reduced genome make it an attractive organism for cell biological study (Matsuzaki et al. 2004). Among several homologous kinetochore proteins identified (Table 1), only CENP-A^{CENH3} has been experimentally characterized to date (Maruyama et al. 2007; Kanesaki et al. 2015). Given its hot and acidic living habitats, it will be interesting to test for potential adaptations of kinetochore components that evolved to cope with such extreme environments.

3.3 Supergroup SAR

The supergroup SAR (Stramenopiles, Alveolates, and Rhizaria: also referred to as Harosa) includes diatoms, ciliates, apicomplexans, and dinoflagellates (discussed later). Ciliates have a somatic macronucleus with highly amplified genes for RNA synthesis as well as several germline micronuclei for genome maintenance. The number of chromosomes in the somatic macronucleus can be as high as 16,000 in some species (Swart et al. 2013). While the germline micronucleus has CENP-A^{CNA1} and segregates its chromosomes accurately, the somatic macronucleus does not have CENP-A^{CNA1} and segregates its chromosomes randomly (Cervantes et al. 2006; Cui and Gorovsky 2006).

Apicomplexans include a number of important human pathogens, including *Plasmodium* and *Toxoplasma* (Francia and Striepen 2014). Several kinetochore proteins have been identified and functionally characterized in *Plasmodium falciparum* and *Toxoplasma gondii* including CENP-A^{CENH3}, CENP-C, and members of the Ndc80 complex (Brooks et al. 2011; Verma and Surolia 2013; Farrell and Gubbels 2014). While the domain architecture appears to be conserved, the *T. gondii* Nuf2 homolog contains a conserved amino acid motif that appears specific to apicomplexan (Farrell and Gubbels 2014). The functional relevance of this motif, however, remains unclear.

3.4 Supergroup CCTH

Very little is known about kinetochores in the supergroup CCTH (Cryptophytes, Centrohelids, Telonemids, and Haptophytes: also called Hacrobia). Cryptophyte algae are thought to have evolved by engulfing a red alga that contained a primary plastid (Tanifuji and Archibald 2014). In the cryptomonad *Guillardia theta*, the secondary plastid has retained the red algal-derived relict nucleus (called nucleomorph) (Curtis et al. 2012). How the nucleomorph genome is maintained during cell division remains unknown. While the nucleomorph genome encodes for a putative CENP-A homolog (Douglas et al. 2001) (Table 1), this protein lacks the hallmark of an extended loop 1 region. It will therefore be necessary to experimentally confirm whether it indeed functions as the centromeric histone variant for the segregation of the nucleomorph genome.

3.5 Supergroup Excavata

Excavata is a group of predominantly flagellated species (Walker et al. 2011; Adl et al. 2012). It is divided into Metamonads and Discoba. A number of human parasites belong to this supergroup, such as *Giardia*, *Trichomonas vaginalis*, *Naegleria fowleri*, and *Trypanosoma brucei*.

Giardia intestinalis (Metamonads) has two histone H3-like molecules that have a longer loop 1. Cytological studies have revealed that only one H3 variant incorporates into centromeres, while the other variant localizes to pericentric heterochromatin (Dawson et al. 2007), underlining the need for experimental approaches to corroborate the identity of the centromeric histone H3 variant. As in *Dictyostelium* (see above), another metamonad *Trichomonas vaginalis* has a CENP-A^{CenH3} protein that does not have a longer loop 1, but its localization pattern is suggestive of centromeric incorporation (Zubácová et al. 2012).

Discoba (also called JEH for Jakobids, Euglenozoa, Heterolobosea) includes *Naegleria*, *Euglena*, and kinetoplastids. Although canonical kinetochore proteins have been identified in *Naegleria gruberi* and *Euglena gracilis* (Table 1), none has been identified in the genome of kinetoplastids.

4 Unconventional Kinetoplastid Kinetochores

Identification of at least a fraction of canonical kinetochore proteins (especially CENP-A and the Ndc80 complex) in diverse eukaryotes led to a notion that all eukaryotes may build the structural core of the kinetochore using a conserved set of kinetochore proteins (Meraldi et al. 2006). However, none of the canonical kinetochore proteins were identified in the genome of kinetoplastids (Lowell and Cross 2004; Berriman et al. 2005), a group of unicellular eukaryotes defined by the presence of kinetoplast (a large structure in the mitochondrion that contains mitochondrial DNA) (Vickerman 1962). They belong to the supergroup Excavata, Discoba group, Euglenozoa. Euglenozoa is a diverse group of flagellates that include euglenids, diplonemids, symbiontids, and kinetoplastids (Walker et al. 2011; Cavalier-Smith 2016).

To uncover the repertoire of kinetoplastid kinetochores, recent studies utilized proteomic and functional approaches and identified 20 kinetochore proteins in *Trypanosoma brucei*, named KKT1–20 (Akiyoshi and Gull 2014; Nerusheva and Akiyoshi 2016). The majority of these proteins are conserved among kinetoplastids, including the free-living *Bodo saltans*. However, obvious orthologs of KKT proteins were not found even in euglenids, which instead have canonical kinetochore proteins (Akiyoshi 2016). The unique KKT-based kinetochores are therefore not conserved across Euglenozoa but are apparently restricted to kinetoplastids. It remains unclear why kinetoplastids possess a unique set of kinetochore proteins (discussed below).

4.1 *Domain Architectures of Kinetoplastid Kinetochore Proteins*

Sequence analyses of kinetoplastid kinetochore proteins have revealed the following conserved domains: a BRCT (BRCA1 C terminus) domain in KKT4, FHA (Forkhead-associated) domain in KKT13, WD40-like domain in KKT15, divergent polo boxes (DPB) in KKT2, KKT3 and KKT20, unique protein kinase domain in KKT2 and KKT3, and CLK (cdc2-like kinase) kinase domain in KKT10 and KKT19. While orthologs of any of the KKT proteins have not been identified in non-kinetoplastid species, the domain architecture and sequence similarity of KKT2, KKT3, and KKT20 suggest that these proteins may share common ancestry with a Polo-like kinase (PLK) (Nerusheva and Akiyoshi 2016). Consistent with this possibility, although the kinase domain of KKT2/3 is apparently unique (Parsons et al. 2005), the next closest kinase domain is that of PLK (Akiyoshi 2016). Furthermore, putative DNA-binding motifs are present in KKT2 and KKT3, suggesting that these proteins likely bind DNA and play a critical role in establishing unique kinetochores in kinetoplastids. Although PLK localizes at the kinetochore in some species, it is not considered to be a structural kinetochore protein in any eukaryote. Substrates of these KKT kinases have yet to be identified.

BRCT, FHA, or CLK-like kinase domains are not present in canonical kinetochore proteins. Domains found in canonical kinetochore proteins such as CH (calponin homology) and RWD (RING finger, WD repeat, DEAD-like helicases) domains have not been identified in KKT proteins. Although KKT proteins do not have similarity to canonical kinetochore proteins at the primary sequence level, high-resolution structural data are necessary to reveal if there is any similarity at the tertiary level.

4.2 *Common Features*

Although components of the core kinetoplastid kinetochore appear to be distinct from canonical kinetochore proteins present in other eukaryotes, various regulatory proteins that are known to be important for chromosome segregation are conserved, including Aurora B, Cyclin/CDK, cohesin, condensin, separase, and the anaphase promoting complex (Berriman et al. 2005; Akiyoshi and Gull 2013). Aurora B apparently localizes at the kinetochore during prometaphase and metaphase in *Trypanosoma brucei* (Li et al. 2008), suggesting that its kinetochore regulatory function may be conserved. It is known that the kinase–phosphatase balance is important for regulating kinetochore functions in other eukaryotes. For example, the KNL1 outer kinetochore protein recruits the PP1 phosphatase (Liu et al. 2010; Rosenberg et al. 2011; Meadows et al. 2011; Espeut et al. 2012). Interestingly, a conserved PP1-binding motif is present in KKT7, suggesting that PP1 may regulate kinetochore functions in kinetoplastids. It is therefore possible that kinetoplastid kinetochores, while being structurally distinct, may still utilize a conserved mechanism for the regulation of kinetochore functions.

4.3 Implications from Kinetoplastid Kinetochores

The discovery of KKT-based kinetochores in kinetoplastids challenged a widely held assumption that the core of the kinetochore would be composed of proteins conserved throughout eukaryotes (e.g., CENP-A and Ndc80). A corollary is that eukaryotic chromosome segregation can be achieved using proteins distinct from CENP-A or Ndc80. Understanding how KKT proteins carry out the conserved kinetochore functions will likely provide important insights into fundamental principles of the kinetochore. It also raises a possibility that there might be as yet different types of kinetochores to be discovered in eukaryotes.

5 Membrane-Embedded Kinetochores

In addition to compositional variations, the subcellular location of kinetochores has also been altered in some lineages. In all eukaryotes, chromosomes are enclosed inside the nuclear envelope during most of the cell cycle. This keeps chromosome-based activities physically separated from the cytoplasm where protein synthesis and metabolic processes take place (Martin and Koonin 2006; Koumandou et al. 2013). This separation necessitates proper nuclear remodeling to be coordinated with the chromosome segregation apparatus. There are mainly three types of mitoses depending on the extent of nuclear envelope breakdown: open, semi-open, and closed (Sazer et al. 2014; Makarova and Oliferenko 2016). In open mitosis, the nuclear envelope breaks down completely during mitosis, facilitating access for cytoplasmic spindle microtubules to chromosomes. Semi-open mitosis involves a partial breakdown of the nuclear envelope, allowing transport of material while keeping chromosomes inside the nucleus. In this case, the spindle assembles either inside or outside of the nucleus. In the latter case, spindle microtubules appear to fenestrate through the nuclear envelope and capture chromosomes that are located inside the nucleus. Finally, in closed mitosis, the nuclear envelope does not break down. To enable capturing of sister chromatids, most eukaryotes with closed mitosis assemble an intranuclear spindle. Some eukaryotes, however, assemble an extranuclear spindle where spindle microtubules are located outside of the nucleus. This type of mitosis, though not very common, is found in some Alveolata (dinoflagellates and Perkinsozoa) and Parabasalids (Trichomonads and Hypermastigia), suggesting that it arose independently. To enable attachments between spindle microtubules and kinetochores, these organisms embed their kinetochores in the nuclear envelope. Below we will summarize the current literature on these organisms as well as their sister species and then discuss potential adaptations and implications from such kinetochores.

5.1 Dinoflagellates

Dinoflagellates are a highly diverse group of flagellates, including photosynthetic free-living and parasitic species (Taylor et al. 2007). They belong to the supergroup SAR, Alveolata group, and their sister groups include Perkinsozoa and Apicomplexa (Saldarriaga et al. 2004) (Fig. 2). Dinoflagellates are characterized by large genome sizes in the range of 1,500 Mbp to 185,000 Mbp (Wisecaver and Hackett 2011). Despite having all core histone genes, histones are not involved in packaging the majority of nuclear DNA (Hackett et al. 2005; Marinov and Lynch 2015). In addition, other basic nuclear proteins including Dinoflagellates/Viral NucleoProteins (DVNPs) and HU-like proteins might substitute major histone functions in some of these organisms (Sala-Rovira et al. 1991; Chan and Wong 2007; Gornik et al. 2012; Talbert and Henikoff 2012; Bachvaroff et al. 2014). Their chromosomes are permanently condensed, showing a characteristic liquid crystalline state even in interphase. Interestingly, some, but not all, dinoflagellates have kinetochores embedded in the nuclear envelope with an extranuclear spindle (Leadbeater and Dodge 1967; Kubai and Ris 1969; Spector and Triemer 1981).

Dinoflagellates are divided into core dinoflagellates, Syndiniales, and early diverging Oxyrrhinales (Fig. 2) (Wisecaver and Hackett 2011). Electron microscopy revealed that kinetochores are embedded in the nuclear envelope in core dinoflagellates [e.g., *Amphidinium* (Oakley and Dodge 1974) and *Crypthecodinium cohnii* (Bhaud et al. 2000) (Fig. 3)] as well as in Syndiniales (e.g., *Syndinium* sp. (Ris and Kubai 1974)). Due to their large genome sizes, genome sequence data are limited in dinoflagellates. In fact, the only dinoflagellate genome sequence available to date is for *Symbiodinium minutum* (Shoguchi et al. 2013), which revealed putative CENP-A and outer kinetochore components (Table 1) as well as a spindle assembly checkpoint protein (Mad3/BubR1: symbB.v1.2.026514.t1). These findings suggest that this organism still utilizes canonical kinetochore components and the spindle checkpoint. Indeed, a microtubule inhibitor nocodazole delayed mitotic exit in *Crypthecodinium cohnii*, showing that the spindle checkpoint is functional in core dinoflagellates (Yeung et al. 2000).

In contrast, a member of the early diverging Oxyrrhinales, *Oxyrrhis marina,* has an intranuclear spindle, and its chromosomes are not attached to the nuclear envelope (Triemer 1982; Gao and Li 1986; Kato et al. 2000). These studies show that the extranuclear spindle is not a ubiquitous feature of dinoflagellates.

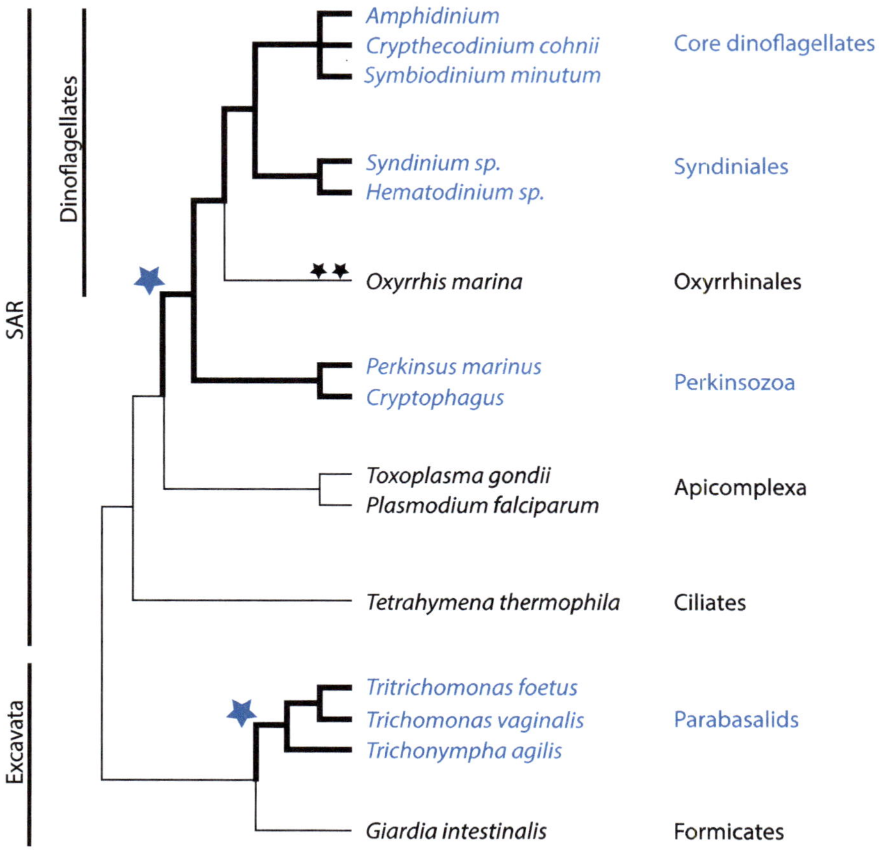

Fig. 2 Membrane-bound kinetochores have independently evolved at least twice. The diagram shows the evolutionary transition to membrane-bound kinetochores in Perkinsozoa and Parabasalids indicated by the *blue star* and *thick* branches. *Two black stars* indicate the reversion to non-membrane-bound kinetochores

5.2 *Perkinsozoa*

Perkinsozoa is one of the closest relatives of dinoflagellates (Fig. 2). Like core dinoflagellates and Syndiniales, Perkinsozoa undergoes a closed mitosis with an extranuclear spindle, suggesting that its kinetochores are embedded in the nuclear envelope (e.g., *Perkinsus marinus* (Perkins 1996) and *Cryptophagus* (Brugerolle 2002)). Unlike dinoflagellates, however, Perkinsozoa has a smaller genome size that is packaged into nucleosomes (58 Mbp in *Perkinsus marinus* (Gornik et al. 2012)), and its chromosomes are not permanently condensed. Taken together, the observations in Perkinsozoa suggest that extranuclear spindles and membrane-embedded

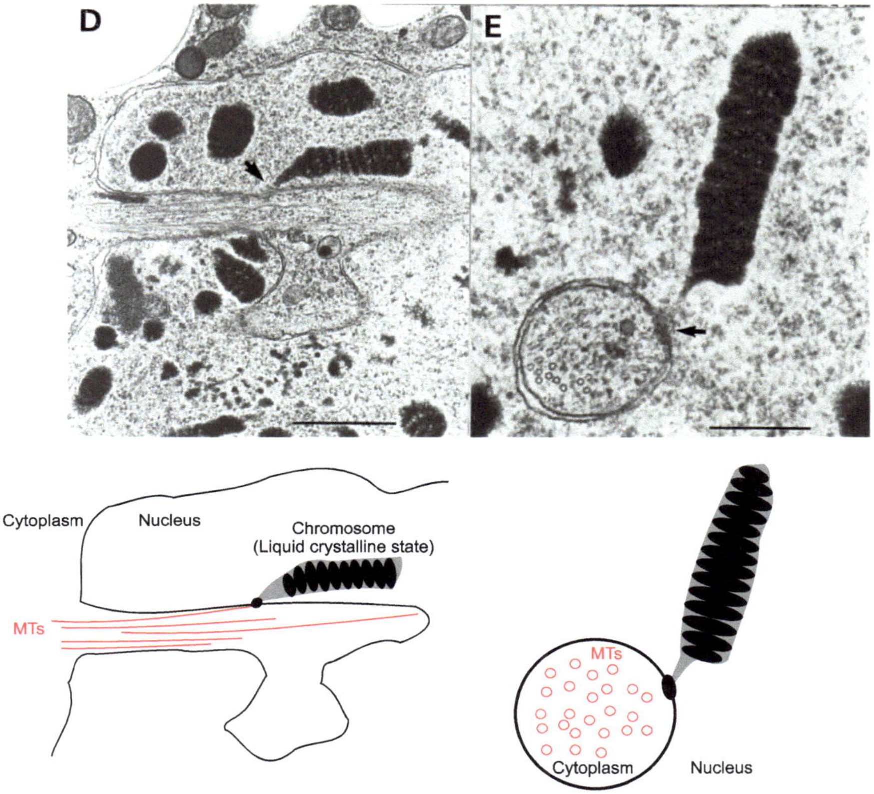

Fig. 3 Membrane-embedded kinetochores in dinoflagellates. *Top* Electron microscopy micrograph of mitotic *Crypthecodinium cohnii* cells. Note that the kinetochore-like structure embedded in the nuclear membrane makes contact with extranuclear spindle microtubules (*arrows*). *Bars* 0.8 μm (*left*), 0.3 μm (*right*). Reproduced from Bhaud et al. (2000) with permission from the Company of Biologists Limited. *Bottom* Simplified schematic of images on top

kinetochores are not necessarily the consequence of an expanded genome or the diminution of packaging histones.

In contrast to Perkinsozoa, its sister group Apicomplexa (e.g., *Plasmodium* and *Toxoplasma gondii*) undergoes a closed mitosis with an intranuclear spindle as is the case for many other species in the SAR supergroup (Francia and Striepen 2014). These observations suggest that the nuclear envelop-embedded kinetochores and the extranuclear spindle appeared at or before the emergence of Perkinsozoa (Cavalier-Smith and Chao 2004) (Fig. 2). Therefore, the intranuclear spindle in *Oxyrrhis* is most likely a derived feature, i.e., back to a more canonical state. The driving forces underlying the switch to the extranuclear spindle or back, however, remain unclear.

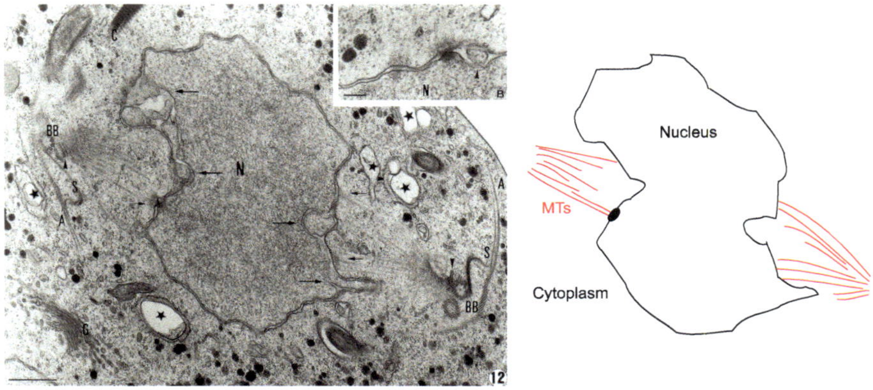

Fig. 4 Bipolar organization of an extranuclear mitotic spindle in Parabasalids. *Left* Electron microscopy image of *Tritrichomonas foetus*. Note that some extranuclear spindle microtubules terminate outside the nuclear membrane. *Bars* 560 and 320 nm (*inset*). Reproduced from Ribeiro et al. (2002) with permission from John Wiley and Sons. *Right* Simplified schematic of the images on *left*

5.3 Parabasalids

Membrane-bound kinetochores have also evolved in Parabasalids that belong to the supergroup Excavata, Metamonads group. They are characterized by a unique parabasal apparatus (Honigberg 1963). Electron microscopy studies showed that Trichomonads (*Tritrichomonas foetus* and *Trichomonas vaginalis*) and Hypermastigia (*Trichonympha agilis*) undergo a closed mitosis with an extranuclear spindle, and have kinetochores embedded in the nuclear envelope (Kubai 1973; Ribeiro et al. 2002) (Fig. 4). As in dinoflagellates and Perkinsozoa, canonical kinetochore proteins are found in the genome of *Trichomonas vaginalis* (Carlton et al. 2007; Zubácová et al. 2012) (Table 1). Because other members of metamonads such as *Giardia* have intranuclear spindles (Sagolla et al. 2006), membrane-embedded kinetochores and extranuclear spindles in Parabasalids appear to be a derived feature that independently evolved in this lineage.

5.4 Implications from Membrane-Bound Kinetochores

The findings of nuclear envelope-embedded kinetochores raise several questions.

It is likely that the change in the location required adaptations of the kinetochore due to the change in biophysical environment. What modifications are necessary to allow kinetochores to be embedded in the nuclear envelope and what are possible consequences? Although the exact position of kinetochores/centromeres within the lipid bilayer of nuclear membranes remains unclear, electron microscopy data indicate that microtubules likely interact with kinetochores in the cytoplasm rather than in the nuclear envelope. This implies that the microtubule-binding domain of the Ndc80 complex is located outside of the nuclear envelope. Other kinetochore proteins that bridge between the Ndc80 complex and CENP-A-containing centromeric chromatin within the nucleus must therefore be embedded within the nuclear envelope. Transmembrane domains have so far not been identified in any of *Symbiodinium minutum* and *Trichomonas vaginalis* kinetochore proteins. It is possible that their kinetochores insert into the lipid bilayer by interacting with other nuclear envelope-embedded components such as the nuclear pore complex as previously suggested (Kubai 1973; Ris and Kubai 1974; Cachon and Cachon 1977; Drechsler and McAinsh 2012).

Another question is how membrane-embedded kinetochores form biorientation and regulate cell cycle progression. Can kinetochores move freely in the nuclear envelope or do they require new membrane synthesis? And once biorientation is achieved, how do nuclear and cytoplasmic environments communicate to promote the transition to anaphase, activating the anaphase promoting complex to disrupt cohesion (in the nucleus) and Cyclin B (in the nucleus or cytoplasm), while coordinating the elongation of spindle microtubules (in the cytoplasm)? Finally, what was the evolutionary driving force that underlies the assembly of kinetochores within the nuclear envelope? To address these unknowns, new tools and model systems need to be developed. Importantly, genetic manipulations have already been established in some dinoflagellates (Te and Lohuis 1998; Radakovits et al. 2010) and *Trichomonas vaginalis* (Delgadillo et al. 1997). Studies on these membrane-embedded kinetochores will likely shed new light onto the diverse mechanism of kinetochore assembly and chromosome segregation in eukaryotes.

6 Speculation of Kinetochores in Early Eukaryotes

Chromosome segregation in the last eukaryotic common ancestor (LECA) was likely driven by tubulin-based polymers because microtubules are a universal feature of the chromosome segregation machinery in all known eukaryotes (McIntosh et al. 2010; Yutin and Koonin 2012; Findeisen et al. 2014). In addition, the LECA likely used condensins to compact chromosomes and cohesins to connect duplicated sister chromatids until anaphase (Nasmyth and Haering 2009; Hirano 2016). Furthermore, the presumed presence of cyclin-dependent kinases and the anaphase promoting complex suggests that chromosome segregation was probably already regulated in the cell cycle dependent manner (Nasmyth 1995; Cavalier-Smith 2010a; Garg and Martin 2016).

In contrast to these components, no obvious ortholog for any of the kinetochore proteins has been identified in prokaryotes, including Lokiarchaeota that is considered to be the closest sister group to eukaryotes (Spang et al. 2015). Therefore, it is unclear whether the LECA utilized canonical kinetochore components, such as CENP-A and Ndc80 that are found in nearly all extant species. It is formally possible that the LECA utilized a KKT-based complex that has later been replaced by canonical kinetochore components in most eukaryotic lineages. This model is consistent with the controversial hypothesis that kinetoplastids might represent the earliest branching eukaryotes (Cavalier-Smith 2010b; Cavalier-Smith 2013; Akiyoshi and Gull 2014). Alternatively, the KKT-based kinetochore may be a derived feature that replaced early eukaryotic kinetochores at some point during the kinetoplastid evolution. A third possibility is that either canonical kinetochores or the KKT-based kinetochores had not yet evolved in the LECA. In this case, chromosome segregation in early eukaryotes might have been similar to the plasmid-partitioning systems found in Bacteria (Gerdes et al. 2010; Reyes-Lamothe et al. 2012) and Archaea (Barillà 2016) where specific DNA elements are recognized by DNA-binding proteins that connect to filament-forming proteins to drive segregation. In such a system, chromosome movement and DNA attachment in the LECA could have been mediated by kinesin or dynein motor proteins. In fact, motor proteins that transport cargo along microtubules and chromokinesins that are capable of connecting chromosomes to microtubules were likely already present in the LECA (Wickstead and Gull 2007; Wickstead et al. 2010).

Kinetochores in all extant eukaryotes are highly complex and consist of many components. Gene duplication likely played an important role in increasing the structural complexity in both canonical and kinetoplastid kinetochores, as evident by the presence of multiple kinetochore proteins that apparently share common ancestry (Schmitzberger and Harrison 2012; Nerusheva and Akiyoshi 2016; Dimitrova et al. 2016; Petrovic et al. 2016). To ensure proper assembly and biorientation of kinetochores, the invention of the Aurora kinase could have been a key evolutionary step that likely had occurred before the emergence of the LECA (Lampson and Cheeseman 2011; Carmena et al. 2012; Hochegger et al. 2013). Error correction by Aurora and direct stabilization of kinetochore-microtubule attachment by tension likely increased the fidelity of chromosome segregation (Akiyoshi et al. 2010; Miller et al. 2016).

7 Conclusions

Most cell biological research over the last several decades has focused on a limited number of model organisms that were selected largely based on historical, not necessarily biological, reasons. Although these studies revealed insights into basic principles of kinetochore organization, a number of differences have been noted even among traditional animal and fungal model organisms. In addition, the unconventional kinetochore in kinetoplastids, the absence of CENP-A in holocentric insects, and nuclear envelope-embedded kinetochores in some eukaryotic lineages all suggest that kinetochores are more plastic than previously thought. The advance of sequencing and genome editing techniques combined with experimental approaches should enable researchers to characterize kinetochores in nontraditional model organisms in a relatively short space of time (Warren 2015; Kobayashi et al. 2015; Gladfelter 2015; Goldstein and King 2016). Insights into kinetochores from diverse species outside of our current catalog of model organisms have a potential to reveal fundamental design and working principles of the eukaryotic segregation machines.

Note to Nomenclature

In different organisms, the centromeric histone H3 variant is referred to with different names (Earnshaw et al. 2013; Talbert and Henikoff 2013). To be consistent with other chapters, we generally refer to the centromeric histone as CENP-A across species. To account for the differences in nomenclature in specific organisms, we donate the superscript of the original name wherever appropriate (for example, CENP-A$^{\mathrm{HCP\text{-}3}}$ for *C. elegans*).

Acknowledgements We thank Paul Talbert, Kim Nasmyth, Geert Kops, and members of the Akiyoshi and Drinnenberg groups for comments on the manuscript. I.A.D. was supported by funds from the CNRS (Atip Avenir) and Institut Curie. B.A. was supported by a Sir Henry Dale Fellowship jointly supported by the Wellcome Trust and the Royal Society (grant number 098403/Z/12/Z), as well as a Wellcome-Beit Prize Fellowship (grant number 098403/Z/12/A).

References

Adl SM, Simpson AGB, Lane CE et al (2012) The revised classification of eukaryotes. J Eukaryot Microbiol 59:429–514. doi:10.1111/j.1550-7408.2012.00644.x

Akiyoshi B (2016) The unconventional kinetoplastid kinetochore: from discovery toward functional understanding. Biochem Soc Trans 44:1201–1217. doi:10.1042/BST20160112

Akiyoshi B, Gull K (2013) Evolutionary cell biology of chromosome segregation: insights from trypanosomes. Open Biol 3:130023. doi:10.1098/rsob.130023

Akiyoshi B, Gull K (2014) Discovery of unconventional kinetochores in kinetoplastids. Cell 156:1247–1258. doi:10.1016/j.cell.2014.01.049

Akiyoshi B, Sarangapani KK, Powers AF et al (2010) Tension directly stabilizes reconstituted kinetochore-microtubule attachments. Nature 468:576–579. doi:10.1038/nature09594

Allshire RC, Karpen GH (2008) Epigenetic regulation of centromeric chromatin: old dogs, new tricks? Nat Rev Genet 9:923–937. doi:10.1038/nrg2466

Bachvaroff TR, Gornik SG, Concepcion GT et al (2014) Dinoflagellate phylogeny revisited: using ribosomal proteins to resolve deep branching dinoflagellate clades. Mol Phylogenet Evol 70:314–322. doi:10.1016/j.ympev.2013.10.007

Barillà D (2016) Driving apart and segregating genomes in archaea. Trends Microbiol. doi:10.1016/j.tim.2016.07.001

Basilico F, Maffini S, Weir JR et al (2014) The pseudo GTPase CENP-M drives human kinetochore assembly. Elife 3:e02978

Berriman M, Ghedin E, Hertz-Fowler C et al (2005) The genome of the African trypanosome *Trypanosoma brucei*. Science 309:416–422. doi:10.1126/science.1112642

Bhaud Y, Guillebault D, Lennon J et al (2000) Morphology and behaviour of dinoflagellate chromosomes during the cell cycle and mitosis. J Cell Sci 113(Pt 7):1231–1239

Biggins S (2013) The composition, functions, and regulation of the budding yeast kinetochore. Genetics 194:817–846. doi:10.1534/genetics.112.145276

Black BE, Cleveland DW (2011) Epigenetic centromere propagation and the nature of CENP-A nucleosomes. Cell 144:471–479. doi:10.1016/j.cell.2011.02.002

Blattner AC, Aguilar-Rodríguez J, Kränzlin M et al (2016) *Drosophila* Nnf1 paralogs are partially redundant for somatic and germ line kinetochore function. Chromosoma. doi:10.1007/s00412-016-0579-4

Brooks CF, Francia ME, Gissot M et al (2011) *Toxoplasma gondii* sequesters centromeres to a specific nuclear region throughout the cell cycle. Proc Natl Acad Sci U S A 108:3767–3772. doi:10.1073/pnas.1006741108

Brugerolle G (2002) *Cryptophagus subtilis*: a new parasite of cryptophytes affiliated with the Perkinsozoa lineage. Eur J Protistol 37:379–390. doi:10.1078/0932-4739-00837

Cachon J, Cachon M (1977) Observations on the mitosis and on the chromosome evolution during the lifecycle of *Oodinium*, a parasitic dinoflagellate. Chromosoma 60:237–251

Carlton JM, Hirt RP, Silva JC et al (2007) Draft genome sequence of the sexually transmitted pathogen *Trichomonas vaginalis*. Science 315:207–212. doi:10.1126/science.1132894

Carmena M, Wheelock M, Funabiki H, Earnshaw WC (2012) The chromosomal passenger complex (CPC): from easy rider to the godfather of mitosis. Nat Rev Mol Cell Biol 13:789–803. doi:10.1038/nrm3474

Carroll CW, Milks KJ, Straight AF (2010) Dual recognition of CENP-A nucleosomes is required for centromere assembly. J Cell Biol 189:1143–1155. doi:10.1083/jcb.201001013

Cavalier-Smith T (2010a) Origin of the cell nucleus, mitosis and sex: roles of intracellular coevolution. Biol Direct 5:7. doi:10.1186/1745-6150-5-7

Cavalier-Smith T (2010b) Kingdoms Protozoa and Chromista and the eozoan root of the eukaryotic tree. Biol Lett 6:342–345. doi:10.1098/rsbl.2009.0948

Cavalier-Smith T (2013) Early evolution of eukaryote feeding modes, cell structural diversity, and classification of the protozoan phyla Loukozoa, Sulcozoa, and Choanozoa. Eur J Protistol 49:115–178. doi:10.1016/j.ejop.2012.06.001

Cavalier-Smith T (2016) Higher classification and phylogeny of Euglenozoa. Eur J Protistol 56:250–276. doi:10.1016/j.ejop.2016.09.003

Cavalier-Smith T, Chao EE (2004) Protalveolate phylogeny and systematics and the origins of Sporozoa and dinoflagellates (phylum Myzozoa nom. nov.). EUR J PROTISTOL 40:185–212. doi:10.1016/j.ejop.2004.01.002

Cervantes MD, Xi X, Vermaak D et al (2006) The CNA1 histone of the ciliate *Tetrahymena thermophila* is essential for chromosome segregation in the germline micronucleus. Mol Biol Cell 17:485–497. doi:10.1091/mbc.E05-07-0698

Chan Y-H, Wong JTY (2007) Concentration-dependent organization of DNA by the dinoflagellate histone-like protein HCc3. Nucleic Acids Res 35:2573–2583. doi:10.1093/nar/gkm165

Cheerambathur DK, Desai A (2014) Linked in: formation and regulation of microtubule attachments during chromosome segregation. Curr Opin Cell Biol 26:113–122. doi:10.1016/j.ceb.2013.12.005

Cheeseman IM (2014) The kinetochore. Cold Spring Harb Perspect Biol 6:a015826. doi:10.1101/cshperspect.a015826

Cheeseman IM, Desai A (2008) Molecular architecture of the kinetochore-microtubule interface. Nat Rev Mol Cell Biol 9:33–46. doi:10.1038/nrm2310

Cheeseman IM, Niessen S, Anderson S et al (2004) A conserved protein network controls assembly of the outer kinetochore and its ability to sustain tension. Genes Dev 18:2255–2268. doi:10.1101/gad.1234104

Cheeseman IM, Chappie JS, Wilson-Kubalek EM, Desai A (2006) The conserved KMN network constitutes the core microtubule-binding site of the kinetochore. Cell 127:983–997. doi:10.1016/j.cell.2006.09.039

Cui B, Gorovsky MA (2006) Centromeric histone H3 is essential for vegetative cell division and for DNA elimination during conjugation in *Tetrahymena thermophila*. Mol Cell Biol 26:4499–4510. doi:10.1128/MCB.00079-06

Curtis BA, Tanifuji G, Burki F et al (2012) Algal genomes reveal evolutionary mosaicism and the fate of nucleomorphs. Nature 492:59–65. doi:10.1038/nature11681

Dawe RK, Reed LM, Yu HG et al (1999) A maize homolog of mammalian CENPC is a constitutive component of the inner kinetochore. Plant Cell 11:1227–1238

Dawson SC, Sagolla MS, Cande WZ (2007) The cenH3 histone variant defines centromeres in *Giardia intestinalis*. Chromosoma 116:175–184. doi:10.1007/s00412-006-0091-3
Delgadillo MG, Liston DR, Niazi K, Johnson PJ (1997) Transient and selectable transformation of the parasitic protist *Trichomonas vaginalis*. Proc Natl Acad Sci U S A 94:4716–4720
Dimitrova YN, Jenni S, Valverde R et al (2016) Structure of the MIND complex defines a regulatory focus for yeast kinetochore assembly. Cell 167(1014–1027):e12. doi:10.1016/j.cell.2016.10.011
Douglas S, Zauner S, Fraunholz M et al (2001) The highly reduced genome of an enslaved algal nucleus. Nature 410:1091–1096. doi:10.1038/35074092
Drechsler H, McAinsh AD (2012) Exotic mitotic mechanisms. Open Biol 2:120140. doi:10.1098/rsob.120140
Drinnenberg IA, deYoung D, Henikoff S, Malik HS (2014) Recurrent loss of CenH3 is associated with independent transitions to holocentricity in insects. Elife 3:e03676. doi:10.7554/eLife.03676
Drinnenberg IA, Henikoff S, Malik HS (2016) Evolutionary turnover of kinetochore proteins: a Ship of Theseus? Trends Cell Biol 26:498–510. doi:10.1016/j.tcb.2016.01.005
Dubin M, Fuchs J, Gräf R et al (2010) Dynamics of a novel centromeric histone variant CenH3 reveals the evolutionary ancestral timing of centromere biogenesis. Nucleic Acids Res 38:7526–7537. doi:10.1093/nar/gkq664
Earnshaw WC (2015) Discovering centromere proteins: from cold white hands to the A, B, C of CENPs. Nat Rev Mol Cell Biol 16:443–449. doi:10.1038/nrm4001
Earnshaw WC, Allshire RC, Black BE et al (2013) Esperanto for histones: CENP-A, not CenH3, is the centromeric histone H3 variant. Chromosome Res 21:101–106. doi:10.1007/s10577-013-9347-y
Espeut J, Cheerambathur DK, Krenning L et al (2012) Microtubule binding by KNL-1 contributes to spindle checkpoint silencing at the kinetochore. J Cell Biol 196:469–482. doi:10.1083/jcb.201111107
Etemad B, Kops GJPL (2016) Attachment issues: kinetochore transformations and spindle checkpoint silencing. Curr Opin Cell Biol 39:101–108. doi:10.1016/j.ceb.2016.02.016
Farrell M, Gubbels M-J (2014) The *Toxoplasma gondii* kinetochore is required for centrosome association with the centrocone (spindle pole). Cell Microbiol 16:78–94. doi:10.1111/cmi.12185
Findeisen P, Mühlhausen S, Dempewolf S et al (2014) Six subgroups and extensive recent duplications characterize the evolution of the eukaryotic tubulin protein family. Genome Biol Evol 6:2274–2288. doi:10.1093/gbe/evu187
Finseth FR, Dong Y, Saunders A, Fishman L (2015) Duplication and adaptive evolution of a key centromeric protein in *Mimulus*, a genus with female meiotic drive. Mol Biol Evol 32:2694–2706. doi:10.1093/molbev/msv145
Foley EA, Kapoor TM (2013) Microtubule attachment and spindle assembly checkpoint signalling at the kinetochore. Nat Rev Mol Cell Biol 14:25–37. doi:10.1038/nrm3494
Francia ME, Striepen B (2014) Cell division in apicomplexan parasites. Nat Rev Micro 12:125–136. doi:10.1038/nrmicro3184
Fukagawa T, Earnshaw WC (2014) The centromere: chromatin foundation for the kinetochore machinery. Dev Cell 30:496–508. doi:10.1016/j.devcel.2014.08.016
Gao XP, Li JY (1986) Nuclear division in the marine dinoflagellate *Oxyrrhis marina*. J Cell Sci 85:161–175
Garg SG, Martin WF (2016) Mitochondria, the cell cycle, and the origin of sex via a syncytial eukaryote common ancestor. Genome Biol Evol 8:1950–1970. doi:10.1093/gbe/evw136
Gerdes K, Howard M, Szardenings F (2010) Pushing and pulling in prokaryotic DNA segregation. Cell 141:927–942. doi:10.1016/j.cell.2010.05.033
Gladfelter AS (2015) How nontraditional model systems can save us. Mol Biol Cell 26:3687–3689. doi:10.1091/mbc.E15-06-0429
Goldstein B, King N (2016) The future of cell biology: emerging model organisms. Trends Cell Biol 26:818–824. doi:10.1016/j.tcb.2016.08.005

Gornik SG, Ford KL, Mulhern TD et al (2012) Loss of nucleosomal DNA condensation coincides with appearance of a novel nuclear protein in dinoflagellates. Curr Biol 22:2303–2312. doi:10.1016/j.cub.2012.10.036

Goshima G, Wollman R, Goodwin SS et al (2007) Genes required for mitotic spindle assembly in *Drosophila* S2 cells. Science 316:417–421. doi:10.1126/science.1141314

Hackett JD, Scheetz TE, Yoon HS et al (2005) Insights into a dinoflagellate genome through expressed sequence tag analysis. BMC Genom 6:80. doi:10.1186/1471-2164-6-80

Hedges SB, Dudley J, Kumar S (2006) TimeTree: a public knowledge-base of divergence times among organisms. Bioinformatics 22:2971–2972. doi:10.1093/bioinformatics/btl505

Henikoff S, Ahmad K, Malik HS (2001) The centromere paradox: stable inheritance with rapidly evolving DNA. Science 293:1098–1102. doi:10.1126/science.1062939

Hirano T (2016) Condensin-based chromosome organization from bacteria to vertebrates. Cell 164:847–857. doi:10.1016/j.cell.2016.01.033

Hochegger H, Hégarat N, Pereira-Leal JB (2013) Aurora at the pole and equator: overlapping functions of Aurora kinases in the mitotic spindle. Open Biol 3:120185. doi:10.1098/rsob.120185

Honigberg BM (1963) Evolutionary and systematic relationships in the flagellate order Trichomonadida Kirby. J Protozool 10:20–63

Hori T, Amano M, Suzuki A et al (2008) CCAN makes multiple contacts with centromeric DNA to provide distinct pathways to the outer kinetochore. Cell 135:1039–1052. doi:10.1016/j.cell.2008.10.019

Hornung P, Maier M, Alushin GM et al (2011) Molecular architecture and connectivity of the budding yeast Mtw1 kinetochore complex. J Mol Biol 405:548–559. doi:10.1016/j.jmb.2010.11.012

Hornung P, Troc P, Malvezzi F et al (2014) A cooperative mechanism drives budding yeast kinetochore assembly downstream of CENP-A. J Cell Biol 206:509–524. doi:10.1083/jcb.201403081

Ishii T, Karimi-Ashtiyani R, Banaei-Moghaddam AM et al (2015) The differential loading of two barley CENH3 variants into distinct centromeric substructures is cell type- and development-specific. Chromosome Res 23:277–284. doi:10.1007/s10577-015-9466-8

Kanesaki Y, Imamura S, Matsuzaki M, Tanaka K (2015) Identification of centromere regions in chromosomes of a unicellular red alga, *Cyanidioschyzon merolae*. FEBS Lett 589:1219–1224. doi:10.1016/j.febslet.2015.04.009

Kato KH, Moriyama A, Itoh TJ et al (2000) Dynamic changes in microtubule organization during division of the primitive dinoflagellate *Oxyrrhis marina*. Biol Cell 92:583–594. doi:10.1016/S0248-4900(00)01106-0

Kawabe A, Nasuda S, Charlesworth D (2006) Duplication of centromeric histone H3 (HTR12) gene in *Arabidopsis halleri* and *A. lyrata*, plant species with multiple centromeric satellite sequences. Genetics 174:2021–2032. doi:10.1534/genetics.106.063628

Kobayashi N, Suzuki Y, Schoenfeld LW et al (2015) Discovery of an unconventional centromere in budding yeast redefines evolution of point centromeres. Curr Biol 25:2026–2033. doi:10.1016/j.cub.2015.06.023

Koumandou VL, Wickstead B, Ginger ML et al (2013) Molecular paleontology and complexity in the last eukaryotic common ancestor. Crit Rev Biochem Mol Biol 48:373–396. doi:10.3109/10409238.2013.821444

Kubai DF (1973) Unorthodox mitosis in *Trichonympha agilis*: kinetochore differentiation and chromosome movement. J Cell Sci 13:511–552

Kubai DF, Ris H (1969) Division in the dinoflagellate *Gyrodinium Cohnii* (schiller). J Cell Biol 40:508–528. doi:10.1083/jcb.40.2.508

Lampson MA, Cheeseman IM (2011) Sensing centromere tension: Aurora B and the regulation of kinetochore function. Trends Cell Biol 21:133–140. doi:10.1016/j.tcb.2010.10.007

Leadbeater B, Dodge JD (1967) An electron microscope study of nuclear and cell division in a dinoflagellate. Arch Mikrobiol 57:239–254

Li Z, Umeyama T, Wang CC (2008) The chromosomal passenger complex and a mitotic kinesin interact with the Tousled-like kinase in trypanosomes to regulate mitosis and cytokinesis. PLoS ONE 3:e3814. doi:10.1371/journal.pone.0003814

Liu D, Vleugel M, Backer CB et al (2010) Regulated targeting of protein phosphatase 1 to the outer kinetochore by KNL1 opposes Aurora B kinase. J Cell Biol 188:809–820. doi:10.1083/jcb.201001006

Liu Y, Petrovic A, Rombaut P et al (2016) Insights from the reconstitution of the divergent outer kinetochore of *Drosophila melanogaster*. Open Biol 6:150236. doi:10.1098/rsob.150236

London N, Biggins S (2014) Signalling dynamics in the spindle checkpoint response. Nat Rev Mol Cell Biol 15:736–747. doi:10.1038/nrm3888

Lowell JE, Cross GAM (2004) A variant histone H3 is enriched at telomeres in *Trypanosoma brucei*. J Cell Sci 117:5937–5947. doi:10.1242/jcs.01515

Makarova M, Oliferenko S (2016) Mixing and matching nuclear envelope remodeling and spindle assembly strategies in the evolution of mitosis. Curr Opin Cell Biol 41:43–50. doi:10.1016/j.ceb.2016.03.016

Malik HS, Henikoff S (2003) Phylogenomics of the nucleosome. Nat Struct Biol 10:882–891. doi:10.1038/nsb996

Malik HS, Henikoff S (2009) Major evolutionary transitions in centromere complexity. Cell 138:1067–1082. doi:10.1016/j.cell.2009.08.036

Marinov GK, Lynch M (2015) Diversity and divergence of dinoflagellate histone proteins. G3 (Bethesda) 6:397–422. doi:10.1534/g3.115.023275

Martin W, Koonin EV (2006) Introns and the origin of nucleus-cytosol compartmentalization. Nature 440:41–45. doi:10.1038/nature04531

Maruyama S, Kuroiwa H, Miyagishima S et al (2007) Centromere dynamics in the primitive red alga *Cyanidioschyzon merolae*. Plant J 49:1122–1129. doi:10.1111/j.1365-313X.2006.03024.x

Matsuzaki M, Misumi O, Shin-I T et al (2004) Genome sequence of the ultrasmall unicellular red alga *Cyanidioschyzon merolae* 10D. Nature 428:653–657. doi:10.1038/nature02398

McIntosh JR (2016) Mitosis. Cold Spring Harb Perspect Biol. doi:10.1101/cshperspect.a023218

McIntosh JR, Volkov V, Ataullakhanov FI, Grishchuk EL (2010) Tubulin depolymerization may be an ancient biological motor. J Cell Sci 123:3425–3434. doi:10.1242/jcs.067611

McKinley KL, Cheeseman IM (2016) The molecular basis for centromere identity and function. Nat Rev Mol Cell Biol 17:16–29. doi:10.1038/nrm.2015.5

Meadows JC, Shepperd LA, Vanoosthuyse V et al (2011) Spindle checkpoint silencing requires association of PP1 to both Spc7 and kinesin-8 motors. Dev Cell 20:739–750. doi:10.1016/j.devcel.2011.05.008

Melters DP, Paliulis LV, Korf IF, Chan SWL (2012) Holocentric chromosomes: convergent evolution, meiotic adaptations, and genomic analysis. Chromosome Res 20:579–593. doi:10.1007/s10577-012-9292-1

Meraldi P, McAinsh AD, Rheinbay E, Sorger PK (2006) Phylogenetic and structural analysis of centromeric DNA and kinetochore proteins. Genome Biol 7:R23. doi:10.1186/gb-2006-7-3-r23

Miller MP, Asbury CL, Biggins S (2016) A TOG protein confers tension sensitivity to kinetochore-microtubule attachments. Cell 165:1–12. doi:10.1016/j.cell.2016.04.030

Müller S, Almouzni G (2014) A network of players in H3 histone variant deposition and maintenance at centromeres. Biochim Biophys Acta 1839:241–250. doi:10.1016/j.bbagrm.2013.11.008

Nasmyth K (1995) Evolution of the cell cycle. Philos Trans R Soc Lond B Biol Sci 349:271–281. doi:10.1098/rstb.1995.0113

Nasmyth K, Haering CH (2009) Cohesin: its roles and mechanisms. Annu Rev Genet 43:525–558. doi:10.1146/annurev-genet-102108-134233

Nerusheva OO, Akiyoshi B (2016) Divergent polo box domains underpin the unique kinetoplastid kinetochore. Open Biol 6:150206. doi:10.1098/rsob.150206

Neumann P, Navrátilová A, Schroeder-Reiter E et al (2012) Stretching the rules: monocentric chromosomes with multiple centromere domains. PLoS Genet 8:e1002777. doi:10.1371/journal.pgen.1002777

Neumann P, Pavlíková Z, Koblížková A et al (2015) Centromeres off the hook: massive changes in centromere size and structure following duplication of *cenh3* gene in *Fabeae* species. Mol Biol Evol 32:1862–1879. doi:10.1093/molbev/msv070

Nicklas RB (1997) How cells get the right chromosomes. Science 275:632–637

Oakley BR, Dodge JD (1974) Kinetochores associated with the nuclear envelope in the mitosis of a dinoflagellate. J Cell Biol 63:322–325

Parsons M, Worthey EA, Ward PN, Mottram JC (2005) Comparative analysis of the kinomes of three pathogenic trypanosomatids: *Leishmania major, Trypanosoma brucei* and *Trypanosoma cruzi*. BMC Genom 6:127. doi:10.1186/1471-2164-6-127

Perkins FO (1996) The structure of *Perkinsus marinus* (Mackin, Owen and Collier, 1950) Levine 1978 with comments on taxonomy and phylogeny of *Perkinsus* spp. J Shellfish Res 6:65–87

Petrovic A, Mosalaganti S, Keller J et al (2014) Modular assembly of RWD domains on the Mis12 complex underlies outer kinetochore organization. Mol Cell 53:591–605. doi:10.1016/j.molcel. 2014.01.019

Petrovic A, Keller J, Liu Y et al (2016) Structure of the MIS12 complex and molecular basis of its interaction with CENP-C at human kinetochores. Cell 167(1028–1040):e15. doi:10.1016/j.cell. 2016.10.005

Przewloka MR, Glover DM (2009) The kinetochore and the centromere: a working long distance relationship. Annu Rev Genet 43:439–465. doi:10.1146/annurev-genet-102108-134310

Przewloka MR, Zhang W, Costa P et al (2007) Molecular analysis of core kinetochore composition and assembly in *Drosophila melanogaster*. PLoS ONE 2:e478. doi:10.1371/ journal.pone.0000478

Przewloka MR, Venkei Z, Glover DM (2009) Searching for *Drosophila* Dsn1 kinetochore protein. Cell Cycle 8:1292–1293. doi:10.4161/cc.8.8.8159

Przewloka MR, Venkei Z, Bolanos-Garcia VM et al (2011) CENP-C is a structural platform for kinetochore assembly. Curr Biol 21:399–405. doi:10.1016/j.cub.2011.02.005

Radakovits R, Jinkerson RE, Darzins A, Posewitz MC (2010) Genetic engineering of algae for enhanced biofuel production. Eukaryot Cell 9:486–501. doi:10.1128/EC.00364-09

Reyes-Lamothe R, Nicolas E, Sherratt DJ (2012) Chromosome replication and segregation in bacteria. Annu Rev Genet. doi:10.1146/annurev-genet-110711-155421

Ribeiro KC, Pereira-Neves A, Benchimol M (2002) The mitotic spindle and associated membranes in the closed mitosis of trichomonads. Biol Cell 94:157–172

Richter MM, Poznanski J, Zdziarska A et al (2016) Network of protein interactions within the *Drosophila* inner kinetochore. Open Biol 6:150238. doi:10.1098/rsob.150238

Ris H, Kubai DF (1974) An unusual mitotic mechanism in the parasitic protozoan *Syndinium* sp. J Cell Biol 60:702–720

Rosenberg JS, Cross FR, Funabiki H (2011) KNL1/Spc105 recruits PP1 to silence the spindle assembly checkpoint. Curr Biol 21:942–947. doi:10.1016/j.cub.2011.04.011

Sagolla MS, Dawson SC, Mancuso JJ, Cande WZ (2006) Three-dimensional analysis of mitosis and cytokinesis in the binucleate parasite *Giardia intestinalis*. J Cell Sci 119:4889–4900. doi:10.1242/jcs.03276

Sala-Rovira M, Geraud ML, Caput D et al (1991) Molecular cloning and immunolocalization of two variants of the major basic nuclear protein (HCc) from the histone-less eukaryote *Crypthecodinium cohnii* (Pyrrhophyta). Chromosoma 100:510–518

Saldarriaga JF, "Max" Taylor FJR, Cavalier-Smith T et al (2004) Molecular data and the evolutionary history of dinoflagellates. EUR J PROTISTOL 40:85–111. doi:10.1016/j.ejop. 2003.11.003

Santaguida S, Musacchio A (2009) The life and miracles of kinetochores. EMBO J 28:2511–2531. doi:10.1038/emboj.2009.173

Sato H, Shibata F, Murata M (2005) Characterization of a Mis12 homologue in *Arabidopsis thaliana*. Chromosome Res 13:827–834. doi:10.1007/s10577-005-1016-3

Sazer S, Lynch M, Needleman D (2014) Deciphering the evolutionary history of open and closed mitosis. Curr Biol 24:R1099–R1103. doi:10.1016/j.cub.2014.10.011

Schittenhelm RB, Heeger S, Althoff F et al (2007) Spatial organization of a ubiquitous eukaryotic kinetochore protein network in *Drosophila* chromosomes. Chromosoma 116:385–402. doi:10.1007/s00412-007-0103-y

Schleiffer A, Maier M, Litos G et al (2012) CENP-T proteins are conserved centromere receptors of the Ndc80 complex. Nat Cell Biol 14:604–613. doi:10.1038/ncb2493

Schmitzberger F, Harrison SC (2012) RWD domain: a recurring module in kinetochore architecture shown by a Ctf19–Mcm21 complex structure. EMBO Rep 13:216–222. doi:10.1038/embor.2012.1

Screpanti E, De Antoni A, Alushin GM et al (2011) Direct binding of Cenp-C to the Mis12 complex joins the inner and outer kinetochore. Curr Biol 21:391–398. doi:10.1016/j.cub.2010.12.039

Shoguchi E, Shinzato C, Kawashima T et al (2013) Draft assembly of the *Symbiodinium minutum* nuclear genome reveals dinoflagellate gene structure. Curr Biol 23:1399–1408. doi:10.1016/j.cub.2013.05.062

Spang A, Saw JH, Jørgensen SL et al (2015) Complex archaea that bridge the gap between prokaryotes and eukaryotes. Nature 521:173–179. doi:10.1038/nature14447

Spector DL, Triemer RE (1981) Chromosome structure and mitosis in the dinoflagellates: an ultrastructural approach to an evolutionary problem. BioSystems 14:289–298

Swart EC, Bracht JR, Magrini V et al (2013) The *Oxytricha trifallax* macronuclear genome: a complex eukaryotic genome with 16,000 tiny chromosomes. PLoS Biol 11:e1001473. doi:10.1371/journal.pbio.1001473

Talbert PB, Henikoff S (2012) Chromatin: packaging without nucleosomes. Curr Biol 22:R1040–R1043. doi:10.1016/j.cub.2012.10.052

Talbert PB, Henikoff S (2013) Phylogeny as the basis for naming histones. Trends Genet 29:499–500. doi:10.1016/j.tig.2013.06.009

Talbert PB, Bayes JJ, Henikoff S (2009) Evolution of centromeres and kinetochores: a two-part fugue. In: De Wulf P, Earnshaw WC (eds) The Kinetochore. Springer, New York, pp 1–37

Tanifuji G, Archibald JM (2014) Nucleomorph comparative genomics. In: Löffelhardt W (ed) Endosymbiosis. Springer, Vienna, pp 197–213

Taylor FJR, Hoppenrath M, Saldarriaga JF (2007) Dinoflagellate diversity and distribution. Biodivers Conserv 17:407–418. doi:10.1007/s10531-007-9258-3

Te MR, Lohuis Miller DJ (1998) Genetic transformation of dinoflagellates (*Amphidinium* and *Symbiodinium*): expression of GUS in microalgae using heterologous promoter constructs. Plant J 13:427–435. doi:10.1046/j.1365-313X.1998.00040.x

ten Hoopen R, Manteuffel R, Dolezel J et al (2000) Evolutionary conservation of kinetochore protein sequences in plants. Chromosoma 109:482–489

Triemer RE (1982) A unique mitotic variation in the marine dinoflagellate *Oxyrrhis marina* (pyrrophyta)1. J Phycol 18:399–411. doi:10.1111/j.1529-8817.1982.tb03202.x

Tromer E, Snel B, Kops GJPL (2015) Widespread recurrent patterns of rapid repeat evolution in the kinetochore scaffold KNL1. Genome Biol Evol 7:2383–2393. doi:10.1093/gbe/evv140

Verma G, Surolia N (2013) *Plasmodium falciparum* CENH3 is able to functionally complement Cse4p and its, C-terminus is essential for centromere function. Mol Biochem Parasitol 192:21–29. doi:10.1016/j.molbiopara.2013.11.002

Vermaak D, Hayden HS, Henikoff S (2002) Centromere targeting element within the histone fold domain of Cid. Mol Cell Biol 22:7553–7561

Vickerman K (1962) The mechanism of cyclical development in trypanosomes of the *Trypanosoma brucei* sub-group: an hypothesis based on ultrastructural observations. Trans R Soc Trop Med Hyg 56:487–495

Walker G, Dorrell RG, Schlacht A, Dacks JB (2011) Eukaryotic systematics: a user's guide for cell biologists and parasitologists. Parasitology 138:1638–1663. doi:10.1017/S0031182010001708

Wang G, He Q, Liu F et al (2011) Characterization of CENH3 proteins and centromere-associated DNA sequences in diploid and allotetraploid *Brassica* species. Chromosoma 120:353–365. doi:10.1007/s00412-011-0315-z

Warren G (2015) In praise of other model organisms. J Cell Biol 208:387–389. doi:10.1083/jcb.201412145

Westermann S, Schleiffer A (2013) Family matters: structural and functional conservation of centromere-associated proteins from yeast to humans. Trends Cell Biol 23:260–269. doi:10.1016/j.tcb.2013.01.010

Westhorpe FG, Straight AF (2013) Functions of the centromere and kinetochore in chromosome segregation. Curr Opin Cell Biol 25:334–340. doi:10.1016/j.ceb.2013.02.001

Westhorpe FG, Straight AF (2015) The centromere: epigenetic control of chromosome segregation during mitosis. Cold Spring Harb Perspect Biol 7:a015818. doi:10.1101/cshperspect.a015818

Wickstead B, Gull K (2007) Dyneins across eukaryotes: a comparative genomic analysis. Traffic 8:1708–1721. doi:10.1111/j.1600-0854.2007.00646.x

Wickstead B, Gull K, Richards TA (2010) Patterns of kinesin evolution reveal a complex ancestral eukaryote with a multifunctional cytoskeleton. BMC Evol Biol 10:110. doi:10.1186/1471-2148-10-110

Wisecaver JH, Hackett JD (2011) Dinoflagellate genome evolution. Annu Rev Microbiol 65:369–387. doi:10.1146/annurev-micro-090110-102841

Yeung PK, New DC, Leveson A et al (2000) The spindle checkpoint in the dinoflagellate *Crypthecodinium cohnii*. Exp Cell Res 254:120–129. doi:10.1006/excr.1999.4749

Yutin N, Koonin EV (2012) Archaeal origin of tubulin. Biol Direct 7:10. doi:10.1186/1745-6150-7-10

Zubácová Z, Hostomská J, Tachezy J (2012) Histone H3 variants in *Trichomonas vaginalis*. Eukaryot Cell 11:654–661. doi:10.1128/EC.00006-12

Quantitative Microscopy Reveals Centromeric Chromatin Stability, Size, and Cell Cycle Mechanisms to Maintain Centromere Homeostasis

Ana Stankovic and Lars E.T. Jansen

Abstract Centromeres are chromatin domains specified by nucleosomes containing the histone H3 variant, CENP-A. This unique centromeric structure is at the heart of a strong self-templating epigenetic mechanism that renders centromeres heritable. We review how specific quantitative microscopy approaches have contributed to the determination of the copy number, architecture, size, and dynamics of centromeric chromatin and its associated centromere complex and kinetochore. These efforts revealed that the key to long-term centromere maintenance is the slow turnover of CENP-A nucleosomes, a critical size of the chromatin domain and its cell cycle-coupled replication. These features come together to maintain homeostasis of a chromatin locus that directs its own epigenetic inheritance and facilitates the assembly of the mitotic kinetochore.

1 CENP-A as the Key Epigenetic Determinant of Active Centromeres

Epigenetic traits are heritable features whose propagation is not solely driven by underlying DNA sequences. Centromeres are chromosomal loci whose propagation depends on such a mechanism. The current consensus in the centromere field is that the centromere-specific histone H3 variant CENP-A lies at the core of a positive epigenetic feedback loop and is sufficient to initiate and propagate centromeres. CENP-A, along with CENP-B and CENP-C were among the first centromere proteins to be identified using antibodies isolated from autoimmune sera from human scleroderma patients (CREST) (Earnshaw and Rothfield 1985). These sera stained proteins at all active centromeres but, importantly, they are absent from an inactive centromere, suggesting a "chromatin based regulation" of the centromere (Earnshaw and Migeon 1985). Soon after its initial discovery CENP-A was found to have histone-like properties and to copurify with core histone proteins (Palmer

A. Stankovic · L.E.T. Jansen (✉)
Instituto Gulbenkian de Ciência, 2780-156 Oeiras, Portugal
e-mail: ljansen@igc.gulbenkian.pt

B.E. Black (ed.), *Centromeres and Kinetochores*, Progress in Molecular and Subcellular Biology 56, DOI 10.1007/978-3-319-58592-5_6

et al. 1987). Subsequent cloning of the gene, confirmed these properties (Sullivan et al. 1994). In 1993, the first human neocentromere was described (Voullaire et al. 1993), a functional centromere located on a deleted derivative of chromosome 10 in human patient samples, lacking typical centromeric sequence as well as the CENP-B protein that binds to those sequences. Indeed, CENP-B knockout mice are viable (Hudson et al. 1998), strengthening the notion that centromeric DNA is not the main driver of centromere positioning. In addition, centromere-specific CENP-A homologues exist in nearly all species analyzed so far (Malik and Henikoff 2003; Talbert et al. 2012), with the exception of kinetoplastids and holocentric insects that do not appear to contain a recognizable CENP-A homologue (Akiyoshi and Gull 2013; Drinnenberg et al. 2014). A remarkable feature of centromeric chromatin is the requirement for its maintenance across the germ line in several, but not all organisms analyzed thus far. In mammals, early work has shown that CENP-A is present in mature bovine sperm, evading protamine deposition (Palmer et al. 1990), suggesting CENP-A may play a transgenerational role in mammals. Indeed, stable paternal transmissions of neocentromeres within human families demonstrate that the position of the centromere is inherited epigenetically at least through the male germ line (Amor et al. 2004; Tyler-Smith et al. 1999). Sperm retained CENP-A was also found in *X. laevis* and *D. melanogaster* (Dunleavy et al. 2012; Milks et al. 2009; Raychaudhuri et al. 2012). In Drosophila, a causative role for CENP-A in germ line centromere maintenance has been shown. Selective removal of the CENP-A homologue [known as CID or cenH3 (Talbert and Henikoff 2013)] from paternal centromeres resulted in successful fertilization but in the selective failure to segregate paternal chromosomes in the zygote, despite normal segregation of maternal chromosomes and the availability of a maternal pool of CID (Raychaudhuri et al. 2012). The transgenerational necessity of CENP-A is not universal. *C. elegans* sperm is devoid of CENP-A which is provided de novo through the maternally deposited pool of CENP-A (Gassmann et al. 2012). Further, during oogenesis, pre-existing CENP-A is removed, and is de novo deposited (Monen et al. 2005).

In proliferating somatic cells, loss of CENP-A is lethal due to the severe defects in chromosome segregation in all species analyzed (Black et al. 2007a; Blower and Karpen 2001; Buchwitz et al. 1999; Fachinetti et al. 2013; Henikoff et al. 2000; Howman et al. 2000; Régnier et al. 2005; Stoler et al. 1995; Talbert et al. 2002) Additionally, CENP-A is sufficient for the recruitment of virtually all known centromere and kinetochore proteins (Barnhart et al. 2011; Carroll et al. 2009; Foltz et al. 2006; Guse et al. 2011; Heun et al. 2006; Liu et al. 2006; Mendiburo et al. 2011; Okada et al. 2006), with the exception of the sequence specific DNA-binding protein CENP-B (Pluta et al. 1992; Voullaire et al. 1993). In a groundbreaking study, (Mendiburo et al. 2011) used *Drosophila* S2 cells to tether CENP-A to a naïve chromatin domain containing Lac operator sequences (using a LacI DNA binding domain), not previously associated with centromere function. Once tethered, CENP-A^{CID}-LacI creates a local nucleosomes pool that is able to recruit virtually all known downstream centromere and kinetochore proteins allowing stable binding of microtubules. Importantly, once formed, this nascent centromere recruited naïve CENP-A^{CID}, not previously associated with this region, even after the initial tether has been lost, indicative of self-propagation of CENP-A^{CID}. Analogous experiments were performed with the CENP-A loading factor HJURP. In this case not only neocentromere formation was observed (Barnhart et al. 2011; Hori et al. 2013) but this centromere was shown to rescue chromosome stability and cell viability after deletion of the endogenous centromere in chicken DT40 cells (Hori et al. 2013). A large network of proteins, termed the constitutive centromere associated network (CCAN), is assembled on the centromere throughout the cell cycle (Cheeseman and Desai 2008; Foltz et al. 2006; Izuta et al. 2006; Okada et al. 2006). Intriguingly, (Hori et al. 2013) found that tethering of the CCAN components CENP-C or CENP-I also initiates centromere formation, indicating that the broader centromere is actively participating in maintenance of a positive epigenetic feedback loop. These experiments provide compelling evidence that CENP-A is central to a positive feedback loop which supports stable inheritance of a centromere structure. A key question that follows is, if CENP-A is the heritable mark of the centromere, how is it itself inherited? Heritable systems, whether genetic or epigenetic, adhere to some basic principles that include (1) the ability to survive through key steps of the cell cycle such as DNA replication, transcription and mitosis, (2) have the capacity to drive template-directed duplication and (3), the duplication of the mark is regulated such that each molecule gives rise to an equal number of copies in synchrony with cell division (see also Gómez-Rodríguez and Jansen 2013). In this chapter we discuss our current understanding of the heritable nature of centromeric chromatin which is the sum of its molecular stability, rates of replenishment and mechanisms that maintain these parameters in balance.

2 CENP-A Nucleosomes Are Stably Propagated at Centromeres Through Mitotic and Meiotic Divisions

Early work indicates that total cellular CENP-A protein exhibits a remarkably long half-life and lives as long as the cell itself, equating $\sim 50\%$ decrease per cell generation (Shelby et al. 1997). The apparent slow turnover required the employment of specific tools to assess protein dynamics. Fluorescence recovery after photobleaching (FRAP) which relies on local, irreversible photobleaching of a fluorophore, followed by subsequent repopulation of a bleached area with unbleached molecules provides information of the local rate of protein turnover. FRAP experiments on budding yeast kinetochores (containing a single microtubule attachment site), revealed that the yeast CENP-A homologue, Cse4 displays very low turnover rates at centromeres except during S phase where all of the preexisting Cse4 nucleosomes are exchanged (Pearson et al. 2004). Cse4 was found to be stable specifically at the centromere, whereas the non-centromeric Cse4 is degraded via ubiquitin-mediated proteolysis (Collins et al. 2004). Stable binding of Cse4 at centromeres was recently confirmed in elegant experiments using a photoconvertible Cse4-tdEos (Wisniewski et al. 2014). Eos, green in the unconverted state can be stably switched to red emission upon short wavelength excitation. Following conversion, Cse4 molecules were found to be stably associated with centromeres until their turnover during DNA replication.

Stability of the fission yeast, kinetochore-bound, CENP-A homologue was demonstrated using, once again, photobleaching of Cnp1-GFP (Coffman et al. 2011), which displayed a similar dynamics as previously described for Cse4 (Pearson et al. 2004). Interestingly, in contrast to the yeasts, holocentric *C. elegans* embryos, characterized by extremely short division times (~15 min), photobleaching of embryonic CeCENP-A-GFP in anaphase in the one-cell embryo results in the complete fluorescence recovery in the next cell division, indicative of complete loss of pre-existing CeCENP-A nucleosomes (Gassmann et al. 2012). Here, sites for CeCENP-A deposition appear to be based on other genomic features rather than pre-existing CENP-A. These regions include those with low transcriptional activity in the parental germ line (Gassmann et al. 2012) and sites of high DNA accessibility (Steiner and Henikoff 2014).

In vertebrate cells, following the initial determination of CENP-A stability with a tagged shut-off allele in human cells (Shelby et al. 1997), a shut-off in the context of a full deletion of the CENP-A gene in chicken DT40 cells (Régnier et al. 2005) revealed that the loss rate of the cellular CENP-A pool is very slow indeed, with the first mitotic defects occurring only after 7–8 cell cycles. Similar results were obtained in human cells after conditional deletion of CENP-A (Fachinetti et al. 2013). The fact that these cells can survive for extended amount of time without continuous supply of fresh CENP-A, strongly suggests that pre-existing CENP-A, once assembled into nucleosomes, remains stably bound to centromeric chromatin. While these studies determined that CENP-A turns over slowly, establishing the actual turnover rate proved difficult to determine. The FRAP methodology is suitable for determining protein dynamics at short timescales such as in organisms which have a short cell division time, but proofs limiting for dissecting protein turnover and replenishment rates at long time intervals. This limitation was surmounted by the use of a fluorescent pulse labeling strategy such as SNAP-tag technology, which allows for pulse labeling and visualization of different cohorts of the same protein within whole cell populations. SNAP is a derivative of a human DNA repair enzyme, O^6-alkylguanine-DNA alkyltransferase (AGT). The endogenous AGT enzyme recognizes O^6-alkylated guanine in DNA, and transfers the alkyl group to a reactive cysteine residue. This self-labeling capacity is exploited in a mutant version of AGT (commonly known as SNAP) which has a high affinity toward synthetically engineered small, cell permeable molecules, such as benzylguanine (BG) (Keppler et al. 2003). The enzymatic reaction between SNAP and its substrate is irreversible, highly efficient and specific. Combining serial labeling of SNAP-tagged proteins with different SNAP substrates enables visualization and fate determination of pre-existing versus newly synthesized pools of the same protein (see Bodor et al. 2012 for extended review). Following of a pulse labeled cohort of CENP-A-SNAP molecules over the course of 48−72 h, demonstrated the stable transmission of CENP-A through mitotic divisions (Bodor et al. 2013; Jansen et al. 2007). The loss rate of this pool was found to equate ~50% during each cell division, consistent with quantitative recycling of old CENP-A during S phase, with no additional turnover (Bodor et al. 2013; Dunleavy et al. 2011; Jansen et al. 2007). This high rate of retention appears to be unique to CENP-A nucleosomes. Similar pulse labeling experiments on H3.1 and H3.3 did not reveal such retention at centromeric chromatin (Bodor et al. 2013; Falk et al. 2016), indicating that the property of stable transmission is linked to CENP-A itself, not the centromeric chromatin environment as a whole. However, histone H4 shows a striking differential stability. In the genome overall its turnover rates are similar to that of H3.1, but at the centromere H4 is retained to the extent of CENP-A (Bodor et al. 2013). CENP-A directly contacts H4 in the prenucleosomal complex as well as within the nucleosome, forming a highly rigid structure (Black et al. 2004, 2007b), likely directly stabilizing H4 at the centromere. The other remaining nucleosome partners, H2A and H2B, like H3.1 and H3.3 do not display any elevated retention at the centromere (Bodor et al. 2013). Hence, CENP-A/H4 forms a stable subnucleosomal complex that represents the epigenetic core of the centromere which is quantitatively maintained throughout multiple cell divisions. The portion of CENP-A that confers its centromere targeting lies within its histone fold domain (HFD), in a subdomain termed CENP-A targeting domain (CATD), consisting of loop1 and the α2-helix (Black et al. 2004). Replacement of the equivalent domain in H3 with that of CENP-A is sufficient to target an $H3^{CATD}$ chimera to centromeres (Black et al. 2004, 2007a) and neocentromeres (Bassett et al. 2010). Importantly, the CATD confers increased conformational rigidity to (CENP-A/H4)2 tetramers as well as to CENP-A nucleosomes (Black et al. 2004, 2007b) and maintains the same loading dynamics as wild-type CENP-A (Bodor et al. 2013). Remarkably, although not all CENP-A properties are reproduced after a genetic substitution by $H3^{CATD}$ (Fachinetti et al. 2013), this chimera retains the capacity to maintain its own centromeric levels over multiple cell cycles, suggesting that the CATD is the critical subdomain responsible for longevity of the CENP-A nucleosome in vivo. Therefore, the CATD emerges as a key molecular determinant discriminating CENP-A from histone H3, and implies that the extreme stability of CENP-A nucleosomes is encoded within CENP-A molecule itself. Recent work however defined CENP-C, a member of CCAN network, as an additional extrinsic factor contributing to CENP-A stability. CENP-C binds directly to chromatin-bound CENP-A, and as a consequence, induces structural changes in conformation of CENP-A nucleosomes. This results in increased rigidity of the CENP-A nucleosome, a feature likely contributing to its stable maintenance at centromeres, since CENP-C depletion causes a rapid loss of CENP-A from the chromatin (Falk et al. 2015).

The most striking example showcasing extreme stability of CENP-A nucleosomes is recent work in female mouse meiosis (Smoak et al. 2016). Like in humans, mouse oocytes are arrested in meiotic prophase I for an extended period of time. CENP-A is readily detected in arrested mouse oocytes. However, no assembly occurs at any appreciable rate. Remarkably, deletion of the CENP-A in early oogenesis has no impact on long-term (~ 1 year) retention of centromeric CENP-A despite the lack of a nascent pool.

3 CENP-A Copy Number and the Size of Centromeric Chromatin

Due to its particularly strong epigenetic nature, centromeres represent an ideal model system for studying the basic principles of epigenetic inheritance. In the case of genetic inheritance, one DNA molecule will give a rise to two, and these will be inherited by two daughter cells. Likewise, a pre-S phase, parental centromere will give rise to two daughter centromeres, in a process that ultimately depends, not only on CENP-A but on a critical number of CENP-A molecules to maintain centromere identity.

3.1 Budding Yeast

Initial biochemical characterization of centromeric chromatin was performed on the non-repetitive point centromere of budding yeast. In contrast to higher eukaryotes, whose centromeres associate with highly repetitive long DNA regions, *S.cerevisiae* centromeres assemble on a unique ~ 125 bp DNA sequence, allowing Chromatin Immunoprecipitation (ChIP) analysis of CENP-A bound domains. This approach found the budding yeast CENP-A homolog, Cse4 to be highly enriched at a single nucleosome position and devoid from the adjacent sequences (Furuyama and Biggins 2007), strongly indicating that budding yeast centromeres harbor a single stably bound Cse4 nucleosome. Since *S. cerevisiae* contains 16 clustered centromeres, bearing two Cse4 molecules per nucleosome, yeast centromere foci have been extensively used as fluorescent standard representing 32 molecules.

Orthogonal methods to determine Cse4 copy number include fluorescence correlation spectroscopy (FCS) measurements of Cse4-EGFP (Shivaraju et al. 2012). FCS provides a measure of protein concentration in solution by determining fluctuations of fluorescence as molecules pass through a sub-femtoliter volume excited by a laser. FCS was used to calibrate cytosolic EGFP fluorescence and applied as standard to estimate the number of Cse4-EGFP molecules at the cluster of 16 centromeres. The results pointed at a single molecule of Cse4 per centromere, a surprisingly low number, which only transiently doubles in anaphase through mitotic exit. However, these changes in fluorescence could be confounded by the higher degree of centromeric chromatin compaction at this stage (Pearson et al.

2001; Wisniewski et al. 2014). Another attempt to count the absolute number of Cse4 (Aravamudhan et al. 2013) used stepwise photobleaching to find ~1.7 molecules at S. *cerevisiae* centromeres.

However, the single Cse4 nucleosome per centromere model was challenged by two contemporaneous studies that combined fluorescence measurements of Cse4-GFP in living cells with established external fluorescent standards. Using *E. coli* EGFP-MotB (~22 molecules per focus) as a fluorescent standard (Coffman et al. 2011), authors reported 8 Cse4 molecules per centromere. In a second study (Lawrimore et al. 2011) multiple fluorescent standards were employed, including single EGFP molecules, rotavirus-like particle-GFP-VLP2/6 (containing 120 EGFP molecules), a stably integrated 4-kb LacO array (containing 102 potential binding sites for LacI-GFP dimers) as well as the GFP-MotB protein from *E.coli*. By combining these standards, the authors obtained a mean number of Cse4 molecules per centromere. Further, centromere dependency on a single nucleosome is also inconsistent with the observation that the amount of Cse4 can be reduced by ~40–60%, without affecting kinetochore-microtubule attachments (Haase et al. 2013). It is possible that, in addition to a single stable positioned Cse4 nucleosome, extra copies are locally bound, e.g., in a chaperone complex near the centromere that would be captured by microscopy-based methods.

The most recent study on this theme (Wisniewski et al. 2014), casts some doubt on previous studies, reporting extracentromeric nuclear localization of Cse4 and impaired budding yeast growth when Cse4 is C-terminally GFP tagged. Normal cell growth can be obtained when Cse4 is internally tagged within its unstructured N-terminal tail. This study reported ~2 molecules of Cse4 per centromere based on ratiometric measurements against TetR-GFP bound to a tetO array. Nevertheless, it is not clear whether the tag interference affects all studies in a similar manner. Even though the precise CENP-A$^{\mathrm{Cse4}}$ remains elusive (if there is indeed a fixed number), there is general consensus that few (≤ 4) nucleosomes are present on budding yeast centromeres.

3.2 *Other Yeasts*

The uncertainties of the Cse4 copy number propagated to attempts to count CENP-A at centromeres of other organisms. Based on Cse4, numbers were determined at centromeres of two other yeast species, *C. albicans* and fission yeast, *S. pombe*, (Joglekar et al. 2008). The authors reported ~5 molecules of CENP-A$^{\mathrm{Cnp1}}$ at fission yeast centromeres and ~8 CaCse4 molecules in *C. albicans*. Taking into account the uncertainty in the budding yeast numbers, Candida features between 8 and 32 molecules of CENP-A$^{\mathrm{CaCse4}}$ per centromere. For fission yeast, the range would be 5–20 molecules per centromere. However, (Coffman et al. 2011) reported that the fission yeast strain used for these comparisons, is probably expressing a competing wildtype Cnp1 resulting in underestimation of Cnp1 numbers based on fluorescence. To readdress these confounded numbers, the authors used a clean genetic substitution of Cnp1 and the bacterial flagellar motor protein MotB, as fluorescent standard (Coffman et al. 2011; Leake et al. 2006), resulting in a much higher estimate of ~226 Cnp1 molecules per centromere. However, it is not clear how more than a hundred nucleosomes would fit a space of the 10 kb central core. Another, super-resolution-based method was used to count Cnp1 based on the photoactivatable protein, mEos2, which converts stochastically from a dark state to a fluorescent state once illuminated with low-intensity light (Lando et al. 2012). Subsequent bleaching ensures that each molecule is counted only once. Potential reactivation of fluorescence (blinking) can lead to double counting of molecules. After correction for blinking effects, ~26 molecules of Cnp1 per centromere were reported. These numbers were corroborated using ChIP coupled to high throughput sequencing (ChIP-seq), identifying ~20 distinct peaks of Cnp1 per centromere on average, placing an upper limit to the Cnp1 centromere occupancy (~20 nucleosomes per centromere). Taken together, it is clear that fission yeast centromeres are defined by a number of CENP-A nucleosomes that is an order of magnitude higher compared to budding yeast, clearly defining a regional centromere.

3.3 Metazoans

The first study carried out in metazoans aiming at establishing a centromeric CENP-A copy number used *Drosophila* imaginal disks carrying CENP-A^{CID}-EGFP as the sole source of CID and, once again, employing budding yeast Cse4-GFP as a standard for 32 fluorescent molecules (Schittenhelm et al. 2010). According to these measurements, 84–336 molecules of CENP-A^{CID} are present per centromere, depending on the budding yeast numbers. Similar studies were performed in vertebrates, in chicken DT40 cells (Johnston et al. 2010; Ribeiro et al. 2010). The Johnston et al. study reported at least 62 molecules (using Cse4 as a fluorescent standard). Ribeiro et al., relied on counting of photoblinking events of a photoconvertible Dronpa CENP-A fusion arriving at 25–40 molecules of CENP-A-Dronpa. As stated by the authors, variation in photoblinking confound the results to some extent. Importantly, both studies were performed in the presence of endogenous CENP-A pools, restricting the results to lower estimates. In human cells, using a 3D imaging strategy combined with a clean genetic replacement of endogenous CENP-A in retinal pigment epithelium (RPE) cells reported ~400 molecules per centromere (Bodor et al. 2014). Centromere-derived YFP-CENP-A signals (the only source of CENP-A in the cell) were measured and compared with total cellular levels. Remarkably, this analysis showed that while CENP-A is enriched at the centromere, on average only 0.44% of cellular CENP-A resides at each centromere. Interestingly, this ratio appeared to be fixed between RPE cell lines expressing variable levels of CENP-A, suggesting this ratio is likely preserved in unmodified, wild type RPE cells. The total cellular pool of CENP-A in wild type RPEs was found to be ~91,000 molecules (as determined by quantitative Western blotting using highly purified CENPA/H4 as a reference), which translates into ~400 molecules of CENP-A per centromere. The results were corroborated by employing the yeast the LacO/LacI-GFP standard (Lawrimore et al. 2011) as well as a statistical method based on the random segregation of CENP-A during DNA replication. Given, the predominantly octameric nature of CENP-A nucleosomes (Black and Cleveland 2011; Hasson et al. 2013), this number converts into ~200 CENP-A nucleosomes in interphase, which are split into ~100 nucleosomes on mitotic centromeres (Fig. 1a). Surprisingly, this number is not uniform across different cell types which can be as low as 50 nucleosomes, still retaining the capacity to form a functional and heritable centromere (Bodor et al. 2014).

The scarcity of CENP-A nucleosomes at the centromere [1 in 25 compared to H3 on average (Bodor et al. 2014) appears to be inconsistent with the stable maintenance of a self-templating positive feedback loop, which typically relies on local cooperativity (Dodd et al. 2007). However, analysis of nucleosome distribution at neocentromeres, where such analysis is possible, shows that CENP-A nucleosomes tend to be organized in clusters, as also found by chromatin fiber analysis (Blower et al. 2002). Within these clusters, individual positions harbor CENP-A with a remarkably high occupancy [up to 80% of total cells (Bodor et al. 2014)], indicative of a strong nucleosome positioning favoring CENP-A. Therefore, strong enrichment of CENP-A nucleosomes coupled with their possible clustering at the centromere likely provides an ample amount of CENP-A nucleosomes sufficient to maintain a positive epigenetic feedback loop (Fig. 1b).

4 The Modularity of CENP-A Dependent Kinetochore Assembly

CENP-A acts as the most upstream component in kinetochore assembly by specifying the point of contact between the DNA and mitotic spindle. CENP-A directs the formation of the constitutive centromere associated network (CCAN) which in turn, during mitosis, recruits a secondary protein complex known as the kinetochore. The kinetochore includes the conserved microtubule-binding KMN network, consisting of the protein KNL1, the Mis12 and Ndc80 complexes (Cheeseman et al. 2004, 2006; DeLuca et al. 2006). Kinetochores serve as a platform for binding of dynamic spindle microtubules which exert poleward pulling forces onto centromeres and separate sister chromatids in opposite direction during anaphase.

Current models for centromere and kinetochore architecture are based on repeated individual subunits, in which the amount of centromere components directly dictates the number of downstream kinetochore proteins, and ultimately the number of microtubule attachment sites. This form of organization was initially proposed in 1991, when islets of proteins recognized via CREST antibodies were identified in a stretched centromeric DNA fiber (Zinkowski et al. 1991). Evidence for such a modular organization is found at the *S. cerevisiae* point centromere.

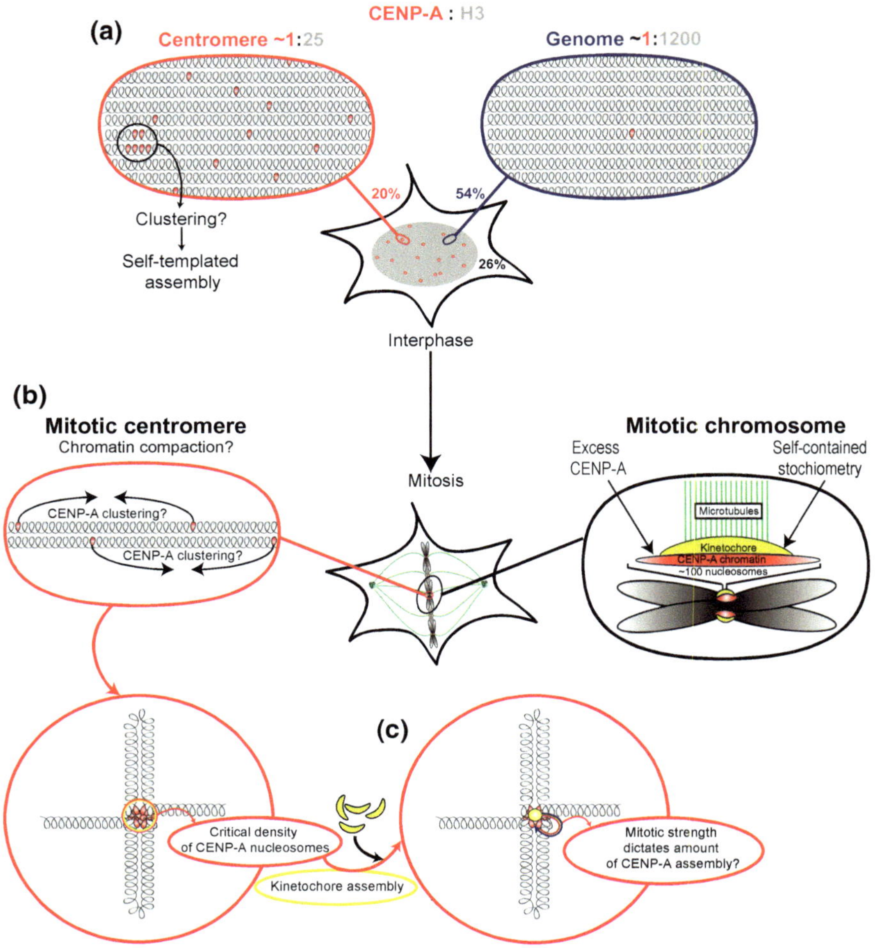

Fig. 1 An integrated view of human centromere architecture. **a** Interphase distribution of CENP-A relative to histone H3 at an average human centromere (*left*) and whole genome level (*right*) adapted from Bodor et al. (2014). **b** Organization of mitotic chromosome in which individual centromeres contain ~100 CENP-A nucleosomes, which is in excess of what is required to nucleate the kinetochore of a fixed size (*right*). Compaction of centromeric chromatin during mitosis possibly leads to clustering of CENP-A nucleosomes, which may reach a critical density of CENP-A nucleosomes for efficient kinetochore assembly (*left* and *bottom*). **c** Normalization of CENP-A levels could be initiated during mitosis through signals dictated by microtubule pulling forces

Joglekar et al. (2006) used endogenous GFP-tagging of the C-termini of kinetochore proteins and compared copy numbers to centromeric Cse4-GFP as a fluorescent standard assuming a single Cse4 nucleosome. They found the proteins forming the interface between centromeric chromatin and the microtubule plus end to be present in specific stoichiometries. For example, 1–2 copies of Mif2p, the yeast CENP-C homolog, 2–3 copies of the COMA complex (containing several CCAN members), 6–7 copies of Mtw1p, the Mis12 homolog and 8 copies of the Ndc80 complex. However, it should be noted that the precise number of kinetochore units could be potentially higher, depending on the actual number of Cse4 molecules at budding yeast centromeres.

Regional centromeres tend to assemble on large stretches of centromeric DNA compared to the budding yeast point centromeres and they are bound by multiple spindle microtubules [ranging from 2 to 4 in fission yeast to ~17 in the case of humans (McEwen et al. 2001; Sagolla et al. 2003)]. Initial studies, focused on the centromeres of fission yeast and *C. albicans* (Joglekar et al. 2008), found a strikingly constant ratio between the amount of centromeric CENP-A nucleosomes, structural components of kinetochore and number of microtubules attached during mitosis. Based on mitotic fluorescent intensities of a multitude of kinetochore components [in a manner analogous to Joglekar et al. (2006)], the authors revealed that, while absolute numbers differ, the number of kinetochore proteins per microtubule attachment are very similar between budding and fission yeast. For both yeasts there are 6–8 molecules of KMN network per kinetochore-microtubule attachment. These findings strongly argue that the regional centromeres of fission yeast are composed of repeated structures reminiscent of the ones existing in budding yeast. This apparent kinetochore architecture extends to certain metazoan species, such are chicken DT40 cells, in which the copy number of CCAN network members (namely CENP-C, CENP-H, CENP-I and CENP-T) is in nearly stoichiometric relation to KMN network members (Mis12, Knl1 and Ndc80), which, once again assemble at ~8 molecules per microtubule (Johnston et al. 2010). However, a direct relationship between the number of centromeric CENP-A nucleosomes and amount of downstream kinetochore components is incompatible with the fact that constitutive overexpression of Cnp1 does not lead to significant changes in the copy number of kinetochore protein (Joglekar et al. 2008). Consistently, in *C. albicans*, the number of CaCse4 nucleosomes is larger than the number of microtubule attachment sites (Joglekar et al. 2008), indicating that the relationship between centromeric chromatin and microtubule attachment sites is less defined. This notion is further supported by the fact that CENP-A depletion in human cells resulting in ~7% of total centromeric (Fachinetti et al. 2013) or ~10% of cellular pool (Liu et al. 2006) had no effect on centromere integrity at least in the short term. Upon partial loss of CENP-A, proteins such as CENP-C and CENP-T remain largely unaffected (Fachinetti et al. 2013). In an extreme case, upon complete acute loss of CENP-A, the centromere remains mitotically functional at least initially, after which failure to propagate the centromere in the next division results in gradual loss of centromere components (Hoffmann et al. 2016). In agreement with the stoichiometric disconnect between centromeric chromatin and the rest of the centromere, altering CENP-A levels in human RPE cell line between 40 and 240% relative to wild type, showed no significant effect on the amount of critical kinetochore proteins (Bodor et al. 2014). These included CENP-C and CENP-T, which are responsible for mitotic recruitment of the KMN network (Gascoigne et al. 2011), as well as the key microtubule binding protein Hec1/NDC80 (Cheeseman et al. 2006; DeLuca et al. 2006). Taken together, these results argue that on a typical human centromere the amount of CENP-A nucleosomes is in excess compared to the critical number necessary to maintain the centromere, which could in part be facilitated through semi-stable self-regulated recruitment of downstream CCAN proteins.

Another insight into the relationship between CENP-A chromatin and the kinetochore comes from overexpression studies. Excess CENP-A results in its mislocalization to non-centromeric sites (Athwal et al. 2015; Heun et al. 2006; Lacoste et al. 2014). Mistargeted CENP-A is not randomly distributed, rather it is enriched at sites of high histone turnover (Athwal et al. 2015; Lacoste et al. 2014). Even at physiological expression levels, CENP-A is present outside the centromere in a surprisingly high amounts. Quantitative fluorescence microscopy methods have estimated that only ~20% of CENP-A is centromeric and about half of all CENP-A is chromatin-bound elsewhere. However, due to the large genome size these CENP-A nucleosomes represent less than one in a thousand nucleosomes, compared to ~50 fold higher enrichment at centromeres (Bodor et al. 2014). Nonetheless, despite their presence in non-centromeric genomic locations, these CENP-A containing nucleosomes do not instigate the formation of the functional centromere (Bodor et al. 2014; Lacoste et al. 2014). It is tempting to speculate that whereas these sporadic genomic CENP-A nucleosomes might have limited capacity to attract some centromeric components, particularly those that directly interact with CENP-A (Gascoigne et al. 2011), the local pool of CENP-A does not reach a critical threshold sufficient to initiate the formation of a functional centromere. Therefore, rather than maintaining a linear relationship between CENP-A nucleosomes and downstream components, the CCAN and the kinetochore, once formed, maintain an internal stoichiometry and become to some extend independent of fluctuation in the centromeric CENP-A pool size.

One curious case in which the levels of centromeric CENP-A appear to dictate the amount of downstream kinetochore proteins has been reported to occur during meiosis in mice (Chmátal et al. 2014). In mammals, during female oogenesis only one out of four meiotic product will give rise to the future gamete. The probability for any allele to be transmitted should, in principle, follow Mendelian rules of inheritance. However, certain "selfish" genomic elements can skew this ratio and are preferentially retained in the mature egg, a process known as meiotic drive. The Chmátal et al. study showed that the amount of kinetochore proteins assembled at the meiotic centromere correlates with the amount of CENP-A nucleosomes. Chromosomes having fewer CENP-A nucleosomes at the centromere relative to the other ones, assembled a lower amount of Hec1/NDC80, which results in its positioning near the cell cortex due to asymmetric microtubule forces within the meiotic spindle resulting in its preferential exclusion to the polar body. The inverse was found for chromosomes with a higher amount of centromeric CENP-A nucleosomes, which were preferentially retained in the mature egg. While the resulting drive is not large, only by 10% from random (Chmátal et al. 2014), at evolutionary timescales, this would have a profound effect on the frequency of a specific chromosome within a population. While in mitosis such inequalities may be equalized by the mitotic checkpoint, this is much weaker during meiosis allowing for centromere discrepancies to evolve.

5 Propagation of Centromeric Chromatin Across Cell Divisions

As outlined above, CENP-A nucleosomes are stably maintained and propagated at mitotic and meiotic centromeres (Bodor et al. 2013; Jansen et al. 2007; Smoak et al. 2016). This unusually slow turnover of CENP-A at each centromere (Falk et al. 2015) has consequences for how the correct levels are maintained across subsequent cell division cycles. New CENP-A histones can either be incorporated at a continuous slow rate to compensate for the twofold reduction during S phase, or alternatively, assembly is restricted to a discrete cell cycle window to control the rate and quantity of assembly. It turns out that, in all species examined thus far, control of CENP-A assembly is maintained by rendering it tightly cell cycle restricted rather than allowing continuous slow assembly. Given the key role of centromeres in mitosis and the fact that CENP-A is lost by twofold during the preceding S phase, it was initially expected that the replenishment of the S-phase diluted pool of CENP-A would occur prior to mitosis (Csink and Henikoff 1998; Shelby et al. 2000). In budding yeast, as outlined above, CENP-A turns over during S-phase (Pearson et al. 2004; Wisniewski et al. 2014). Such turnover appears to be a common feature among unicellular eukaryotes. In an interesting case of the unicellular red algae *Cyanidioschyzon merolae*, CENP-A^{CENH3} is detected at the centromeres only between S-phase and mitosis, and remains undetectable in G1 phase, indicating eviction of CENP-A^{CENH3} (Kanesaki et al. 2015; Maruyama et al. 2007). Upon re-entry into subsequent S-phase, CENP-A^{CENH3} is de novo deposited at regional centromeres of *C. merolae* (Kanesaki et al. 2015). With the exception of these single-celled organisms, CENP-A assembly appears to be uncoupled from DNA replication in metazoans and plants.

In most animal systems examined, a unique pattern of cell cycle-coupled CENP-A replenishment was uncovered where assembly of newly synthesized CENP-A is delayed until mitotic exit, in G1 phase of the next cell cycle, after the primary function of the centromere has been fulfilled. This paradoxical timing of centromeric chromatin assembly was initially discovered in Drosophila and human cells based on steady state fluorescence, FRAP experiments and SNAP-based pulse labeling, respectively (Jansen et al. 2007; Schuh et al. 2007). The SNAP technology has proven extremely useful in dissecting chromatin dynamics (Bergmann et al. 2011; Bodor et al. 2013; Deaton et al. 2016; Dunleavy et al. 2011; Jansen et al. 2007; Prendergast et al. 2011; Ray-Gallet et al. 2011). To assay for the assembly of nascent CENP-A-SNAP specifically, the pre-existing (chromatin bound) pool of CENP-A-SNAP is labeled with a nonfluorescent SNAP substrate (quench). During the ensuing chase period new, unlabeled CENP-A is synthesized which can be fluorescently labeled at a later time point (Bodor et al. 2012). This methodology allows for the visualization of centromeres decorated with nascent CENP-A. G1-restricted assembly of CENP-A in human cells was confirmed by photobleaching experiments of CENP-A-GFP (Hemmerich et al. 2008), and later also found be the conserved in chicken DT40 cells (Silva et al. 2012), and Xenopus (Bernad et al. 2011; Westhorpe et al. 2015). A key question that follows is how CENP-A assembly is coupled to the cell cycle to maintain correct centromere levels. Early work showed that microtubule attachment and checkpoint signaling, two key aspects of mitosis, are not required for subsequent assembly (Jansen et al. 2007; Schuh et al. 2007). Instead, mitotic passage is primarily needed to result in APC-mediated cyclin destruction and concomitant loss of Cdk activity. This notion resulted from experiments demonstrating that selective inhibition of both Cdk1 and Cdk2 (Cdk1/2) in S or G2 phase is sufficient to induce premature, premitotic CENP-A assembly (Silva et al. 2012). CENP-A assembly commences rapidly upon Cdk inactivation, either naturally or artificially. This has led to a model in which all factors necessary for CENP-A loading are present and poised for activity prior to mitotic exit, but are held inactive due to the Cdk1/2 activities in S, G2 and mitosis, when these kinases are active. While CENP-A is the prime candidate regulating propagation of centromeric chromatin, the fact that $H3^{CATD}$ chimera still retained G1-restricted timing of loading to the centromeres argues that external binding factors are likely contributors to cell cycle dependent CENP-A assembly, compared to CENP-A itself (Bodor et al. 2013). Indeed, the CENP-A specific chaperone HJURP is exclusively targeted to G1 centromeres (Dunleavy et al. 2009; Foltz et al. 2009), concurrent with its dephosphorylation on Cdk consensus residues (Müller et al. 2014; Stankovic et al. 2017). Mutation of Cdk responsive residues within HJURP prior to mitotic exit is sufficient to induce limited precocious loading of CENP-A at S and G2 centromeres (Müller et al. 2014; Stankovic et al. 2017). In addition, ectopic targeting of HJURP to centromeres prior to mitotic exit also leads to premature incorporation of CENP-A molecules, suggesting that rather than controlling the interaction interface between CENP-A and HJURP, the negative regulation occurs primarily at the level of localization of the assembly factor (Stankovic et al. 2017). Similarly, Cdk1/2 activities also negatively regulate centromeric localization of another CENP-A assembly factor, the M18 complex. This complex is targeted to centromeres in anaphase of mitosis, prior to the onset of CENP-A deposition, and its activity is necessary for subsequent steps in CENP-A deposition which involves the targeting of HJURP to the centromeres (Barnhart et al. 2011; Fujita et al. 2007). The largest member of the M18 complex, M18BP1 is under Cdk1/2 control, which limits its centromeric recruitment until loss of Cdk1 activity in anaphase (McKinley and Cheeseman 2014; Silva et al. 2012; Stankovic et al. 2017). Interestingly, like HJURP, forced premature recruitment of M18BP1 to the centromeres can overcome negative cell cycle regulation to some extent (McKinley and Cheeseman 2014; Stankovic et al. 2017). A single phosphorylation site at Threonine 653 is key to this control (Stankovic et al. 2017). This latter study showed that simultaneous expression of unphosphorylatable mutant forms of

M18BP1 and HJURP, leads to their premature centromere targeting, resulting in essentially complete reconstitution of CENP-A assembly.

The requirement and sufficiency of these two targets defines a two-step inhibitory mechanism in which Cdk1/2 are directly targeting both assembly factors. This dual level control ultimately allows for a strict cell cycle coupled timing of CENP-A assembly (Fig. 2). Recently, another kinase, Plk1, was shown to act as a positive regulator of CENP-A deposition. Its localization to G1 centromeres and contemporaneous phosphorylation of M18BP1 proved to be important for robust recruitment of the M18 complex to G1 centromeres. Interestingly, Plk1 activity is necessary for both canonical and premature (G2 phase) deposition of CENP-A, indicating the requirement of positive phosphosignaling at all cell cycle stages. Therefore, the strict cell cycle coupling of CENP-A loading is achieved through negative Cdk1- and 2-dependant signals, restricting assembly to G1 while positive signals, such as Plk1 are needed to stimulate assembly (Fig. 2).

While CENP-A assembly is uncoupled from DNA replication in most eukaryotes, in fission yeast and plants, CENP-A assembly occurs in premitotic G2 phase

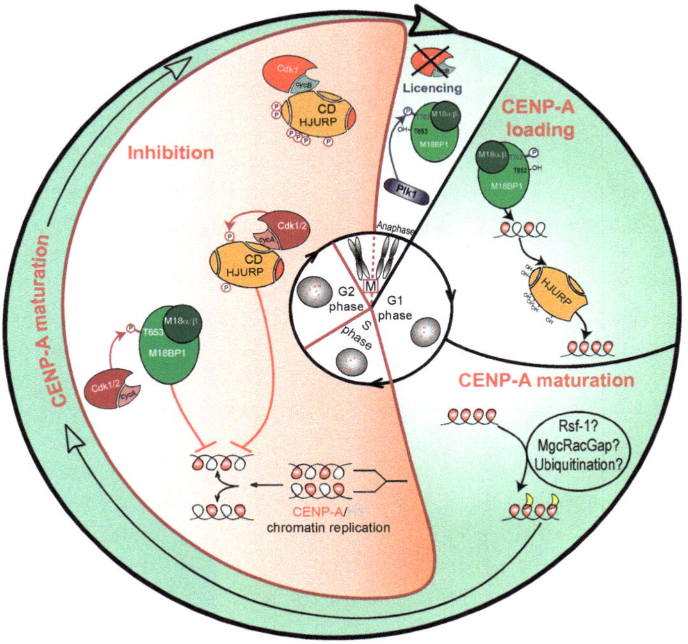

Fig. 2 Overview of mechanisms ensuring cell cycle-coupled CENP-A assembly. CENP-A deposition is restricted to early G1 phase by the Cdk1/2 based phosphorylation of two key loading factors, M18BP1 and HJURP. During mitosis, positive regulation takes place in a form of licensing phosphorylation of M18BP1 by Plk1. Upon mitotic exit, negative regulation is alleviated and CENP-A assembly initiates. An additional step of "maturation" may be necessary in order to stabilize newly loaded pool of CENP-A and to normalize CENP-A levels

(Lando et al. 2012; Lermontova et al. 2006), although the molecular details remain elusive. Another outstanding question is assembly control in Drosophila. While G1 phase is the major cell cycle window where CENP-A assembly occurs (Lidsky et al. 2013; Schuh et al. 2007), in Drosophila somatic cell lines, some degree of assembly also takes place in other phases, notably in mitosis (Lidsky et al. 2013; Mellone et al. 2011). However, in neuroblasts, within the in vivo context of the organism, CENP-A assembly remains G1-restricted (Dunleavy et al. 2012). Rather than indicting a fundamentally different logic of control, these differences likely reflect physiological differences in the efficiency of inhibition by the cell cycle machinery, as artificially achieved in human cells.

In sum, a picture emerges where different mechanisms have evolved all of which tie the CENP-A assembly machinery to the cell cycle. However, the importance of this for the maintenance of centromere structure and function remains largely undefined.

6 Possible Mechanisms to Maintain Homeostasis of CENP-A Levels Across Cell Divisions

The presence of pre-existing, chromatin bound CENP-A nucleosomes is a prerequisite for the stable propagation of centromeric domain. Parental CENP-A nucleosomes direct the incorporation of a nascent CENP-A molecules, which are placed adjacent to the pre-existing ones (Ross et al. 2016). This precise positioning of CENP-A molecules is likely facilitated through interaction between the constitutive centromeric protein CENP-C, which on one hand recognizes chromatin bound CENP-A (Carroll et al. 2010; Kato et al. 2013), and on the other, forms an interaction platform between the M18 licensing complex and centromeric chromatin (Dambacher et al. 2012; Moree et al. 2011; Shono et al. 2015; Westhorpe et al. 2015). This complex in turn recruits the CENP-A specific chaperone HJURP (Nardi et al. 2016; Stellfox et al. 2016; Wang et al. 2014) which deposits newly synthetized CENP-A (Barnhart et al. 2011; Dunleavy et al. 2009; Foltz et al. 2009). These molecular connections likely contribute to a closed positive epigenetic feedback loop where deposition of new CENP-A is ultimately dependent on the previously incorporated pool. However, how the correct CENP-A levels are maintained remains an open question. Too little would render centromeres dysfunctional [e.g. reducing CENP-A levels to 10% is ultimately incompatible with viability of cells (Black et al. 2007a)], while too much CENP-A can potentially lead to neocentromere formation as is the case in Drosophila (Heun et al. 2006; Olszak et al. 2011).

The amount of CENP-A present at the centromeres is in a direct proportion to varying total cellular levels (Bodor et al. 2014) suggesting that the CENP-A loading machinery is not a rate-limiting factor controlling the size of centromeric domain, rather, it is CENP-A itself. The challenge to our understanding of how CENP-A levels are maintained is the fact that the chromatin bound pool does not exchange, rendering it invisible to a classic equilibrium. There is no apparent communication between soluble and centromeric CENP-A. This indicates that cells need some other measure of how much CENP-A is in chromatin and to adjust the assembly accordingly. Given the nature of a positive feedback loop, in the absence of a dynamic equilibrium, individual centromeres would have the potential of reaching extreme values, spinning out of control unless there is a mechanisms to curb the assembly of new CENP-A. In addition, due to the nature of chromatin recycling during DNA replication, CENP-A levels would be increasingly variable. Current evidence indicates that existing centromeric CENP-A is redistributed stochastically during DNA replication. The ratio in pool size between two sister centromeres follows a normal distribution averaging at 50/50 with a certain probability that one daughter centromere inherits a disproportionally larger (or smaller) number of parental CENP-A nucleosomes (Bodor et al. 2014). It is conceivable that there are surveillance mechanisms which would monitor and sense imbalanced number of CENP-A nucleosomes at each centromere. One possibility is that the CENP-A assembly machinery would incorporate a pool of molecules not in a direct relation to the number present in chromatin but load in excess, which has been observed (Jansen et al. 2007; Lagana et al. 2010). In this scenario, the correct amount would be determined in a later "maturation" step, in which the overloaded pool of new CENP-A would be removed from the centromere having an excess of parental CENP-A, whereas those with reduced levels would be stripped to a lesser extent (Fig. 3). Should there be such an eviction mechanism, it would have to allow discrimination between CENP-A marked for instability versus the one which is destined to be stably inherited over cell cycle. Whereas molecular steps allowing eviction of overloaded pool of CENP-A are largely unknown, there are reports of stabilization of nascent CENP-A occurring in G1 (Lagana et al. 2010; Liu and Mao 2016; Perpelescu et al. 2009), suggesting that addition of CENP-A "stabilization" mark would happen prior to DNA synthesis. A recent addition to this theme is the report of ubiquitylation of parental CENP-A as a requirement to recruit nascent CENP-A (Niikura et al. 2016). Centromeric CENP-A levels could also be normalized during S phase passage, in which the mix of parental and G1-loaded pools of CENP-A would be coordinately and preferentially segregated to the grand-daughter centromere which inherited a decreased number of CENP-A molecules from the previous generation. An elegant model has been proposed linking the amount of CENP-A assembly in G1 phase directly to the strength of the centromere in mitosis (Brown and Xu 2009). In this model, weaker centromeres would bind a smaller number of microtubules that would in turn generate a signal driving the assembly of a compensatory number of CENP-A molecules in the subsequent G1 phase (Fig. 1c). One drawback of this model is that it assumes a proportional nature of kinetochore assembly in relation to the number of CENP-A molecules. However, variations of this model could be extended to modular kinetochores (assembled in a fixed rate independently of the number of CENP-A nucleosomes). Assuming nearly equal numbers of microtubules attached to each daughter centromere (due to checkpoint signaling), the signal required to stabilize the amount of CENP-A molecules would come from the tension generated within centromeric chromatin. A speculative idea is that only those CENP-A molecules that are under tension are marked for stability whereas superfluous ones are marked for removal. In this way, over multiple mitotic divisions the number of CENP-A molecules would equalize. Individually or in combination, these mechanisms would have to rely on the presence of a yet to be identified rate limiting factors or a combination of factors that constitute a more stable measure of centromere size. These would need to have a capacity to recognize chromatin-bound pool of CENP-A and contain "counting" properties allowing sensing of the size of CENP-A populated domain. CENP-C, a factor stabilizing CENP-A (Falk et al. 2015) could be one of such factors, limiting CENP-A domain size.

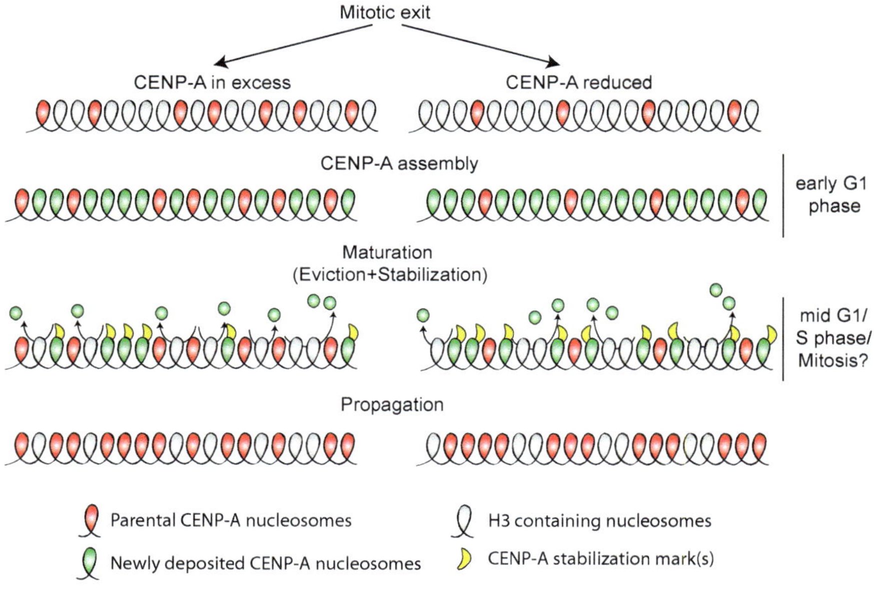

Fig. 3 A model for normalization of CENP-A levels across mitotic divisions. Stochastic redistribution of CENP-A during S-phase may give rise to daughter centromeres having an unequal amount of parental nucleosomes upon mitotic exit. To accommodate for this, an excessive amount of nascent CENP-A is deposited to the centromere in early G1 phase, followed by selective stabilization of a portion of newly loaded CENP-A molecules. This would occur in an inverse proportion to the number of parental nucleosomes: the greater the number of parental nucleosomes is, the smaller the pool of new CENP-A is marked for stability, the remainder of which will be evicted. The combination of these two processes (stabilization and eviction) could encompass the previously proposed "′maturation"′ step of centromeric chromatin

References

Akiyoshi B, Gull K (2013) Evolutionary cell biology of chromosome segregation: insights from trypanosomes. Open Biol 3:130023

Amor DJ, Bentley K, Ryan J, Perry J, Wong L, Slater H, Choo KHA (2004) Human centromere repositioning "in progress". Proc Natl Acad Sci U S A 101:6542–6547

Aravamudhan P, Felzer-Kim I, Joglekar AP (2013) The budding yeast point centromere associates with two Cse4 molecules during mitosis. Curr Biol 23:770–774

Athwal RK, Walkiewicz MP, Baek S, Fu S, Bui M, Camps J, Ried T, Sung M-H, Dalal Y (2015) CENP-A nucleosomes localize to transcription factor hotspots and subtelomeric sites in human cancer cells. Epigenetics Chromatin 8:2

Barnhart MC, Kuich PHJL, Stellfox ME, Ward JA, Bassett EA, Black BE, Foltz DR (2011) HJURP is a CENP-A chromatin assembly factor sufficient to form a functional de novo kinetochore. J Cell Biol 194:229–243

Bassett EA, Wood S, Salimian KJ, Ajith S, Foltz DR, Black BE (2010) Epigenetic centromere specification directs aurora B accumulation but is insufficient to efficiently correct mitotic errors. J Cell Biol 190:177–185

Bergmann JH, Rodríguez MG, Martins NMC, Kimura H, Kelly DA, Masumoto H, Larionov V, Jansen LET, Earnshaw WC (2011) Epigenetic engineering shows H3K4me2 is required for HJURP targeting and CENP-A assembly on a synthetic human kinetochore. EMBO J 30: 328–340

Bernad R, Sánchez P, Rivera T, Rodríguez-Corsino M, Boyarchuk E, Vassias I, Ray-Gallet D, Arnaoutov A, Dasso M, Almouzni G et al (2011) Xenopus HJURP and condensin II are required for CENP-A assembly. J Cell Biol 192:569–582

Black BE, Cleveland DW (2011) Epigenetic centromere propagation and the nature of CENP-A nucleosomes. Cell 144:471–479

Black BE, Foltz DR, Chakravarthy S, Luger K, Woods VL, Cleveland DW (2004) Structural determinants for generating centromeric chromatin. Nature 430:578–582

Black BE, Jansen LET, Maddox PS, Foltz DR, Desai AB, Shah JV, Cleveland DW (2007a) Centromere identity maintained by nucleosomes assembled with histone H3 containing the CENP-A targeting domain. Mol Cell 25:309–322

Black BE, Brock MA, Bédard S, Woods VL, Cleveland DW (2007b) An epigenetic mark generated by the incorporation of CENP-A into centromeric nucleosomes. Proc Natl Acad Sci U S A 104:5008–5013

Blower MD, Karpen GH (2001) The role of Drosophila CID in kinetochore formation, cell-cycle progression and heterochromatin interactions. Nat Cell Biol 3:730–739

Blower MD, Sullivan BA, Karpen GH (2002) Conserved organization of centromeric chromatin in flies and humans. Dev Cell 2:319–330

Bodor DL, Rodríguez MG, Moreno N, Jansen LET (2012) Analysis of protein turnover by quantitative SNAP-based pulse-chase imaging. Curr Protoc Cell Biol 8 (Chapter 8, Unit 8.8)

Bodor DL, Valente LP, Mata JF, Black BE, Jansen LET (2013) Assembly in G1 phase and long-term stability are unique intrinsic features of CENP-A nucleosomes. Mol Biol Cell 24:923–932

Bodor DL, Mata JF, Sergeev M, David AF, Salimian KJ, Panchenko T, Cleveland DW, Black BE, Shah JV, Jansen LET (2014) The quantitative architecture of centromeric chromatin. Elife 3: 1–26

Brown WRA, Xu Z (2009) The "kinetochore maintenance loop"—the mark of regulation? BioEssays 31:228–236

Buchwitz BJ, Ahmad K, Moore LL, Roth MB, Henikoff S (1999) Cell division: a histone-H3-like protein in *C. elegans*. Nature 401:547–548

Carroll CW, Silva MCCC, Godek KM, Jansen LETT, Straight AF (2009) Centromere assembly requires the direct recognition of CENP-A nucleosomes by CENP-N. Nat Cell Biol 11: 896–902

Carroll CW, Milks KJ, Straight AF (2010) Dual recognition of CENP-A nucleosomes is required for centromere assembly. J Cell Biol 189:1143–1155

Cheeseman IM, Desai A (2008) Molecular architecture of the kinetochore-microtubule interface. Nat Rev Mol Cell Biol 9:33–46

Cheeseman IM, Niessen S, Anderson S, Hyndman F, Yates JR, Oegema K, Desai A (2004) A conserved protein network controls assembly of the outer kinetochore and its ability to sustain tension. Genes Dev 18:2255–2268

Cheeseman IM, Chappie JS, Wilson-Kubalek EM, Desai A (2006) The conserved KMN network constitutes the core microtubule-binding site of the kinetochore. Cell 127:983–997

Chmátal L, Gabriel SIII, Mitsainas GP, Martínez-Vargas J, Ventura J, Searle JB, Schultz RM, Lampson MA (2014) Centromere strength provides the cell biological basis for meiotic drive and karyotype evolution in mice. Curr Biol 24:2295–2300

Coffman VC, Wu P, Parthun MR, Wu J-Q (2011) CENP-A exceeds microtubule attachment sites in centromere clusters of both budding and fission yeast. J Cell Biol 195:563–572

Collins KA, Furuyama S, Biggins S (2004) Proteolysis contributes to the exclusive centromere localization of the yeast Cse4/CENP-A Histone H3 variant. *Current Biology* 14, no. 21 (2004): 1968–1972

Csink AK, Henikoff S (1998) Something from nothing: the evolution and utility of satellite repeats. Trends Genet 14:200–204

Dambacher S, Deng W, Hahn M, Sadic D, Fröhlich J, Nuber A, Hoischen C, Diekmann S, Leonhardt H, Schotta G (2012) CENP-C facilitates the recruitment of M18BP1 to centromeric chromatin. Nucleus 3:101–110

Deaton AM, Gómez-Rodríguez M, Mieczkowski J, Tolstorukov MY, Kundu S, Sadreyev RI, Jansen LE, Kingston RE (2016) Enhancer regions show high histone H3.3 turnover that changes during differentiation. Elife 5:e15316

DeLuca JG, Gall WE, Ciferri C, Cimini D, Musacchio A, Salmon ED (2006) Kinetochore microtubule dynamics and attachment stability are regulated by Hec1. Cell 127:969–982

Dodd IB, Micheelsen MA, Sneppen K, Thon G (2007) Theoretical analysis of epigenetic cell memory by nucleosome modification. Cell 129:813–822

Drinnenberg IA, DeYoung D, Henikoff S, Malik HS (2014) Recurrent loss of CenH3 is associated with independent transitions to holocentricity in insects. Elife 3:e03676

Dunleavy EM, Roche D, Tagami H, Lacoste N, Ray-Gallet D, Nakamura Y, Daigo Y, Nakatani Y, Almouzni-Pettinotti G (2009) HJURP is a cell-cycle-dependent maintenance and deposition factor of CENP-A at centromeres. Cell 137:485–497

Dunleavy EM, Almouzni G, Karpen GH (2011) H3.3 is deposited at centromeres in S phase as a placeholder for newly assembled CENP-A in G(1) phase. Nucleus 2:146–157

Dunleavy EM, Beier NL, Gorgescu W, Tang J, Costes SV, Karpen GH (2012) The cell cycle timing of centromeric chromatin assembly in drosophila meiosis is distinct from mitosis yet requires CAL1 and CENP-C. PLoS Biol 10:e1001460

Earnshaw WC, Migeon BR (1985) Three related centromere proteins are absent from the inactive centromere of a stable isodicentric chromosome. Chromosoma 92:290–296

Earnshaw WC, Rothfield N (1985) Identification of a family of human centromere proteins using autoimmune sera from patients with scleroderma. Chromosoma 91:313–321

Fachinetti D, Diego Folco H, Nechemia-Arbely Y, Valente LP, Nguyen K, Wong AJ, Zhu Q, Holland AJ, Desai A, Jansen LET et al (2013) A two-step mechanism for epigenetic specification of centromere identity and function. Nat Cell Biol 15:1056–1066

Falk SJ, Guo LY, Sekulic N, Smoak EM, Mani T, Logsdon GA, Gupta K, Jansen LET, Van Duyne GD, Vinogradov SA et al (2015) CENP-C reshapes and stabilizes CENP-A nucleosomes at the centromere. Science (80)348:699–703

Falk SJ, Lee J, Sekulic N, Sennett MA, Lee T-H, Black BE (2016) CENP-C directs a structural transition of CENP-A nucleosomes mainly through sliding of DNA gyres. Nat Struct Mol Biol 23:204–208

Foltz DR, Jansen LET, Black BE, Bailey AO, Yates JR, Cleveland DW (2006) The human CENP-A centromeric nucleosome-associated complex. Nat Cell Biol 8:458–469

Foltz DR, Jansen LET, Bailey AO, Yates JR, Bassett EA, Wood S, Cleveland DW (2009) Centromere-specific assembly of CENP-a nucleosomes is mediated by HJURP. Cell 137: 472–484
Fujita Y, Hayashi T, Kiyomitsu T, Toyoda Y, Kokubu A, Obuse C, Yanagida M (2007) Priming of centromere for CENP-A recruitment by human hMis18alpha, hMis18beta, and M18BP1. Dev Cell 12:17–30
Furuyama S, Biggins S (2007) Centromere identity is specified by a single centromeric nucleosome in budding yeast. Proc Natl Acad Sci U S A 104:14706–14711
Gascoigne KE, Takeuchi K, Suzuki A, Hori T, Fukagawa T, Cheeseman IM (2011) Induced ectopic kinetochore assembly bypasses the requirement for CENP-A nucleosomes. Cell 145:410–422
Gassmann R, Rechtsteiner A, Yuen KW, Muroyama A, Egelhofer T, Gaydos L, Barron F, Maddox P, Essex A, Monen J et al (2012) An inverse relationship to germline transcription defines centromeric chromatin in *C. elegans*. Nature 484:534–537
Gómez-Rodríguez M, Jansen LET (2013) Basic properties of epigenetic systems: lessons from the centromere. Curr Opin Genet Dev 23:219–227
Guse A, Carroll CW, Moree B, Fuller CJ, Straight AF (2011) In vitro centromere and kinetochore assembly on defined chromatin templates. Nature 477:354–358
Haase J, Mishra PK, Stephens A, Haggerty R, Quammen C, Taylor RM, Yeh E, Basrai MA, Bloom K (2013) A 3D map of the yeast kinetochore reveals the presence of core and accessory centromere-specific histone. Curr Biol 23:1939–1944
Hasson D, Panchenko T, Salimian KJ, Salman MU, Sekulic N, Alonso A, Warburton PE, Black BE (2013) The octamer is the major form of CENP-A nucleosomes at human centromeres. Nat Struct Mol Biol 6:687–695
Hemmerich P, Weidtkamp-Peters S, Hoischen C, Schmiedeberg L, Erliandri I, Diekmann S (2008) Dynamics of inner kinetochore assembly and maintenance in living cells. J Cell Biol 180:1101–1114
Henikoff S, Ahmad K, Platero JS, van Steensel B (2000) Heterochromatic deposition of centromeric histone H3-like proteins. Proc Natl Acad Sci 97:716–721
Heun P, Erhardt S, Blower MD, Weiss S, Skora AD, Karpen GH (2006) Mislocalization of the Drosophila centromere-specific histone CID promotes formation of functional ectopic kinetochores. Dev Cell 10:303–315
Hoffmann S, Dumont M, Barra V, Ly P, Nechemia-Arbely Y, McMahon MA, Hervé S, Cleveland DW, Fachinetti D, Bade D et al (2016) CENP-A is dispensable for mitotic centromere function after initial centromere/kinetochore assembly. Cell Rep 17:2394–2404
Hori T, Shang W-HH, Takeuchi K, Fukagawa T (2013) The CCAN recruits CENP-A to the centromere and forms the structural core for kinetochore assembly. J Cell Biol 200:45–60
Howman EV, Fowler KJ, Newson AJ, Redward S, MacDonald AC, Kalitsis P, Choo KHA (2000) Early disruption of centromeric chromatin organization in centromere protein A (Cenpa) null mice. Proc Natl Acad Sci 97:1148–1153
Hudson DF, Fowler KJ, Earle E, Saffery R, Kalitsis P, Trowell H, Hill J, Wreford NG, de Kretser DM, Cancilla MR et al (1998) centromere protein B null mice are mitotically and meiotically normal but have lower body and testis weights. J Cell Biol 141:309–319
Izuta H, Ikeno M, Suzuki N, Tomonaga T, Nozaki N, Obuse C, Kisu Y, Goshima N, Nomura F, Nomura N et al (2006) Comprehensive analysis of the ICEN (Interphase Centromere Complex) components enriched in the CENP-A chromatin of human cells. Genes Cells 11:673–684
Jansen LET, Black BE, Foltz DR, Cleveland DW (2007) Propagation of centromeric chromatin requires exit from mitosis. J Cell Biol 176:795–805
Joglekar AP, Bouck DC, Molk JN, Bloom KS, Salmon ED (2006) Molecular architecture of a kinetochore-microtubule attachment site. Nat Cell Biol 8:581–585
Joglekar AP, Bouck D, Finley K, Liu X, Wan Y, Berman J, He X, Salmon ED, Bloom KS (2008) Molecular architecture of the kinetochore-microtubule attachment site is conserved between point and regional centromeres. J Cell Biol 181:587–594

Johnston K, Joglekar A, Hori T, Suzuki A, Fukagawa T, Salmon EDD (2010) Vertebrate kinetochore protein architecture: protein copy number. J Cell Biol 189:937–943
Kanesaki Y, Imamura S, Matsuzaki M, Tanaka K (2015) Identification of centromere regions in chromosomes of a unicellular red alga, *Cyanidioschyzon merolae*. FEBS Lett 589:1219–1224
Kato H, Jiang J, Zhou B-R, Rozendaal M, Feng H, Ghirlando R, Xiao TS, Straight AF, Bai Y (2013) A conserved mechanism for centromeric nucleosome recognition by centromere protein CENP-C. Science (80)340:1110–1113
Keppler A, Gendreizig S, Gronemeyer T, Pick H, Vogel H, Johnsson K (2003) A general method for the covalent labeling of fusion proteins with small molecules in vivo. Nat Biotechnol 21:86–89
Lacoste N, Woolfe A, Tachiwana H, Garea AV, Barth T, Cantaloube S, Kurumizaka H, Imhof A, Almouzni G (2014) Mislocalization of the centromeric histone variant CenH3/CENP-A in human cells depends on the chaperone DAXX. Mol Cell 53:631–644
Lagana AA, Dorn JF, De Rop VV, Ladouceur A-M, Maddox AS, Maddox PS (2010) A small GTPase molecular switch regulates epigenetic centromere maintenance by stabilizing newly incorporated CENP-A. Nat Cell Biol 12:1186–1193
Lando D, Endesfelder U, Berger H, Subramanian L, Dunne PD, McColl J, Klenerman D, Carr AM, Sauer M, Allshire RC et al (2012) Quantitative single-molecule microscopy reveals that CENP-ACnp1 deposition occurs during G2 in fission yeast. Open Biol 2:120078
Lawrimore J, Bloom KS, Salmon ED (2011) Point centromeres contain more than a single centromere-specific Cse4 (CENP-A) nucleosome. J Cell Biol 195:573–582
Leake MC, Chandler JH, Wadhams GH, Bai F, Berry RM, Armitage JP (2006) Stoichiometry and turnover in single, functioning membrane protein complexes. Nature 443:355–358
Lermontova I, Schubert V, Fuchs J, Klatte S, Macas J, Schubert I (2006) Loading of Arabidopsis centromeric histone CENH3 occurs mainly during G2 and requires the presence of the histone fold domain. Plant Cell 18:2443–2451
Lidsky PV, Sprenger F, Lehner CF (2013) Distinct modes of centromere protein dynamics during cell cycle progression in Drosophila S2R+ cells. J Cell Sci 126:4782–4793
Liu C, Mao Y (2016) Diaphanous formin mDia2 regulates CENP-A levels at centromeres. J Cell Biol 213:415–424
Liu S-T, Rattner JB, Jablonski SA, Yen TJ (2006) Mapping the assembly pathways that specify formation of the trilaminar kinetochore plates in human cells. J Cell Biol 175:41–53
Malik HS, Henikoff S (2003) Phylogenomics of the nucleosome. Nat Struct Mol Biol 10:882–891
Maruyama S, Kuroiwa H, Miyagishima S, Tanaka K, Kuroiwa T (2007) Centromere dynamics in the primitive red alga *Cyanidioschyzon merolae*. Plant J 49:1122–1129
McEwen BF, Chan GKT, Zubrowski B, Savoian MS, Sauer MT, Yen TJ (2001) CENP-E is essential for reliable bioriented spindle attachment, but chromosome alignment can be achieved via redundant mechanisms in mammalian cells. Mol Biol Cell 12:2776–2789
McKinley KL, Cheeseman IM (2014) Polo-like Kinase 1 licenses CENP-A deposition at centromeres. Cell 158:397–411
Mellone BG, Grive KJ, Shteyn V, Bowers SR, Oderberg I, Karpen GH (2011) Assembly of Drosophila centromeric chromatin proteins during mitosis. PLoS Genet 7:e1002068
Mendiburo MJ, Padeken J, Fülöp S, Schepers A, Heun P (2011) Drosophila CENH3 is sufficient for centromere formation. Science (80)334:686–690
Milks KJ, Moree B, Straight AF (2009) Dissection of CENP-C-directed centromere and kinetochore assembly. Mol Biol Cell 20:4246–4255
Monen J, Maddox PS, Hyndman F, Oegema K, Desai A (2005) Differential role of CENP-A in the segregation of holocentric *C. elegans* chromosomes during meiosis and mitosis. Nat Cell Biol 7:1248–1255
Moree B, Meyer CB, Fuller CJ, Straight AF (2011) CENP-C recruits M18BP1 to centromeres to promote CENP-A chromatin assembly. J Cell Biol 194:855–871
Müller S, Montes de Oca R, Lacoste N, Dingli F, Loew D, Almouzni G, Montes de Oca R, Lacoste N, Dingli F, Loew D et al (2014) Phosphorylation and DNA binding of HJURP

determine its centromeric recruitment and function in CenH3(CENP-A) loading. Cell Report 8:190–203

Nardi IKK, Zasadzińska E, Stellfox MEE, Knippler CMM, Foltz DRR (2016) Licensing of centromeric chromatin assembly through the Mis18α-Mis18β heterotetramer. Mol Cell 61:774–787

Niikura Y, Kitagawa R, Kitagawa K (2016) CENP-A ubiquitylation is inherited through dimerization between cell divisions. Cell Report 15:61–76

Okada M, Cheeseman IM, Hori T, Okawa K, McLeod IX, Yates JR 3rd, Desai A, Fukagawa T (2006) The CENP-H-I complex is required for the efficient incorporation of newly synthesized CENP-A into centromeres. Nat Cell Biol 8:446–457

Olszak AM, van Essen D, Pereira AJ, Diehl S, Manke T, Maiato H, Saccani S, Heun P (2011) Heterochromatin boundaries are hotspots for de novo kinetochore formation. Nat Cell Biol 13:799–808

Palmer DK, O'Day K, Wener MH, Andrews BS, Margolis RL (1987) A 17-kD centromere protein (CENP-A) copurifies with nucleosome core particles and with histones. J Cell Biol 104: 805–815

Palmer DK, O'Day K, Margolis RL (1990) The centromere specific histone CENP-A is selectively retained in discrete foci in mammalian sperm nuclei. Chromosoma 100:32–36

Pearson CG, Maddox PS, Salmon ED, Bloom K (2001) Budding yeast chromosome structure and dynamics during mitosis. J Cell Biol 152:1255–1266

Pearson CG, Yeh E, Gardner M, Odde D, Salmon ED, Bloom K (2004) Stable kinetochore-microtubule attachment constrains centromere positioning in metaphase. Curr Biol 14:1962–1967

Perpelescu M, Nozaki N, Obuse C, Yang H, Yoda K (2009) Active establishment of centromeric CENP-A chromatin by RSF complex. J Cell Biol 185:397–407

Pluta AF, Saitoh N, Goldberg I, Earnshaw WC (1992) Identification of a subdomain of CENP-B that is necessary and sufficient for localization to the human centromere. J Cell Biol 116: 1081–1093

Prendergast L, van Vuuren C, Kaczmarczyk A, Doering V, Hellwig D, Quinn N, Hoischen C, Diekmann S, Sullivan KF (2011) Premitotic assembly of human CENPs-T and -W switches centromeric chromatin to a mitotic state. PLoS Biol 9:e1001082

Raychaudhuri N, Dubruille R, Orsi GA, Bagheri HC, Loppin B, Lehner CF (2012) Transgenerational propagation and quantitative maintenance of paternal centromeres depends on Cid/Cenp-A presence in drosophila sperm. PLoS Biol 10:e1001434

Ray-Gallet D, Woolfe A, Vassias I, Pellentz C, Lacoste N, Puri A, Schultz DC, Pchelintsev NA, Adams PD, Jansen LE et al (2011) Dynamics of histone H3 deposition in vivo reveal a nucleosome gap-filling mechanism for H3.3 to maintain chromatin integrity. Mol Cell 44: 928–941

Régnier V, Vagnarelli P, Fukagawa T, Zerjal T, Burns E, Trouche D, Earnshaw W, Brown W (2005) CENP-A is required for accurate chromosome segregation and sustained kinetochore association of BubR1. Mol Cell Biol 25:3967–3981

Ribeiro SA, Vagnarelli P, Dong Y, Hori T, McEwen BF, Fukagawa T, Flors C, Earnshaw WC (2010) A super-resolution map of the vertebrate kinetochore. J Cell Sci USA 107:10484–10489

Ross JE, Woodlief KS, Sullivan BA (2016) Inheritance of the CENP-A chromatin domain is spatially and temporally constrained at human centromeres. Epigenetics Chromatin 9:20

Sagolla MJ, Uzawa S, Cande WZ (2003) Individual microtubule dynamics contribute to the function of mitotic and cytoplasmic arrays in fission yeast. J Cell Sci 116

Schittenhelm RB, Althoff F, Heidmann S, Lehner CF (2010) Detrimental incorporation of excess Cenp-A/Cid and Cenp-C into Drosophila centromeres is prevented by limiting amounts of the bridging factor Cal1. J Cell Sci 123:3768–3779

Schuh M, Lehner CF, Heidmann S (2007) Incorporation of drosophila CID/CENP-A and CENP-C into centromeres during early embryonic anaphase. Curr Biol 17:237–243

Shelby RD, Vafa O, Sullivan KF (1997) Assembly of CENP-A into centromeric chromatin requires a cooperative array of nucleosomal DNA contact sites. J Cell Biol 136:501–513

Shelby RD, Monier K, Sullivan KF (2000) Chromatin assembly at kinetochores is uncoupled from dna replication. J Cell Biol 151:1113–1118

Shivaraju M, Unruh JRR, Slaughter BDD, Mattingly M, Berman J, Gerton JL, Alber F, Dokudovskaya S, Veenhoff LM, Zhang W et al (2012) Cell-cycle-coupled structural oscillation of centromeric nucleosomes in yeast. Cell 150:304–316

Shono N, Ohzeki J-I, Otake K, Martins NMC, Nagase T, Kimura H, Larionov V, Earnshaw WC, Masumoto H (2015) CENP-C and CENP-I are key connecting factors for kinetochore and CENP-A assembly. J Cell Sci jcs-180786

Silva MCC, Bodor DL, Stellfox ME, Martins NMC, Hochegger H, Foltz DR, Jansen LET (2012) Cdk activity couples epigenetic centromere inheritance to cell cycle progression. Dev Cell 22:52–63

Smoak EM, Stein P, Schultz RM, Lampson MA, Black BE (2016) Long-term retention of Cenp-A nucleosomes in mammalian oocytes underpins transgenerational inheritance of centromere identity. Curr Biol 26:1110–1116

Stankovic A, Guo LY, Mata JF, Bodor DL, Cao X-Y, Bailey AO, Shabanowitz Jeffrey, Hunt DF, Garcia BA, Black BE, Jansen LET (2017) A dual inhibitory mechanism sufficient to maintain cell cycle restricted CENP-A assembly. Mol Cell 65:231–246

Steiner FA, Henikoff S (2014) Holocentromeres are dispersed point centromeres localized at transcription factor hotspots. Elife 3:e02025

Stellfox ME, Nardi IK, Knippler CM, Foltz DR (2016) Differential binding partners of the Mis18α/β YIPPEE domains regulate Mis18 complex recruitment to centromeres. Cell Rep 1–9

Stoler S, Keith KC, Curnick KE, Fitzgerald-Hayes M (1995) A mutation in CSE4, an essential gene encoding a novel chromatin-associated protein in yeast, causes chromosome nondisjunction and cell cycle arrest at mitosis. Genes Dev 9:573–586

Sullivan KF, Hechenberger M, Masri K (1994) Human CENP-A contains a histone H3 related histone fold domain that is required for targeting to the centromere. J Cell Biol 127:581–592

Talbert PB, Henikoff S (2013) Phylogeny as the basis for naming histones. Trends Genet 29: 499–500

Talbert PB, Masuelli R, Tyagi AP, Comai L, Henikoff S (2002) Centromeric localization and adaptive evolution of an arabidopsis histone H3 variant. Plant Cell Online 14:1053–1066

Talbert PB, Ahmad K, Almouzni G, Ausió J, Berger F, Bhalla PL, Bonner WM, Cande WZ, Chadwick BP, Chan SWL et al (2012) A unified phylogeny-based nomenclature for histone variants. Epigenetics Chromatin 5:7

Tyler-Smith C, Gimelli G, Giglio S, Floridia G, Pandya A, Terzoli G, Warburton PE, Earnshaw WC, Zuffardi O (1999) Transmission of a fully functional human neocentromere through three generations. Am J Hum Genet 64:1440–1444

Voullaire LE, Slater HR, Petrovic V, Choo KH (1993) A functional marker centromere with no detectable alpha-satellite, satellite III, or CENP-B protein: activation of a latent centromere? Am J Hum Genet 52:1153–1163

Wang J, Liu X, Dou Z, Chen L, Jiang H, Fu C, Fu G, Liu D, Zhang J, Zhu T (2014) Mitotic regulator mis18 interacts with and specifies the centromeric assembly of molecular chaperone holliday junction recognition protein (HJURP). J Biol Chem 289:8326–8336

Westhorpe FG, Fuller CJ, Straight AF (2015) A cell-free CENP-A assembly system defines the chromatin requirements for centromere maintenance. J Cell Biol jcb-201503132

Wisniewski J, Hajj B, Chen J, Mizuguchi G, Xiao H, Wei D, Dahan M, Wu C (2014) Imaging the fate of histone Cse4 reveals de novo replacement in S phase and subsequent stable residence at centromeres. Elife 3:e02203

Zinkowski RP, Meyne J, Brinkley BR (1991) The centromere-kinetochore complex: a repeat subunit model. J Cell Biol 113:1091–1110

Part II
Epigenetic Contributions to Centromere Identity and Function

Orchestrating the Specific Assembly of Centromeric Nucleosomes

Ewelina Zasadzińska and Daniel R. Foltz

Abstract Centromeres are chromosomal loci that are defined epigenetically in most eukaryotes by incorporation of a centromere-specific nucleosome in which the canonical histone H3 variant is replaced by Centromere Protein A (CENP-A). Therefore, the assembly and propagation of centromeric nucleosomes are critical for maintaining centromere identify and ensuring genomic stability. Centromeres direct chromosome segregation (during mitosis and meiosis) by recruiting the constitutive centromere-associated network of proteins throughout the cell cycle that in turn recruits the kinetochore during mitosis. Assembly of centromere-specific nucleosomes in humans requires the dedicated CENP-A chaperone HJURP, and the Mis18 complex to couple the deposition of new CENP-A to the site of the pre-existing centromere, which is essential for maintaining centromere identity. Human CENP-A deposition occurs specifically in early G1, into pre-existing chromatin, and several additional chromatin-associated complexes regulate CENP-A nucleosome deposition and stability. Here we review the current knowledge on how new CENP-A nucleosomes are assembled selectively at the existing centromere in different species and how this process is controlled to ensure stable epigenetic inheritance of the centromere.

In all eukaryotes centromeres serve as a site of kinetochore formation that facilitates faithful chromosome segregation during cell division. Centromeres in most species are characterized by the presence of unique nucleosomes containing the histone H3

E. Zasadzińska · D.R. Foltz (✉)
Department of Biochemistry and Molecular Genetics, University of Virginia, Charlottesville, VA 22908, USA
e-mail: dfoltz@northwestern.edu

D.R. Foltz
Department of Biochemistry and Molecular Genetics, Northwestern University Feinberg School of Medicine, Chicago, IL 60611, USA

D.R. Foltz
Robert H. Lurie Comprehensive Cancer Center, Northwestern University Feinberg School of Medicine, Chicago, IL 60611, USA

B.E. Black (ed.), *Centromeres and Kinetochores*, Progress in Molecular and Subcellular Biology 56, DOI 10.1007/978-3-319-58592-5_7

variant Centromere Protein A (CENP-A). However, different organisms employ distinct strategies to specify centromere location. Budding yeast contain point centromeres, the location of which is determined by the presence of approximately 150 bp domain with three distinct DNA sequences: CDEI, CDEII, and CDEIII (Clarke and Carbon 1980; Fitzgerald-Hayes et al. 1982). In budding yeast, these sequences are sufficient for the establishment of a functional centromere. However, the wide variation of centromere DNA repeat sequences across species, and indeed the lack of DNA repetitive elements in several species suggest that DNA sequence elements may not be critical for centromere function in higher eukaryotes (Allshire and Karpen 2008; Shang et al. 2010; Wade et al. 2009). Moreover, the existences of neocentromeres and pseudodicentromeric chromosomes (Scott and Sullivan 2013) strongly suggest that centromeres do not depend on the underlying DNA sequence for their inheritance but are epigenetic loci that are stably inhered through epigenetic processes. Indeed, in higher eukaryotes the centromeric chromatin is defined by epigenetic chromatin features, primarily by the presence of a centromere-specific CENP-A histone variant, rather than underlying DNA sequence. CENP-A specification of epigenetic centromeres means that the process of nucleosome assembly is a key event in inheritance of the locus.

All histone H3 variants employ distinct mechanisms, facilitated by histone chaperones, which selectively recognize them upon synthesis and escort to the site of nucleosome assembly (Filipescu et al. 2014). Similarly, CENP-A uses its own specific machinery that orchestrates the spatiotemporal assembly of centromeric chromatin during the cell cycle. In humans new CENP-A incorporation is a multistep mechanism that involves identification of centromeric chromatin for new CENP-A incorporation, deposition of newly synthesized CENP-A/H4 and stabilization of CENP-A nucleosomes. Each of those steps requires the activity of multiple protein factors which work together to ensure that CENP-A nucleosomes are deposited specifically at the centromeric domain, at the correct time and only once per cell cycle. Mechanisms regulating CENP-A incorporation are well conserved across eukaryotes and here we summarize the current knowledge on the processes regulating new CENP-A deposition in different species.

1 Histone Chaperones and Centromere Assembly

Incorporation of histones into the chromatin requires assembly factors or chaperones that work together to facilitate nucleosome deposition (Burgess and Zhang 2013; Ransom et al. 2010). Histone H3 variants use their specific independent chaperone complexes that govern a selective recognition and facilitate their deposition in replication-dependent (H3.1 variant) or replication-independent (H3.3 and CENP-A variants) nucleosome assembly pathways (Sarma and Reinberg 2005; Szenker et al. 2011; Weber and Henikoff 2014). The major histone variant H3.1 is deposited into newly replicated naked DNA during DNA replication via the CAF-1 complex that include p150, p60, and p46/48 (Tagami et al. 2004; Tyler et al. 1999,

2001) The H3.3 variant is regulated by two chaperone complexes distinct from the H3.1 replication-dependent chaperones responsible for H3.1 deposition. The HIRA chaperone is devoted to the genome-wide deposition of histone H3.3 at active and repressed genes (Chow et al. 2005; Goldberg et al. 2010; Lewis et al. 2010; Mito et al. 2005; Szenker et al. 2011; Tagami et al. 2004; Tamura et al. 2009). DAXX also acts as a chaperone for H3.3 and mediates H3.3 deposition at telomeric and pericentric heterochromatin in conjunction with the H3K9-binding protein ATRX (Goldberg et al. 2010; Lewis et al. 2010).

Similar to the other H3 variants, the centromere-specific histone H3 variant CENP-A interacts with a dedicated chaperone prior to deposition into chromatin. Prenucleosomal human CENP-A associates with the Holliday junction recognition protein (HJURP) (Dunleavy et al. 2009; Foltz et al. 2009; Shuaib et al. 2010). HJURP is necessary for incorporation of vertebrate CENP-A into the centromeric chromatin and is recruited to centromeres in early G1, when new CENP-A assembly is occurring (Fig. 1) (Bernad et al. 2011; Dunleavy et al. 2009; Foltz et al. 2009; Jansen et al. 2007). Suppression of HJURP completely abolishes new CENP-A deposition, results in errors in kinetochore assembly and ultimately leads to a high rate of chromosome segregation defects (Dunleavy et al. 2009; Foltz et al. 2009).

The centromere targeting domain (CATD) of CENP-A is sufficient to determine the centromeric deposition of CENP-A. The CATD domain spans loop 1 and the $\alpha2$ helix of CENP-A and when replaced with corresponding domain within canonical H3.1 was demonstrated to confer both HJURP binding and centromeric localization (Black et al. 2007; Foltz et al. 2009). His 104 and Leu112 residues within the CATD C-terminal region, together with either Asn85 or Gln89 within CATD N-terminus, are sufficient to confer HJURP binding, but not sufficient to facilitate centromere incorporation (Bassett et al. 2012).

HJURP specifically recognizes the CATD domain of CENP-A through its N-terminal CENP-A binding domain (Fig. 2). The CENP-A binding domain of HJURP shares homology with the yeast Scm3 proteins that also act as CENP-A (Cse4, Cnp1)-specific chaperone (Figs. 2 and 3) (Camahort et al. 2007; Mizuguchi et al. 2007; Pidoux et al. 2009; Sanchez-Pulido et al. 2009; Stoler et al. 2007; Williams et al. 2009). Although the mechanism of centromere inheritance between budding yeast and humans is very different, both systems are dependent on a CENP-A-specific histone chaperone. HJURP binds CENP-A through the conserved Scm3 domain. A number of residues within yeast Scm3 were proposed to be essential for CENP-A^{Cnp1} incorporation including Leucine 56 and Leucine 73. The fact that those key residues required for CENP-A^{Cnp1} deposition are conserved as hydrophobic amino acids in other eukaryotes, including humans, implies the mechanism by which CENP-A is selectively recognized and deposited at centromeric chromatin by its chaperone is common between yeast and humans (Cho and Harrison 2011; Pidoux et al. 2009).

In contrast to HJURP, which is recruited to centromeres with a refined temporal window when new CENP-A nucleosomes assembly occurs, the fission yeast Scm3 protein remains associated with the centromere through most of the cell cycle (Pidoux et al. 2009). This localization may provide a mechanism to insure the reassembly of

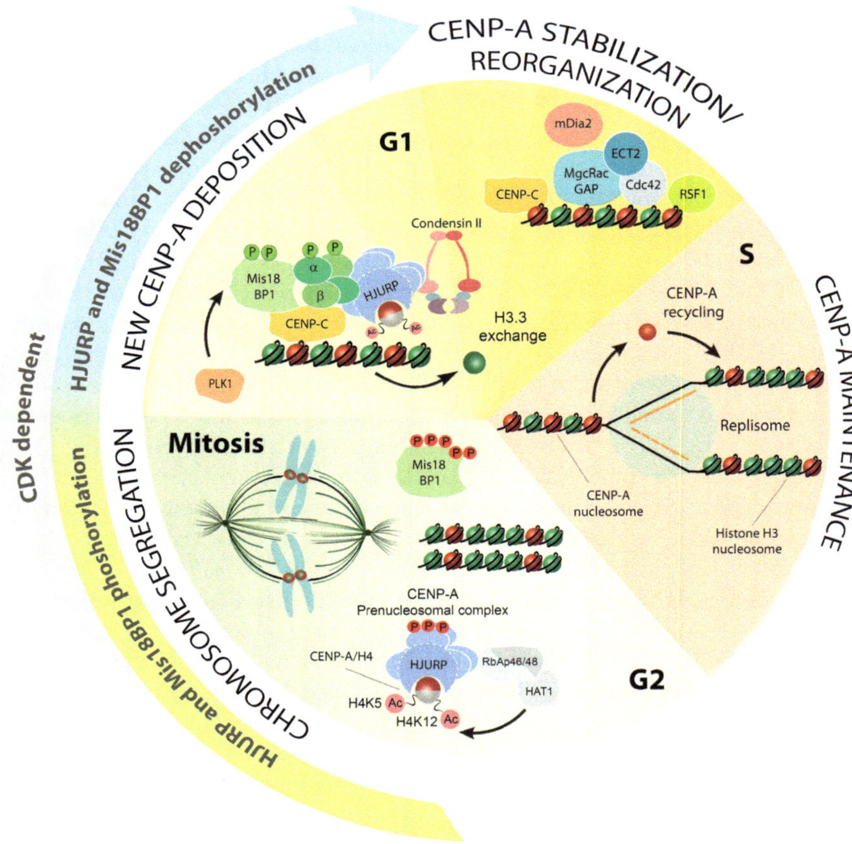

Fig. 1 The model of cell cycle regulated CENP-A deposition in humans. New CENP-A deposition occurs exclusively during early G1 and protein complexes involved are depicted in the model. HJURP binds the newly synthesized CENP-A/H4 complex in prenucleosomal form (Dunleavy et al. 2009; Foltz et al. 2009; Shuaib et al. 2010). H4K5Ac and H4K12Ac histone marks present in the CENP-A prenucleosomal complex are dependent upon RbAp46/48/HAT1 activity and required for CENP-A deposition (Shang et al. 2016). HJURP/CENP-A/H4 localization relies on the Mis18 complex (Barnhart et al. 2011; Fujita et al. 2007). The CENP-A deposition machinery is controlled by CDK activity. Cell cycle regulated and CDK1/CDK2-dependent phosphorylation of Mis18BP1 and HJURP prevents premature CENP-A loading during G2 and mitosis, and dephosphorylation of these proteins occurs prior new CENP-A deposition in G1 (Muller et al. 2014; Silva et al. 2012). G1-coupled and PLK1-mediated phosphorylation of the Mis18 complex promotes its centromeric localization and CENP-A deposition (McKinley and Cheeseman 2014). Mis18BP1 is recruited to centromeres upon its direct interaction with CENP-C (Dambacher et al. 2012). Human Mis18α and Mis18β form a multi subunit complex which is recruited to the centromere through interaction of Mis18α with Mis18BP1 and Mis18β with CENP-C (Nardi et al. 2016; Stellfox et al. 2016). HJURP mediates deposition of CENP-A nucleosomes, and histone H3.3 placeholder is removed from the centromeric chromatin (Dunleavy et al. 2011). Following new CENP-A deposition centromeric nucleosomes are stabilized and protein factors involved in this process are depicted in the model. During DNA replication existing CENP-A nucleosomes are retained across the replication fork (Bodor et al. 2014; Jansen et al. 2007)

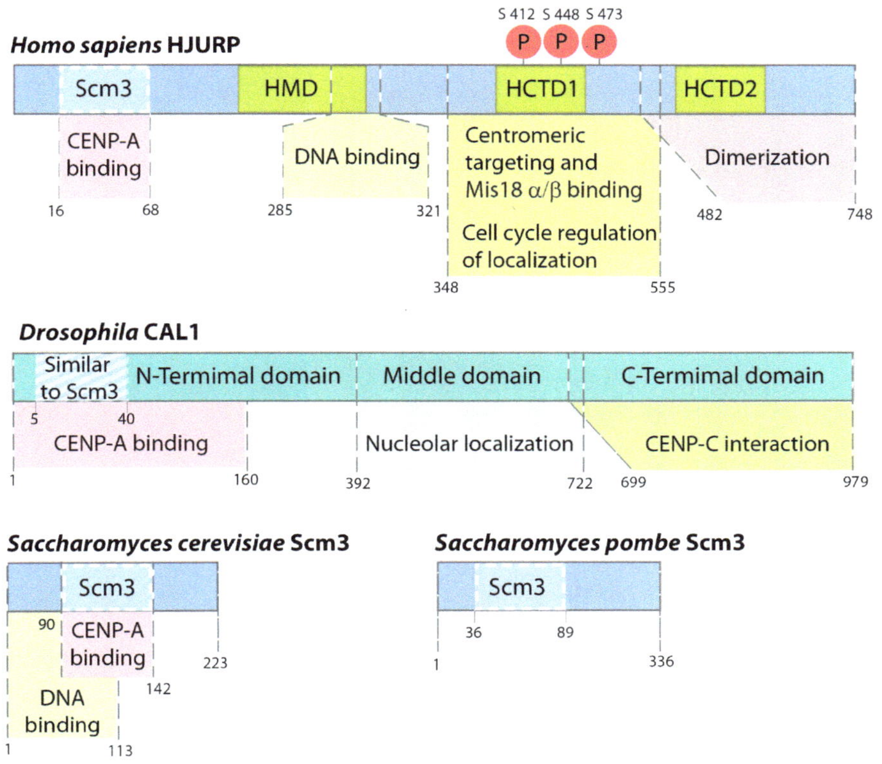

Fig. 2 Molecular organization of CENP-A-specific chaperone in different species. Domains identified within CENP-A chaperones among different species and their roles are depicted (Barnhart et al. 2011; Bassett et al. 2012; Cho and Harrison 2011; Dechassa et al. 2011; Hu et al. 2011; Muller et al. 2014; Sanchez-Pulido et al. 2009; Schittenhelm et al. 2010; Shuaib et al. 2010; Wang et al. 2014; Zasadzinska et al. 2013). The Scm3 domain is conserved among eukaryotes except for the *Drosophila melanogaster* where the similarity was assessed based on both sequence and secondary structure similarity (Phansalkar et al. 2012; Sanchez-Pulido et al. 2009). HMD-HJURP mid domain; HCTD1-HJURP carboxy terminal domain 1; HCTD2-HJURP carboxy terminal domain 2

CENP-A^{Cnp1} in the event of centromeric chromatin disruption, or to block the ubiquitination and degradation of centromeric CENP-A^{Cnp1}. Alternatively Scm3 may provide additional function at the centromere beyond CENP-A^{Cnp1} deposition.

The crystal structures of both yeast Scm3/CENP-A^{Cse4}/H4 and human HJURP-Scm3/CENP-A/H4 complexes demonstrate that the association of CENP-A (Cse4) with its chaperone prevents CENP-A/H4 tetramer formation, and precludes spontaneous DNA interactions by the histone complex in the prenucleosomal form (Cho and Harrison 2011; Hu et al. 2011). Vertebrate HJURP is much larger than its yeast orthologue Scm3, and contains several domains that are absent from both the *S. pombe* and *S. cerevisiae* orthologues (Fig. 3) (Sanchez-Pulido et al. 2009). Similar to *S. cerevisiae* Scm3, human HJURP was demonstrated to mediate an

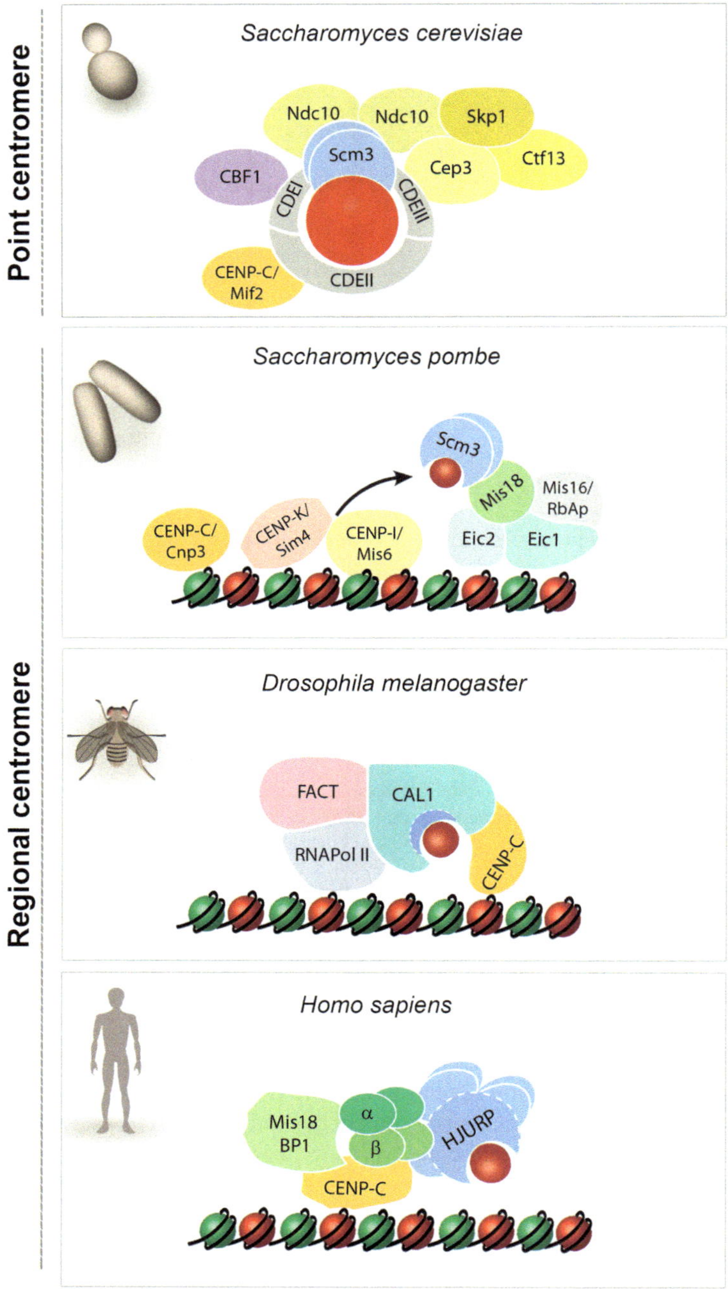
Point centromere
Regional centromere
Saccharomyces cerevisiae
Ndc10
Ndc10
Skp1
Scm3
Cep3
Ctf13
CBF1
CDEI
CDEIII
CENP-C/
Mif2
CDEII
Saccharomyces pombe
Scm3
Mis18
Mis16/
RbAp
CENP-C/
Cnp3
CENP-K/
Sim4
CENP-I/
Mis6
Eic2
Eic1
Drosophila melanogaster
FACT
CAL1
RNAPol II
CENP-C
Homo sapiens
Mis18
BP1
α
β
HJURP
CENP-C

◀**Fig. 3** Protein complexes involved in CENP-A deposition pathway in eukaryotes. The comparison of CENP-A deposition machinery across species. All conserved proteins involved in the CENP-A deposition pathway are colored similarly. Budding yeast point centromeres are specified by unique DNA elements: CDEI, CDEII, and CDEIII (Clarke and Carbon 1980; Fitzgerald-Hayes et al. 1982), which are required for recruitment of DNA-binding proteins as depicted in the model. The regional centromeres in fission yeast and higher eukaryotes are specified by the presence of CENP-A containing nucleosomes. CENP-A incorporation into centromeric chromatin is mediated by its distinct histone chaperone HJURP in vertebrates, Scm3 in yeast and CAL1 in *Drosophila melanogaster* (Barnhart et al. 2011; Bernad et al. 2011; Camahort et al. 2007; Dechassa et al. 2011; Dunleavy et al. 2009; Foltz et al. 2009; Mizuguchi et al. 2007; Pidoux et al. 2009; Shuaib et al. 2010; Stoler et al. 2007; Williams et al. 2009). HJURP and Scm3 share common ancestry, as depicted on the model, and CAL1 shares similarity to Scm3 based on the sequence and secondary structure similarity (Phansalkar et al. 2012; Sanchez-Pulido et al. 2009). CENP-C is conserved in all eukaryotes but its essential role in centromere specification is restricted to higher eukaryotes where it is required for recruitment of the Mis18 complex (Dambacher et al. 2012; Moree et al. 2011). The role of the Mis18 complex in CENP-A deposition pathway is conserved from fission yeast to humans; however, no Mis18 homologue was identified in *Drosophila* (Fujita et al. 2007; Maddox et al. 2007). The fission yeast has only one copy of Mis18 protein and the function of Mis 8BP1 was replaced by the Eic1 protein (Hayashi et al. 2014; Subramanian et al. 2014). Human Mis18 complex is a multisubunit complex composing of Mis18α/β and Mis18BP1 (Fujita et al. 2007; Maddox et al. 2007; Nardi et al. 2016). CENP-A deposition in *Drosophila* requires active transcription mediated by the FACT and RNA Polymerase II (Chen et al. 2015)

interaction with DNA through its "mid" domain (HMD), which is required for new CENP-A deposition (Fig. 2) (Muller et al. 2014; Xiao et al. 2011). It is not known whether in addition to its ability to bind DNA, HJURP also has a capacity to interact with RNA. Given the evidence that RNA plays a role in centromere specification and HJURP recruitment it is an outstanding question that awaits future studies (Bergmann et al. 2011; Quenet and Dalal 2014a).

Centromeric recruitment of HJURP is independent of CENP-A binding and is mediated by the HJURP carboxyl terminal domain 1 (HCTD1) (Fig. 2) (Wang et al. 2014; Zasadzinska et al. 2013). The HJURP carboxyl terminal domain 2 (HCTD2) serves as a homo-dimerization interface and facilitates HJURP self-association, consistent with formation of the budding yeast Scm3/CENP-A^{Cse4}/H4 hexamer and Scm3 self-association in fission yeast. In those species the multimerization mediated by the CENP-A chaperone is required for new CENP-A deposition (Mizuguchi et al. 2004; Pidoux et al. 2009; Wang et al. 2014; Zasadzinska et al. 2013). This evidence provides a mechanism by which prenucleosomal HJURP complex brings two CENP-A molecules to the site of CENP-A deposition consistent with the CENP-A nucleosomes forming an octamer. Alternatively, one HJURP present in the prenucleosomal complex brings newly synthetized CENP-A/H4 heterodimer, and the other HJURP molecule can recognize CENP-A present within centromeric chromatin, consistent with the hemisome hypothesis (Wang et al. 2014; Zasadzinska et al. 2013).

The proposed role of the histone chaperone has been to preclude the stochastic interactions between the histone protein and DNA prior to nucleosomes assembly. Consistent with this idea, the interaction of HJURP with the CENP-A/H4 heterotetramer blocks several key residues along the DNA interface of CENP-A (Cse4) (Cho and Harrison 2011; Hu et al. 2011). In addition, histone chaperones are known to facilitate the assembly of histone subunits into nucleosomes. Both Scm3 and HJURP mediate CENP-A (Cse4) nucleosome assembly in vitro (Barnhart et al. 2011; Camahort et al. 2009; Dechassa et al. 2011; Shivaraju et al. 2011). Much consideration has been given to whether the CENP-A nucleosome adopts non-canonical forms which have been reviewed extensively (Black and Clevel 2011; Quenet and Dalal 2012). Deposition experiments suggest that, while CENP-A may take on varied conformations, the CENP-A chaperone facilitates the formation of octameric nucleosomes with a left-handed wrap of the DNA (Barnhart et al. 2011; Dechassa et al. 2011).

2 Prenucleosomal Posttranslational Modifications and CENP-A Deposition

CENP-A is bound to its chaperone as a heterodimer with histone H4, thus modification of H4 may contribute to CENP-A nucleosome assembly. Indeed, histone H4 is acetylated on K5ac and K12ac within the prenucleosomal complex, and these modifications are necessary for CENP-A deposition (Fig. 1) (Shang et al. 2016).

Human RbAp46 (a.k.a. RBBP7) and RpAp48 (a.k.a. RBBP4) are highly homologous genes whose protein products are present in many chromatin remodeling complexes (Loyola and Almouzni 2004). Mutants of the *S. pombe* homolog of the RbAp proteins, Mis16, cause chromosome segregation defects due to a failure to assemble CENP-A^{Cnp1} nucleosomes (Hayashi et al. 2004). RbAp46/48 co-purified with HJURP in the prenucleosomal CENP-A complex (Dunleavy et al. 2009; Shuaib et al. 2010). A crystal structure of the Mis16-Scm3-CENP-A^{Cnp1}/H4 complex shows that Mis16 contacts both the Scm3 chaperone and histone H4 (An et al. 2015). Depletion of RbAp proteins reduces HJURP recruitment and new CENP-A deposition (Dunleavy et al. 2009; Shang et al. 2016). K5 and K12 acetylation of the histone H4 bound to CENP-A within the prenucleosomal complex are dependent on RbAp48, and these modifications are required for CENP-A deposition in vivo (Fig. 1) (Shang et al. 2016). In the *Xenopus* system H4K5 and H4K12 acetylation marks in prenucleosomal CENP-A complex are dependent upon HAT1 activity (Shang et al. 2016) which is also required for CENP-A deposition in *Drosophila* (Boltengagen et al. 2016). Therefore, a major role of RbAP48 may be the recruitment of the histone acetyltransferase required for modifying Histone H4. What components may read out the presence of H4 acetylation within the assembly pathway is not known.

RbAp46 and RbAp48 depletion results in reduced HJURP protein levels (Dunleavy 2009) and a second role for these proteins may be in regulating the stability of the CENP-A prenucleosomal complex (Mouysset et al. 2015). RbAP46 forms a complex with the CRL4 ubiquitin ligase, a member of the cullin-RING-ligase family, and DDB1 protein (where DDB1 mediates the association of CUL4 with its substrate-specific receptor—RbAP46) (Lee and Zhou 2007; Mouysset et al. 2015). RbAp46 is required for stabilizing CENP-A protein levels and the CRL4-RbAp46 complex activity promotes efficient new CENP-A deposition in humans (Mouysset et al. 2015). This is in contrast to studies in yeast and *Drosophila*, where the association of CENP-A with the SCF E3-ubiquitin ligase complex leads to CENP-A degradation (see below).

Two different posttranslational modifications of human CENP-A are proposed to be important for CENP-A deposition. These are phosphorylation of serine 68 and ubiquitylation of lysine 124 (Niikura et al. 2015; Yu et al. 2015). Both modifications are located outside of the CATD domain that is sufficient for HJURP binding, and situated on the helix $\alpha1$ and helix $\alpha3$ of CENP-A, respectively. However, both are proposed to influence HJURP binding to CENP-A. CENP-A lysine 124 (K124) in humans undergoes mono- and di-ubiquitylation mediated by the CUL4A-RBX1-COPS8 E3 ligase complex (Niikura et al. 2015). Downregulation of any of the CUL4A-RBX1-COPS8 subunits or mutation of Lys124 leads to loss of centromeric CENP-A in mitosis and interphase cells. Mutation of CENP-A lysine 124 weakens the interaction with CENP-A chaperone HJURP.

Phosphorylation at CENP-A-Ser68 is proposed to preclude its interaction with HJURP, negatively regulating new CENP-A deposition. CENP-A Ser68 phosphorylation depends on Cdk1/cyclin B activity during early mitosis and PP1a phosphatase dephosphorylates Ser68 in late mitosis, making CENP-A competent for HJURP binding and new incorporation in the following G1 (Yu 2015). While the phosphomimetic S68Q mutation appears to preclude HJURP binding both in vivo and in vitro (Hu et al. 2011; Yu et al. 2015), Bassett et al. reported that S68Q substitution within CENP-A has no effect on HJURP-mediated targeting and subsequent incorporation into chromatin at non-centromeric sites. Moreover, recombinant CENP-A containing the S68Q mutation forms a complex in vitro with HJURP with similar efficiency when compared to the wild-type form (Bassett et al. 2012). Despite the effects observed in vivo for the S68 and K124 mutations, both mutations are fully able to rescue CENP-A null cells, suggesting that these modifications are not essential for the process of centromere specification and inheritance (Fachinetti 2016).

In budding yeast, Psh1 prevents ectopic localization of CENP-A^{Cse4} (Hewawasam et al. 2010; Ranjitkar et al. 2010). Psh1 is an E3 ubiquitin ligase that was identified as associated with yeast CENP-A^{Cse4} in immunoprecipitation experiments and characterized as a kinetochore and centromere-associated protein. Psh1 regulates CENP-A^{Cse4} levels by ubiquitylating CENP-A^{Cse4} and targeting it for proteolysis, thus preventing its accumulation outside the centromeric chromatin. Psh1 and Scm3 both recognize the CENP-A^{Cse4}-CATD domain; therefore, Scm3 appears to protect CENP-A^{Cse4} from the Psh1-mediated ubiquitination and subsequent degradation (Hewawasam et al. 2010; Ranjitkar et al. 2010). In flies, CENP-A^{CID} directly interacts with the with the F-Box Protein Partner of Paired (Ppa), a variable component of a SCF E3-ubiquitin ligase complex in *Drosophila*.

Ppa binds CENP-A^{CID} through the CATD domain and regulates its stability (Cardozo and Pagano 2004; Moreno-Moreno et al. 2011; Nakayama and Nakayama 2006; Schuh et al. 2007).

3 Coupling Chaperone Recruitment to Existing Centromeres

Human centromeres range from 0.3 to 5 Mbp in size and account for less than 1% of the chromosome (Cleveland et al. 2003). The restriction of centromeres to a single locus ensures the stable inheritance of centromeres by avoiding situations where multiple centromeres on one chromosome could make attachments to opposing poles and result in chromosome breakage during mitosis.

The recruitment of the CENP-A-specific histone chaperone to the existing centromere is an essential step in epigenetic inheritance. Mis18 is a key adapter protein that mediates the recruitment of the CENP-A chaperone to centromeres in several organisms (Figs. 3 and 4), but is absent from organisms with point centromeres. Mis18 was originally identified in a genetic screen in fission yeast to identify genes required for proper chromosome segregation (Hayashi et al. 2004). spMis18 mutants eliminate CENP-A^{Cnp1} incorporation to centromeres and Mis18 directly interacts with Scm3 to determine its recruitment (Pidoux et al. 2009). In humans, Mis18 exists as a complex comprised of Mis18α, Mis18β, and Mis18BP1 proteins (Figs. 3 and 4). The Mis18 complex is essential for the recruitment of HJURP and CENP-A to the centromeric chromatin due to a direct interaction with the HJURP centromere targeting domain within the HCTD1 (Fig. 2) (Barnhart et al. 2011; Fujita et al. 2007; Nardi et al. 2016; Wang et al. 2014). Mis18 proteins do not require HJURP for recruitment, demonstrating that they are upstream components of the pathway (Barnhart et al. 2011; Bernad et al. 2011). Consistent with studies in yeast, depletion of the Mis18 complex subunits in human cells results in a high rate of chromosome segregation defects and loss of centromeric CENP-A (Fujita et al. 2007). The role of the Mis18 proteins in the CENP-A deposition pathway is evolutionarily conserved, as depletion of Mis18BP1 (KNL-2) homologues in *C. elegans* and *Xenopus* also leads to defects in CENP-A deposition in these species, although, as we discussed below, the pathway has undergone several permutations in different organisms (Maddox et al. 2007; Moree et al. 2011).

Since the Mis18 proteins are required for HJURP recruitment, the key question becomes how the Mis18 protein recognizes the existing centromere. CENP-A nucleosomes recruit the CCAN (constitutive centromere-associated network), a multiprotein complex comprised of 16 subunits, present at the centromere throughout the cell cycle, that serves as a structural core for kinetochore assembly during mitosis (Amano et al. 2009; Cheeseman and Desai 2008; Earnshaw et al. 1986; Foltz et al. 2006; Izuta et al. 2006; McKinley and Cheeseman 2016; Nishihashi et al. 2002; Okada et al. 2006; Saitoh et al. 1992; Sugata et al. 1999). The CENP-C component of the CCAN directly recognizes the CENP-A nucleosome

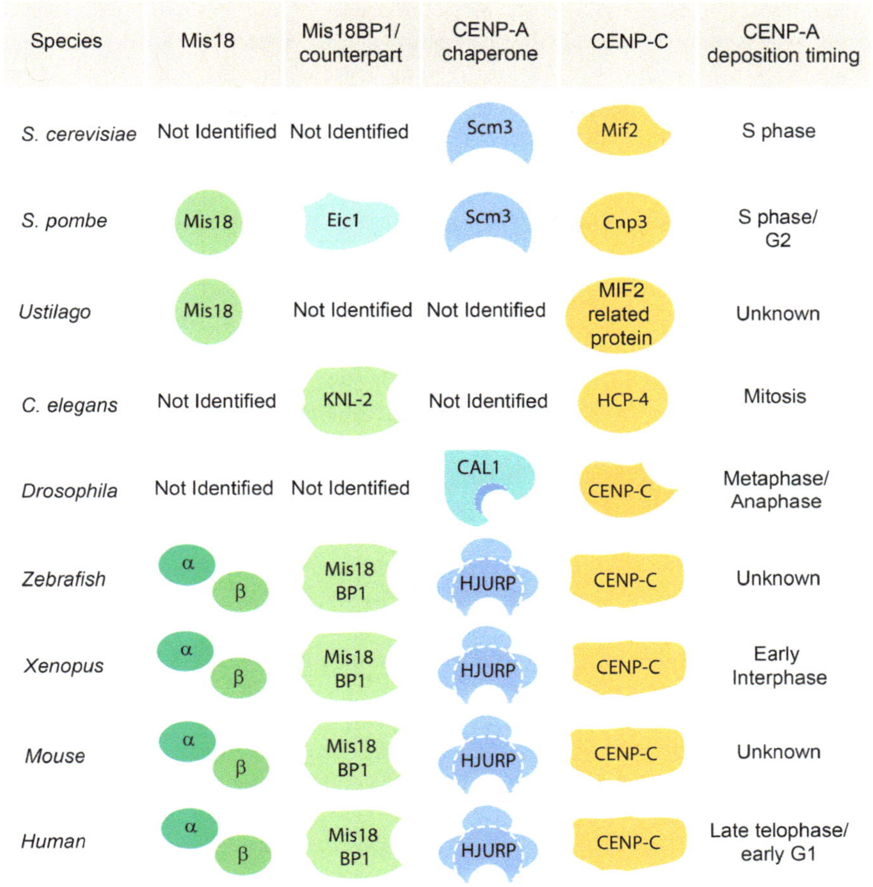

Fig. 4 Conservation of CENP-A deposition factors across species. Table detailing the conserved proteins involved in CENP-A deposition pathway as well as timing of CENP-A deposition in different model organisms (Bernad et al. 2011; Dunleavy et al. 2007; Jansen et al. 2007; Maddox et al. 2007; Mellone et al. 2011; Moree et al.2011; Pearson et al. 2004; Schuh et al. 2007; Takayama et al. 2008)

(Carroll et al. 2010; Guse et al. 2011; Kato et al. 2013). New CENP-A nucleosomes within the alpha satellite DNA are assembled directly adjacent to the existing CENP-A (Ross et al. 2016). CENP-C plays a crucial role in recruiting the proteins required for CENP-A deposition (Figs. 1 and 2), and thus links the existing centromere to the assembly of new CENP-A nucleosomes in early G1. CENP-C interacts directly with two proteins within the Mis18 complex, Mis18BP1 and Mis18β (Dambacher et al. 2012; Moree et al. 2011; Stellfox et al. 2016). CENP-C depletion causes defects in Mis18BP1 and HJURP recruitment and leads to loss of CENP-A chromatin assembly (Dambacher et al. 2012; Moree et al. 2011; Stellfox et al. 2016).

The pivotal role that CENP-C plays in determining the site of centromeric chromatin assembly is exemplified by experiments in chicken DT40 cells, where the endogenous centromere is conditionally removed and the functional kinetochore assembled at an ectopic LacO locus. These experiments show that tethering the LacI-fused HJURP or full-length CENP-C are sufficient to recruit CENP-A in order to establish a functional epigenetic de novo centromere (Hori et al. 2013). Although tethering the CENP-C N-terminus (1–643 aa) in this system is sufficient to recruit microtubule binding proteins and the CPC complex, it fails to incorporate CENP-A nucleosomes (Hori et al. 2013). This is consistent with the identification of the N-terminus of CENP-C as the region of interaction with Mis18BP1 and Mis18β (Dambacher et al. 2012; Moree et al. 2011; Stellfox et al. 2016). In contrast, CENP-C homologues in yeast (Mif2 and Cnp3) are not essential to facilitate CENP-A deposition (Fig. 3) (Meluh and Koshland 1995, 1997; Westermann et al. 2003).

Additional factors in the CCAN also contribute to directing new CENP-A nucleosome deposition. Depletion of the CENP-HIKM complex in chicken cells compromise the incorporation of newly synthesized CENP-A (Okada et al. 2006). Consistent with this observation, fission yeast CENP-I^{Mis6} and CENP-K^{Sim4} are required for CENP-A nucleosome deposition (Fig. 3) (Pidoux et al. 2009; Takahashi et al. 2000). Similar to CENP-C, tethering CENP-I to a non-centromeric site in chicken DT40 cells drives new CENP-A deposition and forms an epigenetic centromere (Hori et alz et al. 2013). This suggests that the CCAN components play a dual role, and are required for both centromere specification in G1 and recruitment of kinetochore components during mitosis.

Budding yeast centromeres are determined by DNA sequence. And although they share a homologous CENP-A chaperone, Scm3, the mechanism by which Scm3 is recruited to centromeres is distinct from epigenetic centromeres (Fig. 3). The centromere-determining elements (CDE) in budding yeast are essential for recruitment of a DNA-binding protein Cbf1 specifically recognizing CDEI and a multisubunit protein complex: CBF3 (centromere binding factor 3), containing Ndc10, Cep3, Ctf13, and Skp1, associated with CDEIII DNA element (Cho and Harrison 2011; Doheny et al. 1993; Goh and Kilmartin 1993; Hyman et al. 1992; Lechner and Carbon 1991; Mizuguchi et al. 2007; Russell et al. 1999; Strunnikov et al. 1995). The CBF3 subunit-Ndc10 is required for the recruitment of the Scm3 chaperone and subsequent deposition of the CENP-A^{Cse4} containing nucleosome (Camahort et al. 2007; Mizuguchi et al. 2007).

4 The Chromatin Landscape Influence on CENP-A Deposition

CENP-A nucleosomes are interspersed with the canonical H3 nucleosomes within the centromeres of flies and humans (Blower et al. 2002). Centromeres were initially thought to be transcriptionally silent loci, a characteristic that is consistent with the posttranslational modifications found in the surrounding pericentric heterochromatin (Peters et al. 2001; Ribeiro et al. 2010; Rice et al. 2003). However, studies in human and *Drosophila*-derived chromatin fibers demonstrated that H3K9me2 and H3K9me3 marks are absent from the CENP-A (CID) domain (Lam et al. 2006; Sullivan and Karpen 2004). Furthermore, histone H3 nucleosomes found interspersed with CENP-A nucleosomes in humans are decorated with histone marks associated with active or poised chromatin, such as H3K4me1/2 and H3K36me2/3 (Bergmann et al. 2011; Sullivan and Karpen 2004). The histone H3K4 trimethylation, associated with actively transcribed regions, is absent from the centromeric core domain in humans and *Drosophila*, but is present at chicken centromeric DNA (Ribeiro et al. 2010; Sullivan and Karpen 2004). Until recently the centrochromatin-localized histones in higher eukaryotes were thought to be hypo-acetylated and lack acetylated marks found generally in euchromatin such as H3K9Ac, H4K5Ac, H4K8Ac, H4K12Ac, or H4K16Ac. However, a recent study documented the presence of H4K5Ac and H4K12Ac within CENP-A containing nucleosomes in chicken and humans (Bailey et al. 2016; Shang et al. 2016; Sullivan and Karpen 2004).

Transcripts from centromeric repeat sequences have been observed in multiple model organisms (Bergmann et al. 2011; Bouzinba-Segard et al. 2006; Carone et al. 2009, 2013; Chan et al. 2011, 2012; Eymery et al. 2009; Hall et al. 2012; Lam et al. 2006; May et al. 2005; Ohkuni and Kitagawa 2011; Quenet and Dalal 2014b; Stimpson and Sullivan 2010; Topp et al. 2004; Wong et al. 2007). Active RNA Polymerase II is recruited to endogenous human centromeres during mitosis and early G1 (Chan et al. 2012; Quenet and Dalal 2014b). Inhibition of RNA-Polymerase-II-mediated transcription in HeLa cells leads to decreased α-satellite transcript levels in mitosis, loss of CENP-C recruitment to endogenous centromeres, and chromosome segregation defects (Chan et al. 2012). The mechanistic role of centromeric transcripts and the act of transcription in centromere function is not yet clear, although histone H3 eviction may be a key aspect.

Utilizing a synthetic human artificial chromosome (HAC), Bergmann et al. demonstrated that the presence of H3K4me2 and transcription events at the centromere play a critical role in CENP-A assembly and centromere function by altering the recruitment of CENP-A deposition machinery (Bergmann et al. 2011). Tethering a lysine-specific demethylase 1 (LSD1) to the HAC centromeric domain leads to removal of H3K4 methylation and results in loss of transcription of α-satellite DNA at this loci. This correlates with loss of HJURP localization, impaired CENP-A deposition, and ultimately leads to loss of kinetochore function (Bergmann et al. 2011).

Biochemical purification of RNA associated with the prenucleosomal CENP-A/HJURP complex identified a 1.3 kb RNA product that co-localizes with a-satellite DNA and CENP-A, and hybridizes to centromeric α-satellite probes, suggesting it originated from α-satellite transcripts (Quenet and Dalal 2014a). Targeting of α-satellite transcripts as well as other centromere-derived RNAs by siRNA in vivo results in reduced CENP-A and HJURP recruitment to the centromere, suggesting that the RNA component partially encoded within α-satellite DNA plays a role in CENP-A deposition pathway (Quenet and Dalal 2014a).

Exactly how RNAs are associated with the CENP-A prenucleosomal complex is still unknown, as well as how this association would contribute mechanistically to CENP-A deposition.

A strong link between CENP-A deposition and transcription was demonstrated in *Drosophila*. Chen et al. using an inducible ectopic centromere approach demonstrated that new CENP-A^{CID} deposition at the ectopic centromere requires transcription (Fig. 3) (Chen et al. 2015). The mass spec analysis of binding partners of the *Drosophila* CENP-A^{CID} chaperone CAL1 in vivo identified two subunits of the FACT complex: Spt16 and SSRP1, both of which physically interact with CAL1. FACT was also previously found associated with centromere in human cells (Foltz et al. 2009; Obuse et al. 2004). FACT is involved in transcription elongation from chromatin templates in vitro and promoting deposition of histone H3.3 nucleosomes in vivo in *Drosophila* (Orphanides et al. 1998). Spt16 and SSRP1 subunits colocalize with CENP-A^{CID} in *Drosophila* cells and downregulation of FACT leads to defects in CENP-A^{CID} recruitment at endogenous centromeres. CAL1 along with FACT facilitates RNA-Polymerase-II-mediated transcription at the site of CENP-A^{CID} deposition, which is required for CENP-A^{CID} incorporation to occur. In support to these findings other groups reported localization of the active form of RNA Polymerase II at endogenous centromeres in *Drosophila* during mitosis, which is coincident with new CENP-A^{CID} deposition timing (Rosic et al. 2014).

In addition to the role of the Mis18 complex in the recognition of the CCAN and direct recruitment of HJURP, the Mis18 complex influences posttranslational modifications within the centromeric chromatin (Kim et al. 2012). Deletion of Mis18 in *S. pombe* leads to increased levels of histone H3 and H4 acetylation at centromeres (Hayashi et al. 2004). In vertebrates, the Mis18 complex influences histone modifications and DNA methylation. Knockout of Mis18α in mice leads to reduced H3K9 and H3K4 methylation and increased acetylation within centromeric repeats (Kim et al. 2012). The de novo methyltransferase enzymes DNMT3a/b are also recruited to centromeres by Mis18α/β (Kim et al. 2012). Downregulation of DNMT3b or Mis18α leads to increased transcription of centromeric repeats (Gopalakrishnan et al. 2009). However, the importance of DNMT3a/b in centromere function is unclear since cells lacking DNTM3a/b are viable (Reviewed in Brown and Robertson 2007).

More recently Mis18BP1 was shown to recruit the KAT7 lysine methyltransferase complex to centromeres (Ohzeki et al. 2016). Disruption of the KAT7 complex leads to reduced CENP-A deposition. KAT7 in conjunction with RSF1 may regulate histone turnover to facilitate new CENP-A deposition in G1. In future work it will be important to determine exactly how the Mis18 complex may integrate multiple downstream chromatin modifying pathways to promote centromere deposition.

5 Licensing of Centromere Assembly

CENP-A incorporation into the centromeric chromatin is cell cycle regulated, although the timing of CENP-A deposition differs across species (Figs. 1 and 4) (Allshire and Karpen 2008; Boyarchuk et al. 2011). Budding yeast CENP-A^{Cse4} incorporation is coincident with DNA replication (Pearson et al. 2004; Wisniewski et al. 2014). Similarly, in fission yeast, CENP-A deposition occurs during early S phase, but also during G2 phase (Takayama et al. 2008). In vertebrates, new CENP-A incorporation is uncoupled from DNA replication and restricted to late telophase/early G1 phase (Bernad et al. 2011; Jansen et al. 2007; Silva et al. 2012).

The process of human CENP-A deposition occurs via a licensing mechanism that restricts deposition to the G1 phase and controls the assembly of CENP-A to ensure that only a limited amount of new CENP-A is assembled in each cell cycle. The timing of CENP-A deposition is restricted to the early G1 phase by inhibition of CENP-A deposition through CDK activity, which is high during S and G2-phase, and drops rapidly following satisfaction of the mitotic checkpoint (Fig. 1) (Silva et al. 2012). Although CENP-A transcript and protein levels accumulate from mid-S phase into G2, CDK1/CDK2-dependent phosphorylation of Mis18BP1 prevents from premature CENP-A loading during this time (Silva et al. 2012). Mis18BP1 dephosphorylation occurs during early G1, coincident with new CENP-A deposition. The PLK1 kinase positively regulates CENP-A deposition. PLK1 phosphorylates the Mis18 complex during G1 to promote its recruitment to centromeres (Fig. 1) (McKinley and Cheeseman 2014). Inhibition of the PLK1 kinase activity abrogates new CENP-A deposition. The opposing functions of PLK1 and CDK1 phosphorylation provide tight temporal control of CENP-A deposition by limiting Mis18 recruitment.

The assembly of CENP-A nucleosomes in G1 is limited by at least two mechanisms. The Mis18 complex forms a conserved multimer (Nardi et al. 2016; Subramanian et al. 2016). Mis18 binds the centromere stably in late telophase. Binding of HJURP to Mis18 disrupts the Mis18 multimer and eliminates the ability of Mis18 to continue to interact with the centromere, essentially removing the signal for HJURP recruitment, and blocking further CENP-A deposition at that site. In addition, the Mis18β subunit undergoes ubiquitylation and degradation by the SCF$^{\beta TrCP}$ E3 ubiquitin ligase, thus degrading the signal for HJURP recruitment to centromeres (Kim et al. 2014).

6 Viva la Difference—Evolutionary Diversity in CENP-A Deposition Pathways

Despite the high degree of conservation between the CENP-A-binding domains within the HJURP and Scm3 chaperones, that spans billions of years of evolution, there is a great variety in the CENP-A deposition pathways across organisms (Figs. 3 and 4). This likely reflects the unique strategies for centromeric chromatin assembly that these organisms employ.

Drosophila species lack a clear HJURP homolog, but an siRNA screen for genes involved in CENP-A^{CID} centromere deposition in *Drosophila* S2 cells identified CAL1 (chromosome alignment defect 1) as a key factor (Erhardt et al. 2008). *Drosophila* CAL1 is a fly-specific protein that functions as a CENP-A^{CID} chaperone (Chen et al. 2014; Erhardt et al. 2008; Goshima et al. 2007). Despite the small similarity to the Scm3 domain of *K.lactis*, based on sequence and secondary structure, CALI does not share common ancestry with yeast Scm3 and human HJURP (Phansalkar et al. 2012; Sanchez-Pulido et al. 2009). CAL1 directly binds to CENP-A^{CID}/H4 dimer and was shown to function as the CENP-A^{CID}-specific assembly factor in fruit flies (Chen et al. 2014). Its depletion in *Drosophila* results in loss of centromeric CENP-A^{CID} localization and is associated with chromosome segregation defects (Chen et al. 2014; Erhardt et al. 2008; Goshima et al. 2007). Both HJURP and CAL1 are sufficient to promote de novo centromere establishment. Tethering HJURP to the chromosome arm or to a naïve alpha satellite array is sufficient to facilitate CENP-A deposition outside of the centromeric chromatin and results in formation of a functional kinetochore at an ectopic site in human cells (Barnhart et al. 2011; Ohzeki et al. 2012). Similarly, targeting CAL1 to an ectopic site was demonstrated to mediate de novo CENP-A^{CID} deposition in *Drosophila,* which leads to formation of a de novo centromere outside of the endogenous centromeric loci (Chen et al. 2014). This de novo centromere is epigenetically maintained and serves as platform for recruitment of a functional kinetochore (Chen et al. 2014). There are several organisms that contain CENP-A nucleosomes for which a functional chaperone has not been identified, including the well-studied nematode *C. elegans* (Fig. 4). *C. elegans* have holocentric chromosomes in which the centromere position may be variable and obfuscate the need for specific targeting of the CENP-A histone variant.

Conservation of the Mis18 complex is also highly variable across species. Species as divergent as *S. pombe* and humans possess Mis18, but in higher eukaryotes the Mis18 gene underwent duplication (Fig. 4). The Mis18 paralogs, termed Mis18α and Mis18β, share about 30% sequence identity, but have diverged in their function in higher eukaryote centromeres (Fujita et al. 2007; Hayashi et al. 2004; Stellfox et al. 2016). The Mis18 complex has not been identified in *Drosophila* or *S. cerevisiae* (Fig. 4). In both cases, these organisms have devised alternative strategies to couple the CENP-A chaperones to the existing centromere. CALI binds CENP-C in *Drosophila* and the Ndc10 complex, which directly recognizes DNA, recruits the Scm3 chaperone in budding yeast (Camahort et al. 2007;

Doheny et al. 1993; Erhardt et al. 2008; Goh and Kilmartin 1993; Jiang et al. 1993; Lechner and Carbon 1991; Mellone et al. 2011; Sorger et al. 1995).

While *S. pombe* possess a Mis18 homolog, it lacks the vertebrate Mis18BP1 orthologue (Fig. 4). The Mis18BP1 function in *S. pombe* is replaced by the Eic1 protein (a.k.a Mis19) (Fig. 3). The Eic1 and Eic2 proteins co-purified with the spMis18 and exhibit a similar temporal pattern of centromeric localization throughout the cell cycle (Hayashi et al. 2014; Subramanian et al. 2014). Eic1 was demonstrated to be essential for the recruitment of the Mis18, Mis16, and Scm3 proteins to the centromere and for CENP-A^{Cnp1} incorporation. However, Eic2 is dispensable for recruitment of CENP-A^{Cnp1} to the centromere. This suggests Eic1 is functionally analogous to the Mis18BP1 subunit in recruitment of CENP-A deposition, although Eic1 is evolutionary distinct and does not share any apparent sequence homology to Mis18BP1 (Hayashi et al. 2014; Subramanian et al. 2014).

7 Centromere Stabilization and Re-organization

The recruitment of CENP-A to centromeres via HJURP and Mis18 is not sufficient for the stability of CENP-A, but requires additional proteins that may potentially reorganize centromeric chromatin to increase stability. These factors include the Rho GTPase MgcRacGAP, the formin protein mDia, and the RSF-1 remodeling complex, and appear to be recruited to centromeres later than Mis18 and HJURP (Fig. 1) (Izuta et al. 2006; Lagana et al. 2010; Liu and Mao 2016; Obuse et al. 2004; Perpelescu et al. 2009).

MgcRacGAP co-purifies with centromeric chromatin and with Mis18BP1 from HeLa cells (Izuta et al. 2006; Lagana et al. 2010; Perpelescu et al. 2009). MgcRacGAP localizes to centromeres in late G1. Although the exact timing between MgcRacGAP recruitment and HJURP recruitment has not been established, it appears that MgcRacGAP is recruited later, after new CENP-A incorporation is accomplished. Depletion of MgcRacGAP or its binding partner, ECT2 (guanine nucleotide exchange factor), results in loss of newly incorporated CENP-A, while existing CENP-A is not affected. This suggests that new and old CENP-A populations during G1 are in some way unique. Furthermore, Cdc42, a small GTPase identified as a target of MgcRacGAP-ECT2 complex, is also recruited to the centromeres during interphase. The Cdc42 activity requires GTPase cycling mediated by MgcRacGAP-ECT2, proposing a GTPase switch implicated in the maturation of the newly deposited CENP-A containing nucleosomes (Lagana et al. 2010). mDia2 is a downstream effector of Rho signaling (Gasman et al. 2003; Lammers et al. 2008). mDia2 depletion leads to defects in new CENP-A deposition. The constitutively active form of mDia2 restores CENP-A levels at the centromere resulting from MgcRacGAP downregulation, consistent with its role downstream of MgcRacGAP in this process. Interestingly, mDia2 depletion leads to prolonged HJURP association with the centromere, suggesting that the processes of HJURP recruitment and MgcRacGAP stabilization are mechanistically linked (Liu and Mao 2016).

The RSF (remodeling and spacing factor), comprised the RSF-1 and SNF2h subunits, has been characterized as an ATP-dependent nucleosome remodeling and spacing factor that together with the FACT complex is implicated in transcription initiation (LeRoy et al. 1998; Orphanides et al. 1998). The RSF complex co-purified with CENP-A nucleosomes prepared from interphase cell extracts (Izuta et al. 2006; Obuse et al. 2004; Perpelescu et al. 2009). RSF centromere localization peaks during the middle of G1 phase. RSF1 can reconstitute and space CENP-A nucleosomes on a naked DNA template, and is required for stability of CENP-A nucleosomes within the centromeric chromatin (Perpelescu et al. 2009). This argues that energy-dependent remodeling events are involved in stabilization of newly deposited CENP-A nucleosomes.

Condensation of centromeric chromatin is a potentially important step in efficient CENP-A deposition. The condensin complexes are involved in ATP-dependent chromosome condensation during mitosis, and are also implicated in centromere establishment in yeast and humans (Hagstrom et al. 2002; Ono et al. 2004; Samoshkin et al. 2009; Wignall et al. 2003; Yong-Gonzalez et al. 2007). Of the two partially overlapping condensin complexes that have been characterized (Condensin I and II) the Condensin II complex is selectivity involved in centromeric chromatin assembly (Barnhart-Dailey et al. 2016; Bernad et al. 2011; Hirano 2005). Downregulation of common components to the Condensin complexes (SMC2 and SMC4) or the Condensin-II-specific subunits (CapH2 and CapD3) leads to reduced assembly of new CENP-A nucleosomes in humans and *Xenopus* extracts (Barnhart-Dailey et al. 2016; Bernad et al. 2011; Samoshkin et al. 2009). CAPH2 was found at human centromeres in early G1, coincident with new CENP-A deposition, and its recruitment is HJURP dependent (Fig. 1) (Barnhart-Dailey et al. 2016).

In chicken cells FACT subunits: SSRP1 and SPT16 co-purified with CENP-A and localize to the centromeric chromatin. FACT interacts with ATP-dependent chromatin remodeling factor CHD1, and the centromeric recruitment of these proteins throughout the cell cycle is dependent upon the CENP-HIKM complex. The downregulation of FACT or CHD1 factors leads to loss of new CENP-A deposition, demonstrating that chromatin remodeling activity of the FACT and CHD1 complexes plays a critical role in CENP-A deposition (Okada et al. 2009). It remains elusive whether the FACT and CHD1 complexes require active transcription in order to fulfill their role in CENP-A incorporation.

8 HJURP, CENP-A, and Cancer

Coordinated up-regulated expression of CENP-A and HJURP mRNA is observed in many cancers, including breast cancer, adenocarcinoma of the colon, gliomas and lipomas, and is a potentially powerful biomarker in several distinct types of cancers (Athwal et al. 2015; Dai et al. 2012; de Tayrac et al. 2011; Hu et al. 2010; Tomonaga et al. 2003; Valente et al. 2013; Wang et al. 2009). Misregulation of CENP-A deposition is a potential mechanism of generating genome instability, as even the missegregation of a single chromosome is sufficient to cause the rearrangement of that chromosome in a process called chromothripsis (Stephens et al. 2011; Zhang et al. 2015). This co-overexpression may be driven by the dysregulation of the transcription factor FoxM1 (Thiru et al. 2014); however, phenotypic consequences of CENP-A overexpression are beginning to emerge. Elevated CENP-A levels in human cells and other organisms result in mistargeting of CENP-A to ectopic sites and can lead to genomic instability (Choi et al. 2011, 2012; Gascoigne et al. 2011; Heun et al. 2006; Mendiburo et al. 2011; Mishra et al. 2011; Van Hooser et al. 2001). Overexpressed CENP-A can form a heterotypic nucleosome containing one copy of CENP-A and one copy of H3.3. The accumulation of CENP-A in the chromosome arms occurs by co-opting the H3.3 chaperone DAXX to mediate its mislocalization (Lacoste et al. 2014).

In budding yeast, the misregualtion of CENP-A^{Cse4} proteolysis results in accumulation of CENP-A^{Cse4} in gene promotors and causes altered gene expression (Hildebrand and Biggins 2016). Accumulation of CENP-A overexpressed in human cells has been observed in CTCF sites, DNAseI hypersensitive sites, and regions of nucleosome turnover, as well as at some key oncognene promoters (Athwal et al. 2015; Lacoste et al. 2014). The question remains whether the overexpression of CENP-A will drive genomic instability in tumors, and if so whether the presence of CENP-A at the ectopic sites alters the transcriptional profile of underlying genes, changes 3D chromatin arrangement due to disruption of CTCF sites, or weakens of endogenous centromere by the redistribution of centromere components is the key event.

9 Conclusions

Centromere specification is essential to ensure genome stability, as defects in centromere establishment can lead to errors in chromosome segregation during mitosis. Because centromeres of many higher eukaryotes are specified epigenetically by the presence of a unique nucleosome containing the histone variant CENP-A, the key step in centromere inheritance is the assembly of new CENP-A nucleosomes. Diverse organisms with unique strategies for centromere inheritance utilize the recruitment of a CENP-A-specific chaperone to ensure the perpetuation of centromere identity. In humans, the centromere deposition machinery is coupled to centromeric proteins that depend on CENP-A for their localization, thus creating an epigenetic mechanism for inheritance of the centromeric locus.

References

Allshire RC, Karpen GH (2008) Epigenetic regulation of centromeric chromatin: old dogs, new tricks? Nat Rev Genet 9:923–937

Amano M, Suzuki A, Hori T, Backer C, Okawa K, Cheeseman IM, Fukagawa T (2009) The CENP-S complex is essential for the stable assembly of outer kinetochore structure. J Cell Biol 186:173–182

An S, Kim H, Cho US (2015) Mis16 Independently Recognizes Histone H4 and the CENP-ACnp1-specific chaperone Scm3sp. J Mol Biol 427:3230–3240

Athwal RK, Walkiewicz MP, Baek S, Fu S, Bui M, Camps J, Ried T, Sung MH, Dalal Y (2015) CENP-A nucleosomes localize to transcription factor hotspots and subtelomeric sites in human cancer cells. Epigenetics Chromatin 8:2

Bailey AO, Panchenko T, Shabanowitz J, Lehman SM, Bai DL, Hunt DF, Black BE, Foltz DR (2016) Identification of the post-translational modifications present in centromeric chromatin. Mol cell proteomics 15:918–931

Barnhart MC, Kuich PH, Stellfox ME, Ward JA, Bassett EA, Black BE, Foltz DR (2011) HJURP is a CENP-A chromatin assembly factor sufficient to form a functional de novo kinetochore. J Cell Biol 194:229–243

Barnhart-Dailey MC, Trivedi P, Stukenberg PT, Foltz DR (2016) HJURP interaction with the condensin II complex during G1 promotes CENP-A deposition. Mol Biol Cell 28(1):54–64

Bassett EA, DeNizio J, Barnhart-Dailey MC, Panchenko T, Sekulic N, Rogers DJ, Foltz DR, Black BE (2012) HJURP uses distinct CENP-A surfaces to recognize and to stabilize CENP-A/histone H4 for centromere assembly. Dev Cell 22:749–762

Bergmann JH, Rodriguez MG, Martins NM, Kimura H, Kelly DA, Masumoto H, Larionov V, Jansen LE, Earnshaw WC (2011) Epigenetic engineering shows H3K4me2 is required for HJURP targeting and CENP-A assembly on a synthetic human kinetochore. EMBO J 30: 328–340

Bernad R, Sanchez P, Rivera T, Rodriguez-Corsino M, Boyarchuk E, Vassias I, Ray-Gallet D, Arnaoutov A, Dasso M, Almouzni G et al (2011) Xenopus HJURP and condensin II are required for CENP-A assembly. J Cell Biol 192:569–582

Black BE, Cleveland DW (2011) Epigenetic centromere propagation and the nature of CENP-A nucleosomes. Cell 144:471–479

Black BE, Brock MA, Bedard S, Woods VL Jr, Cleveland DW (2007) An epigenetic mark generated by the incorporation of CENP-A into centromeric nucleosomes. Proc Natl Acad Sci U S A 104:5008–5013

Blower MD, Sullivan BA, Karpen GH (2002) Conserved organization of centromeric chromatin in flies and humans. Dev Cell 2:319–330

Bodor DL, Mata JF, Sergeev M, David AF, Salimian KJ, Panchenko T, Cleveland DW, Black BE, Shah JV, Jansen LE (2014) The quantitative architecture of centromeric chromatin. Elife 3: e02137

Boltengagen M, Huang A, Boltengagen A, Trixl L, Lindner H, Kremser L, Offterdinger M, Lusser A (2016) A novel role for the histone acetyltransferase Hat1 in the CENP-A/CID assembly pathway in *Drosophila melanogaster*. Nucleic Acids Res 44:2145–2159

Bouzinba-Segard H, Guais A, Francastel C (2006) Accumulation of small murine minor satellite transcripts leads to impaired centromeric architecture and function. Proc Natl Acad Sci U S A 103:8709–8714

Boyarchuk E, Montes de Oca R, Almouzni G (2011) Cell cycle dynamics of histone variants at the centromere, a model for chromosomal landmarks. Curr Opin Cell Biol 23:266–276

Brown KD, Robertson KD (2007) DNMT1 knockout delivers a strong blow to genome stability and cell viability. Nat Genet 39:289–290

Burgess RJ, Zhang Z (2013) Histone chaperones in nucleosome assembly and human disease. Nat Struct Mol Biol 20:14–22

Camahort R, Li B, Florens L, Swanson SK, Washburn MP, Gerton JL (2007) Scm3 is essential to recruit the histone h3 variant cse4 to centromeres and to maintain a functional kinetochore. Mol Cell 26:853–865

Camahort R, Shivaraju M, Mattingly M, Li B, Nakanishi S, Zhu D, Shilatifard A, Workman JL, Gerton JL (2009) Cse4 is part of an octameric nucleosome in budding yeast. Mol Cell 35: 794–805

Cardozo T, Pagano M (2004) The SCF ubiquitin ligase: insights into a molecular machine. Nat Rev Mol Cell Biol 5:739–751

Carone DM, Longo MS, Ferreri GC, Hall L, Harris M, Shook N, Bulazel KV, Carone BR, Obergfell C, O'Neill MJ et al (2009) A new class of retroviral and satellite encoded small RNAs emanates from mammalian centromeres. Chromosoma 118:113–125

Carone DM, Zhang C, Hall LE, Obergfell C, Carone BR, O'Neill MJ, O'Neill RJ (2013) Hypermorphic expression of centromeric retroelement-encoded small RNAs impairs CENP-A loading. Chromosome Res 21:49–62

Carroll CW, Milks KJ, Straight AF (2010) Dual recognition of CENP-A nucleosomes is required for centromere assembly. J Cell Biol 189:1143–1155

Chan FL, Marshall OJ, Saffery R, Won Kim B, Earle E, Choo KH, Wong LH (2012) Active transcription and essential role of RNA polymerase II at the centromere during mitosis. Proc Natl Acad Sci U S A 109:1979–1984

Cheeseman IM, Desai A (2008) Molecular architecture of the kinetochore-microtubule interface. Nat Rev Mol Cell Biol 9:33–46

Chen CC, Dechassa ML, Bettini E, Ledoux MB, Belisario C, Heun P, Luger K, Mellone BG (2014) CAL1 is the drosophila CENP-A assembly factor. J Cell Biol 204:313–329

Chen CC, Bowers S, Lipinszki Z, Palladino J, Trusiak S, Bettini E, Rosin L, Przewloka MR, Glover DM, O'Neill RJ et al (2015) Establishment of centromeric chromatin by the CENP-A assembly factor CAL1 requires FACT-mediated transcription. Dev Cell 34:73–84

Cho US, Harrison SC (2011) Recognition of the centromere-specific histone Cse4 by the chaperone Scm3. Proc Natl Acad Sci U S A 108:9367–9371

Choi ES, Stralfors A, Castillo AG, Durand-Dubief M, Ekwall K, Allshire RC (2011) Identification of noncoding transcripts from within CENP-A chromatin at fission yeast centromeres. J Biol Chem 286:23600–23607

Choi ES, Stralfors A, Catania S, Castillo AG, Svensson JP, Pidoux AL, Ekwall K, Allshire RC (2012) Factors that promote H3 chromatin integrity during transcription prevent promiscuous deposition of CENP-A(Cnp1) in fission yeast. PLoS Genet 8:e1002985

Chow CM, Georgiou A, Szutorisz H, Maia e Silva A, Pombo A, Barahona I, Dargelos E, Canzonetta C, Dillon N (2005) Variant histone H3.3 marks promoters of transcriptionally active genes during mammalian cell division. EMBO Rep 6:354–360

Clarke L, Carbon J (1980) Isolation of a yeast centromere and construction of functional small circular chromosomes. Nature 287:504–509

Cleveland DW, Mao Y, Sullivan KF (2003) Centromeres and kinetochores: from epigenetics to mitotic checkpoint signaling. Cell 112:407–421

Dai Y, Tang Y, He F, Zhang Y, Cheng A, Gan R, Wu Y (2012) Screening and functional analysis of differentially expressed genes in EBV-transformed lymphoblasts. Virol J 9:77

Dambacher S, Deng W, Hahn M, Sadic D, Frohlich J, Nuber A, Hoischen C, Diekmann S, Leonhardt H, Schotta G (2012) CENP-C facilitates the recruitment of M18BP1 to centromeric chromatin. Nucleus 3:101–110

de Tayrac M, Aubry M, Saikali S, Etcheverry A, Surbled C, Guenot F, Galibert MD, Hamlat A, Lesimple T, Quillien V et al (2011) A 4-gene signature associated with clinical outcome in high-grade gliomas. Clin Cancer Res 17:317–327

Dechassa ML, Wyns K, Li M, Hall MA, Wang MD, Luger K (2011) Structure and Scm3-mediated assembly of budding yeast centromeric nucleosomes. Nat Commun 2:313

Doheny KF, Sorger PK, Hyman AA, Tugendreich S, Spencer F, Hieter P (1993) Identification of essential components of the *S. cerevisiae* kinetochore. Cell 73:761–774

Dunleavy EM, Pidoux AL, Monet M, Bonilla C, Richardson W, Hamilton GL, Ekwall K, McLaughlin PJ, Allshire RC (2007) A NASP (N1/N2)-related protein, Sim3, binds CENP-A and is required for its deposition at fission yeast centromeres. Mol Cell 28:1029–1044
Dunleavy EM, Roche D, Tagami H, Lacoste N, Ray-Gallet D, Nakamura Y, Daigo Y, Nakatani Y, Almouzni-Pettinotti G (2009) HJURP is a cell-cycle-dependent maintenance and deposition factor of CENP-A at centromeres. Cell 137:485–497
Dunleavy EM, Almouzni G, Karpen GH (2011) H3.3 is deposited at centromeres in S phase as a placeholder for newly assembled CENP-A in G(1) phase. Nucleus 2:146–157
Earnshaw W, Bordwell B, Marino C, Rothfield N (1986) Three human chromosomal autoantigens are recognized by sera from patients with anti-centromere antibodies. J Clin Invest 77:426–430
Erhardt S, Mellone BG, Betts CM, Zhang W, Karpen GH, Straight AF (2008) Genome-wide analysis reveals a cell cycle-dependent mechanism controlling centromere propagation. J Cell Biol 183:805–818
Eymery A, Callanan M, Vourc'h C (2009) The secret message of heterochromatin: new insights into the mechanisms and function of centromeric and pericentric repeat sequence transcription. Int J Dev Biol 53:259–268
Fachinetti DL, Logsdon GA, Abdullah A, Selzer EB, Cleveland DW, Black BE (2016) CENP-A modifications on Ser68 and Lys124 are dispensable for establishment, maintenance, and long-term function of human centromeres Dev Cell 40(1):104–113
Filipescu D, Muller S, Almouzni G (2014) Histone H3 variants and their chaperones during development and disease: contributing to epigenetic control. Annu Rev Cell Dev Biol 30: 615–646
Fitzgerald-Hayes M, Clarke L, Carbon J (1982) Nucleotide sequence comparisons and functional analysis of yeast centromere DNAs. Cell 29:235–244
Foltz DR, Jansen LE, Black BE, Bailey AO, Yates JR 3rd, Cleveland DW (2006) The human CENP-A centromeric nucleosome-associated complex. Nat Cell Biol 8:458–469
Foltz DR, Jansen LE, Bailey AO, Yates JR 3rd, Bassett EA, Wood S, Black BE, Cleveland DW (2009) Centromere-specific assembly of CENP-a nucleosomes is mediated by HJURP. Cell 137:472–484
Fujita Y, Hayashi T, Kiyomitsu T, Toyoda Y, Kokubu A, Obuse C, Yanagida M (2007) Priming of centromere for CENP-A recruitment by human hMis18alpha, hMis18beta, and M18BP1. Dev Cell 12:17–30
Gascoigne KE, Takeuchi K, Suzuki A, Hori T, Fukagawa M, Cheeseman IM (2011) Induced ectopic kinetochore assembly bypasses the requirement for CENP-A nucleosomes. Cell 145:410–422
Gasman S, Kalaidzidis Y, Zerial M (2003) RhoD regulates endosome dynamics through diaphanous-related formin and Src tyrosine kinase. Nat Cell Biol 5:195–204
Goh PY, Kilmartin JV (1993) NDC10: a gene involved in chromosome segregation in Saccharomyces cerevisiae. J Cell Biol 121:503–512
Goldberg AD, Banaszynski LA, Noh KM, Lewis PW, Elsaesser SJ, Stadler S, Dewell S, Law M, Guo X, Li X et al (2010) Distinct factors control histone variant H3.3 localization at specific genomic regions. Cell 140:678–691
Gopalakrishnan S, Sullivan BA, Trazzi S, Della Valle G, Robertson KD (2009) DNMT3B interacts with constitutive centromere protein CENP-C to modulate DNA methylation and the histone code at centromeric regions. Hum Mol Genet 18:3178–3193
Goshima G, Wollman R, Goodwin SS, Zhang N, Scholey JM, Vale RD, Stuurman N (2007) Genes required for mitotic spindle assembly in Drosophila S2 cells. Science 316:417–421
Guse A, Carroll CW, Moree B, Fuller CJ, Straight AF (2011) In vitro centromere and kinetochore assembly on defined chromatin templates. Nature 477:354–358
Hagstrom KA, Holmes VF, Cozzarelli NR, Meyer BJ (2002) *C. elegans* condensin promotes mitotic chromosome architecture, centromere organization, and sister chromatid segregation during mitosis and meiosis. Genes Dev 16:729–742
Hall LE, Mitchell SE, O'Neill RJ (2012) Pericentric and centromeric transcription: a perfect balance required. Chromosome Res 20:535–546

Hayashi T, Fujita Y, Iwasaki O, Adachi Y, Takahashi K, Yanagida M (2004) Mis16 and Mis18 are required for CENP-A loading and histone deacetylation at centromeres. Cell 118:715–729
Hayashi T, Ebe M, Nagao K, Kokubu A, Sajiki K, Yanagida M (2014) Schizosaccharomyces pombe centromere protein Mis19 links Mis16 and Mis18 to recruit CENP-A through interacting with NMD factors and the SWI/SNF complex. Genes Cells 19:541–554
Heun P, Erhardt S, Blower MD, Weiss S, Skora AD, Karpen GH (2006) Mislocalization of the Drosophila centromere-specific histone CID promotes formation of functional ectopic kinetochores. Dev Cell 10:303–315
Hewawasam G, Shivaraju M, Mattingly M, Venkatesh S, Martin-Brown S, Florens L, Workman JL, Gerton JL (2010) Psh1 is an E3 ubiquitin ligase that targets the centromeric histone variant Cse4. Mol Cell 40:444–454
Hildebrand EM, Biggins S (2016) Regulation of budding yeast CENP-A levels prevents misincorporation at promoter nucleosomes and transcriptional defects. PLoS Genet 12: e1005930
Hirano T (2005) Condensins: organizing and segregating the genome. Curr Biol 15:R265–R275
Hori T, Shang WH, Takeuchi K, Fukagawa T (2013) The CCAN recruits CENP-A to the centromere and forms the structural core for kinetochore assembly. J Cell Biol 200:45–60
Hu Z, Huang G, Sadanandam A, Gu S, Lenburg ME, Pai M, Bayani N, Blakely EA, Gray JW, Mao JH (2010) The expression level of HJURP has an independent prognostic impact and predicts the sensitivity to radiotherapy in breast cancer. Breast Cancer Res 12:R18
Hu H, Liu Y, Wang M, Fang J, Huang H, Yang N, Li Y, Wang J, Yao X, Shi Y et al (2011) Structure of a CENP-A-histone H4 heterodimer in complex with chaperone HJURP. Genes Dev 25:901–906
Hyman AA, Middleton K, Centola M, Mitchison TJ, Carbon J (1992) Microtubule-motor activity of a yeast centromere-binding protein complex. Nature 359:533–536
Izuta H, Ikeno M, Suzuki N, Tomonaga T, Nozaki N, Obuse C, Kisu Y, Goshima N, Nomura F, Nomura N et al (2006) Comprehensive analysis of the ICEN (Interphase Centromere Complex) components enriched in the CENP-A chromatin of human cells. Genes Cells 11:673–684
Jansen LE, Black BE, Foltz DR, Cleveland DW (2007) Propagation of centromeric chromatin requires exit from mitosis. J Cell Biol 176:795–805
Jiang W, Lechner J, Carbon J (1993) Isolation and characterization of a gene (CBF2) specifying a protein component of the budding yeast kinetochore. J Cell Biol 121:513–519
Kato H, Jiang J, Zhou BR, Rozendaal M, Feng H, Ghirlando R, Xiao TS, Straight AF, Bai Y (2013) A conserved mechanism for centromeric nucleosome recognition by centromere protein CENP-C. Science 340:1110–1113
Kim IS, Lee M, Park KC, Jeon Y, Park JH, Hwang EJ, Jeon TI, Ko S, Lee H, Baek SH et al (2012) Roles of Mis18α in epigenetic regulation of centromeric chromatin and CENP-A loading. Mol Cell 46(3):260–273
Kim IS, Lee M, Park JH, Jeon R, Baek SH, Kim KI (2014) βTrCP-mediated ubiquitylation regulates protein stability of Mis18β in a cell cycle-dependent manner. Biochem Biophys Res Commun 443:62–67
Lacoste N, Woolfe A, Tachiwana H, Garea AV, Barth T, Cantaloube S, Kurumizaka H, Imhof A, Almouzni G (2014) Mislocalization of the centromeric histone variant CenH3/CENP-A in human cells depends on the chaperone DAXX. Mol Cell 53:631–644
Lagana A, Dorn JF, De Rop V, Ladouceur AM, Maddox AS, Maddox PS (2010) A small GTPase molecular switch regulates epigenetic centromere maintenance by stabilizing newly incorporated CENP-A. Nat Cell Biol 12:1186–1193
Lam AL, Boivin CD, Bonney CF, Rudd MK, Sullivan BA (2006) Human centromeric chromatin is a dynamic chromosomal domain that can spread over noncentromeric DNA. Proc Natl Acad Sci U S A 103:4186–4191
Lammers M, Meyer S, Kuhlmann D, Wittinghofer A (2008) Specificity of interactions between mDia isoforms and Rho proteins. J Biol Chem 283:35236–35246
Lechner J, Carbon J (1991) A 240 kd multisubunit protein complex, CBF3, is a major component of the budding yeast centromere. Cell 64:717–725

Lee J, Zhou P (2007) DCAFs, the missing link of the CUL4-DDB1 ubiquitin ligase. Mol Cell 26:775–780
LeRoy G, Orphanides G, Lane WS, Reinberg D (1998) Requirement of RSF and FACT for transcription of chromatin templates in vitro. Science 282:1900–1904
Lewis PW, Elsaesser SJ, Noh KM, Stadler SC, Allis CD (2010) Daxx is an H3.3-specific histone chaperone and cooperates with ATRX in replication-independent chromatin assembly at telomeres. Proc Nat Acad Sci U S A 107:14075–14080
Liu, C., and Mao, Y. (2016). Formin-mediated epigenetic maintenance of centromere identity. Small GTPases, 1–6
Loyola A, Almouzni G (2004) Histone chaperones, a supporting role in the limelight. Biochem Biophys Acta 1677:3–11
Maddox PS, Hyndman F, Monen J, Oegema K, Desai A (2007) Functional genomics identifies a Myb domain-containing protein family required for assembly of CENP-A chromatin. J Cell Biol 176:757–763
May BP, Lippman ZB, Fang Y, Spector DL, Martienssen RA (2005) Differential regulation of strand-specific transcripts from Arabidopsis centromeric satellite repeats. PLoS Genet 1:e79
McKinley KL, Cheeseman IM (2014) Polo-like kinase 1 licenses CENP-A deposition at centromeres. Cell 158:397–411
McKinley KL, Cheeseman IM (2016) The molecular basis for centromere identity and function. Nat Rev Mol Cell Biol 17:16–29
Mellone BG, Grive KJ, Shteyn V, Bowers SR, Oderberg I, Karpen GH (2011) Assembly of Drosophila centromeric chromatin proteins during mitosis. PLoS Genet 7:e1002068
Meluh PB, Koshland D (1995) Evidence that the MIF2 gene of *Saccharomyces cerevisiae* encodes a centromere protein with homology to the mammalian centromere protein CENP-C. Mol Biol Cell 6:793–807
Meluh PB, Koshland D (1997) Budding yeast centromere composition and assembly as revealed by in vivo cross-linking. Genes Dev 11:3401–3412
Mendiburo MJ, Padeken J, Fulop S, Schepers A, Heun P (2011) Drosophila CENH3 is sufficient for centromere formation. Science 334:686–690
Mishra PK, Au WC, Choy JS, Kuich PH, Baker RE, Foltz DR, Basrai MA (2011) Misregulation of Scm3p/HJURP causes chromosome instability in *Saccharomyces cerevisiae* and human cells. PLoS Genet 7:e1002303
Mito Y, Henikoff JG, Henikoff S (2005) Genome-scale profiling of histone H3.3 replacement patterns. Nat Genet 37:1090–1097
Mizuguchi G, Shen X, Landry J, Wu WH, Sen S, Wu C (2004) ATP-driven exchange of histone H2AZ variant catalyzed by SWR1 chromatin remodeling complex. Science 303:343–348
Mizuguchi G, Xiao H, Wisniewski J, Smith MM, Wu C (2007) Nonhistone Scm3 and histones CenH3-H4 assemble the core of centromere-specific nucleosomes. Cell 129:1153–1164
Moree B, Meyer CB, Fuller CJ, Straight AF (2011) CENP-C recruits M18BP1 to centromeres to promote CENP-A chromatin assembly. J Cell Biol 194:855–871
Moreno-Moreno O, Medina-Giro S, Torras-Llort M, Azorin F (2011) The F box protein partner of paired regulates stability of Drosophila centromeric histone H3, CenH3(CID). Curr Biol 21:1488–1493
Mouysset J, Gilberto S, Meier MG, Lampert F, Belwal M, Meraldi P, Peter M (2015) CRL4 (RBBP7) is required for efficient CENP-A deposition at centromeres. J Cell Sci 128: 1732–1745
Muller S, Montes de Oca R, Lacoste N, Dingli F, Loew D, Almouzni G (2014) Phosphorylation and DNA binding of HJURP determine its centromeric recruitment and function in CenH3 (CENP-A) loading. Cell Reports 8:190–203
Nakayama KI, Nakayama K (2006) Ubiquitin ligases: cell-cycle control and cancer. Nat Rev Cancer 6:369–381
Nardi IK, Zasadzinska E, Stellfox ME, Knippler CM, Foltz DR (2016) Licensing of centromeric chromatin assembly through the Mis18alpha-Mis18beta heterotetramer. Mol Cell 61:774–787

Niikura Y, Kitagawa R, Ogi H, Abdulle R, Pagala V, Kitagawa K (2015) CENP-A K124 ubiquitylation is required for CENP-A deposition at the centromere. Dev Cell 32:589–603
Nishihashi A, Haraguchi T, Hiraoka Y, Ikemura T, Regnier V, Dodson H, Earnshaw WC, Fukagawa T (2002) CENP-I is essential for centromere function in vertebrate cells. Dev Cell 2:463–476
Obuse C, Yang H, Nozaki N, Goto S, Okazaki T, Yoda K (2004) Proteomics analysis of the centromere complex from HeLa interphase cells: UV-damaged DNA binding protein 1 (DDB-1) is a component of the CEN-complex, while BMI-1 is transiently co-localized with the centromeric region in interphase. Genes Cells 9:105–120
Ohkuni K, Kitagawa K (2011) Endogenous transcription at the centromere facilitates centromere activity in budding yeast. Curr Biol 21:1695–1703
Ohzeki JI, Bergmann JH, Kouprina N, Noskov VN, Nakano M, Kimura H, Earnshaw WC, Larionov V, Masumoto H (2012) Breaking the HAC barrier: histone H3K9 acetyl/methyl balance regulates CENP-A assembly. EMBO J 31:2391–2402
Ohzeki J, Shono N, Otake K, Martins NM, Kugou K, Kimura H, Nagase T, Larionov V, Earnshaw WC, Masumoto H (2016) KAT7/HBO1/MYST2 regulates CENP-A chromatin assembly by antagonizing Suv39h1-mediated centromere inactivation. Dev Cell 37:413–427
Okada M, Cheeseman IM, Hori T, Okawa K, McLeod IX, Yates JR 3rd, Desai A, Fukagawa T (2006) The CENP-H-I complex is required for the efficient incorporation of newly synthesized CENP-A into centromeres. Nat Cell Biol 8:446–457
Okada M, Okawa K, Isobe T, Fukagawa T (2009) CENP-H-containing complex facilitates centromere deposition of CENP-A in cooperation with FACT and CHD1. Mol Biol Cell 20:3986–3995
Ono T, Fang Y, Spector DL, Hirano T (2004) Spatial and temporal regulation of Condensins I and II in mitotic chromosome assembly in human cells. Mol Biol Cell 15:3296–3308
Orphanides G, LeRoy G, Chang CH, Luse DS, Reinberg D (1998) FACT, a factor that facilitates transcript elongation through nucleosomes. Cell 92:105–116
Pearson CG, Yeh E, Gardner M, Odde D, Salmon ED, Bloom K (2004) Stable kinetochore-microtubule attachment constrains centromere positioning in metaphase. Curr Biol 14:1962–1967
Perpelescu M, Nozaki N, Obuse C, Yang H, Yoda K (2009) Active establishment of centromeric CENP-A chromatin by RSF complex. J Cell Biol 185:397–407
Peters AH, O'Carroll D, Scherthan H, Mechtler K, Sauer S, Schofer C, Weipoltshammer K, Pagani M, Lachner M, Kohlmaier A et al (2001) Loss of the Suv39h histone methyltransferases impairs mammalian heterochromatin and genome stability. Cell 107:323–337
Phansalkar R, Lapierre P, Mellone BG (2012) Evolutionary insights into the role of the essential centromere protein CAL1 in Drosophila. Chromosome Res 20:493–504
Pidoux AL, Choi ES, Abbott JK, Liu X, Kagansky A, Castillo AG, Hamilton GL, Richardson W, Rappsilber J, He X et al (2009) Fission yeast Scm3: a CENP-A receptor required for integrity of subkinetochore chromatin. Mol Cell 33:299–311
Quenet D, Dalal Y (2014a) A long non-coding RNA is required for targeting centromeric protein A to the human centromere. Elife 3:e03254
Quenet D, Dalal Y (2014b) A long non-coding RNA is required for targeting centromeric protein A to the human centromere. Elife 3:e03254
Quénet D, Dalal Y (2012) The CENP-A nucleosome: a dynamic structure and role at the centromere. Chromosome Res 20(5):465–479
Ranjitkar P, Press MO, Yi X, Baker R, MacCoss MJ, Biggins S (2010) An E3 ubiquitin ligase prevents ectopic localization of the centromeric histone H3 variant via the centromere targeting domain. Mol Cell 40:455–464
Ransom M, Dennehey BK, Tyler JK (2010) Chaperoning histones during DNA replication and repair. Cell 140:183–195
Ribeiro SA, Vagnarelli P, Dong Y, Hori T, McEwen BF, Fukagawa T, Flors C, Earnshaw WC (2010) A super-resolution map of the vertebrate kinetochore. Proc Natl Acad Sci U S A 107:10484–10489

Rice JC, Briggs SD, Ueberheide B, Barber CM, Shabanowitz J, Hunt DF, Shinkai Y, Allis CD (2003) Histone methyltransferases direct different degrees of methylation to define distinct chromatin domains. Mol Cell 12:1591–1598
Rosic S, Kohler F, Erhardt S (2014) Repetitive centromeric satellite RNA is essential for kinetochore formation and cell division. J Cell Biol 207:335–349
Ross JE, Woodlief KS, Sullivan BA (2016) Inheritance of the CENP-A chromatin domain is spatially and temporally constrained at human centromeres. Epigenetics Chromatin 9:20
Russell ID, Grancell AS, Sorger PK (1999) The unstable F-box protein p58-Ctf13 forms the structural core of the CBF3 kinetochore complex. J Cell Biol 145:933–950
Saitoh H, Tomkiel J, Cooke CA, Ratrie H 3rd, Maurer M, Rothfield NF, Earnshaw WC (1992) CENP-C, an autoantigen in scleroderma, is a component of the human inner kinetochore plate. Cell 70:115–125
Samoshkin A, Arnaoutov A, Jansen LE, Ouspenski I, Dye L, Karpova T, McNally J, Dasso M, Cleveland DW, Strunnikov A (2009) Human condensin function is essential for centromeric chromatin assembly and proper sister kinetochore orientation. PLoS ONE 4:e6831
Sanchez-Pulido L, Pidoux AL, Ponting CP, Allshire RC (2009) Common ancestry of the CENP-A chaperones Scm3 and HJURP. Cell 137:1173–1174
Sarma K, Reinberg D (2005) Histone variants meet their match. Nat Rev Mol Cell Biol 6:139–149
Schittenhelm RB, Althoff F, Heidmann S, Lehner CF (2010) Detrimental incorporation of excess Cenp-A/Cid and Cenp-C into Drosophila centromeres is prevented by limiting amounts of the bridging factor Cal1. J Cell Sci 123:3768–3779
Schuh M, Lehner CF, Heidmann S (2007) Incorporation of Drosophila CID/CENP-A and CENP-C into centromeres during early embryonic anaphase. Curr Biol 17:237–243
Scott KC, Sullivan BA (2013) Neocentromeres: a place for everything and everything in its place. Trends Genet 30(2):66–74
Shang WH, Hori T, Toyoda A, Kato J, Popendorf K, Sakakibara Y, Fujiyama A, Fukagawa T (2010) Chickens possess centromeres with both extended tandem repeats and short non-tandem-repetitive sequences. Genome Res 20:1219–1228
Shang WH, Hori T, Westhorpe FG, Godek KM, Toyoda A, Misu S, Monma N, Ikeo K, Carroll CW, Takami Y et al (2016) Acetylation of histone H4 lysine 5 and 12 is required for CENP-A deposition into centromeres. Nat Commun 7:13465
Shivaraju M, Camahort R, Mattingly M, Gerton JL (2011) Scm3 is a centromeric nucleosome assembly factor. J Biol Chem 286:12016–12023
Shuaib M, Ouararhni K, Dimitrov S, Hamiche A (2010) HJURP binds CENP-A via a highly conserved N-terminal domain and mediates its deposition at centromeres. Proc Nat Acad Sci U S A 107:1349–1354
Silva MC, Bodor DL, Stellfox ME, Martins NM, Hochegger H, Foltz DR, Jansen LE (2012) Cdk activity couples epigenetic centromere inheritance to cell cycle progression. Dev Cell 22:52–63
Sorger PK, Doheny KF, Hieter P, Kopski KM, Huffaker TC, Hyman AA (1995) Two genes required for the binding of an essential Saccharomyces cerevisiae kinetochore complex to DNA. Proc Natl Acad Sci U S A 92:12026–12030
Stellfox ME, Nardi IK, Knippler CM, Foltz DR (2016) Differential binding partners of the Mis18alpha/beta YIPPEE domains regulate Mis18 complex recruitment to centromeres. Cell Rep 15:2127–2135
Stephens PJ, Greenman CD, Fu B, Yang F, Bignell GR, Mudie LJ, Pleasance ED, Lau KW, Beare D, Stebbings LA et al (2011) Massive genomic rearrangement acquired in a single catastrophic event during cancer development. Cell 144:27–40
Stimpson KM, Sullivan BA (2010) Epigenomics of centromere assembly and function. Curr Opin Cell Biol 22:772–780
Stoler S, Rogers K, Weitze S, Morey L, Fitzgerald-Hayes M, Baker RE (2007) Scm3, an essential *Saccharomyces cerevisiae* centromere protein required for G2/M progression and Cse4 localization. Proc Natl Acad Sci U S A 104:10571–10576
Strunnikov AV, Kingsbury J, Koshland D (1995) CEP3 encodes a centromere protein of *Saccharomyces cerevisiae*. J Cell Biol 128:749–760

Subramanian L, Toda NR, Rappsilber J, Allshire RC (2014) Eic1 links Mis18 with the CCAN/Mis6/Ctf19 complex to promote CENP-A assembly. Open Biology 4:140043

Subramanian L, Medina-Pritchard B, Barton R, Spiller F, Kulasegaran-Shylini R, Radaviciute G, Allshire RC, Jeyaprakash AA (2016) Centromere localization and function of Mis18 requires Yippee-like domain-mediated oligomerization. EMBO Rep 17(4):496–507

Sugata N, Munekata E, Todokoro K (1999) Characterization of a novel kinetochore protein, CENP-H. J Biol Chem 274:27343–27346

Sullivan BA, Karpen GH (2004) Centromeric chromatin exhibits a histone modification pattern that is distinct from both euchromatin and heterochromatin. Nat Struct Mol Biol 11:1076–1083

Szenker E, Ray-Gallet D, Almouzni G (2011) The double face of the histone variant H3.3. Cell Res 21:421–434

Tagami H, Ray-Gallet D, Almouzni G, Nakatani Y (2004) Histone H3.1 and H3.3 complexes mediate nucleosome assembly pathways dependent or independent of DNA synthesis. Cell 116:51–61

Takahashi K, Chen ES, Yanagida M (2000) Requirement of Mis6 centromere connector for localizing a CENP-A-like protein in fission yeast. Science 288:2215–2219

Takayama Y, Sato H, Saitoh S, Ogiyama Y, Masuda F, Takahashi K (2008) Biphasic incorporation of centromeric histone CENP-A in fission yeast. Mol Biol Cell 19:682–690

Tamura T, Smith M, Kanno T, Dasenbrock H, Nishiyama A, Ozato K (2009) Inducible deposition of the histone variant H3.3 in interferon-stimulated genes. J Biol Chem 284:12217–12225

Thiru P, Kern DM, McKinley KL, Monda JK, Rago F, Su KC, Tsinman T, Yarar D, Bell GW, Cheeseman IM (2014) Kinetochore genes are coordinately up-regulated in human tumors as part of a FoxM1-related cell division program. Mol Biol Cell 25:1983–1994

Tomonaga T, Matsushita K, Yamaguchi S, Oohashi T, Shimada H, Ochiai T, Yoda K, Nomura F (2003) Overexpression and mistargeting of centromere protein-A in human primary colorectal cancer. Can Res 63:3511–3516

Topp CN, Zhong CX, Dawe RK (2004) Centromere-encoded RNAs are integral components of the maize kinetochore. Proc Natl Acad Sci U S A 101:15986–15991

Tyler JK, Adams CR, Chen SR, Kobayashi R, Kamakaka RT, Kadonaga JT (1999) The RCAF complex mediates chromatin assembly during DNA replication and repair. Nature 402: 555–560

Tyler JK, Collins KA, Prasad-Sinha J, Amiott E, Bulger M, Harte PJ, Kobayashi R, Kadonaga JT (2001) Interaction between the Drosophila CAF-1 and ASF1 chromatin assembly factors. Mol Cell Biol 21:6574–6584

Valente V, Serafim RB, de Oliveira LC, Adorni FS, Torrieri R, Tirapelli DP, Espreafico EM, Oba-Shinjo SM, Marie SK, Paco-Larson ML et al (2013) Modulation of HJURP (Holliday Junction-Recognizing Protein) levels is correlated with glioblastoma cells survival. PLoS ONE 8:e62200

Van Hooser AA, Ouspenski II, Gregson HC, Starr DA, Yen TJ, Goldberg ML, Yokomori K, Earnshaw WC, Sullivan KF, Brinkley BR (2001) Specification of kinetochore-forming chromatin by the histone H3 variant CENP-A. J Cell Sci 114:3529–3542

Wade CM, Giulotto E, Sigurdsson S, Zoli M, Gnerre S, Imsland F, Lear TL, Adelson DL, Bailey E, Bellone RR et al (2009) Genome sequence, comparative analysis, and population genetics of the domestic horse. Science 326:865–867

Wang L, Tang H, Thayanithy V, Subramanian S, Oberg AL, Cunningham JM, Cerhan JR, Steer CJ, Thibodeau SN (2009) Gene networks and microRNAs implicated in aggressive prostate cancer. Can Res 69:9490–9497

Wang J, Liu X, Dou Z, Chen L, Jiang H, Fu C, Fu G, Liu D, Zhang J, Zhu T et al (2014) Mitotic regulator Mis18beta interacts with and specifies the centromeric assembly of molecular chaperone holliday junction recognition protein (HJURP). J Biol Chem 289:8326–8336

Weber CM, Henikoff S (2014) Histone variants: dynamic punctuation in transcription. Genes Dev 28:672–682

Westermann S, Cheeseman IM, Anderson S, Yates JR 3rd, Drubin DG, Barnes G (2003) Architecture of the budding yeast kinetochore reveals a conserved molecular core. J Cell Biol 163:215–222

Wignall SM, Deehan R, Maresca TJ, Heald R (2003) The condensin complex is required for proper spindle assembly and chromosome segregation in Xenopus egg extracts. J Cell Biol 161:1041–1051

Williams JS, Hayashi T, Yanagida M, Russell P (2009) Fission yeast Scm3 mediates stable assembly of Cnp1/CENP-A into centromeric chromatin. Mol Cell 33:287–298

Wisniewski J, Hajj B, Chen J, Mizuguchi G, Xiao H, Wei D, Dahan M, Wu C (2014) Imaging the fate of histone Cse4 reveals de novo replacement in S phase and subsequent stable residence at centromeres. Elife 3:e02203

Wong LH, Brettingham-Moore KH, Chan L, Quach JM, Anderson MA, Northrop EL, Hannan R, Saffery R, Shaw ML, Williams E et al (2007) Centromere RNA is a key component for the assembly of nucleoproteins at the nucleolus and centromere. Genome Res 17:1146–1160

Xiao H, Mizuguchi G, Wisniewski J, Huang Y, Wei D, Wu C (2011) Nonhistone Scm3 binds to AT-rich DNA to organize atypical centromeric nucleosome of budding yeast. Mol Cell 43:369–380

Yong-Gonzalez V, Wang BD, Butylin P, Ouspenski I, Strunnikov A (2007) Condensin function at centromere chromatin facilitates proper kinetochore tension and ensures correct mitotic segregation of sister chromatids. Genes Cells 12:1075–1090

Yu Z, Zhou X, Wang W, Deng W, Fang J, Hu H, Wang Z, Li S, Cui L, Shen J et al (2015) Dynamic phosphorylation of CENP-A at Ser68 orchestrates its cell-cycle-dependent deposition at centromeres. Dev Cell 32:68–81

Zasadzinska E, Barnhart-Dailey MC, Kuich PH, Foltz DR (2013) Dimerization of the CENP-A assembly factor HJURP is required for centromeric nucleosome deposition. EMBO J 32: 2113–2124

Zhang CZ, Spektor A, Cornils H, Francis JM, Jackson EK, Liu S, Meyerson M, Pellman D (2015) Chromothripsis from DNA damage in micronuclei. Nature 522:179–184

Artificial Chromosomes and Strategies to Initiate Epigenetic Centromere Establishment

Evelyne J. Barrey and Patrick Heun

Abstract In recent years, various synthetic approaches have been developed to address the question of what directs centromere establishment and maintenance. In this chapter, we will discuss how approaches aimed at constructing synthetic centromeres have co-evolved with and contributed to shape the theory describing the determinants of centromere identity. We will first review lessons learned from artificial chromosomes created from "naked" centromeric sequences to investigate the role of the underlying DNA for centromere formation. We will then discuss how several studies, which applied removal of endogenous centromeres or over-expression of the centromere-specific histone CENP-A, helped to investigate the contribution of chromatin context to centromere establishment. Finally, we will examine various biosynthetic approaches taking advantage of targeting specific proteins to ectopic sites in the genome to dissect the role of many centromere-associated proteins and chromatin modifiers for centromere inheritance and function. Together, these studies showed that chromatin context matters, particularly proximity to heterochromatin or repetitive DNA sequences. Moreover, despite the important contribution of centromeric DNA, the centromere-specific histone H3-variant CENP-A emerges as a key epigenetic mark to establish and maintain functional centromeres on artificial chromosomes or at ectopic sites of the genome.

1 Introduction

A fundamental task during cell division is to segregate the sister chromosomes faithfully to two daughter cells. This is mediated by a highly specialized region of the chromosome called the centromere, which needs to fulfill two critical functions: first, building a kinetochore, a large multi-protein complex that serves to attach the chromosomes to spindle microtubules in mitosis or meiosis; and second, ensuring

E.J. Barrey · P. Heun (✉)
Wellcome Trust Centre for Cell Biology, The University of Edinburgh, Edinburgh, UK
e-mail: patrick.heun@ed.ac.uk

B.E. Black (ed.), *Centromeres and Kinetochores*, Progress in Molecular and Subcellular Biology 56, DOI 10.1007/978-3-319-58592-5_8

that its identity is passed down through generations. Failure to do so correctly gives rise to cells with an abnormal chromosome number or rearranged chromosomes, which can promote tumor formation, miscarriage, reduced fertility and developmental issues (Hassold and Hunt 2001; Weaver and Cleveland 2007).

Centromeres consist of a complex and dynamic network of components. This includes the underlying DNA sequence and a specific type of centric chromatin that in most eukaryotes contains the centromere-specific histone H3 variant CENP-A in a subset of nucleosomes and a unique mix of histone modifications. The chromatin is further bound by centromere factors that display either constitutive or cell cycle-dependent binding. Recent findings add to the complexity by providing evidence that maintenance and proper function of centromeres also depends on general processes like transcription, chromatin remodeling, and DNA replication.

Although a complete picture of the mechanisms how these various components determine and propagate centromere identity throughout cell division is still missing, considerable progress has been made in the last decades, in large part thanks to biosynthetic approaches. To tease apart the role of all the different players and reduce complexity, many groups have employed a bottom-up approach, consisting in artificially building a new centromere from a selected number of factors, namely defined DNA sequences and centromere-associated proteins. How approaches aimed at constructing synthetic centromeres have co-evolved with and contributed to shape the theory that centromeres in most eukaryotes are not genetically but rather epigenetically defined is the focus of this review.

We start by discussing how the first neocentromeres were created using "naked" centromeric sequences to investigate the role of the underlying DNA in centromere identity. We then review studies that aimed to understand the contribution of chromatin context to centromere establishment by supporting the formation of neocentromeres onto existing chromosomes in an unbiased way. This was achieved either by induced deletion of an endogenous centromere or by over-expression of CENP-A. Finally, we examine various biosynthetic approaches taking advantage of targeting specific proteins to ectopic sites in the genome. This helped dissecting the role of many centromere factors and chromatin modifiers for centromere inheritance and function, with the CENP-A nucleosome emerging as a key centromere mark.

2 Artificial Chromosomes Engineered from Naked Centromeric DNA

The first artificial chromosomes carrying a functional centromere were generated in budding yeast by introducing a centromeric DNA element into a circular vector containing a replication origin (Clarke and Carbon 1980). Subsequently, the first linear minichromosome comprised centromeric DNA, a selectable marker and the Autonomous Replicating Sequence (ARS), to which rDNA repeats containing telomere structures were added (Murray and Szostak 1983). These yeast artificial chromosomes (YACs) were faithfully segregated, and therefore maintained stably through cell divisions. Inducing long-term centromere establishment on artificial chromosomes was first achieved in budding yeast as, in this model system, centromeres are encoded by a short DNA sequence of 125 bp comprising three elements, termed the conserved DNA elements (CDE I, II and III). The budding yeast point centromere is therefore genetically defined unlike the regional centromeres found in most eukaryotes, including fission yeast, flies and humans, as will be discussed further below.

Fission yeast centromeres are composed of a unique central core sequence surrounded by repetitive sequences, which assemble heterochromatin marked by histone H3 dimethylated on Lysine 9 (H3K9me2). To successfully build a functional centromere on transformed naked DNA, both the central core domain and the outer repeats are required (Hahnenberger et al. 1989). It was later found that heterochromatin integrity is essential for establishment of fission yeast centromeres but dispensable for their maintenance. Once the minichromosomes had assembled CENP-A^{cnp1} on the central core domain, they could be propagated in absence of flanking heterochromatin (Folco et al. 2008). It was also hypothesized that pericentric heterochromatin may play a role in ensuring robust cohesion between sister chromatids (Bernard et al. 2001; Nonaka et al. 2002).

Human centromeres have an analogous structure to fission yeast centromeres composed of centromeric chromatin flanked by pericentromeric heterochromatin, but are in general larger and can cover up to five megabases. The core centromere contains an array of higher order repeats of 171 bp α-satellite monomer. Higher order α-satellite DNA sequences from different chromosomes differ slightly, but all chromosomes, apart from the Y-chromosome, contain a 17 bp motif termed CENP-B box, which is specifically recognized and bound by the centromeric protein CENP-B. In contrast, so-called monomeric α-satellite DNA that is found in the pericentromeric region varies greatly in sequence and lack higher order organization (Schueler and Sullivan 2006).

The first generation of Human Artificial Chromosomes (HACs) resulted from cloning an extended array of higher order alphoid DNA containing CENP-B boxes from chromosome 17 into a BAC vector and co-transfecting this BAC into the HT1080 fibrosarcoma cell line together with telomeric sequences and genomic DNA (Harrington et al. 1997). The generated HACs were multimers of several megabases, bound centromere proteins specific for active centromeres and proved to be mitotically stable for up to 6 months in the absence of selection. A second type of HACs was built by introducing monomeric or higher order alphoid DNA from chromosome 21 and a selectable marker into a YAC vector in a recombination-deficient yeast strain. After purification of the modified YAC from yeast cells and transfection into HT1080 cells, multimerization of the introduced DNA led to formation of megabase-sized HACs that were capable of binding CENP-A, CENP-B, CENP-C, and CENP-E without acquiring host DNA. Once again these HACs were stably maintained even in the absence of selection (Ikeno et al. 1998; Masumoto et al. 1998). Importantly, HAC formation was successful only with higher order-DNA-containing YACs but never with the YAC vectors containing monomeric alphoid DNA, suggesting that CENP-B boxes are required for efficient centromere establishment. To further investigate the role of CENP-B boxes in HAC formation, another generation of minichromosomes was developed by mutating the CENP-B boxes comprised in the higher order alphoid array of chromosome 21, so that CENP-B binding is lost. These mutated arrays were introduced into HT1080 cells and assessed for HAC formation. It was concluded that centromere formation can only occur on α-satellite DNA containing functional CENP-B boxes. However, synthetic non-alphoid arrays of DNA containing the CENP-B box were not able to form HACs, suggesting that also other features of higher order alphoid DNA are crucial for centromere establishment (Ohzeki et al. 2002). A separate approach using small synthetic chromosome 17-derived alphoid arrays (16 monomers) with varying CENP-B box density in HT1080 cells confirmed that the CENP-B box is necessary for de novo centromere formation on the input DNA. In agreement with this, increasing the density of CENP-B boxes on alphoid DNA enhances the efficiency of HAC formation (Basu et al. 2005), while decreasing CENP-B boxes density impaired the recruitment of CENP-A and negatively impacted the efficiency of HAC formation.

As large genomic tandem repeat DNA sequences, such as α-satellites, are not entirely uniform and therefore difficult to investigate, the need to build large well-defined synthetic alphoid DNA arrays became obvious. This technical challenge was solved by Larionov and colleagues (Ebersole et al. 2005), who combined rolling circle amplification (RCA) of short defined DNA multimers followed by assembly of the fragment using in vivo homologous recombination in yeast to construct a circular alphoid DNA array of a defined structure of up to 120 kb (Ebersole et al. 2005). This allowed to compare alphoid DNA arrays of various lengths containing wild-type or mutated CENP-B boxes for their ability to promote HAC formation. It was shown that at least 30 kb of alphoid DNA are required to stably maintain CENP-A chromatin and support HAC formation, while fragments larger than 70 kb were surprisingly much less efficient to do so (Okamoto et al. 2007). Constructs engineered with mutant CENP-B boxes did not form a functional centromere, in agreement with findings described above.

The observation that CENP-B boxes are a requirement for functional centromere establishment came as a surprise because the CENP-B gene had been found to be non-essential in mice (Hudson et al. 1998; Perez-Castro et al. 1998; Kapoor et al. 1998). Thus, it was suggested that CENP-B might play a role in establishment rather than in maintenance of centromeres or that its function might be redundant. To further test the involvement of CENP-B in centromere formation, BAC constructs comprising wild-type or CENP-B box-mutated alphoid DNA were transfected into mouse embryonic fibroblasts in the presence or absence of the CENP-B protein and HAC formation was assessed. De novo centromere formation was dependent on both intact CENP-B boxes and expression of the CENP-B protein. When the introduced alphoid DNA did not successfully yield a HAC, an alternative fate was its integration into the genome. In this scenario, CENP-A assembly did not occur and CENP-B was found to induce heterochromatin formation at the chromosomally integrated alphoid array. This suggested that CENP-B exerts a dual antagonistic function promoting centromere formation on DNA that lacks an active centromere, but preventing extra-centromere formation on chromosomes carrying an active centromere (Okada et al. 2007). To date, this is the only DNA sequence-specific binding factor known to influence centromere formation outside of budding yeast. (For additional related information on CENP-B, please refer to the chapter "DNA Sequences in Centromere Formation and Function" by Dumont and Fachinetti.)

It therefore appeared that native centromeric DNA could be sufficient for centromere formation in experimental conditions in yeast and human cells. However, the study of dicentric chromosomes suggests otherwise. Dicentric chromosomes, which result from chromosome fusions or translocations, contain two regions of centromeric sequences (Stimpson et al. 2012). Two different fates have been reported for these dicentric chromosomes, depending on the model system. Typically both centromeric DNA regions are functional, which makes the dicentric chromosome unstable and causes it to break down into two fragments (Hill and Bloom 1989). However, it was observed in flies, plants and mammals that one of the centromere regions can get "inactivated", leaving only one "active" centromere sequence that assembles kinetochores and attaches to microtubules (Earnshaw and Migeon 1985; Sullivan and Schwartz 1995; Agudo et al. 2000; Han et al. 2006). The natural occurrence of these "pseudo-dicentric" chromosomes thus showed that centromeric DNA alone is not sufficient for centromere establishment. Moreover, the clinical discovery of fully functional "neocentromeres" that can arise stochastically within the human genome on non-centromeric DNA argues against a specific primary sequence to be essential for centromere formation (Marshall et al. 2008). It was therefore proposed that centromere specification can be dictated by epigenetic mechanisms (Karpen and Allshire 1997). To investigate epigenetic inheritance of centromeres, several groups have focused on artificially inducing neocentromeres at sites devoid of any native centromeric DNA component.

3 Unbiased Induction of Artificial Neocentromeres

To investigate early events leading to neocentromere formation, centromere disruption assays have been used to drive neocentromere formation in many different model systems (Fig. 1). This was first accomplished in *Drosophila* that stably maintained an X-chromosome-derived minichromosome (Fig. 1a). Two similar minichromosomes that differ in centromere position were subjected to irradiation mutagenesis for fragmentation. This released a piece of the minichromosome with identical sequence that only built a neocentromere and was stably maintained when the fragment was formerly juxtaposed to the centromere of the source minichromosome. This finding supported the hypothesis that centromere identity can spread and is regulated epigenetically (Maggert and Karpen 2001).

In *Schizosaccharomyces pombe*, conditional deletion of an endogenous centromere was achieved by inserting LoxP sites flanking the endogenous centromere

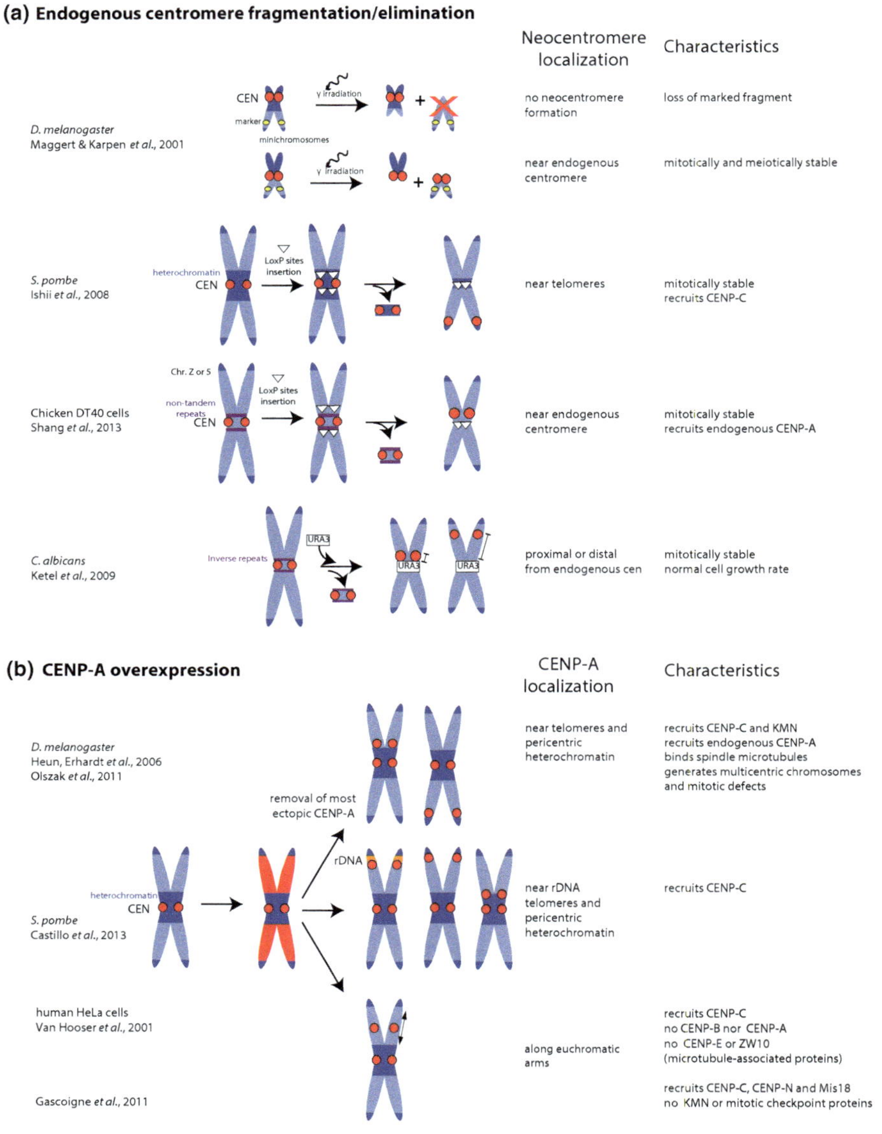

Fig. 1 Unbiased induction of neocentromere formation on endogenous chromosomes. **a** Fragmentation of minichromosomes (*in D. melanogaster*) or elimination of endogenous centromeres was followed by selection for a stabilized neocentromere containing chromosome fragment (in *S. pombe*, DT40 cells, and *C. albicans*). **b** Ectopic kinetochore and neocentromere formation was initiated by over-expression of the CENP-A protein. Active centromeres are marked by CENP-A, depicted as a *red circle*. Heterochromatin is represented as *dark blue* areas at pericentromeres and telomeres, repetitive DNA sequences are shown in purple. LoxP sites are depicted as *white triangles*. rDNA repeats are represented as an *orange mark*

on chromosome 1 (Fig. 1a). Induction of Cre-recombinase resulted in an acentric chromosome and produced rare survivors in which the acentric fragment had either fused to one of the other chromosomes or acquired a neocentromere. Interestingly, neocentromere formation occurred preferentially near telomeres and was significantly reduced, although not abolished, in a heterochromatin-deficient strain. This suggested that chromatin context, such as neighboring heterochromatin found at telomeres, may play a role in centromere formation in this model system (Ishii et al. 2008). The preference for proximity to heterochromatin has also been found for *Drosophila* neocentromeres, as will be discussed further below. In *C. albicans*, replacement of centromere 5 (cen5) with a URA3 gene cassette led to very highly efficient assembly of fully functional neocentromeres of two classes (Fig. 1a). The first class of neocentromeres was directly adjacent to cen5. The second class of neocentromeres appeared 200–450 kb distant from the original cen5, on either side of chromosomal arms (Ketel et al. 2009). Although *C. albicans* lacks heterochromatin components, the neocentromeres characterized in this study revealed interesting properties shared with larger regional centromeres found in other organisms. In particular, neocentromeres tended to form within non-transcribed regions flanked by repetitive sequences, and neocentromeres which moved within a transcribed gene caused silencing of this gene.

To better understand how neocentromeres can form in vertebrates, Fukagawa and colleagues adapted the Cre/Lox-mediated recombination strategy described in *S. pombe* to the chicken DT40 cells chromosome Z and 5 (Shang et al. 2013) (Fig. 1a). This was possible as these chromosomes carry a relatively short centromere, which, unlike most eukaryote centromeres, is assembled onto non-repetitive DNA, allowing insertions of flanking LoxP sites and selection markers. Cre-mediated removal of a 127 kb fragment including the endogenous centromere allowed to isolate clones that retained the Z chromosome despite the loss of its centromere. This protocol yielded 136 colonies, of which 126 acquired a neocentromere, while the remaining clones survived by chromosome fusion events. Importantly, the newly formed centromeres were functionally similar to endogenous centromeres in DT40 cells. ChIP-sequencing analysis on 18 neocentromere cell lines revealed that there is no DNA sequence preference on which neocentromeres assemble. The CENP-A-associated regions of all neocentromeres were similar in size to that of the endogenous Z centromere ($\sim$30 kb), suggesting a mechanism that maintains centromere size constant. Unlike most endogenous centromeres, however, the isolated neocentromeres were not characterized by the presence of flanking heterochromatin, in contrast to what was found in other organisms. Interestingly, it was also observed that neocentromeres could arise at actively transcribed genes, yet establishment of centromeric chromatin was found to be incompatible with housekeeping levels of transcription. Strikingly, for centromere Z and centromere 5 deletions, 76 and 97% of isolated neocentromeres had formed, respectively, at regions adjacent to the original position of the endogenous centromere, again in agreement with findings in *Drosophila* and *C. albicans*. It was thus proposed that low but significant levels of CENP-A detected in the vicinity of endogenous centromeres may nucleate neocentromere formation when the centromere core is removed.

Another approach to induce ectopic formation of neocentromeres has been to over-express the centromere-specific histone H3 variant CENP-A, as this histone is considered to be high in the hierarchy of centromeric proteins (Blower and Karpen 2001) (Fig. 1b). When CENP-A^{cnp1} is moderately over-expressed in fission yeast, it accumulates at pericentric heterochromatin, rDNA repeats and telomeric regions. At telomeric repeats, the Ccq1 protein was shown to play a crucial role by recruiting the methyltransferase Clr4 which leads to formation of heterochromatin that promotes the incorporation of CENP-A ^{cnp1}. However, heterochromatin was not absolutely required for CENP-A ^{cnp1} deposition in these regions, as subtelomeric chromatin still assembled CENP-A ^{cnp1} in heterochromatin-deficient mutants. CENP-A ^{cnp1} also recruited the kinetochore protein CENP-C, suggesting that functional centromeres can be formed at these ectopic sites. (Castillo et al. 2013). In *Drosophila* S2 cells and flies, over-expressing CENP-A^{CID} also led to establishment of functional ectopic centromeres, which attracted key centromere and kinetochore proteins and were sites of microtubule attachments (Fig. 1b). This caused multicentric chromosomes, and consequently chromosome missegregation, cell growth defects and developmental abnormalities. Interestingly, however, not all sites of ectopic CENP-A^{CID} incorporation gave rise to ectopic centromeres, suggesting that additional features of chromatin favor or restrict neocentromere formation (Heun et al. 2006). Indeed, it was later shown that similar to fission yeast, ectopic incorporation of CENP-A^{CID} in *Drosophila* S2 cells occurred primarily in the proximity of telomeres and near pericentric heterochromatin. More specifically, it appeared that transcriptionally silent regions at heterochromatin–euchromatin boundaries behaved as hotspots for CENP-A^{CID} ectopic deposition. These ectopic sites assembled functional kinetochores and bound microtubules, leading to mitotic defects (Olszak et al. 2011). This reoccurring preference for a neocentromere to form within silent domains adjacent to heterochromatin or repetitive DNA observed in *S. pombe*, *C. albicans,* and *Drosophila* is currently not understood. It is possible that subtelomeric heterochromatin mimics the environment of endogenous centromeres in these systems. Moreover, low nucleosome turnover near heterochromatin may help to achieve a critical local concentration of CENP-A needed to seed a neocentromere. Heterochromatin or repetitive DNA may also act as a boundary to prevent invasion of chromatin marks that might antagonize CENP-A deposition.

In contrast, over-expression of CENP-A in human tissue culture cells did not prove to be sufficient for ectopic centromere formation (Fig. 1b), although a subset of kinetochore proteins including CENP-C and CENP-N were recruited to non-centromeric CENP-A foci (Van Hooser et al. 2001; Gascoigne et al. 2011). Human centromeres are more complex than *Drosophila* centromeres as they contain a 16-subunit Constitutive-Centromere-Associated-Network (CCAN, composed of CENP-C/LN/HIKM/PORQU/SXTW), present at the centromere throughout the cell cycle as compared to one conserved member in *Drosophila* (CENP-C). Therefore, additional factors in humans may enable the distinction between correctly localized and mislocalized CENP-A, to avoid complete kinetochore formation and microtubule recognition at mistargeted CENP-A loci. This could also be achieved by post-translational modifications carried by CENP-A or other centromere factors. Interestingly, it was shown recently that CENP-A gets ubiquitinylated on Lysine 124 and it was proposed that ubiquitinylated CENP-A is the actual epigenetic mark for centromeres as over-expression of ubiquitin-fused CENP-A K124R mutants promotes neocentromere formation at ectopic sites in human cells (Niikura et al. 2016). However, the view that CENP-A K124ub is essential for centromere establishment and maintenance has recently been challenged by a combination of gene replacement experiments and ectopic centromere formation assays, which demonstrated that lack of this modification does not affect neither assembly of CENP-A into centromeres nor its ability to nucleate long-term centromere function (Fachinetti et al. 2017).

4 Targeting of Centromere Factors at a Defined Non-centromeric Locus

To overcome unwanted side effects of protein over-expression and potentially titrating less-abundant factors, more controllable tethering techniques were adapted to target centromere proteins of interest to specific ectopic loci using the Lactose repressor (LacI) and Lactose operator (LacO) in various organisms. Originally developed by the Belmont lab to visualize chromosome regions in vivo, these highly efficient systems exploit the strong affinity of the bacterial LacI for its corresponding recognition sequence LacO (Robinett et al. 1996). Insertion of large arrays of operators at an ectopic locus and expression of their respective DNA-binding protein fused to centromere proteins of interest enabled to test the contribution of specific centromere factors in the establishment and maintenance of centromeric chromatin (Fig. 2). This type of strategy presents an important complement to loss-of-function studies as most of these proteins are essential, making it difficult to dissect their role in detail without causing immediate stress or death to the cell.

To address the question of whether CENP-A is sufficient to nucleate centromere formation, a LacI-fused CENP-A^{CID} was tethered to an array of integrated LacO on the arm of chromosome 3 or on a plasmid in *Drosophila* S2 cells (Fig. 2a). The CENP-A^{CID}-tethered ectopic LacO site efficiently recruited inner and outer kinetochore proteins and attracted spindle microtubules, therefore producing chromosome fragmentation and mitotic defects. Endogenous CENP-A^{CID} molecules were assembled at the LacO array following CENP-A^{CID}-LacI targeting, providing evidence that CENP-A serves as a template for its self-propagation. Importantly, a pulse of targeting was sufficient to trigger long-term maintenance of a LacO plasmid, which showed that CENP-A is not only necessary, but also sufficient to initiate centromere establishment and maintenance throughout multiple generations (Mendiburo et al. 2011). In human cells, targeting the CENP-A chaperone HJURP

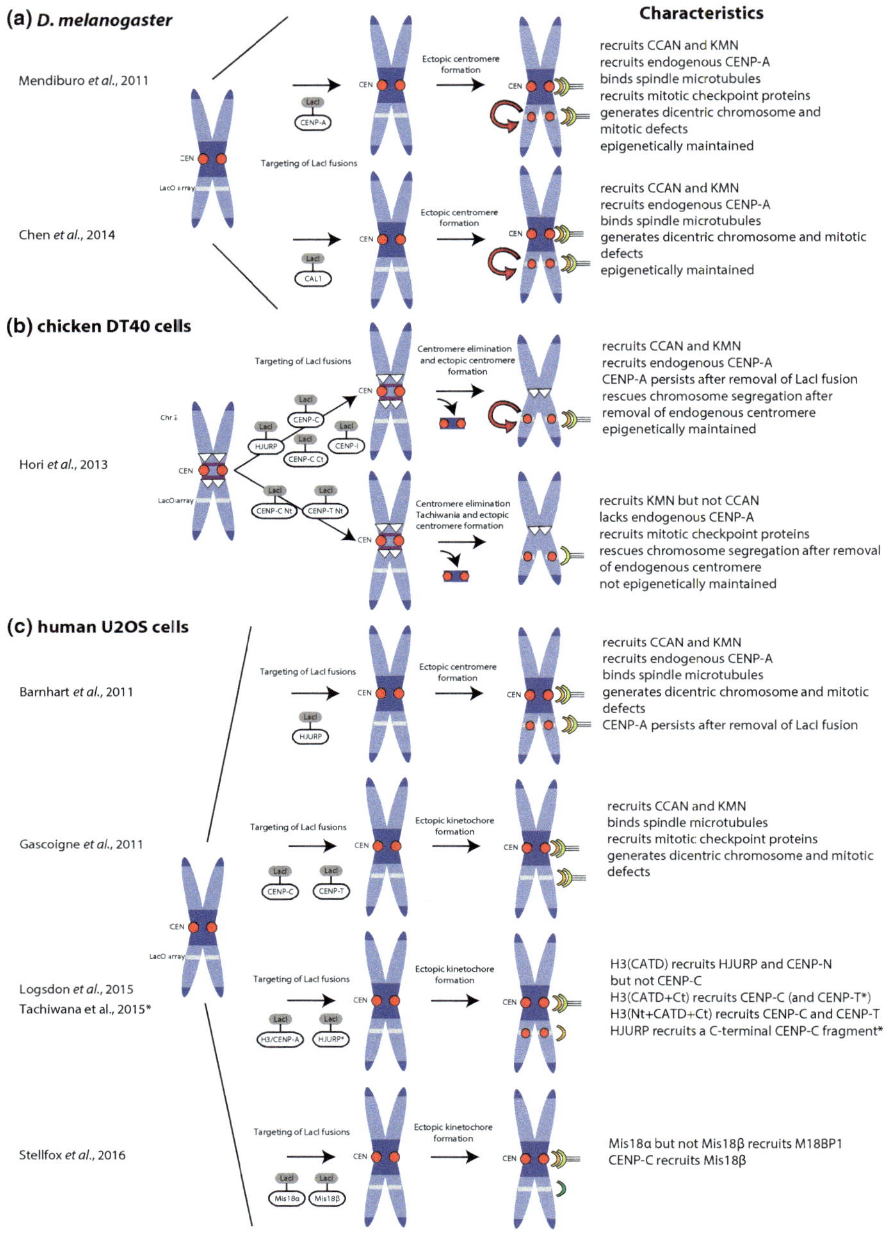
(a) D. melanogaster
Characteristics
Mendiburo et al., 2011
Chen et al., 2014
CEN
LacO array
Targeting of LacI fusions
LacI
CENP-A
CAL1
Ectopic centromere formation
recruits CCAN and KMN
recruits endogenous CENP-A
binds spindle microtubules
recruits mitotic checkpoint proteins
generates dicentric chromosome and mitotic defects
epigenetically maintained
recruits CCAN and KMN
recruits endogenous CENP-A
binds spindle microtubules
generates dicentric chromosome and mitotic defects
epigenetically maintained
(b) chicken DT40 cells
Hori et al., 2013
LacO array
Targeting of LacI fusions
HJURP
CENP-C
CENP-C Ct
CENP-I
CENP-C Nt
CENP-T Nt
Centromere elimination and ectopic centromere formation
Centromere elimination Tachiwana and ectopic centromere formation
recruits CCAN and KMN
recruits endogenous CENP-A
CENP-A persists after removal of LacI fusion
rescues chromosome segregation after removal of endogenous centromere
epigenetically maintained
recruits KMN but not CCAN
lacks endogenous CENP-A
recruits mitotic checkpoint proteins
rescues chromosome segregation after removal of endogenous centromere
not epigenetically maintained
(c) human U2OS cells
Barnhart et al., 2011
Gascoigne et al., 2011
Logsdon et al., 2015
Tachiwana et al., 2015*
Stellfox et al., 2016
Targeting of LacI fusions
HJURP
CENP-C
CENP-T
H3/CENP-A
HJURP*
Mis18α
Mis18β
Ectopic centromere formation
Ectopic kinetochore formation
recruits CCAN and KMN
recruits endogenous CENP-A
binds spindle microtubules
generates dicentric chromosome and mitotic defects
CENP-A persists after removal of LacI fusion
recruits CCAN and KMN
binds spindle microtubules
recruits mitotic checkpoint proteins
generates dicentric chromosome and mitotic defects
H3(CATD) recruits HJURP and CENP-N but not CENP-C
H3(CATD+Ct) recruits CENP-C (and CENP-T*)
H3(Nt+CATD+Ct) recruits CENP-C and CENP-T
HJURP recruits a C-terminal CENP-C fragment*
Mis18α but not Mis18β recruits M18BP1
CENP-C recruits Mis18β

◀**Fig. 2** Targeting of centromere factors to a defined non-centromeric locus (*LacO*). Different approaches are shown for *Drosophila* S2 cells (**a**), chicken DT40 cells (**b**), and human U2OS cells (**c**) to induce and dissect early steps of ectopic kinetochore and neocentromere formation. The LacO array is represented as a gray mark on the chromosome arm. Active centromeres are marked by CENP-A, depicted as a *red circle*. *Orange and yellow arches* correspond to CCAN and KMN proteins, respectively. Microtubules are parallel black lines. The *red arrow* in (**a**) and (**b**) represents epigenetic propagation of the CENP-A mark. In (**b**), LoxP sites are depicted as *white triangles* and the *purple mark* denotes repetitive DNA (non-tandem repeats) at Chromosome Z centromere. The *green arch* in (**c**) depicts Mis18 proteins

was tested using a similar strategy. Binding of a LacI-HJURP fusion to the LacO array in U2OS cells was sufficient to drive CENP-A deposition at this site (Fig. 2c). Using IPTG to disrupt the LacI interaction with the LacO array, a significant subset of CENP-A was found to be retained at the array, suggesting that CENP-A is not solely bound to HJURP in a prenucleosomal form, but that it can also stably incorporate into chromatin. Furthermore, HJURP-deposited CENP-A was able to recruit the CENP-C, CENP-N, CENP-M, and CENP-T inner kinetochore proteins and the outer kinetochore protein Ndc80 at the LacO array. Another piece of evidence that LacI-HJURP targeting gave rise to a functional ectopic centromere at the LacO array came from the observation that the array exhibited a constriction similar to that of endogenous centromeres on mitotic chromosomes, that microtubule ends attach to the LacO array, and that the LacO array containing chromosome is often lagging in anaphase (Barnhart et al. 2011). Together with in vitro reconstitutions using a heterologous system based on *Xenopus*-egg extracts (Guse et al. 2011), the two tethering studies came to the conclusion that CENP-A is a key epigenetic mark sufficient to define centromere identity.

Further efforts were made to study the contribution of other centromere proteins in the establishment of centromeres. In particular, the role of a subset of the 16 CCAN members was tested using ectopic targeting approaches. By replacing the DNA-binding domain of the inner kinetochore proteins CENP-C and CENP-T with LacI and targeting each construct to a LacO array stably integrated on the arm of chromosome 1 in U2OS cells, fully functional ectopic kinetochores capable of recruiting outer kinetochore, spindle assembly checkpoint proteins, and microtubules were generated in a CENP-A-independent manner (Gascoigne et al. 2011) (Fig. 2c).

To investigate CENP-A inheritance in more detail, synthetic neocentromeres were successfully produced in chicken DT40 cells (Fig. 2b). This was accomplished by tethering LacI-fused to HJURP, CENP-I, CENP-T and either full-length, N-terminal or C-terminal fragments of CENP-C to a LacO array, following removal of the endogenous centromere. Upon targeting, all of the constructs induced functional ectopic kinetochores, but only a subset, namely HJURP, full-length CENP-C, CENP-C C-terminus and CENP-I led to recruitment of CENP-A itself to the LacO site. The resulting artificial centromeres were stably inherited even after removal of the LacI fusions, providing evidence that the CCAN plays a crucial role in the epigenetic propagation of centromeres (Hori et al. 2013).

More recently, the LacI/LacO tethering system was used in combination with pulldown assays to dissect interactions involving the Mis18 complex (Fig. 2c). The Mis18 complex is a highly conserved family of proteins, consisting in humans of Mis18α, Mis18β, and the Mis18 binding protein 1 (M18BP1), which are thought to act as a licensing factor to prime centromeric chromatin prior to CENP-A deposition and to recruit the CENP-A-specific chaperone HJURP (Hayashi et al. 2004; Fujita et al. 2007). Wild-type Mis18α, but not wild-type Mis18β, targeted to the LacO array of U2OS cells was sufficient to recruit M18BP1, which suggested that only Mis18α interacts with M18BP1. Moreover, a CENP-C fragment fused to LacI was found to direct the recruitment of Mis18β (Stellfox et al. 2016). Interestingly, this interaction, like the interaction between CENP-C and M18BP1, seemed to occur preferentially in cells that are in the G1 phase of the cell cycle (Stellfox et al. 2016). Considering that the Mis18 complex association with centromeres is the first known step in CENP-A deposition, it seems crucial that multiple redundant contacts are made to stabilize its localization and this was the first line of evidence for this assumption.

Focusing on specific regions of CENP-A essential to direct the first steps of centromere establishment via recruitment of CCAN proteins and CENP-A assembly factors, human CENP-A or various H3/CENP-A chimeras were directly targeted in U2OS LacO cells (Logsdon et al. 2015; Tachiwana et al. 2015) (Fig. 2c). A transient LacI-CENP-A tethering of 48 h was sufficient to recruit CENP-C and HJURP but not endogenous CENP-A to the array. Using a series of H3/CENP-A chimeras, it was concluded that CENP-C recruitment requires both the CENP-A carboxy-terminus as well as the CENP-A targeting domain (CATD). The CATD is known as the minimal domain required to direct CENP-A to centromeres, via an interaction with its chaperone HJURP (Black et al. 2007; Foltz et al. 2009). This extended previous findings that the C-terminal tail of CENP-A is sufficient to bind CENP-C (Guse et al. 2011) and that CENP-C can be maintained at endogenous centromeres only via the C-terminal tail of CENP-A (Fachinetti et al. 2013). The requirements for CENP-C recruitment therefore seem to be different at a de novo centromere compared to an already established centromere. Interestingly, an H3 chimera containing the CATD of CENP-A is unable to recruit CENP-C but can recruit CENP-N, suggesting that the CATD functions in CENP-C recruitment independently of CENP-N. Importantly, CENP-T was found to be recruited to the array upon CENP-A targeting, which also required CENP-C, CENP-N, and a portion of the N-terminal tail of CENP-A (Logsdon et al. 2015). The latter, however, was not found to be required in a similar study using a different U2OS LacO cell line (Tachiwana et al. 2015). In addition, because CENP-C recruitment to CENP-A is dependent on HJURP and because HJURP targeting can recruit a C-terminal CENP-C fragment to the array, it was proposed that HJURP may be involved in CENP-C loading (Tachiwana et al. 2015).

In *Drosophila* S2 cells, targeting of CAL1 to the stably integrated LacO repeats was sufficient to induce inheritable deposition of CENP-A^{CID} and formation of a functional kinetochore at the ectopic site, which led to the proposal that CAL1, is the functional homolog of HJURP in flies (Chen et al. 2014) (Fig. 2a). An additional study using the same experimental setup revealed a link to transcription and the histone chaperone FACT, which allows RNA-polymerase-II-mediated transcription through nucleosome destabilization. Transcription is in turn thought to be critical for eviction of H3.1 and H3.3 placeholders prior to CENP-A deposition (Chen et al. 2015).

Collectively, this set of data showed that artificially targeting key centromere proteins such as CENP-A itself, its assembly factors, and specific CCAN proteins to a non-centromeric site is sufficient to cause this site to be epigenetically marked as a centromere. The following section focuses on recent work aimed at investigating the coordinated functions of both DNA and centromeric or general chromatin factors in centromere establishment.

5 Combining Ectopic Targeting of Chromatin Factors and Centromeric DNA Sequences

One important feature of centromeric chromatin is its unique set of histone modifications. Histone modifications are associated with specific properties of chromatin towards transcriptionally active, "open" euchromatin and mostly inactive, "silent" heterochromatin. Pericentric heterochromatin is characterized by the repressive methylation on lysine 9 of histone H3 marks H3K9me2, H3K9me3, histone H4 trimethylated at Lysine 20 (H4K20me3) and hypoacetylation. In contrast, euchromatin tends to contain the active mark H3K4me2 and is typically hyperacetylated. Centromeric chromatin is distinct from the surrounding heterochromatin in that it contains both CENP-A nucleosomes and H3 nucleosomes with euchromatic marks (H3K4me2) combined with low amounts of H3K9me2/3. However, human and *Drosophila* centromere chromatin lack acetylated histones, which is reminiscent of heterochromatin (Blower et al. 2002; Sullivan and Karpen 2004). To study the chromatin context and the histone marks of centromeric chromatin that are involved in the process of centromere formation, another strategy, combining artificial chromosomes and the tethering technology, was developed. By targeting key chromatin factors onto synthetic centromeric DNA, the characteristics of chromatin that are essential for centromere establishment have been thoroughly investigated (Fig. 3).

As discussed above, fission yeast minichromosomes were successfully generated by transforming vectors containing both the central core domain, on which CENP-A assembles, and the pericentric heterochromatin domain, enriched in H3K9me2. Importantly, targeting the methyltransferase Clr4 next to the central core domain on a minichromosome induces heterochromatin formation and bypasses the need for outer repeat sequences. It was concluded that the main role of outer repeats is to provide RNAi substrates to promote heterochromatin formation (Kagansky et al. 2009).

(a)

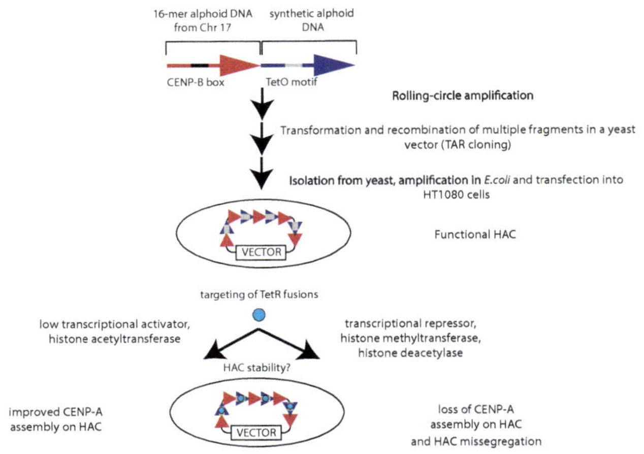

(b)

Protein targeted	Effect on HAC chromatin	Effect on HAC functionality	References
KRAB (transcription silencer)	↑ H3K9me3, HP1 recruitment, ↓ H3K4me2	↓ CENP-A, CENP-B, CENP-C, missegregation	Nakano *et al.*, 2008
KAP1 (transcriptional co-repressor)	↑ H3K9me3	↓ CENP-A, CENP-C, CENP-H, ↓ KMN ; missegregation	Cardinale *et al.*, 2009
LSD1 (lysine demethylase)	↓ H3K4me2, ↓ transcription	↓ HJURP, CENP-A	Bergmann *et al.*, 2011
NFκB	↑ H3K9ac, ↑ transcription	no change in CENP-A	Bergmann *et al.*, 2012
VP16	↑↑ H3K9ac, ↑↑ transcription	↓ CENP-A, CENP-C, missegregation	
p300, PCAF (acetyltransferases), Mis18, HJURP	↑ H3K9ac	↑ CENP-A levels, 1st successful HAC formation in HeLa cells	Ohzeki *et al.*, 2012
Suv39h1	↑ H3K9me3	↓ CENP-A	
Suv39h1 fragment	↑ H3K9me3, ↓H3K4me2	mild ↓ CENP-A	Martins *et al.*, 2016
EZH2	low H3K27me3, ↓ H3K4me2 and H3K36me2	mild ↓ CENP-A, CENP-C, CENP-T	
HP1	↑↑ H3K9me3	↓ CENP-A, missegregation	
PRC1	↑↑ H3K27me3, ↓ H3K4me2	↓ CENP-A, missegregation	
CCAN, KMN, M18BP1, MgcRacGAP, CENP-B	not checked	↑CENP-A	Shono *et al.*, 2015
HMTs, HDACs	not checked	↓ CENP-A	
HATs	not checked	no change in CENP-A	

Fig. 3 Combining ectopic targeting of chromatin factors and centromeric DNA sequences. Approach for targeting different chromatin modifiers to the engineered TetO-HAC in human cells is shown. **a** Schematic representation of the experimental setup. The TetO-HAC was constructed by rolling circle amplification followed by transformation associated recombination (*TAR*) cloning of a fragment made up of two units: a 16-mer alphoid DNA comprising a CENP-B box (*black mark*) derived from chromosome 17, next to synthetic alphoid DNA containing an array of Tet operators (*gray mark*). Following amplification in *E. coli*, the construct was transfected into HT1080 cells and yielded a HAC carrying a functional centromere. Underlying centromeric chromatin was tested for TetO-HAC centromere maintenance and centromeric or chromatin factor-TetR fusions (*blue circle*) were tethered to the TetO-HAC to assess its mitotic stability. **b** Summary of the observed effects of tethering various factors to the TetO-HAC on its chromatin state and on the functionality of its centromere

In human cells, the chromatin context was also carefully investigated using a special HAC designed as a 50 kb array of 2 subunits: a natural alphoid DNA monomer from chromosome 17 containing a CENP-B box adjacent to a synthetic alphoid DNA monomer containing a Tetracycline operator (TetO) in place of the CENP-B box (Fig. 3). This second unit enables targeting of specific proteins fused to the Tetracycline repressor (TetR) within the context of centromeric DNA. It was first shown that this construct (referred to as TetO-HAC thereafter) segregates correctly as an independent chromosome in HT1080 cells and that tethering RFP-TetR or TetR-GFP fusions had no effect on the stability of the HAC (Nakano et al. 2008). Meanwhile, this tool has been successfully used to tether a variety of chromatin factors to modulate and study the role of chromatin context on centromeres (Fig. 3b). By tethering factors involved in transcriptional activation or repression, it was shown that centromere establishment and centromere function are generally favored by low levels of transcription while repressive heterochromatin hinders centromere formation and kinetochore assembly. Indeed, tethering the tTS transcription silencer to the TetO-HAC caused elevated levels of H3K9me3, loss of CENP-A, CENP-B, and CENP-C and missegregation of the TetO-HAC (Nakano et al. 2008). A similar disruption of the TetO-HAC stability was observed when tethering the transcriptional corepressor KAP1 (Cardinale et al. 2009). Conversely, targeting the transcription factor NFκB subunit p65 to the TetO-HAC resulted in low levels of transcription marked by H3K9ac and did not interfere with kinetochore formation. Remarkably, targeting the herpes viral trans-inducing factor VP16 led to higher levels of transcription and increased levels of H3K9ac were accompanied by a CENP-A loading defect and impaired kinetochore function (Bergmann et al. 2012). The lysine-specific demethylase 1 (LSD1) efficiently removed H3K4me2 from the TetO-HAC and led to reduced levels of transcription. Targeting LSD1, Bergmann and colleagues found that CENP-A incorporation was lost due to a decreased recruitment of its chaperone HJURP (Bergmann et al. 2011), confirming that an open chromatin state promotes centromere formation. A complementary study comparing the effects of targeting the acetyltransferase domains of p300 or PCAF versus the methyltransferase Suv39h1 revealed that centromere formation is dependent on a tight balance between H3K9 acetylation and methylation, whereby acetylation promotes centromere assembly in contrast to methylation that blocks it. This finding also explained the long-standing question why HT1080 cells are much more proficient than other commonly used cell lines for centromere establishment on HACs. HeLa cells displayed higher levels of H3K9me3 and robust heterochromatin but lower levels of H3K9ac at their endogenous centromeres than HT1080 cells. Therefore, HeLa cells were more refractory to HAC formation, which could be complemented by targeting the histone acetyltransferases p300 of PCAF (Ohzeki et al. 2012). In contrast with the view that centromere formation is negatively impacted by invasion of silent chromatin, a recent study showed that a TetO-HAC can resist low levels of H3K9me3 (deposited by Suv39h1) or low levels of H3K27me3 (deposited by EZH2). Surprisingly, targeting the methyltransferase EZH2 to the TetO-HAC led to deposition of H3K27me3 but did not significantly affect centromere function.

However, direct tethering of the H3K27me3 reader PRC1 led to centromere inactivation on the TetO-HAC. Analogous observations were made with targeting a truncated fragment of the methyltransferase Suv39h1 versus the H3K9me2/3 reader HP1. This previously unreported ability of synthetic centromeres to resist local transcriptional silencing may help to explain how human centromeres maintain a pseudo-euchromatic signature despite being surrounded by heterochromatin (Martins et al. 2016).

The same construct was also used to screen for proteins involved in CENP-A maintenance at the TetO-HAC, on one hand; and in de novo CENP-A establishment at an ectopic site where the TetO-HAC was randomly integrated, on the other hand (Fig. 3b) (Shono et al. 2015). Factors that induced both de novo centromere establishment and maintenance include CENP-C, CENP-I, and outer kinetochore proteins. Importantly this class of factors depended on CENP-C, which also recruited M18BP1 of the Mis18 complex. The only exception seemed to be CENP-I, which was still able to recruit M18BP1 to the integrated alphoid array even in CENP-C-depleted cells. CENP-N, CENP-T, CENP-B, and MgcRacGAP belonged to a class of factors, which were involved in CENP-A inheritance but not in de novo CENP-A deposition. These are thought to act, not by directly recruiting, but rather by enhancing the stability or activity of CENP-A assembly factors. Histone acetyltransferases including the MYST family, proteins of the Mis18 complex, and histone chaperones involved in transcription, such as SSRP1 or RSF1, were able to promote CENP-A deposition only when CENP-A was over-expressed. It was proposed that these proteins function by changing the chromatin state to allow histone exchange, possibly via transcription. Conversely, most histone methyltransferases and deacetylases, like HDAC and SIRT proteins, were found to decrease CENP-A levels at the integrated HAC, possibly through heterochromatin formation and repression of histone turnover. This is in agreement with previous findings indicating that centromeric chromatin assembly and heterochromatin assembly antagonize each other.

5.1 Concluding Remarks

In recent years, many different synthetic approaches have been successfully undertaken to address the question of how centromeric chromatin is established and maintained. To induce centromere formation, transfecting naïve centromeric sequences has proved to be sufficient in yeast as well as in human cells, albeit with low efficiency. HAC formation assays showed that the CENP-B box embedded in human alphoid DNA is necessary for centromere establishment on naïve input DNA.

To investigate the mechanism of neocentromere formation, one approach has been to remove the endogenous active centromere of a specific chromosome and select for the cells who maintained this chromosome. Used in yeast and chicken cells, this led to the generation of artificial neocentromeres, which formed non-stochastically on the chromosome. To address the same question, an alternative strategy consisted in over-expressing the key centromere epigenetic mark CENP-A. In fission yeast or flies, but not in human cells, this gave rise to functional neo-centromeres at a limited number of preferred sites within the genome. In several model systems, this occurs either in proximity to the former endogenous centromere, potentially mediated by CENP-A spreading, or next to heterochromatin or repetitive DNA for currently unknown reasons. To dissect the molecular pathway by which centromeres are formed and propagated, factors such as CENP-A, CCAN proteins, or CENP-A assembly factors were artificially tethered to non-centromeric sites and efficiently produced functional ectopic centromeres in several model systems including *Drosophila* S2 cells, chicken cells, and human cells. Taken together, these approaches have significantly advanced our understanding of how centromere identity is determined. They suggest that CENP-A lies high in the hierarchy of centromere-associated proteins and that CCAN proteins, and especially CENP-C, play an important role both in kinetochore assembly and in the epigenetic propagation of centromeres. As an outlook, the accumulated knowledge on pathways leading to centromere establishment could be exploited for the design of a new generation of artificial chromosomes, which will be useful tools not only for basic research but also for the development of delivery systems for applications such as gene therapy or cell reprogramming.

References

Agudo M, Abad JP, Molina I et al (2000) A dicentric chromosome of Drosophila melanogaster showing alternate centromere inactivation. Chromosoma 109:190–196

Barnhart MC, Kuich PHJL, Stellfox ME et al (2011) HJURP is a CENP-A chromatin assembly factor sufficient to form a functional de novo kinetochore. J Cell Biol. doi:10.1083/jcb.201012017

Basu J, Stromberg G, Compitello G et al (2005) Rapid creation of BAC-based human artificial chromosome vectors by transposition with synthetic alpha-satellite arrays. Nucleic Acids Res 33:587–596. doi:10.1093/nar/gki207

Bergmann JH, Jakubsche JN, Martins NM et al (2012) Epigenetic engineering: histone H3K9 acetylation is compatible with kinetochore structure and function. J Cell Sci 125:411–421. doi:10.1242/jcs.090639

Bergmann JH, Rodriguez MG, Martins NMC et al (2011) Epigenetic engineering shows H3K4me2 is required for HJURP targeting and CENP-A assembly on a synthetic human kinetochore. EMBO J 30:328–340. doi:10.1038/emboj.2010.329

Bernard P, Maure JF, Partridge JF et al (2001) Requirement of heterochromatin for cohesion at centromeres. Science 294:2539–2542. doi:10.1126/science.1064027

Black BE, Jansen LET, Maddox PS et al (2007) centromere identity maintained by nucleosomes assembled with histone H3 containing the CENP-A targeting domain. Mol Cell. doi:10.1016/j.molcel.2006.12.018

Blower MD, Karpen GH (2001) The role of Drosophila CID in kinetochore formation, cell-cycle progression and heterochromatin interactions. Nat Cell Biol 3:730–739. doi:10.1038/35087045

Blower MD, Sullivan BA, Karpen GH (2002) Conserved organization of centromeric chromatin in flies and humans. Dev Cell 2:319–330

Cardinale S, Bergmann JH, Kelly D et al (2009) Hierarchical inactivation of a synthetic human kinetochore by a chromatin modifier. Mol Biol Cell 20:4194–4204. doi:10.1091/mbc.E09-06-0489

Castillo AG, Pidoux AL, Catania S et al (2013) Telomeric repeats facilitate CENP-ACnp1 Incorporation via telomere binding proteins. PLoS ONE. doi:10.1371/journal.pone.0069673

Chen C-C, Dechassa ML, Bettini E et al (2014) CAL1 is the Drosophila CENP-A assembly factor. J Cell Biol 204:313–329. doi:10.1083/jcb.201305036

Chen CC, Bowers S, Lipinszki Z et al (2015) Establishment of centromeric chromatin by the CENP-A assembly factor CAL1 requires FACT-mediated transcription. Dev Cell. doi:10.1016/j.devcel.2015.05.012

Clarke L, Carbon J (1980) Isolation of a yeast centromere and construction of functional small circular chromosomes. Nature 287:504–509

Earnshaw WC, Migeon BR (1985) Three related centromere proteins are absent from the inactive centromere of a stable isodicentric chromosome. Chromosoma 92:290–296

Ebersole T, Okamoto Y, Noskov VN et al (2005) Rapid generation of long synthetic tandem repeats and its application for analysis in human artificial chromosome formation. Nucleic Acids Res 33:e130. doi:10.1093/nar/gni129

Fachinetti D, Folco HD, Nechemia-Arbely Y et al (2013) A two-step mechanism for epigenetic specification of centromere identity and function. Nat Cell Biol 15:1056–1066. doi:10.1038/ncb2805

Fachinetti D, Logsdon GA, Abdullah A et al (2017) CENP-A modifications on ser68 and Lys124 are dispensable for establishment, maintenance, and long-term function of human centromeres. Dev Cell 40:104–113. doi:10.1016/j.devcel.2016.12.014

Folco HD, Pidoux AL, Urano T, Allshire RC (2008) Heterochromatin and RNAi are required to establish CENP-A chromatin at centromeres. Science 319:94–97. doi:10.1126/science.1150944

Foltz DR, Jansen LET, Bailey AO et al (2009) Centromere-specific assembly of CENP-A nucleosomes Is mediated by HJURP. Cell. doi:10.1016/j.cell.2009.02.039

Fujita Y, Hayashi T, Kiyomitsu T et al (2007) Priming of centromere for CENP-A recruitment by human hMis18, hMis18, and M18BP1. Dev Cell. doi:10.1016/j.devcel.2006.11.002

Gascoigne KE, Takeuchi K, Suzuki A et al (2011) Induced ectopic kinetochore assembly bypasses the requirement for CENP-A nucleosomes. Cell. doi:10.1016/j.cell.2011.03.031

Guse A, Carroll CW, Moree B et al (2011) In vitro centromere and kinetochore assembly on defined chromatin templates. Nature 477:354–358. doi:10.1038/nature10379

Hahnenberger KM, Baum MP, Polizzi CM et al (1989) Construction of functional artificial minichromosomes in the fission yeast Schizosaccharomyces pombe. Proc Natl Acad Sci U S A 86:577–581

Han F, Lamb JC, Birchler JA (2006) High frequency of centromere inactivation resulting in stable dicentric chromosomes of maize. Proc Natl Acad Sci U S A 103:3238–3243. doi:10.1073/pnas.0509650103

Harrington JJ, Van Bokkelen G, Mays RW et al (1997) Formation of de novo centromeres and construction of first-generation human artificial microchromosomes. Nat Genet 15:345–355. doi:10.1038/ng0497-345

Hassold T, Hunt P (2001) To err (meiotically) is human: the genesis of human aneuploidy. Nat Rev Genet 2:280–291. doi:10.1038/35066065

Hayashi T, Fujita Y, Iwasaki O et al (2004) Mis16 and Mis18 are required for CENP-A loading and histone deacetylation at centromeres. Cell. doi:10.1016/j.cell.2004.09.002

Heun P, Erhardt S, Blower MD et al (2006) Mislocalization of the drosophila centromere-specific histone CID promotes formation of functional ectopic kinetochores. Dev Cell. doi:10.1016/j.devcel.2006.01.014

Hill A, Bloom K (1989) Acquisition and processing of a conditional dicentric chromosome in *Saccharomyces cerevisiae*. Mol Cell Biol 9:1368–1370

Hori T, Shang WH, Takeuchi K, Fukagawa T (2013) The CCAN recruits CENP-A to the centromere and forms the structural core for kinetochore assembly. J Cell Biol. doi:10.1083/jcb.201210106
Hudson DF, Fowler KJ, Earle E et al (1998) Centromere protein B null mice are mitotically and meiotically normal but have lower body and testis weights. J Cell Biol 141:309–319
Ikeno M, Grimes B, Okazaki T et al (1998) Construction of YAC-based mammalian artificial chromosomes. Nat Biotechnol 16:431–439. doi:10.1038/nbt0598-431
Ishii K, Ogiyama Y, Chikashige Y et al (2008) Heterochromatin integrity affects chromosome reorganization after centromere dysfunction. Science 321:1088–1091. doi:10.1126/science.1158699
Kagansky A, Folco HD, Almeida R et al (2009) Synthetic heterochromatin bypasses RNAi and centromeric repeats to establish functional centromeres. Science 324:1716–1719. doi:10.1126/science.1172026
Kapoor M, de Montes Oca Luna R, Liu G et al (1998) The cenpB gene is not essential in mice. Chromosoma 107:570–576
Karpen GH, Allshire RC (1997) The case for epigenetic effects on centromere identity and function. Trends Genet 13:489–496
Ketel C, Wang HSW, McClellan M et al (2009) Neocentromeres form efficiently at multiple possible loci in *Candida albicans*. PLoS Genet 5:e1000400. doi:10.1371/journal.pgen.1000400
Logsdon GA, Barrey EJ, Bassett EA et al (2015) Both tails and the centromere targeting domain of CENP-A are required for centromere establishment. J Cell Biol. doi:10.1083/jcb.201412011
Maggert KA, Karpen GH (2001) The activation of a neocentromere in Drosophila requires proximity to an endogenous centromere. Genetics 158:1615–1628
Marshall OJ, Chueh AC, Wong LH, Choo KHA (2008) Neocentromeres: new insights into centromere structure, disease development, and karyotype evolution. Am. J. Hum, Genet
Martins NMC, Bergmann JH, Shono N et al (2016) Epigenetic engineering shows that a human centromere resists silencing mediated by H3K27me3/K9me3. Mol Biol Cell 27:177–196. doi:10.1091/mbc.E15-08-0605
Masumoto H, Ikeno M, Nakano M et al (1998) Assay of centromere function using a human artificial chromosome. Chromosoma 107:406–416
Mendiburo MJ, Padeken J, Fulop S et al (2011) Drosophila CENH3 is sufficient for centromere formation. Science 334:686–690. doi:10.1126/science.1206880
Murray AW, Szostak JW (1983) Construction of artificial chromosomes in yeast. Nature 305:189–193
Nakano M, Cardinale S, Noskov VN et al (2008) Inactivation of a human kinetochore by specific targeting of chromatin modifiers. Dev Cell. doi:10.1016/j.devcel.2008.02.001
Niikura Y, Kitagawa R, Kitagawa K (2016) CENP-A ubiquitylation is inherited through dimerization between cell divisions. Cell Rep 15:61–76. doi:10.1016/j.celrep.2016.03.010
Nonaka N, Kitajima T, Yokobayashi S et al (2002) Recruitment of cohesin to heterochromatic regions by Swi6/HP1 in fission yeast. Nat Cell Biol 4:89–93. doi:10.1038/ncb739
Ohzeki J-I, Bergmann JH, Kouprina N et al (2012) Breaking the HAC barrier: histone H3K9 acetyl/ methyl balance regulates CENP-A assembly. EMBO J 3182:2391–2402. doi:10.1038/emboj.2012.82
Ohzeki J, Nakano M, Okada T, Masumoto H (2002) CENP-B box is required for de novo centromere chromatin assembly on human alphoid DNA. J Cell Biol. doi:10.1083/jcb.200207112
Okada T, Ohzeki J, Nakano M et al (2007) CENP-B controls centromere formation depending on the chromatin context. Cell. doi:10.1016/j.cell.2007.10.045
Okamoto Y, Nakano M, Ohzeki J et al (2007) A minimal CENP-A core is required for nucleation and maintenance of a functional human centromere. EMBO J 26:1279–1291. doi:10.1038/sj.emboj.7601584
Olszak AM, van Essen D, Pereira AJ et al (2011) Heterochromatin boundaries are hotspots for de novo kinetochore formation. Nat Cell Biol 13:799–808. doi:10.1038/ncb2272

Perez-Castro AV, Shamanski FL, Meneses JJ et al (1998) Centromeric protein B null mice are viable with no apparent abnormalities. Dev Biol 201:135–143. doi:10.1006/dbio.1998.9005

Robinett CC, Straight A, Li G et al (1996) In vivo localization of DNA sequences and visualization of large-scale chromatin organization using lac operator/repressor recognition. J Cell Biol 135:1685–1700

Schueler MG, Sullivan BA (2006) Structural and functional dynamics of human centromeric chromatin. Annu Rev Genomics Hum Genet 7:301–313. doi:10.1146/annurev.genom.7.080505.115613

Shang WH, Hori T, Martins NMC et al (2013) Chromosome engineering allows the efficient isolation of vertebrate neocentromeres. Dev Cell. doi:10.1016/j.devcel.2013.02.009

Shono N, Ohzeki J, Otake K et al (2015) CENP-C and CENP-I are key connecting factors for kinetochore and CENP-A assembly. J Cell Sci 128:4572–4587. doi:10.1242/jcs.180786

Stellfox ME, Nardi IK, Knippler CM, Foltz DR (2016) Differential binding partners of the Mis18α/β YIPPEE domains regulate Mis18 complex recruitment to centromeres. Cell Rep. doi:10.1016/j.celrep.2016.05.004

Stimpson KM, Matheny JE, Sullivan BA (2012) Dicentric chromosomes: unique models to study centromere function and inactivation. Chromosom Res 20:595–605. doi:10.1007/s10577-012-9302-3

Sullivan BA, Karpen GH (2004) Centromeric chromatin exhibits a histone modification pattern that is distinct from both euchromatin and heterochromatin. Nat Struct Mol Biol 11:1076–1083. doi:10.1038/nsmb845

Sullivan BA, Schwartz S (1995) Identification of centromeric antigens in dicentric Robertsonian translocations: CENP-C and CENP-E are necessary components of functional centromeres. Hum Mol Genet 4:2189–2197

Tachiwana H, Müller S, Blümer J et al (2015) HJURP involvement in de novo CenH3CENP-A and CENP-C recruitment. Cell Rep. doi:10.1016/j.celrep.2015.03.013

Van Hooser AA, Ouspenski II, Gregson HC et al (2001) Specification of kinetochore-forming chromatin by the histone H3 variant CENP-A. J Cell Sci 114:3529–3542

Weaver BAA, Cleveland DW (2007) Aneuploidy: instigator and inhibitor of tumorigenesis. Cancer Res 67:10103–10105. doi:10.1158/0008-5472.CAN-07-2266

Post-translational Modifications of Centromeric Chromatin

Ana García del Arco and Sylvia Erhardt

Abstract Regulation of chromatin structures is important for the control of DNA processes such as gene expression, and misregulation of chromatin is implicated in diverse diseases. Covalent post-translational modifications of histones are a prominent way to regulate chromatin structure and different chromatin regions bear their specific signature of histone modifications. The composition of centromeric chromatin is significantly different from other chromatin structures and mainly defined by the presence of the histone H3-variant CENP-A. Here we summarize the composition of centromeric chromatin and what we know about its differential regulation by post-translational modifications.

Abbreviations

aa	Amino acid
C. elegans	*Caenorhabditis elegans*
CAL1	Chromosome alignment defect 1
CID	Centromere identifier (*Drosophila*)
Dam	Deoxyadenosine methylase
D. melanogaster	*Drosophila melanogaster*
Dnmt	DNA methyltransferase
HACs	Human artificial chromosomes
HAT B	Histone acetyltransferase B
HDACs	Histone Deacetylases
HFD	Histone fold domain
HJURP	Holliday junction-recognition protein
H3	Histone 3
LSD1	Lysine-specific demethylase 1

A. García del Arco · S. Erhardt
Center for Molecular Biology of Heidelberg (ZMBH), DKFZ-ZMBH Alliance, Heidelberg University, Im Neuenheimer Feld 282, 69120 Heidelberg, Germany

S. Erhardt (✉)
Cell Networks Excellence Cluster, Heidelberg University, Im Neuenheimer Feld 282, 69120 Heidelberg, Germany
e-mail: s.erhardt@zmbh.uni-heidelberg.de

B.E. Black (ed.), *Centromeres and Kinetochores*, Progress in Molecular and Subcellular Biology 56, DOI 10.1007/978-3-319-58592-5_9

PTM	Post-translational modifications
PcG	Polycomb group
Plk	Polo-like kinase
RCC1	Regulator of chromatin condensation 1
Rdx	Roadkill
S. cerevisiae	*Saccharomyces* cerevisiae
S. pombe	*Schizosaccharomyces pombe*
TrxG	Trithorax group
Z. may	*Zea mays*

1 Chromatin

In the nuclei of all eukaryotic cells, genomic DNA is highly folded, constrained, and compacted by both histone and non-histone proteins in a dynamic polymer called chromatin. Chromatin is organized in the dynamic structuring of nucleosomes, which represent the basic repeating unit of the chromatin fiber. Each nucleosome is formed by 146–147 bp of chromosomal DNA tightly wrapped around an octamer of proteins comprising two subunits each of the canonical histones H3, H4, H2A, and H2B, or variants of these histones (Davey et al. 2002; Luger et al. 1997).

Histones are small basic proteins consisting of a globular domain, called the histone fold domain (HFD), and a more flexible and charged NH_2-terminus (histone tail). These flexible N-terminal tails of the four core histones undergo a range of post-translational modifications (PTMs), including acetylation, methylation, phosphorylation, ubiquitination, sumoylation, ribosylation, and many others (Hatakeyama et al. 2016). These covalent modifications reveal a "histone code" that is involved in generating epigenetic information, the code that is 'above' (greek 'epi') the genetic code by influencing the state of chromatin structure and function (Jenuwein and Allis 2001; Kouzarides 2007). The two most distinct chromatin states are euchromatin, which is mainly found on chromosome arms and comprises most of the protein-coding genes. In contrast, heterochromatin is more compact, highly condensed, and most prominently found at telomeres and in vicinity to centromeres (Grewal and Elgin 2007).

The transcriptional consequences of histone modifications are revealed either as a result of the direct biophysical consequences of the modification, or through the catalytic activities of proteins and complexes that recognize and bind these specifically modified histones. For instance, acetylation, which reduces the net positive charge on the nucleosome, results in decreased stability of the histone associations with the negatively charged DNA, facilitating access of chromatin associating factors and promoting, most prominently, transcription. Therefore, by balancing levels and activities of histone acetyltransferases (HATs) and deacetylases (HDACs), which, respectively, add and remove acetyl groups from histones (and other proteins), transcription can be activated or repressed (Bannister and Kouzarides 2011). In contrast, the effects of histone methylation appear to be transmitted more indirectly, as the addition of a methyl group does not modify histone chain charge. Methylated histones rather seem to serve as a recognition platform for protein complexes that bind to chromatin and remodel its compaction, localization, and transcriptional activity (Cohen et al. 2011).

Apart from covalent modifications occurring at histone tails, the DNA sequence can also be chemically modified. A detailed description of DNA modifications is beyond the scope of this chapter, but it is important to mention that there is significant crosstalk between histone PTMs and DNA modifications, and that these two gene expression regulatory phenomena can be dependent on each other. For instance, DNA methylation has been proposed to serve as a template for establishing particular PTMs on newly synthesized histones after DNA replication (Cedar and Bergman 2009). DNA methylation is the classic example of a heritable epigenetic mark, specifically at CpG islands, which are maintained in a semi-conservative manner by the activity of the DNA methyltransferase 1 (Dnmt1) (Jones and Liang 2009). DNA methylation is found symmetrically on parental strands, each daughter strand therefore contains one methylated strand acting as a template for faithfully maintaining the methylation pattern at the newly synthesized DNA strand.

2 Machineries of Histone Post-translational Modifications (PTMs)

The addition and removal of histone PTMs are considered key regulatory processes of chromatin function. Enzymes which catalyze the addition of PTMs to histones are often referred to as "writers." The added modifications are then recognized by so-called "readers" (proteins which are sensitive to the PTMs presented by the histones), and, ultimately, some of them by "erasers" (enzymes which catalyze the removal of these marks) (Falkenberg and Johnstone 2014). The actions of these proteins are crucial for the dynamic regulation of histone modification levels on the chromatin fiber, since most PTMs have a writer/eraser pair with opposing effects. For example, histone acetylation is regulated by the opposing actions of HATs and HDACs, and phosphorylation is regulated by the activity of kinases and phosphatases. The fine regulation of these enzyme activities determine whether histone PTMs contribute to short-term chromatin regulation (immediate functions) or long-term chromatin changes (heritable function). In cellular signaling events, rapid responses to environmental stimuli necessitate a high turnover rate of histone PTMs, whereas defining and maintaining chromatin structures, such as constitutive heterochromatin, throughout the cell cycle or from one cell generation to the next, requires constant PTMs with little dynamics (Turner 2007).

3 PTMs in Epigenetic Inheritance

Epigenetics is generally defined as heritable changes in gene activity and expression that occur without altering the underlying DNA sequence (Bird 2007; Goldberg et al. 2007). Epigenetics is therefore often considered a link between genotype and phenotype. Besides DNA methylation (which is not found in all species), histone PTMs are a major regulatory entity for epigenetic chromatin regulation and function, since they determine heritable differences in chromatin states. Important players at all stages of histone and chromatin regulation (including PTM modulation) are histone chaperones. They associate with newly synthesized histones, help transport them into the nucleus, and associate them to DNA (Avvakumov et al. 2011).

Epigenetic inheritance and the transmission of information beyond the DNA sequence during cell division is crucial for maintaining differential gene expression patterns during development and differentiation, and also for the development of many diseases (Heard and Martienssen 2014). The known association of histone chaperones and chromatin-modifying enzymes to the replication fork highlights that, at least for some of these marks, their inheritance is of crucial importance for cell viability and function.

An epigenetic role for histones was first proposed in 1980 (Stein 1980), but it was almost 20 years later, in 1997, when the first mechanistic understanding of how histone PTMs mediate the inheritance of silenced chromatin was reported, using budding yeast as a model (Sherman and Pillus 1997). More and more examples of histone-mediated inheritance were discovered in higher eukaryotes, including humans, resulting in more effort to unravel the epigenetic roles of PTMs. Today, the information on different epigenetic roles of PTMs is overwhelming and summarized elsewhere (Allis and Jenuwein 2016).

4 Maintenance of PTMs Throughout Cell Cycle

Chromatin in proliferating cells is highly dynamic. During cell cycle, there are two major events involving a global chromatin reconstruction. First, during S-phase, newly synthesized histones must be correctly incorporated with nascent DNA, and histone PTMs must be reestablished (Annunziato 2005). Second, major chromatin remodeling and distributional changes of chromatin-associated factors occur during mitosis; most prominent is the condensation of chromosomes and the phosphorylation of histone H3S10 (Hsu et al. 2000).

In the case of histone modifications, canonical histones are displaced and replaced during S-phase and DNA replication. During DNA replication, chromatin is disassembled prior to the replication fork and reassembled on the two daughter strands. How PTMs of recycled and newly incorporated histones are faithfully passed from one generation to the next is not yet fully understood. Alabert et al.

have recently examined the maintenance of histone PTMs after DNA replication in HeLa cells and propose two modes for the transmission of histone modifications at the replication fork: (i) most of the PTMs on parental old histones are retained and newly synthesized histones collect PTMs to become identical within the same cell cycle; and (ii) gradual modification of both new and parental histones. In the case of the H3K9me3 and H3K27me3 marks on new H3, more than one generation is required to complete the trimethylation. As these marks are therefore diluted in half after replication, a potential loss of cellular memory could result (Apostolou and Hochedlinger 2013). However, this is counterbalanced by the continual modification of both new and old histones.

The results of Alabert et al. suggest that histone PTMs are transmitted at the replication fork and would be in contrast to the observation of complete deletion of PTMs in experiments using *Drosophila* embryos, where the TrxG and PcG histone-modifying enzyme complexes remain associated during replication and reestablish the histone modifications after S-phase (Petruk et al. 2012).

5 Centromeric Chromatin: CENP-A, the Epigenetic Marker for Centromere Identity

The proper segregation of genetic information during cell division is crucial to maintain genomic integrity. Errors in segregation can lead to abnormal chromosome number—known as aneuploidy—which is linked to human disease (Kops et al. 2005). Accurate segregation during mitosis is mediated by the centromere, a chromosomal chromatin region of each sister chromatid that serves as a chromatin foundation for kinetochore formation and as a chromosome attachment to the mitotic microtubule spindle (Przewloka and Glover 2009).

Despite differences in complexity, one common centromere feature is conserved among species: the replacement of the canonical histone H3 with the histone H3 variant CENP-A in a subset of centromeric nucleosomes (Allshire and Karpen 2008). The underlying DNA sequence is fast evolving and not conserved between species (Murphy et al. 2005). With the exception of some yeast species such as *Saccharomyzes cerevisiae* and *Kluyveromyces lactis*, the centromeric DNA sequence alone seems insufficient to confer centromeric identity, and it is, therefore, widely accepted that centromeres are regulated epigenetically (Karpen and Allshire 1997).

CENP-A is structurally similar to the canonical histone H3. The C-terminus contains a globular HFD that shares 62% sequence homology with the HFD of canonical H3 (Sullivan et al. 1994). The HFD of CENP-A, like all histone proteins, consists of three α-helices linked by two loops (Arents et al. 1991). In addition to mediating the interaction with histone H4, CENP-A's HFD contains the critical structural features that are needed to deposit CENP-A to centromeres, i.e., loop1 (L1) and α-helix 2, which build up the CENP-A targeting domain (CATD), a region that is necessary and sufficient to promote centromeric targeting (Black et al. 2004). In contrast to the HFD, the N-terminal tail of CENP-A is very diverse and varies in length between different species as discussed later (Smith 2002). X-ray crystallography has revealed that CENP-A and canonical nucleosomes are structurally very similar, and both types of nucleosomes wrap their DNA in a left-handed manner (Tachiwana et al. 2011). The precise composition of centromeric nucleosomes has been a subject of controversy over the past years, however, most evidence points to an octamer as the predominant centromeric structure (Dunleavy et al. 2013). Differences in the octameric structure of CENP-A-containing nucleosomes may be mediated by the binding of CENP-C, which can reshape these nucleosomes into a more rigid structure (Falk et al. 2015).

6 PTMs of Centromeric Chromatin

Despite the essential role of CENP-A for most centromeres, the chromatin environment created by the presence of specific PTMs on all histone species at centromeres is just as important. In most species, centromeres are organized with a central region that is defined by the presence of CENP-A-containing nucleosomes, surrounded on both sides by flanking heterochromatin (pericentric chromatin) (Blower and Karpen 2001; Blower et al. 2002; Partridge et al. 2000). The PTMs on histones present at human centromeres indicate that centromeric chromatin is neither heterochromatic nor euchromatic. This unique mixture of repressive and permissive histone marks has been termed "centrochromatin" (Sullivan and Karpen 2004).

The PTMs of centromeric chromatin can be sorted in four different categories depending on their localization: PTMs (i) of the pericentromeric region, (ii) of nucleosomes adjacent to CENP-A containing nucleosomes, (iii) of CENP-A containing nucleosomes, and (iv) of CENP-A itself (Fig. 1).

6.1 Modifications of the Pericentromeric Region

As mentioned before, chromatin is generally divided into distinct types: transcriptionally 'open' euchromatin and tightly packaged, 'closed' heterochromatin. It is now clear that this "open" or "closed" classification is an oversimplification, and more precise subdivisions of chromatin structures have been proposed based on specific PTMs and the protein associated with these modifications (Filion et al. 2010). Early studies have shown that centromeres are surrounded by heterochromatin (Lima de Faria 1949), and following studies have found that these pericentromeric regions are indeed containing hypermethylated H3K9 that is dependent on the Suv39h histone methyltransferases as a typical repressive mark of heterochromatin (Peters et al. 2003; Rice et al. 2003). Similar to centromeres, the surrounding

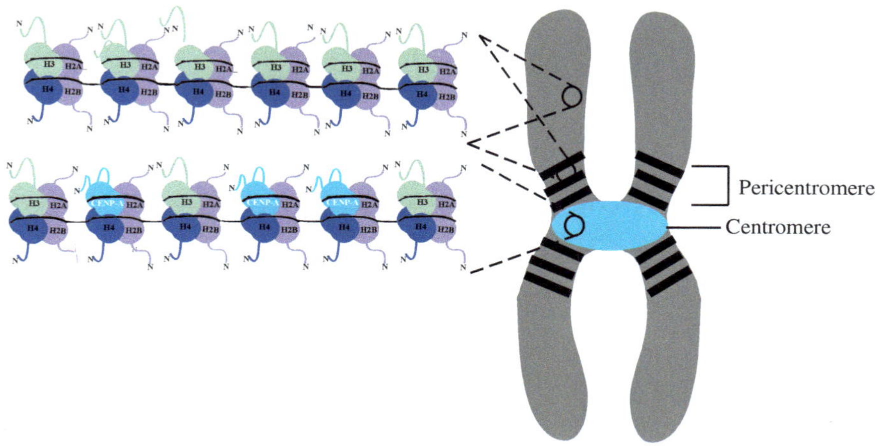

Fig. 1 Different possible localizations of PTMs at centromeric chromatin. A generalized diagram is shown, indicating the composition of chromatin in distinct regions of the chromosome arm

pericentromeric heterochromatin is characterized by hypoacetylated canonical histones. However, in contrast to centromeres, pericentric chromatin is characterized by dimethylation (flies and fission yeast) or trimethylation (in mammalian cells) of H3K9 (Noma et al. 2001). Another repressive mark present at the pericentromeric region is H3K27me3, which may serve slightly different or additional functions for instance a compensatory mechanism for the loss of K9 methylation, causing redistribution of this mark to the region, preserving in this manner the repressive state (Lam et al. 2006). Trimethylation of H4K20 is another marker of constitutive heterochromatin that is present in DNA repetitive regions, focally enriched at pericentric heterochromatin. All these repressive marks index pericentric heterochromatin in a sequential model. After H3K9 and H3K27 methylation occurs, the heterochromatin protein 1 (HP1) binds, further recruiting Suv4-20h enzymes, which trimethylate H4K20, reinforcing the heterochromatic state of pericentromeric chromatin (Schotta et al. 2004). HP1 dissociation from heterochromatin is regulated by the Aurora B phosphorylation of H3S10 in M-phase (Fischle et al. 2005). The levels of HP1 dissociation, however, seem to vary in different model systems. In *Drosophila* cultured cells, for instance, HP1 is still detectable on metaphase chromosomes (Rosic et al. 2014).

The presence of heterochromatin in pericentromeric regions is also required to ensure recruitment of cohesin protein complex, which holds sister chromatids together until anaphase onset (Sakuno et al. 2009; Yamagishi et al. 2008). Moreover, heterochromatin of pericentromeric regions is restricted to a particular portion as shown in *Schizosaccharomyces pombe*, where centromeres are surrounded by chromatin barriers containing tRNA genes (Scott et al. 2006). It is also important to note that there is a correlation between heterochromatin and neocentromere establishment at least in some species: neocentromere formation is reduced in mutants of the *S. pombe* H3K9 methyltransferase Clr4, suggesting that heterochromatin positively influences neocentromere formation (Ishii et al. 2008). Consistently, stable hotspots of overexpressed CENP-A in *Drosophila* cells are preferentially established at euchromatin/heterochromatin boundaries (Olszak et al. 2011).

While repressive marks on the canonical histones H3 or H4 play a role in establishing centromeric chromatin, histone variants (apart from CENP-A) can also do so but are less well understood. For instance, the H3 variant H3.3 replaces the canonical H3.1 in nucleosomes of pericentromeres, telomeres, and regions of active transcription in a replication-independent manner (Chow et al. 2005; Wirbelauer et al. 2005). During mitosis, H3.3S31 is phosphorylated by CHK1 at pericentromeric regions (Hake et al. 2005). H3.3S31ph then spreads along the chromosome arm in lagging or misaligned chromosomes, causing p53-dependent cell cycle arrest. Mitotic H3.3S31ph has been unraveled recently, as a sensor to promote nuclear p53 accumulation in aneuploidy daughter cells, thereby preventing and suppressing aneuploidy (Hinchcliffe et al. 2016).

6.2 Modifications of H3 Nucleosomes Adjacent to CENP-A Containing Nucleosomes

Examination of extended chromatin fibers has revealed that CENP-A nucleosomes occupy discrete domains that are interspersed with chromatin containing canonical histone H3 (Blower et al. 2002; Sullivan and Karpen 2004; Zinkowski et al. 1991). In contrast to the repressive marks at pericentromeres, canonical histone H3 within centromeric chromatin contains some marks that are usually specific for open chromatin, e.g., K36me2 (Bergmann et al. 2011). At the same time, other typical euchromatic modifications, such as acetylation of H3 and H4, are missing, but so are typically silent chromatin marks, such as H3K9me3 (Sullivan and Karpen 2004). The inhibition of HDAC activity by trichostatin A (TSA) leads to hyperacetylated centromeres and chromosome segregation defects in *S. pombe,* and prolonged mitotic arrest in HeLa cells (Shin et al. 2003), suggesting that histone marks are crucial for cell cycle progression and accurate segregation.

Using human artificial chromosomes (HACs), Bergmann et al. found H3K36me as a new centrochromatin modification. This modification is normally associated with transcription elongation, supporting observations that centromeres are transcriptionally active (Bergmann et al. 2011). This study also found that H3K4me2 plays a role in CENP-A maintenance. H3K4me2 depletion at the alphoid^{tetO} centromere of the HAC by tethering the lysine-specific demethylase 1 (LSD1) causes a reduction of CENP-A incorporation as a result of the loss of the CENP-A chaperone HJURP at centromeres, suggesting that this modification is involved in the recruitment of HJURP to centromeres.

A more recent study shows that CENP-A-proximal nucleosomes containing canonical histones are not uniformly modified, but bear complex combinations of PTMs. They confirm the presence of H3K4me2 and H3K36me2/3, and show that these modifications exist in combination with methylation (and some low levels of acetylation) on different lysines of the same histone, predominantly H3K9 (mono-, di-, and trimethylations), and H3K27 (mono, di, and trimethylation) (Bailey et al. 2015). Regulating the balance between H3K9ac, which promotes CENP-A assembly, and H3K9me3, which inhibits it, may be crucial not only for kinetochore assembly but also for genome stability (Ohzeki et al. 2012). In conclusion, the distinct combination of histone modifications associated with centrochromatin distinguishes it from bulk chromatin, thereby creating a chromatin environment crucial for facilitating the centromere function and its propagation.

In addition to establishing a unique chromatin environment, some marks established only during specific processes such as mitosis are also important for centromere function. For instance, the mitotic kinase haspin is responsible for H3T3 phosphorylation and this mark is specifically enriched at H3 nucleosomes of the centromeric core of mitotic chromosomes (Kelly et al. 2010; Wang et al. 2010; Yamagishi et al. 2010) and has been proposed to guarantee proper chromosome congression to the metaphase plate for faithful segregation of sister chromatids during anaphase (Dai and Higgins 2005).

6.3 Modifications of CENP-A Containing Nucleosomes

X-ray crystallography studies showed that human CENP-A and H4 interact to form a heterotetramer (Sekulic et al. 2010; Tachiwana et al. 2011). H4 associated with pre-nucleosomal CENP-A is acetylated in a manner that is essentially identical to H4 in complex with pre-nucleosomal H3. The three predominant acetylation sites of H4 are the α-N-terminus, which is modified constitutively during translation (Hole et al. 2011), and lysines K5 and K12, which are acetylated by histone acetyltransferase B (HAT B) (Chang et al. 1997). Acetylation of H4 at K5 and K12 is found within the pre-nucleosomal CENP-A-H4-HJURP complex and requires RbAp46/48 for its subsequent successful localization of CENP-A to centromeres (Shang et al. 2016).

In contrast to chromatin-associated centromeric H4, pre-nucleosomal CENP-A associated histone H4 lacks K20me (Bailey et al. 2015). H4K20me1 has been reported to be enriched at centromeres and essential for correct kinetochore assembly (Hori et al. 2014). Bailey et al. also found that the most abundant form of centromeric H4 in cycling cells bore H4K20me2. However, as they discussed, H4K20me2 is a common modification within general chromatin and the ubiquitous nature of H4K20me2 makes it unlikely to play a unique role in centromere identity.

In *S. cerevisiae* hypoacetylation of H4K16 at centromeres has been reported to be important for kinetochore function, since its deregulation leads to failures in chromosome segregation (Choy et al. 2011).

6.4 Modifications of CENP-A Itself

The CENP-A N-terminal tail is enriched in arginines and lacks most of the well-characterized lysines of histone H3 that are targets for modification. A divergence is not only found between CENP-A and other canonical histones or other variants, but also between CENP-A orthologs from different species. The N-terminal tails of CENP-A orthologs vary significantly in length (for instance, 20 aa in *S. pombe* versus 200 aa in *Caenorhabditis elegans*) (Smith 2002). Like other histones, CENP-A is also subjected to post-translational modifications. Depending on the modification, the effect will influence CENP-A stability, structure, or positioning. What is special to CENP-A modifications is that many of these PTMs affect the recruitment of kinetochore components. However, for most of the so far discovered PTMs of CENP-A we know very little about their catalysis, dynamics, and function.

Phosphorylation. The CENP-A N-terminus is phosphorylated on S16 and S18 already in prenucleosomal CENP-A, and these marks are important for reliable chromosome portioning during division (Bailey et al. 2013). The phosphorylation state of *Drosophila* CENP-A varies with its subnuclear localizations since mass spectrometry analysis detected cytoplasmic CENP-A peptides in unmodified, mono-, and dephosphorylated form (most prominently at S20ph and S75ph), while nucleoplasmic CENP-A peptides were only detected as unmodified and monophosphorylated (S77ph) peptide (Boltengagen et al. 2015).

Apart from regulating the passage of the newly synthesized protein through different pre-assembly complexes, phosphorylation of CENP-A can impair its deposition at centromeres. CDK1 phosphorylates CENP-A at S68, which interferes with CENP-A binding to its loading factor HJURP and, therefore, with its deposition to centromeric chromatin prior to mitotic exit (Yu et al. 2015; Zhao et al. 2016). At the time of CENP-A loading onto centromeric chromatin this phosphorylation is removed by the phosphatase PP1α. However, in long-term cell survival assays, S68 phosphorylation seems dispensable for CENP-A function and cellular survival, challenging the finding that S68 phosphorylation is necessary for CENP-A recognition by HJURP and therefore faithful loading (Fachinetti et al. 2017).

CENP-A is also phosphorylated by Aurora A and B at S7 and this modification is required for mitotic progression and proper kinetochore function. CENP-A-S7ph is initially established by Aurora A in prophase and this is required for Aurora B restriction to the inner centromere, the maintenance of CENP-A phosphorylation at S7 by Aurora B from late prophase to metaphase, and for recruiting the inner kinetochore protein CENP-C (Goutte-Gattat et al. 2013; Kunitoku et al. 2003). The maize ortholog of CENP-A is phosphorylated at S50 during chromosome segregation, in a temporal pattern very similar to the S7ph in human CENP-A (Zhang et al. 2005).

Methylation. CENP-A N-terminus not only bears phosphorylation sites, but is also α-trimethylated on Gly1 by the N-terminal RCC1 methyltransferase NRMT, though it is unclear how NRMT activity is regulated (Bailey et al. 2013).

The α-N-methylation has previously been reported to facilitate the chromatin localization of the regulator of chromatin condensation 1 (RCC1), a key player in nucleocytoplasmic transport, mitosis, and nuclear envelope assembly (Chen et al. 2007). The addition of three methyl groups on CENP-A implies the introduction of a conformational change that mediates DNA interactions. The conservation of the CENP-A N-terminal motifs among different species and the fact that most of CENP-A nucleosomes carry G1me3 points to the importance of this modification.

In *S. cerevisiae* a single nucleosome defines the centromeric region and therefore constitutes a so-called point centromere (Morey et al. 2004). In the essential N-terminal domain of Cse4 (CENP-A ortholog), R37 is methylated and this modification is proposed to positively regulate the recruitment of the complete kinetochore complex and consequently control proper chromosome segregation (Samel et al. 2012).

Acetylation. Apart from the phosphorylations already discussed that affect CENP-A localization within the cell, CENP-A-K105ac has been described in *Drosophila* cytosolic prenucleosomal CENP-A but not in nuclear extracts (Boltengagen et al. 2015). Therefore, this prenucleosomal modification might be important for its association with chaperones and/or for its import into the nucleus, as has been shown for H4 acetylation at K5 and K12 (Lassallette et al. 2011). Human CENP-A has also been reported to be acetylated at K124 in G1/S-phase-derived cells, a residue located within the HFD closer to the C terminus. Adding an acetyl group to K124 neutralizes the positively charged lysine surface, supposedly loosening the DNA-histone interface and increasing the accessibility of the CENP-A nucleosomal interior to non-histone proteins or to chromatin remodelers (Bui et al. 2012). It was proposed that this CENP-A K124ac functions in "priming" or "blocking" CENP-A K124 for ubiquitylation until the M-phase.

Ubiquitylation. At the same residue as the previously discussed acetylation, CENP-A can be ubiquitylated (K124ub) by the CUL4A-RBX1-COPS8 complex in vivo and in vitro. Acetylation of CENP-A serves as a signal for its deposition at centromeres. The ubiquitylation at this residue occurs in the M and G1 phases and is required for efficient interaction with HJURP to properly localize CENP-A at centromeres and is, therefore, essential for CENP-A loading onto chromatin (Niikura et al. 2015). This study has recently been contradicted by Fachinetti et al. who found no evidence for CENP-A-K124ub to be important for loading or maintenance of CENP-A (Fachinetti et al. 2017). CENP-A mono-ubiquitylation seems epigenetically inherited through dimerization between cell divisions and this inheritance is important for the control of CENP-A deposition and maintenance at centromeres (Niikura et al. 2016). Similar to the human K124ub, mono-ubiquitylation of Drosophila CENP-A by the E3 ligase CUL3/RDX has been reported (Bade et al. 2014). Mono-ubiquitylation stabilizes CENP-A that is bound to its loading factor CAL1. The CAL1 interaction to the ubiquitin machinery mediates the mono-ubiquitylation of CENP-A and, therefore, its accurate loading, securing that chromosomes segregate correctly.

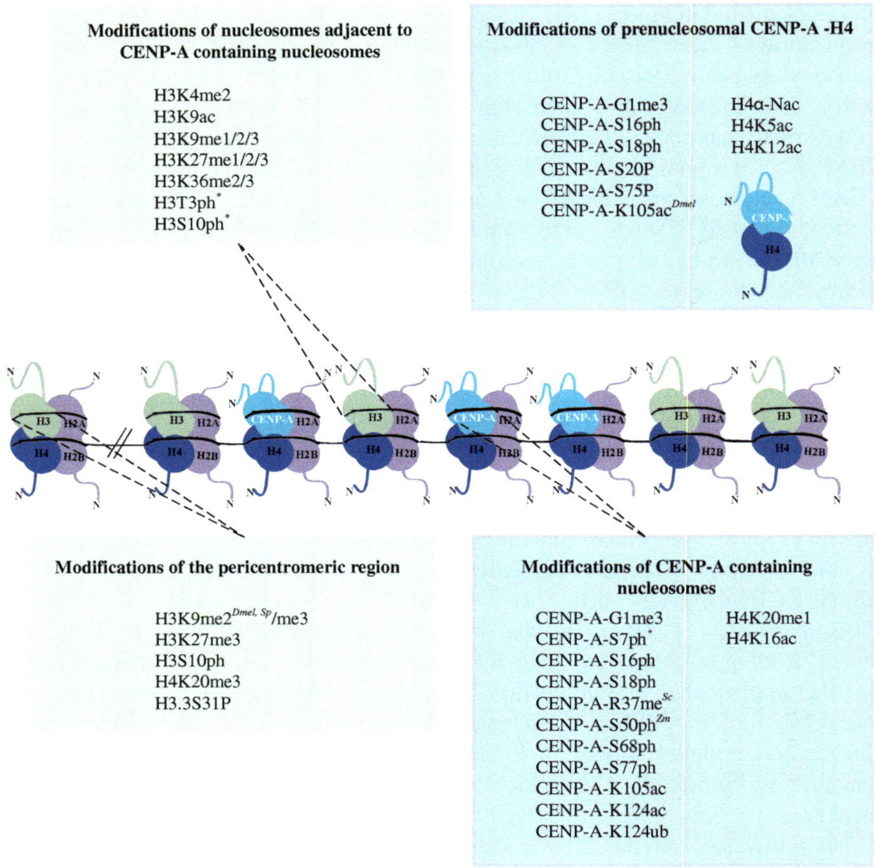

Fig. 2 PTMs of centromeric chromatin. Summary of the PTMs present at histones composing the core centromere and the pericentromeric region, and at CENP-A itself. Mitosis-specific PTMs are labeled with *asterisk* and PTMs specific of certain species are also marked (*Dmel Drosophila melanogaster*, *Sc Saccharomyzes cerevisiae*, *Sp Schizosaccharomyzes pombe*, *Zm Zea mays*)

Mono-ubiquitylation of CENP-A seems to be important for its stability, but ubiquitin is normally used to poly-ubiquitylate proteins, thereby marking them for degradation by the proteasome. In the case of CENP-A, proteolysis of residual, spare, or overproduced CENP-A helps to prevent its spreading into euchromatin in several organisms and restrict loading to centromeric chromatin only (Hewawasam et al. 2010; Moreno-Moreno et al. 2006, 2011; Ranjitkar et al. 2010).

Taken together, post-translational modifications present on centromeric chromatin at different levels impact the prenucleosomal assembly, nuclear import, and the pre-loading states of CENP-A, as well as the formation of centromeric chromatin and kinetochore formation, allowing centromeres to mediate faithful mitosis and meiosis (Fig. 2).

7 Phosphorylation of Non-histone Proteins, Regulatory Effect of Kinetochore Activity

As already mentioned, the interaction between the centromere-attached kinetochore and the microtubules ensures the precise segregation of chromosomes in mitosis as well as in meiosis. For accurate regulation, kinetochore components are also post-translationally modified in large numbers. The most commonly known PTM of kinetochore proteins is phosphorylation. Two prominent kinase families at the kinetochore are Polo and Aurora that phosphorylate many proteins at centromeric regions during the cell cycle. Nevertheless, other kinases such as mps1, haspin, Cdk1, or bub1 are key regulators of mitotic progression that function at or near centromeric chromatin and are essential for correct cell cycle progression in mitosis (Bayliss et al. 2012).

The Polo-like kinases (Plk) compose a family of structurally related Ser/Thr kinases that are highly conserved from yeast to humans. They have multiple cell cycle functions, e.g., coordinating the entry into M-phase by the activation and control of cyclin-dependent kinase 1 (CDK1) (Archambault and Glover 2009). Strikingly, Polo kinases localize to the centromeric and kinetochore regions and have been shown to be required for initial CENP-A deposition in human cells. Plk1, which is the most extensively studied among the four mammalian Plks, is required downstream of CENP-C localization, and its substrate is M18BP1, a subunit of the kinetochore protein family Mis18 complex. The phosphorylation of M18BP1 by Plk1 promotes the centromeric localization of Mis18 complex, thereby licensing centromeres for CENP-A deposition (McKinley and Cheeseman 2016). Plk1 localization has been suggested to be dependent on phosphorylation of inner centromere protein (INCENP) by CDK1. However, it is difficult to rule out whether these modifications directly mediate Plk1 localization to the kinetochore, or whether it is a secondary effect of other aspects in kinetochore assembly. Furthermore, Plk1 seems to act as a sensor of tension at the kinetochore. Plk1 phosphorylates BubR1 at S676 and thereby stabilizes kinetochore–microtubule interactions during mammalian mitosis (Elowe et al. 2007).

A second important family of conserved kinases involved in chromosome segregation is the family of Aurora protein kinases. This Ser/Thr-direct kinase family encompasses Aurora A, B, and C. Aurora B has been implicated in many cell cycle processes, including chromosome condensation, segregation, sister chromatid cohesion, and cytokinesis (Carmena et al. 2009). Aurora B is part of the chromosomal passenger complex (CPC), that it is also composed of INCENP, Borealin, and Survivin. Localization of the CPC is dynamic during mitosis and is an indication of the multiple roles of the CPC. Aurora B, along with CDK1, contributes to sister chromatid's resolution by phosphorylating the cohesion-stabilizing protein Soronin (Nishiyama et al. 2013). CPC also regulates kinetochore to microtubule attachments and activation of the mitotic checkpoint until chromosomes become bi-oriented (Muñoz-Barrera and Monje-Casas 2014). Whether and how these phosphorylated proteins interact with PTMs of centromeric histones or of other enzymes that act on histones is likely to be an area of intense research in the future.

Remarkably, DNA processes can also regulate kinase function. For instance, transcription at the centromere plays an important role in kinetochore assembly, since noncoding RNAs are required for regulating the activation and localization of Aurora B (Blower 2016; Ferri et al. 2009; Jambhekar et al. 2014).

8 Conclusion and Perspectives

In summary, a vast array of PTMs regulate centromeric chromatin and centromere function. Like the rest of the genome, modifications of centromeric chromatin exist in a unique pattern that specifies centromere identity. While previous research on PTMs relied on site-specific antibodies, these methods are replete with technical obstacles. As a result of the progress in protein mass spectrometry, many new aspects of these modifications have been unraveled. Single-cell epigenomic methods are also very rapidly developing, which have the potential to polish our understanding of histone modifications in a more detailed manner. Single-cell DamID could also support genome wide analysis of histone modifications by using Dam fusion with specific histone readers or modifiers.

Improving and refining our knowledge about histone modifications that occur in particular cells at defined moments will improve our understanding of how epigenetic processes crosstalk with one another, and their role in stemness, development, and disease.

Acknowledgements We apologize to those authors whose work we did not cite because of space limitations. We thank Anne-Laure Pauleau, Engin Demirdizen, and Aubry K. Miller for comments on the manuscript and the entire Erhardt lab for fruitful discussions. Work in the Erhardt laboratory is funded by the Deutsche Forschungsgemeinschaft (*EXC81, SFB1036, and ER576/2-2*) and the European Research Council (ERC-*CoG-682496*). AGA is a member of the Hartmut Hoffmann-Berling International Graduate School of Molecular and Cellular Biology.

Glossary

Histone code	It describes the hypothesis that the genetic information encoded in the DNA with a four-letter code is controlled by diverse post-translational modifications of histones which act in combination to provide binding sites for specific regulatory proteins depending on the combinatorial use of histone modifications.
PTM	Post-translational modifications are covalent modifications of proteins catalyzed by enzymes, which occur after proteins translation is completed.

References

Allis CD, Jenuwein T (2016) The molecular hallmarks of epigenetic control. Nat Rev Genet 17:487–500

Allshire RC, Karpen GH (2008) Epigenetic regulation of centromeric chromatin: old dogs, new tricks? Nature reviews. Genetics 9:923–937

Annunziato AT (2005) Split decision: what happens to nucleosomes during DNA replication? J Biol Chem 280:12065–12068

Apostolou E, Hochedlinger K (2013) Chromatin dynamics during cellular reprogramming. Nature 502:462–471

Archambault V, Glover DM (2009) Polo-like kinases: conservation and divergence in their functions and regulation. Nat Rev Mol Cell Biol 10:265–275

Arents G, Burlingame RW, Wang BC, Love WE, Moudrianakis EN (1991) The nucleosomal core histone octamer at 3.1 A resolution: a tripartite protein assembly and a left-handed superhelix. Proc Natl Acad Sci USA 88:10148–10152

Avvakumov N, Nourani A, Cote J (2011) Histone chaperones: modulators of chromatin marks. Mol Cell 41:502–514

Bade D, Pauleau AL, Wendler A, Erhardt S (2014) The E3 ligase CUL3/RDX controls centromere maintenance by ubiquitylating and stabilizing CENP-A in a CAL1-dependent manner. Dev Cell 28:508–519

Bailey AO, Panchenko T, Sathyan KM, Petkowski JJ, Pai P-J, Bai DL, Russell DH, Macara IG, Shabanowitz J, Hunt DF et al (2013) Posttranslational modification of CENP-A influences the conformation of centromeric chromatin. PNAS 110:11827–11832

Bailey AO, Panchenko T, Shabanowitz J, Lehman SM, Bai DL, Hunt DF, Black BE, Foltz DR (2015) Identification of the posttranslational modifications present in centromeric chromatin. Mol Cell Proteomics 15:918–931

Bannister AJ, Kouzarides T (2011) Regulation of chromatin by histone modifications. Cell Res 21:381–395

Bayliss R, Fry A, Haq T, Yeoh S (2012) On the molecular mechanisms of mitotic kinase activation. Open Biol 2:120–136

Bergmann JH, guez MGOMRI, Martins NMC, Kimura H, Kelly DA, Masumoto H, Larionov V, Jansen LET, Earnshaw WC (2011) Epigenetic engineering shows H3K4me2 is required for HJURP targeting and CENP-A assembly on a synthetic human kinetochore. EMBO J 30:328–340

Bird A (2007) Perceptions of epigenetics. Nature 447:396–398

Black BE, Foltz DR, Chakravarthy S, Luger K, Woods VL, Cleveland DW (2004) Structural determinants for generating centromeric chromatin. Nature 430:578–582

Blower MD (2016) Centromeric transcription regulates Aurora-B localization and activation. Cell Rep 15(8):1624–1633

Blower MD, Karpen GH (2001) The role of Drosophila CID in kinetochore formation, cell-cycle progression and heterochromatin interactions. Nat Cell Biol 3:730–739

Blower MD, Sullivan BA, Karpen GH (2002) Conserved organization of centromeric chromatin in flies and humans. Dev Cell 2:319–330

Boltengagen M, Huang A, Boltengagen A, Trixl L, Lindner H, Kremser L, Offterdinger M, Lusser A (2015) A novel role for the histone acetyltransferase Hat1 in the CENP-A/CID assembly pathway in *Drosophila melanogaster*. Nucl Acids Res 1–15

Bui M, Dimitriadis EK, Hoischen C, An E, Quénet D, Giebe S, Nita-Lazar A, Diekmann S, Dalal Y (2012) Cell-cycle-dependent structural transitions in the human CENP-A nucleosome in vivo. Cell 150:317–326

Carmena M, Ruchaud S, Earnshaw WC (2009) Making the Auroras glow: regulation of Aurora A and B kinase function by interacting proteins. Curr Opin Cell Biol 21:796–805

Cedar H, Bergman Y (2009) Linking DNA methylation and histone modification: patterns and paradigms. Nat Rev Genet 10:295–304

Chang L, Loranger SS, Mizzen C, Ernst SG, Allis CD, Annunziato AT (1997) Histones in transit: cytosolic histone complexes and diacetylation of H4 during nucleosome assembly in human cells. Biochemistry 36:469–480
Chen T, Muratore TL, Schaner-Tooley CE, Shabanowitz J, Hunt DF, Macara IG (2007) N-terminal alpha-methylation of RCC1 is necessary for stable chromatin association and normal mitosis. Nat Cell Biol 9:596–603
Chow CM, Georgiou A, Szutorisz H, Maia e Silva A, Pombo A, Barahona I, Dargelos E, Canzonetta C, Dillon N (2005) Variant histone H3.3 marks promoters of transcriptionally active genes during mammalian cell division. EMBO Rep 6:354–360
Choy JS, Acuna R, Au WC, Basrai MA (2011) A role for histone H4K16 hypoacetylation in *Saccharomyces cerevisiae* kinetochore function. Genetics 189:11–21
Cohen I, Poreba E, Kamieniarz K, Schneider R (2011) Histone modifiers in cancer: friends or foes? Genes Cancer 2:631–647
Dai J, Higgins JM (2005) Haspin: a mitotic histone kinase required for metaphase chromosome alignment. Cell Cycle 4:665–668
Davey CA, Sargent DF, Luger K, Maeder AW, Richmond TJ (2002) Solvent mediated interactions in the structure of the nucleosome core particle at 1.9 a resolution. J Mol Biol 319:1097–1113
Dunleavy EM, Zhang W, Karpen GH (2013) Solo or doppio: how many CENP-As make a centromeric nucleosome? Nat Struct Mol Biol 20:648–650
Elowe S, Hummer S, Uldschmid A, Li X, Nigg EA (2007) Tension-sensitive Plk1 phosphorylation on BubR1 regulates the stability of kinetochore microtubule interactions. Genes Dev 21:2205–2219
Fachinetti D, Logsdon GA, Abdullah A, Selzer EB, Cleveland DW, Black BE (2017) CENP-A modifications on Ser68 and Lys124 are dispensable for establishment, maintenance, and long-term function of human centromeres. Dev Cell 40:104–113
Falk SJ, Guo LY, Sekulic N, Smoak EM, Mani T, Logsdon GA, Gupta K, Jansen LET, Van Duyne GD, Vinogradov SA et al (2015) CENP-C reshapes and stabilizes CENP-A nucleosomes at the centromere. Science 348:699–703
Falkenberg KJ, Johnstone RW (2014) Histone deacetylases and their inhibitors in cancer, neurological diseases and immune disorders. Nat Rev Drug Discovery 13:673–691
Ferri F, Bouzinba-Segard H, Velasco G, Hube F, Francastel C (2009) Non-coding murine centromeric transcripts associate with and potentiate Aurora B kinase. Nucleic Acids Res 37:5071–5080
Filion GJ, Bemmel JGV, Braunschweig U, Talhout W, Kind J, Ward LD, Castro JD, Kerkhoven RM, Bussemaker HJ, Steensel BV et al (2010) Systematic protein location mapping reveals five principal chromatin types in Drosophila cells. Cell 143:212–224
Fischle W, Tseng BS, Dormann HL, Ueberheide BM, Garcia BA, Shabanowitz J, Hunt DF, Funabiki H, Allis CD (2005) Regulation of HP1-chromatin binding by histone H3 methylation and phosphorylation. Nature 438:1116–1122
Goldberg AD, Allis CD, Bernstein E (2007) Epigenetics: a landscape takes shape. Cell 128:635–638
Goutte-Gattat D, Shuaib M, Ouararhni K, Gautier T, Skoufias DA, Hamiche A, Dimitrov S (2013) Phosphorylation of the CENP-A amino-terminus in mitotic centromeric chromatin is required for kinetochore function. Proc Natl Acad Sci USA 110:8579–8584
Grewal SIS, Elgin SCR (2007) Transcription and RNA interference in the formation of heterochromatin. Nature 447:399–406
Hake SB, Garcia Ba, Kauer M, Baker SP, Shabanowitz J, Hunt DF, Allis CD (2005) Serine 31 phosphorylation of histone variant H3.3 is specific to regions bordering centromeres in metaphase chromosomes. Proc Natl Acad Sci USA 102:6344–6349
Hatakeyama A, Hartmann B, Travers A, Nogues C, Buckle M (2016) High-resolution biophysical analysis of the dynamics of nucleosome formation. Sci Rep 6:1–14
Heard E, Martienssen RA (2014) Transgenerational epigenetic inheritance: myths and mechanisms. Cell 157:95–109

Hewawasam G, Shivaraju M, Mattingly M, Venkatesh S, Martin-Brown S, Florens L, Workman JL, Gerton JL (2010) Psh1 is an E3 ubiquitin ligase that targets the centromeric histone variant Cse4. Mol Cell 40:444–454

Hinchcliffe EH, Day CA, Karanjeet KB, Fadness S, Langfald A, Vaughan KT, Dong Z (2016) Chromosome missegregation during anaphase triggers p 53 cell cycle arrest through histone H3.3 Ser31 phosphorylation. Nat Cell Biol 18:668–675

Hole K, van Damme P, Dalva M, Aksnes H, Glomnes N, Varhaug JE, Lillehaug JR, Gevaert K, Arnesen T (2011) The human N-Alpha-acetyltransferase 40 (hNaa40p/hNatD) is conserved from yeast and N-terminally acetylates histones H2A and H4. PLoS One 6:1–11

Hori T, Shang WH, Toyoda A, Misu S, Monma N, Ikeo K, Molina O, Vargiu G, Fujiyama A, Kimura H et al (2014) Histone H4 Lys 20 monomethylation of the CENP-A nucleosome is essential for kinetochore assembly. Dev Cell 29:740–749

Hsu J-Y, Sun Z-W, Li X, Reuben M, Tatchell K, Bishop DK, Grushcow JM, Brame CJ, Caldwell JA, Hunt DF et al (2000) Mitotic phosphorylation of histone H3 is governed by Ipl1/Aurora kinase and Glc7/PP1 phosphatase in budding yeast and nematodes. Cell 102:279–291

Ishii K, Ogiyama Y, Chikashige Y, Soejima S, Masuda F, Kakuma T, Hiraoka Y, Takahashi K (2008) Heterochromatin integrity affects chromosome reorganization after centromere dysfunction. Science 321:1088–1091

Jambhekar A, Emerman AB, Schweidenback CT, Blower MD (2014) RNA stimulates Aurora B kinase activity during mitosis. PLoS One 9:e100748

Jenuwein T, Allis CD (2001) Translating the histone code. Science 293:1074–1081

Jones PA, Liang G (2009) Rethinking how DNA methylation patterns are maintained. Nat Rev Genet 10:805–811

Karpen GH, Allshire RC (1997) The case for epigenetic effects on centromere identity and function. Trends Genet: TIG 13:489–496

Kelly AE, Ghenoiu C, Xue JZ, Zierhut C, Kimura H, Funabiki H (2010) Survivin reads phosphorylated histone H3 threonine 3 to activate the mitotic kinase Aurora B. Science 330:235–239

Kops GJPL, Weaver BAA, Cleveland DW (2005) On the road to cancer: aneuploidy and the mitotic checkpoint. Nat Rev Cancer 5:773–785

Kouzarides T (2007) Chromatin modifications and their function. Cell 128:693–705

Kunitoku N, Sasayama T, Marumoto T, Zhang D, Honda S, Kobayashi O, Hatakeyama K, Ushio Y, Saya H, Hirota T (2003) CENP-A phosphorylation by Aurora-A in prophase is required for enrichment of Aurora-B at inner centromeres and for kinetochore function. Dev Cell 5:853–864

Lam AL, Boivin CD, Bonney CF, Rudd MK, Sullivan BA (2006) Human centromeric chromatin is a dynamic chromosomal domain that can spread over noncentromeric DNA. Proc Natl Acad Sci USA 103:4186–4191

Lassallette AE, Mocquard E, Arnaud M-C, Thiriet C, Matera AG (2011) H4 replication-dependent diacetylation and Hat1 promote S-phase chromatin assembly in vivo. Mol Biol Cell 2011 (22):245–255

Lima de Faria A (1949) Genetics, origin and evolution of kinetochores. Hereditas 35:422–444

Luger K, Mader AW, Richmond RK, Sargent DF, Richmond TJ (1997) Crystal structure of the nucleosome core particle at 2.8 A resolution. Nature 389:251–260

McKinley KL, Cheeseman IM (2016) The molecular basis for centromere identity and function. Nat Rev Mol Cell Biol 17:16–29

Moreno-Moreno O, Medina-Gir S, Torras-Llort M, Azorn NF (2011) The F box protein partner of paired regulates stability of drosophila centromeric histone H3, CenH3 CID. Curr Biol 21:1488–1493

Moreno-Moreno O, Torras-Llort MN, Azor NF (2006) Proteolysis restricts localization of CID, the centromere-specific histone H3 variant of Drosophila, to centromeres. Nucl Acids Res 34:6247–6255

Morey L, Barnes K, Chen Y, Fitzgerald-Hayes M, Baker RE (2004) The histone fold domain of Cse4 is sufficient for CEN targeting and propagation of active centromeres in budding yeast. Eukaryot Cell 3:1533–1543

Muñoz-Barrera M, Monje-Casas F (2014) Increased Aurora B activity causes continuous disruption of kinetochore-microtubule attachments and spindle instability. Proc Natl Acad Sci USA 111:E3996–E4005

Murphy WJ, Larkin DM, Everts-van der Wind A, Bourque G, Tesler G, Auvil L, Beever JE, Chowdhary BP, Galibert F, Gatzke L et al (2005) Dynamics of mammalian chromosome evolution inferred from multispecies comparative maps. Science 309:613–617

Niikura Y, Kitagawa R, Kitagawa K (2016) CENP-A ubiquitylation is inherited through dimerization between cell divisions. Cell Rep 15:61–76

Niikura Y, Kitagawa R, Ogi H, Abdulle R, Pagala V, Kitagawa K (2015) CENP-A K124 ubiquitylation is required for CENP-A deposition at the centromere. Dev Cell 32(5):589–603

Nishiyama T, Sykora MM, Mechtler K, Peters J-M (2013) Aurora B and Cdk1 mediate WapI activation and release of acetylated cohesin from chromosomes by phosphorylating Sororin. Proc Natl Acad Sci 110:13404–13409

Noma KI, Allis CD, Grewal SIS (2001) Transitions in distinct histone H3 methylation patterns at the heterochromatin domain boundaries. Science 293:1150–1155

Ohzeki J-I, Bergmann JH, Kouprina N, Noskov VN, Nakano M, Kimura H, Earnshaw WC, Larionov V, Masumoto H (2012) Breaking the HAC barrier: histone H3K9 acetyl/methyl balance regulates CENP-A assembly. EMBO J 31:2391–2402

Olszak AM, van Essen D, Pereira AJ, Diehl S, Manke T, Maiato H, Saccani S, Heun P (2011) Heterochromatin boundaries are hotspots for de novo kinetochore formation. Nat Cell Biol 13:799–808

Partridge JF, Borgstrøm B, Allshire RC (2000) Distinct protein interaction domains and protein spreading in a complex centromere. Genes Dev 14:783–791

Peters AHFM, Kubicek S, Mechtler K, O'Sullivan RJ, Derijck AAHA, Perez-Burgos L, Kohlmaier A, Opravil S, Tachibana M, Shinkai Y et al (2003) Partitioning and plasticity of repressive histone methylation states in mammalian chromatin. Mol Cell 12:1577–1589

Petruk S, Sedkov Y, Johnston DM, Hodgson JW, Black KL, Kovermann SK, Beck S, Canaani E, Brock HW, Mazo A (2012) TrxG and PcG proteins but not methylated histones remain associated with DNA through replication. Cell 150:922–933

Przewloka MR, Glover DM (2009) The kinetochore and the centromere: a working long distance relationship. Annu Rev Genet 43:439–465

Ranjitkar P, Press MO, Yi X, Baker R, MacCoss MJ, Biggins S (2010) An E3 ubiquitin ligase prevents ectopic localization of the centromeric histone H3 variant via the centromere targeting domain. Mol Cell 40:455–464

Rice JC, Briggs SD, Ueberheide B, Barber CM, Shabanowitz J, Hunt DF, Shinkai Y, Allis CD (2003) Histone methyltransferases direct different degrees of methylation to define distinct chromatin domains. Mol Cell 12:1591–1598

Rosic S, Kohler F, Erhardt S (2014) Repetitive centromeric satellite RNA is essential for kinetochore formation and cell division. J Cell Biol 207:335–349

Sakuno T, Tada K, Watanabe Y (2009) Kinetochore geometry defined by cohesion within the centromere. Nature 458:852–858

Samel A, Cuomo A, Bonaldi T, Ehrenhofer-Murray AE (2012) Methylation of CenH3 arginine 37 regulates kinetochore integrity and chromosome segregation. Proc Natl Acad Sci 109:9029–9034

Schotta G, Lachner M, Sarma K, Ebert A, Sengupta R, Reuter G, Reinberg D, Jenuwein T (2004) A silencing pathway to induce H3-K9 and H4-K29 trimethylation at constitutive heterochromatin. Genes Dev 18:1251–1262

Scott KC, Merrett SL, Willard HF (2006) A heterochromatin barrier partitions the fission yeast centromere into discrete chromatin domains. Curr Biol 16:119–129

Sekulic N, Bassett EA, Rogers DJ, Black BE (2010) The structure of (CENP-A-H4)(2) reveals physical features that mark centromeres. Nature 467:347–351

Shang WH, Hori T, Westhorpe FG, Godek KM, Toyoda A, Misu S, Monma N, Ikeo K, Carroll CW, Takami Y et al (2016) Acetylation of histone H4 lysine 5 and 12 is required for CENP-A deposition into centromeres. Nat Commun 7:13465
Sherman JM, Pillus L (1997) An uncertain silence. Trends Genet 13:308–313
Shin H-J, Baek K-H, Jeon A-H, Kim S-J, Jang K-L, Sung Y-C, Kim C-M, Lee C-W (2003) Inhibition of histone deacetylase activity increases chromosomal instability by the aberrant regulation of mitotic checkpoint activation. Oncogene 22:3853–3858
Smith MM (2002) Centromeres and variant histones: what, where, when and why? Curr Opin Cell Biol 14:279–285
Stein WD (1980) The epigenetic address: a model for embryological development. J Theor Biol 82:663–677
Sullivan BA, Karpen GH (2004) Centromeric chromatin exhibits a histone modification pattern that is distinct from both euchromatin and heterochromatin. Nat Struct Mol Biol 11:1076–1083
Sullivan KF, Hechenberger M, Masri K (1994) Human CENP-A contains a histone H3 related histone fold domain that is required for targeting to the centromere. J Cell Biol 127:581–592
Tachiwana H, Kagawa W, Shiga T, Osakabe A, Miya Y, Saito K, Hayashi-Takanaka Y, Oda T, Sato M, Park S-Y et al (2011) Crystal structure of the human centromeric nucleosome containing CENP-A. Nature 476:232–235
Turner BM (2007) Defining an epigenetic code. Nat Cell Biol 9:2–6
Wang F, Dai J, Daum JR, Niedzialkowska E, Banerjee B, Stukenberg PT, Gorbsky GJ, Higgins JM (2010) Histone H3 Thr-3 phosphorylation by haspin positions Aurora B at centromeres in mitosis. Science 330:231–235
Wirbelauer C, Bell O, Schubeler D (2005) Variant histone H3.3 is deposited at sites of nucleosomal displacement throughout transcribed genes while active histone modifications show a promoter-proximal bias. Genes Dev 19:1761–1766
Yamagishi Y, Honda T, Tanno Y, Watanabe Y (2010) Two histone marks establish the inner centromere and chromosome bi-orientation. Science 330:239–243
Yamagishi Y, Sakuno T, Shimura M, Watanabe Y (2008) Heterochromatin links to centromeric protection by recruiting shugoshin. Nature 455:251–255
Yu Z, Zhou X, Wang W, Deng W, Fang J, Hu H, Wang Z, Li S, Cui L, Shen J et al (2015) Dynamic phosphorylation of CENP-A at Ser68 orchestrates its cell-cycle-dependent deposition at centromeres. Dev Cell 32:68–81
Zhang X, Li X, Marshall JB, Zhong CX, Dawe RK (2005) Phosphoserines on maize centromeric histone H3 and histone H3 demarcate the centromere and pericentromere during chromosome segregation. Plant Cell 17:572–583
Zhao H, Bui M, Dalal Y, Papoian GA (2016) Promiscuous histone mis-assembly is actively prevented by chaperones. J Am Chem Soc 138:13207
Zinkowski RP, Meyne J, Brinkley BR (1991) The centromere-kinetochore complex: a repeat subunit model. J Cell Biol 113:1091–1110

Centromere Silencing Mechanisms

Shannon M. McNulty and Beth A. Sullivan

Abstract Centromere function is essential for genome stability and chromosome inheritance. Typically, each chromosome has a single locus that consistently serves as the site of centromere formation and kinetochore assembly. Decades of research have defined the DNA sequence and protein components of functional centromeres, and the interdependencies of specific protein complexes for proper centromere assembly. Less is known about how centromeres are disassembled or functionally silenced. Centromere silencing, or inactivation, is particularly relevant in the cases of dicentric chromosomes that occur via genome rearrangements that place two centromeres on the same chromosome. Dicentrics are usually unstable unless one centromere is inactivated, thereby allowing the structurally dicentric chromosome to behave like one of the monocentric, endogenous chromosomes. The molecular basis for centromere inactivation is not well understood, although studies in model organisms and in humans suggest that both genomic and epigenetic mechanisms are involved. In this chapter, we review recent studies using synthetic chromosomes and engineered or induced dicentrics from various organisms to define the molecular processes that are involved in the complex process of centromere inactivation.

S.M. McNulty · B.A. Sullivan (✉)
Department of Molecular Genetics and Microbiology,
Duke University Medical Center, DUMC 3054, Durham, NC 27710, USA
e-mail: beth.sullivan@duke.edu

S.M. McNulty
e-mail: shannon.mcnulty@duke.edu

S.M. McNulty · B.A. Sullivan
Division of Human Genetics, Duke University Medical Center,
DUMC 3054, Durham, NC 27710, USA

B.E. Black (ed.), *Centromeres and Kinetochores*, Progress in Molecular
and Subcellular Biology 56, DOI 10.1007/978-3-319-58592-5_10

1 Introduction

The centromere is an essential chromosomal region that is defined as the visible primary constriction of eukaryotic chromosomes. It is the locus that segregates chromosomes in mitosis and meiosis and is important for maintaining sister chromatid (mitosis) or homolog (meiosis) attachments prior to mitosis. Each chromosome must contain one, and usually only one, centromere. Chromosomes lacking centromeres are lost during cell division, and those with more than one centromere often undergo anaphase lag and/or chromosome breakage and can trigger genome instability. Defective centromeres are associated with aneuploidy, birth defects, infertility, cancer, and aging.

Centromeres are largely defined epigenetically, by the presence of the histone H3 variant CENP-A that replaces H3 in a subset of centromeric nucleosomes to create a unique type of chromatin that distinguishes the centromere from the rest of the genome (Blower et al. 2002; Palmer et al. 1991). Centromeric chromatin exists as a large domain of interspersed CENP-A and H3 nucleosomes. The location at which CENP-A nucleosomes accumulate is typically restricted to one region per chromosome. This site is marked and inherited as the centromere due to the incorporation of new CENP-A every cell cycle. CENP-A incorporation into nucleosomes triggers the assembly of the constitutive centromere-associated network (CCAN) that connects chromatin within the centromere to the outer kinetochore where microtubule attachments are made.

Much research has been focused on determining the biological processes underlying proper centromere assembly and the achievement of optimal kinetochore architecture that is required for microtubule attachment and chromosome movement toward spindle poles (Izuta et al. 2006; McAinsh et al. 2003; Cheeseman et al. 2008; Suzuki et al. 2015). It has been equally important, though, to determine how centromeres are functionally silenced, or turned off. This is relevant for developing therapeutic approaches to target specific centromeres in different types of diseases. In cases of congenital aneuploidies, such approaches could potentially return a cell to a normal diploid state. Targeting specific, or multiple, centromeres in tumor cells could hold promise for chemotherapies with fewer negative consequences for the normal cells of the patient.

This chapter encompasses current knowledge in the field with respect to mechanisms that silence or inactivate centromeres. We discuss studies from various organisms spanning unicellular yeasts to humans that have focused on endogenous, synthetic, and variant centromeres. These investigations have coordinately increased our knowledge of the factors that are required to disable centromeres.

2 Silencing Endogenous Centromeres

Centromere assembly requires a wide array of proteins to build the many structural and functional levels of the centromere and kinetochore. Recruitment of CENP-A to centromeres is achieved through the cooperative action of the CENP-A chaperone HJURP and the Mis18 complex (Barnhart et al. 2011; Foltz et al. 2009; Hayashi et al. 2004). The 16-subunit complex CCAN is important for creating a platform to link centromeric chromatin of the inner kinetochore to the KMN protein network that connects the kinetochore to spindle microtubules (Cheeseman et al. 2008; Foltz et al. 2006; Hori et al. 2008; McKinley et al. 2015; Nishino et al. 2012, 2013; Suzuki et al. 2015). Genetic deletions or RNAi depletions of almost any of the proteins that contribute to centromeric chromatin or are part of the CCAN and KMN networks will reduce centromere function and inhibit centromere assembly (Carroll et al. 2009; Cheeseman et al. 2008; Fachinetti et al. 2013; Foltz et al. 2006; Hori et al. 2008; Westhorpe and Straight 2013). However, these mutations or protein depletions affect all centromeres equally, leading to cell arrest or death, and have been less useful for understanding the process by which a centromere is progressively stripped of function. A recent study showed that centromere assembly requires at least two steps: (1) establishment of centromere memory by incorporation of CENP-A into chromatin and (2) stabilization of the incorporated CENP-A by coordinated binding to CENP-B and CENP-C (Fachinetti et al. 2013). Centromere silencing, by extension, must involve, to some extent, destabilization of the interactions between CENP-A, CENP-B, and CENP-C, as well as erasure of centromere memory invoked by existing CENP-A to inhibit new CENP-A loading. Understanding how CENP-A is removed and/or prevented from loading at a fully functional centromere has been explored in a variety of contexts, including synthetic centromeres and engineered dicentric chromosomes.

3 Induced Centromere Silencing

In addition to intrinsic mechanisms of centromere silencing, centromeres can also be silenced by ectopically tethering proteins to the centromere domain to alter the level of transcription and the chromatin environment, ultimately inducing kinetochore loss. Studies using this approach have supported a link between specialized centromeric chromatin and kinetochore function, highlighting the importance of a fine balance of euchromatin and heterochromatin and a moderate level of transcription for proper centromere function and maintenance and prevention of centromere inactivation.

3.1 *Human Artificial Chromosomes as Tools to Study Centromere Silencing*

The majority of protein-tethering experiments have utilized human artificial chromosomes (HACs), which are engineered chromosomes that contain a centromere and are faithfully segregated with the endogenous chromosomes over many cell cycles, even in the absence of selection. HACs are advantageous for the study of centromere epigenetics and centromere inactivation because their dysfunction or loss does not affect the stability of the endogenous chromosomes or the viability of mammalian cells at large, allowing the extreme limits of centromere function to be tested. Additionally, changes to the centromere that result in inactivation are easily and rapidly assessed by measuring loss of the entire HAC.

HACs have been created by two methods: telomere-directed truncation, also known as "top-down" assembly, which removes chromosome arms by telomere-mediated truncation to leave only the centromere and the immediately surrounding region (Farr et al. 1991), and by "bottom-up" assembly, which creates de novo chromosomes by the transfection of naked genomic DNA and synthetic alpha satellite DNA capable of centromere formation (Harrington et al. 1997; Ikeno et al. 1998). The first generation HACs constructed with these methods contained centromeres derived from endogenous chromosomes or synthetically assembled alpha satellite sequences and, while useful for determining the minimum sequence requirements to form a viable, faithfully segregating chromosome, they were not amenable to directed epigenetic alteration to study centromere inactivation. The second generation of HACs provided this ability by replacing the CENP-B box sequence of every other monomer of the approximately 100 kb alpha satellite array with a tetracycline operator (tetO) sequence to form alphoid^{tetO}-HACs (Fig. 1a) (Nakano et al. 2008). Using the tetracycline repressor-operator (tetR-tetO) system, specific proteins fused to the tetracycline repressor (tetR) can be tethered to the tetO sequence to induce changes in the level of transcription or chromatin accessibility. These alterations affect only the HAC, as the tetO sequence does not naturally occur within the human genome, allowing epigenetic manipulation of a single, nonessential centromere. Additionally, HAC centromeres assemble centromeric chromatin similar to that observed in endogenous human centromeres (Sullivan and Karpen 2004; Blower et al. 2002), containing a continuous CENP-A and H3K4me2 domain (Fig. 1a) (Lam et al. 2006; Nakano et al. 2008), and are highly stable. HACs closely mimic normal centromeres and are powerful tools for investigating processes of centromere inactivation.

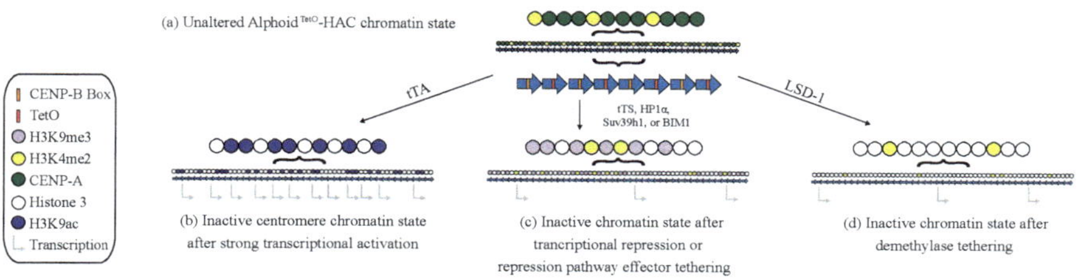

Fig. 1 Silencing of synthetic centromeres. **a** HAC centromeres used in induced inactivation experiments are assembled on synthetic alpha satellite arrays, which have been modified to contain a tet operator sequence in every other monomer. Like endogenous human centromeres, alphoid^{TetO} arrays contain CENP-A and the euchromatic H3K4me2 histone mark. **b** Centromere inactivation is induced by promoting strong transcription through the alphoid^{TetO} array by tethering tTA, which leads to a loss of H3K4me2 and CENP-A, and an influx of H3K9ac marks. **c** Tethering a transcription repressor (tTS) or proteins involved in constitutive heterochromatin and polycomb repression pathways to the alphoid^{TetO} centromere causes H3K9me3 incorporation, inducing a more heterochromatic state with reduced transcription and leading to CENP-A loss and centromere inactivation. **d** Targeting of the demethylase LSD-1 to tetO sequences integrated into the alphoid DNA causes a loss of both H3K4me2 and CENP-A, leading to the gradual inactivation of the centromere, although heterochromatin does not spread into the region

3.2 *Transcriptional Activation of Artificial Centromeres to Induce Silencing*

Although early experiments in budding yeast suggested the incompatibility of centromere function and high levels of transcription (Hill and Bloom 1987), the exact relationship between transcription and centromere activity remained unclear. To directly test the ability of a human centromere to withstand transcription, two transcriptional activators with variable levels of transcriptional induction were tethered to the centromere of an alphoid^{tetO}-HAC. Tethering of tetracycline-controlled activator (tTA), a fusion protein of tetR and a transactivating domain derived from the Herpes simplex virus protein 16 (VP16) (Gossen et al. 1995), increased transcription 150-fold and moderately increased H3K9 acetylation within the alphoid^{tetO} centromere (Fig. 1b) (Nakano et al. 2008; Bergmann et al. 2012). Within 2 days, tTA tethering led to a dramatic reduction in CENP-A loading, centromere inactivation, and loss of the HAC. In contrast, when NF-κB p65, a transcription factor that induces a lower level of transcription, is tethered to the HAC, the centromere remains functional, despite a 10-fold increase in transcription and a moderate increase in H3K9 acetylation (Bergmann et al. 2012). These experiments suggest that, although some level of transcription and a resulting increase in chromatin accessibility is tolerated within the centromere, the plasticity of centromeres is limited, with the combination of strong transcription and hyperacetylation being incompatible with active centromere function.

3.3 Transcriptional Repression of Artificial Centromeres

Centromere inactivation can also be induced by tethering the tetracycline-controlled silencer tTS, a fusion protein of tetR and the KRAB repressor domain of human Kid-1 protein (Witzgall et al. 1994; Freundlieb et al. 1999) to the centromere of an alphoid^{tetO}-HAC to repress transcription (Nakano et al. 2008). tTS tethering results in dramatically reduced levels of alphoid^{tetO} transcripts and loss of H3K4me2 (Fig. 1c). As a result, H3K9me3 levels increased, suggesting heterochromatinization of the alphoid^{tetO} array. Additionally, centromere proteins, including CENP-A, CENP-B, and CENP-C were greatly reduced or undetectable, leading to centromere inactivation and the loss of nearly all HACs within 14 days. This suggests that the CENP-A domain cannot be maintained in a heterochromatic environment completely void of transcription, which is consistent with recent reports of low levels of transcription within endogenous human centromeres (Chan et al. 2012), as well as neocentromeres (Saffery et al. 2003). However, whether the loss of transcription alone or resulting structural changes in the chromatin are responsible for the observed HAC centromere inactivation was unclear. This question was further addressed by additional protein-tethering experiments that changed the chromatin environment of HAC centromeres.

3.4 Chromatin State Switching on Artificial Centromeres

Directly activating and repressing transcription of the alphoid^{tetO} centromere leads to dramatic changes in the chromatin environment, however, the exact players involved in this alteration and, ultimately, in centromere silencing, were unknown. Further experiments to directly tether proteins involved in altering chromatin accessibly have helped narrow down the mechanisms involved in centromere silencing when levels of transcription are altered. Two candidate pathways were investigated for involvement in HAC centromere inactivation: (1) the constitutive heterochromatin repression pathway, which is centered on H3K9me3 histone marks produced by Suv39H1 and transcriptional silencing through the action of HP1α (Saksouk et al. 2015), and (2) the polycomb repression pathway, which involves the methylation of H3K27 and subsequent recruitment of chromatin compaction and transcriptional silencing effectors, such as Polycomb Recruitment Complex 1 (PRC1) (Simon and Kingston 2009). Using HAC-based protein-tethering approaches, there is evidence for involvement of proteins and histone marks in both of these pathways in centromere inactivation.

As described above, tTS tethering to the alphoid^{tetO} centromere leads to heterochromatinization and centromere loss, but could be accomplished by multiple pathways. HP1α, a highly conserved protein involved in transcriptional silencing and heterochromatin formation (Eissenberg and Elgin 2014), was first identified as a major player involved in HAC centromere inactivation following tTS tethering

(Nakano et al. 2008). Unlike tTA, tTS tethering leads to accumulation of HP1α at the kinetochore. HP1α targeting to the alphoid^{tetO} array alone induced centromere protein loss and HAC destabilization comparable to the effects seen with tTS tethering (Fig. 1c). Further evidence of the incompatibility of heterochromatin and centromere function and a role for the constitutive heterochromatin repression pathway in centromere inactivation comes from experiments to tether the histone methyltransferase Suv39h1 to the alphoid^{tetO} array (Ohzeki et al. 2012). Tethering full-length Suv39h1 leads to an increase in H3K9me3 levels and a reduction in loading of new CENP-A, centromere inactivation/HAC destabilization (Fig. 1c). However, tethering a truncated version of Suv36h1 that cannot recruit HP1α (Suv39h1^{SET}) does not lead to centromere inactivation, despite an increase in H3K9me3 within the alphoid^{tetO} array (Martins et al. 2016). Together, these studies suggested that centromere inactivation does not occur due to an increase in H3K9me3 levels alone and may require additional downstream effectors, like HP1α.

Additional HAC tethering experiments explored the succession of events that occurs during centromere inactivation. Tethering KAP1, a downstream effector of tTS involved in the constitutive heterochromatin repression pathway, also led to centromere protein loss and alphoid^{tetO} centromere inactivation (Cardinale et al. 2009). CENP-C and CENP-H were lost more rapidly than CENP-A, suggesting that CENP-A reduction follows the loss of other centromere proteins. These findings led the authors to propose a model in which KAP1 recruits chromatin modifiers, like HP1α, to destabilize CENP-C and CENP-H by weakening their binding to CENP-A or to H3-containing nucleosomes without affecting CENP-A. Only after these proteins are lost does dramatic CENP-A reduction occur.

A recent study has also implicated the polycomb repression pathway in centromere inactivation. The tethering of BIM1, a component of the PRC1 which is involved in reducing chromatin accessibility and blocking transcription, leads to reduced CENP-A and centromere inactivation at levels similar to those observed with Suv39h1, HP1α, and KAP1 tethering (Fig. 1c) (Martins et al. 2016). Interestingly, tethering of the upstream regulator of PRC1, EZH2, is not sufficient to substantially reduce alphoid^{tetO} array centromere protein levels or inactivate HAC centromeres.

Rather than inducing a repressive pathway, centromeres can also be silenced by the removal of H3K4me2 marks, which are interspersed with CENP-A containing nucleosomes in endogenous centromeres (Blower et al. 2002; Sullivan and Karpen 2004). Tethering LSD-1, a lysine-specific demethylase 1 which removes methyl groups from mono- or dimethylated H3K4 and H3K9, to the alphoid^{tetO} array, leads to a rapid loss of H3K4me2 levels, falling below detectable levels within 3 days (Fig. 1d) (Bergmann et al. 2011). Additionally, transcription of the alphoid^{tetO} array was dramatically reduced and the centromere was unable to recruit HJURP or, as a result, incorporate new CENP-A, leading to gradual inactivation. Interestingly, no increase in heterochromatic chromatin marks was observed, indicating that loss of H3K4me2 does not necessarily lead to an influx of H3K9me3, as observed in tTS tethering. These findings suggested transcriptional memory preserved by H3K4me2 might prevent centromere inactivation.

3.5 Induced HAC Centromere Inactivation as a Tool in Gene Therapy and Drug Screening

HACs are promising as gene delivery and expression vectors for potential therapeutic gene therapy use due to their potential to hold multiple kilobases of sequence, faithfully segregate over many cell cycles, and maintain the introduced DNA outside of the genome, preventing issues seen with integrative gene therapy vectors. The ability to conditionally inactivate the centromere of a HAC is also useful because it provides the ability to transiently induce gene expression.

In order to integrate full-length genes for expression, the alphoid^{tetO}-HAC (Cardinale et al. 2009) was modified to contain a loxP cassette for Cre-mediated recombination (Iida et al. 2010). As a proof of principle, an EGFP/HPRT transgene was introduced and demonstrated to robustly and stably express EGFP protein. By tTS tethering-mediated centromere inactivation, the HAC could be removed or "cured". This modified alphoid^{tetO}-HAC was taken one step further to demonstrate the utility of HACs with conditional centromeres as gene therapy vectors. Introduction of HACs containing full-length human genes *NBS1* and *VHL,* known to be mutated in von Hippel-Lindau syndrome and Nijmegen breakage syndrome, respectively, to human cell lines has corrected genetic deficiencies (Kim et al. 2011). These relatively large genes, 55 and 25 kb in length, were individually inserted into the loxP alphoid^{tetO}-HAC (Cardinale et al. 2009; Iida et al. 2010). In both cases, the HACs were maintained in the genome without integration and the protein of interest was expressed from the HAC. A number of physiological tests unique for each syndrome indicated that the HAC-encoded proteins successfully complement the genetic deficiency. Inactivation of the centromere by tTS tethering (Cardinale et al. 2009) was used "cure" the cells of the HAC. This induced centromere inactivation is useful as a control in proof-of-principle experiments and is also a promising tool for gene therapy treatments that require only transient protein expression and to eliminate the concern that the protein may integrate into the genome and lead to mutagenesis, as observed with viral vectors (Hacein-Bey-Abina et al. 2003). Similar proof-of-principle studies using a conditional HAC for gene delivery have expressed other, larger genes, including *BRCA1* (Kononenko et al. 2014). To further improve the efficiency of HAC centromere inactivation for gene function studies and gene therapy uses, an inducible tTA^{VP64} was added to the loxP alphoid^{tetO}-HAC (Kononenko et al. 2015). Previously described methods to cure cells of a HAC required an additional transfection of a tTS or tTA expression construct. By integrating a tTA containing four copies of the Herpes simplex VP16 transactivation domain (Gossen et al. 1995), transcription at the HAC can be strongly induced by doxycycline removal, generating a conditionally self-eliminating HAC.

HACs can also serve as a tool to efficiently identify drugs with the potential to induce chromosome instability (CIN), leading to the loss or gain of chromosomes. Such drugs can be used therapeutically to induce cancer cell death by elevating CIN beyond a tolerable level (Janssen et al. 2009; Silk et al. 2013). To rapidly measure chromosome loss, human cell lines containing an alphoid^{tetO}-HAC expressing EGFP were treated with a spectrum of drugs known to induce CIN and flow cytometry was used to measure HAC loss (observed as loss of EGFP expression) as a readout for elevated chromosome mis-segregation (Lee et al. 2013). This approach was further improved by alteration of the alphoid^{tetO}-HAC to express an EGFP-targeting shRNA and introduction into human cell lines stably expressing EGFP. In this system, loss of the HAC is measured (Kim et al. 2016) by gain of EGFP, as the EGFP is no longer targeted by the HAC-encoded shRNA following an increase in CIN that leads to HAC loss. Using this system, two inhibitors of CENP-E, a centromere-associated motor protein, and one inhibitor of CENP-F, an outer kinetochore protein, were identified as strong inducers of CIN. The potential for these drugs to be used to selectively kill cancer cells by centromere inactivation represents an exciting advance in both centromere biology and cancer therapeutics.

4 Dicentric Chromosomes and Centromere Inactivation

Dicentric chromosomes are the products of genome rearrangement that place two centromeres in proximity on the same chromosome. Dicentrics frequently form during repair of double strand breaks (DSBs) caused during recombination or by DNA damage, resulting in the joining of two different chromosomes or the fusion of sister chromatids. Dicentrics were first described in maize by the cytogeneticist McClintock (1939, 1941). They were observed to be inherently unstable, undergoing multiple rounds of anaphase breakage, bridge formation, and fusion of broken ends (i.e., the BFB cycle), leading to significant genome reshuffling and ongoing instability. Since then, dicentrics have been described in other model organisms where they are also largely unstable. Dicentric chromosomes in *Drosophila melanogaster* undergo mis-segregation and breakage, similar to dicentrics in corn (Novitski 1952). These early studies did not recover the products of dicentric segregation, lending to the conclusion that *Drosophila* are less tolerant of dicentric chromosomes than other organisms. However, more recent studies have captured the fate of experimentally produced dicentrics during *Drosophila* cell division. Most of these dicentrics undergo breakage, with some broken fragments transmitted to offspring (Ahmad and Golic 1998; Titen and Golic 2008). In the budding yeast *Saccharomyces cerevisiae*, dicentric chromosomes are unstable in both meiosis and mitosis, typically undergoing breakage-fusion-bridge cycles and other types of instability that rearrange the genome (Haber et al. 1984; Hill and Bloom 1989; Koshland et al. 1987; Pennaneach and Kolodner 2009). These findings that built on McClintock's original observations perpetuated the long-standing view that most, if not all, dicentrics are unstable.

The concept of stable dicentric chromosomes was not fully appreciated until the first description of human dicentric X chromosomes (dicX). These dicXs were isochromosomes, formed by breakage within the X short arm followed by repair that fused the two sister chromatids of the same chromosome. DicXs occur in ~ 15% of Turner Syndrome patients. Because these patients are viable and have a relatively normal lifespan, their dicXs were proposed to be stable due to inactivation of one centromere (de la Chapelle et al. 1966; Ockey et al. 1966). Subsequently, dicentric Robertsonian translocations (ROBs) have been identified in humans, occurring in every 1000 individuals. Over 90% of ROBs undergo centromere inactivation, stabilizing them so that they are inherited at high frequencies in both mitosis and meiosis (Bandyopadhyay et al. 2002; Niebuhr 1972; Sullivan and Schwartz 1995; Sullivan et al. 1994). As a result of these human cytogenetic findings, our early understanding of centromere inactivation came primarily from observational studies in humans. Cytologically, inactive centromeres lack a constricted appearance on metaphase chromosomes (Daniel 1979; Dewald et al. 1979; Sullivan et al. 1994; Therman et al. 1986). Moreover, nearly all essential centromere and kinetochore proteins and chromosomal and chromatin proteins are absent from inactive centromeres (Craig et al. 2003; Earnshaw and Migeon 1985; Earnshaw et al. 1989; Gascoigne and Cheeseman 2013; Sullivan and Schwartz 1995; Warburton et al. 1997; Page et al. 1995). Beyond these cytological observations, testing mechanisms of centromere inactivation have proven difficult. First, centromere inactivation was not considered to be a major mechanism of dicentric stability in model organisms, so little attention was paid to addressing the molecular basis in model systems. Moreover, until recently, there were no experimental systems to produce de novo dicentric human chromosomes, making it difficult to capture the process of inactivation in human cells where centromere inactivation is common. In the past five years, several labs have created dicentric chromosomes de novo in human cells, allowing the processes that regulate dicentric stabilization to finally be studied in detail. It has been shown recently that centromere inactivation does occur in model organisms, including maize, fission yeast, and *Drosophila* (Han et al. 2006; Sato et al. 2012; Agudo et al. 2000). This finding has galvanized the field to broadly consider mechanisms of centromere inactivation.

4.1 Experimental Models of Dicentric Chromosomes

How inactive centromeres lose centromere proteins, particularly CENP-A that serves as the foundation of the kinetochore and imparts centromere memory, is key to understanding how centromere silencing occurs. Experimentally produced dicentrics in yeasts, *Drosophila*, and mammalian cells have been important tools for uncovering several distinct pathways that render centromeres inactive. A variety of dicentrics have been experimentally created in different organisms (Fig. 2). Some are the results of induced translocations between endogenous chromosomes or integrations of native centromeric DNA at ectopic locations (Stimpson et al. 2010; Nakano et al. 2003; Ohzeki et al. 2016). In budding yeast, a conditional dicentric was created by placing a second copy of the centromere of chromosome III immediately downstream of the inducible GAL1 promoter (Hill and Bloom 1987; Bloom et al. 1989). Fission yeast (*Schizosaccharomyces pombe*) and *Drosophila* dicentrics have been created by Cre-lox or FRT-FLP recombination, respectively (Sato et al. 2012; Ahmad and Golic 1998; Hill and Golic 2015). These engineered chromosomes are the closest physiological models of naturally occurring dicentric chromosomes, since endogenous centromeres/native centromeric DNA are included on the dicentrics. Other experimentally produced dicentrics have been generated by ectopic insertion of lac operator (lacO) or tet operator (tetO) arrays into chromosomal arms (Chen et al. 2015; Mendiburo et al. 2011; Olszak et al. 2011; Gascoigne and Cheeseman 2013; Ho et al. 2014; Gascoigne et al. 2011; Barnhart et al. 2011). These lacO/tetO arrays are platforms to which specific LacI-fusion proteins can be tethered. Although these tethering approaches have been more focused on understanding centromere assembly, they have offered some insight into centromere silencing or inactivation.

4.2 *Centromere Inactivation by Deletion*

Studies in the budding yeast *S. cerevisiae* were among the first to demonstrate both the mitotic instability and fate of dicentric chromosomes. Yeast centromeres all contain the same structural elements CDEI, CDEII, and CDEIII (~ 125 bp) that are absolutely required for proper centromere assembly and function. Some of the first dicentrics in yeast were engineered using nonessential minichromosomes containing two 125 bp CEN regions (Koshland et al. 1987; Mann and Davis 1983; Oertel and Mayer 1984). These dicentric minichromosomes could be stabilized by deletion of one or both CENs or by mutations in CDEII or CDEIII so that they functioned as monocentric structures (Koshland et al. 1987; Mann and Davis 1983). Likewise, engineered linear dicentrics in budding yeast can be stabilized if one centromere is deleted (Pennaneach and Kolodner 2009; Jager and Philippsen 1989; Kramer et al. 1994) (Fig. 3a), although such stabilization is often accompanied by complex rearrangement or translocation of the chromosome.

Stable dicentric chromosomes do not naturally occur at high rates in the fission yeast *S. pombe*. Those that have been produced by end fusions or via replication fork collapse undergo anaphase bridge formation and breakage (Mohebi et al. 2015), suggesting that centromere inactivation does not occur frequently in this organism. However, engineered dicentric chromosomes involving chromosomes I and II were constructed using site-specific or meiotic recombination (Sato et al. 2012). Although nearly all (99%) cells containing dicentrics died, a small number of de novo dicentrics were stably maintained. A fraction (10%) of these dicentrics were observed to delete one centromere, suggesting that budding and fission yeast employ similar modes of centromere inactivation (Fig. 3c).

Dicentrics in humans have not been thought to undergo centromere inactivation by deletion because dicentrics typically maintain two arrays of alpha satellite DNA, even when one is inactivated. However, interpretation of the results has been complicated by the retrospective nature of the studies; the dicentrics were studied after inactivation occurred, so it was difficult or impossible to capture the original

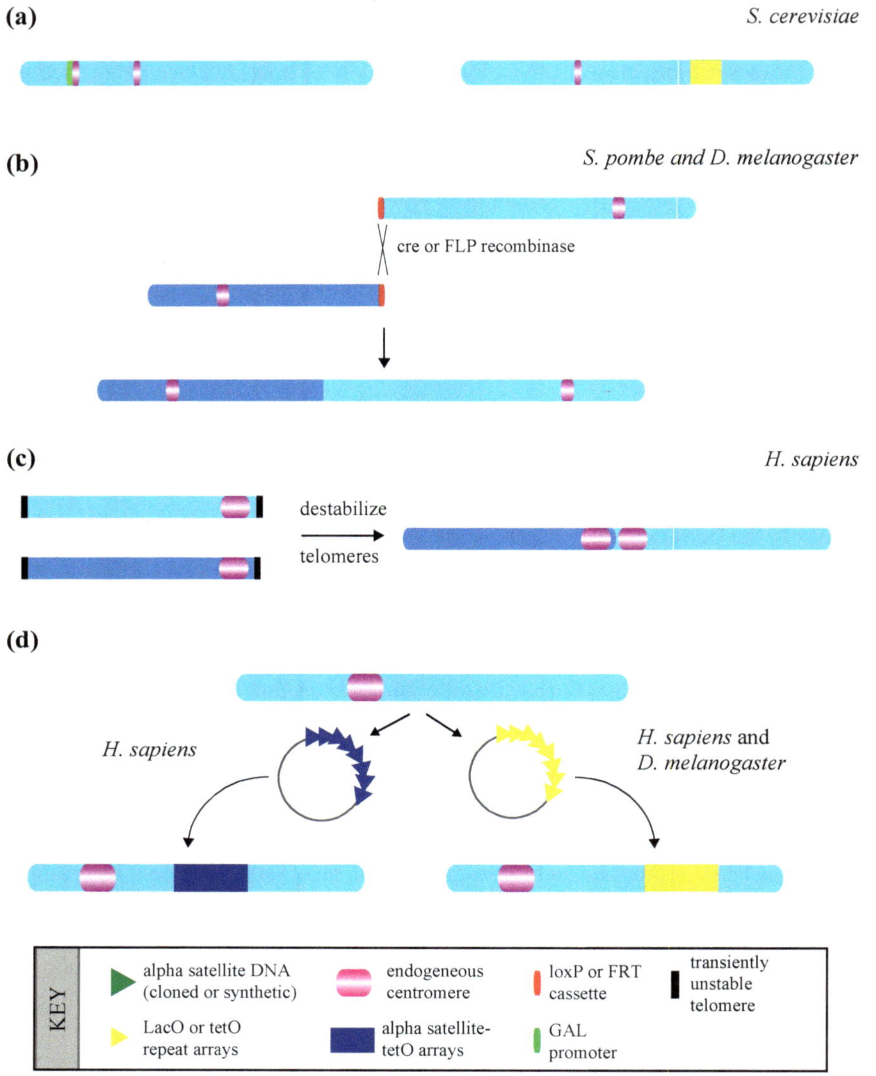

genomic state of the inactivated centromere. However, dicentrics induced by transient telomere dysfunction have permitted alpha satellite arrays involved in dicentric fusions to be studied before and after inactivation (Stimpson et al. 2010). These studies revealed that like budding yeast, centromere inactivation in humans also occurs by centromeric DNA deletion (Fig. 4a), but with one notable difference. Human centromeres form on large, multi-megabase arrays of alpha satellite DNA. The deletions associated with inactivation only remove the portion of satellite DNA associated with centromere proteins, resulting in alpha satellite arrays that are still

◂**Fig. 2** Approaches to making dicentric chromosomes. **a** In the budding yeast *S. cerevisiae*, dicentric chromosomes have been constructed by inserting a copy of endogenous CEN3 downstream of the GAL1 promoter. In other studies, lac operator (LacO) arrays have been inserted into a chromosome arm. LacI-fusions of centromere proteins have been used to assemble a de novo centromere at the ectopic site. **b** Dicentrics in *S. pombe* have been engineered by inserting loxP-selectable marker cassettes at the ends of chromosomes I and II. Upon expression of Cre recombinase, terminal fusions, leading to a dicentric chromosome, were induced. In similar experiments, FRT site were inserted into *Drosophila melanogaster* chromosomes, and dicentrics were generated after expression of FLP recombinase. **c** In humans, controllable destabilization of chromosome ends has been used to temporarily deprotect telomeres, leading to chromosome fusions and resulting in dicentric chromosomes. **d** Ectopic centromeres can be induced in both humans and *Drosophila*. In humans, constructs containing either chimeric alpha satellite DNA-tet operator arrays or lacO arrays alone have been randomly inserted in chromosome arms. TetR- or LacI-fusions of centromere proteins have been driven to the ectopic sites where centromere formation can be induced

present, but truncated. The deletion mechanisms achieve the same result, even though centromeric DNA is retained in humans. This is presumably because centromeric (CENP-A-containing) chromatin encompasses the entire small point centromere of *S. cerevisiae* and the entire central core of the regional *S. pombe* centromere. In humans, centromeric chromatin is only assembled on a portion (30%) of the large repetitive DNA array (Ross et al. 2016; Sullivan et al. 2011). Thus, both small and large eukaryotes appear to undergo centromere silencing by a common, deletion-based mechanism that achieves the same goal—removal of CENP-A nucleosomes.

4.3 *Epigenetic Mechanisms of Centromere Silencing*

Despite the frequency of centromere inactivation by deletion in budding yeast and other eukaryotes, many instances exist in which inactivated centromeres undergo no detectable genomic changes or alterations to the DNA (Higgins et al. 2005; Stimpson et al. 2010; Sato et al. 2012) (Fig. 4b). These observations suggest that sequence-independent, or epigenetic, mechanisms are also involved in centromere inactivation. Combined with the fact that inactive centromeres often differ in morphology from active centromeres (i.e., loss of the primary constriction), chromatin remodeling is thought to be involved in centromere inactivation. Indeed, studies in fission yeast, plants (*Arabidopsis* and maize), and mammals have confirmed that inactive centromeres lack CENP-A and euchromatic histone modifications (i.e., H3K9ac, H3K14ac, H3K4me2, H3K36me2) and other modifications, such as phosphorylated H3 (PH3S10) that are found at active centromeres (Han et al. 2009; Sato et al. 2012; Maloney et al. 2012). Instead, inactive regional centromeres are enriched for heterochromatin-associated epigenetic marks (H3K9me2/3 and H3K27me2/3) (Sato et al. 2012; Zhang et al. 2010; Maloney et al. 2012) (Fig. 4b). Epigenetic DNA changes, such as DNA methylation, have also

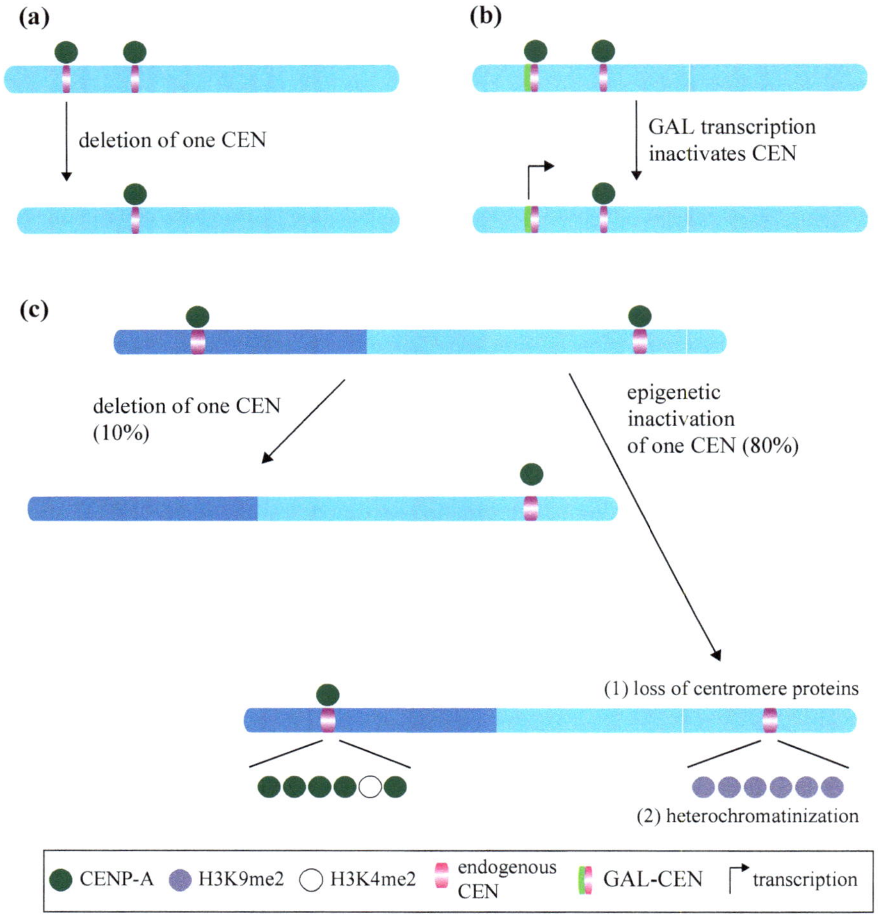

Fig. 3 Centromere silencing mechanisms in yeast. **a** Dicentric chromosomes or circular minichromosomes (not shown) in *S. cerevisiae* are stabilized if one centromere is deleted. The loss of CENP-A nucleosomes at the second centromere reduces the chromosome to a stable, monocentric state. **b** Conditional dicentrics in *S. cerevisiae* have demonstrated that centromeric deletion can be avoided, and centromere inactivation will occur, by driving transcription from the GAL1 promoter. **c** Centromere silencing in *S. pombe* dicentrics occurs in two primary ways. In a small fraction of the dicentrics, one of the centromeres of the dicentric is deleted, reducing the chromosome to a stable monocentric state. However, in 80% of dicentrics, both centromere regions are retained intact, but one undergoes chromatin remodeling through the loss of centromere proteins and subsequent acquisition of heterochromatin

been observed at inactive centromeres (Fu et al. 2012; Koo et al. 2011). Collectively these studies argue that centromere inactivation is accompanied by more compact or distinctly packaged chromatin.

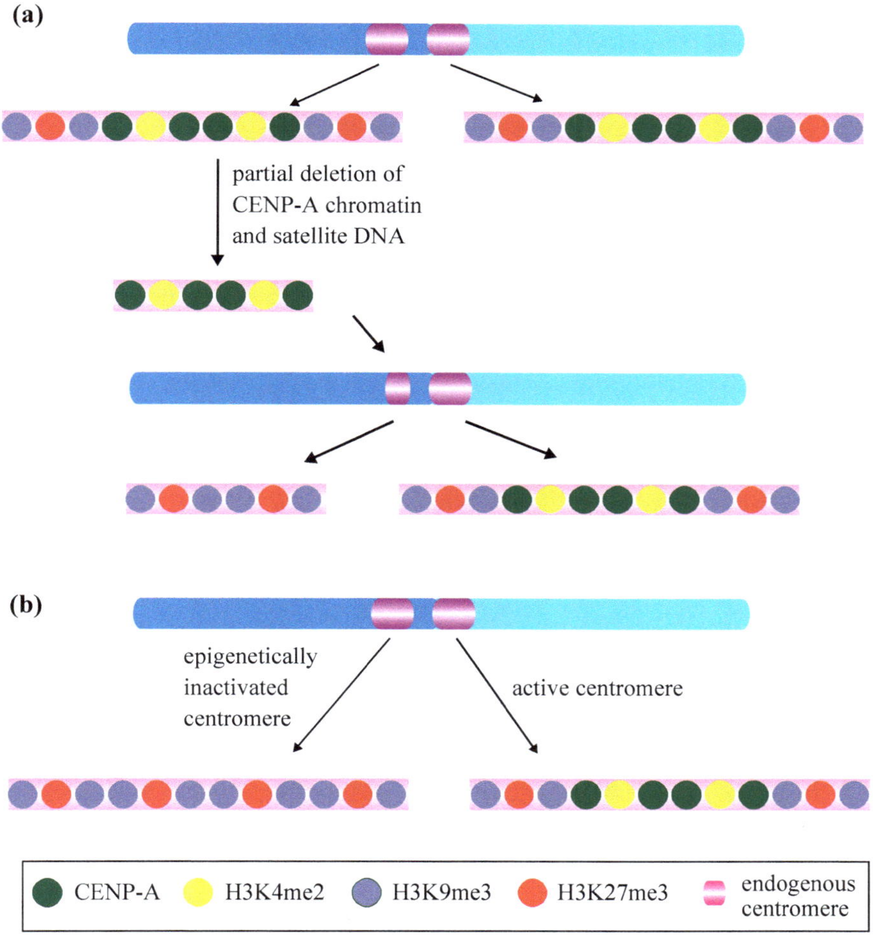

Fig. 4 Centromere inactivation in eukaryotes with large regional centromeres. **a** Half of dicentric human chromosomes induced by transient telomere destabilization undergo centromere inactivation. In approximately 65% of the dicentrics, inactivation occurs by partial deletion of the centromere. The deleted region includes the portion of alpha satellite DNA that contains centromeric (CENP-A chromatin), leaving behind the remainder of the array that is largely enriched for heterochromatin. **b** Most patient-derived and a portion of engineered dicentrics in human cells undergo centromere silencing by epigenetic mechanisms. Inactive centromeres show extensive chromatin remodeling, losing centromeric chromatin and centromere proteins and adopting a heterochromatic state

Still, the important, but elusive question is what triggers epigenetic centromere inactivation. Inactive human centromeres lack centromeric chromatin containing CENP-A and H3K4me2 and are enriched for heterochromatic histone modifications (Stimpson et al. 2010; Maloney et al. 2012). The experimentally engineered chromosome II–III dicentric fusions in *S. pombe* also showed loss of CENP-A^{Cnp1}, enrichment of H3K9me2, and depletion of euchromatic modifications H3K9ac and H3K14ac (Sato et al. 2012; Sato and Saitoh 2013). Inactive centromeres in plant are enriched for both heterochromatic histone modifications and DNA methylation (Han et al. 2006; Stimpson et al. 2012; Fu et al. 2012). Thus, inactive centromeres in several different model systems share centromere protein loss and chromatin reorganization. But, is it the loss of centromere proteins that initiate inactivation or does chromatin remodeling prevent centromere assembly and new CENP-A loading? Experiments from the *S. pombe* engineered dicentric model have shed light on this question. Mutations in key kinetochore components (mis6, mis12) and chromatin assembly factors that license centromeric chromatin for new CENP-A assembly (mis16) increased the number of epigenetically inactivated centromeres up to 50-fold. Moreover, mutations in heterochromatin components did not eliminate epigenetic centromere inactivation, but resulted in a higher rate of centromere reactivation. These studies indicate that loss of kinetochore proteins is an initiating event in centromere inactivation and that heterochromatin formation is a subsequent event that locks in the inactivated state (Fig. 3b). Indeed, a recent study of human chromosomes containing large ectopic arrays of alpha satellite DNA revealed that histone acetylation mediated by the histone acetyltransferase KAT7 was important for initiating centromere assembly while SUV39H1-mediated heterochromatin formation silenced centromere function (Ohzeki et al. 2016). Although it is still not clear if heterochromatin alone is sufficient to silence centromeres in mammals, these results strongly point to loss of centromere proteins and heterochromatin remodeling as important components of centromere inactivation.

4.4 Dicentric Chromosome Centromere Inactivation by Transcriptional Activity

Several studies have reported that centromere assembly involves an RNA component (Volpe et al. 2003; Carone et al. 2009; Chan et al. 2012; Du et al. 2010; Quenet and Dalal 2014; Rosic et al. 2014; Topp et al. 2004; Wong et al. 2007; Bouzinba-Segard et al. 2006; Catania et al. 2015) These noncoding RNAs are thought to recruit CENP-A, establish centromeric chromatin, and/or stabilize the interactions of CENP-C at the centromere (Du et al. 2010; Quenet and Dalal 2014; Catania et al. 2015). From these studies, one might expect that inactivated centromeres would lack transcription. However, as previously described in Sects. 3.2 and 3.3, too much or too little transcription can functionally silence a synthetic centromere. Similar scenarios are observed for endogenous centromeres. In the tammar wallaby model of mammalian centromeres, increased expression of centromeric RNAs disrupts centromere protein recruitment (Carone et al. 2013). Conversely, when centromeric RNAs in *Drosophila* and human cells are depleted by RNAi, centromere assembly and function is impaired (Rosic et al. 2014;

Quenet and Dalal 2014). From these studies, it appears that a discrete level of transcription is important for proper centromere function. However, few studies have focused on dicentric chromosomes to determine the relationship of transcription to an inactivated centromere. The best model has been the *S. cerevisiae* dicentric chromosome containing an inducible GAL1-CEN (Hill and Bloom 1987). When GAL1 transcription was induced with galactose, the second CEN was inactivated, suggesting that increased transcription inhibits centromere function (Fig. 2b). However, a recent study using a *Drosophila* dicentric chromosome model in which the second centromere was activated at an ectopic LacO array showed that transcription was necessary to recruit and maintain CENP-A^{CID} and its chaperone CAL1 at the second centromere (Chen et al. 2015). Thus, the role of centromeric transcription at inactive centromeres, especially on dicentric human chromosomes, remains unclear, and future studies will be important for determining if inactivated centromeres lack transcripts or overproduce them.

4.5 Centromere Inactivation: Permanent or Reversible?

Epigenetic changes to the genome are often reversible, depending on developmental or cell cycle cues. In the cases of epigenetically inactivated centromeres, an open question is if inactive centromeres are permanently inactivated or, alternatively, are dormant and can be reactivated. Isodicentric human chromosomes containing two functional centromeres have been shown to undergo centromere inactivation in the presence of Trichostatin A (TSA), an inhibitor of histone deacetylases (Higgins et al. 2005). However, inactive ectopic alpha satellite arrays can also be reactivated in the presence of TSA (Nakano et al. 2003), suggesting that subtle changes in chromatin state can tip a centromere toward function or silencing. Inactivated centromeres of *S. pombe* dicentrics can reactivate in the absence of heterochromatin (Sato et al. 2012). Thus, some inactive centromeres may exist in a silent, but poised configuration, while others may be more permanently inactivated through the maintenance of a heterochromatic state (Ohzeki et al. 2016; Sato et al. 2012). Given that chromatin organization and transcription intersect in the formation and maintenance of functional centromeres, it is likely that both pathways are perturbed during centromere inactivation. It will important for future studies to determine if inactive "poised" centromeres have a different chromatin and transcriptional signature than permanently inactive centromeres.

5 Conclusion

In the past two decades, enormous progress has been made in defining the DNA and protein components of eukaryotic centromeres, and the order of assembly of the various protein networks. Centromere inactivation does not appear to be simply the reverse of the assembly process. In instances of centromere inactivation by deletion, the entire centromeric chromatin platform is removed, thereby eliminating the memory of where to load new centromere proteins. The event that triggers the deletion event, particularly in dicentric human chromosomes, is largely unknown. Is the centromeric deletion initiated by DSBs that occur during CENP-A loading? Or do centromere breaks occur due to mechanical forces produced by the dicentric twisting on the spindle and centromeres being oriented to the wrong spindle poles? It may be difficult to pinpoint the precise breakpoints at human centromeres because of their large size and homogeneous repetitive sequence, but the role of DSBs could be tested using CRISPR-mediated chromosome breaks.

Although fission yeast studies indicate that kinetochore protein loss is the trigger for epigenetic centromere inactivation, it is not known if centromeres in multicellular eukaryotes are inactivated in the same way. Future studies using many of the synthetic centromere and dicentric models discussed in this chapter would be useful for definitively describing the temporal and molecular steps of centromere inactivation. These studies will underscore similarities and differences in the process of centromere silencing among various organisms and, more importantly, could reconcile divergent data on the role of transcription and noncoding RNAs in centromere assembly and inactivation.

References

Agudo M, Abad JP, Molina I, Losada A, Ripoll P, Villasante A (2000) A dicentric chromosome of Drosophila melanogaster showing alternate centromere inactivation. Chromosoma 109:190–196

Ahmad K, Golic KG (1998) The transmission of fragmented chromosomes in Drosophila melanogaster. Genetics 148:775–792

Bandyopadhyay R, Heller A, Knox-Dubois C, McCaskill C, Berend SA, Page SL, Shaffer LG (2002) Parental origin and timing of de novo Robertsonian translocation formation. Am J Hum Genet 71:1456–1462

Barnhart MC, Kuich PH, Stellfox ME, Ward JA, Bassett EA, Black BE, Foltz DR (2011) HJURP is a CENP-A chromatin assembly factor sufficient to form a functional de novo kinetochore. J Cell Biol 194:229–243

Bergmann JH, Rodriguez MG, Martins NM, Kimura H, Kelly DA, Masumoto H, Larionov V, Jansen LE, Earnshaw WC (2011) Epigenetic engineering shows H3K4me2 is required for HJURP targeting and CENP-A assembly on a synthetic human kinetochore. EMBO J 30:328–340

Bergmann JH, Jakubsche JN, Martins NM, Kagansky A, Nakano M, Kimura H, Kelly DA, Turner BM, Masumoto H, Larionov V, Earnshaw WC (2012) Epigenetic engineering: histone H3K9 acetylation is compatible with kinetochore structure and function. J Cell Sci 125:411

Bloom K, Hill A, Jones E (1989) Conditional dicentric chromosomes in yeast. Prog Clin Biol Res 318:149–158

Blower MD, Sullivan BA, Karpen GH (2002) Conserved organization of centromeric chromatin in flies and humans. Dev Cell 2:319–330

Bouzinba-Segard H, Guais A, Francastel C (2006) Accumulation of small murine minor satellite transcripts leads to impaired centromeric architecture and function. Proc Natl Acad Sci U S A 103:8709–8714

Cardinale S, Bergmann JH, Kelly D, Nakano M, Valdivia MM, Kimura H, Masumoto H, Larionov V, Earnshaw WC (2009) Hierarchical inactivation of a synthetic human kinetochore by a chromatin modifier. Mol Biol Cell 20:4194–4204

Carone DM, Longo MS, Ferreri GC, Hall L, Harris M, Shook N, Bulazel KV, Carone BR, Obergfell C, O'Neill MJ, O'Neill RJ (2009) A new class of retroviral and satellite encoded small RNAs emanates from mammalian centromeres. Chromosoma 118:113–125

Carone DM, Zhang C, Hall LE, Obergfell C, Carone BR, O'Neill MJ, O'Neill RJ (2013) Hypermorphic expression of centromeric retroelement-encoded small RNAs impairs CENP-A loading. Chromosome Res 21:49–62

Carroll CW, Silva MC, Godek KM, Jansen LE, Straight AF (2009) Centromere assembly requires the direct recognition of CENP-A nucleosomes by CENP-N. Nat Cell Biol 11:896–902

Catania S, Pidoux AL, Allshire RC (2015) Sequence features and transcriptional stalling within centromere DNA promote establishment of CENP-A chromatin. PLoS Genet 11:e1004986

Chan FL, Marshall OJ, Saffery R, Kim BW, Earle E, Choo KH, Wong LH (2012) Active transcription and essential role of RNA polymerase II at the centromere during mitosis. Proc Natl Acad Sci U S A 109:1979–1984

Cheeseman IM, Hori T, Fukagawa T, Desai A (2008) KNL1 and the CENP-H/I/K complex coordinately direct kinetochore assembly in vertebrates. Mol Biol Cell 19:587–594

Chen CC, Bowers S, Lipinszki Z, Palladino J, Trusiak S, Bettini E, Rosin L, Przewloka MR, Glover DM, O'Neill RJ, Mellone BG (2015) Establishment of centromeric chromatin by the CENP-A assembly factor CAL1 requires FACT-mediated transcription. Dev Cell 34:73–84

Craig JM, Earle E, Canham P, Wong LH, Anderson M, Choo KH (2003) Analysis of mammalian proteins involved in chromatin modification reveals new metaphase centromeric proteins and distinct chromosomal distribution patterns. Hum Mol Genet 12:3109–3121

Daniel A (1979) Single Cd band in dicentric translocations with one suppressed centromere. Hum Genet 48:85–92

de la Chapelle A, Wennstrom J, Hortling H, Ockey CH (1966) Isochromosome-X in man. I. Hereditas 54:260–276

Dewald GW, Boros SJ, Conroy MM, Dahl RJ, Spurbeck JL, Vitek HA (1979) A tdic(5;15)(p31;p11) chromosome showing variation for constriction in the centromeric regions in a patient with the cri du chat syndrome. Cytogenet Cell Genet 24:15–26

Du Y, Topp CN, Dawe RK (2010) DNA binding of centromere protein C (CENPC) is stabilized by single-stranded RNA. PLoS Genet 6:e1000835

Earnshaw WC, Migeon BR (1985) Three related centromere proteins are absent from the inactive centromere of a stable isodicentric chromosome. Chromosoma 92:290–296

Earnshaw WC, Ratrie H, Stetten G (1989) Visualization of centromere proteins CENP-B and CENP-C on a stable dicentric chromosome in cytological spreads. Chromosoma 98:1–12

Eissenberg JC, Elgin SC (2014) HP1a: a structural chromosomal protein regulating transcription. Trends Genet 30:103–110

Fachinetti D, Folco HD, Nechemia-Arbely Y, Valente LP, Nguyen K, Wong AJ, Zhu Q, Holland AJ, Desai A, Jansen LE, Cleveland DW (2013) A two-step mechanism for epigenetic specification of centromere identity and function. Nat Cell Biol 15:1056–1066

Farr C, Fantes J, Goodfellow P, Cooke H (1991) Functional reintroduction of human telomeres into mammalian cells. Proc Natl Acad Sci U S A 88:7006–7010

Foltz DR, Jansen LE, Black BE, Bailey AO, Yates JR 3rd, Cleveland DW (2006) The human CENP-A centromeric nucleosome-associated complex. Nat Cell Biol 8:458–469

Foltz DR, Jansen LE, Bailey AO, Yates JR 3rd, Bassett EA, Wood S, Black BE, Cleveland DW (2009) Centromere-specific assembly of CENP-a nucleosomes is mediated by HJURP. Cell 137:472–484

Freundlieb S, Schirra-Muller C, Bujard H (1999) A tetracycline controlled activation/repression system with increased potential for gene transfer into mammalian cells. J Gene Med 1:4–12

Fu S, Gao Z, Birchler J, Han F (2012) Dicentric chromosome formation and epigenetics of centromere formation in plants. J Genet Genomics 39:125–130

Gascoigne KE, Cheeseman IM (2013) Induced dicentric chromosome formation promotes genomic rearrangements and tumorigenesis. Chromosome Res 21:407–418
Gascoigne KE, Takeuchi K, Suzuki A, Hori T, Fukagawa T, Cheeseman IM (2011) Induced ectopic kinetochore assembly bypasses the requirement for CENP-A nucleosomes. Cell 145:410–422
Gossen M, Freundlieb S, Bender G, Muller G, Hillen W, Bujard H (1995) Transcriptional activation by tetracyclines in mammalian cells. Science 268:1766–1769
Haber JE, Thorburn PC, Rogers D (1984) Meiotic and mitotic behavior of dicentric chromosomes in *Saccharomyces cerevisiae*. Genetics 106:185–205
Hacein-Bey-abina S, von Kalle C, Schmidt M, McCormack MP, Wulffraat N, Leboulch P, Lim A, Osborne CS, Pawliuk R, Morillon E, Sorensen R, Forster A, Fraser P, Cohen JI, De-Saint-basile G, Alexander I, Wintergerst U, Frebourg T, Aurias A, Stoppa-Lyonnet D, Romana S, Radford-Weiss I, Gross F, Valensi F, Delabesse E, Macintyre E, Sigaux F, Soulier J, Leiva LE, Wissler M, Prinz C, Rabbitts TH, le Deist F, Fischer A, Cavazzana-Calvo M (2003) LMO2-associated clonal T cell proliferation in two patients after gene therapy for SCID-X1. Science 302:415–419
Han F, Lamb JC, Birchler JA (2006) High frequency of centromere inactivation resulting in stable dicentric chromosomes of maize. Proc Natl Acad Sci U S A 103:3238–3243
Han F, Gao Z, Birchler JA (2009) Reactivation of an inactive centromere reveals epigenetic and structural components for centromere specification in maize. Plant Cell 21:1929–1939
Harrington JJ, Bokkelen GV, Mays RW, Gustashaw K, Willard HF (1997) Formation of de novo centromeres and construction of first-generation human artificial microchromosomes. Nat Genet 15:345–355
Hayashi T, Fujita Y, Iwasaki O, Adachi Y, Takahashi K, Yanagida M (2004) Mis16 and Mis18 are required for CENP-A loading and histone deacetylation at centromeres. Cell 118:715–729
Higgins AW, Gustashaw KM, Willard HF (2005) Engineered human dicentric chromosomes show centromere plasticity. Chromosome Res 13:745–762
Hill A, Bloom K (1987) Genetic manipulation of centromere function. Mol Cell Biol 7:2397–2405
Hill A, Bloom K (1989) Acquisition and processing of a conditional dicentric chromosome in *Saccharomyces cerevisiae*. Mol Cell Biol 9:1368–1370
Hill H, Golic KG (2015) Preferential breakpoints in the recovery of broken dicentric chromosomes in *Drosophila melanogaster*. Genetics 201:563–572
Ho KH, Tsuchiya D, Oliger AC, Lacefield S (2014) Localization and function of budding yeast CENP-A depends upon kinetochore protein interactions and is independent of canonical centromere sequence. Cell Rep 9:2027–2033
Hori T, Amano M, Suzuki A, Backer CB, Welburn JP, Dong Y, McEwen BF, Shang WH, Suzuki E, Okawa K, Cheeseman IM, Fukagawa T (2008) CCAN makes multiple contacts with centromeric DNA to provide distinct pathways to the outer kinetochore. Cell 135:1039–1052
Iida Y, Kim JH, Kazuki Y, Hoshiya H, Takiguchi M, Hayashi M, Erliandri I, Lee HS, Samoshkin A, Masumoto H, Earnshaw WC, Kouprina N, Larionov V, Oshimura M (2010) Human artificial chromosome with a conditional centromere for gene delivery and gene expression. DNA Res 17:293–301
Ikeno M, Grimes B, Okazaki T, Nakano M, Saitoh K, Hoshino H, McGill NI, Cooke H, Masumoto H (1998) Construction of YAC-based mammalian artificial chromosomes. Nat Biotechnol 16:431–439
Izuta H, Ikeno M, Suzuki N, Tomonaga T, Nozaki N, Obuse C, Kisu Y, Goshima N, Nomura F, Nomura N, Yoda K (2006) Comprehensive analysis of the ICEN (interphase centromere complex) components enriched in the CENP-A chromatin of human cells. Genes Cells 11:673–684
Jager D, Philippsen P (1989) Stabilization of dicentric chromosomes in *Saccharomyces cerevisiae* by telomere addition to broken ends or by centromere deletion. EMBO J 8:247–254
Janssen A, Kops GJ, Medema RH (2009) Elevating the frequency of chromosome mis-segregation as a strategy to kill tumor cells. Proc Natl Acad Sci U S A 106:19108–19113

Kim JH, Kononenko A, Erliandri I, Kim TA, Nakano M, Iida Y, Barrett JC, Oshimura M, Masumoto H, Earnshaw WC, Larionov V, Kouprina N (2011) Human artificial chromosome (HAC) vector with a conditional centromere for correction of genetic deficiencies in human cells. Proc Natl Acad Sci U S A 108:20048–20053

Kim JH, Lee HS, Lee NC, Goncharov NV, Kumeiko V, Masumoto H, Earnshaw WC, Kouprina N, Larionov V (2016) Development of a novel HAC-based "gain of signal" quantitative assay for measuring chromosome instability (CIN) in cancer cells. Oncotarget 7:14841–14856

Kononenko AV, Bansal R, Lee NC, Grimes BR, Masumoto H, Earnshaw WC, Larionov V, Kouprina N (2014) A portable BRCA1-HAC (human artificial chromosome) module for analysis of BRCA1 tumor suppressor function. Nucleic Acids Res 42:e164

Kononenko AV, Lee NC, Liskovykh M, Masumoto H, Earnshaw WC, Larionov V, Kouprina N (2015) Generation of a conditionally self-eliminating HAC gene delivery vector through incorporation of a tTAVP64 expression cassette. Nucleic Acids Res 43:e57

Koo DH, Han F, Birchler JA, Jiang J (2011) Distinct DNA methylation patterns associated with active and inactive centromeres of the maize B chromosome. Genome Res 21:908–914

Koshland D, Rutledge L, Fitzgerald-Hayes M, Hartwell LH (1987) A genetic analysis of dicentric minichromosomes in *Saccharomyces cerevisiae*. Cell 48:801–812

Kramer KM, Brock JA, Bloom K, Moore JK, Haber JE (1994) Two different types of double-strand breaks in *Saccharomyces cerevisiae* are repaired by similar RAD52-independent, nonhomologous recombination events. Mol Cell Biol 14:1293–1301

Lam AL, Boivin CD, Bonney CF, Rudd MK, Sullivan BA (2006) Human centromeric chromatin is a dynamic chromosomal domain that can spread over noncentromeric DNA. Proc Natl Acad Sci U S A 103:4186–4191

Lee HS, Lee NC, Grimes BR, Samoshkin A, Kononenko AV, Bansal R, Masumoto H, Earnshaw WC, Kouprina N, Larionov V (2013) A new assay for measuring chromosome instability (CIN) and identification of drugs that elevate CIN in cancer cells. BMC Cancer 13:252

Maloney KA, Sullivan LL, Matheny JE, Strome ED, Merrett SL, Ferris A, Sullivan BA (2012) Functional epialleles at an endogenous human centromere. Proc Natl Acad Sci U S A 109:13704–13709

Mann C, Davis RW (1983) Instability of dicentric plasmids in yeast. Proc Natl Acad Sci U S A 80:228–232

Martins NMC, Bergmann JH, Shono N, Kimura H, Larionov V, Masumoto H, Earnshaw WC (2016) Epigenetic engineering shows that a human centromere resists silencing mediated by H3K27me3/K9me3. Mol Biol Cell 27:177

McAinsh AD, Tytell JD, Sorger PK (2003) Structure, function, and regulation of budding yeast kinetochores. Annu Rev Cell Dev Biol 19:519–539

McClintock B (1939) The behaviour of successive nuclear divisions of a chromosome broken at meiosis. Proc Natl Acad Sci U S A 25:405–416

McClintock B (1941) The stability of broken ends of chromosomes in *Zea mays*. Genetics 26:234–282

McKinley KL, Sekulic N, Guo LY, Tsinman T, Black BE, Cheeseman IM (2015) The CENP-L-N complex forms a critical node in an integrated meshwork of interactions at the centromere-kinetochore interface. Mol Cell 60:886–898

Mendiburo MJ, Padeken J, Fulop S, Schepers A, Heun P (2011) Drosophila CENH3 is sufficient for centromere formation. Science 334:686–690

Mohebi S, Mizuno K, Watson A, Carr AM, Murray JM (2015) Checkpoints are blind to replication restart and recombination intermediates that result in gross chromosomal rearrangements. Nat Commun 6:6357

Nakano M, Okamoto Y, Ohzeki J, Masumoto H (2003) Epigenetic assembly of centromeric chromatin at ectopic alpha-satellite sites on human chromosomes. J Cell Sci 116:4021–4034

Nakano M, Cardinale S, Noskov VN, Gassmann R, Vagnarelli P, Kandels-Lewis S, Larionov V, Earnshaw WC, Masumoto H (2008) Inactivation of a human kinetochore by specific targeting of chromatin modifiers. Dev Cell 14:507–522

Niebuhr E (1972) Dicentric and monocentric Robertsonian translocations in man. Humangenetik 16:217–226

Nishino T, Takeuchi K, Gascoigne KE, Suzuki A, Hori T, Oyama T, Morikawa K, Cheeseman IM, Fukagawa T (2012) CENP-T-W-S-X forms a unique centromeric chromatin structure with a histone-like fold. Cell 148:487–501

Nishino T, Rago F, Hori T, Tomii K, Cheeseman IM, Fukagawa T (2013) CENP-T provides a structural platform for outer kinetochore assembly. EMBO J 32:424–436

Novitski E (1952) The genetic consequences of anaphase bridge formation in Drosophila. Genetics 37:270–287

Ockey CH, Wennstrom J, de la Chapelle A (1966) Isochromosome-X in man. II. Hereditas 54:277–292

Oertel W, Mayer M (1984) Structure and mitotic stability of minichromosomes originating in yeast cells transformed with tandem dimers of CEN11 plasmids. Mol Gen Genet 195:300–307

Ohzeki J, Bergmann JH, Kouprina N, Noskov VN, Nakano M, Kimura H, Earnshaw W. C, Larionov V, Masumoto H (2012) Breaking the HAC barrier: histone H3K9 acetyl/methyl balance regulates CENP-A assembly. EMBO J 31:2391

Ohzeki J-I, Shono N, Otake K, Martins NMC, Kugou K, Kimura H, Nagase T, Larionov V, Earnshaw-William C, Masumoto H (2016) KAT7/HBO1/MYST2 regulates CENP-A chromatin assembly by antagonizing Suv39h1-mediated centromere inactivation. Dev Cell 37:413–427

Olszak AM, van Essen D, Pereira AJ, Diehl S, Manke T, Maiato H, Saccani S, Heun P (2011) Heterochromatin boundaries are hotspots for de novo kinetochore formation. Nat Cell Biol 13:799–808

Page SL, Earnshaw WC, Choo KH, Shaffer LG (1995) Further evidence that CENP-C is a necessary component of active centromeres: studies of a dic(X; 15) with simultaneous immunofluorescence and FISH. Hum Mol Genet 4:289–294

Palmer DK, O'Day K, Trong HL, Charbonneau H, Margolis RL (1991) Purification of the centromere-specific protein CENP-A and demonstration that it is a distinctive histone. Proc Natl Acad Sci U S A 88:3734–3738

Pennaneach V, Kolodner RD (2009) Stabilization of dicentric translocations through secondary rearrangements mediated by multiple mechanisms in *S. cerevisiae*. PLoS One 4:e6389

Quenet D, Dalal Y (2014) A long non-coding RNA is required for targeting centromeric protein A to the human centromere. Elife 3:e03254

Rosic S, Kohler F, Erhardt S (2014) Repetitive centromeric satellite RNA is essential for kinetochore formation and cell division. J Cell Biol 207:335–349

Ross JE, Woodlief KS, Sullivan BA (2016) Inheritance of the CENP-A chromatin domain is spatially and temporally constrained at human centromeres. Epigenetics Chromatin 9:20

Saffery R, Sumer H, Hassan S, Wong LH, Craig JM, Todokoro K, Anderson M, Stafford A, Choo KHA (2003) Transcription within a functional human centromere. Mol Cell 12:509–516

Saksouk N, Simboeck E, Dejardin J (2015) Constitutive heterochromatin formation and transcription in mammals. Epigenetics Chromatin 8:3

Sato H, Masuda F, Takayama Y, Takahashi K, Saitoh S (2012) Epigenetic inactivation and subsequent heterochromatinization of a centromere stabilize dicentric chromosomes. Curr Biol 22:658–667

Sato H, Saitoh S (2013) Switching the centromeres on and off: epigenetic chromatin alterations provide plasticity in centromere activity stabilizing aberrant dicentric chromosomes. Biochem Soc Trans 41:1648–1653

Silk AD, Zasadil LM, Holland AJ, Vitre B, Cleveland DW, Weaver BA (2013) Chromosome missegregation rate predicts whether aneuploidy will promote or suppress tumors. Proc Natl Acad Sci U S A 110:4134–4141

Simon JA, Kingston RE (2009) Mechanisms of polycomb gene silencing: knowns and unknowns. Nat Rev Mol Cell Biol 10:697–708

Stimpson KM, Song IY, Jauch A, Holtgreve-Grez H, Hayden KE, Bridger JM, Sullivan BA (2010) Telomere disruption results in non-random formation of de novo dicentric chromosomes involving acrocentric human chromosomes. PLoS Genet 6:e1001061

Stimpson KM, Matheny JE, Sullivan BA (2012) Dicentric chromosomes: unique models to study centromere function and inactivation. Chromosome Res 20:595–605

Sullivan BA, Karpen GH (2004) Centromeric chromatin exhibits a histone modification pattern that is distinct from both euchromatin and heterochromatin. Nat Struct Mol Biol 11:1076–1083

Sullivan BA, Schwartz S (1995) Identification of centromeric antigens in dicentric Robertsonian translocations: CENP-C and CENP-E are necessary components of functional centromeres. Hum Mol Genet 4:2189–2197

Sullivan BA, Wolff DJ, Schwartz S (1994) Analysis of centromeric activity in Robertsonian translocations: implications for a functional acrocentric hierarchy. Chromosoma 103:459–467

Sullivan LL, Boivin CD, Mravinac B, Song IY, Sullivan BA (2011) Genomic size of CENP-A domain is proportional to total alpha satellite array size at human centromeres and expands in cancer cells. Chromosome Res 19:457–470

Suzuki A, Badger BL, Salmon ED (2015) A quantitative description of Ndc80 complex linkage to human kinetochores. Nat Commun 6:8161

Therman E, Trunca C, Kuhn EM, Sarto GE (1986) Dicentric chromosomes and the inactivation of the centromere. Hum Genet 72:191–195

Titen SW, Golic KG (2008) Telomere loss provokes multiple pathways to apoptosis and produces genomic instability in *Drosophila melanogaster*. Genetics 180:1821–1832

Topp CN, Zhong CX, Dawe RK (2004) Centromere-encoded RNAs are integral components of the maize kinetochore. Proc Natl Acad Sci U S A 101:15986–15991

Volpe T, Schramke V, Hamilton GL, White SA, Teng G, Martienssen RA, Allshire RC (2003) RNA interference is required for normal centromere function in fission yeast. Chromosome Res 11:137–146

Warburton PE, Cooke CA, Bourassa S, Vafa O, Sullivan BA, Stetten G, Gimelli G, Warburton D, Tyler-Smith C, Sullivan KF, Poirier GG, Earnshaw WC (1997) Immunolocalization of CENP-A suggests a distinct nucleosome structure at the inner kinetochore plate of active centromeres. Curr Biol 7:901–904

Westhorpe FG, Straight AF (2013) Functions of the centromere and kinetochore in chromosome segregation. Curr Opin Cell Biol 25:334–340

Witzgall R, O'Leary E, Leaf A, Onaldi D, Bonventre JV (1994) The Kruppel-associated box-A (KRAB-A) domain of zinc finger proteins mediates transcriptional repression. Proc Natl Acad Sci U S A 91:4514–4518

Wong LH, Brettingham-Moore KH, Chan L, Quach JM, Anderson MA, Northrop EL, Hannan R, Saffery R, Shaw ML, Williams E, Choo KH (2007) Centromere RNA is a key component for the assembly of nucleoproteins at the nucleolus and centromere. Genome Res 17:1146–1160

Zhang W, Friebe B, Gill BS, Jiang J (2010) Centromere inactivation and epigenetic modifications of a plant chromosome with three functional centromeres. Chromosoma 119:553–563

Centromere Transcription: *Means and Motive*

Zachary Duda, Sarah Trusiak and Rachel O'Neill

Abstract The chromosome biology field at large has benefited from studies of the cell cycle components, protein cascades and genomic landscape that are required for centromere identity, assembly and stable transgenerational inheritance. Research over the past 20 years has challenged the classical descriptions of a centromere as a stable, unmutable, and transcriptionally silent chromosome component. Instead, based on studies from a broad range of eukaryotic species, including yeast, fungi, plants, and animals, the centromere has been redefined as one of the more dynamic areas of the eukaryotic genome, requiring coordination of protein complex assembly, chromatin assembly, and transcriptional activity in a cell cycle specific manner. What has emerged from more recent studies is the realization that the transcription of specific types of nucleic acids is a key process in defining centromere integrity and function. To illustrate the transcriptional landscape of centromeres across eukaryotes, we focus this review on how transcripts interact with centromere proteins, when in the cell cycle centromeric transcription occurs, and what types of sequences are being transcribed. Utilizing data from broadly different organisms, a picture emerges that places centromeric transcription as an integral component of centromere function.

Z. Duda · S. Trusiak · R. O'Neill (✉)
Department of Molecular and Cell Biology, The Institute for Systems Genomics, University of Connecticut, Storrs, CT 06269, USA
e-mail: rachel.oneill@uconn.edu

Z. Duda
e-mail: zachary.duda@uconn.edu

S. Trusiak
e-mail: sarah.trusiak@uconn.edu

B.E. Black (ed.), *Centromeres and Kinetochores*, Progress in Molecular and Subcellular Biology 56, DOI 10.1007/978-3-319-58592-5_11

1 A Centromere Refresher

1.1 Why Centromeres?

Centromeres across eukaryotic lineages range from a relatively small, "point" within a chromosome to sprawling and complex structures that vary in size from 10's of kilobases (KB) to 10's of megabases (MB) (Pluta et al. 1995; Choo 1997, and reviewed in Bayes and Malik 2008; Brown and O'Neill 2014). While the location of the centromere as a single constriction on a chromosome is found broadly across the major clades of eukaryotes, some eukaryotic species do not harbor a distinct centromere; rather, there are multiple nucleating sites across chromosome arms that act as centromeres (termed *holocentricity*) (reviewed in Malik and Henikoff 2009). For example, chromosome segregation in *Caenorhabditis elegans* (and other nematodes) and some insect and plant species is mediated by sites along the entire chromosome.

The diversity in the complexity, density, and distribution of centromere forms across species lies in contrast to the uniform requisite function for the centromere: *to serve as the site of kinetochore assembly and spindle attachment during meiosis and mitosis*. In essence, the proper functioning of centromeres is a requirement for faithful segregation of a chromosome complement. Any failure in this function has catastrophic consequences for the cell, such as chromosome breakage, and/or loss and cellular breakdown (reviewed in Holland and Cleveland 2009); and, consequently has devastating consequences for the organism, such as infertility, loss of cell cycle control, and aberrant proliferation.

1.2 Why NOT Centromeres?

Despite the deep phylogenetic conservation of centromere function—*to mediate kinetochore formation and spindle attachment*—the diversity of centromere forms across species has presented a unique challenge in understanding the components that delineate centromere functionality as well as defining the minimal required elements for centromere integrity. For example, the "point centromeres" of the budding yeast, *Saccharomyces cerevisiae* (Fishel et al. 1988), consist of a 125-bp nucleotide sequence that supports centromere function (Meluh et al. 1998) without the requirement for any other complex repeat structures. The centromeres of the filamentous fungi *Neurospora* are 175–300 KB and harbor AT-rich, degenerate transposons (reviewed in Smith et al. 2012) whose sequences have been ravaged by a genome defense mechanism known as RIP ('repeat induced point mutation') (Smith et al. 2011). Many plants, including maize and grasses, carry satellites and transposons throughout their regional centromeres (Neumann et al. 2011; Gent and Dawe 2012). Like fungi, there does not appear to be any pattern to the repeat structure that defines the functional centromere core in most plants.

The diversity of centromere forms is not only restricted to species-specific genomic arrangements, as several species (e.g., orangutan, horse, chicken) carry a chromosome complement wherein some centromeres are characterized largely by repetitive DNA (satellites and transposable elements) while others are seemingly devoid of a highly repetitive structure (Piras et al. 2010; Shang et al. 2010; Locke et al. 2011). Thus, attempts to reconcile these diverse centromere forms with a generalizable model for centromere function across traditional eukaryotic model organisms (yeast, human, mouse, *Arabidopsis*, maize) have been largely unsuccessful as no simple rule appears to apply to even the majority of centromere types. Large-scale genome sequencing projects for several model species initially showed promise in capturing the DNA landscape of regional centromeres in species with diverse karyotypes [e.g., human (Schueler et al. 2001), *Arabidopsis* (Copenhaver et al. 1999; Kumekawa et al. 2000; Hosouchi et al. 2002), rice (Yan and Jiang 2007), wallaby (Renfree et al. 2011), and gibbon (Carbone et al. 2014)]. However, the highly repetitive nature of such centromeres, composed of expansive arrays of simple satellites [ranging in size anywhere from only 0.2 KB to more than 28 MB (Melters et al. 2013)] and other highly repeated sequences, such as transposable elements, has remained a hurdle in defining genomic maps for complex, regional centromeres. As a consequence, complex eukaryotic centromeres have, to date, remained on the "black list" (Miga et al. 2015) of regions refractive to mapping and assembly techniques (Altemose et al. 2014).

Emerging sequence techniques that afford long-range sequence information (e.g., long-read sequencers capable of sequencing >100 KB of contiguous DNA, such as PacBio and Oxford Nanopore; and, synthetic long-read sequencers such 10X Genomics) offer the potential to overcome the technical challenges of dealing with highly repeated regions of genomes. However, the overall scale of the total repeat regions that encompass the functional centromeres within model systems that are subject to genome sequencing efforts is *orders of magnitude* greater than the long-read sequencing capabilities and has left centromere regions in genome assemblies without the foundation of a linear genetic map in most cases, particularly human and mouse. Confounding this sequencing challenge is the sheer number of centromeres that must be tackled for assembly within any given genome —one per chromosome in a diploid cell—as each centromere contains a unique genomic sequence structure.

2 Centromeric DNA: A Descriptor or Determinant?

Studies aimed at identifying the primary sequence associated with functional centromeric chromatin have revealed a lack of conservation of centromeric sequences, even among closely related species. Thus, the genomic component of eukaryotic centromeres is relatively rapidly evolving despite its conserved role in chromosome segregation (Henikoff et al. 2001). A remarkable computational effort has led to the production of graphical models of human centromere sequences (Miga et al. 2014;

Miga 2015; Rosenbloom et al. 2015), bypassing the need for strict linear assembly in the assessment of nascent genetic content. These "maps" do not delineate the order of sequences within any given centromere, yet reveal the diversity of satellites within and among centromeres, supporting earlier work demonstrating that while satellite higher order repeats (HORs) are homogenized through processes such as molecular drive and concerted evolution (Dover et al. 1982), some satellites are in fact distinct amongst different chromosomes (for an example, see Miga et al. 2014).

Defying another common misconception that each chromosome has only one location that can serve as a functional centromere, several human chromosomes have multiple HORs that act as functional centromeric epialleles (Maloney et al. 2012). Within any given chromosome, only one of these epialleles functions as the active centromere, raising the possibility of heterozygotes for different epialleles on the same chromosome pair. As the quality of sequencing and gap-filling for the human genome has increased, novel annotation workflows have also uncovered retroelements scattered throughout active centromere regions across all human chromosomes, within HORs and between epialleles (Miga et al. 2014; Rosenbloom et al. 2015). Indeed, co-option of repetitive elements, including tandem duplications, may be a general aspect of centromere ontogenesis across eukaryotes (Dawe 2003; Wong and Choo 2004; Chueh et al. 2009; O'Neill and Carone 2009; Brown and O'Neill 2010).

Most multicellular eukaryotic centromeres harbor a similar, characteristic repeat structure highly enriched in species-specific satellites (e.g., α satellites in human and minor satellites (miSAT) in mouse). The functional impact of these satellites with respect to kinetochore assembly remains less clear, however, based on multiple lines of evidence. Several studies highlight that centromeric satellites are not sufficient to form kinetochores. Placing an array of satellites in a cell is not the only requisite to form stable artificial chromosomes in all cases (Nakano et al. 2003). In fact, dicentric chromosomes often retain their satellite array but this array no longer forms functional centromeric chromatin (Warburton et al. 1997). Thus, the presence of satellite DNA alone is not the primary determinant for recruiting centromeric histones. As both ectopic centromeres in abnormal chromosomes (e.g., mini- and marker chromosomes, B chromosomes, neocentromeres) and newly formed centromeres that have only recently become fixed within a species (e.g., evolutionary new centromeres, ENC) are often devoid of satellite DNA, the absence of satellite DNA suggests such repeated DNA is also not required for centromere formation (Lo et al. 2001a; Alonso et al. 2007; Hasson et al. 2013).

While the canonical structure of species-specific satellites, and higher order arrays of groups of satellites, is neither sufficient nor required to facilitate centromere assembly, it is a pervasive feature among eukaryotic centromeres (Brown and O'Neill 2014; Plohl et al. 2014). While the fact that centromeres can form and act on genomic regions devoid of satellite DNA has lent support to the notion that centromere identity is likely under epigenetic control (Karpen and Allshire 1997; Henikoff et al. 2001). The contributions such types of genomic sequence have on defining the functional capacity of centromeric chromatin assembly and evolutionary stability of centromeres cannot be discounted. As exemplified in studies of human neocentromeres, DNA satellites alone are not required to attract centromere proteins to ectopic centromeres (e.g., Lo et al. 2001b). In such cases, another type of repeat found in most complex centromeres, retrotransposons, are found to bind the defining centromeric histone, CENP-A, and define the functional centromere (Chueh et al. 2009). These selfish entities may be the progenitors of satellite arrays (e.g., Macas et al. 2009) that experience accretion and diminution as either monomers or large homogenous arrays following centromere stabilization and fixation in a population. Just as the acquisition of repeat expansions may be linked to the ontogeny of a fixed, stable centromere within a species, the primary establishment of a new centromere may be the result of a seeding event from retroelement(s) that progressively generate novel satellites (Dawe 2003; O'Neill et al. 2004; Brown and O'Neill 2010).

Despite the challenges in delineating a finite sequence demarcating centromere functionality across species, the protein cascade that leads to faithful centromere assembly each cell cycle is more clearly defined. The pivotal event is the loading of the centromere specific H3, CENP-A (Fig. 1a), which occurs in late telophase/early G1 in most organisms (Dunleavy et al. 2009) [n.b. in *S. pombe*, this occurs in S phase (Dunleavy et al. 2007)]. During replication in S phase, the levels of CENP-A are diluted to 1/2 as H3.3 is assembled into centromeric chromatin as a placeholder (Dunleavy et al. 2011). In human, HJURP (holliday junction recognition protein) associates with CENP-A in pre-nucleosomal complexes (Mellone et al. 2009; Foltz et al. 2009; Dunleavy et al. 2009) and chaperones newly synthesized CENP-A to centromeric chromatin following mitosis (telophase/early G1) (Foltz et al. 2009; Dunleavy et al. 2009) when CENP-A loading occurs (Jansen et al. 2007). After mitosis, new CENP-A loading is also facilitated by a priming mechanism involving protein complexes such as hMis18 (Fujita et al. 2007) that prepares the centromeric nucleosome for CENP-A loading (Mellone et al. 2009; Dunleavy et al. 2009). These proteins serve as the pinnacle of the DNA-chromatin interface, yet many other proteins are involved in the coordinated assembly of the kinetochore (described in Parts I and IV of this book).

3 Active Transcription at Centromeres—*Breaking Down Common Myths and Legends*

Challenging another classical description of a eukaryotic centromere as a heterochromatin-rich and transcriptionally inactive region, centromeres are in fact characterized by a complex suite of different chromatin marks supporting active transcription and the production of centromeric noncoding RNAs required for proper centromere formation and function (Wong et al. 2007; Carone et al. 2009, 2013; Ting et al. 2011; Hall et al. 2012; Quenet and Dalal 2014). The chromatin encompassing the centromere core, referred to as "centrochromatin", is distinct from that of pericentromeres and contains histone modifications associated with transcriptionally active chromatin (Sullivan and Karpen 2004; Eymery et al. 2009; Gopalakrishnan et al. 2009; Bergmann et al. 2011, 2012). CENP-A nucleosomes within centrochromatin are interspersed with modified histones, histone H3 methylation, and dimethylation of lysine 4 and di- and trimethylation of lysine 36 of histone H3 (H3K4me1, H3K4me2, H3K36me2, and H3K36me3). These modified histones are not only permissive to transcription, but differentiate centrochromatin from its neighboring pericentromere, a region that, while also characterized by a high density of repeats, is defined by histone modifications typically associated with transcriptional silencing [(Gopalakrishnan et al. 2009; Roadmap Epigenomics et al. 2015): di- and trimethylation of lysine residues 9 and 27 of histone H3 (H3K9me2, H3K9me3,H3K27me2, and H3K27me3)].

Despite the remarkably different chromatin environments that define the peri- and centromere, active transcription has been detected from both regions in many organisms (Carone et al. 2009; Ugarkovic 2005; Eymery et al. 2009; Brown et al. 2012; Gent and Dawe 2012; Hall et al. 2012; Biscotti et al. 2015; Koo et al. 2016; Rosic and Erhardt 2016). Moreover, a clear balance in transcriptional output from each region is required to maintain chromosome stability (Hall et al. 2012). The types of sequences found to produce transcripts within centromeres includes the same sequences represented in the genomic foundation of a centromere: satellites, retroelements and in some cases active genes located within the boundaries of centrochromatin (e.g., Nagaki et al. 2004).

While prevalent in complex eukaryotic centromeres, the importance of these retroelement and satellite-derived transcripts to centromere function is only recently becoming apparent; chromosome missegregation has been associated with aberrant centromere transcription in animals and satellite RNA has been implicated in the assembly of centromere components CENP-A and -C, in *Drosophila*, plants, mouse and human (Mejía 2002; Bergmann et al. 2011; Ting et al. 2011; Carone et al. 2013; Quenet and Dalal 2014; Leung et al. 2015).

4 Genome Engineering to Tease Apart the Transcriptional Framework of the Centromere

Advances in techniques that allow manipulation of DNA and its nascent chromatin have been used synergistically to create and modify artificial centromere constructs within living cells. For example, alpha satellite arrays from human have been isolated and, when placed in HT1080 cells, form functional human artificial chromosomes (HACs) (Harrington et al. 1997; Ikeno et al. 1998; Grimes and Monaco 2005; Lam et al. 2006; Maloney et al. 2012). Focused studies of HACs have shown that active transcription at the centromere is essential to their stable propagation (Okamoto et al. 2007; Nakano et al. 2008; Bergmann et al. 2011, 2012; Molina et al. 2016). DNA constructs that form stable HACs incorporate selectable marker genes (i.e., *neo* and *bsr*) under strong, constitutive promoters juxtaposed to the alphoid arrays. The ability of the resulting HAC to assemble a functional kinetochore and survive cell division was found to be reliant not only on the overall number of satellites but also on the transcriptional activity of these marker genes (Okamoto et al. 2007).

Human artificial chromosomes modified to carry tetO transcriptional regulatory sequences within alphoid arrays were manipulated to increase or decrease transcriptional output in attempts to define the activity for proper centromere function (Nakano et al. 2008). Switching off transcription from the tetO dramatically diminished propagation of the HACs, but upregulating transcription with tet activators had a similar effect, indicating a balanced level of transcription is a requisite for proper centromere function. Further modifications of HACs by tethering a lysine-specific demethylase (LSD1) to alphoid arrays showed that depletion of H3K4me2 from HAC centromeric chromatin results in a loss of satellite transcription and concomitant reduction in local assembly of newly synthesized CENP-A (Bergmann et al. 2011). HACs targeted to increase H3K9 acetylation, a mark permissive to transcription, showed no effect on kinetochore formation despite such a dramatic change in chromatin state. However, when this chromatin change is coupled with a dramatic increase in transcription, rapid centromere inactivation through loss of CENP-A loading results (Bergmann et al. 2012).

Recently, a study using an inducible ectopic centromere system in *Drosophila* showed that CENP-A assembly by the *Drosophila* CENP-A chaperone, CAL1, requires RNA pol II mediated transcription of nascent DNA (Chen et al. 2015). In this ectopic centromere system, transcription is mediated by CAL1's binding partner, the chromatin remodeling complex FACT (facilitates chromatin transcription) and targets an artificial array of lacO sequences, indicating that the passage of RNA polymerase is required for CENP-A chromatin establishment rather than sequence-specific transcripts (Chen et al. 2015). A study of the primary centromere core sequence in *S. pombe* necessary and sufficient for CENP-A assembly was conducted wherein the core sequence was shuffled to create a de novo sequence with the same AT content and nucleosome positioning (Catania et al. 2015). This new construct was not able to effectively establish CENP-A chromatin, indicating some sequence features are required for centromere integrity. Notably, the core sequence is actively transcribed via multiple putative transcription start sites, implicating its ability to facilitate transcription (albeit stalled transcription, see below) as a defining feature of this centromere-competent sequence (Catania et al. 2015). As demonstrated by these studies, centromere integrity requires tight control of centromere transcription, suggesting that centromeric DNA sequence identity may not be an absolute requirement, but the ability to facilitate transcription and act as a fundamentally stable and immutable regulatory element(s) is needed.

5 The "How?" of Centromeric RNA: Centromere Transcripts and Protein Interactions

Centromeric RNAs have been hypothesized to perform diverse functions, including establishing and maintaining pericentromeric heterochromatin and recruiting kinetochore proteins to the centrochromatin core. Recent studies have focused on the transcription of the most prevalent centromeric sequence, satellites, with respect to centromere function, however, the identity and functional roles of satellite transcripts in diverse organisms have not been fully elucidated. Several recent studies highlighted below further support the growing evidence that transcription is an integral part of the centromere chromatin assembly cascade; how, when, and what types of transcripts impact centromere assembly are emerging areas of focus in the centromere biology field.

As the centromere is a tightly regulated network of protein and nucleic acid interactions, noncoding transcripts may only directly interact with a subset of this multi-protein network and yet, indirectly impact the function of many centromere and kinetochore proteins when these transcripts are mis-regulated. CENP-C, CENP-A, HJURP, and certain members of the chromosomal passenger complex (CPC) have been implicated as the centromere proteins that directly associate with, or bind to, noncoding RNAs (Wong et al. 2007; Ferri et al. 2009; Du et al. 2010; Carone et al. 2013; Quenet and Dalal 2014; Rosic et al. 2014; Blower 2016).

CENP-A: The first indication that CENP-A can interact with noncoding RNA was discovered in a human neocentromere; LINE-1 elements within the CENP-A binding region of a neocentromere on 10q25 are actively transcribed into a non-coding RNA that incorporates with CENP-A chromatin (Chueh et al. 2009). While less evidence exists for a direct association of centromeric RNA and CENP-A or HJURP, aberrant expression of these transcripts distinctly perturbs CENP-A localization and loading (Fig. 2). For example, overexpression of noncoding RNA from the centromeric retrotransposon KERV in tammar wallaby disrupts proper CENP-A loading into centromeres in late telophase (Carone et al. 2013). A recent study in human showed a more direct contact between CENP-A and RNA when an alpha satellite related RNA sequence was pulled down with the soluble CENP-A/HJURP complex using RNA immunoprecipitation (RIP) (Quenet and Dalal 2014). While this specific noncoding RNA sequence does not match the alpha satellite consensus, nor any other alpha satellite higher order repeat sequence, and any known repeated elements in the assembled or unassembled contigs of the human genome, DNA FISH showed it may reside in only a subset of chromosomes in the human karyotype (Quenet and Dalal 2014). Quenet and Dalal (2014) complemented their study with an in-silico prediction of potential RNA binding sites in CENP-A and HJURP, finding that 79 out of 140 residues in CENP-A and 286 out of 748 residues in HJURP had a capacity for RNA binding. Intriguingly, the entirety of the CENP-A N-terminal tail was predicted to carry RNA-binding capacity (Quenet and Dalal 2014) (Fig. 1a). CENP-A's N-terminal tail is the most rapidly evolving portion of CENP-A

(Henikoff et al. 2001; Malik and Henikoff 2001), and while it is known to be required for CENP-A stabilization at the centromere (Logsdon et al. 2015), its exact function remains elusive. Given a putative role in RNA interaction, it is possible the vast differences in amino acid sequence and overall length of the N-terminal region among species (Henikoff et al. 2001) could be to enable permissive interaction with a variety of transcripts that emanate from the rapidly evolving, underlying DNA.

CENP-C: CENP-C contains an experimentally validated, distinct RNA binding domain (Wong et al. 2007; Du et al. 2010) (Fig. 1b). Interestingly, the RNA binding domain of CENP-C shares homology to the RNA binding hinge domain region of the pericentromeric heterochromatin proteins HP1 alpha, beta, and gamma (Muchardt et al. 2002; Du et al. 2010). In human, CENP-C associates with single-stranded (ss) alpha satellite transcripts both in vitro and in vivo, and is lost from centromeres upon alpha satellite depletion along with the CPC proteins, INCENP and Survivin (Wong et al. 2007). DNA binding of maize CENP-C is stabilized by a ssRNA in vitro, although this stabilization appears to be independent of the ssRNA sequence (Du et al. 2010) (Fig. 2). This permissive binding in maize is in contrast to human CENP-C that showed a preferential association with alpha satellite ssRNA in competition assays with tRNA, rRNA, and mouse pericentric satellite (Wong et al. 2007). In *Drosophila*, the X chromosome specific satellite,

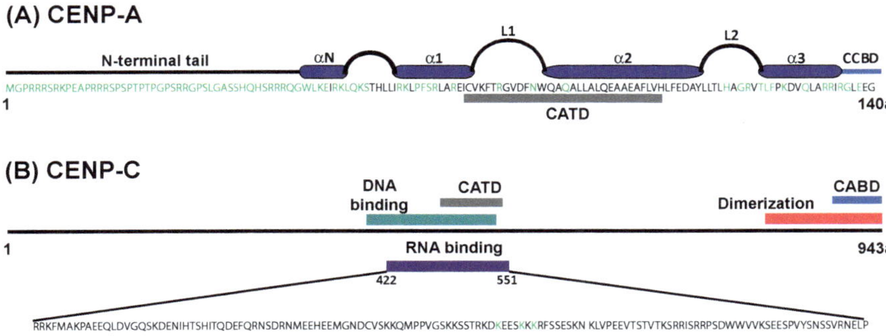

Fig. 1 RNA binding domains of CENP-A and CENP-C. **a** Linear depiction of the complete CENP-A protein domain structure (Regnier et al. 2003) with amino acids that comprise each domain shown underneath. Amino acids in *green* were computationally predicted to have an RNA binding capability by Quenet and Dalal (2014). Most of the potential RNA interaction capability lies in the N-terminal tail region [*a* Alpha helix, *L* Loop region, *CATD* Centromere targeting domain (Black et al. 2004), *CCBD* CENP-C binding domain (Carroll et al. 2010)]. **b** Linear depiction of the CENP-C protein. The RNA binding domain experimentally validated by Wong et al. (2007) is located between amino acids 422 and 551, the sequence of which is shown below. Amino acids in *green* are most critical to RNA binding (Wong et al. 2007). The RNA binding domain of CENP-C overlaps with both the DNA binding domain (*aqua*) and the CATD (*gray*). Note that Wong et al. (2007) also found evidence for a second RNA binding domain between 552 and 943, but did not isolate the exact region. [DNA binding domain (Yang et al. 1996; Sugimoto et al. 1997; Cohen et al. 2008; Schueler et al. 2010), CATD (Yang et al. 1996), Dimerization domain (Sugimoto et al. 1997), *CABD* CENP-A binding domain (Trazzi et al. 2009)]

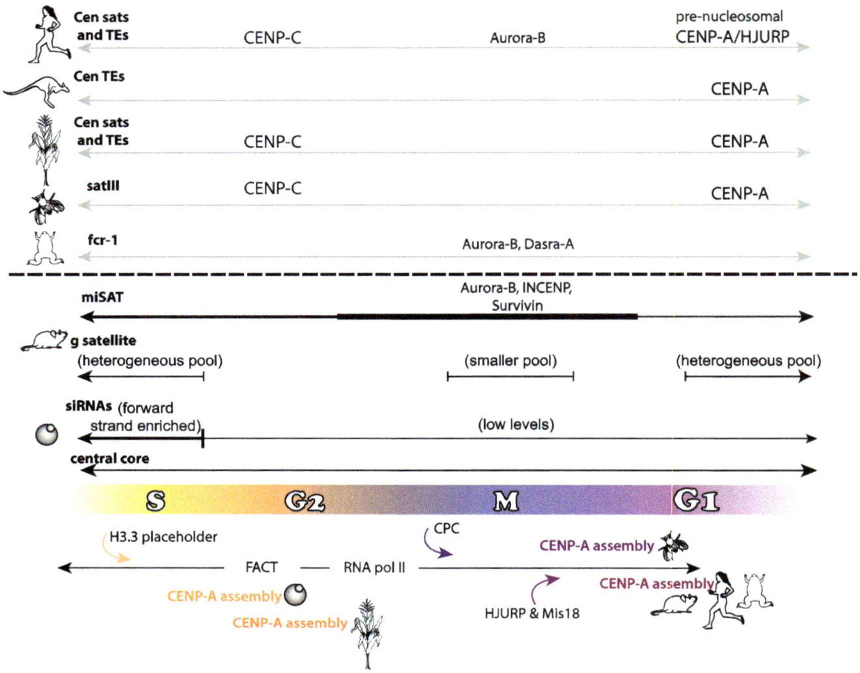

Fig. 2 Cell cycle variation of centromere transcription. Overview of when in the cell cycle centromeric transcripts have been identified in different model species in relation to critical assembly events defining centromere integrity. The cell cycle is indicated (*color*). *Top* The type of transcript for each species [from *top*, human (Wong et al. 2007; Chueh et al. 2009; Quenet and Dalal 2014), wallaby (Carone et al. 2013), plants (Topp et al. 2004; Koo et al. 2016), and reviewed in Gent and Dawe (2012), Drosophila (Rosic et al. 2014), frog (Blower 2016), mouse (Lu and Gilbert 2007; Ferri et al. 2009), yeast (Chen et al. 2008; Choi et al. 2011; Catania et al. 2015)] is indicated. Below the *dashed line* are the transcripts where the timing of transcription is known. A *thinner bar* represents a lower level of transcription while a *thicker bar* represents a higher level. *Gray lines* (above the *dashed line*) indicate that while transcripts have been identified, it is not known when in the cell cycle transcription is initiated. Above each line are the protein associations known for transcripts. *Bottom* The timing of protein cascade components relative to the cell cycle. *Black bar* indicates constitutive association with the centromere (FACT, RNA Pol II). Relevant timing of loading of the H3.3 placeholder, CPC recruitment and CENP-A loading components are indicated. Specific CENP-A assembly times are indicated for each group of species

SAT III, is actively transcribed into long noncoding transcripts that localize to centromeres and associate with CENP-C (Rosic et al. 2014) (Fig. 2). Upon CENP-C depletion, SAT III RNA signal is greatly reduced at centromeres, implying a similar interaction between CENP-C and RNA in *Drosophila*, as in human and maize. When depleting SAT III RNA levels, both newly synthesized CENP-C and CENP-A showed a reduction in centromeric signal that was also observed to cascade up through the kinetochore proteins (Rosic et al. 2014). SAT III-depleted cells also suffered errors in mitosis, including lagging chromosomes and micronuclei formation; notably, all chromosomes were susceptible to mitotic defects, indicating that SAT III RNAs, despite originating from the X chromosome, can act in trans to target the autosomes (Rosic et al. 2014). Lagging chromosomes with reduced CENP-C signal were also prevalent in human cells after RNA pol II inhibition (Chan et al. 2012).

The CPC: Ostensibly, a single centromeric noncoding transcript does not have to bind to just a single protein. When considering the fact that CENP-A assembly is facilitated by a chaperone, HJURP, and that CENP-A and CENP-C are both required for centromere integrity, it is probable that these transcripts contact multiple proteins that are associated with one another. Such RNA interactions may even serve to tether protein complexes together, or to scaffold these complexes to other components of the surrounding chromatin environment. The multi-protein interaction between Aurora-B, Dasra-a/Borealin, Survivin, and Incenp composes the CPC. The CPC aids in mitosis as a phosphorylating agent at chromosome arms, the inner centromere, and mitotic spindles (reviewed in (Carmena et al. 2012). While at the inner centromere, the CPC plays a key role in bipolar spindle attachment by acting as a "sensor" of connection between the centromere and the spindle (Lampson and Cheeseman 2011). Both cen-RNAs and spindle-enriched RNAs are known to congregate with the CPC (Ferri et al. 2009; Ideue et al. 2014; Jambhekar et al. 2014) (Fig. 2). In fact, there is direct binding of RNA to the CPC and this interaction is responsible for inner centromere localization (Jambhekar et al. 2014; Blower 2016), and is required for CPC activation (Blower 2016). Despite the fact that multiple proteins form the CPC, Ferri et al. (2009) showed that miSAT transcripts in mouse are a key partner with CPC proteins Aurora-B, INCENP and Survivin at the onset of mitosis. In *Xenopus* extracts, however, RNA binding was identified that is required for CPC localization, but this binding is specific only to the proteins Aurora-B and Dasra-A, and not INCENP, Survivin, and XMAP215 (Blower 2016).

In *Xenopus,* among the Aurora-B binding RNAs is a $\sim$170 nt centromeric transcript (fcr1, frog centromeric repeat1) that, similar to the sat III RNA in *Drosophila* (Rosic et al. 2014), is only found on a subset of CENP-A defined centromeres within the karyotype (Edwards and Murray 2005). The active transcription of fcr1 is required for Aurora-B localization to the inner centromere of mitotic chromosomes and may act initially on the fcr-1-native chromosomes before diffusing to other centromeres (Blower 2016) (Fig. 2).

6 The "When?" of Centromere Transcription: It is an Around the Clock Job

The emergence of studies on RNA, transcription, and centromere function since the pivotal studies in yeast (Volpe et al. 2003), plants (Topp et al. 2004) and human neocentromeres (Wong et al. 2007) has led to accumulating evidence that transcription is a requirement for centromere function and cell stability. However, the timing of this transcription and a delineation of whether specific transcript sequences, or simply the act of transcription itself, are required for centromere integrity are not known. Studies in several model systems have begun to highlight the intricacies of transcriptional events at the centromere throughout the cell cycle, with a particular emphasis on mitotic transcription (Fig. 2).

Cell Cycle Phase G1: Late telophase/early G1 is the pivotal time in mammalian cells when CENP-A is actively loaded into centromeric chromatin. Thus, the impact of active transcription at this point in the cell cycle may have direct bearing on the ability of CENP-A to assemble functional centromeric chromatin. The 1.3 KB human centromeric transcript described above (Quenet and Dalal 2014) is transcribed by RNA pol II from late telophase into early G1, coincident with the timing of CENP-A deposition by its chaperone HJURP (Fig. 2). This RNA transcript was found to interact with these proteins, suggesting its capacity to aid in CENP-A nucleosome assembly. The transcription of one of two groups of transcripts that emanate from mouse pericentromeric gamma satellites was detected in late G1 and proceeded through mid-S phase (Lu and Gilbert 2007) (Fig. 2). This species of RNA did not show a discrete size range and transcription of this species decreased at a time coincident with the replication of pericentric heterochromatin. This RNA class is a large, heterogeneous group of gamma satellite transcripts whose transcriptional timing may simply be the result of cryptic transcription (Lu and Gilbert 2007). However, given that the appearance of these satellite transcripts is CDK (cyclin dependent kinase)-dependent, and thus is only in cells committed to proliferation, the transcripts may be required for heterochromatin reassembly at the replication fork (Lu and Gilbert 2007).

Cell Cycle Phase S and G2: Double-stranded RNAs are actively transcribed from the pericentric repeats dh and dg from within the centromeres of the yeast, *Shizosaccharomyces pombe*, and are subsequently processed into small interfering RNAs (siRNAs) (Volpe et al. 2002, 2003). These siRNAs are bound to a complex of proteins (the RNA-induced initiation of transcriptional gene silencing, RITS) and result in targeted H3 lysine-9 methylation through RNA interference (Volpe et al. 2002, 2003). Moreover, the disruption of RNAi components compromises heterochromatin assembly (Volpe et al. 2002) and CENP-A deposition (Folco et al. 2008), linking a small RNA component to centromere function.

The forward strand of centromeric repeats is transcribed in S phase in *S. pombe*, thought to be the major initiating point for siRNA production (Chen et al. 2008). siRNA levels are stably detected throughout the cell cycle but increased in S/G2, coincident with RITS complex accumulation and transcript processing (Fig. 2). Similar to that proposed for gamma satellites in mouse, these transcripts may be the result of cryptic or spurious transcription yet are required for the appropriate establishment of heterochromatin (Chen et al. 2008).

While active CENP-A loading occurs in late telophase/early G1 in mammals, H3.3 is loaded into centrochromatin during S phase, likely as a placeholder for CENP-A replenishment after mitosis (Dunleavy et al. 2011). Thus, transcription could be involved in the eviction of the placeholder (Catania et al. 2015; Chen et al.

2015; Chen and Mellone 2016). A recent study in *S. pombe* showed that RNA pol II stalls at centromeric DNA and that the level of stalling is directly related to the level of subsequent CENP-A nucleosome assembly. In this yeast species, CENP-A assembly occurs in S phase and G2 (Takahashi et al. 2005; Dunleavy et al. 2007; Takayama et al. 2008); an increase in RNA pol II stalling, and concomitantly permissive but "low-quality" transcription, may lead to increased CENP-A chromatin through either increased eviction rates for placeholder H3 or through demarcation of a specific environment conducive to efficient CENP-A assembly (Catania et al. 2015).

The timing of detectable increases in RNAPII stalling in S phase is coincident with DNA replication. Centromere transcription at this phase of the cell cycle [forward strands in *S. pombe* (Chen et al. 2008) and pericentromeric satellites in mouse (Lu and Gilbert 2007)] may position replication forks and RNA pol II to collide more often, increasing the rate of RNA pol II stalling (reviewed in Brown et al. 2012). RNA pol II stalling and collisions subsequently increase the generation of large, stable R-loop formation (Reddy et al. 2011). While the RNA–DNA hybrids present in R-loops are typically small and transient, it is notable that the transcriptional framework of the centromere may present an increase in stable R-loops in S phase since increases in R-loops are linked to phosphorylation of H3S10, a marker of subsequent entry into mitosis (M phase) (Castellano-Pozo et al. 2013; Oestergaard and Lisby 2016). Chen et al (2015) showed that FACT is required for CENP-A assembly and FACT has been previously shown to both travel in a complex with the CENP-A chaperone (Foltz et al. 2006) and localize to centromeres throughout the cell cycle (Okada et al. 2009). The presence of FACT at centromeres and in complex with key centromere assembly components supports hypotheses that FACT is an essential part of the chromatin remodeling involved in facilitating CENP-A nucleosome assembly. For example, FACT may be required to destabilize nucleosomes (Hondele and Ladurner 2013) and subsequently facilitate the transcription of centromere sequences preceding assembly of new CENP-A nucleosomes (Chen et al. 2015; Chen and Mellone 2016). Recently, FACT was found to bind the inner kinetochore proteins of the CENP-T/W complex and thus may promote the CENP-T/W deposition at centromeres (Prendergast et al. 2016). In fungi it appears that FACT is necessary to prevent spurious, ectopic incorporation of CENP-A rather than performing a function in primary CENP-A assembly at the centromere (Deyter and Biggins 2014). FACT is known to solve R-loop and replication mediated conflicts in both human and yeast (Herrera-Moyano et al. 2014). Furthermore, low levels of central core transcripts are detected in yeast cells due to an increase in RNA pol II stalling (Catania et al. 2015). We thus propose that FACT may also be present at centromeres to resolve the resulting R-loops prior to progression into mitosis. The multiple, possible roles for FACT in CENP-A assembly are not mutually exclusive, rather are an example of the dynamic state of the centromere during different phases of the cell cycle.

Cell Cycle Phase Mitosis: Whether or not centromeric transcription occurs during mitosis has been hotly debated since the majority of transcription factors and

RNA polymerases are not associated with chromosomes during mitosis (Gottesfeld and Forbes 1997). However, several pieces of evidence suggest centromeric transcription occurs during M phase of the cell cycle (Liu 2016), indicating persistent transcription during mitosis may serve to further distinguish the centromere from chromosome arms. First, RNA pol II is present at kinetochores in M phase (Chan et al. 2012). Second, transcription run-on assays have shown that RNA pol II is capable of transcribing centromeres of mitotic chromosomes (Liu 2016). Third, inhibition of RNA pol II by alpha-amanitin reduces centromeric cohesion and CENP-C localization (Chan et al. 2012; Liu et al. 2015). The defective cohesion following RNA polymerase inhibition was caused by a mislocalization of Sgo1, a protein found at the inner centromere that protects cohesin during mitosis (Liu et al. 2015). Collectively, these data suggest that transcription and/or the transcripts themselves may play a functional role in a time-specific manner (i.e., restricted to specific phases of the cell cycle).

A long noncoding RNA was recently identified as actively transcribed from the centromeres of *Xenopus* egg extracts during mitosis; moreover, these transcripts serve a functional role via binding Aurora-B, a component of the CPC, and are required for normal kinetochore-centromere attachment (Blower 2016). *Xenopus* egg extracts have also led to the discovery that RNA processing assists in kinetochore and spindle assembly (Grenfell et al. 2016); inhibition of the spliceosome in egg extracts leads to an accumulation of long centromeric transcripts and a failure to efficiently recruit CENP-A, CENP-C and NDC80. What this study shows is that transcription is not only active during mitosis, further supporting the growing body of evidence indicating this occurs, but that transcripts undergo processing during this phase, contradicting the theory that RNA processing is repressed during mitosis (Shin and Manley 2002).

Centromere transcript processing is a recurring theme observed for a broad set of centromere transcripts, although the relationship of the processing machinery and/or processed RNA products to centromere integrity is less clear (with the notable exception of *S. pombe*, see above). Early work in mouse cells showed that a loss of DICER activity, the RNAse III enzyme that facilitates small RNA processing, results in an accumulation of larger satellite transcripts (Kanellopoulou et al. 2005). This finding implies that when DICER is available, these larger satellite transcripts are not detected as they are processed into smaller RNAs. The implication that DICER is involved in this RNA processing would also indicate these small RNAs are <40 nt based on the catalytic activity of the enzyme (MacRae et al. 2007). Other sizes of centromeric RNAs have been uncovered that are likely independent of DICER. For example, small RNAs have been detected for the maize centromere satellite CentC (Du et al. 2010) and from the wallaby centromeric retroelement KERV (Carone et al. 2009). Both of these small RNAs were also found to participate in the centromere assembly cascade: CentC associated with CENP-C directly (Du et al. 2010); a reduction in KERV small RNAs resulted in a loss of CENP-A assembly in late telophase (Carone et al. 2013) (Fig. 2). However, any connection between these types of processed, small RNA transcripts and the necessary RNA processing machinery observed in *Xenopus* is unexplored; likewise the timing of transcription for these processed RNAs is currently unknown.

7 Conclusion

Mounting evidence suggests that RNA species and the act of transcription itself is required for the recruitment and/or establishment of centromere and kinetochore proteins. Thus, it is clear that transcription at the centromere, and in neighboring pericentromeric heterochromatin, is functionally distinct yet critical throughout the entirety of the cell cycle. Studies are now beginning to reveal that centromeric transcripts and accompanying chromatin changes are required for different components of the centromere assembly cascade at different points in the cell cycle. Over the last decade, RNA species derived from the centromeric regions of many model species have been uncovered, as have some of their interacting partners. Closer examination of these transcripts, and indeed of the subregions of the centromere previously considered devoid of transcriptional activity, has made it clear that both the act of transcription itself and the resulting transcripts are critical to ensuring proper CENP-A assembly and faithful chromosome segregation. In the same manner that the comparative approach revealed that centromeres evolve rapidly and are established through an epigenetic framework, the use of diverse eukaryotic systems will afford the development of a model to describe key remaining questions, such as: how do specific transcripts mediate centromere function in cis and/or in trans? is splicing or RNA processing a requisite in forming functional transcripts across different cell cycles and among different species? and, how does this transcriptional landscape impact centromere evolution in both a phylogenetic and disease context? In fact, one of the reasons the myth of the centromere as "silent chromatin" prevailed for so long is that centromere transcripts have been difficult to capture and characterize. As highlighted herein, it is the very reason these transcripts are difficult to capture (e.g., RNA pol II stalling, RNA processing, protein-RNA binding) that holds the key to how centromere transcription, and their transcripts, likely function in maintaining centromere integrity.

Acknowledgements Sincerest apologies to all of the scientists' work that could not be covered in this review for lack of space. Thank you to J. Brown, M. O'Neill and B. Mellone for critical review of the manuscript. RJO and ZD are supported by NSF 1613803.

References

Alonso A, Fritz B, Hasson D, Abrusan G, Cheung F, Yoda K, Radlwimmer B, Ladurner AG, Warburton PE (2007) Co-localization of CENP-C and CENP-H to discontinuous domains of CENP-A chromatin at human neocentromeres. Genome Biol 8(7):R148

Altemose N, Miga KH, Maggioni M, Willard HF (2014) Genomic characterization of large heterochromatic gaps in the human genome assembly. PLoS Comput Biol 10(5):e1003628. doi:10.1371/journal.pcbi.1003628

Bayes JJ, Malik HS (2008) The evolution of centromeric DNA sequences. Encycl Life Sci 1–8

Bergmann JH, Rodriguez MG, Martins NM, Kimura H, Kelly DA, Masumoto H, Larionov V, Jansen LE, Earnshaw WC (2011) Epigenetic engineering shows H3K4me2 is required for

HJURP targeting and CENP-A assembly on a synthetic human kinetochore. EMBO J 30 (2):328–340. doi:10.1038/emboj.2010.329 emboj2010329 [pii]

Bergmann JH, Jakubsche JN, Martins NM, Kagansky A, Nakano M, Kimura H, Kelly DA, Turner BM, Masumoto H, Larionov V, Earnshaw WC (2012) Epigenetic engineering: histone H3K9 acetylation is compatible with kinetochore structure and function. J Cell Sci 125(Pt 2):411–421. doi:10.1242/jcs.090639 jcs.090639 [pii]

Biscotti MA, Canapa A, Forconi M, Olmo E, Barucca M (2015) Transcription of tandemly repetitive DNA: functional roles. Chromosome Res 23(3):463–477. doi:10.1007/s10577-015-9494-4

Black BE, Foltz DR, Chakravarthy S, Luger K, Woods VL Jr, Cleveland DW (2004) Structural determinants for generating centromeric chromatin. Nature 430(6999):578–582

Blower MD (2016) Centromeric transcription regulates Aurora-B localization and activation. Cell Rep 15(8):1624–1633. doi:10.1016/j.celrep.2016.04.054

Brown JD, O'Neill RJ (2010) Chromosomes, conflict, and epigenetics: chromosomal speciation revisited. Annu Rev Genomics Hum Genet 11:291–316. doi:10.1146/annurev-genom-082509-141554

Brown JD, O'Neill RJ (2014) The evolution of centromeric DNA sequences. In: eLS, Wiley, USA. doi:10.1002/9780470015902.a0020827.pub2

Brown JD, Mitchell SE, O'Neill RJ (2012) Making a long story short: noncoding RNAs and chromosome change. Heredity (Edinb) 108(1):42–49. doi:10.1038/hdy.2011.10410.1038 hdy2011104 [pii]

Carbone L, Harris RA, Gnerre S, Veeramah KR, Lorente-Galdos B, Huddleston J, Meyer TJ, Herrero J, Roos C, Aken B, Anaclerio F, Archidiacono N, Baker C, Barrell D, Batzer MA, Beal K, Blancher A, Bohrson CL, Brameier M, Campbell MS, Capozzi O, Casola C, Chiatante G, Cree A, Damert A, de Jong PJ, Dumas L, Fernandez-Callejo M, Flicek P, Fuchs NV, Gut I, Gut M, Hahn MW, Hernandez-Rodriguez J, Hillier LW, Hubley R, Ianc B, Izsvak Z, Jablonski NG, Johnstone LM, Karimpour-Fard A, Konkel MK, Kostka D, Lazar NH, Lee SL, Lewis LR, Liu Y, Locke DP, Mallick S, Mendez FL, Muffato M, Nazareth LV, Nevonen KA, O'Bleness M, Ochis C, Odom DT, Pollard KS, Quilez J, Reich D, Rocchi M, Schumann GG, Searle S, Sikela JM, Skollar G, Smit A, Sonmez K, ten Hallers B, Terhune E, Thomas GW, Ullmer B, Ventura M, Walker JA, Wall JD, Walter L, Ward MC, Wheelan SJ, Whelan CW, White S, Wilhelm LJ, Woerner AE, Yandell M, Zhu B, Hammer MF, Marques-Bonet T, Eichler EE, Fulton L, Fronick C, Muzny DM, Warren WC, Worley KC, Rogers J, Wilson RK, Gibbs RA (2014) Gibbon genome and the fast karyotype evolution of small apes. Nature 513(7517):195–201. doi:10.1038/nature13679

Carmena M, Wheelock M, Funabiki H, Earnshaw WC (2012) The chromosomal passenger complex (CPC): from easy rider to the godfather of mitosis. Nat Rev Mol Cell Biol 13 (12):789–803. doi:10.1038/nrm3474

Carone DM, Longo MS, Ferreri GC, Hall L, Harris M, Shook N, Bulazel KV, Carone BR, Obergfell C, O'Neill MJ, O'Neill RJ (2009) A new class of retroviral and satellite encoded small RNAs emanates from mammalian centromeres. Chromosoma 118(1):113–125. doi:10.1007/s00412-008-0181-5

Carone DM, Zhang C, Hall LE, Obergfell C, Carone BR, O'Neill MJ, O'Neill RJ (2013) Hypermorphic expression of centromeric retroelement-encoded small RNAs impairs CENP-A loading. Chromosome Res 21(1):49–62. doi:10.1007/s10577-013-9337-0

Carroll CW, Milks KJ, Straight AF (2010) Dual recognition of CENP-A nucleosomes is required for centromere assembly. J Cell Biol 189(7):1143–1155. doi:10.1083/jcb.201001013 jcb.201001013 [pii]

Castellano-Pozo M, Santos-Pereira JM, Rondon AG, Barroso S, Andujar E, Perez-Alegre M, Garcia-Muse T, Aguilera A (2013) R loops are linked to histone H3 S10 phosphorylation and chromatin condensation. Mol Cell 52(4):583–590. doi:10.1016/j.molcel.2013.10.006

Catania S, Pidoux AL, Allshire RC (2015) Sequence features and transcriptional stalling within centromere DNA promote establishment of CENP-A chromatin. PLoS Genet 11(3):e1004986. doi:10.1371/journal.pgen.1004986

Chan FL, Marshall OJ, Saffery R, Won Kim B, Earle E, Choo KH, Wong LH (2012) Active transcription and essential role of RNA polymerase II at the centromere during mitosis. Proc Natl Acad Sci USA 109(6):1979–1984. doi:10.1073/pnas.1108705109 1108705109 [pii]
Chen CC, Mellone BG (2016) Chromatin assembly: journey to the CENter of the chromosome. J Cell Biol 214(1):13–24. doi:10.1083/jcb.201605005
Chen ES, Zhang K, Nicolas E, Cam HP, Zofall M, Grewal SI (2008) Cell cycle control of centromeric repeat transcription and heterochromatin assembly. Nature 451(7179):734–737. doi:10.1038/nature06561
Chen CC, Bowers S, Lipinszki Z, Palladino J, Trusiak S, Bettini E, Rosin L, Przewloka MR, Glover DM, O'Neill RJ, Mellone BG (2015) Establishment of centromeric chromatin by the CENP-A assembly factor CAL1 requires FACT-mediated transcription. Dev Cell 34(1):73–84. doi:10.1016/j.devcel.2015.05.012
Choi ES, Stralfors A, Castillo AG, Durand-Dubief M, Ekwall K, Allshire RC (2011) Identification of noncoding transcripts from within CENP-A chromatin at fission yeast centromeres. J Biol Chem 286(26):23600–23607
Choo K (1997) The centromere. Oxford University Press, Oxford
Chueh AC, Northrop EL, Brettingham-Moore KH, Choo KH, Wong LH (2009) LINE retrotransposon RNA is an essential structural and functional epigenetic component of a core neocentromeric chromatin. PLoS Genet 5(1):e1000354. doi:10.1371/journal.pgen.1000354
Cohen RL, Espelin CW, De Wulf P, Sorger PK, Harrison SC, Simons KT (2008) Structural and functional dissection of Mif2p, a conserved DNA-binding kinetochore protein. Mol Biol Cell 19(10):4480–4491. doi:10.1091/mbc.E08-03-0297
Copenhaver GP, Nickel K, Kuromori T, Benito MI, Kaul S, Lin X, Bevan M, Murphy G, Harris B, Parnell LD, McCombie WR, Martienssen RA, Marra M, Preuss D (1999) Genetic definition and sequence analysis of Arabidopsis centromeres. Science 286(5449):2468–2474
Dawe RK (2003) RNA interference, transposons, and the centromere. Plant Cell 15(2):297–301
Deyter GM, Biggins S (2014) The FACT complex interacts with the E3 ubiquitin ligase Psh1 to prevent ectopic localization of CENP-A. Genes Dev 28(16):1815–1826. doi:10.1101/gad.243113.114
Dover GA, Strachan T, Coen ES, Brown SD (1982) Molecular drive. Science 218(4577):1069
Du Y, Topp CN, Dawe RK (2010) DNA binding of centromere protein C (CENPC) is stabilized by single-stranded RNA. PLoS Genet 6(2):e1000835. doi:10.1371/journal.pgen.1000835
Dunleavy EM, Pidoux AL, Monet M, Bonilla C, Richardson W, Hamilton GL, Ekwall K, McLaughlin PJ, Allshire RC (2007) A NASP (N1/N2)-related protein, Sim3, binds CENP-A and is required for its deposition at fission yeast centromeres. Mol Cell 28(6):1029–1044
Dunleavy EM, Roche D, Tagami H, Lacoste N, Ray-Gallet D, Nakamura Y, Daigo Y, Nakatani Y, Almouzni-Pettinotti G (2009) HJURP is a cell-cycle-dependent maintenance and deposition factor of CENP-A at centromeres. Cell 137(3):485–497. doi:10.1016/j.cell.2009.02.040 S0092-8674(09)00254-2 [pii]
Dunleavy EM, Almouzni G, Karpen GH (2011) H3.3 is deposited at centromeres in S phase as a placeholder for newly assembled CENP-A in G(1) phase. Nucleus 2(2):146–157. doi:10.4161/nucl.2.2.15211
Edwards NS, Murray AW (2005) Identification of Xenopus CENP-A and an associated centromeric DNA repeat. Mol Biol Cell 16(4):1800–1810. doi:10.1091/mbc.E04-09-0788
Eymery A, Callanan M, Vourc'h C (2009) The secret message of heterochromatin: new insights into the mechanisms and function of centromeric and pericentric repeat sequence transcription. Int J Dev Biol 53(2–3):259–268. doi:10.1387/ijdb.082673ae
Ferri F, Bouzinba-Segard H, Velasco G, Hube F, Francastel C (2009) Non-coding murine centromeric transcripts associate with and potentiate Aurora B kinase. Nucleic Acids Res 37 (15):5071–5080. doi:10.1093/nar/gkp529
Fishel B, Amstutz H, Baum M, Carbon J, Clarke L (1988) Structural organization and functional analysis of centromeric DNA in the fission yeast *Schizosaccharomyces pombe*. Mol Cell Biol 8(2):754–763

Folco HD, Pidoux AL, Urano T, Allshire RC (2008) Heterochromatin and RNAi are required to establish CENP-A chromatin at centromeres. Science 319(5859):94–97. doi:10.1126/science.1150944 319/5859/94 [pii]

Foltz DR, Jansen LE, Black BE, Bailey AO, Yates JR 3rd, Cleveland DW (2006) The human CENP-A centromeric nucleosome-associated complex. Nat Cell Biol 8(5):458–469

Foltz DR, Jansen LE, Bailey AO, Yates JR 3rd, Bassett EA, Wood S, Black BE, Cleveland DW (2009) Centromere-specific assembly of CENP-a nucleosomes is mediated by HJURP. Cell 137(3):472–484. doi:10.1016/j.cell.2009.02.039 S0092-8674(09)00253-0 [pii]

Fujita Y, Hayashi T, Kiyomitsu T, Toyoda Y, Kokubu A, Obuse C, Yanagida M (2007) Priming of centromere for CENP-A recruitment by human hMis18alpha, hMis18beta, and M18BP1. Dev Cell 12(1):17–30. doi:10.1016/j.devcel.2006.11.002 S1534-5807(06)00507-7 [pii]

Gent JI, Dawe RK (2012) RNA as a structural and regulatory component of the centromere. Annu Rev Genet 46:443–453. doi:10.1146/annurev-genet-110711-155419

Gopalakrishnan S, Sullivan BA, Trazzi S, Della Valle G, Robertson KD (2009) DNMT3B interacts with constitutive centromere protein CENP-C to modulate DNA methylation and the histone code at centromeric regions. Hum Mol Genet 18(17):3178–3193

Gottesfeld JM, Forbes DJ (1997) Mitotic repression of the transcriptional machinery. Trends Biochem Sci 22(6):197–202

Grenfell AW, Heald R, Strzelecka M (2016) Mitotic noncoding RNA processing promotes kinetochore and spindle assembly in Xenopus. J Cell Biol 214(2):133–141. doi:10.1083/jcb.201604029

Grimes BR, Monaco ZL (2005) Artificial and engineered chromosomes: developments and prospects for gene therapy. Chromosoma 114(4):230–241

Hall LE, Mitchell SE, O'Neill RJ (2012) Pericentric and centromeric transcription: a perfect balance required. Chromosome Res 20(5):535–546. doi:10.1007/s10577-012-9297-9

Harrington JJ, Van Bokkelen G, Mays RW, Gustashaw K, Willard HF (1997) Formation of de novo centromeres and construction of first-generation human artificial microchromosomes. Nat Genet 15(4):345–355

Hasson D, Panchenko T, Salimian KJ, Salman MU, Sekulic N, Alonso A, Warburton PE, Black BE (2013) The octamer is the major form of CENP-A nucleosomes at human centromeres. Nat Struct Mol Biol 20(6):687–695. doi:10.1038/nsmb.2562

Henikoff S, Ahmad K, Malik H (2001) The centromere paradox: stable inheritance with rapidly evolving DNA. Science 293(5532):1098–1102. doi:10.1126/science.1062939

Herrera-Moyano E, Mergui X, Garcia-Rubio ML, Barroso S, Aguilera A (2014) The yeast and human FACT chromatin-reorganizing complexes solve R-loop-mediated transcription-replication conflicts. Genes Dev 28(7):735–748. doi:10.1101/gad.234070.113

Holland AJ, Cleveland DW (2009) Boveri revisited: chromosomal instability, aneuploidy and tumorigenesis. Nat Rev Mol Cell Biol 10(7):478–487. doi:10.1038/nrm2718

Hondele M, Ladurner AG (2013) Catch me if you can: how the histone chaperone FACT capitalizes on nucleosome breathing. Nucleus 4(6):443–449. doi:10.4161/nucl.27235

Hosouchi T, Kumekawa N, Tsuruoka H, Kotani H (2002) Physical map-based sizes of the centromeric regions of Arabidopsis thaliana chromosomes 1, 2, and 3. DNA Res 9(4):117–121

Ideue T, Cho Y, Nishimura K, Tani T (2014) Involvement of satellite I noncoding RNA in regulation of chromosome segregation. Genes Cells 19(6):528–538. doi:10.1111/gtc.12149

Ikeno M, Grimes B, Okazaki T, Nakano M, Saitoh K, Hoshino H, McGill NI, Cooke H, Masumoto H (1998) Construction of YAC-based mammalian artificial chromosomes. Nat Biotechnol 16(5):431–439

Jambhekar A, Emerman AB, Schweidenback CT, Blower MD (2014) RNA stimulates Aurora B kinase activity during mitosis. PLoS ONE 9(6):e100748. doi:10.1371/journal.pone.0100748

Jansen L, Black B, Foltz D, Cleveland D (2007) Propagation of centromeric chromatin requires exit from mitosis. J Cell Biol 176(6):795–805. doi:10.1083/jcb.200701066

Kanellopoulou C, Muljo S, Kung A, Ganesan S, Drapkin R, Jenuwein T, Livingston D, Rajewsky K (2005) Dicer-deficient mouse embryonic stem cells are defective in differentiation and centromeric silencing. Genes Dev 19(4):489–501. doi:10.1101/gad.1248505

Karpen GH, Allshire RC (1997) The case for epigenetic effects on centromere identity and function. Trends Genet 13(12):489–496

Koo DH, Zhao H, Jiang J (2016) Chromatin-associated transcripts of tandemly repetitive DNA sequences revealed by RNA-FISH. Chromosome Res. doi:10.1007/s10577-016-9537-5

Kumekawa N, Hosouchi T, Tsuruoka H, Kotani H (2000) The size and sequence organization of the centromeric region of arabidopsis thaliana chromosome 5. DNA Res 7(6):315–321

Lam AL, Boivin CD, Bonney CF, Rudd MK, Sullivan BA (2006) Human centromeric chromatin is a dynamic chromosomal domain that can spread over noncentromeric DNA. Proc Nat Acad Sci USA 103(11):4186–4191

Lampson MA, Cheeseman IM (2011) Sensing centromere tension: Aurora B and the regulation of kinetochore function. Trends Cell Biol 21(3):133–140. doi:10.1016/j.tcb.2010.10.007

Leung W, Shaffer CD, Reed LK, Smith ST, Barshop W, Dirkes W, Dothager M, Lee P, Wong J, Xiong D, Yuan H, Bedard JE, Machone JF, Patterson SD, Price AL, Turner BA, Robic S, Luippold EK, McCartha SR, Walji TA, Walker CA, Saville K, Abrams MK, Armstrong AR, Armstrong W, Bailey RJ, Barberi CR, Beck LR, Blaker AL, Blunden CE, Brand JP, Brock EJ, Brooks DW, Brown M, Butzler SC, Clark EM, Clark NB, Collins AA, Cotteleer RJ, Cullimore PR, Dawson SG, Docking CT, Dorsett SL, Dougherty GA, Downey KA, Drake AP, Earl EK, Floyd TG, Forsyth JD, Foust JD, Franchi SL, Geary JF, Hanson CK, Harding TS, Harris CB, Heckman JM, Holderness HL, Howey NA, Jacobs DA, Jewell ES, Kaisler M, Karaska EA, Kehoe JL, Koaches HC, Koehler J, Koenig D, Kujawski AJ, Kus JE, Lammers JA, Leads RR, Leatherman EC, Lippert RN, Messenger GS, Morrow AT, Newcomb V, Plasman HJ, Potocny SJ, Powers MK, Reem RM, Rennhack JP, Reynolds KR, Reynolds LA, Rhee DK, Rivard AB, Ronk AJ, Rooney MB, Rubin LS, Salbert LR, Saluja RK, Schauder T, Schneiter AR, Schulz RW, Smith KE, Spencer S, Swanson BR, Tache MA, Tewilliager AA, Tilot AK, VanEck E, Villerot MM, Vylonis MB, Watson DT, Wurzler JA, Wysocki LM, Yalamanchili M, Zaborowicz MA, Emerson JA, Ortiz C, Deuschle FJ, DiLorenzo LA, Goeller KL, Macchi CR, Muller SE, Pasierb BD, Sable JE, Tucci JM, Tynon M, Dunbar DA, Beken LH, Conturso AC, Danner BL, DeMichele GA, Gonzales JA, Hammond MS, Kelley CV, Kelly EA, Kulich D, Mageeney CM, McCabe NL, Newman AM, Spaeder LA, Tumminello RA, Revie D, Benson JM, Cristostomo MC, DaSilva PA, Harker KS, Jarrell JN, Jimenez LA, Katz BM, Kennedy WR, Kolibas KS, LeBlanc MT, Nguyen TT, Nicolas DS, Patao MD, Patao SM, Rupley BJ, Sessions BJ, Weaver JA, Goodman AL, Alvendia EL, Baldassari SM, Brown AS, Chase IO, Chen M, Chiang S, Cromwell AB, Custer AF, DiTommaso TM, El-Adaimi J, Goscinski NC, Grove RA, Gutierrez N, Harnoto RS, Hedeen H, Hong EL, Hopkins BL, Huerta VF, Khoshabian C, LaForge KM, Lee CT, Lewis BM, Lydon AM, Maniaci BJ, Mitchell RD, Morlock EV, Morris WM, Naik P, Olson NC, Osterloh JM, Perez MA, Presley JD, Randazzo MJ, Regan MK, Rossi FG, Smith MA, Soliterman EA, Sparks CJ, Tran DL, Wan T, Welker AA, Wong JN, Sreenivasan A, Youngblom J, Adams A, Alldredge J, Bryant A, Carranza D, Cifelli A, Coulson K, Debow C, Delacruz N, Emerson C, Farrar C, Foret D, Garibay E, Gooch J, Heslop M, Kaur S, Khan A, Kim V, Lamb T, Lindbeck P, Lucas G, Macias E, Martiniuc D, Mayorga L, Medina J, Membreno N, Messiah S, Neufeld L, Nguyen SF, Nichols Z, Odisho G, Peterson D, Rodela L, Rodriguez P, Rodriguez V, Ruiz J, Sherrill W, Silva V, Sparks J, Statton G, Townsend A, Valdez I, Waters M, Westphal K, Winkler S, Zumkehr J, DeJong RJ, Hoogewerf AJ, Ackerman CM, Armistead IO, Baatenburg L, Borr MJ, Brouwer LK, Burkhart BJ, Bushhouse KT, Cesko L, Choi TY, Cohen H, Damsteegt AM, Darusz JM, Dauphin CM, Davis YP, Diekema EJ, Drewry M, Eisen ME, Faber HM, Faber KJ, Feenstra E, Felzer-Kim IT, Hammond BL, Hendriksma J, Herrold MR, Hilbrands JA, Howell EJ, Jelgerhuis SA, Jelsema TR, Johnson BK, Jones KK, Kim A, Kooienga RD, Menyes EE, Nollet EA, Plescher BE, Rios L, Rose JL, Schepers AJ, Scott G, Smith JR, Sterling AM, Tenney JC, Uitvlugt C, VanDyken RE, VanderVennen M, Vue S, Kokan NP, Agbley K, Boham SK, Broomfield D, Chapman K, Dobbe A, Dobbe I, Harrington W, Ibrahem M, Kennedy A, Koplinsky CA, Kubricky C, Ladzekpo D, Pattison C, Ramirez RE Jr, Wande L, Woehlke S, Wawersik M, Kiernan E, Thompson JS, Banker R, Bartling JR, Bhatiya CI,

Boudoures AL, Christiansen L, Fosselman DS, French KM, Gill IS, Havill JT, Johnson JL, Keny LJ, Kerber JM, Klett BM, Kufel CN, May FJ, Mecoli JP, Merry CR, Meyer LR, Miller EG, Mullen GJ, Palozola KC, Pfeil JJ, Thomas JG, Verbofsky EM, Spana EP, Agarwalla A, Chapman J, Chlebina B, Chong I, Falk IN, Fitzgibbons JD, Friedman H, Ighile O, Kim AJ, Knouse KA, Kung F, Mammo D, Ng CL, Nikam VS, Norton D, Pham P, Polk JW, Prasad S, Rankin H, Ratliff CD, Scala V, Schwartz NU, Shuen JA, Xu A, Xu TQ, Zhang Y, Rosenwald AG, Burg MG, Adams SJ, Baker M, Botsford B, Brinkley B, Brown C, Emiah S, Enoch E, Gier C, Greenwell A, Hoogenboom L, Matthews JE, McDonald M, Mercer A, Monsma N, Ostby K, Ramic A, Shallman D, Simon M, Spencer E, Tomkins T, Wendland P, Wylie A, Wolyniak MJ, Robertson GM, Smith SI, DiAngelo JR, Sassu ED, Bhalla SC, Sharif KA, Choeying T, Macias JS, Sanusi F, Torchon K, Bednarski AE, Alvarez CJ, Davis KC, Dunham CA, Grantham AJ, Hare AN, Schottler J, Scott ZW, Kuleck GA, Yu NS, Kaehler MM, Jipp J, Overvoorde PJ, Shoop E, Cyrankowski O, Hoover B, Kusner M, Lin D, Martinov T, Misch J, Salzman G, Schiedermayer H, Snavely M, Zarrasola S, Parrish S, Baker A, Beckett A, Belella C, Bryant J, Conrad T, Fearnow A, Gomez C, Herbstsomer RA, Hirsch S, Johnson C, Jones M, Kabaso R, Lemmon E, Vieira CM, McFarland D, McLaughlin C, Morgan A, Musokotwane S, Neutzling W, Nietmann J, Paluskievicz C, Penn J, Peoples E, Pozmanter C, Reed E, Rigby N, Schmidt L, Shelton M, Shuford R, Tirasawasdichai T, Undem B, Urick D, Vondy K, Yarrington B, Eckdahl TT, Poet JL, Allen AB, Anderson JE, Barnett JM, Baumgardner JS, Brown AD, Carney JE, Chavez RA, Christgen SL, Christie JS, Clary AN, Conn MA, Cooper KM, Crowley MJ, Crowley ST, Doty JS, Dow BA, Edwards CR, Elder DD, Fanning JP, Janssen BM, Lambright AK, Lane CE, Limle AB, Mazur T, McCracken MR, McDonough AM, Melton AD, Minnick PJ, Musick AE, Newhart WH, Noynaert JW, Ogden BJ, Sandusky MW, Schmuecker SM, Shipman AL, Smith AL, Thomsen KM, Unzicker MR, Vernon WB, Winn WW, Woyski DS, Zhu X, Du C, Ament C, Aso S, Bisogno LS, Caronna J, Fefelova N, Lopez L, Malkowitz L, Marra J, Menillo D, Obiorah I, Onsarigo EN, Primus S, Soos M, Tare A, Zidan A, Jones CJ, Aronhalt T, Bellush JM, Burke C, DeFazio S, Does BR, Johnson TD, Keysock N, Knudsen NH, Messler J, Myirski K, Rekai JL, Rempe RM, Salgado MS, Stagaard E, Starcher JR, Waggoner AW, Yemelyanova AK, Hark AT, Bertolet A, Kuschner CE, Parry K, Quach M, Shantzer L, Shaw ME, Smith MA, Glenn O, Mason P, Williams C, Key SC, Henry TC, Johnson AG, White JX, Haberman A, Asinof S, Drumm K, Freeburg T, Safa N, Schultz D, Shevin Y, Svoronos P, Vuong T, Wellinghoff J, Hoopes LL, Chau KM, Ward A, Regisford EG, Augustine L, Davis-Reyes B, Echendu V, Hales J, Ibarra S, Johnson L, Ovu S, Braverman JM, Bahr TJ, Caesar NM, Campana C, Cassidy DW, Cognetti PA, English JD, Fadus MC, Fick CN, Freda PJ, Hennessy BM, Hockenberger K, Jones JK, King JE, Knob CR, Kraftmann KJ, Li L, Lupey LN, Minniti CJ, Minton TF, Moran JV, Mudumbi K, Nordman EC, Puetz WJ, Robinson LM, Rose TJ, Sweeney EP, Timko AS, Paetkau DW, Eisler HL, Aldrup ME, Bodenberg JM, Cole MG, Deranek KM, DeShetler M, Dowd RM, Eckardt AK, Ehret SC, Fese J, Garrett AD, Kammrath A, Kappes ML, Light MR, Meier AC, O'Rouke A, Perella M, Ramsey K, Ramthun JR, Reilly MT, Robinett D, Rossi NL, Schueler MG, Shoemaker E, Starkey KM, Vetor A, Vrable A, Chandrasekaran V, Beck C, Hatfield KR, Herrick DA, Khoury CB, Lea C, Louie CA, Lowell SM, Reynolds TJ, Schibler J, Scoma AH, Smith-Gee MT, Tuberty S, Smith CD, Lopilato JE, Hauke J, Roecklein-Canfield JA, Corrielus M, Gilman H, Intriago S, Maffa A, Rauf SA, Thistle K, Trieu M, Winters J, Yang B, Hauser CR, Abusheikh T, Ashrawi Y, Benitez P, Boudreaux LR, Bourland M, Chavez M, Cruz S, Elliott G, Farek JR, Flohr S, Flores AH, Friedrichs C, Fusco Z, Goodwin Z, Helmreich E, Kiley J, Knepper JM, Langner C, Martinez M, Mendoza C, Naik M, Ochoa A, Ragland N, Raimey E, Rathore S, Reza E, Sadovsky G, Seydoux MI, Smith JE, Unruh AK, Velasquez V, Wolski MW, Gosser Y, Govind S, Clarke-Medley N, Guadron L, Lau D, Lu A, Mazzeo C, Meghdari M, Ng S, Pamnani B, Plante O, Shum YK, Song R, Johnson DE, Abdelnabi M, Archambault A, Chamma N, Gaur S, Hammett D, Kandahari A, Khayrullina G, Kumar S, Lawrence S, Madden N, Mandelbaum M, Milnthorp H, Mohini S, Patel R, Peacock SJ, Perling E,

Quintana A, Rahimi M, Ramirez K, Singhal R, Weeks C, Wong T, Gillis AT, Moore ZD, Savell CD, Watson R, Mel SF, Anilkumar AA, Bilinski P, Castillo R, Closser M, Cruz NM, Dai T, Garbagnati GF, Horton LS, Kim D, Lau JH, Liu JZ, Mach SD, Phan TA, Ren Y, Stapleton KE, Strelitz JM, Sunjed R, Stamm J, Anderson MC, Bonifield BG, Coomes D, Dillman A, Durchholz EJ, Fafara-Thompson AE, Gross MJ, Gygi AM, Jackson LE, Johnson A, Kocsisova Z, Manghelli JL, McNeil K, Murillo M, Naylor KL, Neely J, Ogawa EE, Rich A, Rogers A, Spencer JD, Stemler KM, Throm AA, Van Camp M, Weihbrecht K, Wiles TA, Williams MA, Williams M, Zoll K, Bailey C, Zhou L, Balthaser DM, Bashiri A, Bower ME, Florian KA, Ghavam N, Greiner-Sosanko ES, Karim H, Mullen VW, Pelchen CE, Yenerall PM, Zhang J, Rubin MR, Arias-Mejias SM, Bermudez-Capo AG, Bernal-Vega GV, Colon-Vazquez M, Flores-Vazquez A, Gines-Rosario M, Llavona-Cartagena IG, Martinez-Rodriguez JO, Ortiz-Fuentes L, Perez-Colomba EO, Perez-Otero J, Rivera E, Rodriguez-Giron LJ, Santiago-Sanabria AJ, Senquiz-Gonzalez AM, delValle FR, Vargas-Franco D, Velazquez-Soto KI, Zambrana-Burgos JD, Martinez-Cruzado JC, Asencio-Zayas L, Babilonia-Figueroa K, Beauchamp-Perez FD, Belen-Rodriguez J, Bracero-Quinones L, Burgos-Bula AP, Collado-Mendez XA, Colon-Cruz LR, Correa-Muller AI, Crooke-Rosado JL, Cruz-Garcia JM, Defendini-Avila M, Delgado-Peraza FM, Feliciano-Cancela AJ, Gonzalez-Perez VM, Guiblet W, Heredia-Negron A, Hernandez-Muniz J, Irizarry-Gonzalez LN, Laboy-Corales AL, Llaurador-Caraballo GA, Marin-Maldonado F, Marrero-Llerena U, Martell-Martinez HA, Martinez-Traverso IM, Medina-Ortega KN, Mendez-Castellanos SG, Menendez-Serrano KC, Morales-Caraballo CI, Ortiz-DeChoudens S, Ortiz-Ortiz P, Pagan-Torres H, Perez-Afanador D, Quintana-Torres EM, Ramirez-Aponte EG, Riascos-Cuero C, Rivera-Llovet MS, Rivera-Pagan IT, Rivera-Vicens RE, Robles-Juarbe F, Rodriguez-Bonilla L, Rodriguez-Echevarria BO, Rodriguez-Garcia PM, Rodriguez-Laboy AE, Rodriguez-Santiago S, Rojas-Vargas ML, Rubio-Marrero EN, Santiago-Colon A, Santiago-Ortiz JL, Santos-Ramos CE, Serrano-Gonzalez J, Tamayo-Figueroa AM, Tascon-Penaranda EP, Torres-Castillo JL, Valentin-Feliciano NA, Valentin-Feliciano YM, Vargas-Barreto NM, Velez-Vazquez M, Vilanova-Velez LR, Zambrana-Echevarria C, MacKinnon C, Chung HM, Kay C, Pinto A, Kopp OR, Burkhardt J, Harward C, Allen R, Bhat P, Chang JH, Chen Y, Chesley C, Cohn D, DuPuis D, Fasano M, Fazzio N, Gavinski K, Gebreyesus H, Giarla T, Gostelow M, Greenstein R, Gunasinghe H, Hanson C, Hay A, He TJ, Homa K, Howe R, Howenstein J, Huang H, Khatri A, Kim YL, Knowles O, Kong S, Krock R, Kroll M, Kuhn J, Kwong M, Lee B, Lee R, Levine K, Li Y, Liu B, Liu L, Liu M, Lousararian A, Ma J, Mallya A, Manchee C, Marcus J, McDaniel S, Miller ML, Molleston JM, Diez CM, Ng P, Ngai N, Nguyen H, Nylander A, Pollack J, Rastogi S, Reddy H, Regenold N, Sarezky J, Schultz M, Shim J, Skorupa T, Smith K, Spencer SJ, Srikanth P, Stancu G, Stein AP, Strother M, Sudmeier L, Sun M, Sundaram V, Tazudeen N, Tseng A, Tzeng A, Venkat R, Venkataram S, Waldman L, Wang T, Yang H, Yu JY, Zheng Y, Preuss ML, Garcia A, Juergens M, Morris RW, Nagengast AA, Azarewicz J, Carr TJ, Chichearo N, Colgan M, Donegan M, Gardner B, Kolba N, Krumm JL, Lytle S, MacMillian L, Miller M, Montgomery A, Moretti A, Offenbacker B, Polen M, Toth J, Woytanowski J, Kadlec L, Crawford J, Spratt ML, Adams AL, Barnard BK, Cheramie MN, Eime AM, Golden KL, Hawkins AP, Hill JE, Kampmeier JA, Kern CD, Magnuson EE, Miller AR, Morrow CM, Peairs JC, Pickett GL, Popelka SA, Scott AJ, Teepe EJ, TerMeer KA, Watchinski CA, Watson LA, Weber RE, Woodard KA, Barnard DC, Appiah I, Giddens MM, McNeil GP, Adebayo A, Bagaeva K, Chinwong J, Dol C, George E, Haltaufderhyde K, Haye J, Kaur M, Semon M, Serjanov D, Toorie A, Wilson C, Riddle NC, Buhler J, Mardis ER, Elgin SC (2015) Drosophila muller f elements maintain a distinct set of genomic properties over 40 million years of evolution. G3 (Bethesda) 5 (5):719–740. doi:10.1534/g3.114.015966

Liu H (2016) Insights into centromeric transcription in mitosis. Transcription 7(1):21–25. doi:10.1080/21541264.2015.1127315

Liu H, Qu Q, Warrington R, Rice A, Cheng N, Yu H (2015) Mitotic transcription installs Sgo1 at centromeres to coordinate chromosome segregation. Mol Cell 59(3):426–436. doi:10.1016/j.molcel.2015.06.018

Lo AW, Craig JM, Saffery R, Kalitsis P, Irvine DV, Earle E, Magliano DJ, Choo KH (2001a) A 330 kb CENP-A binding domain and altered replication timing at a human neocentromere. EMBO J 20(8):2087–2096

Lo AW, Magliano DJ, Sibson MC, Kalitsis P, Craig JM, Choo KH (2001b) A novel chromatin immunoprecipitation and array (CIA) analysis identifies a 460-kb CENP-A-binding neocentromere DNA. Genome Res 11(3):448–457

Locke DP, Hillier LW, Warren WC, Worley KC, Nazareth LV, Muzny DM, Yang SP, Wang Z, Chinwalla AT, Minx P, Mitreva M, Cook L, Delehaunty KD, Fronick C, Schmidt H, Fulton LA, Fulton RS, Nelson JO, Magrini V, Pohl C, Graves TA, Markovic C, Cree A, Dinh HH, Hume J, Kovar CL, Fowler GR, Lunter G, Meader S, Heger A, Ponting CP, Marques-Bonet T, Alkan C, Chen L, Cheng Z, Kidd JM, Eichler EE, White S, Searle S, Vilella AJ, Chen Y, Flicek P, Ma J, Raney B, Suh B, Burhans R, Herrero J, Haussler D, Faria R, Fernando O, Darre F, Farre D, Gazave E, Oliva M, Navarro A, Roberto R, Capozzi O, Archidiacono N, Della Valle G, Purgato S, Rocchi M, Konkel MK, Walker JA, Ullmer B, Batzer MA, Smit AF, Hubley R, Casola C, Schrider DR, Hahn MW, Quesada V, Puente XS, Ordonez GR, Lopez-Otin C, Vinar T, Brejova B, Ratan A, Harris RS, Miller W, Kosiol C, Lawson HA, Taliwal V, Martins AL, Siepel A, Roychoudhury A, Ma X, Degenhardt J, Bustamante CD, Gutenkunst RN, Mailund T, Dutheil JY, Hobolth A, Schierup MH, Ryder OA, Yoshinaga Y, de Jong PJ, Weinstock GM, Rogers J, Mardis ER, Gibbs RA, Wilson RK (2011) Comparative and demographic analysis of orang-utan genomes. Nature 469 (7331):529–533. doi:10.1038/nature09687

Logsdon GA, Barrey EJ, Bassett EA, DeNizio JE, Guo LY, Panchenko T, Dawicki-McKenna JM, Heun P, Black BE (2015) Both tails and the centromere targeting domain of CENP-A are required for centromere establishment. J Cell Biol 208(5):521–531. doi:10.1083/jcb.201412011

Lu J, Gilbert DM (2007) Proliferation-dependent and cell cycle regulated transcription of mouse pericentric heterochromatin. J Cell Biol 179(3):411–421

Macas J, Koblizkova A, Navratilova A, Neumann P (2009) Hypervariable 3′ UTR region of plant LTR-retrotransposons as a source of novel satellite repeats. Gene 448(2):198–206. doi:10.1016/j.gene.2009.06.014

MacRae IJ, Zhou K, Doudna JA (2007) Structural determinants of RNA recognition and cleavage by Dicer. Nat Struct Mol Biol 14(10):934–940. doi:10.1038/nsmb1293

Malik HS, Henikoff S (2001) Adaptive evolution of Cid, a centromere-specific histone in Drosophila. Genetics 157(3):1293–1298

Malik H, Henikoff S (2009) Major evolutionary transitions in centromere complexity. Cell 138 (6):1067–1082

Maloney KA, Sullivan LL, Matheny JE, Strome ED, Merrett SL, Ferris A, Sullivan BA (2012) Functional epialleles at an endogenous human centromere. Proc Natl Acad Sci USA 109 (34):13704–13709. doi:10.1073/pnas.1203126109

Mejía J (2002) Efficiency of de novo centromere formation in human artificial chromosomes. Genomics 79(3):297–304. doi:10.1006/geno.2002.6704

Mellone BG, Zhang W, Karpen GH (2009) Frodos found: behold the CENP-a “Ring” bearers. Cell 137(3):409–412. doi:10.1016/j.cell.2009.04.035 S0092-8674(09)00461-9 [pii]

Melters DP, Bradnam KR, Young HA, Telis N, May MR, Ruby JG, Sebra R, Peluso P, Eid J, Rank D, Garcia JF, DeRisi JL, Smith T, Tobias C, Ross-Ibarra J, Korf I, Chan SW (2013) Comparative analysis of tandem repeats from hundreds of species reveals unique insights into centromere evolution. Genome Biol 14(1):R10. doi:10.1186/gb-2013-14-1-r10

Meluh PB, Yang P, Glowczewski L, Koshland D, Smith MM (1998) Cse4p is a component of the core centromere of Saccharomyces cerevisiae. Cell 94(5):607–613

Miga KH (2015) Completing the human genome: the progress and challenge of satellite DNA assembly. Chromosome Res. doi:10.1007/s10577-015-9488-2

Miga KH, Newton Y, Jain M, Altemose N, Willard HF, Kent WJ (2014) Centromere reference models for human chromosomes X and Y satellite arrays. Genome Res 24(4):697–707. doi:10.1101/gr.159624.113

Miga KH, Eisenhart C, Kent WJ (2015) Utilizing mapping targets of sequences underrepresented in the reference assembly to reduce false positive alignments. Nucleic Acids Res. doi:10.1093/nar/gkv671

Molina O, Vargiu G, Abad MA, Zhiteneva A, Jeyaprakash AA, Masumoto H, Kouprina N, Larionov V, Earnshaw WC (2016) Epigenetic engineering reveals a balance between histone modifications and transcription in kinetochore maintenance. Nat Commun 7:13334. doi:10.1038/ncomms13334

Muchardt C, Guilleme M, Seeler JS, Trouche D, Dejean A, Yaniv M (2002) Coordinated methyl and RNA binding is required for heterochromatin localization of mammalian HP1alpha. EMBO Rep 3(10):975–981. doi:10.1093/embo-reports/kvf194 kvf194 [pii]

Nagaki K, Cheng Z, Ouyang S, Talbert PB, Kim M, Jones KM, Henikoff S, Buell CR, Jiang J (2004) Sequencing of a rice centromere uncovers active genes. Nat Genet 36(2):138–145

Nakano M, Okamoto Y, Ohzeki J, Masumoto H (2003) Epigenetic assembly of centromeric chromatin at ectopic alpha-satellite sites on human chromosomes. J Cell Sci 116(Pt 19):4021–4034

Nakano M, Cardinale S, Noskov VN, Gassmann R, Vagnarelli P, Kandels-Lewis S, Larionov V, Earnshaw WC, Masumoto H (2008) Inactivation of a human kinetochore by specific targeting of chromatin modifiers. Dev Cell 14(4):507–522

Neumann P, Navratilova A, Koblizkova A, Kejnovsky E, Hribova E, Hobza R, Widmer A, Dolezel J, Macas J (2011) Plant centromeric retrotransposons: a structural and cytogenetic perspective. Mob DNA 2(1):4. doi:10.1186/1759-8753-2-4

Oestergaard VH, Lisby M (2016) Transcription-replication conflicts at chromosomal fragile sites-consequences in M phase and beyond. Chromosoma. doi:10.1007/s00412-016-0617-2

Okada M, Okawa K, Isobe T, Fukagawa T (2009) CENP-H-containing complex facilitates centromere deposition of CENP-A in cooperation with FACT and CHD1. Mol Biol Cell 20 (18):3986–3995. doi:10.1091/mbc.E09-01-0065

Okamoto Y, Nakano M, Ohzeki J, Larionov V, Masumoto H (2007) A minimal CENP-A core is required for nucleation and maintenance of a functional human centromere. EMBO J 26 (5):1279–1291

O'Neill RJ, Carone DM (2009) The role of ncRNA in centromeres: a lesson from marsupials. Prog Mol Subcell Biol 48:77–101. doi:10.1007/978-3-642-00182-6_4

O'Neill RJ, Eldridge MD, Metcalfe CJ (2004) Centromere dynamics and chromosome evolution in marsupials. J Hered 95(5):375–381. doi:10.1093/jhered/esh063 95/5/375 [pii]

Piras FM, Nergadze SG, Magnani E, Bertoni L, Attolini C, Khoriauli L, Raimondi E, Giulotto E (2010) Uncoupling of satellite DNA and centromeric function in the genus Equus. PLoS Genet 6(2):e1000845. doi:10.1371/journal.pgen.1000845

Plohl M, Meštrović N, Mravinac B (2014) Centromere identity from the DNA point of view. Chromosoma 123(4):313–325. doi:10.1007/s00412-014-0462-0

Pluta AF, Mackay AM, Ainsztein AM, Goldberg IG, Earnshaw WC (1995) The centromere: hub of chromosomal activities. Science 270(5242):1591–1594

Prendergast L, Muller S, Liu Y, Huang H, Dingli F, Loew D, Vassias I, Patel DJ, Sullivan KF, Almouzni G (2016) The CENP-T/-W complex is a binding partner of the histone chaperone FACT. Genes Dev 30(11):1313–1326. doi:10.1101/gad.275073.115

Quenet D, Dalal Y (2014) A long non-coding RNA is required for targeting centromeric protein A to the human centromere. eLife 3:e03254. doi:10.7554/eLife.03254

Reddy K, Tam M, Bowater RP, Barber M, Tomlinson M, Nichol Edamura K, Wang YH, Pearson CE (2011) Determinants of R-loop formation at convergent bidirectionally transcribed trinucleotide repeats. Nucleic Acids Res 39(5):1749–1762. doi:10.1093/nar/gkq935

Regnier V, Novelli J, Fukagawa T, Vagnarelli P, Brown W (2003) Characterization of chicken CENP-A and comparative sequence analysis of vertebrate centromere-specific histone H3-like proteins. Gene 316:39–46

Renfree MB, Papenfuss AT, Deakin JE, Lindsay J, Heider T, Belov K, Rens W, Waters PD, Pharo EA, Shaw G, Wong ES, Lefevre CM, Nicholas KR, Kuroki Y, Wakefield MJ, Zenger KR, Wang C, Ferguson-Smith M, Nicholas FW, Hickford D, Yu H, Short KR, Siddle HV, Frankenberg SR, Chew KY, Menzies BR, Stringer JM, Suzuki S, Hore TA, Delbridge ML, Mohammadi A, Schneider NY, Hu Y, O'Hara W, Al Nadaf S, Wu C, Feng ZP, Cocks BG, Wang J, Flicek P, Searle SM, Fairley S, Beal K, Herrero J, Carone DM, Suzuki Y, Sagano S, Toyoda A, Sakaki Y, Kondo S, Nishida Y, Tatsumoto S, Mandiou I, Hsu A, McColl KA, Landsell B, Weinstock G, Kuczek E, McGrath A, Wilson P, Men A, Hazar-Rethinam M, Hall A, Davies J, Wood D, Williams S, Sundaravadanam Y, Muzny DM, Jhangiani SN, Lewis LR, Morgan MB, Okwuonu GO, Ruiz SJ, Santibanez J, Nazareth L, Cree A, Fowler G, Kovar CL, Dinh HH, Joshi V, Jing C, Lara F, Thornton R, Chen L, Deng J, Liu Y, Shen JY, Song XZ, Edson J, Troon C, Thomas D, Stephens A, Yapa L, Levchenko T, Gibbs RA, Cooper DW, Speed TP, Fujiyama A, Graves JA, O'Neill RJ, Pask AJ, Forrest SM, Worley KC (2011) Genome sequence of an Australian kangaroo, Macropus eugenii, provides insight into the evolution of mammalian reproduction and development. Genome Biol 12(8): R81. doi:10.1186/gb-2011-12-8-r81 gb-2011-12-8-r81 [pii]

Roadmap Epigenomics C, Kundaje A, Meuleman W, Ernst J, Bilenky M, Yen A, Heravi-Moussavi A, Kheradpour P, Zhang Z, Wang J, Ziller MJ, Amin V, Whitaker JW, Schultz MD, Ward LD, Sarkar A, Quon G, Sandstrom RS, Eaton ML, Wu YC, Pfenning AR, Wang X, Claussnitzer M, Liu Y, Coarfa C, Harris RA, Shoresh N, Epstein CB, Gjoneska E, Leung D, Xie W, Hawkins RD, Lister R, Hong C, Gascard P, Mungall AJ, Moore R, Chuah E, Tam A, Canfield TK, Hansen RS, Kaul R, Sabo PJ, Bansal MS, Carles A, Dixon JR, Farh KH, Feizi S, Karlic R, Kim AR, Kulkarni A, Li D, Lowdon R, Elliott G, Mercer TR, Neph SJ, Onuchic V, Polak P, Rajagopal N, Ray P, Sallari RC, Siebenthall KT, Sinnott-Armstrong NA, Stevens M, Thurman RE, Wu J, Zhang B, Zhou X, Beaudet AE, Boyer LA, De Jager PL, Farnham PJ, Fisher SJ, Haussler D, Jones SJ, Li W, Marra MA, McManus MT, Sunyaev S, Thomson JA, Tlsty TD, Tsai LH, Wang W, Waterland RA, Zhang MQ, Chadwick LH, Bernstein BE, Costello JF, Ecker JR, Hirst M, Meissner A, Milosavljevic A, Ren B, Stamatoyannopoulos JA, Wang T, Kellis M (2015) Integrative analysis of 111 reference human epigenomes. Nature 518(7539):317–330. doi:10.1038/nature14248

Rosenbloom KR, Armstrong J, Barber GP, Casper J, Clawson H, Diekhans M, Dreszer TR, Fujita PA, Guruvadoo L, Haeussler M, Harte RA, Heitner S, Hickey G, Hinrichs AS, Hubley R, Karolchik D, Learned K, Lee BT, Li CH, Miga KH, Nguyen N, Paten B, Raney BJ, Smit AF, Speir ML, Zweig AS, Haussler D, Kuhn RM, Kent WJ (2015) The UCSC genome browser database: 2015 update. Nucleic Acids Res 43 (Database issue):D670–681. doi:10.1093/nar/gku1177

Rosic S, Erhardt S (2016) No longer a nuisance: long non-coding RNAs join CENP-A in epigenetic centromere regulation. Cell Mol Life Sci 73(7):1387–1398. doi:10.1007/s00018-015-2124-7

Rosic S, Kohler F, Erhardt S (2014) Repetitive centromeric satellite RNA is essential for kinetochore formation and cell division. J Cell Biol 207(3):335–349. doi:10.1083/jcb.201404097

Schueler MG, Higgins AW, Rudd MK, Gustashaw K, Willard HF (2001) Genomic and genetic definition of a functional human centromere. Science 294(5540):109–115

Schueler MG, Swanson W, Thomas PJ, Green ED (2010) Adaptive evolution of foundation kinetochore proteins in primates. Mol Biol Evol 27(7):1585–1597. doi:10.1093/molbev/msq043 msq043 [pii]

Shang WH, Hori T, Toyoda A, Kato J, Popendorf K, Sakakibara Y, Fujiyama A, Fukagawa T (2010) Chickens possess centromeres with both extended tandem repeats and short non-tandem-repetitive sequences. Genome Res 20(9):1219–1228. doi:10.1101/gr.106245.110

Shin C, Manley JL (2002) The SR protein SRp38 represses splicing in M phase cells. Cell 111 (3):407–417

Smith KM, Phatale PA, Sullivan CM, Pomraning KR, Freitag M (2011) Heterochromatin is required for normal distribution of Neurospora crassa CenH3. Mol Cell Biol 31(12):2528–2542. doi:10.1128/MCB.01285-10

Smith KM, Galazka JM, Phatale PA, Connolly LR, Freitag M (2012) Centromeres of filamentous fungi. Chromosome Res 20(5):635–656. doi:10.1007/s10577-012-9290-3

Sugimoto K, Kuriyama K, Shibata A, Himeno M (1997) Characterization of internal DNA-binding and C-terminal dimerization domains of human centromere/kinetochore autoantigen CENP-C in vitro: role of DNA-binding and self-associating activities in kinetochore organization. Chromosome Res 5(2):132–141

Sullivan BA, Karpen GH (2004) Centromeric chromatin exhibits a histone modification pattern that is distinct from both euchromatin and heterochromatin. Nat Struct Mol Biol 11(11):1076–1083

Takahashi K, Takayama Y, Masuda F, Kobayashi Y, Saitoh S (2005) Two distinct pathways responsible for the loading of CENP-A to centromeres in the fission yeast cell cycle. Philos Trans R Soc Lond B Biol Sci 360 (1455):595–606; discussion 606–597. doi:10.1098/rstb.2004.1614

Takayama Y, Sato H, Saitoh S, Ogiyama Y, Masuda F, Takahashi K (2008) Biphasic incorporation of centromeric histone CENP-A in fission yeast. Mol Biol Cell 19(2):682–690. doi:10.1091/mbc.E07-05-0504

Ting DT, Lipson D, Paul S, Brannigan BW, Akhavanfard S, Coffman EJ, Contino G, Deshpande V, Iafrate AJ, Letovsky S, Rivera MN, Bardeesy N, Maheswaran S, Haber DA (2011) Aberrant overexpression of satellite repeats in pancreatic and other epithelial cancers. Science 331(6017):593–596. doi:10.1126/science.1200801

Topp CN, Zhong CX, Dawe RK (2004) Centromere-encoded RNAs are integral components of the maize kinetochore. Proc Nat Acad Sci USA 101(45):15986–15991

Trazzi S, Perini G, Bernardoni R, Zoli M, Reese JC, Musacchio A, Della Valle G (2009) The C-terminal domain of CENP-C displays multiple and critical functions for mammalian centromere formation. PLoS ONE 4(6):e5832. doi:10.1371/journal.pone.0005832

Ugarkovic D (2005) Functional elements residing within satellite DNAs. EMBO Rep 6(11):1035–1039

Volpe TA, Kidner C, Hall IM, Teng G, Grewal SI, Martienssen RA (2002) Regulation of heterochromatic silencing and histone H3 lysine-9 methylation by RNAi. Science 297 (5588):1833–1837. doi:10.1126/science.1074973 1074973 [pii]

Volpe T, Schramke V, Hamilton GL, White SA, Teng G, Martienssen RA, Allshire RC (2003) RNA interference is required for normal centromere function in fission yeast. Chromosome Res 11(2):137–146

Warburton PE, Cooke CA, Bourassa S, Vafa O, Sullivan BA, Stetten G, Gimelli G, Warburton D, Tyler-Smith C, Sullivan KF, Poirier GG, Earnshaw WC (1997) Immunolocalization of CENP-A suggests a distinct nucleosome structure at the inner kinetochore plate of active centromeres. Curr Biol 7(11):901–904

Wong LH, Choo KH (2004) Evolutionary dynamics of transposable elements at the centromere. Trends Genet 20(12):611–616

Wong LH, Brettingham-Moore KH, Chan L, Quach JM, Anderson MA, Northrop EL, Hannan R, Saffery R, Shaw ML, Williams E, Choo KA (2007) Centromere RNA is a key component for the assembly of nucleoproteins at the nucleolus and centromere. Genome Res 17(8):1146–1160. doi:10.1101/gr.6022807

Yan H, Jiang J (2007) Rice as a model for centromere and heterochromatin research. Chromosome Res 15(1):77–84

Yang CH, Tomkiel J, Saitoh H, Johnson DH, Earnshaw WC (1996) Identification of overlapping DNA-binding and centromere-targeting domains in the human kinetochore protein CENP-C. Mol Cell Biol 16(7):3576–3586

Part III
The Role of DNA Sequence in Centromere Identity and Function

The Promises and Challenges of Genomic Studies of Human Centromeres

Karen H. Miga

Abstract Human centromeres are genomic regions that act as sites of kinetochore assembly to ensure proper chromosome segregation during mitosis and meiosis. Although the biological importance of centromeres in genome stability, and ultimately, cell viability are well understood, the complete sequence content and organization in these multi-megabase-sized regions remains unknown. The lack of a high-resolution reference assembly inhibits standard bioinformatics protocols, and as a result, sequence-based studies involving human centromeres lag far behind the advances made for the non-repetitive sequences in the human genome. In this chapter, I introduce what is known about the genomic organization in the highly repetitive regions spanning human centromeres, and discuss the challenges these sequences pose for assembly, alignment, and data interpretation. Overcoming these obstacles is expected to issue a new era for centromere genomics, which will offer new discoveries in basic cell biology and human biomedical research.

1 Introduction

Our understanding of the human genome remains incomplete, with an estimated $\sim 10\%$ of bases, representing highly repetitive sequences spanning centromeres and the acrocentric short arms, that are omitted from genome-wide studies of cellular function and human health (Eichler et al. 2004). The lack of sequence description for the millions of bases in each centromere-assigned assembly gap has barred studies aimed to better understand the influence of the underlying genomic structure on proper establishment and maintenance of inner kinetochore proteins. Progress in long-read sequencing technologies and avant-garde computational strategies enable high-resolution genomic studies to advance in these uncharted regions of the human genome (English et al. 2012; Jain et al. 2015; Miga et al. 2014). In spite of the

K.H. Miga (✉)
Center for Biomolecular Science and Engineering,
University of California, Santa Cruz, CA, USA
e-mail: khmiga@soe.ucsc.edu

B.E. Black (ed.), *Centromeres and Kinetochores*, Progress in Molecular
and Subcellular Biology 56, DOI 10.1007/978-3-319-58592-5_12

implicit challenges, the promise of human centromere genomics is expected to lead to breakthroughs in our understanding of genome biology, stability, and evolution. In this chapter, I describe those sequences that are expected to be in the "dark side", or centromere and heterochromatin-assigned gaps, of the human reference assemblies and discuss efforts to incorporate these regions into high-throughput sequence-based analyses. These advancements are expected to enable functional characterization of human centromeres, thereby, focusing studies on how centromeric sequence structure contributes to our understanding of chromosome biology and human disease.

2 The Dark Side of the Human Genome: A Genomic Model for Centromeres

Expansive satellite arrays, recent transposon insertions, and enrichment of segmentally duplicated sequences define the millions of bases that encompass each human centromeric region (Rudd and Willard 2004; She et al. 2004; Eichler et al. 2004). Previous experimental efforts aimed to characterize repeat abundance and organization in these regions have issued a general genomic model to guide centromere studies (Hayden 2012; Rudd et al. 2003). Here, I discuss what is generally known about repeat classes and long-range sequence organization in human centromeric regions and how this genomic structure may influence centromere identity and function.

2.1 Alpha Satellite DNA Repeat Organization

Human centromeric regions are marked by the enrichment of a primate-specific AT-rich tandem repeat, known as alpha satellite (Manuelidis and Wu 1978; Manuelidis 1976). Individual alpha satellite repeats are defined by a fundamental ~171 bp unit, or monomer, that offers substantial sequence diversity between copies compared genome-wide (Waye and Willard 1987; Alexandrov et al. 2001; Choo et al. 1991). This intrinsic divergence has enabled experimental studies to characterize local monomer organization within each centromeric region. Alpha satellite DNAs are commonly organized into large, often megabase-sized, arrays (Wevrick and Willard 1989). Each array is defined by a highly represented higher order repeat (HOR) structure, or a repeating unit composed of a collection of divergent alpha satellite monomers (illustrated in Fig. 1) (Willard and Waye 1987). The monomer composition within each repeat unit varies between distinct arrays, enabling array sequence characterization to take place in a chromosome-specific manner (Willard and Waye 1987; Willard 1985; Alexandrov et al. 1993).

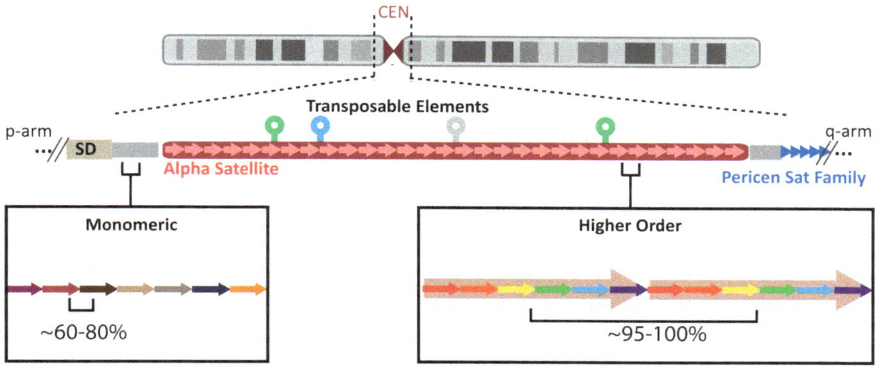

Fig. 1 Genomic model of sequences spanning sites of human centromeres. Human centromeric regions are predominantly defined by one or more multi-megabase-sized HOR arrays of alpha satellite (indicated here as a single array in *red*). HORs in this example array are composed of six highly divergent monomers (*red* through *purple*), yet sequence alignments (indicated by *gray boxes*) between adjacent HORs reveal that they are 95–100% identical. HOR arrays are highly homogenized with few sites of sequence variation, HOR structural rearrangement, inversion, and/or sites of transposable element insertions, shown here as lollipop diagrams with different colors representing different repeat classes. Monomeric alpha satellite DNA, blocks of *gray*, are commonly found directly adjacent to a HOR array. Unlike HORs, monomeric arrays represent ordered head-to-tail copies of the highly divergent 171 bp fundament repeat unit with limited local sequence homology (shown in the example box below to be ~60–80%) or multi-monomeric structure. Segmental duplications are enriched in centric transition regions (indicated as a tan box labeled "S.D."). Additionally, on a subset of chromosomes additional pericentromeric satellite families are found adjacent to alpha satellite sequences (shown here as blue tandem repeats)

Efforts to characterize and map HOR repeats revealed that human centromeric regions were commonly defined by the presence of one or more array (Wevrick and Willard 1991; Alexandrov et al. 2001; Vissel and Choo 1991; Rudd et al. 2006). Each array is composed of hundreds, if not thousands, of copies of a given HOR in tandem (Wevrick and Willard 1989). In contrast to the sequence divergence observed between alpha satellite monomer genome-wide, HORs within a single array are almost entirely identical to one another with only a few sites of sequence variation. Individual arrays are observed to expand and contract in length via unequal crossing-over and/or conversion, and as a consequence, sites of sequence variation are observed to change expand and contract within regional domains (Warburton et al. 1992) (Fig. 2a). This genomic organization results in an array composed of rapidly evolving domains, whereas repeat variants are expected to fluctuate in array position and frequency, introducing a novel source of sequence variation in the human population (Oakey and Tyler-Smith 1990; Warburton et al. 1991). Indeed, arrays on haploid human chromosomes X and Y demonstrated a tenfold difference in satellite DNA array lengths and classified satellite sequence variants enriched in individuals from distinct populations (Miga et al. 2014) (Fig. 2b).

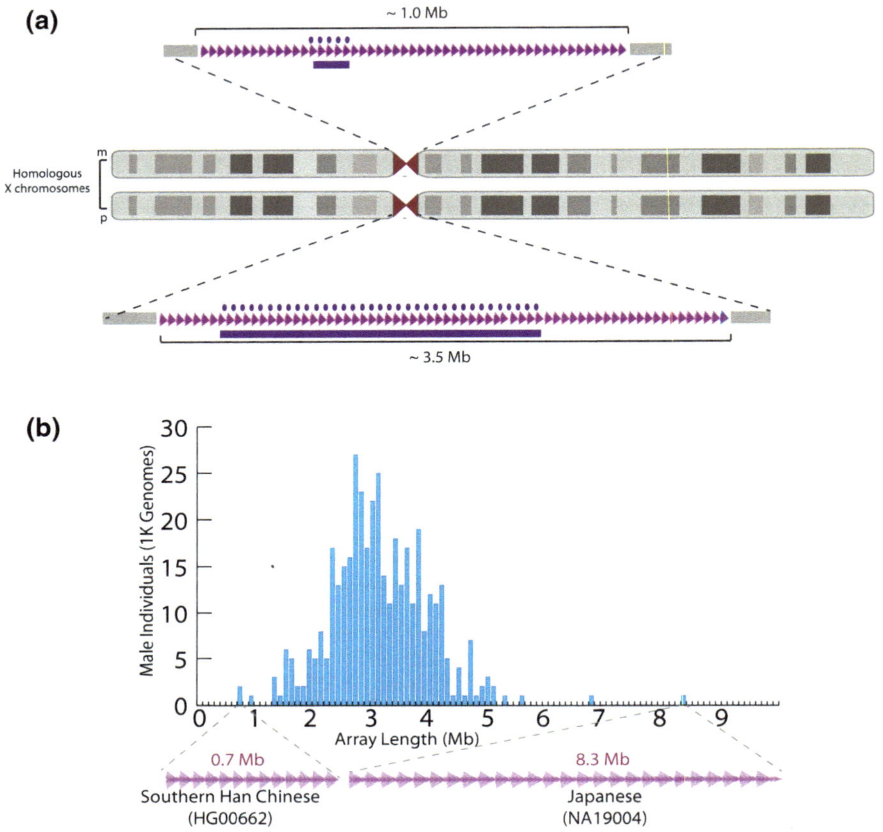

Fig. 2 Array evolution and length variation. HOR arrays vary in the proportion of HOR variants and overall array length in the population, as well as between homologous chromosomes in a single individual (shown in **a**, for example, DXZ1 arrays present on homologous X chromosomes). HOR variants, depicted as *purple circles*, are expected to expand and contract in the array by mechanisms involved in sequence conversion resulting in sequence model where domains (*purple band*) of identical repeats may vary. As indicated in **b** CEN X (DXZ1) haploid array length distribution for male individuals from 1000 Genome Project. Array length is observed to have a tenfold range, shown min: 0.7 Mb (Southern Han Chinese HG00662) to max: 8.3 Mb (Japanese NA19004). (Miga et al. 2014)

Sequence-based studies between distinct HOR arrays in the human genome support the hypothesis that sequence conversion rates are higher within a single array than between two distinct arrays, even when found on the same chromosome (Warburton and Willard 1995). High stringency fluorescence in situ hybridization (FISH) conditions are useful in mapping individual HOR arrays to a single chromosome (Willard 1985). In contrast, under low stringency conditions, subsets of HOR arrays are observed to cross-hybridize, establishing a metric of inter-array relatedness of five suprachromosomal satellite subfamilies (Alexandrov et al. 1993,

1988; Waye and Willard 1987). Cataloging chromosome-assignment and shared sequence homologies of HOR repeats was critical in creating FISH probes to individual chromosomes useful in cytogenetic analyses (Willard 1985), promoting physical mapping efforts (Mahtani and Willard 1998; Wevrick and Willard 1991), and generating HOR-specific sequence libraries (Miga et al. 2014; Durfy and Willard 1987). These collective studies were critical to developing the current genomic model, and will continue to guide studies aimed to improve sequence maps in centromeric regions.

In addition to HORs, arrangements of highly divergent alpha satellite ~ 171 bp repeats, known as "monomeric", are commonly found in transition regions that span hundreds of kilobases adjacent to each array (Alexandrov et al. 1993; Rudd and Willard 2004; Hayden et al. 2013) (Fig. 3a). Sequence comparisons of collections of monomeric repeats found on several chromosomes have revealed that subsets of repeats form evolutionary blocks, thereby promoting a model of alpha satellite sequence evolve by shifting from homogenization states to non-homogenization monomeric states at the edge of the array (Schueler et al. 2005; Shepelev et al. 2015).

With respect to centromere function, monomeric DNAs are not observed to co-localize with inner kinetochore proteins in normal cell lines. Furthermore, of the select monomeric sequences tested by human artificial chromosome (HAC) assays, all have been shown to be insufficient for de novo kinetochore recruitment (Hayden et al. 2013). This is in contrast to studies of HOR arrays, which are observed by immunofluorescence and DNA fluorescence in situ hybridization (immuno-FISH) to associate with inner kinetochore proteins (Spence et al. 2002; Blower et al. 2002). Additionally, centromeres are functionally sensitive to HOR array deletion and/or rearrangement (Tyler-Smith et al. 1993). Finally, particular HOR repeats are competent for de novo centromere formation and stably maintained as functioning HACs (Harrington et al. 1997; Masumoto et al. 1998). In general, HOR alpha satellite DNAs are credited with centromere identity (Schueler et al. 2001), however in a given genome only a fraction of a HOR array is expected to associate with inner kinetochore proteins (Sullivan et al. 2011). Furthermore, not all HOR arrays in a given genome are expected to be "active", in that only a subset interacts with inner kinetochore proteins (Vafa and Sullivan 1997). For example, the centromeric region on chromosome 17 contains at least two distinct HOR arrays yet in most genomes studied to date only a portion of one array (i.e., either D17Z1 or D17Z1B), is considered active (Maloney et al. 2012) (Fig. 3a). Repeats from each array on chromosome 17 provide evidence for centromere competency by exhibiting de novo centromere formation in HAC studies; therefore, although both HORs are *competent* for centromere function only one is *functioning* as an active centromere in a particular genome (Hayden et al. 2013; Maloney et al. 2012) (Fig. 3b). These results support the idea that centromeres are likely not defined strictly by the underlying sequence or epigenetics, but rather certain sequences and/or genomic organization offer favorable sites, or opportunity for kinetochore assembly.

Experimental efforts to study centromere competent alpha satellite sequences and genomic structure have been restricted to HAC assays and ChIP-seq studies

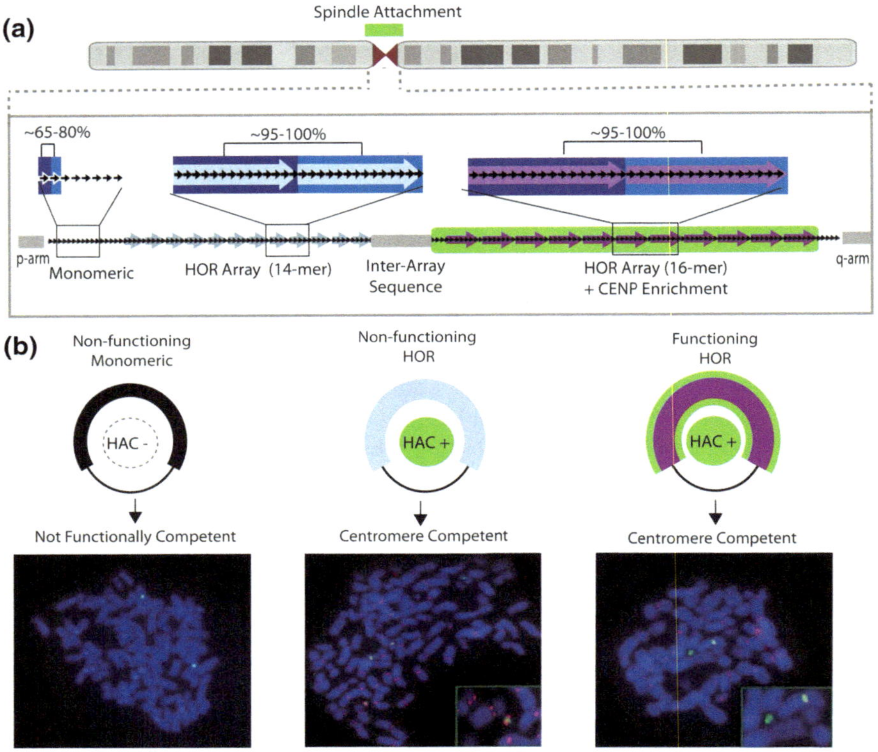

Fig. 3 Defining centromere competent sequences in the human genome. Each human centromere-assigned assembly gap is expected to have one or more HOR array, as indicated in the schematic as two distinct HORs (*light blue*, 14-mer and *purple*, 16-mer). Although HORs within an array are expected to be highly similar to one another, i.e., 95–100%, HORs between arrays often present sufficient sequence divergence to ensure both array and chromosome sequence specificity. Monomeric alpha satellite DNAs are expected to flank HOR arrays and exhibit limited sequence homology with local repeats. Although alpha satellite is credited as a centromeric satellite family, not all alpha satellite DNAs in a given centromeric region bind to inner kinetochore proteins. Green highlighting indicates that the purple HOR array is "functioning" as an active centromere as that particular HOR is determined by ChIP-seq or IF-FISH colocalization to bind inner kinetochore proteins. As expected, HORs from a functioning array are determined by HAC assay to be competent for de novo centromere formation (indicated by HAC+ in *green*). However, one can also observe nonfunctioning higher order repeat to be centromere competent, while nonfunctioning monomeric units are not observed to be centromere competent (marked by HAC−). *Bottom* corresponding FISH images for HAC assay where CENP-A is shown in *red* and the BAC construct is shown in *green* (Hayden et al. 2013)

with known inner kinetochore proteins. Human artificial chromosome studies have been useful in identifying candidate sequence features that influence de novo centromere formation (Harrington et al. 1997). In particular, HORs that are enriched with a 17 bp binding motif for centromere protein B, or CENP-B box, have been shown in HAC assays to be important for centromere establishment (Masumoto et al. 1998; Ohzeki et al. 2002). It is likely that additional HOR sequence features that are antagonistic to nucleosome stability or favor centromere protein binding will influence centromere competency. HAC assays are highly laborious, therefore researchers have only tested to a small number of sequences, which ultimately result in insufficient data to apply robust statistically meaningful surveys of sequence function (Basu et al. 2005; Harrington et al. 1997). In contrast, ChIP-Seq maps of centromere protein A (CENP-A), a specialized histone that serves as the key epigenetic mark needed to ensure kinetochore assembly, provide a broad survey of centromere competent sequences (Warburton et al. 1997; Hayden et al. 2013; Vafa and Sullivan 1997). Acknowledging variability of HOR arrays it has been shown in studies involving HOR arrays specific to chromosome 17 that repeat abundance and variant composition might single out one array as active and the other as inactive (Aldrup-MacDonald et al. 2016). Further, increased rates of array mutation, i.e., transposable element interruption, expansion of repeat variants, and HOR rearrangements in the array, are likely to influence whether or not an array is functioning as a centromere. Therefore studies aimed to understand centromere genomics will need to extend experimental methods to determine libraries of competent centromere sequences and potential genomic configurations for function.

2.2 Exploring the Functional Impact of Transposable Elements

Human centromeric regions are known to contain not only satellite DNAs, but also recent mobile element insertions. Array sequence libraries from the HOR, for example, both long and short interspersed repeat elements (LINEs and SINEs) have been characterized throughout both monomeric and HOR arrays, yet the functional impact of these transposable elements (TE) is an area of active research (Prades et al. 1996; Schueler et al. 2001, 2005). Transposable element insertions are expected to be rare in HOR arrays, where centromere proteins are observed to bind (Santos et al. 2000; Schueler et al. 2001). These findings could support the idea that TE interruptions are selected against to prevent centromere inactivation (Malik and Henikoff 2002). When evaluating read libraries of the X chromosome HOR array (DXZ1), only six insertion sites were identified in an estimated 3.8 Mb sized haploid array (Miga et al. 2014; Levy et al. 2007) (Fig. 4a). Furthermore, when evaluating alpha satellite reference models in GRCh38, LINE and SINEs appear to be the most prevalent transposable element insertion in HOR arrays (Fig. 4b).

Although the functional role of monomeric TEs is unknown, transposable element insertions in other genomes have indicated roles in strand-specific transcription, chromatin remodeling, and specific interactions with centromere proteins that are expected to contribute to identity (Chueh et al. 2009; May et al. 2005; Gent et al. 2011). Indeed, neocentromere studies of a full length LINE element on 10q25 have offered evidence that TE transcripts associate with CENP-A and may influence

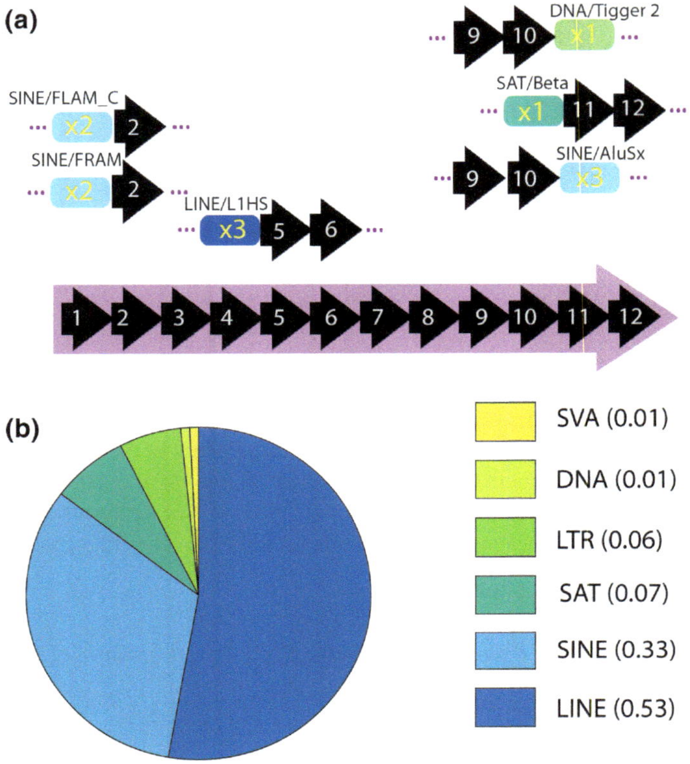

Fig. 4 Transposable elements present in HOR arrays. Transposable elements are expected to be rare in HOR arrays. Shown in **a** only a few examples of interspersed non-satellite sequences are observed in the DXZ1array. The relative position is shown relative to the HOR 12-mer consensus. Repeat class/family and estimated copy number match previously published data (Miga et al. 2014). **b** A study of transposable elements in HOR arrays genome-wide provide evidence for a high proportion of LINEs and SINEs relative to other subfamilies of repeats (53 and 33% of total TE insertions respectively)

centromere function (Chueh et al. 2009). Further analyses designed to explore both functional and evolutionary trends with transposable elements will depend on a more comprehensive library of all insertion sites within human centromeric regions.

2.3 *Exploring the Influence of Pericentromeric Satellite DNAs*

Alpha satellite is commonly found adjacent to other pericentromeric satellite families, including classical human satellite families (HSAT I, II, and III) and beta

Table 1 Survey of the centromeric and pericentromeric satellite families in the human genome representing roughly 5% of the HuRef genome (Levy et al. 2007)

Satellite family name	GenBank accession	Fundamental repeat unit (bp)	% Genome
Alpha satellite	X07685 (consensus)	171	2.58
HSAT 2, 3	X03460 (HSAT2); X03457 (HSAT3)	5	1.42
rDNA	U13369	42,999	0.67
Gamma satellite	X68545 (clone:50E1)	220	0.13
HSAT I	X03463 (clone: B2A)	42	0.12
Beta satellite	JN194202 (clone: pAH45)	68	0.02
ACRO1	KF726396	147	0.01
CER	KF651980	96	0.008
D20S16	U10479	98	0.0003

and gamma satellites, each defined by their respective individual sequence composition and evolution (Lee et al. 1997; Warburton et al. 2008) (Table 1). Unlike alpha satellite, which is present on every centromeric region, pericentromeric satellite families are variable in both their overall abundance and chromosome distribution (Lee et al. 1997; Rudd et al. 2003). Only limited progress has been made characterizing sequence variation or chromosome organization of either new or existing satellite families in our genome, even with the wealth of publicly available genome sequences. Human satellites II and III (HSAT 2,3) are the second most abundant satellite families in the human genome (Altemose et al. 2014; Miga et al. 2015). Cytogenetic studies paired with high-resolution genomic studies using flow-sorted chromosome libraries assigned HSAT 2,3 satellite arrays to at least 13 pericentromeric regions (Altemose et al. 2014; Prosser et al. 1986; Tagarro et al. 1994). Unlike alpha satellite, many HSAT 2,3 sequences in the genome are highly similar between chromosomal subsets, with only a subset revealing sufficient sequence variation to ensure chromosome-specific mapping (Nakahori et al. 1986; Schwarzacher-Robinson et al. 1988). The simple repeat structures observed in HSAT's canonical pentameric (CATTC)n sequence definition (Prosser et al. 1986) challenge efforts to characterize the satellite sequences into higher order repeats. However, efforts to screen through comprehensive HSAT 2 and 3 libraries obtained from whole genomic datasets reveal that these complex combinations of simple repeats can be resolved in the human assembly into larger repeat units of varying size and composition (Altemose et al. 2014). The limited characterization of pericentromeric satellite families in general likely reflects the of functional data, as the cellular role of these satellite families is obscure, with only limited information regarding their respective epigenetic states under normal and stressed conditions (Eymery et al. 2010; Sengupta et al. 2009; Valgardsdottir et al. 2008).

2.4 *Enrichment of Segmental Duplications at Centric Transitions*

Segmental duplications, or genomic sequences that span at least ~1 kb that share high sequence identity among chromosomally distributed copies, are enriched at the majority of centromeric transitions (She et al. 2004; Bailey et al. 2002). Although the satellite-rich, heterochromatic regions spanning human centromeres are gene-poor, the presence of segmental duplications introduced a "euchromatic-like" landscape that is capable of generating novel transcripts in the vicinity of centromeres (She et al. 2004). Segmentally duplicated sequences are observed to exist within the satellite-rich regions on Yq11 and 21p (Lyle et al. 2007; Kirsch et al. 2005). Using patterns of sequence variation produced by population admixture, additional unplaced non-satellite sequences (representing in one study, eight new segmental supplications and over a dozen protein-coding genes) were assigned to pericentromeric/centromeric regions (Genovese et al. 2013a, b). Further, efforts to assign unmapped assemblies found in close proximity to each HSat2,3 subfamilies localized over 1 Mb of additional, satellite-associated sequences to particular chromosomes (Altemose et al. 2014). It remains unknown what functional role, if any, these sequences may play in human centromeric regions. Notably in primate evolution, segmental duplications are enriched at ancient centromere locations and are a common genomic feature at sites of centromere position reuse (Ventura et al. 2001).

In summary, the genomic model of centromeric sequence content and organization is constructed largely from experimentally determined localization and characterization of HOR satellite arrays and sequences embedded and/or directly adjacent to each array (Hayden 2012). This proposed 1000-foot view of sequence organization in regions known to contain sites of active centromeres is foundational for guiding genomic assembly efforts in the human genome. It is important to note, however, that sequence characterization is far from complete, and that it is likely that extensive genomic studies will discover both satellite and non-satellite DNAs that are currently missing from the assembly and/or specific to only a small group of individuals in the population. Discovery of novel human centromeric sequences will rely on studies aimed to explore sequence content in the unmapped and unassembled portion of whole genome sequencing projects (Miga et al. 2014, 2015; Genovese et al. 2013b). Ultimately, efforts to improve upon this model will benefit from both increasing genomic characterization and expanding the number of individual genomes to best represent the range of sequence variability.

3 Efforts to Advance Genome Informatics in Centromere Regions

High-resolution sequence-based studies are largely inhibited due to the lack of a genomic reference assembly and the difficulty generating meaningful short-read alignments to highly repetitive DNAs. Instead of improving computational tools to include centromeric regions, most software development and computational resources have been dedicated to methods to filter out repetitive reads and ignore annotation of repetitive regions in the genome (Miga et al. 2015; Li 2014). This limited ability to detect functional elements in centromeric DNAs presents a clear and urgent gap in human genetics and the biomedical sciences. Efforts to resolve this limitation will require development of a new suite of computational tools that are designed to fully incorporate centromeric sequences into functional annotation procedures. However, major biological and computational obstacles are expected to challenge the advance of centromere genome informatics. Overcoming these barriers will greatly improve our understanding of centromere biology, challenge current models of genome-wide regulation, and ultimately, revitalize a new field in centromere genomics.

3.1 Constructing a Reference Map of Centromeric Regions

Standard overlap-layout-consensus assembly fails in highly repetitive centromere regions due to the insufficient number of unique bases capable of ensuring a correct linear arrangement of repeats (Li et al. 2012; Miller et al. 2010). As a result, the entire multi-megabase-sized regions that span sites of active centromeres are excluded from chromosome assemblies (Rudd et al. 2003; Hayden et al. 2013). Efforts to construct a linear assembly of a single HOR array will rely on informative sequence variants that offer a unique "anchors" to correctly order repeats (Luce et al. 2006). Doing so will require high confidence in base quality to ensure that anchor sites are not sequencing errors. In addition to quality, assembly efforts will require sequencing of high molecular DNA, and will likely be faced with the challenge of spanning long stretches of highly homogenized portions of the array sequences that lack informative variants (Roizes 2006; Warburton et al. 1992). Fortunately, single molecule sequencing technologies have shown potential to resolve repetitive regions that span tens of kilobases, and with protocol improvements and advancements in sequencing chemistry it will be possible to achieve these ultra-long-read lengths in the near future (Jain et al. 2015; English et al. 2012). In lieu of a single high quality sequence that spans the entire multi-megabase-sized array, efforts to generate a true linear reference of an HOR array will ultimately require some form of sequence assembly across a large number of repeats. Success will likely rely on both the availability long, high quality reads that offer sufficient informative sites to confidently permit standard overlap-layout-consensus methods, and finally experimental validation to support the final linear prediction.

Alpha satellite arrays on homologous chromosomes are expected to vary considerably in length and repeat variants. This issues a major challenge to efforts aimed to generate a single, haploid assembled contig for each centromeric region. Further, a small proportion of HOR arrays assigned to more than one chromosome are known to share high sequence similarity (Vissel and Choo 1991). For example, a HOR dimer has been observed to share high sequence identity with at least three arrays assigned to chromosomes 1, 5, and 19 (Carine et al. 1989; Alexandrov et al. 2001). Therefore, efforts to adequate represent sequences from diploid centromeric arrays is complicated not only by the challenge of representing two homologous chromosomes each with inherent repeat variability, but also by the inability to separate HOR sequence libraries from individual chromosomes that share similar arrays (Altemose et al. 2014; Vissel and Choo 1991). It is possible to address the challenge of diploid assembly using haploid genomes (i.e. human hydatidiform moles or gamete DNA (Wang et al. 2012; Chaisson et al. 2015)), and/or isolation of sequence libraries from individual human chromosomes (Bentley et al. 2008), however, such efforts are highly laborious and are not sufficient to broaden centromere genomics studies to include a large number of individuals.

In addition to anticipated hurdles associated with genome assembly in diploid cells, it is possible that prevalent somatic array rearrangement, expansion, and/or contraction of repeat variants within the population of cells used to generate sequence libraries could further complicate efforts to generate a true HOR array reference. Although the frequency of introduced array variation in a given population of cells is unknown, such rearrangements have already been reported in cells exposed to stress where reduced heterochromatin and hypomethylation is expected to promote satellite sequence instability (Peng and Karpen 2007). Therefore, maps of centromeric regions that assume a true linear assembly may require single-cell sequence characterization or an assessment of centromere sequence stability in the population of cells used to generate the reference map.

Although correct sequence representation is necessary to study the genomic role in centromere identity and function, scientific progress may be able to bypass the need for strict linear assembly. For example, studies using unassembled whole genome sequence data have been successful in predicting HOR structure (Macas et al. 2010; Alkan et al. 2007) as well as reporting repeat structure and variation within a single HOR array (Miga et al. 2014; Altemose et al. 2014). Catalogs of HOR repeat variants have been previously represented in a sequence graph, where individual variants within each monomer involved in the higher order repeat are grouped, defining a single "node", and the organization adjacent monomers on the unassembled sequence reads represent the local ordering, or "edges"(Miga et al. 2014) (Fig. 5). Probabilistic traversal of these graphs results in a linear reference model of observed array sequence composition (Miga et al. 2014). Although to date no group has successfully assembled a complete human centromere, the last release of the official reference genome (GRCh38) contained a representative sequence for each centromere by sampling sequence graphs of HOR arrays from a single individual. These data provide a first look into the basic sequence structure and extent of HOR repeat variation when pooling array data across centromeric regions of homologous chromosomes.

Representing centromeric sequence data as a sequence graph may present an unconventional method to bypass the need for linear assembly. This data structure

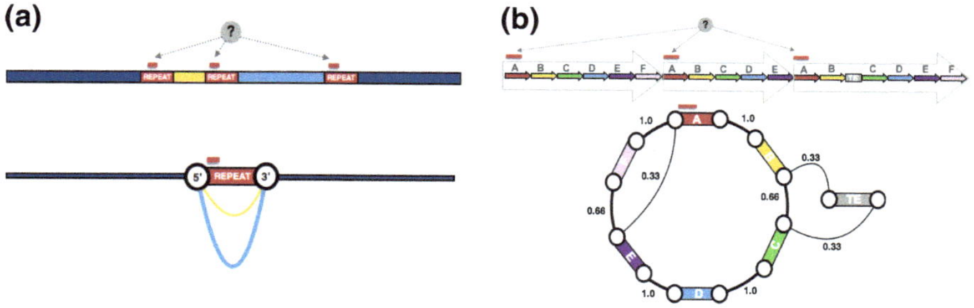

Fig. 5 Illustration of repeat sequence representation in genome graph and improvement of unambiguous short-read alignments. As shown in **a** collapsed representation of three exact repeat copies into a sequence graph, in which three identical repeat copies (shown in *red*) are collapsed into one region noting 5′ to 3′ orientation. This form of sequence data compression reduces multi-mapping issues to a single element allowing for non-ambiguous mapping to the graph. Similarly in centromeric HOR sequences, shown in **b** exact copies of individual monomers within each HOR can be represented as a single graph element or node, with edges providing local, or adjacent sequence information. In this example, the HOR repeat (shown as A–F) is rearranged in the second (deletion of F monomer) and third (insertion of TE element) repeat unit. The graph structure is able to represent these alterations in repeat structure. This data structure is able to represent all forms of repeat structure and variation. Additionally, like the example in (**a**) multi-mapping reads are able to map unambiguously to a single "A" repeat monomer in the graph

offers key advantages over true linear assembly. First, this allows explicit representation of true sequence organization in the genome graph, resulting in a more comprehensive and less biased representation of what is known about centromere structure. In this format, more extensive haplotype information offered by long, high quality sequence reads could be represented as a set of paths in the graph (Eriksson et al. 2008; Stephens et al. 2001). Second, the graph format offers a practical method to store and compare sequence data across a large number of individuals (Nguyen et al. 2015). Indeed graphical data for the X and Y chromosome have been used to study differences in repeat content and abundance between 400 individual genomes, demonstrating that population-based data can be studied without the need for exact linear assembly (Miga et al. 2014). Notably, this proposed data structure for centromeric DNAs aligns with the next anticipated version of the human reference genome, known as the "Human Variation Genome Map" (HVGM) (Nguyen et al. 2015), which also utilizes a graph-based format to better represent all common human sequence variation. Including graphical models of satellite DNA arrays in the HVGM will potentially streamline common mapping, annotation and analysis software operating on a graph structure to also operate on repeats (Paten et al. 2014; Novak et al. 2015). As a result, satellite DNAs could be included in a global platform for medical research and basic research in the life sciences.

3.2 Unambiguous Alignment to Multi-mapping Repeats

Genomic and epigenetic studies will need to incorporate high-throughput short-read functional datasets [e.g., chromatin-immunoprecipitation sequencing (ChIP-Seq) and RNA sequencing (RNA-seq)] to explore models of centromere identity and function. This goal presents a new bioinformatics challenge, in that identical repeats within centromeric regions provide multiple, best scoring alignments (Hayden et al. 2013). Current standard mapping procedures in high-throughput genomic studies eliminate such multi-mapping reads in an effort to optimize algorithmic efficiency (Encode Project Consortium 2012). To address the alignment problem, it is necessary to develop, optimize, and employ a suite of mapping and functional annotation tools to include repeated DNAs. Several existing strategies are designed to include multi-mapping reads and ultimately include repetitive DNAs in genomic studies. For example, assignment of a multi-mapping read can be made probabilistically using the collective mapping location information (Hashimoto et al. 2009), or by normalizing multi-mapping assignments based on the estimates of copy number obtained from the reference assembly (Mortazavi et al. 2008). Alternatively, data compression algorithms, e.g., the use of suffix arrays (a sorted array of all suffixes of a given string), alignment searches are able to identify exact genomic matches with little computational overhead, facilitating an accurate alignment of reads that map to multiple genomic loci (Li and Durbin 2009; Dobin et al. 2013). Reads are considered ‘multi-mapping’ if the best alignment score defines two or more distinct locations in the genome. In many cases, these repeated sequences are ignored from downstream analyses or one of the alignment sites is selected at random (Li and Durbin 2009). Specific to the challenge of short-read mapping to repetitive DNA, a sequence graph is able to collapse multiple exact copies of a repetitive element into a single, representative sequence element so that mapping can be made unambiguously (Novak et al. 2015).

In addition to the bioinformatics challenge aimed to improve mapping and peak enrichment protocols, functional studies must also overcome the difficulties involved in data interpretation. Assuming that the centromeric reference map is complete and representative of the individual genome from whom the functional datasets were obtained, it may be possible to confidently map a sequence to the reference. However, if that particular sequence provides an exact match to a repeat found to occupy the majority of an array, it will be impossible to assume a strict location of a given enrichment peak and/or transcript. Although the true enrichment of a given protein only spans a subset of repeats the exact matching within the array is expected to report all possible sites spanning the majority of the repeats. This problem intensifies when mapping data obtained from a diploid individual where the ordering and prevalence of HOR variants are expected to vary between homologous chromosomes. Efforts to bypass this ambiguity could introduce, or engineer unique markers throughout the array, however it is unclear if such sequence modifications would alter functionality or be readily removed (or expanded in frequency) by sequence conversion. Alternatively, several studies have mapped reads to a single repeat consensus to summarize read alignment profiles, proving that although enrichment signal cannot be mapped uniquely to any one, or a subset of repeats, the general trends offer valuable biological information (Erliandri et al. 2014; Jacobs et al. 2014; Hasson et al. 2013). Additionally, by bypassing the strict need to assign enrichment patterns to a precise site in the array, it is also possible to study alignment trends relative to all possible repeat variants to detect correlated signals between two or more functional datasets (Hayden et al. 2013). Such attempts to advance both the centromeric reference maps and alignment strategies are expected to ultimately shift our view of centromere genomics from the rough generic model of sequence organization to a new era of high-resolution sequence-based functional studies.

4 Future Perspectives: Emerging Genomic and Epigenomic Studies

Efforts to advance genome informatics and optimize long-range sequencing technologies will improve sequence characterization of centromeric regions to promote the use of clinical, epigenetic, and population-based genomic datasets to address the question of satellite DNA biology and centromere function. Importantly, such improvements in our understanding of centromeric sequence content and chromosome-specific organization will be extremely useful in releasing sequence-based and technology-based experimental resources to accelerate tools in genome research and applications in human health. These foundational genomic resources and tools will enable researchers to quantify the extent of centromere sequence variability in the human population and across multi-generational pedigrees. It remains unknown how much variation in the genomic and genetic definition of human centromeres exists if one extensively sampled individuals in the population. Further, researchers will begin to ask if this new source of sequence variation is associated with centromere competency, i.e., are there particular genomic features such as array size, frequency of HOR repeat variants, increased number of TE interruptions, that could positively or negatively predict centromere function? Additionally, are there genomic features that are associated with the increased rates of aneuploidy and chromosome instability? Such mechanistic findings are expected to drive parallel paths in basic cell biology, satellite sequence evolution and lead to the identification of novel disease variants in human biomedical research.

References

Alexandrov IA, Medvedev LI, Mashkova TD, Kisselev LL, Romanova LY, Yurov YB (1993) Definition of a new alpha satellite suprachromosomal family characterized by monomeric organization. Nucleic Acid Res 21(9):2209–2215

Aldrup-MacDonald ME, Kuo ME, Sullivan LL, Chew K, Sullivan BA (2016) Genomic variation within alpha satellite DNA influences centromere location on human chromosomes with metastable epialleles. Genome Res 26(10):1301–1311

Alexandrov IA, Mitkevich SP, Yurov YB (1988) The phylogeny of human chromosome specific alpha satellites. Chromosoma 96(6):443–453

Alexandrov I, Kazakov A, Tumeneva I, Shepelev V, Yurov Y (2001) Alpha-satellite DNA of primates: old and new families. Chromosoma 110(4):253–266

Alkan C, Ventura M, Archidiacono N, Rocchi M, Sahinalp SC, Eichler EE (2007) Organization and evolution of primate centromeric DNA from whole-genome shotgun sequence data. PLoS Comput Biol 3(9):1807–1818

Altemose N, Miga KH, Maggioni M, Willard HF (2014) Genomic characterization of large heterochromatic gaps in the human genome assembly. PLoS Comput Biol 10(5):e1003628

Bailey JA, Gu Z, Clark RA, Reinert K, Samonte RV, Schwartz S, Adams MD, Myers EW, Li PW, Eichler EE (2002) Recent segmental duplications in the human genome. Science 297 (5583):1003–1007

Basu J, Stromberg G, Compitello G, Willard HF, Van Bokkelen G (2005) Rapid creation of BAC-based human artificial chromosome vectors by transposition with synthetic alpha-satellite arrays. Nucleic Acid Res 33(2):587–596

Bentley DR, Balasubramanian S, Swerdlow HP, Smith GP, Milton J, Brown CG et al et al (2008) Accurate whole human genome sequencing using reversible terminator chemistry. Nature 456(7218):53–59

Blower MD, Sullivan BA, Karpen GH (2002) Conserved organization of centromeric chromatin in flies and humans. Dev Cell 2(3):319–330

Carine K, Jacquemin-Sablon A, Waltzer E, Mascarello J, Scheffler IE (1989) Molecular characterization of human minichromosomes with centromere from chromosome 1 in human-hamster hybrid cells. Somat Cell Mol Genet 15(5):445–460

Chaisson MJ, Huddleston J, Dennis MY, Sudmant PH, Malig M, Hormozdiari F et al (2015) Resolving the complexity of the human genome using single-molecule sequencing. Nature 517 (7536):608–611

Choo KH, Vissel B, Nagy A, Earle E, Kalitsis P (1991) A survey of the genomic distribution of alpha satellite DNA on all the human chromosomes, and derivation of a new consensus sequence. Nucleic Acid Res 19(6):1179–1182

Chueh AC, Northrop EL, Brettingham-Moore KH, Choo KH, Wong LH (2009) LINE retrotransposon RNA is an essential structural and functional epigenetic component of a core neocentromeric chromatin. PLoS Genet 5(1):e1000354

Dobin A, Davis CA, Schlesinger F, Drenkow J, Zaleski C, Jha S, Batut P, Chaisson M, Gingeras TR (2013) STAR: ultrafast universal RNA-seq aligner. Bioinformatics 29(1):15–21

Durfy SJ, Willard HF (1987) Molecular analysis of a polymorphic domain of alpha satellite from the human X chromosome. Am J Hum Genet 41(3):391–401

Eichler EE, Clark RA, She X (2004) An assessment of the sequence gaps: unfinished business in a finished human genome. Nat Rev Genet 5(5):345–354

Encode Project Consortium (2012) An integrated encyclopedia of DNA elements in the human genome. Nature 489(7414):57–74

English AC, Richards S, Han Y, Wang M, Vee V, Qu J et al (2012) Mind the gap: upgrading genomes with Pacific biosciences RS long-read sequencing technology. PLoS ONE 7(11): e47768

Eriksson N, Pachter L, Mitsuya Y, Rhee SY, Wang C, Gharizadeh B et al (2008) Viral population estimation using pyrosequencing. PLoS Comput Biol 4(5):e1000074

Erliandri I, Fu H, Nakano M, Kim JH, Miga KH, Liskovykh M et al (2014) Replication of alpha-satellite DNA arrays in endogenous human centromeric regions and in human artificial chromosome. Nucleic Acid Res 42(18):11502–11516

Eymery A, Souchier C, Vourc'h C, Jolly C (2010) Heat shock factor 1 binds to and transcribes satellite II and III sequences at several pericentromeric regions in heat-shocked cells. Exp Cell Res 316(11):1845–1855

Genovese G, Handsaker RE, Li H, Altemose N, Lindgren AM, Chambert K, Pasaniuc B, Price AL, Reich D, Morton CC, Pollak MR, Wilson JG, McCarroll SA (2013a) Using population admixture to help complete maps of the human genome. Nat Genet 45(4):406–414, 414e1-2

Genovese G, Handsaker RE, Li H, Kenny EE, McCarroll SA (2013b) Mapping the human reference genome's missing sequence by three-way admixture in Latino genomes. Am J Hum Genet 93(3):411–421

Gent JI, Schneider KL, Topp CN, Rodriguez C, Presting GG, Dawe RK (2011) Distinct influences of tandem repeats and retrotransposons on CENH3 nucleosome positioning. Epigenetics Chromatin 4:3

Harrington JJ, Van Bokkelen G, Mays RW, Gustashaw K, Willard HF (1997) Formation of de novo centromeres and construction of first-generation human artificial microchromosomes. Nat Genet 15(4):345–355

Hashimoto T, de Hoon MJ, Grimmond SM, Daub CO, Hayashizaki Y, Faulkner GJ (2009) Probabilistic resolution of multi-mapping reads in massively parallel sequencing data using MuMRescueLite. Bioinformatics 25(19):2613–2614

Hasson D, Panchenko T, Salimian KJ, Salman MU, Sekulic N, Alonso A, Warburton PE, Black BE (2013) The octamer is the major form of CENP-A nucleosomes at human centromeres. Nat Struct Mol Biol 20(6):687–695

Hayden KE (2012) Human centromere genomics: now it's personal. Chromosome Res 20(5): 621–633

Hayden KE, Strome ED, Merrett SL, Lee HR, Rudd MK, Willard HF (2013) Sequences associated with centromere competency in the human genome. Mol Cell Biol 33(4):763–772

Jacobs FM, Greenberg D, Nguyen N, Haeussler M, Ewing AD, Katzman S, Paten B, Salama SR, Haussler D (2014) An evolutionary arms race between KRAB zinc-finger genes ZNF91/93 and SVA/L1 retrotransposons. Nature 516(7530):242–245

Jain M, Fiddes IT, Miga KH, Olsen HE, Paten B, Akeson M (2015) Improved data analysis for the MinION nanopore sequencer. Nat Methods 12(4):351–356

Kirsch S, Weiss B, Miner TL, Waterston RH, Clark RA, Eichler EE et al (2005) Interchromosomal segmental duplications of the pericentromeric region on the human Y chromosome. Genome Res 15(2):195–204

Lee C, Wevrick R, Fisher RB, Ferguson-Smith MA, Lin CC (1997) Human centromeric DNAs. Hum Genet 100(3–4):291–304

Levy S, Sutton G, Ng PC, Feuk L, Halpern AL, Walenz BP et al (2007) The diploid genome sequence of an individual human. PLoS Biol 5(10):e254

Li H (2014) Toward better understanding of artifacts in variant calling from high-coverage samples. Bioinformatics 30(20):2843–2851

Li H, Durbin R (2009) Fast and accurate short read alignment with Burrows-Wheeler transform. Bioinformatics 25(14):1754–1760

Li Z, Chen Y, Mu D, Yuan J, Shi Y, Zhang H et al (2012) Comparison of the two major classes of assembly algorithms: overlap-layout-consensus and de-bruijn-graph. Brief Funct Genomics 11(1):25–37

Luce AC, Sharma A, Mollere OS, Wolfgruber TK, Nagaki K, Jiang J et al (2006) Precise centromere mapping using a combination of repeat junction markers and chromatin immunoprecipitation-polymerase chain reaction. Genetics 174(2):1057–1061

Lyle R, Prandini P, Osoegawa K, ten Hallers B, Humphray S, Zhu B et al (2007) Islands of euchromatin-like sequence and expressed polymorphic sequences within the short arm of human chromosome 21. Genome Res 17(11):1690–1696

Macas J, Neumann P, Novak P, Jiang J (2010) Global sequence characterization of rice centromeric satellite based on oligomer frequency analysis in large-scale sequencing data. Bioinformatics 26(17):2101–2108

Mahtani MM, Willard HF (1998) Physical and genetic mapping of the human X chromosome centromere: repression of recombination. Genome Res 8(2):100–110

Malik HS, Henikoff S (2002) Conflict begets complexity: the evolution of centromeres. Curr Opin Genet Dev 12(6):711–718

Maloney KA, Sullivan LL, Matheny JE, Strome ED, Merrett SL, Ferris A, Sullivan BA (2012) Functional epialleles at an endogenous human centromere. Proc Natl Acad Sci USA 109 (34):13704–13709

Manuelidis L (1976) Repeating restriction fragments of human DNA. Nucleic Acid Res 3(11):3063–3076

Manuelidis L, Wu JC (1978) Homology between human and simian repeated DNA. Nature 276(5683):92–94

Masumoto H, Ikeno M, Nakano M, Okazaki T, Grimes B, Cooke H, Suzuki N (1998) Assay of centromere function using a human artificial chromosome. Chromosoma 107(6–7):406–416

May BP, Lippman ZB, Fang Y, Spector DL, Martienssen RA (2005) Differential regulation of strand-specific transcripts from Arabidopsis centromeric satellite repeats. PLoS Genet 1(6):e79

Miga KH, Newton Y, Jain M, Altemose N, Willard HF, Kent WJ (2014) Centromere reference models for human chromosomes X and Y satellite arrays. Genome Res 24(4):697–707

Miga KH, Eisenhart C, Kent WJ (2015) Utilizing mapping targets of sequences underrepresented in the reference assembly to reduce false positive alignments. Nucleic Acid Res gkv671

Miller JR, Koren S, Sutton G (2010) Assembly algorithms for next-generation sequencing data. Genomics 95(6):315–327

Mortazavi A, Williams BA, McCue K, Schaeffer L, Wold B (2008) Mapping and quantifying mammalian transcriptomes by RNA-Seq. Nat Methods 5(7):621–628

Nakahori Y, Mitani K, Yamada M, Nakagome Y (1986) A human Y-chromosome specific repeated DNA family (DYZ1) consists of a tandem array of pentanucleotides. Nucleic Acid Res 14(19):7569–7580

Nguyen N, Hickey G, Zerbino DR, Raney B, Earl D, Armstrong J et al (2015) Building a pan-genome reference for a population. J Comput Biol 22(5):387–401

Novak AM, et al (2015) Canonical, stable, general mapping using context schemes. Bioinformatics btv435

Oakey R, Tyler-Smith C (1990) Y chromosome DNA haplotyping suggests that most European and Asian men are descended from one of two males. Genomics 7(3):325–330

Ohzeki J, Nakano M, Okada T, Masumoto H (2002) CENP-B box is required for de novo centromere chromatin assembly on human alphoid DNA. J Cell Biol 159(5):765–775

Paten B, Novak A, Haussler D (2014) Mapping to a reference genome structure. arXiv preprint arXiv: 1404.5010

Peng JC, Karpen GH (2007) H3K9 methylation and RNA interference regulate nucleolar organization and repeated DNA stability. Nat Cell Biol 9(1):25–35

Prades C, Laurent AM, Puechberty J, Yurov Y, Roizes G (1996) SINE and LINE within human centromeres. J Mol Evol 42(1):37–43

Prosser J, Frommer M, Paul C, Vincent PC (1986) Sequence relationships of three human satellite DNAs. J Mol Biol 187(2):145–155

Roizes G (2006) Human centromeric alphoid domains are periodically homogenized so that they vary substantially between homologues. Mechanism and implications for centromere functioning. Nucleic Acid Res 34(6):1912–1924

Rudd MK, Willard HF (2004) Analysis of the centromeric regions of the human genome assembly. Trends Genet 20(11):529–533

Rudd MK, Schueler MG, Willard HF (2003) Sequence organization and functional annotation of human centromeres. Cold Spring Harb Symp Quant Biol 68:141–149

Rudd MK, Wray GA, Willard HF (2006) The evolutionary dynamics of alpha-satellite. Genome Res 16(1):88–96

Santos FR, Pandya A, Kayser M, Mitchell RJ, Liu A, Singh L et al (2000) A polymorphic L1 retroposon insertion in the centromere of the human Y chromosome. Hum Mol Genet 9(3):421–430

Schueler MG, Higgins AW, Rudd MK, Gustashaw K, Willard HF (2001) Genomic and genetic definition of a functional human centromere. Science 294(5540):109–115

Schueler MG, Dunn JM, Bird CP, Ross MT, Viggiano L, Program NCS et al (2005) Progressive proximal expansion of the primate X chromosome centromere. Proc Natl Acad Sci USA 102(30):10563–10568

Schwarzacher-Robinson T, Cram LS, Meyne J, Moyzis RK (1988) Characterization of human heterochromatin by in situ hybridization with satellite DNA clones. Cytogenet Cell Genet 47(4):192–196

Sengupta S, Parihar R, Ganesh S (2009) Satellite III non-coding RNAs show distinct and stress-specific patterns of induction. Biochem Biophys Res Commun 382(1):102–107

She X, Horvath JE, Jiang Z, Liu G, Furey TS, Christ L et al (2004) The structure and evolution of centromeric transition regions within the human genome. Nature 430(7002):857–864

Shepelev VA, Uralsky LI, Alexandrov AA, Yurov YB, Rogaev EI, Alexandrov IA (2015) Annotation of suprachromosomal families reveals uncommon types of alpha satellite organization in pericentromeric regions of hg38 human genome assembly. Genom Data 5:139–146

Spence JM, Critcher R, Ebersole TA, Valdivia MM, Earnshaw WC, Fukagawa T et al (2002) Co-localization of centromere activity, proteins and topoisomerase II within a subdomain of the major human X alpha-satellite array. EMBO J 21(19):5269–5280

Stephens M, Smith NJ, Donnelly P (2001) A new statistical method for haplotype reconstruction from population data. Am J Hum Genet 68(4):978–989

Sullivan LL, Boivin CD, Mravinac B, Song IY, Sullivan BA (2011) Genomic size of CENP-A domain is proportional to total alpha satellite array size at human centromeres and expands in cancer cells. Chromosome Res 19(4):457–470

Tagarro I, Fernandez-Peralta AM, Gonzalez-Aguilera JJ (1994) Chromosomal localization of human satellites 2 and 3 by a FISH method using oligonucleotides as probes. Hum Genet 93(4):383–388

Tyler-Smith C, Oakey RJ, Larin Z, Fisher RB, Crocker M, Affara NA, Ferguson-Smith MA, Muenke M, Zuffardi O, Jobling MA (1993) Localization of DNA sequences required for human centromere function through an analysis of rearranged Y chromosomes. Nat Genet 5 (4):368–375

Vafa O, Sullivan KF (1997) Chromatin containing CENP-A and alpha-satellite DNA is a major component of the inner kinetochore plate. Curr Biol 7(11):897–900

Valgardsdottir R, Chiodi I, Giordano M, Rossi A, Bazzini S, Ghigna C, Riva S, Biamonti G (2008) Transcription of Satellite III non-coding RNAs is a general stress response in human cells. Nucleic Acid Res 36(2):423–434

Ventura M, Archidiacono N, Rocchi M (2001) Centromere emergence in evolution. Genome Res 11(4):595–599

Vissel B, Choo KH (1991) Four distinct alpha satellite subfamilies shared by human chromosomes 13, 14 and 21. Nucleic Acid Res 19(2):271–277

Wang J, Fan HC, Behr B, Quake SR (2012) Genome-wide single-cell analysis of recombination activity and de novo mutation rates in human sperm. Cell 150(2):402–412

Warburton PE, Willard HF (1995) Interhomologue sequence variation of alpha satellite DNA from human chromosome 17: evidence for concerted evolution along haplotypic lineages. J Mol Evol 41(6):1006–1015

Warburton PE, Greig GM, Haaf T, Willard HF (1991) PCR amplification of chromosome-specific alpha satellite DNA: definition of centromeric STS markers and polymorphic analysis. Genomics 11(2):324–333

Warburton PE, Wevrick R, Mahtani MM, Willard HF (1992) Pulsed-field and two-dimensional gel electrophoresis of long arrays of tandemly repeated DNA: analysis of human centromeric alpha satellite. Methods Mol Biol 12:299–317

Warburton PE, Cooke CA, Bourassa S, Vafa O, Sullivan BA, Stetten G et al (1997) Immunolocalization of CENP-A suggests a distinct nucleosome structure at the inner kinetochore plate of active centromeres. Curr Biol 7(11):901–904
Warburton PE, Hasson D, Guillem F, Lescale C, Jin X, Abrusan G (2008) Analysis of the largest tandemly repeated DNA families in the human genome. BMC Genom 9:533
Waye JS, Willard HF (1987) Nucleotide sequence heterogeneity of alpha satellite repetitive DNA: a survey of alphoid sequences from different human chromosomes. Nucleic Acid Res 15(18):7549–7569
Wevrick R, Willard HF (1989) Long-range organization of tandem arrays of alpha satellite DNA at the centromeres of human chromosomes: high-frequency array-length polymorphism and meiotic stability. Proc Natl Acad Sci USA 86(23):9394–9398
Wevrick R, Willard HF (1991) Physical map of the centromeric region of human chromosome 7: relationship between two distinct alpha satellite arrays. Nucleic Acid Res 19(9):2295–2301
Willard HF (1985) Chromosome-specific organization of human alpha satellite DNA. Am J Hum Genet 37(3):524–532
Willard HF, Waye JS (1987) Hierarchical order in chromosome-specific human alpha satellite DNA. Trends Genet 3:192–198

DNA Sequences in Centromere Formation and Function

M. Dumont and D. Fachinetti

Abstract Faithful chromosome segregation during cell division depends on the centromere, a complex DNA/protein structure that links chromosomes to spindle microtubules. This chromosomal domain has to be marked throughout cell division and its chromosomal localization preserved across cell generations. From fission yeast to human, centromeres are established on a series of repetitive DNA sequences and on specialized centromeric chromatin. This chromatin is enriched with the histone H3 variant, named CENP-A, that was demonstrated to be the epigenetic mark that maintains centromere identity and function indefinitely. Although centromere identity is thought to be exclusively epigenetic, the presence of specific DNA sequences in the majority of eukaryotes and of the centromeric protein CENP-B that binds to these sequences, suggests the existence of a genetic component as well. In this review, we will highlight the importance of centromeric sequences for centromere formation and function, and discuss the centromere DNA sequence/CENP-B paradox.

1 Introduction

In eukaryotic cells, the centromere constitutes the site for sister chromatid attachment to the mitotic spindle and is the foundation for kinetochore assembly during meiosis and mitosis to ensure faithful chromosome segregation. The centromere is comprised of a complex of specific DNA sequences and proteins. At the DNA level, it is characterized by a series of long arrays of highly similar tandem repeats that are not conserved between species, called in human alpha-satellite DNAs (Manuelidis 1978; Vissel and Choo 1987). The DNA wraps around histones to form nucleosomes that are enriched with the histone H3-variant CENP-A, giving

M. Dumont · D. Fachinetti (✉)
Institut Curie, PSL Research University, CNRS, UMR 144, 26 rue d'Ulm, 75005 Paris, France
e-mail: daniele.fachinetti@curie.fr

B.E. Black (ed.), *Centromeres and Kinetochores*, Progress in Molecular and Subcellular Biology 56, DOI 10.1007/978-3-319-58592-5_13

rise to a unique centromeric chromatin (Palmer et al. 1987)—although the nature of its composition is still a matter of debate (Black and Cleveland 2011).

While the preservation of centromere identity and inheritance were initially thought to be tightly linked to the highly repetitive DNA sequence (Grady et al. 1992), the discovery in 1993 of human neocentromeres (Voullaire et al. 1993)—new centromeres formed at an ectopic site deprived of satellite arrays—led researchers to revisit this dogma. Neocentromeres that are formed in euchromatic sites outside the α-satellite DNA allow kinetochore assembly (specifically and uniquely at the new site), show relative normal chromosome segregation—although with reduced fidelity (Bassett et al. 2010; Fachinetti et al. 2015)—and, importantly, are maintained across generations [at least two generations (Amor et al. 2004)]. Therefore, the discovery of neocentromeres and of pseudo-dicentric chromosomes—stable chromosomes with one active centromere while another is silenced without apparent sequence rearrangements (Sullivan and Willard 1998)—indicates that centromere identity can be sequence-independent and specified by one or more epigenetic factors.

In the last decade, several reports suggested that CENP-A acted as an epigenetic mark for centromere specification (Allshire and Karpen 2008). It was demonstrated that centromere position is specifically maintained by the centromeric chromatin bound by CENP-A in flies (Mendiburo et al. 2011), chickens (Hori et al. 2013), and humans (Fachinetti et al. 2013). CENP-A is conserved between species and during evolution (Earnshaw and Rothfield 1985; Palmer et al. 1987; Sullivan 1994; Stoler et al. 1995; Buchwitz et al. 1999; Takahashi 2000; Malik and Henikoff 2001) and is essential for cell viability (Howman et al. 2000; Régnier et al. 2003; Fachinetti et al. 2013). CENP-A is enriched at centromeric regions and interspersed with canonical histone H3 nucleosomes (Blower et al. 2002). However, quantitative measurement of CENP-A molecules revealed that half of it is deposited outside centromeres (Bodor et al. 2014). For these reasons, it was proposed that the conservation of a proper stoichiometry of CENP-A molecules present at centromere might be necessary for maintaining centromere position across generations (Bodor et al. 2014). At centromeres, CENP-A primarily binds to a specific type of α-satellite sequence [α-I; (Ando et al. 2002)], but how CENP-A is precisely distributed at the centromeric regions is unknown, partially due to the repetitive nature of DNA sequences that precludes accurate investigation. Studies on neocentromeres revealed that a CENP-A domain occupies ~90–100 kb (Alonso et al. 2007, 2010; Hasson et al. 2013), although this length might not reflect the actual size of a CENP-A domain at native, repetitive human centromeres. However, studies in chicken cells suggested that CENP-A domains at induced-neocentromeres and at nonrepetitive native centromeres are of comparable size [~40 kb; (Shang et al. 2010, 2013)].

Despite centromere identity being exclusively determined through an epigenetic mechanism is a widely accepted concept, DNA sequences are also likely to play a fundamental role in centromere biology as suggested by their enrichment in the majority of eukaryotes. In this review, we highlight the importance of centromeric DNA sequences and DNA sequence-dependent binding proteins in mediating centromere identity and function.

2 The Arrangement of Centromeric DNA Sequence

In eukaryotes, two types of centromere configurations exist: "point centromeres" found in budding yeast and "regional centromeres" found in most of the model organisms that include the pathogenic yeast *Candida albicans* and the fission yeast *Schizosaccharomyces pombe*. Point centromeres are characterized by a single nucleosome that covers 125 bp of DNA sequence composed of three Conserved DNA Elements (CDE I/II/III) (Clarke and Carbon 1980). The peculiarity of the point centromere is that centromere position and function relies on the centromeric DNA sequence, as conditional deletion or even a single CDEIII point mutation leads to chromosome mis-segregation (McGrew et al. 1986; Ng and Carbon 1987) [mutation of CDEI or II are known to be more permissive; (Cumberledge and Carbon 1987)]. Indeed, deletion of centromeric sequence in budding yeast never drives neocentromere development, indicating that DNA sequence is indeed essential for centromere formation and function, in contrast to other yeasts such as *Candida albicans* (Ketel et al. 2009).

Although centromeric DNA sequences have diverged drastically across evolution in the larger regional centromeres, the overall composition of the centromere is conserved. Centromeres are normally established on highly repetitive DNA arrays such as satellite DNAs and transposable elements (Willard and Waye 1987; Grady et al. 1992; Schueler et al. 2001; Cleveland et al. 2003; Plohl et al. 2014). This tandemly repetitive organization at centromere is found in hundreds of plant and animal species (Melters et al. 2013) and reveals a similar mode of evolution of centromeric DNA sequences even for species that diverged 50 million years ago.

In human, α-satellite DNA represents ~3% of the genome (Hayden et al. 2013) and is organized in a head-to-tail tandem repeat of single AT-rich 171 bp monomers (Waye and Willard 1986). The association of adjacent monomers can also form higher order repeat (HOR) units in which blocks of multiple repeats—discovered by the periodicity of restriction endonuclease cleavage sites on genomic DNA (Wu and Manuelidis 1980)—can form a larger domain (for further details see review by K.H. Miga in this issue). HORs are usually flanked by relatively short stretches of divergent monomeric satellites (Alexandrov et al. 2001; Rudd and Willard 2004). This HOR organization is conserved in all great ape species and was also recently uncovered in small apes (Koga et al. 2014), although they evolved rapidly compared to monomeric α-satellites (Rudd et al. 2006; Shepelev et al. 2009). A HOR is itself repeated constantly hundreds to thousands of times spanning between 0.34 and 6 kb and altogether these HOR repeats can give rise to megabase-size centromeres (0.3–5 Mb) (Fig. 1). Individual monomers share 50–70% sequence identity while the HORs, due to an overall homogenous α-satellite array, can exceed up to 95% identity (Koga et al. 2014 and see also review from Aldrup-MacDonald et al. 2016). However, variation in the number of tandem monomers can be observed and creates a chromosome specificity (Willard 1985; Vissel and Choo 1987) with some chromosomes that possess several different α-satellite (alphoid) DNA subfamilies within their centromeres (Choo et al. 1991; Wevrick and Willard 1991; Alexandrov et al. 2001).

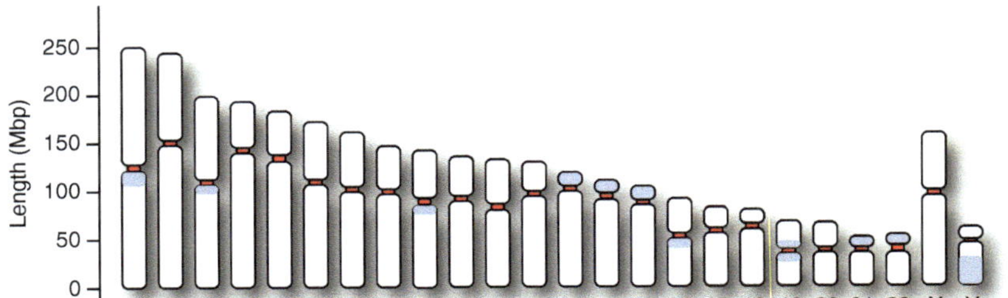

Chr N°	1	2	3	4	5	6	7	8	9	10	11	12	13	14	15	16	17	18	19	20	21	22	X	Y
Chr Type	M	sM	M	sM	sM	sM	sM	sM	sM	sM	sM	sM	Ac	Ac	Ac	M	sM	sM	M	M	Ac	Ac	sM	Ac
Cen Size (Mb)	2.9	2.1	0.8	2	3.4	1.2	2.8	1.8	2.1	1.9	3.3	2.4	ND	ND	2.6	1.9	3.8	5.3	2.6	2.3	ND	ND	3.8	0.3
Σ B-box/M (%)	40	48	36	54	32	37	44	42	44	45	35	45	ND	ND	51	47	46	39	39	38	ND	ND	30	0
N° HOR array	5	1	3	1	6	1	2	1	1	5	2	2	5	1	3	2	3	6	1	6	2	5	1	1

Fig. 1 The human chromosomes and their peri/centromeres. Schematic representation of human chromosomes. Chromosome number and size are shown. *Light blue* color indicates regions with a high level of heterochromatin. Chromosome (Chr) type: *M* metacentric, *sM* sub-metacentric and *Ac* acrocentric. The approximate centromere (Cen) size in each chromosome is also indicated in megabase pairs (Mbs) and is calculated considering all the different high-order repeats (HOR) arrays present at each individual centromere (*N* number). The sum of the frequency of CENP-B boxes (*TTCG****A**CGGG*), in case of several HOR arrays, and number of variants of HOR arrays for each chromosome is indicated based on reference models from one individual genome using HuRef Sanger reads on GRCh38 reference models (Karen Miga personal communication). The frequency of CENP-B boxes/monomers is showed in % and it is calculated by summing the total number of monomers for each chromosome and the number of monomers containing CENP-B. CENP-B boxes were included if they had no more than 2 mismatches from the 17 bp consensus sequences. Data for acrocentric chromosomes cannot be determined (ND). Chromosomes 1, 15 and 19 contain a redundant array that cannot be solved. These data are based on a model (they present sequence variants from an averaging of diploid arrays); it is likely that the indicated frequency of CENP-B boxes and size may change given different cell line or individual genome of interest

At the level of the HOR, additional variation can be present on the same chromosome due to size polymorphism, as was described for the human chromosome 17 (Waye and Willard 1986). In this particular case, it was shown that the two HORs behave as centromeric epialleles with both HORs being capable of recruiting proteins required for kinetochore formation (Maloney et al. 2012). The mechanism that controls the choice of alpha-satellites to be the active form of the centromere is unknown, although recent studies proposed that genomic variation in size and sequences within the HOR is a negative driver of epiallele formation (Aldrup-MacDonald et al. 2016).

3 The Role of Centromeric DNA Sequences

The existence of neocentromeres devoid of repetitive sequences (Depinet et al. 1997) strongly suggested that DNA sequences are neither sufficient nor essential for maintaining centromere position and function. Indeed, nonrepetitive centromeres are found in orangutans (Locke et al. 2011), on one chromosome in horses (Wade et al. 2009) (see also review by Elena Giuliotto in this issue for more information), three chromosomes in chickens (Shang et al. 2010) and in five chromosomes of potato (Gong et al. 2012). However, it was hypothesized that evolutionary new centromeres (ENCs) are "centromeres in progress" and further incorporation of repetitive DNA is then required for full centromere maturation (Marshall et al. 2008). In support of this idea, evolutionary studies in macaque revealed that all ENCs harbor large arrays of alpha-satellite DNA (Ventura et al. 2007). Thus, if alpha-satellites are required to stabilize ENCs and all human centromeres contain these repetitive sequences, what are their main functions?

3.1 Preservation of a Unique Chromosomal Architecture

One possibility is that the existence and conservation of higher order structures might be required to preserve and/or maintain a particular DNA topology. The architecture of centromeres has only recently emerged from studies in flies (Garavís et al. 2015) or using human BACs (containing from 135 to 190 kb of the centromeric regions of the chromosomes 8 and 17) with *Xenopus* egg extracts (Aze et al. 2016). Both studies revealed the presence of specific secondary structures such as DNA loops [by electron microscopy (Aze et al. 2016)] or four antiparallel motifs [by NMR (Garavís et al. 2015)]. The latter study suggested that centromeric sequences were mainly selected throughout evolution for their ability to form secondary structures rather than for the DNA sequence itself, implying that monomer length might be important for HOR organization. In this regard, different organisms have similar lengths of repeating units that approximately correspond to one (171 bp-long α-satellite repeat in human, 120 bp in mouse, 178 bp in Arabidopsis, 156 bp in maize) or two (340 bp repeat in pig, 359 bp-long SAT III in Drosophila) nucleosomal units. The fact that HORs might play a role in centromere folding was hypothesized by Rosandić et al. (2008). Using a computational method, they generated different centromere structural models for the distribution of the HORs in the three known 30-nm DNA fiber models (a crossed-linker model, a solenoid model, and a helical-ribbon model), in which the characteristic geometrical pattern of the DNA folding allows the recognition of specific microtubules.

CENP-B might also play a crucial role in centromere organization mediated by its unique DNA binding capacity and/or by epigenetic modifications on centromeric DNA (see paragraph CENP-B and DNA methylation).

3.2 Heterochromatin Formation

Another proposed role of repetitive sequences at the centromere is to promote pericentromeric heterochromatin and, simultaneously, to create a separated environment for centromeres. Indeed, only a fraction of alpha-satellites constitutes the functional centromere with the binding of CENP-A only occurring at one type of alphoid DNA (Ando et al. 2002). In fission yeast, this distinct separation between peri/centromeric regions is enforced by tRNA genes that flank the centromeric regions that physically act as a barrier to prevent pericentromeric heterochromatin expansion into centromeres (Scott et al. 2006; Noma et al. 2006) or centromere migration. Pericentromeric heterochromatin has a distinguished epigenetic signature with the heterochromatin protein HP1 bound to histone H3 Lysine 9 trimethylation [H3K9me3, for more information see review by Almouzni and Probst (2011)] and is required for the assembly of centromeric cohesion (Bernard et al. 2001; Bailis et al. 2003). Interestingly, it has been demonstrated that heterochromatin has a direct impact on centromeres itself by promoting de novo centromere formation in fission yeast (Folco et al. 2008) or by favoring ectopic CENP-A deposition at heterochromatic sites in flies (Olszak et al. 2011).

3.3 Centromere-Derived Transcripts

The centromere was long considered "junk DNA" without transcriptional activity; only in the last decade has centromere transcription been appreciated to be essential for maintaining centromere integrity in various species. Transcription of centromeric repetitive sequences that produces non coding RNAs (ncRNAs), distinguished in satellite I, II, and III RNAs, was uncovered in mammals (for more information see review Rošić and Erhardt 2016). For example, active RNA polymerase II surprisingly localizes at the mitotic kinetochore and its inhibition was found to affect CENP-C binding and consequently give rise to mitotic error (Chan et al. 2012). Further, the satellite I RNA was shown to co-immuno-precipitate with two components of the mitotic Chromosome Passenger Complex (CPC) (Aurora B and INCENP), although its binding occurs mainly in interphase (Ideue et al. 2014). This ncRNA controls the localization and activity of the CPC and consequently regulates chromosome segregation. Moreover, of particular interest for this review, a 1.3 kb centromere-long transcript was identified to physically interact with the CENP-A/HJURP pre-assembly complex in vivo at early G1 (Quénet and Dalal 2014). Targeted destruction of these centromeric transcripts (by shRNA) resulted in a decrease of CENP-A at centromeres and severe mitotic defects. Similar findings were observed in flies, in which an RNA derived from X chromosome-specific SAT III—curiously with a similar size to the cenRNA observed in human (Quénet and Dalal 2014)—binds centromeric regions and is required for successful mitosis by maintaining correct binding of CENP-A and CENP-C (Rošić et al. 2014). Additional suggested roles for the centromeric transcripts are in the stabilization of the higher order chromatin structure at the centromere, as well as to act as a scaffold for chromatin remodeling complexes, although these mechanisms are not determined yet (see reviews Allshire and Karpen 2008; Rošić and Erhardt 2016).

3.4 *CENP-B Boxes and CENP-B*

Enclosed in the alpha-satellite DNA, the CENP-B box is a short sequence that is the binding site for the centromere protein CENP-B (Masumoto et al. 1989). The involvement of CENP-B boxes in centromere location and function has been studied for many years, and CENP-B, together with the CENP-B box, may therefore constitute a link between the genetic and epigenetic centromere identities.

4 Evolution of the CENP-B Boxes

Two families of α-satellite DNA are distinguished in primates (Romanova et al. 1996). Type A monomeric units (also called alphoid type II) were found in the centromeric region of lower primates and are designated as the "old" family. This monomer diverged in the "new" family of primates, the descendants of the last ancestor of the great apes, and became organized in several monomer units named the alphoid type I DNA that contains CENP-B boxes and gained the ability to bind CENP-B (Alexandrov et al. 2001). The CENP-B box consists of a 17 bp motif that is found at regular intervals in the alphoid DNA (Ikeno et al. 1994) of all the human chromosomes, except for the Y chromosome, and at varying frequencies between the chromosomes (Masumoto et al. 1989; Rosandić et al. 2006) (Fig. 1). For example, analysis of the centromere region of the chromosome 21 revealed the occurrence of two different types of alphoid DNAs (α21-I and α21-II), differing by the nature of the HOR units and the frequency of the CENP-B boxes (Ikeno et al. 1994). The α21-I displays CENP-B boxes in almost every other monomer unit whereas CENP-B boxes were found on the α21-II locus at a low frequency (every 100 monomer repeat unit). The authors proposed a model in which the binding of CENP-B to the alphoid type I DNA forms more compact folding for centromere function while the α21-II might play a role of "supporting area". A further study performed on all the human chromosomes using fluorescence in situ hybridization (FISH) probes and chromosome-specific α-satellite DNA probes (Lo et al. 1999) revealed a relatively common occurrence of low-alphoid DNA centromeres at chromosome 21, called a "low-alphoid" chromosome. While the centromere function of this chromosome does not seem to be impaired, the proper segregation of chromosome 21 compared to the others remains to be investigated.

Variability in the occurrence of CENP-B boxes in vertebrates has been observed during evolution. A computational method showed a strong conservation of the CENP-B box motifs in all of the mammalian consensus sequences, with horses and dogs showing the greatest variability and complexity in centromere organization, and, in rare cases such as the platypus, CENP-B boxes were found outside the satellite DNA (Alkan et al. 2011). An initial study performed in 1995 (Haaf et al. 1995) using FISH probes against CENP-B boxes revealed the absence of this sequence in the chromosomes of Old and New World monkeys (the separation between these two groups arises from the migration of the New World Monkeys from Africa to South America ~30 million years ago), prosimians and gibbons, while it was detected in humans (except on the Y chromosome) and great apes (such as chimpanzee and bonobo). Paradoxically, no or few (100–1000-fold fewer than in human) CENP-B boxes were detected in African green monkeys (AGM), despite the presence of a biochemically functional CENP-B protein and despite their being phylogenetically closer to humans than mice (*Mus musculus*)—a species in which CENP-B boxes were found (Goldberg et al. 1996). In contrast, a parallel study by Okazaki and colleagues (Yoda et al. 1996) demonstrated the existence of CENP-B boxes in the alpha-satellite of AGM and binding of CENP-B at metaphase chromosomes in AGM cells, although at lower levels compared to human centromeres. This controversy was recently resolved by comparing the latest sequence analysis of AGM genome using paired-end sequencing (Warren et al. 2015) with the human datasets; a ~240-fold enrichment of CENP-B boxes in human versus AGM was observed despite ~10-fold less alphoid sequences (Sivakanthan Kasinathan and Steven Henikoff, personal communication). This demonstrates that CENP-B boxes are indeed present in AGM although at low frequency. The reason for this observed reduction is unknown, but suggests additional roles of CENP-B in centromere function other than via its DNA binding domain.

Two recent studies have reported the presence of CENP-B boxes in the α-satellite DNA of three New World monkey species (Suntronpong et al. 2016; Kugou et al. 2016). These binding sites were demonstrated to be functional, as CENP-B was found at centromeric regions (marked by CENP-A), although not on all chromosomes. The repeat unit size in which the CENP-B boxes are found is twice that of humans and great apes (340–350 bp instead of the 171 bp). In addition, the location and the sequence direction of the CENP-B boxes in the studied primates (marmosets, squirrel monkeys, and tamarins) is also different. This is likely due to divergent evolution from a common ancestor or different mutations that occurred after lineage divergence, although the intervals between CENP-B boxes are nearly equivalent. It is of interest to understand why an additional CENP-B box is not found in hominid. The authors suggest that this is due to a limit of the maximum amount of CENP-B boxes in a single repeat unit, taking into consideration that CENP-B boxes along the CENP-B box-positive alpha-satellite DNAs are directly associated with HOR structures in humans (Kugou et al. 2016). Several other hypotheses have been suggested to explain those differences, including a variability in the CENP-B DNA affinity (with binding to type II or type III satellite sequences rather than to a common CENP-B box), an evolution of the DNA binding domain sequence that is able to recognize CENP-B protein variants [although this seems unlikely since no *CENP-B* paralogs were found in mammals (Marshall and Choo 2012)], or that centromere function preceded the satellite repeat accumulation during evolution. In the latter case, the acquisition of satellite tandem repeats is thought to have occurred subsequently to contribute to centromere stability and function.

The occurrence of CENP-B boxes outside of vertebrates remains under debate. Analysis of the structure and the organization of satellite DNAs, however, revealed the presence of CENP-B box-like sequences in other organisms such as in *Xenopus*

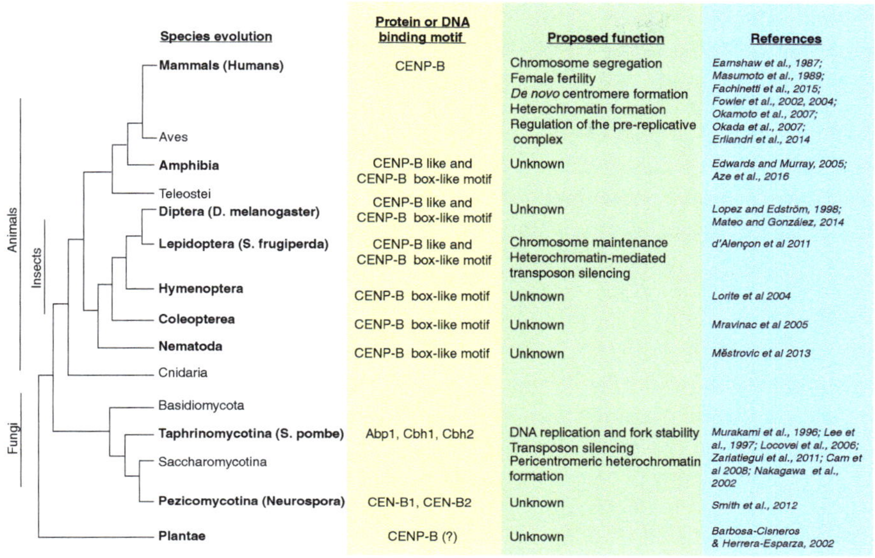

Fig. 2 Phylogeny of CENP-B homologs or CENP-B box-like motifs in animals and fungi. For more details on the different CENP-B-like proteins, the existence of CENP-B-like binding domains and their biological function see the main text. Species in *bold* are the ones in which CENP-B-like or CENP-B box-like have been described. See also Drinnenberg et al. (2016)

(Edwards and Murray 2005), dipterans (López and Edström 1998), lepidopterans (d'Alençon et al. 2011), plants (Barbosa-Cisneros and Herrera-Esparza 2002) and recently also in nematodes (Meštrović et al. 2013) with a similarity of 10–12 out of the 17 nucleotides (Fig. 2).

5 Evolution of CENP-B, the CENP-B Box Binding Protein

Characterized for the first time in 1987 using serum from a patient with anticentromere antibodies (ACA), human CENP-B is a 80 kDa protein that was revealed to be present at varying amounts between different chromosomes and completely absent from the Y chromosome centromere and neocentromeres (Earnshaw et al. 1987; Depinet et al. 1997; Tanaka et al. 2001; Miga et al. 2014). CENP-B amino acid (a.a.) sequence is highly conserved within mammals with ~80% of a.a. homology (92% between human and mice) (Sullivan and Glass 1991). Unlike the single evolutionary origin of other known kinetochore components, CENP-B-like proteins have originated multiple times from a domestication of a major class of transposases, the *pogo*-like family (Kipling and Warburton 1997). In addition to humans, other mammalian species also acquired CENP-B at their kinetochores and are likely derived from a single domestication event. However, in contrast to other *bona fide* transposases encoded by other *pogo*-like family found in human (Smit and Riggs 1996), CENP-B lost its capacity to act as a transposase due to a.a. mutations within the excision catalytic domain—although the nature of the sequence mutation does not exclude that it retains nickase activity (see below).

Despite this high conservation in evolution, it is unclear if other domestication events of CENP-B-like proteins occurred outside mammals with the exception of fission yeast. In *S. pombe* three different CENP-B-like proteins have been found (see below) with sequence homology of 25–50% (Murakami et al. 1996; Halverson et al. 1997; Lee et al. 1997). Similar to mammalian cases, the fission yeast homologs have lost the transposase motif, but retain the NH2-terminal and central domain of pogo-like transposases and localization at peri/centromeric regions (Casola et al. 2008).

Additional domestication events of putative CENP-B-like proteins were also described in the holocentric chromosomes of Lepidoptera *S. frugiperda* (d'Alençon et al. 2011) and in *D. melanogaster* (Mateo and González 2014), the latter of which conserved the same structure of the DNA binding domain observed in human (34% overall identity with CENP-B), yet their localization and role at centromeres have not been demonstrated. Contrasting studies are found regarding the existence of a CENP-B homolog in amphibians; while no ortholog was found in *X. tropicalis* (Casola et al. 2008) or in *Xenopus laevis* (A. Straight, personal communication), a protein with homology to CENP-B was identified by quantitative mass spectrometry and reported to bind human BACs in *Xenopus laevis* egg extract (Aze et al. 2016). This latter finding is in accordance with the discovery of a CENP-A-associated CENP-B box sequence [although with low similarity to the human ones (Edwards and Murray 2005)]. However, further studies are needed to validate this protein as a true *CENP-B* ortholog.

CENP-B binds to the CENP-B box sequence via its N-terminal region (1–129 a.a.) with an equilibrium constant for binding of $6.3 \times 10^8\ M^{-1}$ (Masumoto et al. 1989; Muro et al. 1992; Yoda et al. 1992). The crystal structure of the complex between CENP-B and DNA sequence has been determined at 2.5 Å resolution although a 21 bp-CENP-B box was used in their model [a slightly longer sequence that includes the conventional 17 bp sequence, probably for an easier purification; (Tanaka et al. 2001)]. Four well-defined CENP-B regions were identified in the first 129 a.a. region of CENP-B: the N-terminal arm, domain 1 [a CENP-B mutant of a.a. 10–25 failed to bind to CENP-B boxes (Yoda et al. 1992)], a linker loop and the domain 2. Both domains 1 and 2 have a helix-turn-helix motif that binds adjacent to the major grooves of DNA and is similar to that of proteins with known functions such as transposases. CENP-B forms a homodimer through its carboxy terminus domain that might allow the bundling of two distant CENP-B boxes—and that might be involved in centromere structure (Tawaramoto et al. 2003). Two CENP-B monomers interact via their hydrophobic surfaces involving Van der Waals contacts and hydrogen bonds and form a unique four-helix bundle structure that is asymmetrical and antiparallel as solved by the crystal structure of the dimerization domain at 1.65 Å resolution (Tawaramoto et al. 2003).

DNA sequence recognition by CENP-B involves nine "core nucleotides" of the 17 bp-CENP-B box [the underlined nucleotides: (Py)TTCGTTGGAA(Pu)CGGGA (Masumoto et al. 1993)] and the binding induces a "kink-straight-kink" bend of the DNA by ~60° that might be relevant to distinguish the centromeric chromatin structure from other sites of the genome (Tanaka et al. 2001). Indeed, using in vitro nucleosome reconstitution, the CENP-B dimer was proposed to bind two CENP-B boxes and fold the alphoid DNA to induce nucleosome positioning between two CENP-B boxes that are present every other monomer (Yoda et al. 1998). In this view, two nucleosomes are expected to be between two CENP-B/DNA complexes. Recently, high-resolution chromatin immunoprecipitation (ChIP) followed by clustering of sequence data revealed an enrichment of a dimer of 340 bp α-satellite containing two adjacent CENP-A nucleosomes on either side of a CENP-B box (Henikoff et al. 2015). Further, high-resolution CENP-A mapping at all native human centromeres (except the Y) revealed phasing of CENP-A between CENP-B boxes (Y. Nechemia-Arbely, K.H. Miga and D.W. Cleveland, personal communication) as observed for HORs of the X chromosome, with CENP-A-nucleosome positions peaking between the CENP-B boxes (Hasson et al. 2013). Surprisingly, phasing of CENP-A nucleosomes on α-satellite was also observed on HORs of the Y chromosome, on non-functional CENP-B boxes (Hasson et al. 2013) or following genetic depletion of *CENP-B* (Y. Nechemia-Arbely, K.H. Miga and D.W. Cleveland personal communication), demonstrating the existence of a CENP-B-independent nucleosome phasing.

6 Loading of Nascent CENP-B at Centromeric Regions

The timing and the molecular control of CENP-B loading at the centromere are currently not well understood. A pioneering approach was done using fluorescence recovery at photobleached centromeres (FRAP) on ectopically expressed GFP-tagged centromeric proteins to reveal that CENP-B is highly dynamic during G1- and S-phases (the complete pool of CENP-B turns over at centromeres within 1 h during these phases) to then be stably bound to kinetochore during G2 and M (Hemmerich et al. 2008). This increased CENP-B stability in G2 requires its dimerization domain and might be involved in mediating higher order chromatin structure by nucleosome positioning (Yoda et al. 1998) or kinetochore stability by stabilizing CENP-C (Fachinetti et al. 2015). Tachiwana and colleagues (Tachiwana et al. 2013) found that the human Nap1, an acidic histone chaperone, stimulates CENP-B binding to CENP-A or H3 mono-nucleosome containing CENP-B boxes in vitro. Remarkably, Nap1 also has an opposite activity on CENP-B; its interaction with CENP-B inhibits its non-specific binding to non-centromeric DNA, similarly to its previously described inhibitory activity for non-nucleosomal histone–DNA interaction (Andrews et al. 2010). How Nap1 achieves these contrasting functions is not clear, although its highly acidic core is likely to be involved.

Recently, ADA3 (alteration/deficiency in activation 3)—a component of the transcriptional activator and histone acetyltransferase (HAT) complexes that control faithful mitotic progression (Orpinell et al. 2010)—was shown in mice to interact with α-satellite DNA and CENP-B both in vitro and in vivo (Mohibi et al. 2015). Conditional KO mice rescued with a ADA3 mutant unable to bind CENP-B undergo cell proliferation arrest and show an increase in chromosome segregation defects and abrogation of centromeric CENP-B recruitment. Altogether, the authors conclude that ADA3 binding to centromeric regions is required for CENP-B loading. Although this study provided new insights into CENP-B localization, how ADA3 regulates CENP-B loading is not known.

The binding of CENP-B to DNA was also recently proposed to be auto regulated via post-translational modifications. Mass spectrometry analysis of a transiently transfected tagged CENP-B revealed that the N-terminal region of CENP-B is trimethylated by the α-N-methyltransferase NRMT at every stage of the cell cycle (Dai et al. 2013). Interestingly, the N-terminal methylation of CENP-B enhances its binding to CENP-B boxes, likely via an increase of electrostatic interactions with the DNA phosphate group, and increases under stress conditions such as cell density or heat shock.

7 CENP-B and Its Role in Centromere Function: From CENP-A Deposition to Chromosome Segregation

For many years, the biological role of CENP-B has been enigmatic and still now remains poorly defined. Sequence homology studies have linked CENP-B to *pogo*-like transposases and Tigger elements (Kipling and Warburton 1997; Casola et al. 2008; Mateo and González 2014) and suggest a putative 3′ nicking activity that would involve CENP-B in the evolution of the α-satellite DNA. This, together with activity of the CENP-B dimerization domain, might contribute to the high-order structure of the centromeres by promoting recombination hotspots and therefore by modifying the expansion of α-satellites (Kipling and Warburton 1997). Indeed, many in vivo studies throughout the years arrived to the same conclusion that CENP-B is not an essential factor in centromere function. Nevertheless, other indications from concurrent studies attributed several roles for CENP-B in centromere function, including chromosome segregation. What role, then, does CENP-B play at the centromere, and why has the protein been so conserved throughout mammalian evolution?

CENP-B can be found on inactive centromeres of stable dicentric chromosomes or chromosome-containing neocentromeres, and CENP-B boxes (and consequently CENP-B) are absent from the male Y chromosome and neocentromeres, even though kinetochore formation is functional. However, centromere composition of the Y chromosome or neocentromere-containing chromosomes are compromised (Fachinetti et al. 2015) leading to high rate of mis-segregation of the Y chromosome in tissue culture cells (Hoffmann et al. 2016; Ly et al. 2016). In addition, the loss of the Y in peripheral blood cells (Forsberg et al. 2014) has been linked to non-hematopoietic tumors in elderly men and can act as a predictive biomarker in oncogenesis (Forsberg et al. 2014; Dumanski et al. 2015; Noveski et al. 2016).

The idea that CENP-B has no main role in centromere function was initially suggested when three independent studies conducted in mice concluded that mutations in *CENP-B* do not affect viability (Hudson et al. 1998; Perez-Castro et al. 1998; Kapoor et al. 1998). However, Hudson et al. reported lower body weight, reduced testicles size and sperm content in *CENP-B* KO mice, although these phenotypes were not found by the two other studies. The work by the A. Choo lab was followed up by two additional studies: Fowler et al. reported the first significant phenotype of *CENP-B* gene disruption in female null mice showing abnormal uterine epithelium (where CENP-B is highly expressed) and reproductive dysfunction (Fowler et al. 2000). Interestingly, uterine epithelial tissue is known to be highly mitotically active (Epifanova 1958). The authors suggested a direct implication of CENP-B in human uterine pathologies related to altered fertility. A further study refined the role of CENP-B in female reproductive performance in which they described the loss of fertility over a number of generations due to a decline in endometrial glands' numbers (Fowler et al. 2004), although the molecular mechanisms involved are undefined. However, it is important to note that aneuploidy rate was not examined by the three teams who produced *CENP-B* null mice. Recognizing that remarkable somatic aneuploidy can indeed be generated in mice without a strong phenotype or significant shortening of life span (Michel et al. 2001; Weaver et al. 2007; Jeganathan et al. 2007), it is possible that mice deficient in CENP-B develop somatic aneuploidy. Indeed, in vitro analysis on mouse embryonic fibroblasts (MEFs) derived from one of the *CENP-B*-KO mice studies (Kapoor et al. 1998) showed an increased rate of chromosome mis-segregation and kinetochore defects following CENP-B loss (Fachinetti et al. 2015). Further analysis in vivo would be required to test this crucial issue.

Due to the unexpected results of the in vivo *CENP-B* KO and to the existence of a closely related family of single copy genes named Tigger-derived (TIGD) (Smit and Riggs 1996), it was also proposed that putative *CENP-B* paralogs might be redundant for centromere function, as was observed in fission yeast (Irelan et al. 2001). This hypothesis was recently weakened following immuno-localization on three different TIGD proteins that revealed no centromere binding, therefore strongly suggesting that CENP-B acts alone without functionally redundant partners (Marshall and Choo 2012).

7.1 CENP-B and Its Direct Role in Chromosome Segregation

In terms of centromere localization, analysis of metaphase chromosomes using immuno-electron microscopy revealed that the majority of CENP-B is absent from the outer kinetochore plate, but rather it is located just beneath it into the α-satellite-rich heterochromatin domain (Cooke et al. 1990). These results are in contrast to a previous finding that analyzed the distribution of centromeric proteins using ACA staining (Brenner et al. 1981). However, recently, more reports indicate direct interaction between CENP-B and inner centromeric components such as CENP-A and CENP-C as described below.

First evidences that implicate CENP-B in centromere function come from H. Masumoto's lab. Using a yeast two-hybrid assay for screening of a cDNA library, his team revealed an unexpected interaction between CENP-B acidic residues (368–599 a.a.) and two different Mif2 homology domains of CENP-C (Suzuki et al. 2004). These interactions were confirmed using an in vitro GST/HIS pulldown assay on purified recombinant proteins (Fachinetti et al. 2015). In vivo data showed that overexpression of a truncated CENP-B (containing only its DNA binding domain or lacking the CENP-C interacting sites) drives a cell cycle delay in metaphase by altering CENP-C assembly at centromeres (Suzuki et al. 2004). Accordingly, *CENP-B* gene depletion in both human or mouse cell lines caused a two-fold reduction of the level of centromere-bound CENP-C, that in turn affected the stability of the kinetochore complex leading to an increase in chromosome mis-segregation and aneuploidy (Fachinetti et al. 2015). Recently, using rapid depletion of endogenous CENP-A in human cells, it was shown that CENP-B binding to alphoid DNA is necessary and sufficient for kinetochore anchoring via CENP-C and for maintenance of chromosome segregation fidelity in the first mitosis on CENP-A-depleted centromeres (Hoffmann et al. 2016). Similarly, CENP-B via the CENP-A amino-terminal tail was demonstrated to maintain faithful chromosome segregation (except for the Y chromosome; Ly et al. 2016) in cells in which the CENP-A/CENP-C interaction was abolished (Fachinetti et al. 2013). Indeed, removal of CENP-B led to mitotic failure and cell lethality following complete replacement of CENP-A with a chimeric variant in which the carboxy-terminal tail—the CENP-C-recruiting site—was substituted with the corresponding region of histone H3. Finally, base-pair resolution genomic analysis for mapping of centromeric proteins revealed that two CENP-C molecules dimerize over a CENP-B box flanked by two CENP-A nucleosomes and that CENP-T, essential for kinetochore formation (Foltz et al. 2006; Wood et al. 2016), resides on top of this CENP-B box (Henikoff et al. 2015; Thakur and Henikoff 2016).

All of these findings implicate CENP-B as an important member of the centromere complex modulating centromere function that, together with CENP-A chromatin, acts as a major link from the DNA to the kinetochore to mediate successful chromosome segregation (Fig. 3). It is tantalizing to hypothesize that a correlation might exist between the number of CENP-B binding sites, respectively, present at the centromere of each chromosome and the fidelity of their segregation during mitosis.

In support of the above-mentioned hypothesis is the theory that the "stronger" centromeres preferentially segregate to the egg due to higher amount of centromere proteins that might attract more spindle microtubules during the female asymmetric meiosis. This model—meiotic drive—could explain the positive evolution of the centromere, in response to the modification of the DNA satellite sequence (Malik and Henikoff 2001; Henikoff et al. 2001). In other words, the evolution of alphoid DNA toward a higher number of CENP-B boxes would allow higher deposition of CENP-A and CENP-C at the centromere through binding to CENP-B, and hence reinforce the centromere. Indeed, the absence of this female-specific selection at the male chromosome could justify the absence of CENP-B boxes only at this centromere sequence. This model of the “stronger centromeres” retained in the egg was recently supported by comparing a standard laboratory mouse strain with wild mice harboring a conserved Robertsonian (Rb) fusion (Chmátal et al. 2014). In this approach they observed that in a particular genetic background Rb metacentric fusion recruited more kinetochore protein (Hec1) and, consequently, are preferentially transmitted. Further, this model was supported by the comparison of the centromeric histone H3 evolution in several clades with asymmetric meiosis and clades with symmetric meiosis (no meiotic drive pressure) using codon-substitution models (Zedek and Bureš 2016). This study revealed that the clades with asymmetric meiosis underwent histone H3 evolution more frequently in order to avoid the negative consequences of the meiotic drive.

7.2 De Novo *Centromere Formation*

Other than directly controlling the stability of the kinetochore complex before or during mitosis, previous work mainly using artificial chromosomes placed CENP-B in modulating centromere formation. The first eukaryotic artificial chromosome vectors were engineered in budding yeast (Yeast Artificial Chromosome, YAC) in the beginning of the 1980s (Murray and Szostak 1983). They consisted in a linear plasmid carrying the essential and minimal elements such as DNA origins of replication, the functional centromere DNA components and telomeres. The segregation of linear YACs is highly dependent on their size, with a decrease in frequency of loss with increasing length, and on the position of the centromere (Murray and Szostak 1986). Artificial chromosome vectors were subsequently developed in bacteria (Bacterial Artificial Chromosome, BAC) (Woo et al. 1994) and in mammalian/human systems (Mammalian Artificial Chromosome and Human Artificial Chromosome, MAC and HAC) (Harrington et al. 1997; Ikeno et al. 1998). Although HACs have a lower mitotic stability than endogenous chromosomes, it was shown that the centromere of a HAC behaves as its natural counterpart during mitosis in terms of chromosome alignment, chromatid cohesion and assembly of the kinetochore-dependent checkpoint (Tsuduki et al. 2006).

Of interest for this review, artificial chromosomes were used to study the molecular mechanism necessary to form de novo centromere assembly. Masumoto and colleagues defined the necessity of having the type I alphoid DNA and CENP-B boxes on these mini-chromosomes (using genomic DNA derived from a mouse/human chromosome 21) to form functional centromeres (Ikeno et al. 1998; Masumoto et al. 1998; Ohzeki et al. 2002). By testing four different synthetic

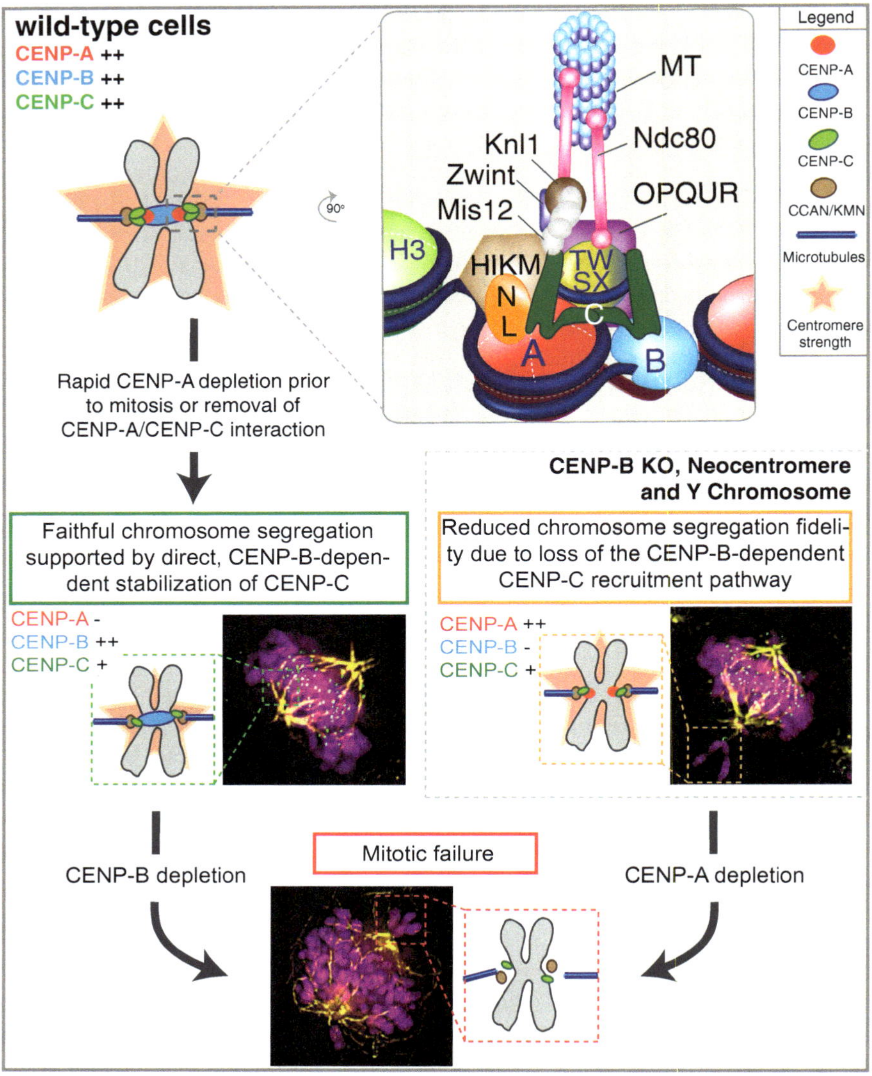

Fig. 3 CENP-B supports kinetochore formation and contributes to the maintenance of chromosome segregation fidelity. CENP-B directly interacts with CENP-A (via CENP-A amino-terminal tail) and CENP-C to maintain kinetochore nucleation (see schematic of the centromere/kinetochore complex). CENP-B is essential and sufficient to stabilize a fraction of CENP-C (in *green*) that allows correct chromosome segregation even under condition in which CENP-A cannot interact with CENP-C (e.g., following rapid CENP-A depletion or in a CENP-A mutant lacking its carboxy-terminal tail), although CENP-C level is decreased and the rate of chromosome segregation errors increased compared to wild type condition. Centromeres that do not have CENP-B (e.g., neocentromeres, Y chromosome or in CENP-B-depleted cells) show reduced CENP-C and chromosome segregation fidelity. However, complete mitotic failure is observed only when both CENP-A and CENP-B are depleted. See the main text for more details

repetitive sequences, combining alphoid or non-alphoid DNA with (or without) point mutations in CENP-B boxes (to create non-functional CENP-B boxes), they found that only the association of the alphoid DNA type I sequence with a functional CENP-B box led to MAC formation with de novo centromere chromatin assembly [measured by scoring for CENP-A assembly; (Ohzeki et al. 2002)] (Fig. 4). Further, both alphoid DNA length and density of CENP-B boxes directly impacted the efficiency to form functional HACs (Okamoto et al. 2007). A BAC containing 60 kb of satellite DNA (30 kb length as a minimum functional core) with five functional CENP-B boxes/11-monomer repeating units promotes centromere nucleation (27% success rate of HAC formation in HT1080 cells) and spreading/maintenance of CENP-A nucleosome. Along the same lines, artificial chromosomes containing α-satellite DNA from chromosome Y, devoid of CENP-B boxes, are unstable and circularize (Taylor et al. 1996), and are unable to form de novo centromeres in HACs (Mejía et al. 2002).

The requirement for CENP-B boxes for centromere establishment was also confirmed in mouse cells using BACs constituted of 60 kb of human alphoid DNA (Okada et al. 2007). These evidences also provided proof of the importance of CENP-B itself for CENP-A deposition and HAC formation. Further, an additional role for CENP-B was proposed in the prevention of extra-centromere formation when a centromere is already established. In chromosomally integrated alphoid DNA, CENP-B binding enhances H3K9me3 and subsequently CpG methylation (except at its binding site) leading to an inactive chromatin state at these ectopic sites, therefore preventing CENP-A incorporation (Fig. 4). These molecular mechanisms controlled by CENP-B might have implications in centromere silencing such as in pseudo-dicentric chromosomes in which one centromere has been inactivated. The role of histone modifications in centromere formation was further investigated using HACs (Bergmann et al. 2011; Ohzeki et al. 2012; for more information, see the Chapter "Artificial Chromosomes and Strategies to Initiate Epigenetic Centromere Establishment" by Barrey and Heun).

Based on these findings, another intriguing possibility is that CENP-B interacts with CENP-A nucleosomes to promote its incorporation into the chromatin. Indeed, in vitro GST-binding assays showed that CENP-B interacts with the first 29 a.a. of the amino-terminal tail of CENP-A, but not with a CENP-A mutant lacking the N-tail (Fachinetti et al. 2015) or with a H3 (in complex with H4) chimera containing only the CATD domain of CENP-A (Fujita et al. 2015). It was proposed that interactions via the first 29 a.a. of CENP-A are required to stabilize CENP-B at centromeric regions, as removal of the N-tail reduces CENP-B levels at centromere (Fachinetti et al. 2015), similar to what was observed in the complete absence of CENP-A (Fachinetti et al. 2013). In vitro reconstitution of CENP-A or H3 nucleosomes demonstrated that CENP-B forms a stable complex preferentially with CENP-A nucleosomes and that the proximal position of CENP-B boxes increases its binding (Fujita et al. 2015). Moreover, in human cells, ChIP analysis and quantitative PCR on alphoid DNA BAC revealed that a CENP-B box mutant has a

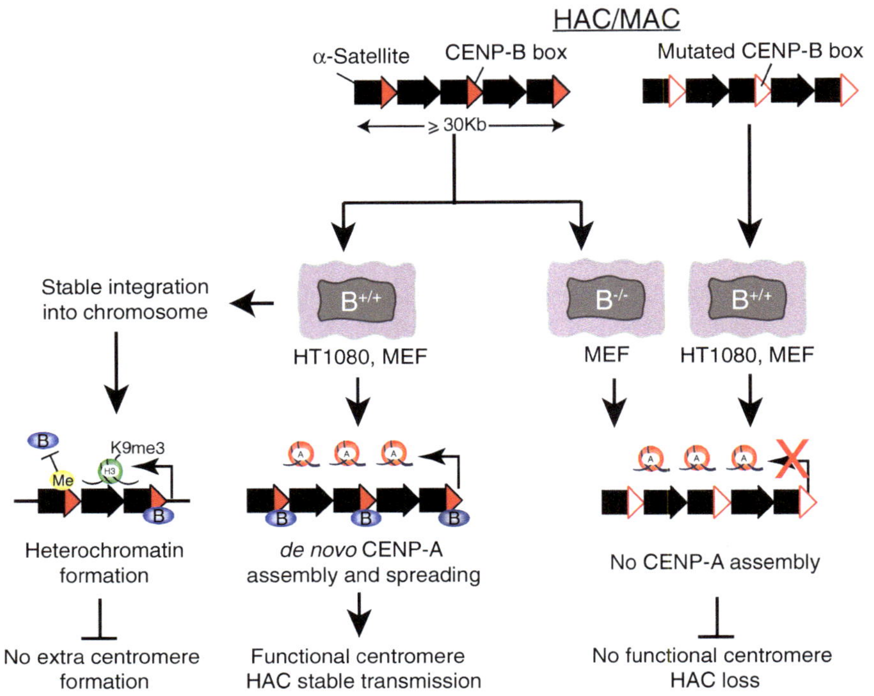

Fig. 4 De novo centromere formation is controlled by CENP-B and DNA sequence. A certain length of alpha-satellite DNA sequence containing CENP-B boxes is required for CENP-B binding and subsequent for CENP-A assembly for Human/Mouse Artificial Chromosome (HAC/MAC) formation and stability. HAC containing a DNA sequence with mutation in CENP-B boxes is unstable and lost. Additionally, when the HAC is integrated into the chromosome that contain already a functional centromere, CENP-B promotes Histone H3-K9 trimethylation and DNA methylation to prevent new CENP-B binding and extra-centromere formation. See the main text for more details

lower retention capacity of CENP-A nucleosome to incorporate into chromatin (Fujita et al. 2015). This suggests that CENP-B binding to the CENP-B box is involved in the stabilization of a pre-assembled CENP-A nucleosome, although CENP-B depletion in cells only slightly affects centromeric CENP-A levels (Fachinetti et al. 2015). How CENP-A/CENP-B interaction is mediated and controlled are still open questions. It is likely that the string of acidic subdomains of CENP-B (Earnshaw et al. 1987) could capture the amino-terminal tail of histones that are normally basic domains.

Altogether, these studies may question the exclusive epigenetic nature of the centromere; despite the fact that centromere identity is epigenetically determined—therefore without DNA sequence commitment—CENP-B could be a missing key element for proper centromere function.

8 CENP-B and Its Implication in DNA Replication

The molecular function of CENP-B had also been associated to DNA replication with several studies performed on the three CENP-B homologs found in fission yeast, Ars binding protein 1 (Abp1) and CENP-B homologs 1 and 2 (cbh1/2) (Halverson et al. 1997; Lee et al. 1997; Baum and Clarke 2000; Irelan et al. 2001). Using a yeast two-hybrid system, Abp1 has been shown to interact with the DNA replication protein Cdc23 (Locovei et al. 2006). In particular, Abp1 has been shown to be important specifically in the initiation of DNA replication since G1-cells depleted for Abp1 delayed entrance in S-phase. In support of this, fission yeast CENP-B homolog Abp1 regulates fork directionality of mating-type switching (Aguilar-Arnal et al. 2008). Even further, the Martienssen group proposes that in fission yeast CENP-B plays a role in replication fork stability; indeed, CENP-B was demonstrated to co-localize with the DNA binding factor switch-activating protein 1 (Sap1) at the long terminal repeat (LTR) of retrotransposons (Zaratiegui et al. 2011). While Sap1 blocks the progression of the replication forks [as previously described for the ribosomal DNA to ensure fork directionality (Krings and Bastia 2005)], CENP-B might raise the Sap1-dependent barrier to promote progression of the replication fork and to stabilize it. It is tantalizing to speculate that CENP-B could control replication fork progression and stability to prevent chromosome breaks and rearrangements also in higher eukaryotes, especially considering that peri or centromeric regions are enriched for retrotransposons.

Evidence supporting an involvement of CENP-B in DNA replication in humans was recently found using HAC or analyzing HORs of chromosome 5 or X (Erliandri et al. 2014). By modulating the chromatin context, CENP-B was suggested to contribute to regulating the replication of alpha-satellite repeats, where DNA replication of centromeric regions initiated. Indeed, some α-satellite DNA blocks that are preferentially enriched in CENP-B boxes are recognized by the replication machinery and are involved in the establishment of pre-replication complexes, in which CENP-B (likely indirectly via heterochromatin formation) might negatively regulate the assembly of multiple origins on these arrays. Further studies are required in this direction, but these new discoveries open a perspective to consider CENP-B as a "structural factor" that may influence chromatin conformation, through its dimerization, for the regulation of the assembly of the pre-replication complex, in line with its role in nucleosome positioning. This could explain why the patterns of CENP-B distribution at centromeres are also regulated by DNA methylation.

9 DNA Methylation and Centromere

Epigenetic modifications such as DNA methylation of cytosine are known to regulate heterochromatin formation (Okano et al. 1999; Xu et al. 1999; Bachman et al. 2001). Within the monomer of the alpha-satellites three CpG sites have been found and curiously, two of them are within the CENP-B box (Choo et al. 1991). Interestingly, some studies reported that CpG dinucleotide methylation of the CENP-B boxes regulates the binding of CENP-B. Indeed, global reduction of CpG methylation (including the alphoid DNA) using 5-aza-2′-deoxycytidine in mouse cells leads to a redistribution of CENP-B binding to a bigger domain than usual (Mitchell et al. 1996).

In agreement with these results, an in vitro complex-reconstitution assay of CENP-B DNA binding domain (1–129) with methylated or unmethylated CENP-B box DNAs revealed that CpG methylation of both sites decreases CENP-B affinity for CENP-B box DNA almost to no DNA specificity (Tanaka et al. 2004). Indeed, the addition of a methyl group was proposed to cause a steric clash with the side chain of residues 44 (threonine) and 125 (arginine) of the DNA binding domain of CENP-B (Tanaka et al. 2001, 2004). Similar results were confirmed in MEFs using HAC formation in which it was demonstrated that methylation of only one residue reduced CENP-B binding (Okada et al. 2007). The role of this epigenetic regulation and how it may be directly relevant for centromere biology still needs to be tested. It is likely that methylated CENP-B boxes are required to limit the available sites for CENP-B to confine the active centromere. In this regard, chromosomally integrated HACs, but not stable extrachromosomal HACs, showed a CENP-B-dependent hypermethylation, suggesting that DNA methylation negatively regulates CENP-B binding to CENP-B boxes and consequently inhibits CENP-A incorporation (Okada et al. 2007) (see above). Alternatively, a link between CENP-B box methylation and CENP-B binding has been proposed in the context of centromeric heterochromatin formation involving RNAi machinery (Tanaka et al. 2004). In this view, CENP-B binding to DNA might inhibit the production of RNA transcripts in proximity of CENP-A nucleosomes. Additionally, the transcription of alpha-satellite in CpG methylated CENP-B boxes is required to maintain heterochromatin formation (as in fission yeast) or to support centromere function. Further, DNA methylation at centromeres has been described as a regulator of centromere transcription levels that might influence the mitotic chromosomes' dynamics (Fuks 2005).

The role of DNA methylation at centromeric regions goes beyond the regulation of CENP-B binding. For example, the DNA methyl transferase DNMT3B, responsible for de novo DNA methylation, was demonstrated to localize at pericentromeric and centromeric regions (Okano et al. 1999; Bachman et al. 2001; Chen et al. 2004) and its centromeric localization to be dependent on a direct interaction with CENP-C via its C-terminal domain, mainly during mitosis (Gopalakrishnan et al. 2009). This interaction was initially identified with yeast two-hybrid screening and subsequently in vivo by co-immunoprecipitation of transiently transfected components. The authors also proposed a cooperative interaction between CENP-C and DNMT3B necessary to reinforce their centromeric binding and to maintain the correct epigenetic marks (such as H3K4me2/3, H3K9me2/3 and H3K27me3) and ensure proper chromosome segregation.

A similar cooperative interaction was observed between Mis18α [a subunit of the Mis18 complex required for centromere priming and CENP-A deposition (Fujita et al. 2007)] and DNMT3A/B in MEFs (Kim et al. 2012). Surprisingly, despite CENP-C playing a role in M18BP1 (Moree et al. 2011) and DNMT3B (Gopalakrishnan et al. 2009) recruitment to centromere, it is not necessary for Mis18α and DNMT3A/B interaction.

Another link between DNA methylation and centromere came from depletion of DNMT3A/B in mice that led to deregulation of mitotic recombination at centromeric regions with changes in the length of the repeat units (Jaco et al. 2008). This correlation between DNA methylation and centromere function/stability can at least partially explain the phenotype observed in patients with a rare disorder due to DNMT3B mutation, the Immunodeficiency, Centromeric region instability, and Facial anomalies (ICF) syndrome (Wijmenga et al. 1998; Okano et al. 1999). Some ICF patients exhibit hypomethylation of peri/centromeric regions, DNA breaks, and increase in chromosome decondensation and chromosome segregation errors (Ehrlich et al. 2008).

Altogether, these evidences implicate DNA methylation (directly or indirectly) as an important player for maintenance of centromere identity and function. However, more studies are required to understand its exact role in these processes.

10 Other Centromeric Proteins with DNA Binding Activity

As discussed above, CENP-B is the only known CENP with a DNA binding domain specific for a centromeric sequence (the CENP-B box). However, DNA binding domains have been found in other components of the CCAN, although without apparent DNA sequence specificity. For example, by expressing a series of truncated versions in *E. coli*, it was demonstrated that CENP-C has a DNA binding domain within its central domain (398–498 a.a.) (Sugimoto et al. 1994, 1997), further confirmed using in vitro assays (422–537 a.a. that might contain two independent DNA binding domains) (Yang et al. 1996) and in vivo (Politi et al. 2002). However, no DNA binding specificity was identified in vitro (Yang et al. 1996).

On the contrary, Della Valle and colleagues demonstrated that CENP-C is preferentially associated with α-satellites and comparison of CENP-B/DNA and CENP-C/DNA complexes revealed that they recognize the same types of alpha-satellite DNA but of different centromere domains (Politi et al. 2002). The discrepancy with the in vitro data might be due to missing cooperative interactions that are, on the contrary, present in vivo. If and to what extent the interaction of CENP-C with DNA other than its interaction with CENP-A (Carroll et al. 2010) is required for its deposition at centromeres are uncertain. However, the DNA binding activity of CENP-C suggests an intriguing correlation with centromere specification, possibly in cooperation with CENP-B. This idea is supported by in vivo evidence that, using an ectopic artificial system to recruit centromere components (LacO/I and TetO/R), shows that CENP-C can recruit CENP-A—via an interaction with the Mis18 complex—to form an ectopic and functional kinetochore (Hori et al. 2013; Shono et al. 2015).

DNA binding domains were also identified for the CENP-T-W-S-X complex (Nishino et al. 2012). Indeed, contact with DNA sequence has been shown to be essential to form a functional kinetochore in chicken DT40 cells. High-resolution structural analysis demonstrated that CENP-T-W-S-X form a nucleosome-like structure involving DNA binding sites, although whether this structure exists in vivo remains undetermined and no DNA sequence specificity was identified.

11 Concluding Remarks

In the last two decades, researchers mainly concentrated their efforts in identifying the epigenetic mechanism for centromere maintenance and function. However, the existence of an evolutionarily conserved α-satellite DNA sequence and a conserved consensus sequence (CENP-B box) for a specific DNA binding protein (CENP-B) emphasizes the importance of a genetic contribution to centromere function. Many different possible roles of centromeric DNA sequences have emerged.

Alphoid DNA has been implicated in creating a unique DNA topology at centromeric regions possibly required to facilitate recombination between repeats (McFarlane and Humphrey 2010) that might have consequences in centromere specification during meiosis. Certainly, centromeric DNA contains unique features such as accumulation of topological constrains [e.g., ultrafine anaphase bridges; (Baumann et al. 2007; Chan et al. 2007)] and a unique DNA replication machinery that relies on one DNA repair mechanism (Aze et al. 2016). Further, transcription is known to exist and to play a major role in centromere maintenance and function (see review Rošić and Erhardt 2016). Formation of heterochromatin at the pericentromeric regions might also be dependent on repetitive sequences (for review see Bierhoff et al. 2014; Biscotti et al. 2015). Finally, α-satellites are binding sites for a range of centromeric proteins, from CENP-A to sequence-specific DNA binding proteins such as CENP-B, to other putative DNA binding proteins such as CENP-C and CENP-T, and are sites of epigenetic marks such as DNA methylation. CENP-B binding to centromeric regions is also required to sustain centromere formation and kinetochore assembly and to control DNA replication. All these aforementioned roles of DNA sequences (and CENP-B) in centromere biology highlight the importance of studying the mechanisms of de novo centromere formation and maintenance in the native context of the centromere.

Acknowledgements The authors would like to thank I. Drinnenberg (Institut Curie, Paris), H. Masumoto (Kazusa, JP), S. McClelland (BCI, UK), K.H. Miga (UCSC, US), C. Bartocci (Institut Curie, Paris), P. Ly (Ludwig, UCSD, La Jolla), A. Musacchio (Max Planck Dortmund, Germany) and all members of the Fachinetti lab for helpful suggestions and discussion. We also thank Y. Nechemia-Arbely and D.W. Cleveland (Ludwig, UCSD, La Jolla), S. Kasinathan and S. Henikoff (FHCRC, Seattle, US), A. Straight (Stanford, US) and V. Costanzo (IFOM, Italy) for sharing unpublished results or observations. We apologize to our colleagues whose important contributions could not be cited due to space constraints. D.F. receives salary support from the CNRS. D.F. has received support by Labex «CelTisPhyBio», the Institut Curie and the ATIP-Avenir 2015 program. This work has also received support under the program "Investissements d'Avenir," launched by the French Government and implemented by ANR with the references ANR-10-LABX-0038 and ANR-10-IDEX-0001-02 PSL.

References

Aguilar-Arnal L, Marsellach F-X, Azorín F (2008) The fission yeast homologue of CENP-B, Abp1, regulates directionality of mating-type switching. EMBO J 27:1029–1038. doi:10.1038/emboj.2008.53

Aldrup-MacDonald ME, Kuo ME, Sullivan LL et al (2016) Genomic variation within alpha satellite DNA influences centromere location on human chromosomes with metastable epialleles. Genome Res 26:1301–1311. doi:10.1101/gr.206706.116

Alexandrov I, Kazakov A, Tumeneva I et al (2001) Alpha-satellite DNA of primates: old and new families. Chromosoma 110:253–266. doi:10.1007/s004120100146

Alkan C, Cardone MF, Catacchio CR et al (2011) Genome-wide characterization of centromeric satellites from multiple mammalian genomes. Genome Res 21:137–145. doi:10.1101/gr.111278.110

Allshire RC, Karpen GH (2008) Epigenetic regulation of centromeric chromatin: old dogs, new tricks? Nat Rev Genet 9:923–937. doi:10.1038/nrg2466

Almouzni G, Probst AV (2011) Heterochromatin maintenance and establishment: lessons from the mouse pericentromere. Nucleus 2:332–338. doi:10.4161/nucl.2.5.17707

Alonso A, Fritz B, Hasson D et al (2007) Co-localization of CENP-C and CENP-H to discontinuous domains of CENP-A chromatin at human neocentromeres. Genome Biol 8: R148. doi:10.1186/gb-2007-8-7-r148

Alonso A, Hasson D, Cheung F, Warburton PE (2010) A paucity of heterochromatin at functional human neocentromeres. Epigenetics Chromatin 3:6. doi:10.1186/1756-8935-3-6

Amor DJ, Bentley K, Ryan J et al (2004) Human centromere repositioning in progress. Proc Natl Acad Sci USA 101:6542–6547. doi:10.1073/pnas.0308637101

Ando S, Yang H, Nozaki N et al (2002) CENP-A, -B, and -C chromatin complex that contains the I-type alpha-satellite array constitutes the prekinetochore in HeLa cells. Mol Cell Biol 22:2229–2241. doi:10.1128/MCB.22.7.-2229-2241.2002

Andrews AJ, Chen X, Zevin A et al (2010) The histone chaperone Nap1 promotes nucleosome assembly by eliminating nonnucleosomal histone DNA interactions. Mol Cell 37:834–842. doi:10.1016/j.molcel.2010.01.037

Aze A, Sannino V, Soffientini P et al (2016) Centromeric DNA replication reconstitution reveals DNA loops and ATR checkpoint suppression. Nat Cell Biol 18:684–691. doi:10.1038/ncb3344

Bachman KE, Rountree MR, Baylin SB (2001) Dnmt3a and Dnmt3b are transcriptional repressors that exhibit unique localization properties to heterochromatin. J Biol Chem 276:32282–32287. doi:10.1074/jbc.M104661200

Bailis JM, Bernard P, Antonelli R et al (2003) Hsk1-Dfp1 is required for heterochromatin-mediated cohesion at centromeres. Nat Cell Biol 5:1111–1116. doi:10.1038/ncb1069

Barbosa-Cisneros O, Herrera-Esparza R (2002) CENP-B is a conserved gene among vegetal species. Genet Mol Res 1:241–245

Bassett EA, Wood S, Salimian KJ et al (2010) Epigenetic centromere specification directs aurora B accumulation but is insufficient to efficiently correct mitotic errors. J Cell Biol 190:177–185. doi:10.1083/jcb.201001035

Baum M, Clarke L (2000) Fission yeast homologs of human CENP-B have redundant functions affecting cell growth and chromosome segregation. Mol Cell Biol 20:2852–2864

Baumann C, Körner R, Hofmann K, Nigg EA (2007) PICH, a centromere-associated SNF2 family ATPase, is regulated by Plk1 and required for the spindle checkpoint. Cell 128:101–114. doi:10.1016/j.cell.2006.11.041

Bergmann JH, Rodríguez MG, Martins NMC et al (2011) Epigenetic engineering shows H3K4me2 is required for HJURP targeting and CENP-A assembly on a synthetic human kinetochore. EMBO J 30:328–340. doi:10.1038/emboj.2010.329

Bernard P, Maure JF, Partridge JF et al (2001) Requirement of heterochromatin for cohesion at centromeres. Science 294:2539–2542. doi:10.1126/science.1064027

Bierhoff H, Postepska-Igielska A, Grummt I (2014) Noisy silence: non-coding RNA and heterochromatin formation at repetitive elements. Epigenetics 9:53–61. doi:10.4161/epi.26485

Biscotti MA, Canapa A, Forconi M et al (2015) Transcription of tandemly repetitive DNA: functional roles. Chromosome Res 23:463–477. doi:10.1007/s10577-015-9494-4

Black BE, Cleveland DW (2011) Epigenetic centromere propagation and the nature of CENP-A nucleosomes. Cell 144:471–479. doi:10.1016/j.cell.2011.02.002

Blower MD, Sullivan BA, Karpen GH (2002) Conserved organization of centromeric chromatin in flies and humans. Dev Cell 2:319–330

Bodor DL, Mata JF, Sergeev M et al (2014) The quantitative architecture of centromeric chromatin. eLife 3:819–826. doi:10.7554/eLife.02137

Brenner S, Pepper D, Berns MW et al (1981) Kinetochore structure, duplication, and distribution in mammalian cells: analysis by human autoantibodies from scleroderma patients. J Cell Biol 91:95–102

Buchwitz BJ, Ahmad K, Moore LL et al (1999) A histone-H3-like protein in C. elegans. Nature 401:547–548. doi:10.1038/44062

Cam HP, Noma K-I, Ebina H et al (2008) Host genome surveillance for retrotransposons by transposon-derived proteins. Nature 451:431–436. doi:10.1038/nature06499

Carroll CW, Milks KJ, Straight AF (2010) Dual recognition of CENP-A nucleosomes is required for centromere assembly. J Cell Biol 189:1143–1155. doi:10.1083/jcb.201001013

Casola C, Hucks D, Feschotte C (2008) Convergent domestication of pogo-like transposases into centromere-binding proteins in fission yeast and mammals. Mol Biol Evol 25:29–41. doi:10.1093/molbev/msm221

Chan KL, North PS, Hickson ID (2007) BLM is required for faithful chromosome segregation and its localization defines a class of ultrafine anaphase bridges. EMBO J 26:3397–3409. doi:10.1038/sj.emboj.7601777

Chan FL, Marshall OJ, Saffery R et al (2012) Active transcription and essential role of RNA polymerase II at the centromere during mitosis. Proc Natl Acad Sci USA 109:1979–1984. doi:10.1073/pnas.1108705109

Chen T, Tsujimoto N, Li E (2004) The PWWP domain of Dnmt3a and Dnmt3b is required for directing DNA methylation to the major satellite repeats at pericentric heterochromatin. Mol Cell Biol 24:9048–9058. doi:10.1128/MCB.24.20.9048-9058.2004

Chmátal L, Gabriel SI, Mitsainas GP et al (2014) Centromere strength provides the cell biological basis for meiotic drive and karyotype evolution in mice. Curr Biol 24:2295–2300. doi:10.1016/j.cub.2014.08.017

Choo KH, Vissel B, Nagy A et al (1991) A survey of the genomic distribution of alpha satellite DNA on all the human chromosomes, and derivation of a new consensus sequence. Nucleic Acid Res 19:1179–1182

Clarke L, Carbon J (1980) Isolation of a yeast centromere and construction of functional small circular chromosomes. Nature 287:504–509

Cleveland DW, Mao Y, Sullivan KF (2003) Centromeres and kinetochores: From epigenetics to mitotic checkpoint signaling. Cell 112:407–421

Cooke CA, Bernat RL, Earnshaw WC (1990) CENP-B: a major human centromere protein located beneath the kinetochore. J Cell Biol 110:1475–1488

Cumberledge S, Carbon J (1987) Mutational analysis of meiotic and mitotic centromere function in *Saccharomyces cerevisiae*. Genetics 117:203–212

Dai X, Otake K, You C et al (2013) Identification of novel α-n-methylation of CENP-B that regulates its binding to the centromeric DNA. J Proteome Res 12:4167–4175. doi:10.1021/pr400498y

d'Alençon E, Nègre N, Stanojcic S et al (2011) Characterization of a CENP-B homolog in the holocentric *Lepidoptera Spodoptera frugiperda*. Gene 485:91–101. doi:10.1016/j.gene.2011.06.007

Depinet TW, Zackowski JL, Earnshaw WC et al (1997) Characterization of neo-centromeres in marker chromosomes lacking detectable alpha-satellite DNA. Hum Mol Genet 6:1195–1204

Drinnenberg IA, Henikoff S, Malik HS (2016) Evolutionary turnover of kinetochore proteins: a ship of theseus? Trends Cell Biol 26:498–510. doi:10.1016/j.tcb.2016.01.005

Dumanski JP, Rasi C, Lönn M et al (2015) Mutagenesis. Smoking is associated with mosaic loss of chromosome Y. Science 347:81–83. doi:10.1126/science.1262092

Earnshaw WC, Rothfield N (1985) Identification of a family of human centromere proteins using autoimmune sera from patients with scleroderma. Chromosoma 91:313–321

Earnshaw WC, Sullivan KF, Machlin PS et al (1987) Molecular cloning of cDNA for CENP-B, the major human centromere autoantigen. J Cell Biol 104:817–829

Edwards NS, Murray AW (2005) Identification of xenopus CENP-A and an associated centromeric DNA repeat. Mol Biol Cell 16:1800–1810. doi:10.1091/mbc.E04-09-0788

Ehrlich M, Sanchez C, Shao C et al (2008) ICF, an immunodeficiency syndrome: DNA methyltransferase 3B involvement, chromosome anomalies, and gene dysregulation. Autoimmunity 41:253–271. doi:10.1080/08916930802024202

Epifanova OI (1958) The mitotic behavior of the epithelium of the uterus at different stages of the sexual cycle in mice. Bull Exp Bio Med 46:1399–1402. doi:10.1007/BF00845190

Erliandri I, Fu H, Nakano M et al (2014) Replication of alpha-satellite DNA arrays in endogenous human centromeric regions and in human artificial chromosome. Nucleic Acids Res 42:11502–11516. doi:10.1093/nar/gku835

Fachinetti D, Folco HD, Nechemia-Arbely Y et al (2013) A two-step mechanism for epigenetic specification of centromere identity and function. Nat Cell Biol 15:1–13. doi:10.1038/ncb2805

Fachinetti D, Han JS, McMahon MA et al (2015) DNA sequence-specific binding of CENP-B enhances the fidelity of human centromere function. Dev Cell 33:314–327. doi:10.1016/j.devcel.2015.03.020

Folco HD, Pidoux AL, Urano T, Allshire RC (2008) Heterochromatin and RNAi are required to establish CENP-A chromatin at centromeres. Science 319:94–97. doi:10.1126/science.1150944

Foltz DR, Jansen LET, Black BE et al (2006) The human CENP-A centromeric nucleosome-associated complex. Nat Cell Biol 8:458–469. doi:10.1038/ncb1397

Forsberg LA, Rasi C, Malmqvist N et al (2014) Mosaic loss of chromosome Y in peripheral blood is associated with shorter survival and higher risk of cancer. 46:624–628. doi:10.1038/ng.2966 (Nature Publishing Group)

Fowler KJ, Hudson DF, Salamonsen LA et al (2000) Uterine dysfunction and genetic modifiers in centromere protein B-deficient mice. Genome Res 10:30–41. doi:10.1101/gr.10.1.30

Fowler KJ, Wong LH, Griffiths BK et al (2004) Centromere protein b-null mice display decreasing reproductive performance through successive generations of breeding due to diminishing endometrial glands. Reproduction 127:367–377. doi:10.1530/rep.1.00102

Fujita Y, Hayashi T, Kiyomitsu T et al (2007) Priming of centromere for CENP-A recruitment by human hMis18α, hMis18β, and M18BP1. Dev Cell 12:17–30. doi:10.1016/j.devcel.2006.11.002

Fujita R, Otake K, Arimura Y et al (2015) Stable complex formation of CENP-B with the CENP-A nucleosome. Nucleic Acids Res 43:4909–4922. doi:10.1093/nar/gkv405

Fuks F (2005) DNA methylation and histone modifications: teaming up to silence genes. Curr Opin Genet Dev 15:490–495. doi:10.1016/j.gde.2005.08.002
Garavís M, Méndez-Lago M, Gabelica V et al (2015) The structure of an endogenous Drosophila centromere reveals the prevalence of tandemly repeated sequences able to form i-motifs. 5:13307. doi:10.1038/srep13307 (Nature Publishing Group)
Goldberg IG, Sawhney H, Pluta AF et al (1996) Surprising deficiency of CENP-B binding sites in African green monkey alpha-satellite DNA: implications for CENP-B function at centromeres. Mol Cell Biol 16:5156–5168
Gong Z, Wu Y, Koblízková A et al (2012) Repeatless and repeat-based centromeres in potato: implications for centromere evolution. Plant Cell 24:3559–3574. doi:10.1105/tpc.112.100511
Gopalakrishnan S, Sullivan BA, Trazzi S et al (2009) DNMT3B interacts with constitutive centromere protein CENP-C to modulate DNA methylation and the histone code at centromeric regions. Hum Mol Genet 18:3178–3193. doi:10.1093/hmg/ddp256
Grady DL, Ratliff RL, Robinson DL et al (1992) Highly conserved repetitive DNA sequences are present at human centromeres. Proc Natl Acad Sci USA 89:1695–1699
Haaf T, Mater AG, Wienberg J, Ward DC (1995) Presence and abundance of CENP-B box sequences in great ape subsets of primate-specific alpha-satellite DNA. J Mol Evol 41:487–491
Halverson D, Baum M, Stryker J et al (1997) A centromere DNA-binding protein from fission yeast affects chromosome segregation and has homology to human CENP-B. J Cell Biol 136:487–500
Harrington JJ, Van Bokkelen G, Mays RW et al (1997) Formation of de novo centromeres and construction of first-generation human artificial microchromosomes. Nat Genet 15:345–355. doi:10.1038/ng0497-345
Hasson D, Panchenko T, Salimian KJ et al (2013) The octamer is the major form of CENP-A nucleosomes at human centromeres. Nat Struct Mol Biol 20:687–695. doi:10.1038/nsmb.2562
Hayden KE, Strome ED, Merrett SL et al (2013) Sequences associated with centromere competency in the human genome. Mol Cell Biol 33:763–772. doi:10.1128/MCB.01198-12
Hemmerich P, Weidtkamp-Peters S, Hoischen C et al (2008) Dynamics of inner kinetochore assembly and maintenance in living cells. J Cell Biol 180:1101–1114. doi:10.1083/jcb.200710052
Henikoff S, Ahmad K, Malik HS (2001) The centromere paradox: stable inheritance with rapidly evolving DNA. Science 293:1098–1102. doi:10.1126/science.1062939
Henikoff JG, Thakur J, Kasinathan S, Henikoff S (2015) A unique chromatin complex occupies young α-satellite arrays of human centromeres. Sci Adv 1:e1400234. doi:10.1126/sciadv.1400234
Hoffmann S, Dumont M, Barra V et al (2016) CENP-A is dispensable for mitotic centromere function after initial centromere/kinetochore assembly. Cell Rep. doi:10.1016/j.celrep.2016
Hori T, Shang W-H, Takeuchi K, Fukagawa T (2013) The CCAN recruits CENP-A to the centromere and forms the structural core for kinetochore assembly. J Cell Biol 200:45–60. doi:10.1083/jcb.201210106
Howman EV, Fowler KJ, Newson AJ et al (2000) Early disruption of centromeric chromatin organization in centromere protein A (Cenpa) null mice. Proc Natl Acad Sci USA 97:1148–1153
Hudson DF, Fowler KJ, Earle E et al (1998) Centromere protein B null mice are mitotically and meiotically normal but have lower body and testis weights. J Cell Biol 141:309–319
Ideue T, Cho Y, Nishimura K, Tani T (2014) Involvement of satellite I noncoding RNA in regulation of chromosome segregation. Genes Cells 19:528–538. doi:10.1111/gtc.12149
Ikeno M, Masumoto H, Okazaki T (1994) Distribution of CENP-B boxes reflected in CREST centromere antigenic sites on long-range α-satellite DNA arrays of human chromosome 21. Hum Mol Genet 3:1245–1257. doi:10.1093/hmg/3.8.1245
Ikeno M, Grimes B, Okazaki T et al (1998) Construction of YAC-based mammalian artificial chromosomes. Nat Biotechnol 16:431–439. doi:10.1038/nbt0598-431
Irelan JT, Gutkin GI, Clarke L (2001) Functional redundancies, distinct localizations and interactions among three fission yeast homologs of centromere protein-B. Genetics 157:1191–1203

Jaco I, Canela A, Vera E, Blasco MA (2008) Centromere mitotic recombination in mammalian cells. J Cell Biol 181:885–892. doi:10.1083/jcb.200803042

Jeganathan K, Malureanu L, Baker DJ et al (2007) Bub1 mediates cell death in response to chromosome missegregation and acts to suppress spontaneous tumorigenesis. J Cell Biol 179:255–267. doi:10.1083/jcb.200706015

Kapoor M, de Oca Montes, Luna R, Liu G et al (1998) The cenpB gene is not essential in mice. Chromosoma 107:570–576

Ketel C, Wang HSW, McClellan M et al (2009) Neocentromeres form efficiently at multiple possible loci in Candida albicans. PLoS Genet 5:e1000400. doi:10.1371/journal.pgen.1000400

Kim IS, Lee M, Park KC et al (2012) Roles of Mis18α in epigenetic regulation of centromeric chromatin and CENP-A loading. Mol Cell 46:260–273. doi:10.1016/j.molcel.2012.03.021

Kipling D, Warburton PE (1997) Centromeres, CENP-B and tigger too. Trends Genet 13:141–145

Koga A, Hirai Y, Terada S et al (2014) Evolutionary origin of higher-order repeat structure in alpha-satellite DNA of primate centromeres. DNA Res 21:407–415. doi:10.1093/dnares/dsu005

Krings G, Bastia D (2005) Sap1p binds to Ter1 at the ribosomal DNA of *Schizosaccharomyces pombe* and causes polar replication fork arrest. J Biol Chem 280:39135–39142. doi:10.1074/jbc.M508996200

Kugou K, Hirai H, Masumoto H, Koga A (2016) Formation of functional CENP-B boxes at diverse locations in repeat units of centromeric DNA in new world monkeys. 6:27833. doi:10.1038/srep27833 (Nature Publishing Group)

Lee JK, Huberman JA, Hurwitz J (1997) Purification and characterization of a CENP-B homologue protein that binds to the centromeric K-type repeat DNA of *Schizosaccharomyces pombe*. Proc Natl Acad Sci USA 94:8427–8432

Lo AW, Liao GC, Rocchi M, Choo KH (1999) Extreme reduction of chromosome-specific alpha-satellite array is unusually common in human chromosome 21. Genome Res 9:895–908

Locke DP, Hillier LW, Warren WC et al (2011) Comparative and demographic analysis of orang-utan genomes. Nature 469:529–533. doi:10.1038/nature09687

Locovei AM, Spiga M-G, Tanaka K et al (2006) The CENP-B homolog, Abp1, interacts with the initiation protein Cdc23 (MCM10) and is required for efficient DNA replication in fission yeast. Cell Div 1:27. doi:10.1186/1747-1028-1-27

López CC, Edström JE (1998) Interspersed centromeric element with a CENP-B box-like motif in Chironomus pallidivittatus. Nucleic Acids Res 26:4168–4172

Lorite P, Carrillo JA, Tinaut A, Palomeque T (2004) Evolutionary dynamics of satellite DNA in species of the Genus Formica (Hymenoptera, Formicidae). Gene 332:159–168. doi:10.1016/j.gene.2004.02.049

Ly P, Teitz LS, Kim DH et al (2016) Selective Y centromere inactivation triggers chromosome shattering in micronuclei and repair by canonical NHEJ. Nat Cell Biol. doi:10.1038/ncb3450

Malik HS, Henikoff S (2001) Adaptive evolution of Cid, a centromere-specific histone in Drosophila. Genetics 157:1293–1298

Maloney KA, Sullivan LL, Matheny JE et al (2012) Functional epialleles at an endogenous human centromere. Proc Natl Acad Sci USA 109:13704–13709. doi:10.1073/pnas.1203126109

Manuelidis L (1978) Chromosomal localization of complex and simple repeated human DNAs. Chromosoma 66:23–32

Marshall OJ, Choo KHA (2012) Putative CENP-B paralogues are not present at mammalian centromeres. Chromosoma 121:169–179. doi:10.1007/s00412-011-0348-3

Marshall OJ, Chueh AC, Wong LH, Choo KHA (2008) Neocentromeres: new insights into centromere structure, disease development, and karyotype evolution. Am J Hum Genet 82:261–282. doi:10.1016/j.ajhg.2007.11.009

Masumoto H, Masukata H, Muro Y et al (1989) A human centromere antigen (CENP-B) interacts with a short specific sequence in alphoid DNA, a human centromeric satellite. J Cell Biol 109:1963–1973. doi:10.1083/jcb.109.5.1963

Masumoto H, Yoda K, Ikeno M et al (1993) Properties of CENP-B and its target sequence in a satellite DNA. Chromosome segregation and aneuploidy. (NATO ASI series). Springer, Berlin, pp 31–43. doi:10.1007/978-3-642-84938-1_3

Masumoto H, Ikeno M, Nakano M et al (1998) Assay of centromere function using a human artificial chromosome. Chromosoma 107:406–416

Mateo L, González J (2014) Pogo-like transposases have been repeatedly domesticated into CENP-B-related proteins. Genome Biol Evol 6:2008–2016. doi:10.1093/gbe/evu153

McFarlane RJ, Humphrey TC (2010) A role for recombination in centromere function. Trends Genet 26:209–213. doi:10.1016/j.tig.2010.02.005

McGrew J, Diehl B, Fitzgerald-Hayes M (1986) Single base-pair mutations in centromere element III cause aberrant chromosome segregation in *Saccharomyces cerevisiae*. Mol Cell Biol 6:530–538. doi:10.1128/MCB.6.2.530

Mejía JE, Alazami A, Willmott A et al (2002) Efficiency of de novo centromere formation in human artificial chromosomes. Genomics 79:297–304. doi:10.1006/geno.2002.6704

Melters DP, Bradnam KR, Young HA et al (2013) Comparative analysis of tandem repeats from hundreds of species reveals unique insights into centromere evolution. Genome Biol 14:R10. doi:10.1186/gb-2013-14-1-r10

Mendiburo MJ, Padeken J, Fulop S et al (2011) Drosophila CENH3 is sufficient for centromere formation. Science 334:686–690. doi:10.1126/science.1206880

Meštrović N, Pavlek M, Car A et al (2013) Conserved DNA Motifs, Including the CENP-B Box-like, are possible promoters of satellite DNA array rearrangements in nematodes. PLoS ONE 8:e67328. doi:10.1371/journal.pone.0067328

Michel LS, Liberal V, Chatterjee A et al (2001) MAD2 haplo-insufficiency causes premature anaphase and chromosome instability in mammalian cells. Nature 409:355–359. doi:10.1038/35053094

Miga KH, Newton Y, Jain M et al (2014) Centromere reference models for human chromosomes X and Y satellite arrays. Genome Res 24:697–707. doi:10.1101/gr.159624.113

Mitchell AR, Jeppesen P, Nicol L et al (1996) Epigenetic control of mammalian centromere protein binding: does DNA methylation have a role? J Cell Sci 109(Pt 9):2199–2206

Mohibi S, Srivastava S, Wang-France J et al (2015) Alteration/deficiency in activation 3 (ADA3) protein, a cell cycle regulator, associates with the centromere through CENP-B and regulates chromosome segregation. J Biol Chem 290:28299–28310. doi:10.1074/jbc.M115.685511

Moree B, Meyer CB, Fuller CJ, Straight AF (2011) CENP-C recruits M18BP1 to centromeres to promote CENP-A chromatin assembly. J Cell Biol 194:855–871. doi:10.1083/jcb.201106079

Mravinac B, Ugarković E, Franjević D, Plohl M (2005) Long inversely oriented subunits form a complex monomer of Tribolium brevicornis satellite DNA. J Mol Evol 60:513–525. doi:10.1007/s00239-004-0236-z

Murakami Y, Huberman JA, Hurwitz J (1996) Identification, purification, and molecular cloning of autonomously replicating sequence-binding protein 1 from fission yeast *Schizosaccharomyces pombe*. Proc Natl Acad Sci USA 93:502–507

Muro Y, Masumoto H, Yoda K et al (1992) Centromere protein B assembles human centromeric alpha-satellite DNA at the 17-bp sequence, CENP-B box. J Cell Biol 116:585–596

Murray AW, Szostak JW (1983) Construction of artificial chromosomes in yeast. Nature 305:189–193

Murray AW, Szostak JW (1986) Construction and behavior of circularly permuted and telocentric chromosomes in *Saccharomyces cerevisiae*. Mol Cell Biol 6:3166–3172

Nakagawa H, Lee J-K, Hurwitz J et al (2002) Fission yeast CENP-B homologs nucleate centromeric heterochromatin by promoting heterochromatin-specific histone tail modifications. Genes Dev 16:1766–1778. doi:10.1101/gad.997702

Ng R, Carbon J (1987) Mutational and in vitro protein-binding studies on centromere DNA from Saccharomyces cerevisiae. Mol Cell Biol 7:4522–4534. doi:10.1128/MCB.7.12.4522

Nishino T, Takeuchi K, Gascoigne KE et al (2012) CENP-T-W-S-X forms a unique centromeric chromatin structure with a histone-like fold. Cell 148:487–501. doi:10.1016/j.cell.2011.11.061

Noma K-I, Cam HP, Maraia RJ, Grewal SIS (2006) A role for TFIIIC transcription factor complex in genome organization. Cell 125:859–872. doi:10.1016/j.cell.2006.04.028
Noveski P, Madjunkova S, Sukarova Stefanovska E et al (2016) Loss of Y chromosome in peripheral blood of colorectal and prostate cancer patients. PLOS One 11:e0146264. doi:10.1371/journal.pone.0146264
Ohzeki J-I, Nakano M, Okada T, Masumoto H (2002) CENP-B box is required for de novo centromere chromatin assembly on human alphoid DNA. J Cell Biol 159:765–775. doi:10.1083/jcb.200207112
Ohzeki J-I, Bergmann JH, Kouprina N et al (2012) Breaking the HAC Barrier: Histone H3K9 acetyl/methyl balance regulates CENP-A assembly. EMBO J 31:2391–2402. doi:10.1038/emboj.2012.82
Okada T, Ohzeki J-I, Nakano M et al (2007) CENP-B controls centromere formation depending on the chromatin context. Cell 131:1287–1300. doi:10.1016/j.cell.2007.10.045
Okamoto Y, Nakano M, Ohzeki J-I et al (2007) A minimal CENP-A core is required for nucleation and maintenance of a functional human centromere. EMBO J 26:1279–1291. doi:10.1038/sj.emboj.7601584
Okano M, Bell DW, Haber DA, Li E (1999) DNA methyltransferases Dnmt3a and Dnmt3b are essential for de novo methylation and mammalian development. Cell 99:247–257
Olszak AM, van Essen D, Pereira AJ et al (2011) Heterochromatin boundaries are hotspots for de novo kinetochore formation. Nat Cell Biol 13:799–808. doi:10.1038/ncb2272
Orpinell M, Fournier M, Riss A et al (2010) The ATAC acetyl transferase complex controls mitotic progression by targeting non-histone substrates. EMBO J 29:2381–2394. doi:10.1038/emboj.2010.125
Palmer DK, O'Day K, Wener MH et al (1987) A 17-kD centromere protein (CENP-A) copurifies with nucleosome core particles and with histones. J Cell Biol 104:805–815
Perez-Castro AV, Shamanski FL, Meneses JJ et al (1998) Centromeric protein B null mice are viable with no apparent abnormalities. Dev Biol 201:135–143. doi:10.1006/dbio.1998.9005
Plohl M, Meštrović N, Mravinac B (2014) Centromere identity from the DNA point of view. Chromosoma 123:313–325. doi:10.1007/s00412-014-0462-0
Politi V, Perini G, Trazzi S et al (2002) CENP-C binds the alpha-satellite DNA in vivo at specific centromere domains. J Cell Sci 115:2317–2327
Quénet D, Dalal Y (2014) A long non-coding RNA is required for targeting centromeric protein A to the human centromere. eLife 3:e03254. doi:10.7554/eLife.03254
Régnier V, Novelli J, Fukagawa T et al (2003) Characterization of chicken CENP-A and comparative sequence analysis of vertebrate centromere-specific histone H3-like proteins. Gene 316:39–46. doi:10.1016/S0378-1119(03)00768-6
Romanova LY, Deriagin GV, Mashkova TD et al (1996) Evidence for selection in evolution of alpha satellite DNA: the central role of CENP-B/pJ alpha binding region. J Mol Biol 261:334–340. doi:10.1006/jmbi.1996.0466
Rosandić M, Paar V, Basar I et al (2006) CENP-B box and pJalpha sequence distribution in human alpha satellite higher-order repeats (HOR). Chromosome Res 14:735–753. doi:10.1007/s10577-006-1078-x
Rosandić M, Glunčić M, Paar V, Basar I (2008) The role of alphoid higher order repeats (HORs) in the centromere folding. J Theor Biol 254:555–560. doi:10.1016/j.jtbi.2008.06.012
Rošić S, Erhardt S (2016) No longer a nuisance: long non-coding RNAs join CENP-A in epigenetic centromere regulation. Cell Mol Life Sci 73:1387–1398. doi:10.1007/s00018-015-2124-7
Rošić S, Köhler F, Erhardt S (2014) Repetitive centromeric satellite RNA is essential for kinetochore formation and cell division. J Cell Biol 207:335–349. doi:10.1083/jcb.201404097
Rudd MK, Willard HF (2004) Analysis of the centromeric regions of the human genome assembly. Trends Genet 20:529–533. doi:10.1016/j.tig.2004.08.008
Rudd MK, Wray GA, Willard HF (2006) The evolutionary dynamics of alpha-satellite. Genome Res 16:88–96. doi:10.1101/gr.3810906

Schueler MG, Higgins AW, Rudd MK et al (2001) Genomic and genetic definition of a functional human centromere. Science 294:109–115. doi:10.1126/science.1065042

Scott KC, Merrett SL, Willard HF (2006) A heterochromatin barrier partitions the fission yeast centromere into discrete chromatin domains. Curr Biol 16:119–129. doi:10.1016/j.cub.2005.11.065

Shang W-H, Hori T, Toyoda A et al (2010) Chickens possess centromeres with both extended tandem repeats and short non-tandem-repetitive sequences. Genome Res 20:1219–1228. doi:10.1101/gr.106245.110

Shang W-H, Hori T, Martins NMC et al (2013) Chromosome engineering allows the efficient isolation of vertebrate neocentromeres. Dev Cell 24:635–648. doi:10.1016/j.devcel.2013.02.009

Shepelev VA, Alexandrov AA, Yurov YB, Alexandrov IA (2009) The evolutionary origin of man can be traced in the layers of defunct ancestral alpha satellites flanking the active centromeres of human chromosomes. PLoS Genet 5:e1000641. doi:10.1371/journal.pgen.1000641

Shono N, Ohzeki J-I, Otake K et al (2015) CENP-C and CENP-I are key connecting factors for kinetochore and CENP-A assembly. J Cell Sci 128:4572–4587. doi:10.1242/jcs.180786

Smit AF, Riggs AD (1996) Tiggers and DNA transposon fossils in the human genome. Proc Natl Acad Sci USA 93:1443–1448

Smith KM, Galazka JM, Phatale PA et al (2012) Centromeres of filamentous fungi. Chromosome Res 20:635–656. doi:10.1007/s10577-012-9290-3

Stoler S, Keith KC, Curnick KE, Fitzgerald-Hayes M (1995) A mutation in CSE4, an essential gene encoding a novel chromatin-associated protein in yeast, causes chromosome nondisjunction and cell cycle arrest at mitosis. Genes Dev 9:573–586. doi:10.1101/gad.9.5.573

Sugimoto K, Yata H, Muro Y, Himeno M (1994) Human centromere protein C (CENP-C) is a DNA-binding protein which possesses a novel DNA-binding motif. J Biochem 116:877–881

Sugimoto K, Kuriyama K, Shibata A, Himeno M (1997) Characterization of internal DNA-binding and C-terminal dimerization domains of human centromere/kinetochore autoantigen CENP-C in vitro: role of DNA-binding and self-associating activities in kinetochore organization. Chromosome Res 5:132–141

Sullivan KF (1994) Human CENP-A contains a histone H3 related histone fold domain that is required for targeting to the centromere. J Cell Biol 127:581–592. doi:10.1083/jcb.127.3.581

Sullivan KF, Glass CA (1991) CENP-B is a highly conserved mammalian centromere protein with homology to the helix-loop-helix family of proteins. Chromosoma 100:360–370

Sullivan BA, Willard HF (1998) Stable dicentric X chromosomes with two functional centromeres. Nat Genet 20:227–228. doi:10.1038/3024

Suntronpong A, Kugou K, Masumoto H et al (2016) CENP-B box, a nucleotide motif involved in centromere formation, occurs in a New World monkey. Biol Lett 12:20150817. doi:10.1098/rsbl.2015.0817

Suzuki N, Nakano M, Nozaki N et al (2004) CENP-B interacts with CENP-C domains containing Mif2 regions responsible for centromere localization. J Biol Chem 279:5934–5946. doi:10.1074/jbc.M306477200

Tachiwana H, Miya Y, Shono N et al (2013) Nap1 regulates proper CENP-B binding to nucleosomes. Nucleic Acids Res 41:2869–2880. doi:10.1093/nar/gks1464

Takahashi K (2000) Requirement of Mis6 centromere connector for localizing a CENP-A-like protein in fission yeast. Science 288:2215–2219. doi:10.1126/science.288.5474.2215

Tanaka Y, Nureki O, Kurumizaka H et al (2001) Crystal structure of the CENP-B protein-DNA complex: the DNA-binding domains of CENP-B induce kinks in the CENP-B box DNA. EMBO J 20:6612–6618. doi:10.1093/emboj/20.23.6612

Tanaka Y, Kurumizaka H, Yokoyama S (2004) CpG methylation of the CENP-B box reduces human CENP-B binding. FEBS J 272:282–289. doi:10.1111/j.1432-1033.2004.04406.x

Tawaramoto MS, Park SY, Tanaka Y et al (2003) Crystal structure of the human centromere protein B (CENP-B) dimerization domain at 1.65-A resolution. J Biol Chem 278:51454–51461. doi:10.1074/jbc.M310388200

Taylor SS, Larin Z, Tyler-Smith C (1996) Analysis of extrachromosomal structures containing human centromeric alphoid satellite DNA sequences in mouse cells. Chromosoma 105:70–81

Thakur J, Henikoff S (2016) CENPT bridges adjacent CENPA nucleosomes on young human α-satellite dimers. Genome Res 26:1178–1187. doi:10.1101/gr.204784.116

Tsuduki T, Nakano M, Yasuoka N et al (2006) An artificially constructed de novo human chromosome behaves almost identically to its natural counterpart during metaphase and anaphase in living cells. Mol Cell Biol 26:7682–7695. doi:10.1128/MCB.00355-06

Ventura M, Antonacci F, Cardone MF et al (2007) Evolutionary formation of new centromeres in macaque. Science 316:243–246. doi:10.1126/science.1140615

Vissel B, Choo KH (1987) Human alpha satellite DNA–consensus sequence and conserved regions. Nucleic Acids Res 15:6751–6752

Voullaire LE, Slater HR, Petrovic V, Choo KH (1993) A functional marker centromere with no detectable alpha-satellite, satellite III, or CENP-B protein: activation of a latent centromere? Am J Hum Genet 52:1153–1163

Wade CM, Giulotto E, Sigurdsson S et al (2009) Genome sequence, comparative analysis, and population genetics of the domestic horse. Science 326:865–867. doi:10.1126/science.1178158

Warren WC, Jasinska AJ, García-Pérez R et al (2015) The genome of the vervet (*Chlorocebus aethiops sabaeus*). Genome Res 25:1921–1933. doi:10.1101/gr.192922.115

Waye JS, Willard HF (1986) Structure, organization, and sequence of alpha satellite DNA from human chromosome 17: evidence for evolution by unequal crossing-over and an ancestral pentamer repeat shared with the human X chromosome. Mol Cell Biol 6:3156–3165

Weaver BAA, Silk AD, Montagna C et al (2007) Aneuploidy acts both oncogenically and as a tumor suppressor. Cancer Cell 11:25–36. doi:10.1016/j.ccr.2006.12.003

Wevrick R, Willard HF (1991) Physical map of the centromeric region of human chromosome 7: relationship between two distinct alpha satellite arrays. Nucleic Acids Res 19:2295–2301

Wijmenga C, van den Heuvel LP, Strengman E et al (1998) Localization of the ICF syndrome to chromosome 20 by homozygosity mapping. Am J Hum Genet 63:803–809. doi:10.1086/302021

Willard HF (1985) Chromosome-specific organization of human alpha satellite DNA. Am J Hum Genet 37:524–532

Willard HF, Waye JS (1987) Chromosome-specific subsets of human alpha satellite DNA: analysis of sequence divergence within and between chromosomal subsets and evidence for an ancestral pentameric repeat. J Mol Evol 25:207–214

Woo SS, Jiang J, Gill BS et al (1994) Construction and characterization of a bacterial artificial chromosome library of Sorghum bicolor. Nucleic Acids Res 22:4922–4931

Wood L, Booth DG, Vargiu G et al (2016) Auxin/AID versus conventional knockouts: distinguishing the roles of CENP-T/W in mitotic kinetochore assembly and stability. Open Biol 6:150230. doi:10.1098/rsob.150230

Wu JC, Manuelidis L (1980) Sequence definition and organization of a human repeated DNA. J Mol Biol 142:363–386. doi:10.1016/0022-2836(80)90277-6

Xu GL, Bestor TH, Bourc'his D et al (1999) Chromosome instability and immunodeficiency syndrome caused by mutations in a DNA methyltransferase gene. Nature 402:187–191. doi:10.1038/46052

Yang CH, Tomkiel J, Saitoh H et al (1996) Identification of overlapping DNA-binding and centromere-targeting domains in the human kinetochore protein CENP-C. Mol Cell Biol 16:3576–3586

Yoda K, Kitagawa K, Masumoto H et al (1992) A human centromere protein, CENP-B, has a DNA binding domain containing four potential alpha helices at the NH2 terminus, which is separable from dimerizing activity. J Cell Biol 119:1413–1427

Yoda K, Nakamura T, Masumoto H et al (1996) Centromere protein B of African green monkey cells: gene structure, cellular expression, and centromeric localization. Mol Cell Biol 16:5169–5177

Yoda K, Ando S, Okuda A et al (1998) In vitro assembly of the CENP-B/alpha-satellite DNA/core histone complex: CENP-B causes nucleosome positioning. Genes Cells 3:533–548

Zaratiegui M, Vaughn MW, Irvine DV et al (2011) CENP-B preserves genome integrity at replication forks paused by retrotransposon LTR. Nature 469:112–115. doi:10.1038/nature09608

Zedek F, Bureš P (2016) CenH3 evolution reflects meiotic symmetry as predicted by the centromere drive model. 6:33308. doi:10.1038/srep33308 (Nature Publishing Group)

The Unique DNA Sequences Underlying Equine Centromeres

Elena Giulotto, Elena Raimondi and Kevin F. Sullivan

Abstract Centromeres are highly distinctive genetic loci whose function is specified largely by epigenetic mechanisms. Understanding the role of DNA sequences in centromere function has been a daunting task due to the highly repetitive nature of centromeres in animal chromosomes. The discovery of a centromere devoid of satellite DNA in the domestic horse consolidated observations on the epigenetic nature of centromere identity, showing that entirely natural chromosomes could function without satellite DNA cues. Horses belong to the genus *Equus* which exhibits a very high degree of evolutionary plasticity in centromere position and DNA sequence composition. Examination of horses has revealed that the position of the satellite-free centromere is variable among individuals. Analysis of centromere location and composition in other *Equus* species, including domestic donkey and zebras, confirms that the satellite-less configuration of centromeres is common in this group which has undergone particularly rapid karyotype evolution. These features have established the equids as a new mammalian system in which to investigate the molecular organization, dynamics and evolutionary behaviour of centromeres.

1 Introduction

Centromeres are the enigmatic loci that specify the site of kinetochore formation during mitosis and meiosis. Unravelling the structure and functional organization of the centromere has been a daunting task due to the surprising variety of DNA sequences and strategies involved in centromere formation, known as the centromere paradox (Henikoff et al. 2001). How does a rapidly evolving DNA

E. Giulotto (✉) · E. Raimondi
Dipartimento di Biologia e Biotecnologie, Università di Pavia,
Via Ferrata 1, 27100 Pavia, Italy
e-mail: elena.giulotto@unipv.it

K.F. Sullivan
National University of Ireland Galway, University Road, Galway, Ireland

B.E. Black (ed.), *Centromeres and Kinetochores*, Progress in Molecular and Subcellular Biology 56, DOI 10.1007/978-3-319-58592-5_14

compartment support the highly conserved mechanism of chromosome segregation? As it has become apparent that centromere determination depends largely on the epigenetic mechanism of CENP-A deposition, the question of how centromere chromatin is organized to deliver centromere function has sharpened (Karpen and Allshire 1997; Ekwall 2007; Panchenko and Black 2009). Understanding how DNA supports this mechanism is a central goal of current work.

Centromeric DNA in most metazoans is comprised of tandemly repeated arrays of satellite DNA, typified by alpha satellite DNA in primates (Willard 1991), generally flanked by other classes of repetitive DNAs and segmental duplications (She et al. 2004). With a repeat size of 171 bp, alpha satellite forms higher order repeats (HOR) of 2 to over 16 monomers which are themselves repeated in arrays ranging from ~200 kb to >5 Mb. These properties of alpha satellite domains are characteristic of metazoan centromere organization, although the sequences that comprise centromere repeats are highly variable in evolution (Alkan et al. 2011). One consequence of this molecular structure is that it has been exceedingly difficult to determine the sequence of genomic centromere domains, which are not fully assembled in current whole genome assemblies (Hayden et al. 2013). Without a primary sequence based map it is difficult to investigate the molecular organization of centromeric chromatin, which dictates the functional properties of the locus.

Unusual behaviour of metazoan centromeres with respect to the DNA they are associated with emerged from two lines of investigation. The first was the discovery of analphoid neocentromeres in human patients, centromeres in novel locations that lack alpha satellite DNA (Voullaire et al. 1993; Tyler-Smith et al. 1999; Marshall et al. 2008). Coupled with the observation that alpha satellite domains on certain stable dicentric chromosomes were inactive in binding centromere proteins and in mitosis (Earnshaw et al. 1989), neocentromeres showed that alpha satellite DNA is neither necessary nor sufficient for centromere function. Precise localization of a neocentromere derivative of human chromosome 10 allowed for identification of the DNA sequences supporting centromere function at this locus, revealing it to be identical with the normal sequence, completely devoid of alpha satellite DNA (Saffery et al. 2000; Barry et al. 2000). The observation that neocentromeres are heritable for multiple generations (Tyler-Smith et al. 1999; Amor et al. 2004) proved that these loci are fully functional centromeres.

The second set of observations that challenged thinking about stable, DNA sequence directed centromeres in vertebrate organisms are those identifying evolutionary centromere repositioning (Rocchi et al. 2012). Using comparative hybridization of ordered BAC clones to examine the evolutionary history of primate karyotypes it was shown that centromeres moved on chromosomes without change in marker order (Montefalcone et al. 1999). In other words, centromere movement occurred without any detectable chromosome rearrangement. These repositioned loci are known as evolutionarily new centromeres or ENCs. Centromere repositioning is a significant contributor to karyotype evolution. The macaque possesses nine ENCs (Ventura et al. 2007) while the human lineage possesses five ENCs (Stanyon et al. 2008). Despite the relative youth of these centromeres, these primate ENCs were all associated with alpha satellite DNA, suggesting that alpha satellite becomes associated with primate centromeres as a normal part of their development. The molecular mechanisms associated with this process are currently obscure, although detailed analysis of alpha satellites in human chromosomes indicates that recombination events repeatedly insert alpha satellite DNA in the core of the centromere over evolutionary time (Shepelev et al. 2009). Interestingly, on human chromosome 17 the centromeric function can be linked to two different alpha satellite families (Maloney et al. 2012), giving rise to functional alleles.

Taken together, these series of observations have documented a remarkable plasticity in centromere identity with respect to its chromosomal substrate. As the centromere has emerged as a locus whose identity is specified by epigenetic mechanisms, understanding the role of this unique mechanism in the evolutionary behaviour of chromosomes has become a key issue. As well, the satellite-free centromeres provide a unique sequence substrate amenable to molecular analysis, providing a view into the centromere that has been obscured by the difficulty to dissect satellite DNA based centromeres. In this respect, the horse and related species of the young genus *Equus* have emerged as an important mammalian system for investigation of the molecular architecture and evolution of the centromere.

2 Why the Genus *Equus*? from Horse Passion to Novel Scientific Findings

The reasons why a researcher works on a project can be very different and it is out of the scope of this book to discuss this matter in detail, however, we would like to briefly describe the history of our scientific interest in equid centromeres. Indeed, everything started from the passion for horses of one of us. Very little was known on the horse genome at that time and the initial decision to work on the genomics of these animals was dictated by the desire to interact with them not only in the leisure time but also at work. These studies brought to the characterization of different types of repetitive DNAs (Anglana et al. 1996a, b; Scocchi et al. 1999; Nergadze et al. 2010; Raimondi et al. 2011; Santagostino et al. 2015) and to the cloning of the major centromeric satellite DNA families. Surprisingly, in spite of several cloning attempts and FISH mapping experiments, we were unable to identify satellite repeats at the centromere of one chromosome pair, horse chromosome 11 (ECA11) (Wade et al. 2009). Even more surprising was the observation that in other species of the genus *Equus* satellite DNA seemed to be missing at several centromeres while present at a number of chromosome ends. These early observations were extended and included in more recent publications (Piras et al. 2009, 2010).

At that time, although the absence of satellite DNA had been already observed at some clinical human neocentromeres (Warburton et al. 2000; Amor and Choo 2002), the hypothesis that satellite-free centromeres might constitute a stable component of mammalian karyotypes seemed bizarre. This hypothesis was only supported by cytogenetic data: the possibility that short stretches of tandem repeats, undetectable by FISH, might be present at the ECA11 centromere could not be excluded. Sequencing data were needed but the horse genome was far from being assembled. At the same time, studies at a cytogenetic level had revealed the phenomenon of evolutionary centromere repositioning (Montefalcone et al. 1999). A large body of evidence suggested that the genus *Equus* evolved very rapidly and we wondered whether, in these fast evolving karyotypes, the peculiar localization of satellite DNA might be related to centromere repositioning events that occurred in recent evolutionary times. We discovered that centromere repositioning had been exceptionally frequent during the evolution of this genus (Carbone et al. 2006; Piras et al. 2010) and one of the repositioning events involved exactly ECA11, thus supporting our prediction. When the international collaboration, Horse Genome Project (http://www.uky.edu/Ag/Horsemap/), started to sequence the horse genome, we were able to take advantage of the assembly being produced, (Wade et al. 2009), to prove that the centromere of this chromosome was completely devoid of extended arrays of tandem repeats demonstrating that a normal mammalian centromere can exist without satellite DNA. By extending this analysis to other individuals, we demonstrated that the centromeric function is not coupled to a specific sequence, but can slide within a relatively wide region (Purgato et al. 2015) giving raise to "epialleles". We also showed that, in other equid species, a number of centromeres are devoid of satellite DNA (Piras et al. 2009, 2010 and unpublished results). The satellite-less equid centromeres represent a new and powerful model system: they are present in all individuals of a given species and can therefore be used as an ideal tool to study the normal mammalian centromere. The non-repetitive nature of these centromeres is allowing us to study the mechanisms of centromere movement during evolution and to dissect, at the molecular level, the factors determining the epigenetic architecture and plasticity of centromeric chromatin.

3 Rapid Karyotype Evolution in the Genus *Equus*

The order *Perissodactyla* includes three extant families: *Tapiridae, Rhinocerotidae* and *Equidae*. Paleontological and molecular evidences suggest that the extant perissodactyl suborders diverged about 56–54 million years ago (Springer et al. 2003). *Tapiridae* and *Rhinocerotidae* belong to the suborder *Ceratomorpha*, while the family *Equidae* belongs to the *Hippomorpha* suborder. The only extant species of the *Equidae* family belong to the genus *Equus*.

Phylogenetic analyses, based on interspecific chromosome painting, allowed the reconstruction of the hypothetic perissodactyl ancestral karyotype, which comprises 74–78 chromosomes, the ambiguity in chromosome number being due to the polymorphic state of some perissodactyl chromosomes or to breakpoint reuse and fusion/fission events (Yang et al. 2003). Following the radiation from the common ancestor, the karyotypes of *Ceratomorpha* remained quite stable and similar to what can be observed in the hypothetical ancestral karyotype, living species showing a prevalence of acrocentric chromosomes. On the contrary, *Equus* karyotypes underwent an evolutionary acceleration after the divergence from the common ancestor.

The karyotypes of the living *Equus* species are characterized by the presence of a variable number of meta- and submetacentric chromosomes derived from fusions between ancestral acrocentric elements (Trifonov et al. 2008). Indeed, equids are considered a representative example of quickly radiating organisms; the eight living species of the genus *Equus* comprise: two horses (*E. caballus* and *E. przewalskii*), two Asiatic asses (*E. kiang* and *E. hemionus*), one African ass (*E. asinus*) and three zebras (*E. grevyi, E. burchelli* and *E. zebra*) (Steiner et al. 2012). An analysis based on the evolution of globin genes suggested that these species shared a common ancestor about 2–3 million years ago and the extant species emerged about 1 million years ago, that is in a very short evolutionary time (Oakenfull and Clegg 1998); recently, genome sequence data from DNA of living and fossil samples suggested that the divergence of the *Equus* lineages from the common ancestor occurred 4.0–4.5 million years ago (Orlando et al. 2013). This recent divergence explains the high degree of morphological similarity and capacity to interbreed. Strikingly, equid karyotypes differ extensively (Ryder et al. 1978; Yang et al. 2003). The variation involves both the structure and the number of chromosomes, which ranges from 32 in *E. zebra* to 66 in *E. przewalskii*. Cross-species chromosome painting confirmed the great karyotype variability of this genus (Trifonov et al. 2008). Data about the chromosomal architecture in different equid karyotypes indicate that the rate of chromosome changes ranges from 2.92 to 22.2 rearrangements per million years, a nearly 80-fold increase compared to the ancient *Ceratomorpha*, making the equid karyotype evolution one of the most rapid observed among mammals (Musilova et al. 2007, 2013; Trifonov et al. 2008).

A further evidence of the evolutionary plasticity of equid karyotype emerged from our molecular cytogenetic analysis of centromere repositioning (Carbone et al. 2006). We investigated the position of the centromere, with respect to flanking markers, in the horse, in the donkey, and in the Burchelli's zebra. The results of these early studies showed that at least eight centromere repositioning events occurred in the genus *Equus*. Surprisingly, at least five of these events arose in the donkey after its divergence from the zebra, which took place approximately 1 million years ago (Carbone et al. 2006). It must be noted that the number of ENCs detected in the paper by Carbone and co-workers was underestimated because cell lines from only three species were investigated and appropriate probes for some of the chromosomes were missing at that time. In order to identify other possible ENCs, we subsequently investigated the evolutionary history of horse chromosome 5q in seven species belonging to the genus *Equus* (*E. caballus, E. asinus, E. burchelli, E. grevyi, E. z. hartmannae, E. h. onager and E. kiang*) (Piras et al. 2009). We could identify two further centromere repositioning events involving donkey chromosome 16 and Burkelli's zebra chromosome 17, respectively. We concluded that the phenomenon of centromere repositioning played a key role in the rapid karyotype evolution of the equids, pointing to these species as a model system for the analysis of neocentromere formation and of centromere evolution.

4 Satellite DNA and Centromere Function are Uncoupled in the Genus *Equus*

As mentioned above, while analysing the distribution of highly repetitive DNA sequences in the genome of four equid species (horse—*E. caballus*, donkey—*E. asinus*, Grevyi's zebra—*E. grevyi,* and Burchelli's zebra—*E. burchelli*), we were intrigued by the peculiarity of the picture that emerged (Fig. 1). It appeared at a glance that one centromere in the horse, and a number of centromeres in the donkey and in the two zebras were devoid of repetitive DNA. On the other hand, in some chromosomes missing repetitive DNA sequences at the centromere, these were unexpectedly present at one or both telomeric termini. Details of this analysis are shown in Fig. 1. The arrows in the figure mark the chromosomes whose centromeres appear satellite-free at the FISH resolution level, red arrows indicating those chromosomes missing satellite DNA at the centromere, but having repetitive DNA sequences located at one or both telomeres. Since our FISH data strongly suggested that, in equid species, centromere function is uncoupled from satellite DNA, we performed immuno-FISH experiments on metaphase chromosomes of the four species using satellite DNA as FISH probe and an anti-CENP-A antibody to unequivocally localize centromere function. In all the species, it was definitely demonstrated that the functional centromere coincides with the primary constriction in satellite-containing as well as in satellite-free centromeres (Wade et al. 2009; Piras et al. 2010 and unpublished results).

The absence of satellite repeats at some centromeres and their presence at terminal positions are in agreement with our previous observation that several centromere repositioning events occurred during the evolution of the *Equidae* (Carbone et al. 2006; Piras et al. 2009); in this scenario, these evolutionarily recent events would have generated new centromeres that, at present, are still "immature" and have not yet acquired the satellite sequence complexity typical of the vertebrate centromeres described to date. Conversely, the presence of satellite DNA at terminal positions in meta- and submetacentric chromosomes, may be interpreted as the trace, left over by centromere repositioning events, of ancient, now inactive, terminal centromeres. In fact, comparative analyses performed using painting probes suggested that the ancestral *Perissodactyla* karyotype was probably composed of acrocentric chromosomes (Trifonov et al. 2008).

According to our experimental results, we proposed the model presented in Fig. 2, where a hypothetical ancestral acrocentric chromosome (Fig. 2a) gives rise to ENCs. The *Equus* ENCs are in a still "immature" stage (Fig. 2b or c), while the previously described ENCs of other orders have acquired satellite DNA reaching "maturity" (Fig. 2d). The persistence of satellite DNA at some inactivated centromere sites (Fig. 2b) could be a fossil relic or may be maintained by selective pressure. On the other hand, the loss of satellite sequences at some inactivated centromeres, such as the one of ECA11, could be the consequence of recombination events eliminating functionally irrelevant sequences. It remains to be established why mature centromeres possess satellite sequences considering that in the genus

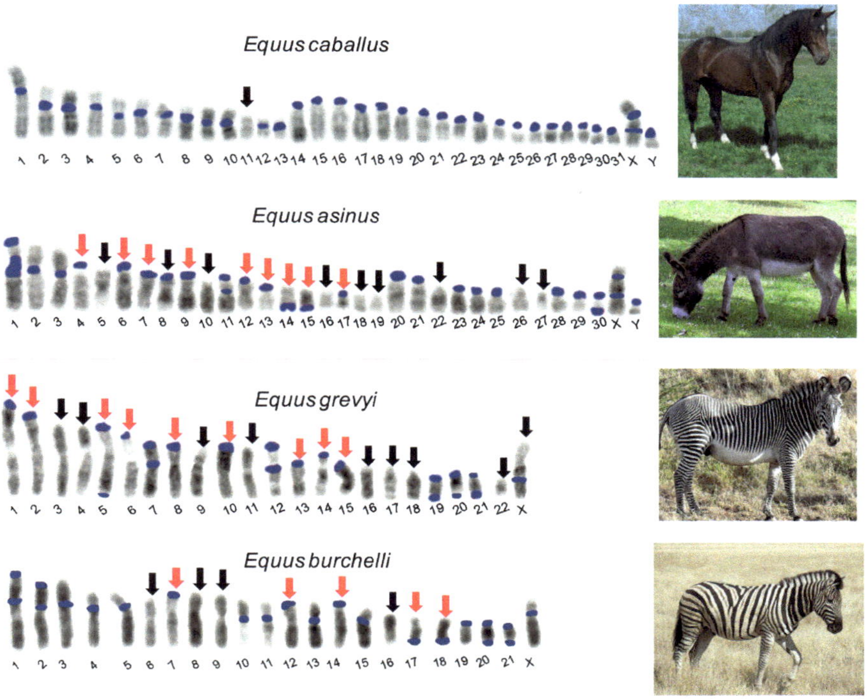

Fig. 1 Schematic representation of the FISH signals obtained after fluorescence in situ hybridization with genomic DNA, detecting total satellite DNA, on metaphase chromosomes from horse, donkey, Grevy's, and Burchelli's zebras. *Black arrows* point to chromosomes completely lacking any hybridization signal, *red arrows* point to chromosomes on which satellite DNA signal is lacking at the centromere while present at one or both non-centromeric ends (Modified from Piras et al. 2010)

Equus some centromeres can stably function in their absence. The mechanisms responsible for the accumulation of tandem repeats and the role, if any, of such repeats during the maturation of ENCs remain obscure.

The complex evolution of satellite sequence distribution in the genus *Equus*, is in agreement with the instability and exceptional plasticity of the karyotype of these species (Ryder et al. 1978; Yang et al. 2003; Trifonov et al. 2008; Musilova et al. 2013). In fact, the centromeric function and the position of satellite DNA turned out to be often uncoupled. Satellite-less centromeres arose from two different evolutionary events: fusions between ancestral acrocentric chromosomes and centromere repositioning. The latter event is unexpectedly frequent in this genus and has occurred independently of the acquisition of satellite DNA. This observation supports the hypothesis that large blocks of satellite repeats are not necessarily required for the stability of centromeres. According to this view, satellite repeats may colonize new centromeres at a later stage giving rise to "mature" centromeres

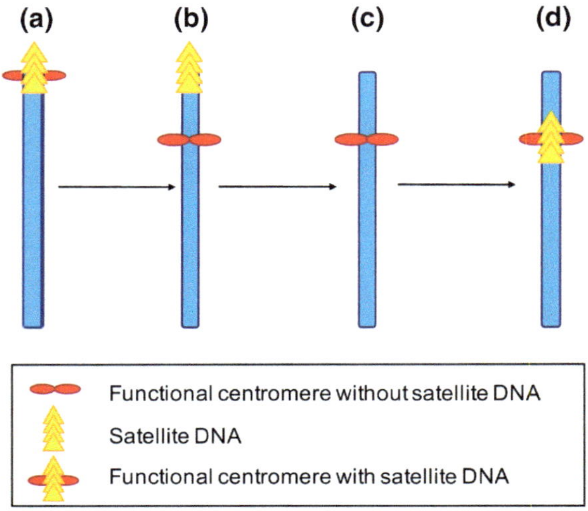

Fig. 2 Four-step mechanism for the formation of evolutionarily new centromeres in equid species. **a** In the ancestral acrocentric chromosome satellite DNA (*yellow*) and the functional centromere (*red*) coincide. **b** The submetacentric chromosome derived from centromere repositioning maintained satellite DNA sequences (*yellow*) at the terminal ancestral position, while the neocentromere (*red*) is devoid of repetitive sequences. **c** The submetacentric chromosome derived from (**b**) lost the terminal satellite sequences. **d** Fully "mature" submetacentric chromosome carrying satellite DNA (*yellow*) at the neocentromere (*red*) (Modified from Piras et al. 2010)

according to the pathway schematized in Fig. 2, as has been observed in primate karyotype evolution (Ventura et al. 2007).

In the horse, the presence of satellite-based together with a satellite-less centromere makes this species a particularly suitable model for studies on the role of centromeric tandem repeats. Using two colour FISH on stretched chromosomes and on combed DNA fibres, the physical relations between the major horse satellite DNA families (37cen, 2PI, and EC137) at satellite-based centromeres were investigated (Nergadze et al. 2014). The 37cen sequence consists of a 221 bp repeat which is 93% identical to the horse major satellite family independently identified by Wijers et al. (1993) and by Sakagami et al. (1994). The 2PI sequence, is formed by 23 bp repeated units, belongs to the e4/1 family described by Broad et al. (1995a, b) and shares 83% identity with it. Finally, the EC137 satellite is composed of 137 bp long units. Sequence analysis demonstrated that EC137 consensus has no significant similarity neither with the two previously described *Equus* satellites, nor with equid repetitive sequences deposited in RepBase. Comparative in silico analysis demonstrated that all the three satellite DNA families are equid specific.

We demonstrated that, in the horse genome, 37cen is the most represented satellite DNA family; its localization coincides with the primary constriction on all chromosomes except ECA11 and, when it is particularly abundant, spreads in the pericentromere. On the contrary, the 2PI and EC137 sequences are much less abundant, EC137 being the less represented family, and their localization is mostly pericentromeric, with partial overlap with 37cen. These results suggest that, in the horse, the 37cen satellite may be a functional centromeric satellite while the 2PI and the EC137 sequences may represent accessory pericentromeric elements. Moreover, while analysing mechanically stretched chromosome preparations (Nergadze et al. 2014), we noted that the 2PI sequence is often present in pericentromeric uncoiled regions pointing to a role of this satellite DNA family in driving the pericentromeric heterochromatin supercoiling which is needed for the correct architectural organization of the centromere core (Blower et al. 2002). The analysis of the arrangement of the three horse satellite DNA families on combed DNA fibres revealed that small arrays of the 2PI and of the EC137 satellites (ranging in size from 2–8 kb) are strictly intermingled and immerged within blocks of the 37cen sequence extending for hundreds of kilobases. This organization indicates that satellite DNA sequence interchanges are a frequent occurrence in the highly plastic horse genome.

The functional role of the 37cen satellite DNA sequence at horse centromeres was definitely confirmed by ChIP-seq and high resolution immune-FISH experiments (Cerutti et al. 2016). Blotting experiments and ChIP-seq on chromatin immunoprecipitated with an anti-CENP-A antibody demonstrated that, in the horse, the 37cen satellite binds CENP-A. Sequence analysis showed that the 37cen sequence bound by CENP-A is GC-rich with 221 bp units organized in a head-to-tail fashion.

Immuno-FISH on stretched chromosomes and chromatin fibres showed that the extension of the 37cen satellite DNA stretches is variable and is not related to the organization of CENP-A binding domains. Interestingly, the horse shares with other species a similar molecular organization of centromeres, relying on CENP-A blocks of variable length immersed in long satellite DNA stretches (Blower et al. 2002).

Finally, we analysed, by means of RNA-seq, the transcriptome profile of a horse fibroblast cell line in order to search for 37cen transcripts. The results proved that the centromeric satellite 37cen is transcriptionally active. Emerging evidence suggests that satellite transcripts may act both in *cis* and in *trans* (Bergmann et al. 2011; Quénet and Dalal 2014). Therefore, in the horse system, it is tempting to speculate that 37cen RNA may play a role, not only at satellite-based centromeres, but also at the satellite-less centromere of chromosome 11.

5 Discovery of the First Natural Satellite-Less Centromere

As mentioned above, the centromere of ECA11 was the only one in the horse lacking any hybridization signal in FISH experiments in which the two major horse satellites (Fig. 3a) or total horse genomic DNA (Fig. 1) were used as probes. To test, at the sequence level, whether satellite DNA was completely missing at this centromere, we localized the primary constriction of ECA11 by performing two and three colour hybridization experiments on horse metaphase spreads with a panel of BAC clones (Fig. 3b). Taking advantage of the ongoing horse genome sequence assembly, we then identified a 2.7 Mb region that should contain the centromeric function and prepared an array covering this region to be further analysed by ChIP-on-chip. The array was hybridized with DNA purified from chromatin immuno-precipitated with antibodies against CENP-A or CENP-C. With both antibodies, two peaks spanning about 135 and 100 kb, respectively, separated by a 165 kb region, were identified whereas no hybridization was observed in the flanking sequences. The 400 kb region comprising the two peaks did not show any peculiar feature (Fig. 3c): no protein coding genes, normal levels of interspersed repetitive elements, no evidence of accumulation of L1 transposons (Chueh et al. 2009) or KERV-1 elements (Carone et al. 2009), which were previously hypothesized to influence ENC formation. The absence of extended tandem repeat arrays demonstrated that our initial hypothesis was correct and that a normal, stable and functional mammalian centromere can be totally deprived of satellite DNA. Using a similar approach, a satellite-free ENC was then identified in orangutan, where it is present in a heterozygous state (Locke et al. 2011). An additional interesting observation was that the genomic region comprising the centromere of ECA11 was contained in a large conserved syntenic region in many mammalian species, strongly supporting the notion that the centromeric function is unrelated to DNA sequence. We proposed that the ECA11 centromere is evolutionarily young and, although functional and stable in all horses, did not yet acquire all the marks typical of mammalian centromeres. We were able for the first time to capture a mammalian ENC in an immature state, as suggested by the model sketched in Fig. 2.

6 The Satellite-Free Horse Centromere Is Wobbling Around a 500 kb Region

The identification of two well-defined peaks of CENP-A binding, in the first individual analysed, (Wade et al. 2009) posed another key question: does each one of the two homologous chromosomes 11 contain both centromeric domains or a single CENP-A binding domain (Fig. 4)? To address this question, we localized the CENP-A binding domains in five additional individuals and found a surprising result: one or two protein binding domains can be present in a single individual; moreover, the localization of each peak varies among individuals within a 500 kb long region (Purgato et al. 2015) (Fig. 5). These results suggest that the centromere function is not coupled to a specific sequence, but can slide within a relatively wide region. Using a SNP-based approach and immune-FISH experiments on single chromatin fibres (Fig. 6), we then demonstrated that the two binding domains for centromeric proteins, described in the first horse (Wade et al. 2009), correspond to the localization of the centromeric function on the two homologous chromosomes and a similar situation is common in the horse population where we observed a surprisingly high positional variation giving rise to multi-allelic epigenetic polymorphism. In the five horses analysed, we identified at least seven functional alleles

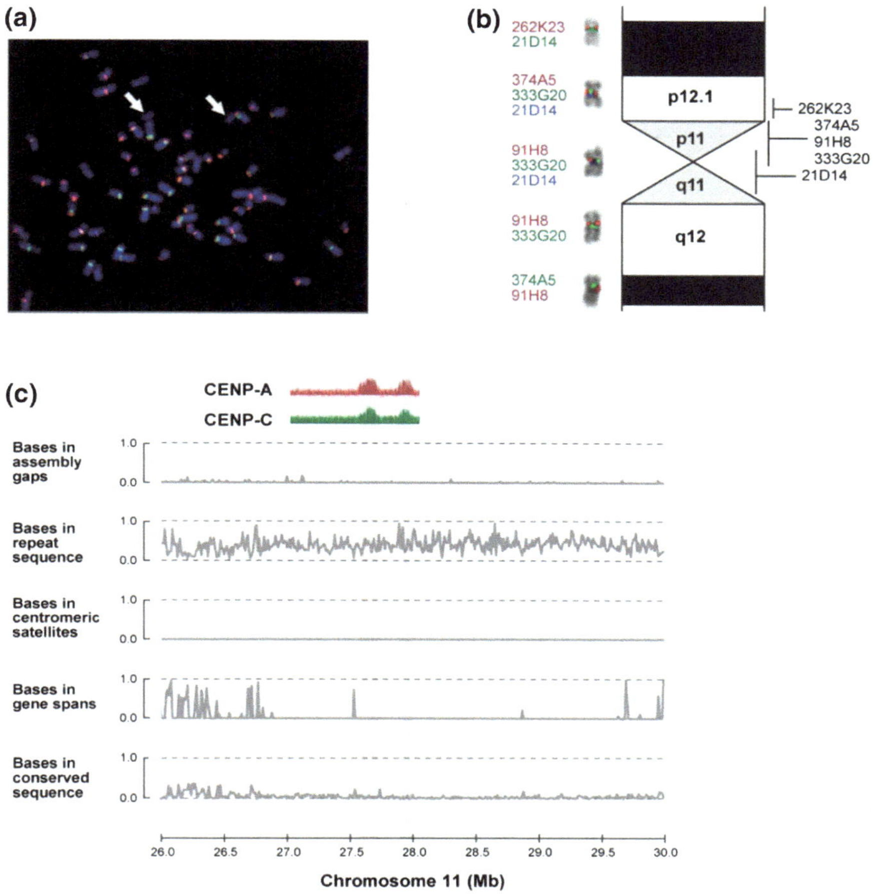

Fig. 3 Localization and sequence analysis of the ECA11 centromere. **a** Hybridization of the two major horse satellite sequences on horse chromosomes; the 23 bp repeat (2PI) is labelled in *red* and the 221 bp repeat (37cen) is labelled in *green*. All centromeres are labelled with one or both satellite probes except chromosome 11 (*arrows*). **b** Schematic representation of the cytogenetic localization of the primary constriction within a 2.7 Mb region identified using hybridization with BAC clones from the CHORI library; numbers correspond to names of informative BAC clones. **c** Bioinformatic analysis of the sequence comprising the primary constriction of ECA11; the two regions of 136 and 99 kb, respectively, bound by CENP-A and CENP-B are indicated on the *top* of this panel (From Wade et al. 2009, Supporting online material)

(epialleles) scattered in the 500 kb region. It is notable that, integrating across the individuals, CENP-A can be localized across the entire 500 kb region. At a molecular level, these results reveal a mobility of CENP-A nucleosome arrays, a property that could be related to the evolutionary mobility of centromeres.

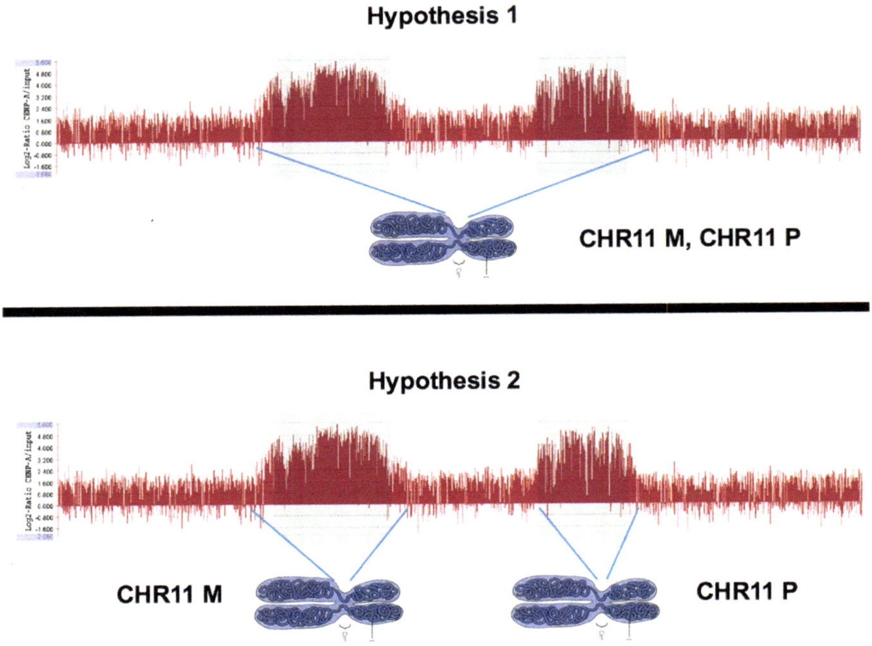

Fig. 4 Two hypotheses for the presence of two CENP-A/C binding domains on horse chromosome 11. Hypothesis 1: Both homologous chromosomes 11 (CHR11 M and CHR11P) contain two CENP-A binding peaks; Hypothesis 2: On one chromosome 11 (CHR11 M) only the 135 kb peak is present (epiallele 1) while the 95 kb domain (epiallele 2) is localized on the other homolog (CHR11P)

A possible explanation of these findings is that, although the centromere function is epigenetically determined, satellite DNA may provide positional stability to this domain along the chromosome. In this regard, it is important to note that, in the ECA11 centromere region, we did not find any evidence for the presence of CENP-B boxes (the consensus sequence binding CENP-B) (data not shown). It has been recently proposed that CENP-B, which directly binds to satellite DNA, may play a role in the stabilization of the centromere (Fachinetti et al. 2015). The lack of specific CENP-B binding within the satellite-less centromeres may contribute to the sliding of CENP-A domains in the horse. In other words, a role of CENP-B and satellite DNA may be specifically related to suppression of the intrinsic mobility of the CENP-A domain on DNA. In the chicken DT40 cell line, when neocentromeres were experimentally induced by chromosome engineering, they preferentially formed close to the original centromere (Shang et al. 2013). This observation is in agreement with our results on ECA11 suggesting that epigenetic marks required for "centromerization" are present around the centromere. Indeed, we recently found heterochromatin marks in a wide region surrounding the ECA11 neocentromere (R. Gamba, M. Corbo, F. Piras, E. Giulotto, unpublished observation).

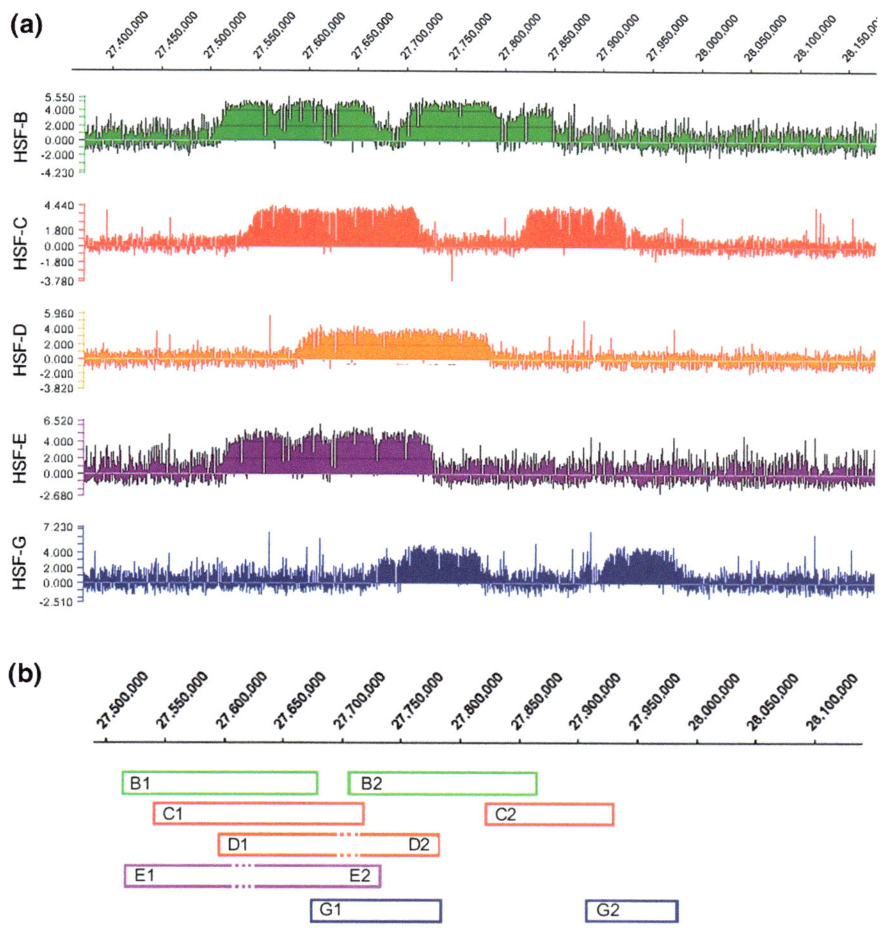

Fig. 5 Variable position of the centromere of horse chromosome 11. **a** Using an anti-CENP-A antibody, DNA obtained by chromatin immuno-precipitation from five horse fibroblast cultures was hybridized to a tiling array covering the centromere region. **b** Peak positions are represented as *boxes*. Epiallele identification was obtained by combining ChIP-on-chip, SNP and fibre FISH results. Sequence coordinates refer to the horse EquCab2.0 (2007) sequence assembly, as reported by the UCSC genome browser (http://genome.ucsc.edu). Alleles are designated by the letter of the horse they derive from, followed by '1' or '2' to distinguish the two variants. *Dotted lines* represent the region of overlap of the two binding domains in the reference sequence. Therefore, at least seven different centromeric domains were identified (Modified from Purgato et al. 2015)

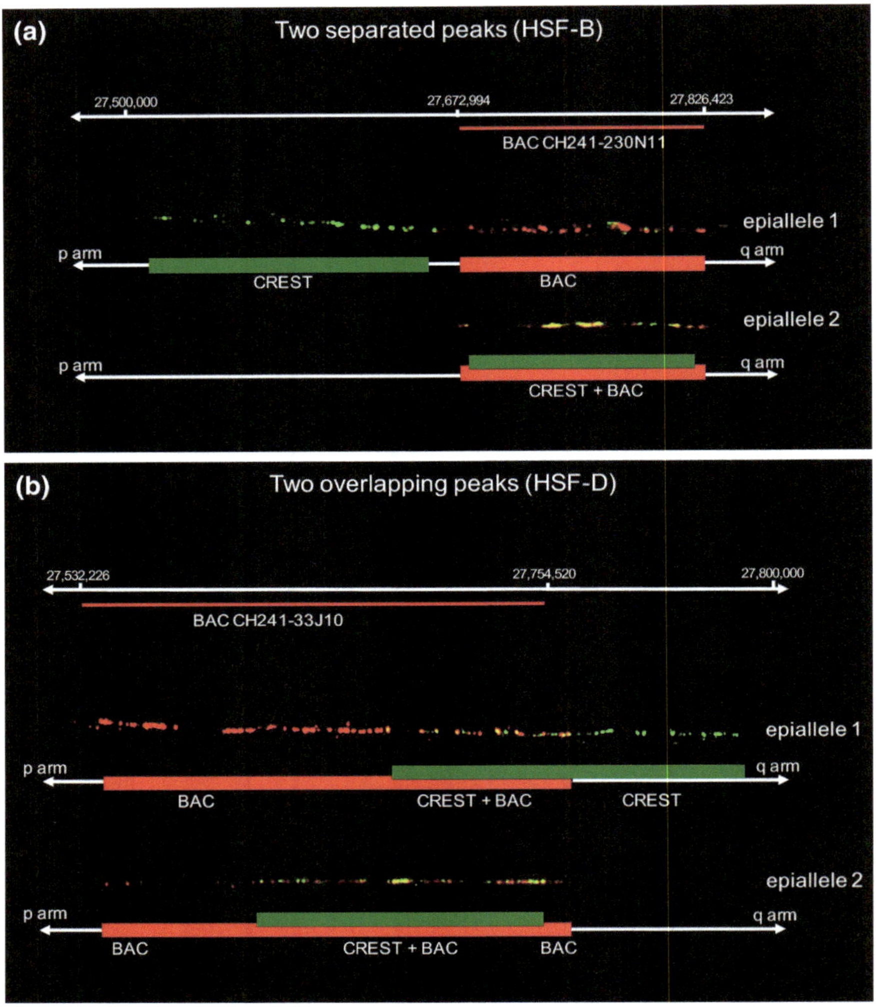

Fig. 6 Single molecule analysis of centromeric epialleles on chromatin fibres by immune-FISH. Organization pattern of functional alleles in horse HSF-B (**a**), displaying two separated ChIP-on-chip peaks (see Fig. 5a), and in horse HSF-D (**b**), displaying two overlapping ChIP-on-chip peaks (see Fig. 5a). At the *top* of each panel, the coordinates of the regions occupied by the centromeric domains are reported, and BAC coverage is represented by a *red line*. CREST immuno-staining is *green* labelled while the BAC FISH signals are *red* labelled. Under each fibre image, a schematic representation is depicted with *green rectangles* corresponding to centromeric domains and *red rectangles* indicating BAC hybridization (From Purgato et al. 2015)

7 Concluding Remarks and Perspectives

The body of evidence indicating that centromeres are primarily epigenetically determined loci was dramatically confirmed by the discovery of a natural satellite-less centromere in the domestic horse, demonstrating that such a centromere configuration is fully compatible with all functional requirements for cell division. Subsequent identification of such "unique sequence" centromeres in vertebrates such as orangutan (Locke et al. 2011) and chicken and non-vertebrates (Gong et al. 2012; Sanyal et al. 2004) has revealed that this type of centromeric DNA configuration is not unique in natural karyotypes. Thus it is clear that centromere identity is not dependent on any particular type of DNA sequence, but rather is a result of epigenetic factors, exemplified by CENP-A, that are stably associated with the locus (Karpen and Allshire 1997; Panchenko and Black 2009). In examination of inter-individual variation of centromere position discussed above it appears that centromeres can form over a 500 kb span of DNA with no exclusion of sequences, demonstrating clearly that DNA sequence per se has little influence on the ability of a centromere to form at a locus (Purgato et al. 2015). Rather, this fluidity of centromere association with DNA focuses attention on the mechanisms that establish and maintain CENP-A and associated proteins at or near their site of origin during DNA replication and cellular propagation. With the identification of CENP-B's role in stabilizing CENP-A at human centromeres, a possible function for CENP-B as a suppressor of the intrinsic dynamics of the epigenetic pathway has come to light. Thus, satellite DNAs may function to stabilize centromeres at a specific locus, perhaps maintaining functional identity between homologous chromosomes that otherwise differ in centromere position. Given the high frequency of satellite-less centromeres in the equids and the utility of unique sequence substrates for chromatin profiling, the horse and related species emerge as an important model system for dissecting the molecular structure of the centromere in a mammal. As well, the rapid karyotype evolution in the equids, coupled to very high rates of centromere repositioning, provides a mammalian system in which the evolution of centromeres may be investigated definitively, towards a deeper understanding of both their molecular architecture and function and their roles in shaping karyotypic evolution and speciation.

Acknowledgements The laboratories were supported by grants from Consiglio Nazionale delle Ricerche (CNR-Progetto Bandiera Epigenomica), from Ministero dell'Istruzione dell'Università e della Ricerca (MIUR-PRIN) and from Science Foundation Ireland (12/A/1370).

References

Alkan C, Cardone MF, Catacchio CR et al (2011) Genome-wide characterization of centromeric satellites from multiple mammalian genomes. Genome Res 21:137–145

Amor DJ, Choo KHA (2002) Neocentromeres: role in human disease, evolution, and centromere study. Am J Hum Genet 71:695–714

Amor DJ, Bentley K, Ryan J et al (2004) Human centromere repositioning "in progress". Proc Natl Acad Sci USA 101:6542–6547

Anglana M, Bertoni L, Giulotto E (1996a) Cloning of a polymorphic sequence from the nontranscribed spacer of horse rDNA. Mamm Genome 7:539–541

Anglana M, Vigoni MT, Giulotto E (1996b) Four horse genomic fragments containing minisatellites detect highly polymorphic DNA fingerprints. Anim Genet 27:286

Barry AE, Bateman M, Howman EV et al (2000) The 10q25 neocentromere and its inactive progenitor have identical primary nucleotide sequence: further evidence for epigenetic modification. Genome Res 10:832–838

Bergmann JH, Rodríguez MG, Martins NMC et al (2011) Epigenetic engineering shows H3K4me2 is required for HJURP targeting and CENP-A assembly on a synthetic human kinetochore: H3K4me2 and kinetochore maintenance. EMBO J 30:328–340

Blower MD, Sullivan BA, Karpen GH (2002) Conserved organization of centromeric chromatin in flies and humans. Dev Cell 2:319–330

Broad TE, Ede AJ, Forrest JW et al (1995a) Families of tandemly repeated DNA elements from horse: cloning, nucleotide sequence, and organization. Genome 38:1285–1289

Broad TE, Forrest JW, Lewis PE et al (1995b) Cloning of a DNA repeat element from horse: DNA sequence and chromosomal localization. Genome 38:1132–1138

Carbone L, Nergadze SG, Magnani E et al (2006) Evolutionary movement of centromeres in horse, donkey, and zebra. Genomics 87:777–782

Carone DM, Longo MS, Ferreri GC et al (2009) A new class of retroviral and satellite encoded small RNAs emanates from mammalian centromeres. Chromosoma 118(1):113–125

Cerutti F, Gamba R, Mazzagatti A et al (2016) The major horse satellite DNA family is associated with centromere competence. Mol Cytogenet 9:35

Chueh AC, Northrop EL, Brettingham-Moore KH et al (2009) LINE retrotransposon RNA is an essential structural and functional epigenetic component of a core neocentromeric chromatin. PLoS Genet 5(1):e1000354

Earnshaw WC, Ratrie H, Stetten G (1989) Visualization of centromere proteins CENP-B and CENP-C on a stable dicentric chromosome in cytological spreads. Chromosoma 98:1–12

Ekwall K (2007) Epigenetic control of centromere behavior. Annu Rev Genet 41:63–81

Fachinetti D, Han JS, McMahon MA et al (2015) dna sequence-specific binding of CENP-B enhances the fidelity of human centromere function. Dev Cell 33:314–327

Gong Z, Wu Y, Koblízková A, Torres GA et al (2012) Repeatless and repeat-based centromeres in potato: implications for centromere evolution. Plant Cell 24(9):3559–3574

Hayden KE, Strome ED, Merrett SL et al (2013) Sequences associated with centromere competency in the human genome. Mol Cell Biol 33:763–772

Henikoff S, Ahmad K, Malik HS (2001) The centromere paradox: stable inheritance with rapidly evolving DNA. Science 293:1098–1102

Karpen GH, Allshire RC (1997) The case for epigenetic effects on centromere identity and function. Trends Genet TIG 13:489–496

Locke DP, Hillier LW, Warren WC et al (2011) Comparative and demographic analysis of orang-utan genomes. Nature 469:529–533

Maloney KA, Sullivan LL, Matheny JE et al (2012) Functional epialleles at an endogenous human centromere. Proc Natl Acad Sci USA 109:13704–13709

Marshall OJ, Chueh AC, Wong LH, Choo KHA (2008) Neocentromeres: new insights into centromere structure, disease development, and karyotype evolution. Am J Hum Genet 82:261–282

Montefalcone G, Tempesta S, Rocchi M, Archidiacono N (1999) Centromere repositioning. Genome Res 9:1184–1188
Musilova P, Kubickova S, Zrnova E et al (2007) Karyotypic relationships among *Equus grevyi*, *Equus burchelli* and domestic horse defined using horse chromosome arm-specific probes. Chromosome Res 15:807–813
Musilova P, Kubickova S, Vahala J, Rubes J (2013) Subchromosomal karyotype evolution in Equidae. Chromosome Res 21:175–187
Nergadze SG, Lupotto M, Pellanda P et al (2010) Mitochondrial DNA insertions in the nuclear horse genome. Anim Genet 41(Suppl 2):176–185
Nergadze SG, Belloni E, Piras FM et al (2014) Discovery and comparative analysis of a novel satellite, EC137, in horses and other equids. Cytogenet Genome Res 144:114–123
Oakenfull EA, Clegg JB (1998) Phylogenetic relationships within the genus *Equus* and the evolution of alpha and theta globin genes. J Mol Evol 47:772–783
Orlando L, Ginolhac A, Zhang G et al (2013) Recalibrating *Equus* evolution using the genome sequence of an early Middle Pleistocene horse. Nature 499(7456):74–78
Panchenko T, Black BE (2009) The epigenetic basis for centromere identity. Springer, Berlin, pp 1–32
Piras FM, Nergadze SG, Poletto V et al (2009) Phylogeny of horse chromosome 5q in the genus *Equus* and centromere repositioning. Cytogenet Genome Res 126:165–172
Piras FM, Nergadze SG, Magnani E et al (2010) Uncoupling of satellite DNA and centromeric function in the genus *Equus*. PLoS Genet 6:e1000845
Purgato S, Belloni E, Piras FM et al (2015) Centromere sliding on a mammalian chromosome. Chromosoma 124:277–287
Quénet D, Dalal Y (2014) A long non-coding RNA is required for targeting centromeric protein A to the human centromere. Elife 3:e03254
Raimondi E, Piras FM, Nergadze SG et al (2011) Polymorphic organization of constitutive heterochromatin in *Equus asinus* (2n = 62) chromosome 1. Hereditas 148:110–113
Rocchi M, Archidiacono N, Schempp W et al (2012) Centromere repositioning in mammals. Heredity 108:59–67
Ryder OA, Epel NC, Benirschke K (1978) Chromosome banding studies of the Equidae. Cytogenet Cell Genet 20:332–350
Saffery R, Irvine DV, Griffiths B et al (2000) Human centromeres and neocentromeres show identical distribution patterns of >20 functionally important kinetochore-associated proteins. Hum Mol Genet 9:175–185
Sakagami M, Hirota K, Awata T, Yasue H (1994) Molecular cloning of an equine satellite-type DNA sequence and its chromosomal localization. Cytogenet Cell Genet 66:27–30
Santagostino M, Khoriauli L, Gamba R et al (2015) Genome-wide evolutionary and functional analysis of the equine repetitive element 1: an insertion in the myostatin promoter affects gene expression. BMC Genet 16:126
Sanyal K, Baum M, Carbon J (2004) Centromeric DNA sequences in the pathogenic yeast Candida albicans are all different and unique. Proc Natl Acad Sci USA 101(31):11374–11379
Scocchi M, Bontempo D, Boscolo S et al (1999) Novel cathelicidins in horse leukocytes(1). FEBS Lett 457:459–464
Shang W-H, Hori T, Martins NMC et al (2013) Chromosome engineering allows the efficient isolation of vertebrate neocentromeres. Dev Cell 24:635–648
She X, Horvath JE, Jiang Z et al (2004) The structure and evolution of centromeric transition regions within the human genome. Nature 430:857–864
Shepelev VA, Alexandrov AA, Yurov YB, Alexandrov IA (2009) The evolutionary origin of man can be traced in the layers of defunct ancestral alpha satellites flanking the active centromeres of human chromosomes. PLoS Genet 5(9):e1000641
Springer MS, Murphy WJ, Eizirik E, O'Brien SJ (2003) Placental mammal diversification and the Cretaceous-Tertiary boundary. PNAS 100:1056–1061
Stanyon R, Rocchi M, Capozzi O et al (2008) Primate chromosome evolution: ancestral karyotypes, marker order and neocentromeres. Chromosome Res 16:17–39

Steiner CC, Mitelberg A, Tursi R, Ryder OA (2012) Molecular phylogeny of extant equids and effects of ancestral polymorphism in resolving species-level phylogenies. Mol Phylogenet Evol 65:573–581
Trifonov VA, Stanyon R, Nesterenko AI et al (2008) Multidirectional cross-species painting illuminates the history of karyotypic evolution in Perissodactyla. Chromosome Res 16:89–107
Tyler-Smith C, Gimelli G, Giglio S et al (1999) Transmission of a fully functional human neocentromere through three generations. Am J Hum Genet 64:1440–1444
Ventura M, Antonacci F, Cardone MF et al (2007) Evolutionary formation of new centromeres in macaque. Science 316:243–246
Voullaire LE, Slater HR, Petrovic V, Choo KH (1993) A functional marker centromere with no detectable alpha-satellite, satellite III, or CENP-B protein: activation of a latent centromere? Am J Hum Genet 52:1153–1163
Wade CM, Giulotto E, Sigurdsson S et al (2009) Genome sequence, comparative analysis, and population genetics of the domestic horse. Science 326:865–867
Warburton PE, Dolled M, Mahmood R et al (2000) Molecular cytogenetic analysis of eight inversion duplications of human chromosome 13q that each contain a neocentromere. Am J Hum Genet 66:1794–1806
Wijers ER, Zijlstra C, Lenstra JA (1993) Rapid evolution of horse satellite DNA. Genomics 18:113–117
Willard HF (1991) Evolution of alpha satellite. Curr Opin Genet Dev 1:509–514
Yang F, Fu B, O'Brien PCM et al (2003) Karyotypic relationships of horses and zebras: results of cross-species chromosome painting. Cytogenet Genome Res 102:235–243

Part IV
The Functions and Mechanisms of Centromeres and Kinetochores in Healthy Meiotic and Mitotic Divisions and their Role in Human Disease

Centromere Dynamics in Male and Female Germ Cells

Elaine M. Dunleavy and Caitríona M. Collins

Abstract In sexually reproducing organisms the germ line is the cellular lineage that gives rise to gametes. All germ cells originate from germline stem cells that divide asymmetrically to generate gonial pre-cursors, which are amplified in number by mitotic divisions, undergo meiosis and eventually differentiate into mature gametes (haploid eggs and sperm). Information transmitted with gametes is inherited by offspring, and potentially by subsequent generations, instructing in organismal development and beyond. Meiosis comprises one round of DNA replication, followed by two rounds of chromosome segregation; homologous chromosomes segregate in the first division (meiosis I) and sister chromatids segregate in the second division (meiosis II). Important mechanistic features of meiosis occur in substages of prophase I and are critical for genetic recombination, including pairing and synapsis of homologous chromosomes (at leptotene and zygotene), crossing-over (at pachytene), and the appearance of chiasmata (at diplotene/diakinesis). Another unique feature of meiosis is the altered centromere/kinetochore geometry at metaphase I, such that sister kinetochores face the same spindle pole (mono-orientation) and stay together at anaphase I. This chapter reviews centromere dynamics in germ cells, focusing on centromere function and assembly in meiotic cell cycles, as well as centromere inheritance in zygotes. Centromeres are functionally defined by the presence of the histone H3 variant CENP-A, the epigenetic determinant of centromere identity. In most eukaryotes, it is well established that CENP-A function is essential for chromosome segregation in mitosis. CENP-A function in meiosis is less well understood and emerging insights into the differential regulation of meiotic and mitotic CENP-A are discussed.

E.M. Dunleavy (✉) · C.M. Collins
Centre for Chromosome Biology, Biomedical Sciences, National University of Ireland, Galway, Ireland
e-mail: elaine.dunleavy@nuigalway.ie

B.E. Black (ed.), *Centromeres and Kinetochores*, Progress in Molecular and Subcellular Biology 56, DOI 10.1007/978-3-319-58592-5_15

1 CENP-A Function in Germ Cells

CENP-A function is essential for kinetochore assembly in mitosis (Fachinetti et al. 2013; Black et al. 2007; Takahashi et al. 2000; Blower and Karpen 2001; Buchwitz et al. 1999; Howman et al. 2000; Ravi et al. 2010; Stoler et al. 1995). CENP-A function in meiosis is less well understood and is currently unknown in many organisms, including humans. One of the first investigations into CENP-A function in meiosis was performed in the nematode *Caenorhabditis elegans*. Surprisingly, its function appears to be dispensable in worm meiosis, or at least it is not required at mitotic levels (Monen et al. 2005). Fixed and time-lapse imaging of CENP-A-depleted gonads showed that meiotic chromosome segregation was not perturbed, yet chromosome segregation completely failed in CENP-A-depleted mitotic cells. The lack of a requirement for CENP-A in worm meiosis most likely relates to the fact that this organism is holocentric. As CENP-A is incorporated throughout the length of holocentric chromosomes, they lack a single site to co-orient sister chromatids and to protect cohesins against degradation in the first meiotic division (Marques and Pedrosa-Harand 2016). To overcome this problem, *C. elegans* oocytes have departed from the typical kinetochore-driven mode of chromosome segregation to one in which microtubules between chromosomes mediate separation (Dumont et al. 2010), perhaps excluding a need for CENP-A. In contrast, in the fruit fly *Drosophila melanogaster*, CENP-A is required for centromere function in meiosis, at least in males. RNAi knockdown of CENP-A in pre-meiotic gonial cells in testes revealed meiotic chromosome missegregation events, including uneven nuclear segregation in meiosis I and II (Dunleavy et al. 2012). Moreover, depletion or mutation of key CENP-A assembly factors Chromosome alignment defect 1 (CAL1) and Centromeric protein-C (CENP-C) give rise to defects in meiotic centromere function in both male and female flies (Dunleavy et al. 2012; Raychaudhuri et al. 2012; Kwenda et al. 2016; Unhavaithaya and Orr-Weaver 2013). In the thale cress *Arabidopsis thaliana*, RNAi knockdown of CENP-A to a level sufficient for mitosis leads to partial plant sterility, with lagging chromosomes in meiosis I and II, and gametes (pollen spores) have micronuclei (Lermontova et al. 2011). Indeed, meiosis-specific functions for CENP-A are best illustrated in plants, in which alterations to CENP-A can give rise to haploid progeny (Ravi and Chan 2010). Therefore, aside from worm, the consensus finding from model organisms examined so far is that CENP-A is functionally required for meiosis.

1.1 Germ-Cell-Specific Functions of the CENP-A N-Terminus

The CENP-A N-terminus shares no similarity between eukaryotes, it is highly divergent and is rapidly evolving (Malik and Henikoff 2003) (Fig. 1). First

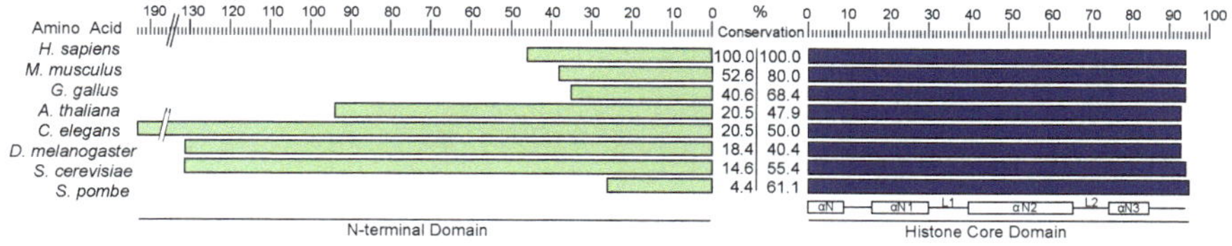

Fig. 1 The CENP-A N-terminal domain is highly divergent in both length and amino acid sequence. Amino acids 1–46 and 47–140 of human CENP-A represent the N-terminal and histone core domain of CENP-A, respectively. Percent identity relative to human CENP-A was determined using TCoffee (Notredame et al. 2000)

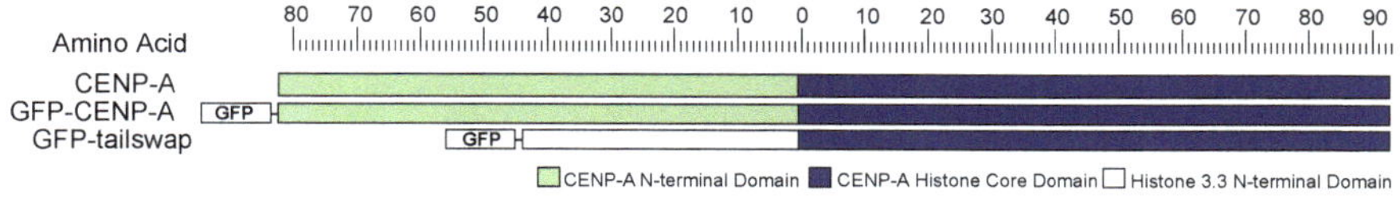

Fig. 2 In *A. thaliana*, GFP-tailswap plants express a GFP-tagged CENP-A transgene in which the N-terminus of CENP-A is replaced with the N-terminus of histone H3.3 (Ravi et al. 2010)

experiments investigating CENP-A domains critical for centromere specification showed that the centromere-targeting domain (CATD) within the C-terminal histone core domain is both necessary and sufficient for CENP-A deposition and function (Black et al. 2004, 2007; Fachinetti et al. 2013). At this point, the function of the CENP-A N-terminus was not known and it was presumed to be largely dispensable for CENP-A deposition. More recent long-term viability assays in human cultured cells and fission yeast, in which centromere function can be rescued by the expression of chimeric CENP-A/H3 transgenes, now indicate functional requirements for the CENP-A N-terminus in addition to the CATD (Fachinetti et al. 2013). In both organisms, the CENP-A N-terminus supports long-term cellular viability and in human cells it directs CENP-B binding to reinforce kinetochore function. Additional centromere establishment assays in human cells revealed that the CENP-A N-terminus is required for the initial recruitment of the kinetochore proteins CENP-C and CENP-T (Logsdon et al. 2015). Studies carried out in *A. thaliana* first demonstrated an unexpected functional requirement for the CENP-A N-terminus in plant germ cells. Ravi and colleagues isolated a *cenh3* (plant CENP-A) null mutation by ethylmethanesulfonate (EMS) mutagenesis, which was embryo lethal as expected. This CENP-A-lethal mutation could be rescued by the expression of an N-terminally GFP-tagged CENP-A (Ravi et al. 2010). Surprisingly, expression of a modified, GFP-tagged version of CENP-A in which its N-terminus is replaced with that of histone H3.3 (so-called 'GFP-tailswap', Fig. 2) complemented lethality in *cenh3* null plants, but the resulting plants were largely sterile (Ravi et al. 2010; Ravi and Chan 2010). Similar experiments, instead performed using fluorescently tagged CENP-A transgenes completely lacking its N-terminus, showed that 'tailless' CENP-A localises to mitotic centromeres (Lermontova et al. 2006; Ravi et al. 2010). Yet, in line with

GFP-tailswap plants, plants over expressing the YFP-tagged tailless transgene were sterile or partially sterile (Lermontova et al. 2011). Sterility in tailless and GFP-tailswap plants was attributed to meiotic abnormalities in gametes. GFP-tailswap plants were defective in homolog disjunction in meiosis I, in chromosome alignment at metaphase II and pollen spores display micronuclei (Lermontova et al. 2011; Ravi et al. 2011). YFP-tagged tailless plants displayed lagging chromosomes in meiosis II also resulting in pollen spores with micronuclei (Lermontova et al. 2011). These findings point to critical roles for the CENP-A N-terminus in meiotic chromosome segregation.

In their analyses of transgenic plants expressing modified versions of CENP-A, Ravi and Chan made a second striking observation. GFP-tailswap plants could be crossed to wild-type plants; however, viable offsprings were haploid and only contained the wild-type set of chromosomes (Ravi and Chan 2010). This phenomenon termed centromere-mediated genome elimination has been exploited for accelerated plant breeding purposes, but has also fuelled research into CENP-A function in germ cells and zygotes in a number of plant species. To explain this uniparental chromosome loss, current models propose that gamete chromosomes with modified CENP-A are 'weaker' than those with wild-type CENP-A and either fail to interact with the mitotic spindle or lag in early embryo mitoses (Karimi Ashtiyani et al. 2015; Ravi and Chan 2010). A similar mechanism was also proposed for genome elimination in barley hybrids (Sanei et al. 2011). Plant breeders are now positioned to carry out targeted mutagenesis screens to identify additional CENP-A alterations that might give rise to haploid plants. For example, in *A. thaliana*, a single point mutation within the centromere-targeting domain (CATD) of CENP-A impairs its localisation to both mitotic and meiotic centromeres and can give rise to haploid embryos (Karimi-Ashtiyani et al. 2015), albeit at a lower efficiency than crosses with the GFP-tailswap. This exact point mutation impairs CENP-A localisation in barley and sugar beet (Karimi-Ashtiyani et al. 2015), raising the possibility that this method for haploid plant induction could be extended to other species. In summary, plant breeding experiments strongly support a germ-cell-specific role for CENP-A and its N-terminus; parents with defective or modified versions of CENP-A generate chromosome loss both in meiocytes and resulting progeny.

A third observation of GFP-tagged CENP-A localisation in plants demonstrated an unexpected role for the N-terminus in meiotic CENP-A dynamics. GFP-tailswap and GFP- or YFP-tailless CENP-As fail to localise to meiotic centromeres; GFP-tailswap was only faintly visible at centromeres during early prophase I (leptotene and zygotene) and was not detected beyond the start of pachytene, nor for remaining phases of meiosis I and II (Ravi et al. 2011; Lermontova et al. 2011). Therefore, it is likely that plants expressing modified or truncated CENP-A are sterile due to a specific failure in meiotic CENP-A retention at centromeres that gives rise to chromosome segregation defects in gametes. Remarkably, the small percent of GFP-tailswap or tailless gametes that survive meiosis were competent to reload CENP-A in subsequent mitotic (gametophytic) divisions of pollen spores that occur in plants (Ravi et al. 2011; Schubert et al. 2014). These findings point to de novo CENP-A assembly after meiotic exit, possibly due to the re-availability of mitotic CENP-A assembly factors. While experiments with GFP-tailswap plants have been extremely informative, it is important to note that presence of the GFP-tag alone compromises CENP-A function (Ravi et al. 2011). Complementation assays with untagged, mutated versions of CENP-A might prove more biologically relevant for future studies (Maheshwari et al. 2015).

Interestingly, despite its hyper-variability among divergent organisms, the CENP-A N-terminus of flies and plants harbour blocks of conserved amino acid motifs (Torras-Llort et al. 2009; Malik et al. 2002; Maheshwari et al. 2015). The N-terminus of CENP-As from the Drosophila clade harbour three conserved arginine-rich domains (Malik et al. 2002) (Fig. 3), whereas at least two conserved blocks were identified in nearly all plant CENP-As analysed ranging from green algae to flowering plants (Maheshwari et al. 2015) (Fig. 4). The function of such conserved sequence blocks is largely unknown, but suggest functional specialisation. Indeed, evolutionarily divergent plant CENP-As can complement mitotic and meiotic functions in an *A. thaliana cenh3* null background (Maheshwari et al. 2015), suggesting that despite sequence differences in N-terminal tails, critical functional features are conserved. Further investigations into requirements of the CENP-A N-terminus and its potential conserved blocks in other organisms might reveal germ-cell-specific functions, for example in meiotic CENP-A assembly or maintenance.

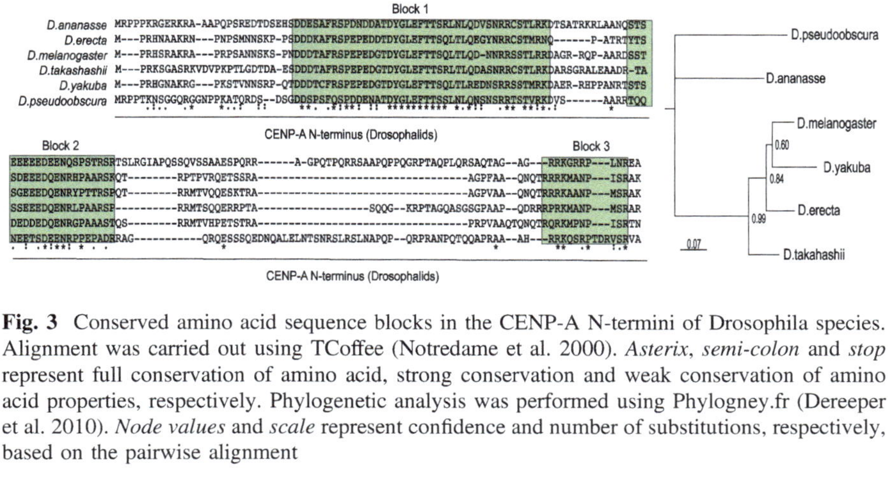

Fig. 3 Conserved amino acid sequence blocks in the CENP-A N-termini of Drosophila species. Alignment was carried out using TCoffee (Notredame et al. 2000). *Asterix*, *semi-colon* and *stop* represent full conservation of amino acid, strong conservation and weak conservation of amino acid properties, respectively. Phylogenetic analysis was performed using Phylogney.fr (Dereeper et al. 2010). *Node values* and *scale* represent confidence and number of substitutions, respectively, based on the pairwise alignment

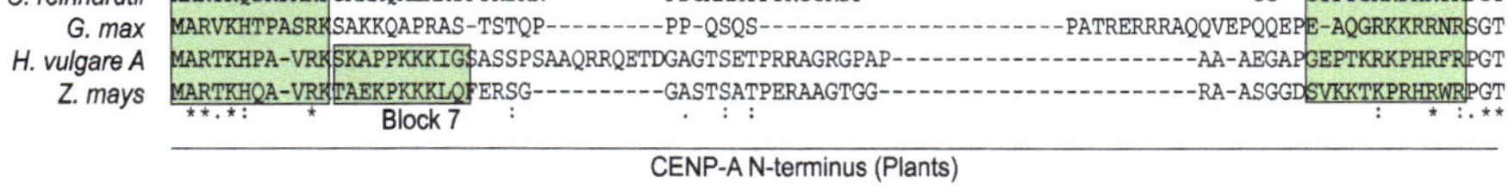

Fig. 4 Conserved amino acid sequence blocks in the CENP-A N-termini of highly divergent plant species. Alignment was performed as shown in Fig. 3

2 Cell Cycle Assembly Timing of Meiotic CENP-A

In the last decade, investigations into the cell cycle timing of centromere assembly in mitosis have proven critical to understanding mechanisms of centromere specification and function. In human cultured cells, seminal first studies showed that CENP-A is assembled at late telophase and early G1 phase (Jansen et al. 2007; Hemmerich et al. 2008) or at anaphase in syncytial divisions in fly embryos (Schuh et al. 2007). Critically, these experiments highlighted the stability of pre-existing CENP-A nucleosomes incorporated at centromeres that are stably retained through mitotic cell cycles. CENP-A assembly in G1 phase was unexpected as it indicated that centromeres are competent for kinetochore assembly and chromosome segregation with half the total amount of CENP-A. Different from metazoans, CENP-A assembly occurs in G2 phase in plants and fission yeast (Lando et al. 2012; Lermontova et al. 2006, 2007). The significance of pre-divisional CENP-A assembly in most organisms and post-divisional CENP-A assembly in others is currently not clear, but could have mechanistic implications for centromere function and kinetochore assembly at chromosome segregation. What is clear is that new CENP-A deposition should occur at least once per cell cycle to ensure centromere propagation. Given that the meiotic cell cycle comprises two rounds of nuclear division, investigations into meiotic CENP-A assembly have focused on determining if CENP-A is replenished in both divisions, or in only one division, or not at all, i.e., is the pre-meiotic CENP-A level sufficient to support both meiotic divisions?

Earliest investigations into the cell cycle timing of meiotic CENP-A assembly were conducted in *C. elegans*. Fixed and live analysis of oocytes showed that meiotic CENP-A was dynamic (Monen et al. 2005), a result that was unexpected given the stability of CENP-A nucleosomes in mitosis (Buchwitz et al. 1999). Surprisingly, CENP-A was not detected on prophase I chromosomes at early pachytene and was first detected at late pachytene/diplotene (Fig. 5). Unusual CENP-A localisation dynamics in this system might again be related to the finding that CENP-A is largely dispensable for holocentric worm meiosis (Monen et al. 2005). Intriguingly, CENP-A removal in early pachytene and reloading by late diplotene coincides with the timing of key recombination events in prophase I. An unexpected drop in CENP-A signal was also observed between meiosis I and II; however, its functional importance was not tested.

Meiotic CENP-A assembly in prophase I was also observed in *D. melanogaster* males (Dunleavy et al. 2012; Raychaudhuri et al. 2012) (Fig. 5). Quantitation of endogenous and GFP-tagged CENP-A levels from fixed and live testes revealed an increase in CENP-A intensity between early and late prophase I (Dunleavy et al. 2012; Raychaudhuri et al. 2012). CENP-A assembly in prophase I was also observed in Drosophila females (between zygotene and diplotene) (Dunleavy et al. 2012). Unlike Drosophila males that lack conventional features of meiotic prophase I, Drosophila females carry out chromosome synapsis and homologous recombination, indicating that CENP-A assembly at this time is unlikely to be a peculiarity

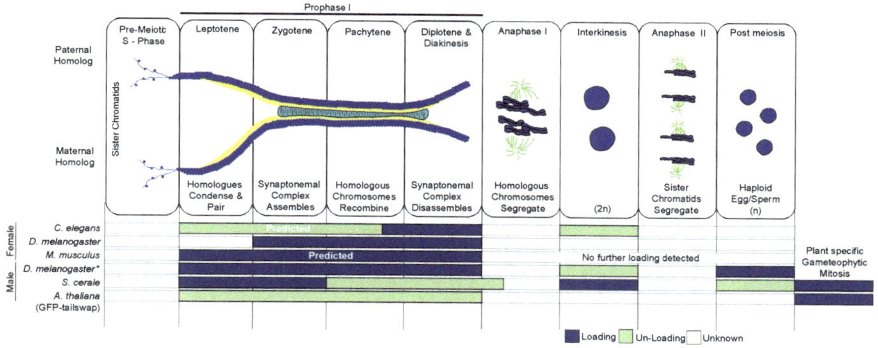

Fig. 5 Chromosome and CENP-A dynamics during meiosis in *C. elegans* (Monen et al. 2005), *D. melanogaster* (Dunleavy et al. 2012; Raychaudhuri et al. 2012), *M. musculus* (Smoak et al. 2016), *S. cerale* (Schubert et al. 2014) and *A. thaliana* (Ravi et al. 2011). *Synapsis and recombination are absent in Drosophila male meiosis

of male fruit flies. In contrast to mitosis, in which the majority of CENP-A is assembled in minutes to hours (Schuh et al. 2007; Jansen et al. 2007; Hemmerich et al. 2008), meiotic CENP-A assembly dynamics appear to be slow in flies. Prophase I lasts over 90 h in Drosophila males and gradual increments in CENP-A assembly were measured between early and late substages (Dunleavy et al. 2012; Raychaudhuri et al. 2012). Similar to findings in worm, an unexpected drop in CENP-A intensity of greater than half was also measured between meiosis I and II (Dunleavy et al. 2012), but again the significance of this drop is not clear. Intriguingly, the localisation of CAL1, the major CENP-A assembly factor in flies, inversely correlates with the dynamics of CENP-A deposition in prophase I. Centromeric CAL1 level is reduced in early prophase I and is undetectable at late prophase I, the time when CENP-A assembly reaches its peak (Dunleavy et al. 2012; Raychaudhuri et al. 2012). It is possible that CAL1 is gradually removed from centromeres once meiotic CENP-A assembly is complete. An additional phase of CENP-A assembly was measured on spermatids post-meiosis II (Dunleavy et al. 2012). Given that neither CAL1 nor CENP-C is detected at centromeres at this time, it is still unclear if either centromere assembly factor is specifically required for this second loading phase.

Consistent with findings in worms and flies, measurements of CENP-A immuno-fluorescent intensities in meiocytes of the rye plant *Secale cerale* also revealed a first major phase of CENP-A assembly in early prophase I (Schubert 2014) (Fig. 5). Additional unloading and loading events were also measured between late prophase and metaphase I, and at interkinesis, respectively. An unexpected drop in CENP-A intensity, comparable to the drop reported in Drosophila males, was also reported after anaphase, in this instance in tetrad pollen nuclei immediately after the second meiotic division. Taken together, unusual CENP-A dynamics including loading and unloading events appear to be a common feature of meiotic centromere assembly in worm, flies and plants (Fig. 5). Analysis in *A. thaliana* of GFP-tailswap localisation dynamics has also added to current understandings of such meiotic CENP-A loading and unloading events. Strikingly, GFP-tailswap mutants fail to localise to meiotic centromeres (Ravi et al. 2011). Failure to detect the GFP-tailswap was first observed from leptotene in early prophase I, the time when meiotic pairing between homologs initiates (Fig. 5). One hypothesis that might explain such a loss is that CENP-A is gradually removed during prophase I, supporting the notion that numbers and types of CENP-A molecules are normally subject to a quality check at discrete meiotic substages. Given that GFP-tailswap plants express an N-terminally modified version of the CENP-A, it is possible that the N-terminus is a substrate for unloading and might normally direct a quality check (discussed in Sect. 2.1).

CENP-A assembly in early prophase I might also be conserved in mammals. A recent investigation in mouse oocytes failed to detect GFP-CENP-A assembly in prophase I arrested oocytes or upon maturation of oocytes to meiosis II (Smoak et al. 2016) (Fig. 5). This result suggests that sufficient CENP-A for meiotic centromere function is already assembled prior to prophase I arrest at diplotene. Presumably, a round of CENP-A assembly occurs in the last mitosis before entry into meiosis, but additional CENP-A assembly events in early prophase I prior to diplotene remain an open possibility. It is clear that CENP-A incorporated prior to the prophase I arrest is extremely stable, as arrested oocytes in which *cenp-a* was knocked out retain 70% of the pre-existing CENP-A protein one year later (Smoak et al. 2016). Moreover, when bred, the mice were fertile and could support early embryogenesis, reinforcing a model in which CENP-A assembled prior to diplotene is sufficient for meiotic centromere function. However, given that a 30% reduction in CENP-A was measured, some CENP-A was lost after one year and a low level of CENP-A turnover cannot be excluded. Similar investigations into CENP-A assembly and maintenance dynamics in mammalian testes are currently lacking, but will be important to corroborate findings in Drosophila and to identify common features of centromere assembly pathways in males. In summary, while in mitosis CENP-A is stably incorporated at centromeres during the cell cycle, its localisation in meiosis is dynamic, with both assembly and disassembly events reported in most model organisms examined so far.

2.1 Significance of Unusual CENP-A Dynamics in Meiotic Prophase I

With the exception of plants and fission yeast, most eukaryotes assemble CENP-A after chromosome segregation in mitosis (Valente et al. 2012). An emerging theme from investigations into the temporal control of meiotic CENP-A is its assembly before chromosome segregation, in the first meiotic division (Fig. 5). Why switch between post-divisional loading in mitosis to pre-divisional loading in meiosis? One hypothesis is that pre-divisional loading in meiosis I might relate to unique mechanistic features of prophase I that result in genetic recombination. In rye and female fruit flies, CENP-A assembly initiates in early prophase (Schubert et al.

2014; Dunleavy et al. 2012) and likely coincides with synaptonemal complex assembly. In worm, CENP-A assembly initiates later in prophase I, at late pachytene (Monen et al. 2005), when synapsis is complete and crossing-over takes place. In mouse oocytes, any CENP-A assembly likely occurs prior to diplotene (Smoak et al. 2016), the stage at which chiasmata that mark sites of genetic crossover are clearly visible. Therefore, on the one hand, CENP-A assembly could be coupled to the initiation or completion of homolog pairing, chromosomal synapsis or homologous recombination. This is unlikely to be the case in Drosophila males, however, which assemble CENP-A in prophase I despite a lack of synapsis and homologous recombination (Meyer 1960). On the other hand, CENP-A and its associated kinetochore complex may be incompatible with aspects of homologous recombination that necessitate its removal and subsequent reloading.

A second hypothesis is that pre-divisional CENP-A assembly in prophase I prepares the centromere/kinetochore for the mono-orientation of sister chromatids at metaphase I. It is possible that the absolute number of CENP-A nucleosomes is critical to build the kinetochore for the first meiotic division and this number requires adjustment at this cell cycle time. In addition to assembly, evidence for the selective removal of a modified version of CENP-A in early prophase I in plants (Ravi et al. 2011), raises the possibility that CENP-A is turned over at this cell cycle stage. It is possible that CENP-A disassembly removes imperfect or not correctly modified versions of CENP-A before the first meiotic division, as part of a quality control step. Fluorescent recovery after photobleaching (FRAP) experiments with functional fluorescently tagged CENP-A are currently lacking in any organism, but might confirm CENP-A turnover and dynamics in prophase I. Findings that meiotic CENP-A assembly is dependent on an intact CENP-A N-terminus in plants (Ravi et al. 2011) supports models in which the N-terminus directs CENP-A removal. It is possibly subject to meiosis-specific post-translational modifications, or it interacts with a meiosis-specific chaperone/assembly/disassembly factor, or it directs protein-folding activities that instruct CENP-A stability. Finally, the selective CENP-A drop in interkinesis after meiosis I, so far observed in flies and worms, might reflect CENP-A loss due to the reconfiguration of centromeres/kinetochores from a side-to-side to a back-to-back orientation for sister chromatid segregation in meiosis II (Watanabe 2012). The second major phase of CENP-A assembly after meiosis II, so far observed in Drosophila males, is post-divisional and more comparable to the assembly triggered by mitotic exit, for example in human cells (Jansen et al. 2007; Silva et al. 2012). Here additional CENP-A assembly might compensate for excess removal in prophase or anaphase I, or ensure that a sufficient amount of CENP-A is present on mature gametes for epigenetic inheritance in the zygote.

3 Transgenerational Inheritance of CENP-A

In most organisms studied so far, it is apparent that sufficient CENP-A for the two rounds of nuclear division in meiosis is assembled prior to the end of prophase I, with additional CENP-A assembly events immediately after meiotic exit in some organisms (Fig. 5). However, in order to epigenetically specify centromere identity in the next generation, CENP-A must also be maintained on gamete chromatin. Females face the challenge of CENP-A maintenance on egg chromatin arrested at prophase I, which lasts for months to years depending on the species (Von Stetina and Orr-Weaver 2011). Remarkably, a first quantitative analysis of CENP-A protein in mouse oocytes confirms its long-lived stability for at least one year with little turnover (Smoak et al. 2016). Similar analysis of CENP-A stability in arrested human oocytes is of interest, although technically challenging. Male gametes face a different challenge; mature spermatozoa must retain CENP-A despite the dramatic removal and exchange of most other histones for protamines during differentiation. Early immuno-fluorescent studies of fully differentiated bovine spermatozoa first confirmed that CENP-A is retained in discrete nuclear foci in males (Palmer et al. 1990). Moreover, CENP-A was identified as one of the most abundant proteins on bovine sperm, which investigators exploited for its purification and sequencing (Palmer et al. 1991). Subsequent studies confirmed CENP-A localisation on mature sperm in diverse organisms such as Xenopus, plants and flies (Zeitlin et al. 2005; Raychaudhuri et al. 2012; Dunleavy et al. 2012; Ingouff et al. 2010), indicating that CENP-A maintenance on male gametes is likely of wide functional importance for epigenetic centromere inheritance.

One of the most extensive investigations into the inheritance of CENP-A from one generation to the next was performed in Drosophila males (Raychaudhuri et al. 2012). These experiments provide support for a template-governed centromere inheritance and assembly model, in which pre-existing CENP-A nucleosomes direct the deposition of new CENP-A nucleosomes (Figs. 6 and 7). First, Raychaudhuri and colleagues crossed male flies expressing only a GFP-tagged copy of CENP-A to wild-type females and observed dilution of GFP-CENP-A by half in each of the early embryonic cell cycles 1–3 (Fig. 6). This result indicates that each cell cycle pre-existing CENP-A on paternal chromosomes is diluted by unlabeled, maternally supplied CENP-A and is in line with the dilution of pre-existing CENP-A by newly synthesised CENP-A in mitotic cell cycles (Jansen et al. 2007). Next, the authors use a genetic approach to generate flies harbouring sperm in which CENP-A was degraded and no longer detectable at centromeres. In embryos generated from 'CENP-A-degraded' sperm, paternal chromosomes did not assemble new CENP-A

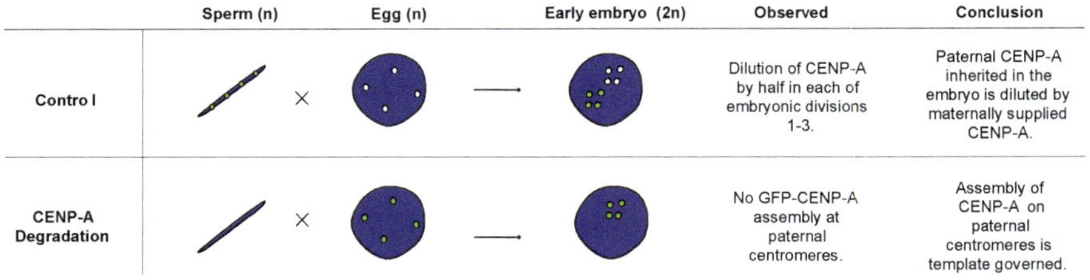

Fig. 6 This experiment suggests that CENP-A inherited on paternal chromosomes acts as a template for subsequent CENP-A loading in the embryo (Raychaudhuri et al. 2012)

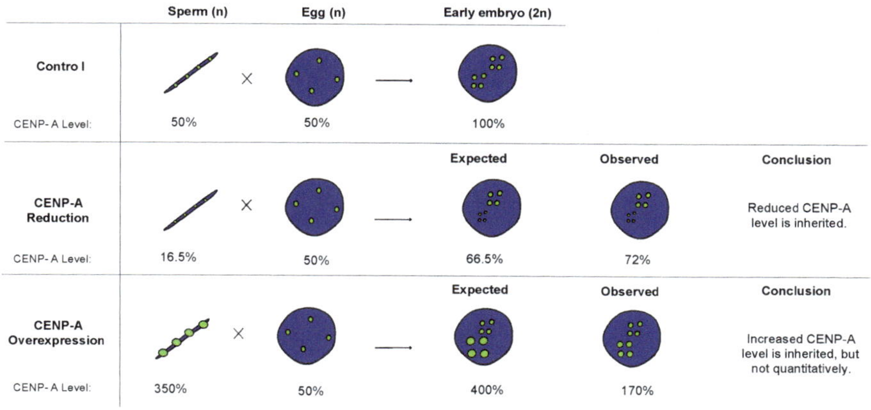

Fig. 7 These experiments suggest that altered centromeric CENP-A levels on sperm are inherited by the embryo (Raychaudhuri et al. 2012)

and were lost during the initial embryonic cell cycles (Fig. 6). This result demonstrates that a minimal amount of pre-existing paternal CENP-A is required to direct the assembly of maternal CENP-A. In a second set of experiments, the authors use additional genetic approaches to manipulate CENP-A levels on sperm and track whether high or low CENP-A levels are inherited in the next generation (Fig. 7). They report two major findings. (i) When males with reduced CENP-A on sperm were crossed to control females, resulting progeny had reduced total CENP-A in embryonic nuclei, as well as adult sperm nuclei. Given that half the chromosomes were of paternal origin, the observed drop in total CENP-A level was in line with the expected drop. (ii) When males with a higher CENP-A level on sperm were crossed to control females, resulting progeny had a higher CENP-A level on chromosomes in embryonic nuclei. While the observed CENP-A increase was lower than expected, it is likely that CENP-A is not quantitatively maintained as levels of CENP-A and its assembly factor CAL1 are limiting (Schittenhelm et al. 2010). Alternatively, it is possible that high CENP-A levels at centromeres can gradually revert back to normal levels. Both sets of experiments support template-driven epigenetic memory at sperm centromeres and indicate that CENP-A levels are not reset, at least not in the next generation. Intriguingly, in RNAi experiments reduced CENP-A levels were also measured in wing imaginal discs and mature adult sperm, indicating that quantitative changes in CENP-A at centromeres were maintained beyond embryogenesis. It will be insightful to now investigate CENP-A inheritance in subsequent generations, i.e., in adult tissues and sperm from grandsons and great grandsons. It will also be of interest to determine if mechanisms for resetting CENP-A at centromeres could exist, for example to counterbalance unequal CENP-A inheritance on paternal and maternal homologs.

In striking contrast to many eukaryotes, CENP-A is not retained on mature sperm in *C. elegans,* arguing against a template-governed centromere inheritance mechanism in worms. Indeed, photobleaching experiments show that GFP-tagged CENP-A is turned over in embryo cell cycles and progeny generated from CENP-A-depleted oocytes fertilised with wild-type sperm do not inherit any CENP-A (Gassmann et al. 2012). Instead genome-wide chromatin immunoprecipitation of CENP-A and hybridisation to a tiling microarray (ChIP-chip) experiments show that CENP-A position in zygotes is linked to transcription (Gassmann et al. 2012). Genes transcribed in the germline, or in embryos, are refractory to CENP-A incorporation, whereas genes silent in embryos are permissive for CENP-A incorporation. The link between de novo centromere specification and transcriptional silencing might be a unique feature of worm holocentric chromosomes. Indeed, an increasing body of evidence supports transcription-coupled CENP-A assembly in somatic cell mitosis in other organisms (Chan and Wong 2012). A second scenario, which argues against a template-driven model for centromere inheritance, comes from a report of de novo CENP-A assembly in zygotic divisions in *A. thaliana* (Ingouff et al. 2010). In this study, analysis of plants expressing a GFP-tagged CENP-A revealed that CENP-A is present on mature sperm, but is absent on mature eggs (Ingouff et al. 2010). Analysis of the earliest divisions post-fertilisation revealed that paternal GFP-CENP-A foci are undetectable after karyogamy (fusion of male and female genomes), and are only detectable at the 16-cell stage of development (Ingouff 2010). These results suggest that CENP-A is assembled de novo in the zygote without pre-existing CENP-A as a guide, the mechanism of which is currently unknown. However, it is possible that a residual amount of CENP-A beyond the limits of detection is retained on egg chromatin or after karyogamy that is sufficient to direct new CENP-A deposition.

A common feature of plant and worm germ cells is the absence of CENP-A on one mature gamete (Table 1); CENP-A is absent from plant eggs (Ingouff et al. 2010), while CENP-A is absent from worm sperm (Gassmann et al. 2012). Alternative mechanisms to assemble CENP-A in zygotes might exist to counterbalance unequal CENP-A loading between male and female gametes; if CENP-A is absent, reduced or not loaded on one gamete, the organism can still reset CENP-A deposition in the zygote. It is also possible that one gamete is more sensitive to CENP-A reduction or modification. For example, GFP-tailswap plants are mostly male sterile; measured fertility was 3.5% for pollen, but 68.5% for ovules (Ravi

Table 1 CENP-A maintenance on mature gametes and the major mechanism of CENP-A assembly in early embryonic cell cycles in *D. melanogaster*, *C. elegans* and *A. thaliana*

Species	CENP-A on mature sperm	CENP-A on mature egg	CENP-A assembly in early embryos	References
D. melanogaster	✓	✓	Template-governed	Raychaudhuri et al. (2012)
C. elegans	✗	✓	de novo	Gassmann et al. (2012)
A. thaliana	✓	✗	de novo	Ingouff et al. (2010)

et al. 2011). This bias in sterility between sexes might reflect different dependencies for meiotic CENP-A and its assembly pathways in males and females.

4 Functions, Assembly and Inheritance of CENP-C in Germ Cells

CENP-C is a conserved primary component of the constitutive centromere-associated network and can directly interact with CENP-A chromatin (Carroll et al. 2009, 2010; Foltz et al. 2006; Falk et al. 2015). CENP-C is essential for mitosis and cell viability in diverse organisms (Meluh and Koshland 1995; Heeger et al. 2005; Kalitsis et al. 1998; Kwon et al. 2007), which has hampered investigations into potential roles in meiosis. However, the targeted isolation of point mutants in *cnp3* (fission yeast CENP-C) that leave mitotic functions intact, provide evidence for meiosis-specific functions (Tanaka et al. 2009). Specifically, C-terminal mutations in CENP-C perturb Moa1 (monopolar attachment 1) recruitment, a meiosis-specific protein exclusively required for mono-orientation of kinetochores in meiosis I (Tanaka et al. 2009). Interestingly, Moa1 and its functional equivalent in mammals Meikin share no significant sequence homology, yet Meikin was discovered through its interaction with CENP-C in mouse testes (Kim et al. 2015). Similarly in *D. melanogaster*, the isolation of a C-terminal point mutation in *cenp-C* that renders flies sterile, but leaves mitotic functions mostly intact, has aided dissection of CENP-C's roles in meiosis. In females, functional CENP-C is required for meiotic centromere clustering, pairing and chromosome segregation (Unhavaithaya and Orr-Weaver 2013). In males, CENP-C is also required for meiotic chromosome segregation, with additional roles in meiotic CENP-A assembly and the timely release of CAL1 and CENP-A from nucleoli (Kwenda et al. 2016). Additional, separation-of-function mutations in other organisms are likely to reveal further meiosis-specific roles of CENP-C. Remarkably, CENP-C appears to be dispensable for worm meiosis, in line with findings reported for CENP-A (Monen et al. 2005).

Compared to CENP-A, CENP-C assembly dynamics in either mitosis or meiosis are less well characterised. Unexpectedly, although CENP-C directly binds to CENP-A nucleosomes, its dynamics do not always follow those of CENP-A. In human mitotic cells in culture, quantitative FRAP and Fluorescence Correlation Spectroscopy (FCS) experiments show that CENP-C undergoes dynamic exchange in both G1 and G2 phase, but not S phase (Hemmerich et al. 2008). This result contrasts findings for CENP-A in this system, which only recovers in early G1 phase. In mouse meiosis, CENP-C-GFP can assemble in oocytes arrested at prophase I or upon maturation, yet CENP-A-GFP cannot (Smoak et al. 2016), further highlighting differential CENP-C and CENP-A dynamics in mammals. In other organisms, CENP-C assembly dynamics appear to align more closely with those of CENP-A. In worm oocytes, like CENP-A, CENP-C is first detected at centromeres in prophase I, at late pachytene/diplotene stages (Monen et al. 2005). In flies, also in line with CENP-A, CENP-C is assembled at anaphase of mitosis in embryonic divisions (Schuh et al. 2007) and at prophase I in meiosis in spermatocytes (Kwenda et al. 2016). Yet, unlike CENP-A, CENP-C level on spermatids drops off after the second meiotic division in male flies (Dunleavy et al. 2012; Raychaudhuri et al. 2012). Intriguingly, the timing of CENP-C 'removal' inversely correlates with an increase in CENP-A intensity at this stage, raising the possibility that novel factors participate in this second phase of CENP-A assembly.

One common feature between frogs, worm and flies is the absence of CENP-C on mature sperm (Milks et al. 2009; Gassmann et al. 2012; Raychaudhuri et al. 2012; Dunleavy et al. 2012). Additionally, CENP-C was not detected on plant meiocytes (Ravi et al. 2011). Therefore, it is unlikely that CENP-C is a mark of paternal centromere identity in the next generation, an epigenetic function attributed instead to CENP-A. Indeed, both in vitro chromatin assembly experiments in Xenopus (Milks et al. 2009) and in vivo dynamics of GFP-tagged CENP-C in Drosophila early embryos (Raychaudhuri et al. 2012), confirm de novo CENP-C assembly from a maternal pool supplied in the egg cytoplasm. Thus, in zygotes, CENP-C assembly is most likely specified by pre-existing or newly assembled CENP-A.

5 Centromere Structure/Function Roles in Homolog Pairing, Clustering and Synapsis in Prophase I

Distinguishing and critical features of meiotic prophase I include the pairing of homologous chromosomes, the assembly of the synaptonemal complex between homologs (synapsis) and homologous recombination leading to genetic exchange, as well as the formation of chiasma that hold homologs together until anaphase of meiosis (Watanabe 2012; Cahoon and Hawley 2016) (Fig. 5). Accumulating evidence suggests centromeres might play structural roles at very early stages of meiotic prophase I. In budding yeast, assembly of the synaptonemal complex component Zip1 in early zygotene initiates at paired homologous centromeres (Tsubouchi et al. 2008). In Drosophila females, assembly of C(3)G, the Zip1 equivalent in flies, is also first detected at centromeres (Tanneti et al. 2011; Takeo et al. 2011). In this system, centromere clustering is coincident with (Tanneti et al. 2011; Takeo et al. 2011) or immediately prior to (Christophorou et al. 2013) the initiation of meiosis. Intriguingly, mutations in *cenp-C* and *cal1* disrupt centromere clustering and homolog pairing in zygotene, as well as the retention of C(3)G at centromeres (Unhavaithaya and Orr-Weaver 2013). In this context, clustered and paired centromeres might serve as an effective structural platform to build and maintain functional associations between chromosomes. It is possible that CENP-A assembly at this time could reinforce such associations.

6 Current and Future Perspectives

Centromere dynamics in germ cells is a relatively new and exciting research field. Investigations into mechanisms of germ cell centromere function, assembly and maintenance in fly, plant, worm and frog model systems are certainly proving fruitful. A common emerging theme is that problems in centromere dynamics in germ cells can give rise to defective gametes, potentially resulting in aneuploidy in the next generation. Yet, experiments in more complex mammalian systems are currently lacking, limited by the accessibility to germ cells, as well as lack of genetic tools for germ cell-specific manipulations and appropriate in vitro culture systems. For example, centromere dynamics and functions in human germ cells are relatively unexplored, but are likely to prove important with clinical relevance for fertility and ageing. Major unresolved themes and future research questions include the following:

(i) Identifying key players and mechanisms of centromere specification and function in germ cells: Investigations so far have highlighted unexpected differences in CENP-A requirements and assembly dynamics in meiosis compared to mitosis. Given that in many organisms meiotic CENP-A assembles in prophase I, a cell cycle time when cyclin-dependent-kinase (CDK) activity is high, molecular signals must differ from mitosis (Silva et al. 2012). How is CENP-A assembly coupled to the meiotic cell cycle? What is the cell cycle timing of CENP-A assembly in male and female meiosis in mammals and in other species? Do mitotic chaperones and assembly factors function in meiosis or are meiosis-specific chaperones required? Undoubtedly, the targeted generation of separation-of-function mutations via the clustered regularly interspaced short palindromic repeats (CRISPR)/Cas9 system will aid the dissection of meiosis-specific functions of essential genes. Additionally, genetic screens in model organisms and biochemical purifications of centromere proteins from meiotic cells will identify critical interactions and regulators. It will also be of interest to determine mechanisms and consequences of CENP-A disassembly in meiosis.

(ii) Identifying the function of the CENP-A N-terminus in germ cells: What is the meiosis-specific role of the CENP-A N-terminus in plants? Are the N-terminal conserved sequence blocks, identified in plants and flies, important for this function? Outside of plants, is the CENP-A N-terminus required for meiosis in other organisms? To answer these questions, the targeted deletion of the motifs by the CRISPR/Cas9 system in transgenic plants and animals is sure to prove informative. Is the CENP-A N-terminus differentially modified in meiosis? Do meiosis-specific CENP-A assembly and maintenance factors interact with the N-terminus? For this, protocols that facilitate precise sorting of meiotic cells improved strategies for biochemical purifications of germ cells from tissues, as well as the development of more sensitive mass spectrometry approaches to identify proteins from reduced amounts of material are critical. Does the CENP-A N-terminus direct its removal in prophase I? FRAP experiments with truncated CENP-A transgenes could give insight into potential phases of meiotic CENP-A turnover and provide evidence to support a CENP-A quality control check before chromosome segregation in meiosis I. For this, conditions that enhance the viability of germ cells in tissues for high-resolution time-lapse imaging are key.

(iii) Identifying mechanisms that govern the transgenerational inheritance of CENP-A: In most organisms, CENP-A epigenetically marks the position of the centromere on male and female gametes, which likely determines centromere position and function in every cell in the next generation. Therefore, it is critical that this epigenetic information is transmitted with fidelity from one generation to the next. On one hand, evidence from flies strongly supports a template-driven mechanism for CENP-A transmission. It is now important to confirm if this model holds true through multiple generations and in other organisms and if so, which molecular mechanisms determine the quantitative, template-driven inheritance of CENP-A. On the other hand, it is important to consider the possibility that de novo CENP-A assembly in zygotes might occur in other species aside from plant and worm. If so, is CENP-A sufficient to nucleate centromere establishment in this context? It is also important to consider whether the balance of CENP-A inherited with male and female gametes is critical. For example, what is the relative contribution of egg and sperm chromatin to centromere identity in progeny and subsequent generations? What determines dependencies on specific CENP-A assembly pathways in male and female gametes in a given organism? Investigations into germ cell centromere dynamics in appropriate model organisms with short generational times and accessibility to both male and female gametes are key to the success of such transgenerational studies. Finally it will be important to determine how CENP-A resists protamine exchange during sperm differentiation.

References

Black BE, Foltz DR, Chakravarthy S, Luger K, Woods VL, Cleveland DW (2004) Structural determinants for generating centromeric chromatin. Nature 430:578–582

Black BE, Jansen LE, Maddox PS, Foltz DR, Desai AB, Shah JV, Cleveland DW (2007) Centromere identity maintained by nucleosomes assembled with histone H3 containing the CENP-A targeting domain. Mol Cell 25:309–322

Blower MD, Karpen GH (2001) The role of Drosophila CID in kinetochore formation, cell-cycle progression and heterochromatin interactions. Nat Cell Biol 3:730–739

Buchwitz BJ, Ahmad K, Moore LL, Roth MB, Henikoff S (1999) A histone-H3-like protein in *C. elegans*. Nature 401:547–548

Cahoon CK, Hawley RS (2016) Regulating the construction and demolition of the synaptonemal complex. Nat Struct Mol Biol 23:369–377

Carroll CW, Milks KJ, Straight AF (2010) Dual recognition of CENP-A nucleosomes is required for centromere assembly. J Cell Biol 189:1143–1155

Carroll CW, Silva MC, Godek KM, Jansen LE, Straight AF (2009) Centromere assembly requires the direct recognition of CENP-A nucleosomes by CENP-N. Nat Cell Biol 11:896–902

Chan FL, Wong LH (2012) Transcription in the maintenance of centromere chromatin identity. Nucleic Acids Res 40:11178–11188

Christophorou N, Rubin T, Huynh JR (2013) Synaptonemal complex components promote centromere pairing in pre-meiotic germ cells. PLoS Genet 9:e1004012

Dereeper A, Audic S, Claverie JM, Blanc G (2010) BLAST-EXPLORER helps you building datasets for phylogenetic analysis. BMC Evol Biol 10:18

Dumont J, Oegema K, Desai A (2010) A kinetochore-independent mechanism drives anaphase chromosome separation during acentrosomal meiosis. Nat Cell Biol 12:894–901

Dunleavy EM, Beier NL, Gorgescu W, Tang J, Costes SV, Karpen GH (2012) The cell cycle timing of centromeric chromatin assembly in Drosophila meiosis is distinct from mitosis yet requires CAL1 and CENP-C. PLoS Biol 10:e1001460

Fachinetti D, Folco HD, Nechemia-Arbely Y, Valente LP, Nguyen K, Wong AJ, Zhu Q, Holland AJ, Desai A, Jansen LE, Cleveland DW (2013) A two-step mechanism for epigenetic specification of centromere identity and function. Nat Cell Biol 15:1056–1066

Falk SJ, Guo LY, Sekulic N, Smoak EM, Mani T, Logsdon GA, Gupta K, Jansen LE, van Duyne GD, Vinogradov SA, Lampson MA, Black BE (2015) Chromosomes. CENP-C reshapes and stabilizes CENP-A nucleosomes at the centromere. Science 348:699–703

Foltz DR, Jansen LE, Black BE, Bailey AO, Yates JR, Cleveland DW (2006) The human CENP-A centromeric nucleosome-associated complex. Nat Cell Biol 8:458–469

Gassmann R, Rechtsteiner A, Yuen KW, Muroyama A, Egelhofer T, Gaydos L, Barron F, Maddox P, Essex A, Monen J, Ercan S, Lieb JD, Oegema K, Strome S, Desai A (2012) An inverse relationship to germline transcription defines centromeric chromatin in *C. elegans*. Nature 484:534–537

Heeger S, Leismann O, Schittenhelm R, Schraidt O, Heidmann S, Lehner CF (2005) Genetic interactions of separase regulatory subunits reveal the diverged Drosophila Cenp-C homolog. Genes Dev 19:2041–2053

Hemmerich P, Weidtkamp-Peters S, Hoischen C, Schmiedeberg L, Erliandri I, Diekmann S (2008) Dynamics of inner kinetochore assembly and maintenance in living cells. J Cell Biol 180:1101–1114

Howman EV, Fowler KJ, Newson AJ, Redward S, Macdonald AC, Kalitsis P, Choo KH (2000) Early disruption of centromeric chromatin organization in centromere protein A (Cenpa) null mice. Proc Natl Acad Sci U S A 97:1148–1153

Ingouff M, Rademacher S, Holec S, Soljić L, Xin N, Readshaw A, Foo SH, Lahouze B, Sprunck S, Berger F (2010) Zygotic resetting of the HISTONE 3 variant repertoire participates in epigenetic reprogramming in Arabidopsis. Curr Biol 20:2137–2143

Jansen LE, Black BE, Foltz DR, Cleveland DW (2007) Propagation of centromeric chromatin requires exit from mitosis. J Cell Biol 176:795–805

Kalitsis P, Fowler KJ, Earle E, Hill J, Choo KH (1998) Targeted disruption of mouse centromere protein C gene leads to mitotic disarray and early embryo death. Proc Natl Acad Sci U S A 95:1136–1141

Karimi-Ashtiyani R, Ishii T, Niessen M, Stein N, Heckmann S, Gurushidze M, Banaei-Moghaddam AM, Fuchs J, Schubert V, Koch K, Weiss O, Demidov D, Schmidt K, Kumlehn J, Houben A (2015) Point mutation impairs centromeric CENH3 loading and induces haploid plants. Proc Natl Acad Sci U S A 112:11211–11216

Kim J, Ishiguro K, Nambu A, Akiyoshi B, Yokobayashi S, Kagami A, Ishiguro T, Pendas AM, Takeda N, Sakakibara Y, Kitajima TS, Tanno Y, Sakuno T, Watanabe Y (2015) Meikin is a conserved regulator of meiosis-I-specific kinetochore function. Nature 517:466–471

Kwenda L, Collins CM, Dattoli AA, Dunleavy EM (2016) Nucleolar activity and CENP-C regulate CENP-A and CAL1 availability for centromere assembly in meiosis. Development 143:1400–1412

Kwon MS, Hori T, Okada M, Fukagawa T (2007) CENP-C is involved in chromosome segregation, mitotic checkpoint function, and kinetochore assembly. Mol Biol Cell 18: 2155–2168

Lando D, Endesfelder U, Berger H, Subramanian L, Dunne PD, McColl J, Klenerman D, Carr AM, Sauer M, Allshire RC, Heilemann M, Laue ED (2012) Quantitative single-molecule microscopy reveals that CENP-A(Cnp1) deposition occurs during G2 in fission yeast. Open Biol 2:120078

Lermontova I, Fuchs J, Schubert V, Schubert I (2007) Loading time of the centromeric histone H3 variant differs between plants and animals. Chromosoma 116:507–510

Lermontova I, Koroleva O, Rutten T, Fuchs J, Schubert V, Moraes I, Koszegi D, Schubert I (2011) Knockdown of CENH3 in Arabidopsis reduces mitotic divisions and causes sterility by disturbed meiotic chromosome segregation. Plant J 68:40–50

Lermontova I, Schubert V, Fuchs J, Klatte S, Macas J, Schubert I (2006) Loading of Arabidopsis centromeric histone CENH3 occurs mainly during G2 and requires the presence of the histone fold domain. Plant Cell 18:2443–2451

Logsdon GA, Barrey EJ, Bassett EA, Denizio JE, Guo LY, Panchenko T, Dawicki-Mckenna JM, Heun P, Black BE (2015) Both tails and the centromere targeting domain of CENP-A are required for centromere establishment. J Cell Biol 208:521–531

Maheshwari S, Tan EH, West A, Franklin FC, Comai L, Chan SW (2015) Naturally occurring differences in CENH3 affect chromosome segregation in zygotic mitosis of hybrids. PLoS Genet 11:e1004970

Malik HS, Henikoff S (2003) Phylogenomics of the nucleosome. Nat Struct Biol 10:882–891

Malik HS, Vermaak D, Henikoff S (2002) Recurrent evolution of DNA-binding motifs in the Drosophila centromeric histone. Proc Natl Acad Sci U S A 99:1449–1454

Marques A, Pedrosa-Harand A (2016) Holocentromere identity: from the typical mitotic linear structure to the great plasticity of meiotic holocentromeres. Chromosoma 125(4):669–81

Meluh PB, Koshland D (1995) Evidence that the MIF2 gene of Saccharomyces cerevisiae encodes a centromere protein with homology to the mammalian centromere protein CENP-C. Mol Biol Cell 6:793–807

Meyer GF (1960) Proceedings of the European regional conference on electron microscopy. Die Nederlandse Verening voor Electronmicroscopic Delft

Milks KJ, Moree B, Straight AF (2009) Dissection of CENP-C-directed centromere and kinetochore assembly. Mol Biol Cell 20:4246–4255

Monen J, Maddox PS, Hyndman F, Oegema K, Desai A (2005) Differential role of CENP-A in the segregation of holocentric *C. elegans* chromosomes during meiosis and mitosis. Nat Cell Biol 7:1248–1255

Notredame C, Higgins DG, Heringa J (2000) T-Coffee: a novel method for fast and accurate multiple sequence alignment. J Mol Biol 302:205–217

Palmer DK, O'Day K, Margolis RL (1990) The centromere specific histone CENP-A is selectively retained in discrete foci in mammalian sperm nuclei. Chromosoma 100:32–36

Palmer DK, O'Day K, Trong HL, Charbonneau H, Margolis RL (1991) Purification of the centromere-specific protein CENP-A and demonstration that it is a distinctive histone. Proc Natl Acad Sci U S A 88:3734–3738

Ravi M, Chan SW (2010) Haploid plants produced by centromere-mediated genome elimination. Nature 464:615–618

Ravi M, Kwong PN, Menorca RM, Valencia JT, Ramahi JS, Stewart JL, Tran RK, Sundaresan V, Comai L, Chan SW (2010) The rapidly evolving centromere-specific histone has stringent functional requirements in *Arabidopsis thaliana*. Genetics 186:461–471

Ravi M, Shibata F, Ramahi JS, Nagaki K, Chen C, Murata M, Chan SW (2011) Meiosis-specific loading of the centromere-specific histone CENH3 in *Arabidopsis thaliana*. PLoS Genet 7: e1002121

Raychaudhuri N, Dubruille R, Orsi GA, Bagheri HC, Loppin B, Lehner CF (2012) Transgenerational propagation and quantitative maintenance of paternal centromeres depends on Cid/Cenp-A presence in Drosophila sperm. PLoS Biol 10:e1001434

Sanei M, Pickering R, Kumke K, Nasuda S, Houben A (2011) Loss of centromeric histone H3 (CENH3) from centromeres precedes uniparental chromosome elimination in interspecific barley hybrids. Proc Natl Acad Sci U S A 108:E498–E505

Schittenhelm RB, Althoff F, Heidmann S, Lehner CF (2010) Detrimental incorporation of excess Cenp-A/Cid and Cenp-C into Drosophila centromeres is prevented by limiting amounts of the bridging factor Cal1. J Cell Sci 123:3768–3779

Schubert V, Lermontova I, Schubert I (2014) Loading of the centromeric histone H3 variant during meiosis-how does it differ from mitosis? Chromosoma 123:491–497

Schuh M, Lehner CF, Heidmann S (2007) Incorporation of Drosophila CID/CENP-A and CENP-C into centromeres during early embryonic anaphase. Curr Biol 17:237–243

Silva MC, Bodor DL, Stellfox ME, Martins NM, Hochegger H, Foltz DR, Jansen LE (2012) Cdk activity couples epigenetic centromere inheritance to cell cycle progression. Dev Cell 22:52–63

Smoak EM, Stein P, Schultz RM, Lampson MA, Black BE (2016) Long-term retention of CENP-A nucleosomes in Mammalian oocytes underpins transgenerational inheritance of centromere identity. Curr Biol 26:1110–1116

Stoler S, Keith KC, Curnick KE, Fitzgerald-Hayes M (1995) A mutation in CSE4, an essential gene encoding a novel chromatin-associated protein in yeast, causes chromosome nondisjunction and cell cycle arrest at mitosis. Genes Dev 9:573–586

Takahashi K, Chen ES, Yanagida M (2000) Requirement of Mis6 centromere connector for localizing a CENP-A-like protein in fission yeast. Science 288:2215–2219

Takeo S, Lake CM, Morais-De-sá E, Sunkel CE, Hawley RS (2011) Synaptonemal complex-dependent centromeric clustering and the initiation of synapsis in Drosophila oocytes. Curr Biol 21:1845–1851

Tanaka K, Chang HL, Kagami A, Watanabe Y (2009) CENP-C functions as a scaffold for effectors with essential kinetochore functions in mitosis and meiosis. Dev Cell 17:334–343

Tanneti NS, Landy K, Joyce EF, McKim KS (2011) A pathway for synapsis initiation during zygotene in Drosophila oocytes. Curr Biol 21:1852–1857

Torras-Llort M, Moreno-Moreno O, Azorín F (2009) Focus on the centre: the role of chromatin on the regulation of centromere identity and function. EMBO J 28:2337–2348

Tsubouchi T, Macqueen AJ, Roeder GS (2008) Initiation of meiotic chromosome synapsis at centromeres in budding yeast. Genes Dev 22:3217–3226

Unhavaithaya Y, Orr-Weaver TL (2013) Centromere proteins CENP-C and CAL1 functionally interact in meiosis for centromere clustering, pairing, and chromosome segregation. Proc Natl Acad Sci U S A 110:19878–19883

Valente LP, Silva MC, Jansen LE (2012) Temporal control of epigenetic centromere specification. Chromosome Res 20:481–492

von Stetina JR, Orr-Weaver TL (2011) Developmental control of oocyte maturation and egg activation in metazoan models. Cold Spring Harb Perspect Biol 3:a005553

Watanabe Y (2012) Geometry and force behind kinetochore orientation: lessons from meiosis. Nat Rev Mol Cell Biol 13:370–382

Zeitlin SG, Patel S, Kavli B, Slupphaug G (2005) Xenopus CENP-A assembly into chromatin requires base excision repair proteins. DNA Repair (Amst) 4:760–772

Cell Biology of Cheating—Transmission of Centromeres and Other Selfish Elements Through Asymmetric Meiosis

Lukáš Chmátal, Richard M. Schultz, Ben E. Black and Michael A. Lampson

Abstract Mendel's First Law of Genetics states that a pair of alleles segregates randomly during meiosis so that one copy of each is represented equally in gametes. Whereas male meiosis produces four equal sperm, in female meiosis only one cell, the egg, survives, and the others degenerate. Meiotic drive is a process in which a selfish DNA element exploits female meiotic asymmetry and segregates preferentially to the egg in violation of Mendel's First Law, thereby increasing its transmission to the offspring and frequency in a population. In principle, the selfish element can consist either of a centromere that increases its transmission via an altered kinetochore connection to the meiotic spindle or a centromere-like element that somehow bypasses the kinetochore altogether in doing so. There are now examples from eukaryotic model systems for both types of meiotic drive. Although meiotic drive has profound evolutionary consequences across many species, relatively little is known about the underlying mechanisms. We discuss examples in various systems and open questions about the underlying cell biology, and propose a mechanism to explain biased segregation in mammalian female meiosis.

L. Chmátal · R.M. Schultz · M.A. Lampson (✉)
Department of Biology, University of Pennsylvania, 433 South University Ave, Philadelphia, PA 19104, USA
e-mail: lampson@sas.upenn.edu

L. Chmátal
e-mail: chmatal@wi.mit.edu

R.M. Schultz
e-mail: rschultz@sas.upenn.edu

B.E. Black
Department of Biochemistry and Biophysics, University of Pennsylvania, 422 Curie Blvd, Philadelphia, PA 19104-6059, USA
e-mail: blackbe@mail.med.upenn.edu

L. Chmátal
Whitehead Institute for Biomedical Research, Cambridge, MA 02142, USA

B.E. Black (ed.), *Centromeres and Kinetochores*, Progress in Molecular and Subcellular Biology 56, DOI 10.1007/978-3-319-58592-5_16

1 Introduction

Asymmetric meiotic division creates an opportunity for cheating when selfish elements increase their likelihood of transmission to the functional gamete. In cases where mechanisms of such preferential segregation are understood, they are mediated by repetitive DNA elements that ultimately interact with microtubules (MTs). The first example of such preferential transmission was observed by Marcus Rhoades in maize (Rhoades 1942), and the repetitive DNA elements were termed 'neocentromeres' because of their ability to drive segregation at cell division. It should be noted that because maize neocentromeres use a special connection to MTs (Yu et al. 1997), they are distinct from "neocentromeres" subsequently described in many other eukaryotic species that recruit a conventional kinetochore-mediated MT connection. The term "meiotic drive" was later introduced to emphasize the key role of asymmetric female meiosis (Sandler and Novitski 1957). Over time, the meaning of "meiotic drive" has been extended to include other forms of transmission ratio distortion that are not strictly a consequence of asymmetric meiosis (meiotic drive sensu lato, Fig. 1), but could be a result of post-meiotic processes (e.g., gamete competition, post-fertilization selection) (Lyttle 1991; Pardo-Manuel de Villena and Sapienza 2001a). Here, we use the term meiotic drive in its originally defined meaning, as depending on the mechanics of asymmetric female meiosis (meiotic drive sensu stricto, Fig. 1).

Meiotic drive violates Mendel's First Law, and although the mechanisms of meiotic chromosome segregation underlying Mendelian genetics have been extensively studied, the cell biology of meiotic drive is relatively unexplored. There are many open questions regarding how selfish elements exploit the meiotic chromosome segregation machinery to maximize their own propagation. We discuss examples of meiotic drive in various systems, to illustrate the breadth of the phenomenon, starting with examples (e.g., plant neocentromeres) that do not depend on typical centromeres. The mechanisms are unknown in most cases and raise intriguing questions about the underlying cell and molecular biology. We also discuss the concept of "centromere drive", for which cell biological models are more intuitive but many open questions remain, and propose a mechanism to explain biased segregation in mammalian female meiosis. Our proposed mechanism incorporates the apparent paradox that exists because centromeres in most eukaryotes are defined epigenetically (i.e., not by a particular DNA sequence; (Black and Cleveland 2011), yet centromere drive is also thought to involve changes in the underlying DNA that strengthen MT connections (Henikoff et al. 2001). In considering the applicability of our model for mammalian centromeres to other systems, one should ask the following questions: *In which cases does the drive mechanism rely on repeat expansion or other changes to the DNA sequence related to modifying a conventional, kinetochore-building centromere? And in which cases does it relate to creating something altogether different (e.g., the abnormal, specialized connections seen with maize neocentromeres, detailed immediately below)*?

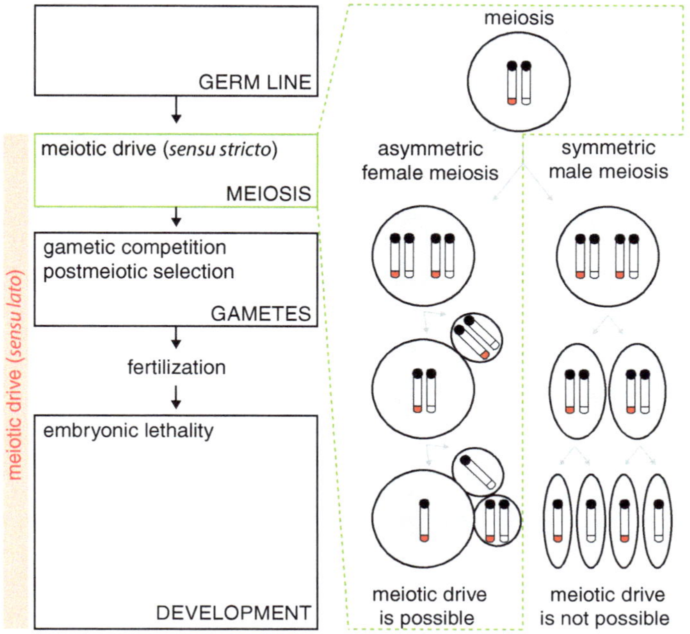

Fig. 1 Meiotic drive terminology. Meiotic drive sensu stricto refers to selfish DNA elements (*red part* of chromosome) exploiting asymmetric female meiosis by preferential transmission to the viable gamete, whereas other gametes degenerate as polar bodies. Meiotic drive sensu lato includes changes in allele frequency due to other mechanisms, such as gametic competition or embryonic lethality (Sandler et al. 1959). *Black circles* indicate centromeres. Adapted from Pardo-Manuel de Villena and Sapienza (2001a)

2 Abnormal Chromosome 10 in Maize

The first experimental evidence of meiotic drive arose from observations of abnormal chromosome 10 (Ab10) in maize (*Zea mays*). In contrast to normal chromosome 10 (N10), Ab10 contains an extra DNA segment that includes regions of euchromatin, an inverted portion of N10, and a repetitive DNA sequence (knob). By following a knob-linked genetic marker, Rhoades showed that Ab10 preferentially segregates to the surviving gamete (megaspore) during female meiosis and proposed a model to explain the phenomenon (Rhoades 1942, 1952). In this model, after recombination between the Ab10 and N10 chromosomes, Ab10 knob activity results in a shifted chromosome position toward meiotic spindle poles in anaphase I, and this position is maintained through meiosis II. Ab10 is, therefore, more likely to segregate to the basal cell that later becomes the megaspore, whereas the other cells degenerate (Fig. 2). Several lines of experimental evidence support Rhoades's model (Rhoades and Vilkomerson 1942; Dawe and Cande 1996; Yu et al. 1997). First, Ab10 knobs act as neocentromeres in that they bind MTs, although they do

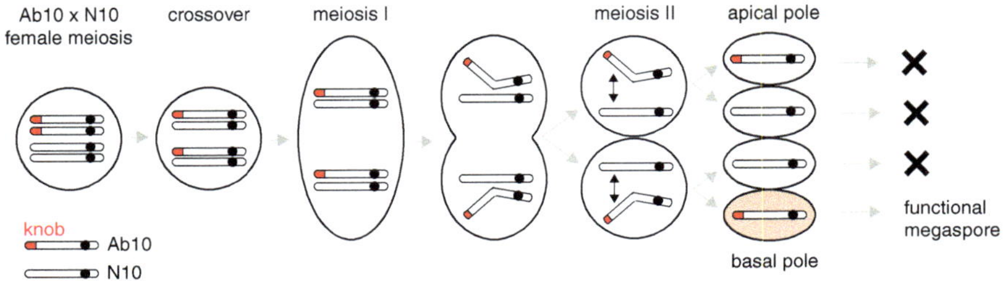

Fig. 2 Ab10 chromosome in maize. Abnormal chromosome 10 (Ab10) with an extra DNA sequence (knob, *red*) recombines with normal chromosome 10 (N10) in meiosis I. Neocentromere activity of Ab10 knobs leads to faster movement to the cell poles and preferential Ab10 position maintained through meiosis II, so that Ab10 chromosomes are transmitted into apical and basal cells. Ab10 transmits preferentially to the next generation because only the basal cell develops to a functional gamete whereas the others degenerate

not assemble typical kinetochores as shown by the absence of the major structural component CENP-C (Dawe et al. 1999). The knobs also lack nucleosomes containing the histone H3 variant, CENP-A, that specify centromere location on typical chromosomes in maize and other eukaryotes. Second, the knob drives faster movement of Ab10 to the spindle poles in anaphase, compared to N10, suggesting that the knob binds a minus-end-directed microtubule motor. Third, in telophase I the knobs are positioned peripherally, on the poleward side of the nucleus. If this localization is maintained into meiosis II, through mechanisms that are unclear, the knobs would end up in the outer megaspores. Multiple other knobs have also been identified on other maize chromosomes and drive in the presence of Ab10, likely by a similar mechanism (Longley 1945; Buckler et al. 1999).

More than seven decades after Rhoades's original model, many outstanding questions still remain. If the knob binds molecular motors, what motors are involved, and how is the interaction mediated? Why is the knob active only in meiosis and not in mitosis? How is drive suppressed by the *Smd1* (suppressor of meiotic drive 1 locus) mutation that was identified using transposon mutagenesis to isolate maize mutants with reduced meiotic drive (Dawe and Cande 1996)? Another challenge is to extend the conclusions from cytological experiments performed in male gametes to biased segregation in the female germline.

3 B Chromosomes

B chromosomes, detected in numerous plants, fungi, and animals (Jones 1995; Burt and Trivers 2006), represent dispensable DNA elements in addition to the standard (A) chromosomes. Although B chromosomes evolved from A chromosomes, they are highly heterochromatic and mostly genetically inactive and act as independent parasitical units, increasing their frequency in the population (Östergren 1945; Jones and Rees 1982). B chromosome drive refers to preferential transmission that can occur as either a pre-meiotic, meiotic, or post-meiotic process, mediated by various mechanisms in males and females (Hewitt 1976; Jones 1991; Banaei-Moghaddam et al. 2012). In an example of meiotic B chromosome drive in grasshopper (*Myrmeleotettix maculates*), biased transmission into the egg during female meiosis is mediated by an asymmetric meiotic spindle, with the egg side of the spindle estimated as approximately three times longer than the polar body side (Hewitt 1976; Jones 1991). A simple model is that B chromosome positioning on such a meiotic spindle is random, so the chromosome more likely attaches to the larger, egg side of the spindle. The mechanistic bases are unknown for both elements of the drive model: B chromosome attachment to the spindle and meiotic spindle asymmetry.

4 Driving Loci in Mouse

Several non-centromere driving loci have been identified in mouse. One example is a homogeneously staining region (HSR) of long-range repeats of about 100 kb each, found on chromosome 1 in remote natural populations of *Mus musculus musculus* (Traut et al. 1984; Yukimenko and Korobitsyna 1988; Agulnik et al. 1990, 1993a, b, c; Sabantsev et al. 1993). An inversion (*In*) splits the HSR sequence into two distinct loci (Fig. 3a), and females heterozygous for the inverted HSR repeat (In/+) preferentially transmit *In* over the normal chromosome 1 ($\sim$85% vs. 15%). Because HSR is far from the centromere, it frequently segregates from the normal chromosome 1 in MII due to recombination, which implies that drive occurs in MII (Fig. 3b) (Ruvinsky 1995). Another example of meiotic drive at MII is the Ovum mutant locus (*Om*) mapped to mouse chromosome 11 (Pardo-Manuel de Villena et al. 2000; Wu et al. 2005). Intriguingly, MII drive in both cases depends on the genotype of the sperm, for example only if the In/+ MII egg is fertilized by sperm with the normal version of chromosome 1 lacking the HSR.

A third example of a driving locus, with as high as 94% preferential transmission in heterozygous females, is *R2d2* (Responder to drive on chromosome 2) (Didion et al. 2015), a massive copy number expansion formed by 36 units of repetitive DNA. Drive depends on *R2d2* expansion, as a strain with only one unit of this repetitive sequence does not exhibit drive. In all three cases of driving loci in mouse, the underlying mechanisms are completely unknown. Outstanding questions include whether any of the loci exhibit neocentromere activity (similar to Ab10), possible contributions of genes present in the HSR or in the *R2d2* cluster, and how the sperm contributes to drive in MII. Future studies in these drive systems may provide new insight into meiotic chromosome segregation.

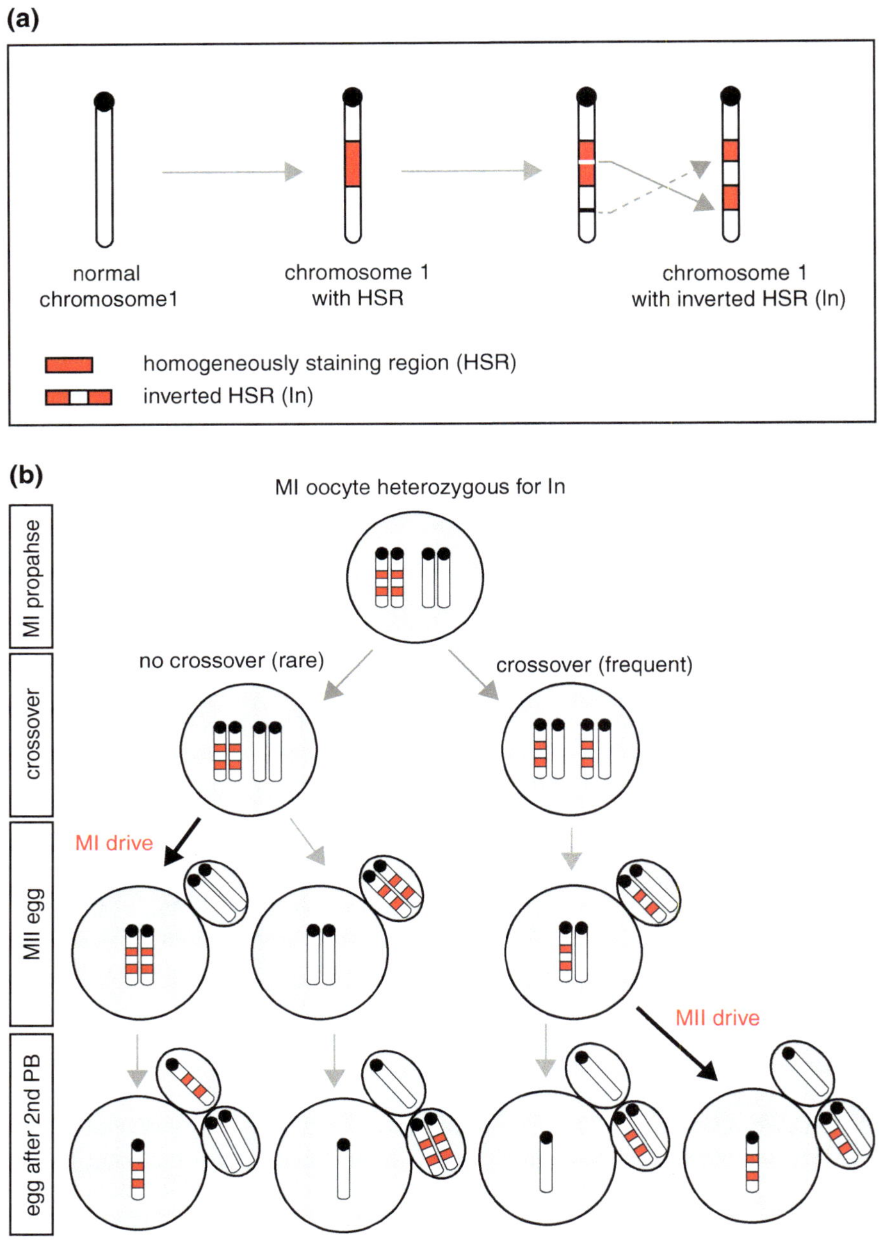

Fig. 3 Abnormal chromosome 1 in mouse. **a** Normal chromosome 1 and its abnormal forms with the homogeneously staining region (HSR) or the inverted HSR (*In*). **b** Chromosome 1 with the inverted HSR drives in MI when no crossover occurs (rare) and in MII when *In* recombines with normal chromosome 1 (frequent). Adapted from Ruvinsky (1995)

5 Univalent X Chromosome Segregation

Female mice with only one X chromosome in their karyotype (XO) are fertile (unlike human XO) and preferentially produce XX females (60%) rather than XO females (40%) when crossed to normal males (Cattanach 1962; Kaufman 1972). These findings imply that the single X chromosome is preferentially retained in female MI oocytes. Visualizing the X chromosome in MII eggs from XO females shows that 30% of the oocytes segregate the single X sister chromatids equationally in MI with no drive, and the remaining 70% of oocytes segregate the complete X univalent preferentially to the egg rather than to the polar body (2:1 bias) (Fig. 4a) (LeMaire-Adkins and Hunt 2000). How biased segregation of X univalents is achieved is unclear, but the mechanism may be similar to centromere drive as discussed below.

Caenorhabditis elegans with an additional X chromosome (XX + X) produce ~70% normal ova with a single X, and only 30% defective ova with two X chromosomes, suggesting preferential segregation of the extra X to the polar body (Hodgkin et al. 1979). Consistent with these results, X univalents are preferentially eliminated during MI in *him-8* mutants in which chiasmata fail to form between the two X chromosomes in a normal XX karyotype (Cortes et al. 2015). The biased univalent segregation during anaphase I is likely due its abnormal position on the MI spindle, closer to the cortex, and the position of the contractile ring separating the ovum and the first polar body. The contractile ring forms preferentially between the lagging univalent chromosome and the egg spindle pole, so the univalent ends up in the polar body (Fig. 4b). This mechanism is not specific for univalent X chromosomes but likely applies to all univalents (Cortes et al. 2015). How lagging univalents influence the position of the contractile ring is unclear.

6 Centromere-Associated Drive in Yellow Monkey-Flowers (*Mimulus guttatus*)

In natural populations of yellow monkey-flower (*Mimulus guttatus*), the D allele exhibits almost complete (98%) transmission advantage in interspecific crosses between *M. guttatus* and *Mylohyus nasutus* (Fishman and Willis 2005). This biased transmission strongly correlates with the presence of an extra centromere-associated DNA repeat domain (Fishman and Saunders 2008). Preferential transmission of the D allele (58%) is also detected within *M. guttatus* but is counterbalanced by male infertility, as *M. guttatus* homozygous for the D allele suffer from significantly reduced pollen viability, compared to other genotypic classes. Thus, the balance between transmission advantage of the D allele through females and pollen inviability in males leads to a D allele polymorphism in *M. guttatus* populations.

Overall the observations in monkey-flower support the centromere drive hypothesis (Henikoff et al. 2001; Malik and Henikoff 2009). In this model,

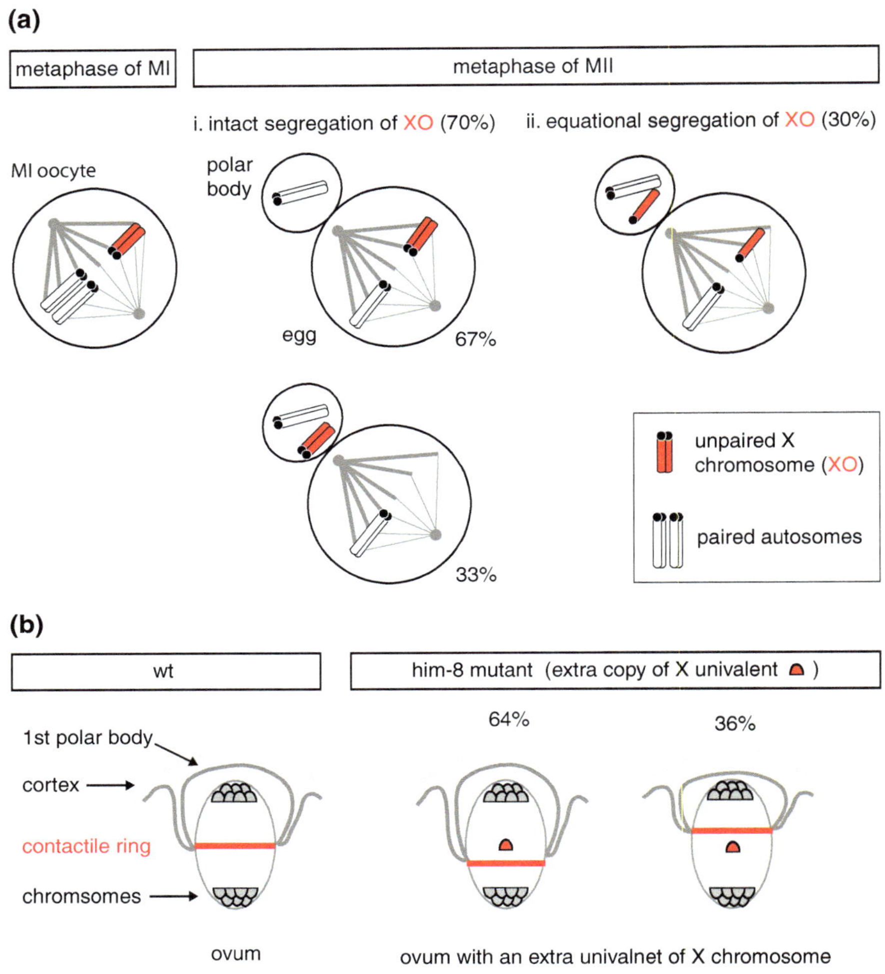

Fig. 4 XO chromosome drive in mouse and ***C. elegans***. **a** In XO mouse oocytes, an unpaired X chromosome is either preferentially retained in the egg as an intact univalent during MI (*i*) or the two sister chromatids segregate equationally in MI (*ii*). **b** A univalent X chromosome is preferentially expelled to the polar body during MI in *C. elegans* due to its lagging position at the metaphase plate and the positioning of the contractile ring. Adapted from LeMaire-Adkins and Hunt (2000) and Cortes et al. (2015)

"stronger" centromeres with expanded repetitive sequences recruit more kinetochore proteins and are more likely to remain in the egg during female meiosis, which would provide a selective advantage driving centromere expansion. However, imbalances between centromeres may also be associated with a fitness cost, such as reduced male fertility, creating selection pressure to equalize kinetochores by changing the centromere proteins binding to the expanded repeats.

Such opposite selection forces may explain the paradox of rapidly evolving centromere sequences and proteins that bind to them, despite conserved centromere function (Henikoff et al. 2001).

Although the observations in monkey-flower are largely consistent with the centromere drive hypothesis, several fundamental questions remain. *What are the consequences of the centromere expansion for kinetochore function, how do chromosomes with the D allele achieve preferential segregation, how does centromere strength relate to the typical epigenetic determinants of centromere identity, and what is the cause of reduced male fertility?* Future mechanistic studies may yield exciting insights into the cell biology of centromere drive, particularly given the magnitude of transmission bias in this system.

7 Robertsonian Fusion Chromosomes in Mouse

Robertsonian (Rb) fusions are common chromosomal rearrangements formed by two telocentric chromosomes (centromere at the end) joining at their centromeres to create one metacentric chromosome (internal centromere) (Fig. 5a) (White et al. 2010). Because they occur with high frequency in the germline compared to other rearrangements (Evans et al. 1978; Jacobs et al. 1992), their preferential accumulation through female meiosis can lead to massive karyotype change from a predominantly telocentric to a predominantly metacentric chromosome constitution. Western house mouse (*Mus musculus domesticus*) is an example of such karyotype divergence (Gropp et al. 1969; Piálek et al. 2005). The karyotype typically consists of all telocentric chromosomes ($2N = 40$), but numerous natural mouse populations have fixed multiple different Rb fusions within 10^2–10^5 years, reducing their chromosome numbers to almost a half in some cases (e.g., $2N = 22$) (Piálek et al. 2005; Garagna et al. 2014).

Fixation of Rb fusions can be explained by meiotic drive. When a new Rb fusion forms in the germline and is present in the heterozygous state, it pairs with the homologous telocentric chromosomes to form a trivalent in MI (Fig. 5a). Biased segregation of the Rb fusion can in principle drive karyotype change in a population (Fig. 5b) (Pardo-Manuel de Vill ena and Sapienza 2001b). According to this model, Rb fusions segregate preferentially to the egg in natural populations that have changed karyotype by accumulating metacentric Rb fusions, and preferentially to the polar body in other populations that have remained telocentric. Such biased transmission of Rb fusions is consistent with karyotype change in numerous other mammalian species (Buckland and Evans 1978; Pardo-Manuel de Villena and Sapienza 2001b; Aniskin et al. 2006; Mao et al. 2008; White et al. 2010) and with preferential retention of Rb fusions in the egg in humans, which tends to maintain the fusions in the female germline and increases the risk of producing aneuploid eggs (de Villena and Sapienza 2001).

Consistent with the centromere drive hypothesis (Henikoff et al. 2001), preferential transmission of Rb fusions correlates with the strength of the fusion

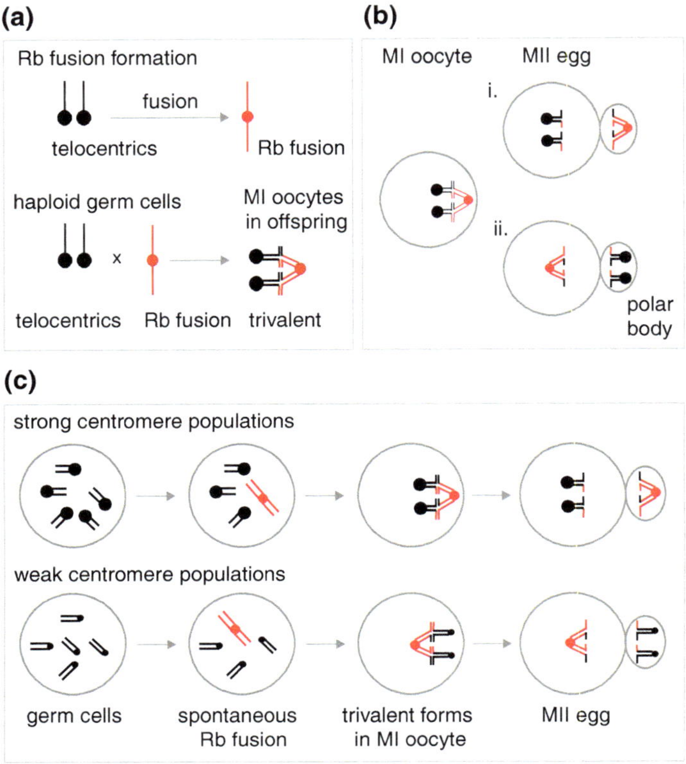

Fig. 5 Rb fusion chromosomes in mouse. **a** Rb fusion and trivalent formation. **b** Two possible outcomes of trivalent segregation in MI. **c** Model of Rb fusion drive in different directions in populations with strong or weak centromere backgrounds (see text for details). Centromere strength is indicated by the size of the circle (*red* Rb fusion, *black* telocentric). Adapted from Chmátal et al. (2014)

centromere relative to centromeres of the homologous unfused telocentrics, as determined in several ways (Chmátal et al. 2014). First, in an oocyte heterozygous for a single Rb fusion, kinetochore proteins are enriched at centromeres of the telocentric chromosomes that preferentially remain in the egg, relative to the homologous metacentric fusion that preferentially segregates to the polar body. These kinetochore proteins include CENP-A and the major MT-binding protein NDC80/HEC1. Second, in a natural metacentric population that accumulated Rb fusions (CHPO, $2n = 26$), the fusion centromeres are enriched for these same kinetochore proteins relative to the telocentric chromosomes. Biased segregation cannot be measured in CHPO because the metacentrics are present in the homozygous state, but it is likely that these fusions were subject to drive to preferentially remain in the egg as they accumulated in the population. Third, CHPO centromeres overall recruit less NDC80/HEC1 compared to a standard lab strain such as CF-1. Because CHPO contains telocentrics as well as metacentrics, a CHPO × CF-1 cross generates asymmetric bivalents with different levels of NDC80/HEC1 on the two sides of the bivalent. These bivalents are positioned off-center on the MI spindle, indicating that the larger kinetochores bind more MTs. Together, these data establish correlations between kinetochore size as measured by the abundance of MT-binding kinetochore proteins, MT-binding capacity, and preferential retention of the chromosome in the egg; centromere strength can be considered to reflect all three characteristics.

Based on these results, we proposed a model for how Rb fusions can exhibit drive in opposite directions to either maintain a telocentric karyotype or accumulate metacentrics, based on relative centromere strength (Fig. 5c). If centromere strength varies between populations, the strength of a newly formed Rb fusion centromere relative to the homologous telocentrics may depend on the background in which the fusion occurs. In this model, fusions arising on a strong centromere background would tend to have weaker centromeres than the homologous telocentrics, so that the fusions would preferentially segregate to the polar body and disappear from the population. Conversely, fusions arising on a weak centromere background would be stronger than the telocentrics and accumulate in the population because they are preferentially retained in the egg. This model is supported by kinetochore protein (NDC80/HEC1) staining in mice from various geographical regions in Europe, which suggests that natural metacentric populations (like CHPO) generally have weaker centromeres, which would have predisposed them to accumulate metacentrics (Chmátal et al. 2014).

8 Mechanistic Models of Meiotic Drive—Weaker and Stronger Centromeres

In general, models for meiotic drive acting on homologous chromosomes include functional heterozygosity of the drive locus, which influences chromosome interactions with the meiotic spindle, and asymmetry in meiosis with respect to cell fate (Rhoades 1952; Pardo-Manuel de Villena and Sapienza 2001a; Henikoff et al. 2001). In the most intuitive case the drive locus is the centromere, and centromeres of different strengths form kinetochores that interact differently with MTs. Alternatively, the drive locus exhibits neocentromere activity, independent of the normal centromere, as shown for Ab10. Many outstanding questions remain about the cell biology of various meiotic drive systems. For example, what is the molecular basis of Ab10 neocentromere activity, and how do other non-centromere drive loci (such as *R2d2*) influence MT interactions? Furthermore, the centromere drive hypothesis suggests that centromere strength depends on centromere DNA, but it is unclear how DNA sequence can influence kinetochore function, given that centromeres are typically specified epigenetically (Black and Cleveland 2011). In particular, it is not yet clear how centromeres may drive in individuals within a species where a centromere has moved to a location lacking centromere repeats, or within an entire species, like orangutan (Locke et al. 2011), horse (Wade et al. 2009), or chicken (Shang et al. 2010) where an 'evolutionary new centromere' has formed on one or more chromosomes. There is some evidence that human neocentromeres may be weaker by virtue of faulty mitotic error correction (Bassett et al. 2010) or through reduced recruitment of the constitutive centromere-associated network (CCAN) of proteins (Fachinetti et al. 2015).

It is clearly important to distinguish kinetochore-independent drive models, where expansion of repetitive DNA elements at a non-centromere locus generates and expands an interaction with the spindle (e.g., maize knob neocentromeres) (Fig. 6a), from those where centromere strength is increased by somehow building a larger kinetochore (e.g., mouse centromeres) (Fig. 6b). Because there is a strong epigenetic component to centromere identity that would impact the latter form of drive, proteins that define a functional centromere are likely involved in both generating a stronger centromere and as candidates to evolve to counteract the proposed detrimental consequences of imbalances between stronger and weaker centromeres (Henikoff et al. 2001).

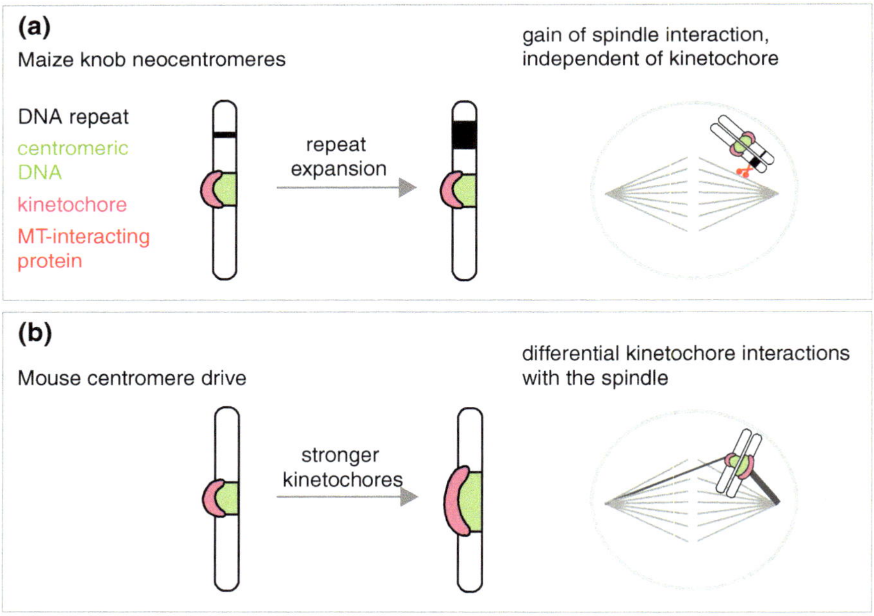

Fig. 6 Knob neocentromeres and expanded mammalian centromeres interact differently with meiotic spindle MTs. **a** Knob repeat expansion with kinetochore-independent interactions with MTs leading to a preferential position on the spindle. **b** Centromere expansion to achieve higher levels of kinetochore proteins and preferential orientation on the spindle. In each case the modified chromosome is shown on the spindle pairing with its homologous partner. Sister chromatids are not shown for simplicity

Focus has logically fallen on the CENP-A nucleosomes that mark centromere location and the proteins immediately proximal to it. A high local concentration of CENP-A nucleosomes recruit the constitutive centromere-associated network (CCAN) of 16 proteins, some of which are involved in the epigenetic recruitment of new CENP-A at each cell cycle to propagate centromere identity and/or the assembly of the mitotic kinetochore (McKinley and Cheeseman 2016). At least two CCAN components, CENP-C and CENP-N, also help maintain centromere identity by driving a nucleosome structural transition that stabilizes CENP-A at the centromere (CENP-C) (Falk et al. 2015, 2016) and fastening CENP-A to the DNA (CENP-N) (Guo et al. 2017).

The amino acid changes in CENP-A, relative to its canonical counterpart histone H3, distinguish CENP-A chromatin from bulk chromatin. Thus, it is worth considering the molecular basis for the epigenetic features that may participate in centromere drive. The CENP-A N-terminal 'tail' has been proposed to interact with CENP-B (Fachinetti et al. 2015), a protein that directly binds to a specific sequence found in typical mammalian centromere DNA repeats (Masumoto et al. 1989). The N-terminal tail also appears to have a CENP-B-independent role in centromere function (Folco et al. 2015; Logsdon et al. 2015). The histone-fold domain harbors the CENP-A targeting domain (CATD) that is sufficient for centromere targeting (Black et al. 2004) through recognition by its specific histone chaperone, HJURP (Foltz et al. 2009). The CATD also drives three structural and dynamic features particular to centromeric nucleosomes. First, the CATD contains hydrophobic stitches that stabilize the $(CENP\text{-}A/H4)_2$ heterotetramer (Black et al. 2004; Sekulic et al. 2010; Bassett et al. 2012). Second, the CATD drives an atypically shaped nucleosome (Sekulic et al. 2010; Falk et al. 2015, 2016) that only achieves its full stability after the CCAN component, CENP-C, drives a structural transition that rigidifies internal inter-histone connections (Falk et al. 2015, 2016). Third, the CATD includes residues that generate a bulged loop L1 that extends from the histone surface of the nucleosome (Sekulic et al. 2010; Tachiwana et al. 2011) and recruits the CCAN component, CENP-N (Carroll et al. 2009; Fang et al. 2015; Guo et al. 2017). In addition, the CATD and the C-terminal tail of CENP-A recruits the CCAN component CENP-C (Carroll et al. 2010; Kato et al. 2013; Logsdon et al. 2015; Tachiwana et al. 2015; Westhorpe et al. 2015). Outside of the CATD, a divergent helix (the αN helix) at the connection to the final turn of nucleosomal DNA has changes relative to H3 that lead to looser wrapping at this particular part of the nucleosome (Conde e Silva et al. 2007; Panchenko et al. 2011; Tachiwana et al. 2011; Hasson et al. 2013; Roulland et al. 2016). Ongoing research is aimed at understanding the importance of each of these distinguishing features of centromeric nucleosomes in producing the epigenetic mark that defines centromere location.

In all likely models for centromere propagation, new CENP-A deposition at the centromere involves nascent CENP-A chromatin assembly once per cell cycle at or near the site of preexisting CENP-A nucleosomes. The special stability of this chromatin (Bodor et al. 2013; Falk et al. 2015) then maintains the centromere over the demandingly long timescales required in mammalian biology (Smoak et al. 2016). It is very likely that some of the molecular features described above play prominent roles in centromere drive mechanisms in diverse eukaryotes, contributing to differences in centromere strength.

9 Mechanistic Models of Meiotic Drive-Spindle Asymmetry

In many cases, drive models also depend on functional asymmetry in the meiotic spindle that biases the segregation of bivalents or trivalents in MI or sister chromosomes in MII (Pardo-Manuel de Villena and Sapienza 2001a; Henikoff et al. 2001). Meiotic spindle asymmetry has been reported in grasshopper (Hewitt 1976), as well as examples in several other organisms (Crowder et al. 2015). We also observe asymmetry within the MI spindle in mouse oocytes, with MTs closer to the cortex more stable than those farther from the cortex (our unpublished results) (Fig. 7a). How spindle asymmetry is regulated is not known, but possible mechanisms include: (1) regulation of MT dynamics by an unknown cortical signal, or (2) asymmetric distribution of spindle pole proteins, as observed in previous studies (Carabatsos et al. 2000; Shuda et al. 2009; Meng et al. 2004; Michaut et al. 2005). The first mechanism simplifies the problem in that it would also explain how the asymmetric spindle orients relative to the cortex.

Based on our observations of spindle asymmetry in mouse oocytes and preferential retention of stronger centromeres in the egg, we speculate on models for meiotic drive in mammalian female meiosis. Two mechanisms, which are not

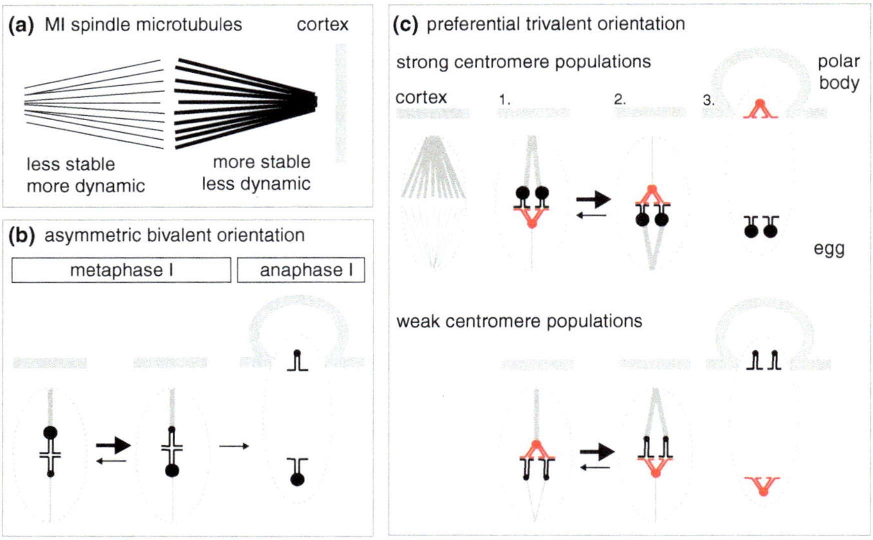

Fig. 7 Model for meiotic drive in mouse oocytes. **a** An asymmetric MI spindle with more stable MTs oriented toward the oocyte cortex. **b** Orientation of an asymmetric bivalent on the asymmetric MI spindle. The weaker centromere preferentially orients toward the more stable MTs on the cortical side of the spindle. **c** Model for preferential Rb fusion segregation in strong and weak centromere populations, based on trivalent orientation on the asymmetric MI spindle. Centromere strength is indicated by the size of the circle (*red* Rb fusion, *black* telocentric)

mutually exclusive, could explain preferential orientation of weaker centromeres toward more stable MTs at the cortical pole and stronger centromere toward less stable (more dynamic) MTs at the egg pole (Fig. 7b). First, more dynamic MTs from the egg pole may initially capture the stronger centromeres, which provide a larger target with more MT-binding proteins at their kinetochores. Weaker centromeres would subsequently bind MTs from the cortical pole to establish tension across the bivalent. Second, the bivalent may sample both orientations, and one is preferred because it is relatively more stable than the other. Under this model the interaction of the weak centromere with more dynamic MTs is labile and will tend to re-orient, and the interaction of the weak centromere with more stable MTs is preferred. This trial-and-error mechanism, in which the preferred configuration is selectively stabilized, is analogous to the long-standing model for how correct, bi-oriented attachments are stabilized by tension (Nicklas 1997), but in this case the outcome is biased orientation. Further observations from live imaging of chromosome dynamics during MI may test these models.

The same models can explain the preferential orientation and segregation of a trivalent. The stronger centromeres may attach first to the more dynamic egg pole, and/or the interaction of weak centromeres with more dynamic MTs may be unstable. As a result, the Rb fusion centromere orients preferentially to the egg pole if it is stronger relative to the homologous telocentrics, or preferentially to the cortical pole if it is relatively weaker (Fig. 7c).

10 Conclusion

Chromosomal rearrangements are frequent events involving chromosomal fusion, fission (chromosomal splitting) or translocations. Their role in speciation via meiotic drive was proposed nearly 50 years ago: *"It may be that the very few chromosomal rearrangements which play a critical role in speciation through the ability to generate powerful isolating mechanisms are precisely those which happen to possess a segregational advantage in the female meiosis"* (White 1968). Under that model karyotype of a given species is not fixed but can change over time. If karyotype changes between populations by preferential transmission of a chromosomal rearrangement (such as an Rb fusion) through female meiosis, meiotic abnormalities in the hybrids would generate a reproduction barrier, promoting speciation (Hauffe et al. 2012; Shurtliff 2013). Chromosomal reorganizations are also a major mechanism of reproductive isolation in *Saccharomyces cerevisiae* (Hou et al. 2014) and contribute to karyotype evolution in higher plants (Jones 1998). In addition, meiotic drive of Rb fusions can explain the bimodal distribution of mammalian karyotypes: most species have either predominantly telocentric or predominantly metacentric karyotypes. A similar bimodal distribution of karyotypes is found in fish (Molina et al. 2014). Transmission advantage of either chromosomal fusions (metacentrics) or fissions (telocentrics) through female meiosis predicts the biased accumulation of a given chromosomal rearrangement, which would shape the karyotype in one direction or the other, depending on the direction of drive.

There is a deleterious effect on male carriers in several meiotic drive systems. For example, Ab10 exhibits reduced pollen fitness, probably due to late replication of the knob sequence that extends the cell cycle and leads to mitotic abnormalities in microsporogenesis (Fluminhan and Kameya 1997), and the overall frequency in natural populations is low ($\sim$ 14%) (Buckler et al. 1999). In monkey-flower, pollen grains from males homozygous for the D allele have reduced fitness (Fishman and Saunders 2008). Similarly, male mice with Rb fusion chromosomes in the heterozygous state, which form trivalents, tend to have decreased fitness due to higher incidence of chromosome nondisjunction during meiosis (Manieu et al. 2014; Green 1981). Such deleterious effects may be frequently linked to drive in systems that have been studied because these deleterious effects prevent complete fixation of a selfish element; otherwise, there would be no meiotic drive to measure.

The various systems exhibiting meiotic drive share one common aspect: an asymmetric meiotic division. Distinct gamete architecture is one of the strongest differences between sexes (Gorelick et al. 2016). Male gametes are typically small, abundant and autonomously moving elements, whereas female gametes are large, stockpiled, stationary cells that are limited in number. An elegant way to achieve such desired egg morphology is combining the cytoplasm from several cells while expelling the redundant DNA, by dividing asymmetrically. Alternatively, asymmetry in meiosis could be explained by selfish DNA elements competing for their transmission. In this model, the DNA elements would transform the architecture of gametogenesis to trigger the elimination of gametes that are not transmitting them (Malik and Henikoff 2009). On the other hand, evolution of asymmetric meiosis could have been driven by the opposite force, to eliminate selfish elements. Regardless of the evolutionary force driving the asymmetry in meiosis, it was evolutionarily successful as it arose independently several times (Malik and Henikoff 2009), and it seems inevitable that selfish elements will exploit the opportunity to increase their chances of survival.

We call attention to a study (Iwata-Otsubo et al. 2017) published during the production of this book. This study provides evidence that amplified satellite repeats act as selfish elements in female meiosis.

References

Agulnik SI, Agulnik AI, Ruvinsky AO (1990) Meiotic drive in female mice heterozygous for the HSR inserts on chromosome 1. Genet Res 55:97–100

Agulnik SI, Adolph S, Winking H, Traut W (1993a) Zoogeography of the chromosome 1 of the house mouse (*Mus musculus*). Hereditas 39–46

Agulnik SI, Sabantsev ID, Orlova GV, Ruvinsky AO (1993b) Meiotic drive on aberrant chromosome 1 in the mouse is determined by a linked distorter. Genet Res 61:91–96

Agulnik SI, Sabantsev ID, Ruvinsky AO (1993c) Effect of sperm genotype on chromatid segregation in female mice heterozygous for aberrant chromosome 1. Genet Res 61:97–100

Aniskin VM, Benazzou T, Biltueva L et al (2006) Unusually extensive karyotype reorganization in four congeneric Gerbillus species (Muridae: Gerbillinae). Cytogenet Genome Res 112:131–140. doi:10.1159/000087525

Banaei-Moghaddam AM, Schubert V, Kumke K et al (2012) Nondisjunction in favor of a chromosome: the mechanism of rye B chromosome drive during pollen mitosis. Plant Cell 24:4124–4134. doi:10.1105/tpc.112.105270

Bassett EA, Wood S, Salimian KJ et al (2010) Epigenetic centromere specification directs aurora B accumulation but is insufficient to efficiently correct mitotic errors. J Cell Biol 190:177–185. doi:10.1083/jcb.201001035

Bassett EA, DeNizio J, Barnhart-Dailey MC et al (2012) HJURP uses distinct CENP-A surfaces to recognize and to stabilize CENP-A/histone H4 for centromere assembly. Dev Cell 22:749–762. doi:10.1016/j.devcel.2012.02.001

Black BE, Cleveland DW (2011) Epigenetic centromere propagation and the nature of CENP-A nucleosomes. Cell 144:471–479. doi:10.1016/j.cell.2011.02.002

Black BE, Foltz DR, Chakravarthy S et al (2004) Structural determinants for generating centromeric chromatin. Nature 430:578–582. doi:10.1038/nature02766

Bodor DL, Valente LP, Mata JF et al (2013) Assembly in G1 phase and long-term stability are unique intrinsic features of CENP-A nucleosomes. Mol Biol Cell 24:923–932. doi:10.1091/mbc.E13-01-0034

Buckland RA, Evans HJ (1978) Cytogenetic aspects of phylogeny in the Bovidae. I. G-banding. Cytogenet Cell Genet 21:42–63

Buckler ES, Phelps-Durr TL, Buckler CS et al (1999) Meiotic drive of chromosomal knobs reshaped the maize genome. Genetics 153:415–426

Burt A, Trivers R (2006) Genes in Conflict. Harvard University Press

Carabatsos MJ, Combelles CM, Messinger SM, Albertini DF (2000) Sorting and reorganization of centrosomes during oocyte maturation in the mouse. Microsc Res Tech 49:435–444. doi:10.1002/(SICI)1097-0029(20000601)49:5<435:AID-JEMT5>3.0.CO;2-H

Carroll CW, Silva MCC, Godek KM et al (2009) Centromere assembly requires the direct recognition of CENP-A nucleosomes by CENP-N. Nat Cell Biol 11:896–902. doi:10.1038/ncb1899

Carroll CW, Milks KJ, Straight AF (2010) Dual recognition of CENP-A nucleosomes is required for centromere assembly. J Cell Biol 189:1143–1155. doi:10.1083/jcb.201001013

Cattanach BM (1962) XO mice. Genet Res 487–490

Chmátal L, Gabriel SI, Mitsainas GP et al (2014) Centromere strength provides the cell biological basis for meiotic drive and karyotype evolution in mice. Curr Biol 24:2295–2300. doi:10.1016/j.cub.2014.08.017

Cortes DB, McNally KL, Mains PE, McNally FJ (2015) The asymmetry of female meiosis reduces the frequency of inheritance of unpaired chromosomes. Elife. doi:10.7554/eLife.06056

Crowder ME, Strzelecka M, Wilbur JD et al (2015) A comparative analysis of spindle morphometrics across metazoans. Curr Biol 25:1542–1550. doi:10.1016/j.cub.2015.04.036

Dawe RK, Cande WZ (1996) Induction of centromeric activity in maize by suppressor of meiotic drive 1. Proc Natl Acad Sci USA 93:8512–8517

Dawe RK, Reed LM, Yu HG et al (1999) A maize homolog of mammalian CENPC is a constitutive component of the inner kinetochore. Plant Cell 11:1227–1238

de Villena F, Sapienza C (2001) Transmission ratio distortion in offspring of heterozygous female carriers of Robertsonian translocations. Hum Genet 108:31–36. doi:10.1007/s004390000437

Didion JP, Morgan AP, Clayshulte AM-F et al (2015) A multi-megabase copy number gain causes maternal transmission ratio distortion on mouse chromosome 2. PLoS Genet 11:1004850. doi:10.1371/journal.pgen.1004850

Evans JA, Canning N, Hunter AG et al (1978) A cytogenetic survey of 14,069 newborn infants. III. An analysis of the significance and cytologic behavior of the Robertsonian and reciprocal translocations. Cytogenet Cell Genet 20:96–123

Fachinetti D, Han JS, McMahon MA et al (2015) DNA sequence-specific binding of CENP-B enhances the fidelity of human centromere function. Dev Cell 33:314–327. doi:10.1016/j.devcel.2015.03.020

Falk SJ, Guo LY, Sekulic N et al (2015) Chromosomes. CENP-C reshapes and stabilizes CENP-A nucleosomes at the centromere. Science 348:699–703. doi:10.1126/science.1259308

Falk SJ, Lee J, Sekulic N et al (2016) CENP-C directs a structural transition of CENP-A nucleosomes mainly through sliding of DNA gyres. Nat Struct Mol Biol 23:204–208. doi:10.1038/nsmb.3175

Fang J, Liu Y, Wei Y et al (2015) Structural transitions of centromeric chromatin regulate the cell cycle-dependent recruitment of CENP-N. Genes Dev 29:1058–1073. doi:10.1101/gad.259432.115

Fishman L, Saunders A (2008) Centromere-associated female meiotic drive entails male fitness costs in monkeyflowers. Science 322:1559–1562. doi:10.1126/science.1161406

Fishman L, Willis JH (2005) A novel meiotic drive locus almost completely distorts segregation in mimulus (monkeyflower) hybrids. Genetics 169:347–353. doi:10.1534/genetics.104.032789

Fluminhan A, Kameya T (1997) Involvement of knob heterochromatin in mitotic abnormalities in germinating aged seeds of maize. Genome 40:91–98

Folco HD, Campbell CS, May KM et al (2015) The CENP-A N-tail confers epigenetic stability to centromeres via the CENP-T branch of the CCAN in fission yeast. Curr Biol 25:348–356. doi:10.1016/j.cub.2014.11.060

Foltz DR, Jansen LET, Bailey AO et al (2009) Centromere-specific assembly of CENP-A nucleosomes is mediated by HJURP. Cell 137:472–484. doi:10.1016/j.cell.2009.02.039

Garagna S, Page J, Fernandez-Donoso R et al (2014) The Robertsonian phenomenon in the house mouse: mutation, meiosis and speciation. Chromosoma 123:529–544. doi:10.1007/s00412-014-0477-6

Gorelick R, Fraser D, Mansfield M et al (2016) Abrupt shortening of bird W chromosomes in ancestral Neognathae. Biol J Linn Soc 119:488–496. doi:10.1111/bij.12832

Green MC (1981) Genetic variants and strains of the laboratory mouse. Fischer, New York, pp 339–343

Gropp A, Tettenborn U, von Lehman E (1969) Chromosomenuntersuchungen bei der Tabakmaus (*M. poschiavinus*) und bei den Hybriden mit der Laboratorumsmaus. Experientia 25:875–876

Guo LY, Allu PK, Zandarashvili L et al (2017) Centromeres are maintained by fastening CENP-A to DNA and directing an arginine anchor-dependent nucleosome transition. Nat Commun 8:15775. doi:10.1038/ncomms15775

Hasson D, Panchenko T, Salimian KJ et al (2013) The octamer is the major form of CENP-A nucleosomes at human centromeres. Nat Struct Mol Biol 20:687–695. doi:10.1038/nsmb.2562

Hauffe HC, Giménez MD, Searle JB (2012) Chromosomal hybrid zones in the house mouse. In: Evolution of the house mouse. Cambridge University Press, pp 407–430

Henikoff S, Ahmad K, Malik HS (2001) The centromere paradox: stable inheritance with rapidly evolving DNA. Science 293:1098–1102. doi:10.1126/science.1062939

Hewitt GM (1976) Meiotic drive for B-chromosomes in the primary oocytes of *Myrmeleotettix maculatus* (Orthopera: Acrididae). Chromosoma 56:381–391

Hodgkin J, Horvitz HR, Brenner S (1979) Nondisjunction mutants of the nematode *Caenorhabditis elegans*. Genetics 91:67–94

Hou J, Friedrich A, de Montigny J, Schacherer J (2014) Chromosomal rearrangements as a major mechanism in the onset of reproductive isolation in *Saccharomyces cerevisiae*. Curr Biol 24:1153–1159. doi:10.1016/j.cub.2014.03.063

Iwata-Otsubo A, Dawicki-McKenna JM, Akera T et al (2017) Expanded satellite repeats amplify a discrete CENP-A nucleosome assembly site on chromosomes that drive in female meiosis. Curr Biol 27:2365–2373. doi:10.1016/j.cub.2017.06.069

Jacobs PA, Browne C, Gregson N et al (1992) Estimates of the frequency of chromosome abnormalities detectable in unselected newborns using moderate levels of banding. J Med Genet 29:103–108

Jones R (1991) B-chromosome drive. Am Nat 430–442

Jones R (1995) B chromosome in plants. New Phytol 131:411–434

Jones K (1998) Robertsonian fusion and centric fission in karyotype evolution of higher plants. Bot Rev 64:273–289

Jones R, Rees H (1982) B Chromosomes. Academic Press, London

Kato H, Jiang J, Zhou B-R et al (2013) A conserved mechanism for centromeric nucleosome recognition by centromere protein CENP-C. Science (80-)340:1110–1113. doi:10.1126/science.1235532

Kaufman MH (1972) Non-random segregation during mammalian oogenesis. Nature 238:465–466

LeMaire-Adkins R, Hunt PA (2000) Nonrandom segregation of the mouse univalent X chromosome: evidence of spindle-mediated meiotic drive. Genetics 156:775–783

Locke DP, Hillier LW, Warren WC et al (2011) Comparative and demographic analysis of orang-utan genomes. Nature 469:529–533. doi:10.1038/nature09687

Logsdon GA, Barrey EJ, Bassett EA et al (2015) Both tails and the centromere targeting domain of CENP-A are required for centromere establishment. J Cell Biol 208:521–531. doi:10.1083/jcb.201412011

Longley AE (1945) Abnormal segregation during megasporogenesis in Maize. Genetics 30:100–113

Lyttle TW (1991) Segregation distorters. Annu Rev Genet 25:511–557. doi:10.1146/annurev.ge.25.120191.002455

Malik HS, Henikoff S (2009) Major evolutionary transitions in centromere complexity. Cell 138:1067–1082. doi:10.1016/j.cell.2009.08.036

Manieu C, González M, López-Fenner J et al (2014) Aneuploidy in spermatids of Robertsonian (Rb) chromosome heterozygous mice. Chromosome Res 22:545–557. doi:10.1007/s10577-014-9443-7

Mao X, Nie W, Wang J et al (2008) Comparative cytogenetics of bats (Chiroptera): the prevalence of Robertsonian translocations limits the power of chromosomal characters in resolving interfamily phylogenetic relationships. Chromosome Res 16:155–170. doi:10.1007/s10577-007-1206-2

Masumoto H, Masukata H, Muro Y et al (1989) A human centromere antigen (CENP-B) interacts with a short specific sequence in alphoid DNA, a human centromeric satellite. J Cell Biol 109:1963–1973. doi:10.1083/jcb.109.5.1963

McKinley KL, Cheeseman IM (2016) The molecular basis for centromere identity and function. Nat Rev Mol Cell Biol 17:16–29. doi:10.1038/nrm.2015.5

Meng X, Fan H, Zhong Z et al (2004) Localization of gamma-tubulin in mouse eggs during meiotic maturation, fertilization, and early embryonic development. J Reprod Dev 50:97–105

Michaut MA, Williams CJ, Schultz RM (2005) Phosphorylated MARCKS: a novel centrosome component that also defines a peripheral subdomain of the cortical actin cap in mouse eggs. Dev Biol 280:26–37 S0012-1606(05)00007-2[pii]10.1016/j.ydbio.2005.01.006

Molina WF, Martinez PA, Bertollo LAC, Bidau CJ (2014) Evidence for meiotic drive as an explanation for karyotype changes in fishes. Mar Genomics 15:29–34. doi:10.1016/j.margen.2014.05.001

Nicklas RB (1997) How cells get the right chromosomes. Science (80-)275:632–637

Östergren G (1945) Parasitic nature of extra fragment chromosomes. Bot Not 157–163

Panchenko T, Sorensen TC, Woodcock CL et al (2011) Replacement of histone H3 with CENP-A directs global nucleosome array condensation and loosening of nucleosome superhelical termini. Proc Natl Acad Sci 108:16588–16593. doi:10.1073/pnas.1113621108

Pardo-Manuel de Villena F, Sapienza C (2001a) Nonrandom segregation during meiosis: the unfairness of females. Mamm Genome 12:331–339. doi:10.1007/s003350040003

Pardo-Manuel de Villena F, Sapienza C (2001b) Female meiosis drives karyotypic evolution in mammals. Genetics 159:1179–1189

Pardo-Manuel de Villena F, de la Casa-Esperon E, Briscoe TL, Sapienza C (2000) A genetic test to determine the origin of maternal transmission ratio distortion. Meiotic drive at the mouse Om locus. Genetics 154:333–342

Piálek J, Hauffe HC, Searle JB (2005) Chromosomal variation in the house mouse. Biol J Linn Soc 84:535–563

Rhoades MM (1942) Preferential egregation in maize. Genetics 27:395–407

Rhoades M (1952) Preferential segregation in maize. In: Gowen J (ed) Heterosis. Iowa College Press, Ames, pp 66–80

Rhoades MM, Vilkomerson H (1942) On the anaphase movement of chromosomes. Proc Natl Acad Sci U S A 28:433–436

Roulland Y, Ouararhni K, Naidenov M et al (2016) The flexible ends of CENP-A nucleosome are required for mitotic fidelity. Mol Cell 63:674–685. doi:10.1016/j.molcel.2016.06.023

Ruvinsky A (1995) Meiotic drive in female mice: an essay. Mamm Genome 6:315–320

Sabantsev I, Spitsin O, Agulnik S, Ruvinsky A (1993) Population dynamics of aberrant chromosome 1 in mice. Heredity (Edinb) 70(Pt 5):481–489

Sandler L, Novitski E (1957) Meiotic drive as an evolutionary force. Am Nat 91:105. doi:10.1086/281969

Sandler L, Hiraizumi Y, Sandler I (1959) Meiotic drive in natural populations of drosophila melanogaster. I. the cytogenetic basis of segregation-Distortion. Genetics 44:233–250

Sekulic N, Bassett EA, Rogers DJ, Black BE (2010) The structure of (CENP-A-H4)(2) reveals physical features that mark centromeres. Nature 467:347–351. doi:10.1038/nature09323

Shang WH, Hori T, Toyoda A et al (2010) Chickens possess centromeres with both extended tandem repeats and short non-tandem-repetitive sequences. Genome Res 20:1219–1228. doi:10.1101/gr.106245.110

Shuda K, Schindler K, Ma J et al (2009) Aurora kinase B modulates chromosome alignment in mouse oocytes. Mol Reprod Dev 76:1094–1105. doi:10.1002/mrd.21075

Shurtliff QR (2013) Mammalian hybrid zones: a review. Mamm Rev 43:1–21. doi:10.1111/j.1365-2907.2011.00205.x

Silvan CE, Black BE, Sivolob A et al (2007) CENP-A-containing nucleosomes: easier disassembly versus exclusive centromeric localization. J Mol Biol 370:555–573. doi:10.1016/j.jmb.2007.04.064

Smoak EM, Stein P, Schultz RM et al (2016) Long-term retention of CENP-A nucleosomes in mammalian oocytes underpins transgenerational inheritance of centromere identity. Curr Biol 26:1110–1116. doi:10.1016/j.cub.2016.02.061

Tachiwana H, Kagawa W, Shiga T et al (2011) Crystal structure of the human centromeric nucleosome containing CENP-A. Nature 476:232–235. doi:10.1038/nature10258

Tachiwana H, Müller S, Blümer J et al (2015) HJURP involvement in de novo CenH3CENP-A and CENP-C recruitment. Cell Rep 11:22–32. doi:10.1016/j.celrep.2015.03.013

Traut W, Winking H, Adolph S (1984) An extra segment in chromosome 1 of wild *Mus musculus*: a C-band positive homogeneously staining region. Cytogenet Cell Genet 38:290–297

Wade CM, Giulotto E, Sigurdsson S et al (2009) Genome sequence, comparative analysis, and population genetics of the domestic horse. Science (80-)326:865–867

Westhorpe FG, Fuller CJ, Straight AF (2015) A cell-free CENP-A assembly system defines the chromatin requirements for centromere maintenance. J Cell Biol 209:789–801. doi:10.1083/jcb.201503132

White MJ (1968) Models of speciation. New concepts suggest that the classical sympatric and allopatric models are not the only alternatives. Science 159:1065–1070

White TA, Bordewich M, Searle JB (2010) A network approach to study karyotypic evolution: the chromosomal races of the common shrew (*Sorex araneus*) and house mouse (*Mus musculus*) as model systems. Syst Biol 59:262–276. doi:10.1093/sysbio/syq004

Wu G, Hao L, Han Z et al (2005) Maternal transmission ratio distortion at the mouse Om locus results from meiotic drive at the second meiotic division. Genetics 170:327–334. doi:10.1534/genetics.104.039479

Yu HG, Hiatt EN, Chan A et al (1997) Neocentromere-mediated chromosome movement in maize. J Cell Biol 139:831–840. doi:10.1083/jcb.139.4.831

Yukimenko LV, Korobitsyna KV (1988) A rare variant of the chromosome 1 in house mouse: occurence of two extra heterochromatin segments. Genetica 376–378

Biophysics of Microtubule End Coupling at the Kinetochore

Ekaterina L. Grishchuk

Abstract The main physiological function of mitotic kinetochores is to provide durable attachment to spindle microtubules, which segregate chromosomes in order to partition them equally between the two daughter cells. Numerous kinetochore components that can bind directly to microtubules have been identified, including ATP-dependent motors and various microtubule-associated proteins with no motor activity. A major challenge facing the field is to explain chromosome motions based on the biochemical and structural properties of these individual kinetochore components and their assemblies. This chapter reviews the molecular mechanisms responsible for the motions associated with dynamic microtubule tips at the single-molecule level, as well as the activities of multimolecular ensembles called couplers. These couplers enable persistent kinetochore motion even under load, but their exact composition and structure remain unknown. Because no natural or artificial macro-machines function in an analogous manner to these molecular nano-devices, understanding their underlying biophysical mechanisms will require conceptual advances.

Key terminology

Microtubule end-tracking ability of the kinetochore (or isolated protein) to move with a dynamic microtubule end;

Microtubule end conversion transition from microtubule wall binding to microtubule-end attachment and subsequent microtubule end-tracking;

E.L. Grishchuk (✉)
Department of Physiology, Perelman School of Medicine, University of Pennsylvania, Philadelphia, PA 19104, USA
e-mail: gekate@mail.med.upenn.edu

B.E. Black (ed.), *Centromeres and Kinetochores*, Progress in Molecular and Subcellular Biology 56, DOI 10.1007/978-3-319-58592-5_17

Microtubule end coupler	a mobile protein-mediated link between a dynamic microtubule end and a cargo (chromosome or microbead);
Load-bearing	ability of microtubule end coupling to persist under tension.

1 The Kinetochore as a Versatile Molecular Machine

Kinetochores are multicomponent molecular assemblies that are capable of various modes of microtubule-dependent motility (reviewed in Mitchison 1988; Rieder and Salmon 1998; McIntosh et al. 2002). Elucidating how the underlying connections are established, maintained, and regulated is crucial for our understanding of normal cell division.

1.1 Microtubule Wall-to-End Transition

Early in mitosis, kinetochores often bind to the walls of spindle microtubules (Hayden et al. 1990; Tanaka et al. 2005; Magidson et al. 2011). The initial binding is facilitated by an expansion of the outer kinetochore layer, called the corona, which is rich in dynein and the kinetochore-localized kinesin CENP-E (Pfarr et al. 1990; Steuer et al. 1990; Cooke et al. 1997; Yao et al. 1997; Putkey et al. 2002; Wan et al. 2009). These mechanochemical enzymes use ATP to drive kinetochore motion along the microtubule. This mode of kinetochore motility is similar to the transport of other intracellular organelles containing motors of opposite polarities, e.g., during axonal or intraflagellar transport. In many cell types, mitotic kinetochores simultaneously contain dynein, a minus-end–directed transporter that moves the chromosomes to the spindle poles, and CENP-E kinesin, a plus-end-directed motor that helps to gather chromosomes at the spindle equator (Fig. 1a).

Many questions persist regarding the coordination of these opposing activities, and the role of a "polar wind," a microtubule-dependent force acting on the chromosome arms and pushing chromosomes away from the spindle poles (Ke et al. 2009; Cheerambathur et al. 2013; Barisic et al. 2014). Collectively, these forces ultimately transport kinetochores to the spindle midzone, where the microtubule plus ends are located. Not all chromosomes in a typical mammalian cell undergo long-distance transport along microtubule walls before coming into contact with the plus ends, and many of the translocations that do occur are probably too short to be detected (Magidson et al. 2011). Furthermore, some microtubule ends

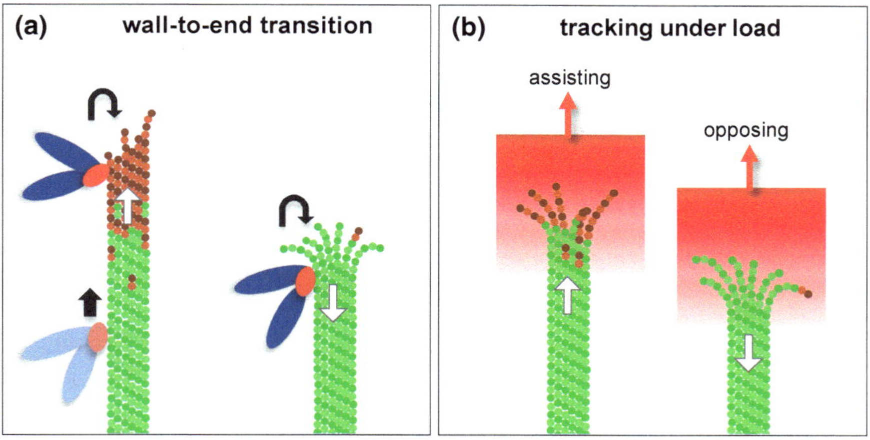

Fig. 1 Modes of kinetochore motility. Each cartoon shows a microtubule plus end interacting with a kinetochore (in *red*). In the microtubules, GTP-bound tubulin dimers are shown in brown and GDP-tubulin dimers in *green*. *Black arrows* indicate kinetochore motion, and *white arrows* indicate direction of microtubule dynamics. **a** Pathways that bring a laterally attached kinetochore in contact with the plus end of a dynamic microtubule. *Left* Plus-end-directed kinetochore-localized kinesin CENP-E transports the kinetochore, dragging it along the microtubule wall. *Right* When the microtubule depolymerizes, its plus end reaches the wall-bound kinetochore. The kinetochore rotates to assume microtubule end-on attachment due to the forces exerted by other spindle microtubules (not shown), which orient the sister kinetochore pair coaxially with the spindle (e.g., see Zaytsev and Grishchuk 2015). For simplicity, the kinetochore is drawn as a small oval, whereas in reality the size of the vertebrate kinetochore is 10–20-fold larger than the diameter of the microtubule; chromosome arms (*blue*) are also depicted on a much smaller scale. **b** In the end-on attachment mode, the kinetochore tracks the dynamic microtubule ends bidirectionally and processively. In metaphase, force acting from the sister kinetochore (*upward red arrow*) assists the motion of the kinetochore when it moves at the polymerizing end, but exerts a load when the kinetochore reverses its motion and tracks the end of depolymerizing microtubule. The fine structure of the kinetochore-bound microtubule ends is not known, but the ends of polymerizing and depolymerizing kinetochore microtubules appear quite similar, with protofilament flare smaller than that in freely depolymerizing microtubules, but larger than that in freely polymerizing microtubules in vitro (McIntosh et al. 2008)

become captured by kinetochores via either a direct encounter or following depolymerization of the laterally bound microtubule (Fig. 1a). Through these various mechanisms, the predominantly lateral microtubule attachments are gradually replaced by attachment to microtubule ends, often referred to as "end-on" binding, a prerequisite for normal segregation (reviewed in Tanaka 2010; Cheerambathur and Desai 2014). No other essential cellular cargo undergoes such a dramatic wall-to-end transition, and the biophysical mechanisms underlying this dynamic process remain poorly understood.

1.2 Bidirectional Processive Tracking of the Dynamic Microtubule Ends

The mechanism by which kinetochores maintain their persistent association with microtubule ends also remains to be elucidated (Inoué and Salmon 1995; Maiato et al. 2004). Textbook images of the kinetochores, e.g., in PtK1 cells, show kinetochore "plates," but these dense structures are seen only on a subset of kinetochores after conventional chemical fixation (Rieder 1982; O'Connell et al. 2012). Methods that preserve cellular structures more accurately, such as rapid freezing and freeze substitution, reveal subtler plates resembling a fibrous meshwork (McEwen et al. 1998; Dong et al. 2007; McIntosh et al. 2008, 2013). Microtubule plus ends become embedded within this meshwork while exhibiting a conspicuous flaring of linear tubulin arrays called protofilaments (Fig. 1).

Remarkably, the attachments between the embedded ends and kinetochore are not static, but instead exhibit dynamic instability (Mitchison et al. 1986; Skibbens et al. 1993), although with different rates and transition frequencies than unbound microtubule ends. As tubulin subunits are added or lost from these ends, the kinetochores move concomitantly (reviewed in Rieder and Salmon 1998; Maiato et al. 2004; Cheeseman and Desai 2008; Santaguida and Musacchio 2009), a behavior referred to as microtubule tip- or end-tracking. Thus, during microtubule end conversion, the kinetochores first transit from wall-to-end binding, then continuously track the dynamic microtubule ends. Kinetochore tracking is bidirectional during chromosome oscillations in metaphase, whereas during anaphase the tip-tracking motion is almost exclusively toward the spindle poles. Strikingly, the fine structure of kinetochore–microtubule ends differs only slightly during these stages (McIntosh et al. 2008). This observation is contrary to expectations based *on* in vitro studies with purified tubulin, in which the structures of polymerizing and depolymerizing ends are dramatically different. At the kinetochore, both polymerizing and depolymerizing microtubule ends contain flaring protofilaments (Fig. 1b). As described later in this chapter, several molecular mechanisms could in principle explain bidirectional kinetochore tracking of dynamic microtubule ends, but many questions remain regarding the identity of the molecules involved and their respective contributions.

1.3 Load-Bearing by Tracking Kinetochores

In cells, kinetochore–microtubule connections are maintained even in the presence of opposing forces (Fig. 1b), as evidenced by chromatin stretching between bi-oriented sister kinetochores in metaphase (reviewed in Rieder and Salmon 1998; Rago and Cheeseman 2013). The magnitudes of the forces at these stretched kinetochores, and those that move in anaphase, remain unknown. Because chromosome motions at the ends of kinetochore microtubules in metaphase and anaphase are relatively slow (1–3 μm/min), the force required to move a typically sized mammalian chromosome in the end-on configuration is very small, on the order of 1 pN (Nicklas 1965; Taylor 1965; Alexander and Rieder 1991). Nonetheless, some estimates suggest that mitotic kinetochores experience forces up to 10 pN (reviewed in Asbury et al. 2011), whereas other studies indicate that the physiological load at the kinetochore may reach hundreds of pN (Alexander and Rieder 1991; Ye et al. 2016). On the scale of intracellular transport, this is a very large force; by comparison, the CENP-E kinesin stall force is 5 pN (Yardimci et al. 2008). Micromanipulation studies in grasshopper spermatocytes suggest that end-on attached kinetochores can move even under loads of 400–600 pN (Nicklas 1983). Analogous experiments in vertebrate cells are lacking because of the difficulty of applying calibrated forces in these more fragile cells (Skibbens and Salmon 1997). Although laser trapping of segregating chromosomes in different cell types shows that they can be stalled by forces on the order of 2–10 pN (Ferraro-Gideon et al. 2013), the direct manipulation of chromosomes with a powerful laser beam is likely to be harmful. Kinetochore particles isolated from budding yeast can stay attached to dynamic microtubule ends under the forces of up to 10 pN (Akiyoshi et al. 2010). Thus, although direct measurements of average physiological loads and maximal (stall) forces in live dividing cells are still lacking, it is plausible that larger kinetochores, such as those of human cells, which connect to 15–20 microtubules, are built to withstand significant forces.

The ability of kinetochores to continue their motions despite significant opposing loads may constitute a fail-safe mechanism for preventing chromosome loss, which could result from resistance due to cellular obstacles or from counteraction by improperly attached (merotelic) microtubules in anaphase (Cimini et al. 2004). The specific molecular mechanisms that govern kinetochore motility under load remain largely unknown (reviewed by Inoué and Salmon 1995; Joglekar et al. 2010). Accordingly, the overarching goal of this field is not only to identify the key components, but also to understand the underlying biophysical mechanisms, which are far from trivial. Indeed, the coexistence of the two major properties of the kinetochore (the processive tracking and persistence under force) is counterintuitive, because tracking implies mobility (i.e., constant dissolution of existing kinetochore–microtubule bonds and formation of new bonds), whereas stability of attachment is most feasibly ensured by static bonds (reviewed in Mitchison 1988; Inoué and Salmon 1995; Grishchuk et al. 2012). Understanding of such intricate biophysical relationships has been facilitated by a combination of in vitro reconstitutions and quantitative experimental and theoretical approaches, as summarized briefly in the following sections.

2 Molecular Mechanisms of Microtubule Tip-Tracking

Classical experiments identified the kinetochore as the major site of force generation for chromosome motion (Nicklas 1989; Khodjakov and Rieder 1996). However, subsequent analyses revealed that kinetochore-localized microtubule-dependent motors are largely dispensable for chromosome motions (Grishchuk and McIntosh 2006; Tanaka et al. 2007), implying that the dynamics of end-on attached microtubules are the primary driver of chromosome motility (reviewed in Inoué and Salmon 1995; McIntosh et al. 2010). In this sense, the kinetochore is not a motor per se, but rather a complex macromolecular device that couples chromosomes to the ends of microtubules that do the actual work of translocation. One possible explanation of the ability of kinetochores to track microtubule ends is that it results directly from the activities of various kinetochore-localized microtubule-binding proteins that are capable of tracking at a single-molecule level. Such molecules could create individual mobile molecular bonds between kinetochores and microtubules, enabling processive kinetochore motions. Alternatively, kinetochore tip-tracking could emerge from the collective behavior of multiple molecules that are individually incapable of tracking (or track poorly) at the single-molecule level. Therefore, it is important to identify the tip-tracking abilities of all kinetochore-localized microtubule binders and their multi-molecular assemblies, elucidate the underlying biophysical mechanisms, and determine the respective contribution of these mechanisms to kinetochore tracking.

2.1 *Affinity-Based Microtubule Tip-Tracking*

2.1.1 Polymerizing Microtubule Ends

The EB proteins are a well-characterized example of affinity-based tracking of polymerizing microtubule tips (reviewed in Akhmanova and Steinmetz 2010). These proteins discriminate between different types of microtubule lattices, as demonstrated by their strong preference for tubulin polymerized with non-hydrolysable GTP analogs in vitro (Zanic et al. 2009; Maurer et al. 2011). Because microtubule tips grow by addition of tubulins bound to GTP that is later hydrolyzed to GDP, the nascent microtubule wall is rich in GTP-tubulin (Desai and Mitchison 1997). When microtubules polymerize in the presence of soluble fluorescently labeled EB proteins, their growing ends are highlighted and appear as moving "comets," whereas the shortening ends are not labeled. Individual EB molecules bind to the tip region very transiently (0.05–0.8 s) (Bieling et al. 2008; Chen et al. 2014). Although a single EB molecule is capable of diffusive motions on the microtubule wall, the transient nature of this interaction prevents the individual molecule from moving processively (see below). Hence, tip-tracking by EB is merely apparent, rather than reflecting actual motion, and consequently represents

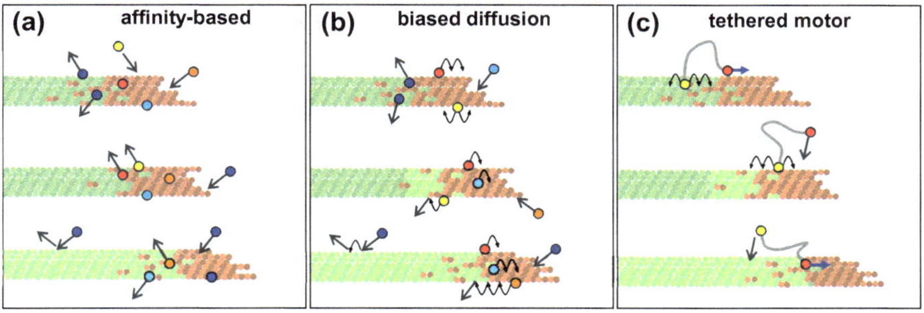

Fig. 2 Mechanisms of single-molecule tracking of polymerizing microtubule tip. In each panel, elongating microtubule end is depicted with three consecutive configurations. *Straight arrows* indicate molecular binding/unbinding events; *curved arrows* correspond to diffusional hopping. **a** MAP molecules are shown in different colors, so that they can be traced easily on the microtubule lattice. In the affinity mechanism, individual MAPs do not move relative to the lattice, but binding of multiple molecules leads to continuous decoration of the growing microtubule end. **b** In contrast to tracking that is based on affinity alone, in this mechanism the individual molecules change their position on the microtubule lattice due to diffusion. The outcome of this diffusion is biased by affinity: MAP molecules that happen to diffuse in the direction of microtubule elongation will retain their binding to the microtubule for a longer time due to their preference for the biochemical or structural features of tubulins at the tip. **c** A motor domain (*red circle*) walks processively to the plus end (*blue arrow*), where it dissociates, while the tail (*yellow circle*) behaves as a diffusing MAP with a limited residency time. When these two domains are tethered together, as in full-length CENP-E, the molecule can track microtubule ends processively and bidirectionally. For simplicity, only a single chain of this homodimeric motor is shown

treadmilling (Fig. 2a). The conspicuous microtubule comets seen in the presence of soluble EB result from multiple individual EB molecules quickly binding and unbinding to the moving microtubule tip.

Transient microtubule interactions can lead to lasting attachments when multiple molecules are involved, even if the binding of individual molecules is stochastic and uncoordinated (Zaytsev et al. 2013). Thus, although binding by a single EB molecule is too transient to lead to processive tracking, multiple EB molecules connected to a surface or scaffold could in principle provide a loose “glue” on the microtubule tip capable of sustaining the motion of small objects. For example, in vitro, short actin filaments can move with growing microtubule ends in the presence of EB and a linking protein that binds both EB and actin (Preciado López et al. 2014). In mitotic cells, EB proteins do not constitutively associate with the kinetochore, but they are seen at the ends of growing kinetochore-bound microtubules (Tirnauer et al. 2002; Armond et al. 2015). It is conceivable that in this capacity, multiple EB molecules binding to a more permanent kinetochore protein, e.g., Ska1 (Thomas et al. 2016), could contribute to kinetochore tracking at polymerizing microtubule ends.

A similar affinity-based mechanism may facilitate tracking with polymerizing microtubule tips by the bona fide kinetochore microtubule-associated proteins (MAPs). For example, the Dam1 heterodecameric protein complex, which is persistently associated with metaphase kinetochores in yeast, prefers to bind GMPCPP-containing microtubule walls in vitro (Westermann et al. 2005). Not surprisingly, small Dam1 oligomers can move with the growing microtubule ends (Lampert et al. 2010), although it remains unclear whether these motions are processive and how they depend on the number of Dam1 subunits. Microbeads coated with Dam1 heterodecamers travel continuously with growing microtubule ends (Asbury et al. 2006), indicating that multiple scaffold-bound Dam1 complexes can readily sustain processive tip-tracking. Likewise, the major microtubule-binding kinetochore component, Ndc80 complex (reviewed in Cheeseman and Desai 2008), can support polymerization-driven motion of microbeads (Powers et al. 2009), but tracking of growing microtubule tips by single Ndc80 molecules is very poor (Lampert et al. 2010). Although Ndc80 has not been reported to have preference for the GTP-containing microtubule lattice, its ability to support the polymerization-driven motion of the microbeads suggests that it recognizes some feature of the growing microtubule tip. Even a small difference in binding affinity will be amplified when multiple bead-bound molecules interact with the microtubule tip, explaining the processive tracking.

In addition, the TOG domain protein XMAP215 has also been reported to move with the growing microtubule tips at a single-molecule level (Brouhard et al. 2008; Widlund et al. 2011) and can sustain the motion of beads (Trushko et al. 2013). Tip-tracking by single XMAP215 molecules is not highly processive, and is thought to involve XMAP215 catalytic activity specifically at the microtubule tip (Kerssemakers et al. 2006; Ayaz et al. 2014). Interestingly, the yeast homolog of XMAP215, Stu2, does not autonomously follow growing microtubule tips in mitotic cells; instead, it is delivered to this location by kinesin-dependent transport (Gandhi et al. 2011). Stu2 localization at the kinetochore is very transient, with a half-life less than 50 s (Aravamudhan et al. 2014). Despite this dynamicity, kinetochore particles purified from Stu2-depleted yeast cells have been reported to interact with MT ends in vitro differently than the wild type particles (Miller et al. 2016), so a minor population of Stu2 must be binding strongly to the kinetochores. This stably bound Stu2 has been proposed to mediate complex tension-sensitive and microtubule dynamics-sensitive responding of yeast kinetochore (Miller et al. 2016).

Thus, many candidate molecules could potentially contribute to affinity-based recognition of growing microtubule ends at the kinetochore. It remains unclear, however, whether the main role of these MAPs is to provide end-tracking motion that pushes the kinetochore away from the pole (so-called AP motion). In vertebrate cells, a kinetochore moving with polymerizing microtubule ends is thought to be dragged by the pulling force acting through its poleward-moving sister (Skibbens et al. 1993; Waters et al. 1996). In this view, the role of kinetochore MAPs with high affinity for the polymerizing microtubule tip would be to provide molecular "friction" that prevents the AP kinetochore from slipping from the growing end (Maddox et al. 2003; Dumont et al. 2012), rather than to actively transport this kinetochore via tip-tracking mechanisms.

2.1.2 Depolymerizing Microtubule Ends

On the other hand, affinity-based tracking of depolymerizing microtubule ends is much less well understood. Two MAPs that might move via this motility mechanism are the human kinetochore proteins Ska1 and CENP-F, which can track depolymerizing microtubule ends in vitro (Schmidt et al. 2012; Volkov et al. 2015). CENP-F exhibits a weak preference for protofilament curls, which decorate the ends of shortening microtubules, whereas human Ska1 has similar affinities to tubulins in microtubule walls and curls. It remains to be seen whether tip-tracking by these proteins is truly affinity-driven, or instead occurs by a biased-diffusion mechanism, as described in the following sections. Individual protofilament curls are highly transient structures, with a lifetime of less than 0.1 s (assuming protofilament length of five dimers and a microtubule depolymerization rate of 25 μm/min). Thus, only proteins with a strong preference for GDP-tubulin curls should be capable of apparent tracking of shortening ends in vitro. The lifetime of the protofilament curls is much shorter than that of the GTP-tubulins in the growing tip. Thus, it is challenging to study depolymerization-driven motions at a single-molecule level, and consequently this type of non-processive tracking remains to be examined.

2.2 *Biased-Diffusion Tracking of Polymerizing Microtubule Ends*

Single molecules of most examined MAPs can diffuse on the microtubule wall (reviewed in Cooper and Wordeman 2009; Grishchuk et al. 2012; Reithmann et al. 2015). Diffusion along a polymer is not unique to microtubules, and has also been observed for actin-binding and DNA-binding proteins. The underlying biophysical mechanisms of microtubule-dependent diffusion have not been elucidated, but this phenomenon is likely to rely on the presence of multiple microtubule-binding sites within the diffusing protein molecule. Importantly, thermal motions are not directional and cannot by themselves lead to tip-tracking. However, they can be biased in the direction of microtubule dynamics via various mechanisms. At the growing microtubule tip, such bias is provided by an increased affinity for this structure. All kinetochore MAPs described in the previous section diffuse on the microtubule wall to some extent, so their tip-tracking may involve these diffusive motions. As pointed out above, a nondiffusing protein can lead only to apparent tip-tracking, whereas true processivity requires the molecule to move along the microtubule. Diffusing molecules can exhibit more persistent tracking because they can "hop" from one binding site to the next at the microtubule tip without fully dissociating from it (Fig. 2b). Such tracking cannot be highly processive because eventually a diffusing molecule will hop by chance in the wrong direction, away from the high-affinity sites at the tip, quickly losing its microtubule attachment.

No general theory has yet been developed to predict how the rate of MAP diffusion and its affinity for different features at the microtubule tip vs. wall will affect the processivity of tracking by single and multiple molecules. Vertebrate kinetochore represents a spatially distributed ensemble of various MAPs capable of microtubule binding and diffusion, probably 20 or more such molecules per microtubule end (Lawrimore et al. 2011; Suzuki et al. 2015). The weakly processive motion of a single molecule should be greatly enhanced when many such molecules are bunched together, enabling persistent association of the kinetochore with the growing microtubule ends. However, the rate of collective diffusion is expected to be much slower than mobility of a single molecule (Volkov et al. 2013). Given the large number of kinetochore MAPs, such as Ndc80, CENP-F, Ska1, Knl1, TOG domain proteins, microtubule-binding domains of CENP-E kinesin and dynein, and other factors (Cheeseman and Desai 2008; Nagpal and Fukagawa 2016), it is surprising that microtubule ends can polymerize while bound to the kinetochores of isolated mammalian chromosomes (Mitchison and Kirschner 1985; Hunt and McIntosh 1998). A rigorous understanding of this phenomenon will require thorough quantification of the diffusion rates and microtubule residency times of all kinetochore MAPs, as well as application of advanced mathematical models capable of incorporating these kinetic features and the mechanical properties of the MAPs and their kinetochore receptors.

2.3 *Biased-Diffusion Tracking of Depolymerizing Microtubule Ends*

2.3.1 Single Molecules

Tracking with the depolymerizing microtubule end can also take place by biased diffusion, driven by thermal energy in association with the unidirectionality of tubulin disassembly. In this case, the bias arises as a direct consequence of microtubule shortening. For successful operation of this mechanism, it is imperative that when a MAP molecule diffusing along a microtubule encounters the end, its probability of detachment is low. Mathematical models that assume this property—e.g., the "burnt bridge" model, in which the molecule's detachment from the end is simply prohibited (Mai et al. 2001)—can recapitulate end-tracking. Because this postulate does not rely directly on a specific mechanochemical pathway of microtubule disassembly, this type of tip-tracking is possible for any shortening polymer, not just microtubules. The mechanisms that could prevent a diffusing molecule from falling off the microtubule end are not well understood. Such property, however, has been reported for the microtubule-binding domain in the tail of CENP-E kinesin (Gudimchuk et al. 2013). When random motions bring the purified CENP-E Tail to the microtubule end, the Tail falls off only infrequently, and most of the time bounces off the end and continues to diffuse, as if the microtubule end constituted a reflective barrier. This remarkable behavior has also been noticed for Ndc80 protein (Powers et al. 2009). While it is likely that other MAPs may also be "reflected" by the microtubule tip, it remains unclear whether this will turn out to be a general feature of all microtubule diffusers.

Even if a MAP is capable of both wall diffusion and bouncing off the tip, such a MAP will not necessarily exhibit processive tracking at a single-molecule level. This is because successful tip-tracking requires an intricate balance between the rate of MAP diffusion and the rate of tubulin dissociation from the shortening microtubule tip. For example, the CENP-E Tail does not track the depolymerizing tip because its diffusion is too fast (1 $\mu m^2/s$) relative to microtubule shortening. After the Tail molecule bounces off the tip, it moves ahead of it much more rapidly than tubulin disassembly (Fig. 3). Thus, while the Tail's random motion is biased by microtubule depolymerization, this molecule spends most of the time diffusing on the wall and only a small fraction of time at the microtubule end (Gudimchuk et al. 2013). A molecule diffusing on microtubule wall ten times slower, e.g., Dam1 heterodecamer or Ndc80 protein ($\sim$0.1 $\mu m^2/s$) (Gestaut et al. 2008; Grishchuk et al. 2008b; Powers et al. 2009; Volkov et al. 2013; Zaytsev et al. 2015), will not move far on the microtubule wall before the shortening polymer end catches up. Thereafter, such MAP molecule will exhibit directed motion, staying close to the shortening tip, because its slow diffusion becomes rate-limiting for microtubule disassembly (Grishchuk et al. 2012). MAP molecules that diffuse even more slowly will simply block microtubule depolymerization until either the molecule dissociates or the terminal tubulin falls off, taking the MAP molecule with it (Fig. 3). Thus, successful tracking also depends on how long the diffusing molecule can remain associated with the microtubule. For all MAPs studied to date, the duration of diffusive motion is brief: Dam1 heterodecamer remains bound to microtubules for only 2 s (Gestaut et al. 2008), whereas Ndc80 kinetochore protein detaches in less than 0.5–1 s (Powers et al. 2009; Zaytsev et al. 2015). Because kinetochore–microtubules disassemble at 1–2 μm/min, a kinetochore would move with one such molecule for only $\sim$40 nm, the length of five tubulin dimers, before disconnecting from the microtubule. In vitro, microtubules depolymerize much faster, so small oligomers of Dam1 can move for longer distances. Despite having a similar rate of diffusion, single Ndc80 molecules are poor trackers of shortening microtubule ends (Powers et al. 2009; Umbreit et al. 2012; Schmidt et al. 2012).

2.3.2 Multiple Molecules Attached to a Microbead

As discussed in regard to polymerization-dependent tracking, motions at the shortening microtubule end are also expected to be much more processive when multiple MAP molecules are involved. Coupling by multiple molecules is usually tested in vitro using microbeads randomly coated with a MAP. This approach has revealed that numerous MAPs, not just those localized to kinetochores, can support transport of beads with microtubule disassembly to varying extents (reviewed in McIntosh et al. 2010). One of the best proteins for bead coupling is Dam1, which

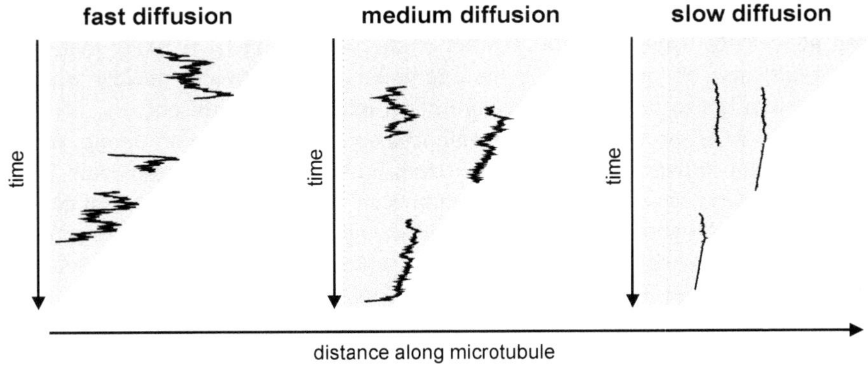

Fig. 3 Quantitative features of the single-molecule tracking of depolymerizing microtubule end. Drawings represent kymograph-like plots for three microtubules (in *grey*, *horizontal arrow* points to the plus-end direction). With time, each microtubule shortens, so the position of the disassembling plus end corresponds to the rightmost side of the *grey triangle*. The slopes of these lines correspond to depolymerization velocities. Each panel also shows changes in position versus time for three diffusing molecules (*black lines*). All molecules have roughly similar residency times, but the rates of diffusion are different in each panel. Rapidly diffusing molecules (diffusion coefficient $\sim 1\ \mu m^2/s$, leftmost kymograph) spend too little time at the microtubule tip, so they do not really track it. A moderate rate of diffusion ($\sim 0.1\ \mu m^2/s$) is optimal for tip-tracking in vitro, because such molecules can move significant distances, staying in close vicinity with the tip. With slower diffusion, however, microtubule depolymerization becomes severely inhibited because the molecular hops are too infrequent. Drawings are based on calculations in Grishchuk et al. (2012)

depending on soluble Dam1 concentration forms small oligomeric patches, microtubule-encircling rings, or stabilizing spirals (Westermann et al. 2005). When soluble Dam1 is present, the Dam1-coated beads "slide" along microtubules, just as expected if the beads were transported by the bead-bound Dam1 ring encircling the microtubule (Grishchuk et al. 2008b). However, when soluble Dam1 is not included in the assay buffer, Dam1-coated beads can still follow the shortening microtubule ends, but under these conditions they roll on the microtubule surface. Because this complex motion is likely to reflect the biased rotational diffusion of the beads (Peskin and Oster 1995), it represents a poor model for coupling at mitotic chromosomes. Furthermore, small non-protein particles with polyvalent positive bonds can also diffuse on the microtubule surface (Minoura et al. 2010) and follow shortening microtubule tips (unpublished observation). Apparently, the negative charges associated with tubulin globular domains and tails, which collectively form a relatively large interaction surface on the multiprotofilament microtubule, can support multivalent interactions with the positively charged molecular clusters. Processive motion of these clusters during microtubule depolymerization could arise through translational or rotational diffusive motions, or represent their combination. Such transport may not be physiological, so caution should be exercised when interpreting depolymerization coupled motility of protein clusters/aggregates, especially under experimental conditions that slow the rate of microtubule disassembly or use buffers with lower than physiological ionic strength. In this regard, a recently developed procedure for clustering kinetochore proteins in a highly controlled manner with the help of origami scaffolds is very promising (Verma et al. 2015). Such precise and quantitative approaches should help overcome the significant technical limitations of randomly coated microbeads, advancing experimental studies of microtubule end-tracking by molecular ensembles with defined compositions.

2.4 *Tethered Motor Mechanism of Bidirectional Tip-Tracking*

This unusual, ATP-dependent mode of tracking has been proposed for kinetochore kinesin CENP-E, which has two microtubule-binding domains: one with motor activity at the N-terminus of the molecule, and one with weak microtubule-binding affinity at the C-terminal tail (Liao et al. 1994; Wood et al. 1997; Kim et al. 2008; Espeut et al. 2008). In vitro studies have shown that this dimeric motor can walk along microtubule walls much faster than the rate at which microtubules polymerize (Yardimci et al. 2008), allowing it to catch up with the growing tip. Afterward, CENP-E does not dissociate, but instead continues to move processively at the rate of microtubule elongation (Gudimchuk et al. 2013). When the microtubule switches into depolymerization, CENP-E also reverses direction and tracks with the shortening end. Although these motions are not highly processive, lasting only for tens of seconds, they can be sustained by a single molecule. This is remarkable because it demonstrates that an individual kinetochore component can possess intrinsic abilities analogous to the physiological behavior of an intact kinetochore: motion along the microtubule wall, transition into tip-binding mode, and bidirectional tip-tracking.

This bidirectional tip-tracking does not rely on the increased CENP-E affinity for microtubule tips, but is instead an emergent behavior of the distinct activities of its two microtubule-binding domains (Fig. 2c). None of these domains exhibit a preference for polymerizing or depolymerizing tips, and neither can track these dynamic ends on their own. When the CENP-E molecule reaches a microtubule plus end, the CENP-E motor domains fall off the tip. Because they are tethered to the microtubule wall by the tail domains of this molecule, the motor domains rebind quickly. The tethering via CENP-E tails is very short (0.5 s), but it is sufficient for rebinding that is estimated to occur within a millisecond. Thus, CENP-E molecule tracks microtubule tips, both assembling and disassembling, by repeating the cycles of plus-end-directed walking, motor domains dissociation, and rebinding. As a tip-tracking protein, CENP-E kinesin can potentially provide bidirectional mobile links between the kinetochore and microtubules (Gudimchuk et al. 2013; Shrestha and Draviam 2013; Vitre et al. 2014), thereby enhancing the ability of chromosomes to follow microtubule ends (Lombillo et al. 1995).

Other plus-end-directed kinesins, e.g., Kif18A and Kip3, can associate processively with the growing microtubule tips in vitro, assisted by the microtubule-binding activity of the C-terminal tails (Mayr et al. 2011; Su et al. 2011; Weaver et al. 2011). However, these kinesins fail to track with microtubule disassembly. This is likely to be due to the slow diffusion of their tail domains. The tail of Kif18A, which diffuses 100-times slower than the tail of CENP-E, is expected to dissociate rapidly from shortening microtubule end, because the end catches up repeatedly with a molecule diffusing slowly in front of the wave of tubulin depolymerization. Each of these encounters may cause stochastic loss of the terminal tubulin dimer together with the bound molecule, so a slowly diffusing tail is not an effective tether for the motor domains on the shortening microtubule. It remains to be seen if the tethered motor mechanism is involved in tracking by these and other kinesin motors of the polymerizing microtubule ends.

3 Load-Bearing Coupling to Dynamic Microtubule Ends

In cells, kinetochore–microtubule connections persist despite the stochasticity of tubulin dynamics even in the presence of variable forces, both assisting and opposing. Load-bearing (processive motion under the force that acts oppositely to the vector of motion, Fig. 1b) is not a single-molecule phenomenon, and instead requires operation of multi-molecular ensembles, called kinetochore couplers. Load-bearing by couplers moving with a polymerizing microtubule is inherently limited because the microtubule buckles when its growing tip experiences a resisting force (Dogterom and Yurke 1997). Nonetheless, couplers based on multiple independent EB binders are strong enough to bear forces that can sustain microtubule bending for ~100 times longer than the binding time of a single EB molecule (Chen et al. 2014; Doodhi et al. 2014). The assisting force, which acts in the same direction as the vector of microtubule dynamics (Fig. 1b), can be applied at the growing microtubule end coupled to a microbead coated with purified proteins or yeast kinetochore particles (Akiyoshi et al. 2010). The latter can maintain persistent attachment to the assembling tips for tens of minutes under an assisting force of 2–6 pN. Such coupling is thought to rely on the affinity-biased and force-biased diffusion of the bead-bound molecular components, but the underlying theory has not yet been developed. Also, possible contributions from more complex phenomena, such as force-induced modification of the MAPs' hopping rates or attachment times, as well as the bead's rotational diffusion, have not yet been examined.

By contrast, load-bearing by couplers moving at depolymerizing microtubule ends has been investigated both theoretically and experimentally, and these results are covered in more detail in the following sections. Although the ultimate role of such couplers is to capitalize on the unidirectionality of microtubule disassembly, the energy for transporting a load can come from two sources: thermal motions (biased-diffusion mechanism) or changes in tubulin conformation (power stroke mechanism). The exact composition and structure of the kinetochore couplers is not yet known, but theoretical analyses have deepened our understanding of the physically possible coupling mechanisms (reviewed in Grishchuk et al. 2012). Sections below summarize these findings to emphasize that the requirements for processive tracking with no load can be readily met by couplers with different designs, moving via either the biased diffusion or power stroke mechanism. However, large load-bearing can be achieved only by couplers that can capture energy from tubulin power strokes. Moreover, the biomechanical properties of such couplers must be finely tuned to enable them to move at the force-generating microtubule end without detaching. Advanced experimental approaches are needed to test these models and reconstitute the couplers acting at kinetochores in different cell types.

3.1 Load-Bearing by Couplers Moving via the Biased-Diffusion Mechanism

3.1.1 Single Molecules

Biased-diffusion tracking of depolymerizing microtubule tip is fundamentally similar to the work of other "thermal ratchets" (Peskin et al. 1993; Reimann 2002; Mogilner and Oster 2003). Force for MAP motion is generated by thermal fluctuation, and the role of microtubule disassembly is to rectify the resulting motion. In this mechanism, the ultimate role of GTP hydrolysis is to control the rate and location of depolymerization, but it does not directly provide the energy for this motion. A single MAP molecule tracking a dynamic microtubule tip by the biased-diffusion mechanism is a poor vehicle for cargo transport. This is because a molecule can be mobile only if its binding energy is low enough for thermal fluctuation to cause the molecule to hop from one binding site to another. Thus, forces that exceed thermal (0.5–1 pN, assuming 4–8 nm step size), should easily detach such a molecule from the microtubule. To prevent this undesirable outcome, it is usually assumed that the energy required to detach a molecule completely from the microtubule is greater than the energy needed for the molecule to hop. Indeed, some MAPs diffuse on microtubules for much longer than would be expected based on the frequency of hops (Powers et al. 2009), indicating that the assumption of two different energies is reasonable. In this case, however, another problem arises, because instead of detaching, the load will pull the mobile molecule along the microtubule until it reaches the end. As discussed above for diffusing tip-tracking molecules, in order to avoid the loss of the MAP molecule bound to the terminal tubulin subunit, the rate of microtubule depolymerization must decrease and become limited by the rate of the MAP's diffusion (Fig. 3). Thus, load-bearing by a diffusing MAP is limited by the requirement for a fluctuation in thermal energy that can overcome the opposing force and advance the molecule (and its load) away

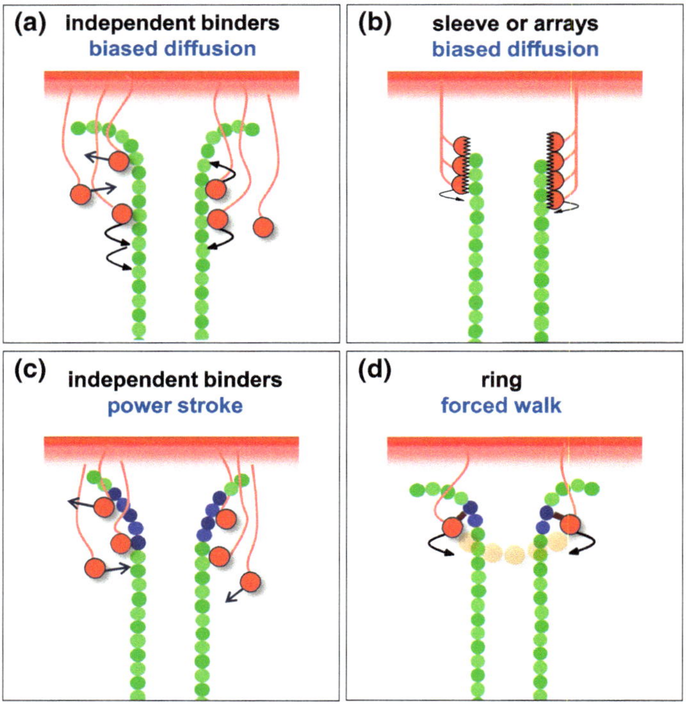

from the tip. Because thermal energy on average generates a small force, even a relatively small load of <1 pN can block tip-tracking of a single MAP, so the biased-diffusion mechanism of tip-tracking has an inherent limit on the load it can carry at physiological velocities (Grishchuk et al. 2012). This conclusion is consistent with the general tendency of thermal ratchets to perform poorly relative to power stroke–dependent motors (Wagoner and Dill 2016).

3.1.2 Multiple Independent Binders

Load capacity should obviously improve if multiple MAP molecules are involved. The simplest such coupler could consist of multiple "independent binders" (Fig. 4a) (Zaytsev et al. 2013), similar to the EB-dependent coupler for polymerizing microtubule tips, considered earlier in this review. In this model, each MAP hops randomly along the microtubule wall in steps of 4 or 8 nm (step size is dictated by the spacing of the binding sites on the microtubule wall), biased by the moving reflective boundary (i.e., the shortening end). It is challenging to predict the load that can be borne by this multimolecular ensemble without a detailed mathematical

◂**Fig. 4** Designs of the most popular models for multimolecular couplers for the depolymerizing microtubule end. **a** In the "independent binders" coupler moving by biased-diffusion mechanism, each MAP binds and diffuses on the microtubule stochastically and independently, analogous to the behavior of molecules in Fig. 2b. The reflection of diffusing molecules at the tip is thought to occur due to some specific feature of bending protofilaments. For example, Ndc80 molecules have weaker affinity for bent tubulin protofilaments than to the tubulins in the cylindrical wall (Alushin et al. 2010; Schmidt et al. 2012). The major concern with this coupling design is that a large pulling load will collect all MAPs at the affinity boundary, where they will either detach or block tubulin disassembly. A corresponding theoretical analysis of these effects, as well as their possible remedies, has not yet been performed. **b** Schematics corresponding to the cross section of a cylindrical sleeve, which completely encircles the microtubule wall, but also illustrates the oligomeric arrays, i.e., groups of three to five MAP subunits with strong lateral bonds that are all bound to the same protofilament. The "teeth" in each MAP symbolize the 0.6-nm step size for coordinated diffusion, the characteristic feature of Hill's design. Also, in this design, the biased motion of the coupler under load arises only in the presence of the "overhangs," i.e., the parts of the cylinder or the arrays that are not bound to the microtubule. Consequently, Hill's design does not function when tubulin ram's horns are present, and this coupler cannot follow polymerizing microtubule ends. **c** The design of the "independent binders" coupler moving via the power stroke mechanism is highly similar to that shown in panel A. It differs only in that the affinity of the MAPs for the microtubule wall is stronger, so they have no or little diffusion. This distinction is similar to that between the affinity-based and biased-diffusion-based mechanisms of tracking of growing microtubule tips (Fig. 2). Thus, in the coupler in *panel A*, the individual MAPs can move processively by thermal hopping, whereas in the power stroke dependent coupler shown here the individual MAPs are not processive. Instead, they bind to straight tubulin in the microtubule wall, mediate pulling on the fibril when the attached tubulin is curling (in *blue*), and dissociate after the tubulin becomes completely bent. **d** Ring couplers can move via different mechanisms. This drawing shows the predicted position of the ring, moving as described in the forced-walk model (Efremov et al. 2007). In this model, the ring binds strongly to the microtubule wall and does not diffuse. The ring moves (*curved arrows*) only when the tubulins at the base of the curls (in *blue*) push on the flexible linkers (*dark bars*). The linkers connect ring subunits with the microtubule wall, both in the models and in the real Dam1 ring (Wang et al. 2007)

model that considers tubulin subunit dynamics and the quantitative characteristics of the MAP's diffusion, its residency time at the microtubule wall and tip, and other details. One estimate for 13 independent MAPs that diffuse similarly to Ndc80 but without detaching from the microtubule wall, suggests that tracking would stall at 6–7 pN (Grishchuk et al. 2012). However, the time of Ndc80 diffusion on the microtubule wall is brief (<1 s) (Powers et al. 2009; Zaytsev et al. 2015). Although with no load these transient individual interactions can sustain lasting attachment of the coupler to the microtubule (Zaytsev et al. 2013), when load is applied the "independent binders" coupler is likely to detach before the stalling force is reached, so its ability to bear a load is probably even smaller. In vitro, Ndc80 molecules conjugated randomly to beads can sustain only 1–2 pN at the ends of depolymerizing microtubules (Powers et al. 2009), consistent with the idea that under these conditions Ndc80 molecules operate independently, coupling the bead to dynamic microtubule via biased diffusion.

3.1.3 Sleeve and Oligomeric Arrays

Another possibility is that at the kinetochore the Ndc80 molecules do not operate as independent binders, but instead move as in Hill's model for a cylindrically shaped "sleeve" coupler (Hill 1985; Powers et al. 2009). Hill and Kirschner pioneered the theoretical investigation of end-on coupling mechanisms (Hill and Kirschner 1982), and Hill was the first to introduce the biased-diffusion mechanism to explain kinetochore load-bearing (Hill 1985; reviewed and analyzed in Mogilner and Oster 1996; Joglekar and Hunt 2002; Shtylla and Keener 2011; Grishchuk et al. 2012). There is a common misconception in the mitosis field that all biased-diffusion couplers move in the same fashion as the Hill's sleeve (Fig. 4b), but this is not so. Rather, there are multiple designs, and accordingly multiple underlying molecular and physical mechanisms that could explain how the biased diffusion of multiple MAPs couples the kinetochore to the disassembling microtubule tip. Indeed, in the coupler with independent binders, considered in previous paragraph (Fig. 4a), the individual MAPs bind and unbind the microtubule wall stochastically and diffuse in an uncoordinated manner. By contrast, in the sleeve coupler the individual MAPs do not operate independently, but hop synchronously in the same direction. As a result, the sleeve's tracking is biased by its motion toward the thermodynamic energy minimum, which is characterized by the maximal number of microtubule-bound MAPs. On the other hand, for the independent binders coupler the entropic component plays a significant role, and at steady-state the number of microtubule-bound MAPs is less than maximal, as dictated by their molecular kinetic constants (Zaytsev et al. 2013). Thus, although both types of couplers exhibit biased diffusion and rectify thermal motions, their underlying mechanisms are very different.

The sleeve design, as proposed many years ago by Hill, deviates in many ways from our current knowledge regarding the molecular and structural biology of the kinetochore (discussed in Efremov et al. 2007), and most researchers agree that such a coupling design is unlikely. It is certainly not a good model for Ndc80, which binds the microtubule wall with 4 nm spacing (Alushin et al. 2010; Zaytsev et al. 2015). In the sleeve model of Ndc80 coupling, the molecular clusters are assumed to diffuse with 0.6 nm step (Powers et al. 2009), while using the physiologically accurate step size (4 nm) should significantly decrease the predicted load-bearing by such a coupler. Another proposed structural arrangement for Ndc80 involves the oligomeric arrays of three to five Ndc80 molecules (Alushin et al. 2010). Such oriented "high-affinity" cluster is thought to maintain microtubule attachment while diffusing along a microtubule protofilament (reviewed in Alushin and Nogales 2011; Tooley and Stukenberg 2011). However, fluorescence analysis of Ndc80 molecules in vitro revealed that this protein has little tendency to form such diffusing arrays (Zaytsev et al. 2015), and the degree of cooperativity implied by electron microscopy studies, which gave rise to this model, appears to be unrealistically high (Zaytsev et al. 2013). Moreover, Ndc80 is not a very fast diffuser, and if three to five Ndc80 molecules were clustered together, such an array would diffuse so slow that it would not be capable of keeping up with the rate of microtubule dynamics at the kinetochore (Zaytsev et al. 2015). Thus, although

Ndc80 remains the main candidate responsible for load-bearing end-on kinetochore attachment in vertebrate cells, the exact design of the Ndc80-containing coupler and whether it moves by the biased-diffusion mechanism is controversial.

3.2 *Microtubule Depolymerization as a Powerful Motor*

It remains unknown how strong the end-on coupler needs to be in order to safely segregate mitotic chromosomes in different cell types, largely because, as pointed out above, the magnitude of forces acting at the kinetochore is not known with certainty. If the load were relatively small, not exceeding 10–15 pN, a well-designed multi-molecular coupler moving via the biased-diffusion mechanism would probably be sufficient to safely transport mitotic chromosomes. However, a different alternative is needed in organisms in which moving chromosomes may encounter large opposing forces or if large pulling forces are needed to assist proper bi-orientation of sister chromatids. Perhaps cells rely on microtubules to transport chromosomes because microtubules are unusual polymers in that they permit not only the biased-diffusion mechanism for processive motility but also a mechanism that can generate force that exceeds the force from thermal fluctuations (reviewed in Inoué and Salmon 1995). Depolymerizing microtubules generate force, and can work as depolymerization motor, thanks to the specific pathway by which tubulin adds to and leaves the microtubule end (reviewed in Desai and Mitchison 1997; Nogales 2001). Each microtubule is like a loaded spring because, during polymerization, only GTP-bound tubulin assembles at the tip and the GTP becomes hydrolyzed. Some of the chemical energy liberated from GTP hydrolysis is stored in conformational strain in the microtubule wall (Caplow et al. 1994; Alushin et al. 2014). During depolymerization of the GDP-containing microtubule wall, linear strands of tubulin, called protofilaments, curl to form "ram's horns" (Mandelkow et al. 1991). This curling protofilaments have been proposed to be capable of delivering a power stroke (Koshland et al. 1988), thereby moving the chromosomes (reviewed in McIntosh et al. 2010). If all the energy from hydrolyzed GTP were channeled into mechanical stroke, the thermodynamically maximal force that depolymerizing microtubule could generate is $\sim$80 pN (Molodtsov et al. 2005). This force is large enough to explain why chromosome motion is not blocked by the experimental application of hundreds of pN (Nicklas 1983), and to rationalize the large force estimate obtained by using fluorescent sensors at the kinetochores in mammalian cells (Ye et al. 2016). Such a force-producing mechanism could also explain how shortening microtubules move chromosomes in purified systems in vitro (Coue et al. 1991). In contrast to the biased-diffusion mechanism, the tip-tracking coupler does not have to wait for thermal fluctuations to overcome the force field of an opposing load, making transport of large loads at physiological velocities possible. However, in the power stroke mechanism, the kinetochore-associated couplers must not only ensure the processivity of tracking, but should also be capable of capturing the energy from microtubule depolymerization. The final sections of this review describe two different ways in which this could be achieved: via couplers with non-processive elements, or via a ring-shaped processive coupler.

3.3 Load-Bearing by a Coupler with Non-processive Binding Elements

Force from bending protofilaments has been shown to act on a micron-size streptavidin-coated bead stably attached to a wall of biotinylated microtubule (Grishchuk et al. 2005). This finding is interesting because biotin–streptavidin attachment is so strong that on the intact microtubule wall the bead is completely immobile, and the force from laser tweezers cannot rupture the underlying bonds. Very strong bonds do not necessarily ensure processivity, and streptavidin bead fails to move with the shortening microtubule end, even in the absence of load. However, as the microtubule depolymerizes at the bead attachment site, the bead exhibits a small jerk in the direction of microtubule shortening, just as expected from the shape of curving protofilaments. Because this motion can be observed in the absence of soluble proteins and nucleotides, the pull must have been generated by conformational changes that took place at the shortening microtubule end as it passed by the bead. This "single-shot" force is only a fraction of a pN, which is much smaller than the rupture force for the bead from the microtubule. This is because these forces are different in nature, and rupture does not represent the force with which the microtubule pulls on the associated coupler. "Single-shot" force exerted by bending protofilaments on the laterally attached micron-size bead is so small, that such a bead would be a terrible coupler from the standpoint of chromosome motion.

Interestingly, both experiments and calculations indicate that if the bead were much smaller, similar in size to a protein molecule, the force it could transduce would have been much greater (Grishchuk et al. 2008a). Moreover, many such "mini-beads" bunched together in a multiprotein coupler could potentially transduce even larger force, since they would have experienced the jerks from all bending protofilaments, not just the one or two protofilaments that interact with the laterally attached microbead. Moreover, with a different design, strong static bonds may become advantageous. This idea was tested with a mathematical model in which multiple MAPs were placed at the ends of elongated fibrils (Fig. 4c) (McIntosh et al. 2008). This design is analogous to the "independent binders" coupler, which was discussed earlier in this chapter as a candidate for biased-diffusion-based coupling. It could be converted into a coupler for the power stroke mechanism by assuming that MAP binding to tubulins in the cylindrical microtubule wall is fairly strong, so under a load they stay attached to the protofilaments long enough to experience the initial stages of their bending. However, to recycle these MAPs, allowing them to rebind microtubule wall and undergo new single-shot pulls, the MAPs must detach fairly quickly from the fully bent tubulins in the ram's horns. This aspect of the power stroke mechanism is frequently misunderstood, because it seems intuitive that bending protofilament generates the power stroke all along its length. However, the main stroke is exerted by the tubulins that undergo conformational changes, i.e., first two or three tubulins at the base of the curl (Molodtsov et al. 2005). The rest of the curl contains tubulin in the fully bent state, so it represents the "exhaust" of the depolymerization motor. For this reason, the non-processive coupler must first attach to tubulin in the wall of the microtubule, and this mechanism will not work if the MAP's only attachment is to the fully bent (i.e., low-energy) form of tubulin. Modeling shows that, indeed, under the power strokes from curling tubulins (shown in blue in Fig. 4), each wall-bound MAP in the coupler can pull transiently on the fibril, whose other end is attached to a cargo. Even though individual MAPs do not diffuse on the microtubule surface, their collective jerks can move the cargo processively against a significant load. In silico, such a coupler becomes stalled at only $\sim$70 pN, so this transporting mechanism with depolymerizing microtubule end is both feasible and potentially very powerful (McIntosh et al. 2008). This coupler could be adapted to move with the polymerizing microtubule tip by the affinity-based mechanism, as in Fig. 2a, but such model has not yet been developed.

Indirect evidence that force at the kinetochores is generated by protofilament power strokes is provided by electron tomography of mammalian kinetochores, which has revealed slender fibrils connected with the curved protofilaments at the end-on attached microtubules (McIntosh et al. 2008, 2013). The fibril-bound protofilaments observed in these studies were slightly straighter than in the typical ram's horns (Fig. 4c), suggesting that they were pulling against a significant load. Currently, however, there is no direct evidence that kinetochore couplers work through high-affinity non-processive binders. The molecular identity of the kinetochore fibril remains unknown, but many fibrillar MAPs could be involved (reviewed in Cheeseman and Desai 2008; Nagpal and Fukagawa 2016). The main candidates are Ndc80 and CENP-F proteins, each comprising a microtubule-binding domain and a highly elongated fibrillar domain. However, when coupled to the microbeads, they can sustain only a small force of several pN (Powers et al. 2009; Volkov et al. 2015), presumably because these MAPs have moderate microtubule-binding affinity. Thus, other kinetochore MAPs would need to be involved in order to realize successfully this coupling mechanism.

Obviously, other coupling designs could be developed based on the ensemble of nondiffusing MAPs, and the movement of such couplers should not necessarily rely on the energy from protofilament power strokes. In one study (Civelekoglu-Scholey et al. 2013), the coupling is hypothesized to involve a viscoelastic protein that does not diffuse and has long residency time and complex force-dependent detachment kinetics. In the "flexible" coupler model (Keener and Shtylla 2014), the MAPs are arranged as in Hill's sleeve, but the sleeve is not rigid, and the nondiffusive MAPs bind to the microtubule wall independently. In the future, this field will undoubtedly see more theoretical studies that analyze the performance of various coupling designs in the context of biophysical properties of specific kinetochore MAPs.

3.4 Ring-Shaped Couplers

3.4.1 Theoretical Motility Mechanisms

Mitchison and colleagues were the first to suggest that a microtubule-encircling ring coupler would be advantageous because it could move processively along the microtubule wall without detaching (Koshland et al. 1988; Mitchison 1988). Ring coupling is particularly interesting because in theory it permits motion by both the thermally driven biased-diffusion mechanism and the diffusion-free mechanism, in which the motion is due to protofilament power strokes (Mitchison 1988; Efremov et al. 2007). This is because the mechanism of motion is not defined by the coupler's geometry so much as by microtubule-binding affinity, as discussed for the "independent binders" designs. If the ring subunits bind the microtubule wall weakly enough to permit the ring's diffusion, its "random walk" (Pearson 1905; Howard 2002) will be biased by the flared protofilaments at the end of the shortening microtubule, assuming that these structures are persistent. As expected for all biased-diffusion couplers, calculations have shown that the opposing load pulls the low-affinity ring to the microtubule tip (Efremov et al. 2007), where the ring will either resist further depolymerization or detach quickly if the protofilament curls are lost, e.g., through stochastic fluctuations or straightening under a load.

To force the loaded ring to move away from the terminal tip subunits and permit microtubule disassembly, tubulin power strokes are necessary; accordingly, this mode of tip-tracking has been termed the "forced walk" (Efremov et al. 2007). The primary distinction between the biased-walk and forced-walk mechanisms is related not to the presence or absence of the "conformational wave" of tubulin bending (Koshland et al. 1988), but rather to the role it plays in the ring's motion (Mitchison 1988). In the biased-diffusion mechanism, the "conformational wave" forms a reflective barrier, whereas in the forced walk this conformational change represents a force-producing element. As the name "forced walk" implies, in this mechanism the MAP elements of the ring coupler move processively under the influence of the depolymerization force. This contrasts with the operation of the independent binders coupler in Fig. 4c, which is also moved by power strokes. In that design, however, the motions of the individual MAPs are not processive, so this variation of the "independent binders" coupler does not operate via the forced-walk mechanism. Hopefully, this consideration demonstrates the diversity and multitude of possible coupling designs, as well as the need to rigorously analyze them via quantitative modeling. The important lesson from theoretical studies thus far is that the competing constraints on the molecular parameters of each design arise from the nontrivial requirement of processivity of tracking under a large load at the force-generating end of the polymer.

Theoretical analyses of ring coupling have allowed evaluating the relative benefits of various mechanisms of operation (Efremov et al. 2007). One important finding relates to the ability of the power stroke mechanism to work in conjunction with rings that bind to the microtubule wall much more strongly than is permissive for ring's diffusion, so such rings cannot be moved by the biased-diffusion mechanism. Stronger ring binding is highly beneficial for chromosome segregation because ring's adhesion represents a back-up mechanism to prevent ring's slipping from the microtubule end that is not protected by the protofilament flare, which can be temporarily lost due to a stochastic fluctuation or disassembly pausing (Zakharov et al. 2015). Such protection would come at the expense of maximum load-bearing by the ring coupler, because the energy from power strokes will now have to be spent moving the wall-binding ring, not just the load. Obviously, if the binding is excessive, the internal friction of this coupler may become so strong that the depolymerization motor will not be able to move such a ring, and microtubule disassembly will be blocked until the ring falls off. Thus, the back-up mechanism to prevent ring's slipping and load-bearing impose competing constraints on the strength of ring's binding to microtubule wall. When it is optimal, the power stroke mechanism can transport larger load and provide better stability of end-attachment than the biased-diffusion mechanism. On the other hand, load-bearing by such a ring is smaller than that by a coupler with non-processive independent binders (as in Fig. 4c) because the latter do not "walk," and therefore such a coupler has no internal friction (McIntosh et al. 2008). Other models for a ring coupler have also been proposed (Liu and Onuchic 2006; Armond and Turner 2010; Vichare et al. 2013), so theoretical investigations evaluating different ring coupling designs are ongoing.

3.4.2 The Yeast Dam1 Ring

The ring coupler is also unusual and exciting because it can be studied experimentally using Dam1 protein complex from the yeast kinetochore, which oligomerizes spontaneously to form microtubule-encircling rings in vitro (Miranda et al. 2005; Westermann et al. 2005). The electron-dense core of the Dam1 ring is separated from the outer microtubule wall by a 3–6 nm gap, spanned by flexible linkages (Miranda et al. 2007; Wang et al. 2007). Such large ring diameter is optimally suited for the power stroke mechanism because it maximizes force transduction by the ring (Molodtsov et al. 2005). Although Dam1 heterodecamers, the individual subunits of the ring, diffuse well on microtubule walls (Westermann et al. 2006; Gestaut et al. 2008; Grishchuk et al. 2008b), their collective microtubule affinity is so strong that the 16-subunit ring diffuses extremely slowly (Volkov et al. 2013); for alternative view on ring's diffusion see (Ramey et al. 2011). If such slow ring had to move via biased diffusion, the rate of microtubule disassembly would have decreased below 0.1 μm/min. On a side note, this behavior is significantly different from that of the sleeve coupler, because the latter was designed to always track at the same velocity by adjusting the size of the overhangs; ring, however, has all subunits bound to the microtubule wall. Real Dam1 ring appears to be able to track the shortening microtubule ends in vitro much faster than would be expected for the biased-diffusion mechanism, at 7–10 μm/min (Grishchuk et al. 2008a). Collectively, this evidence points to the forced-walk mechanism of yeast Dam1 ring tracking. Such nondiffusive coupler with strong microtubule adhesion would be particularly useful in budding yeast, in which each kinetochore is attached to only one microtubule; consequently, loss of this attachment cannot be permitted.

Additional evidence for the forced-walk mechanism of yeast Dam1 ring tracking came from measurements of the load-bearing by this coupler. The relatively strong microtubule wall binding by Dam1 ring, as deduced from its negligible diffusion, implies that it should be able to carry ~40 pN load (Efremov et al. 2007). The initial experiments, however, found that microtubule depolymerization force transduced by Dam1 did not exceed 5 pN (Asbury et al. 2006; Grishchuk et al. 2008a). This is much larger than the force measured with streptavidin beads, but much smaller than predicted by the forced-walk model. This discrepancy could be explained by differences in the geometry of load application between theory and experiment. The experimental measurements were carried out with Dam1 rings attached laterally to the microbeads, whereas the theoretical prediction was for suspended rings (Fig. 4d), such as those that are thought to operate at yeast kinetochores (Gonen et al. 2012). Indeed, when Dam1 was linked to beads with the help of elongated fibrillar tethers, force transduction increased dramatically (Volkov et al. 2013). This force still has not reached the theoretical maximum, presumably because experimental rings can rupture, whereas theoretical rings are rupture-free. Weak ring integrity may limit load-bearing capacity at the kinetochore, or may simply reflect the inadequacy of in vitro systems that use purified components. One important outcome from experiments using the suspended Dam1 ring is that its load-bearing appears to be similar to that of isolated kinetochore particles tracking depolymerizing microtubule ends (Akiyoshi et al. 2010; Volkov et al. 2013). Therefore, it is possible that the suspended Dam1 ring moving via the forced-walk mechanism represents the main force-transducing unit of yeast kinetochore. Elongated tethers are highly important for ring's load-bearing because they allow the load to become aligned along the microtubule axis, thereby minimizing the lever-arm effect that limits force transduction by the wall-bound beads (Grishchuk et al. 2005). Long tethers also permit distributing the load evenly among the microtubule protofilaments. Although ring-dependent coupling may be unique to the yeast kinetochore, fibrillar elements will undoubtedly be found to play essential roles in load-bearing by kinetochore couplers in other cell types.

Acknowledgements The author thanks F. Ataullakhanov for critical reading of the manuscript, and M. Godzi for technical assistance with preparing the manuscript and figures. Funding for this work was provided by the National Institutes of Health grant GM-R01098389. Numerical calculations in Fig. 3 were carried out with support from Russian Science Foundation (grant # 16-14-00-224). E.L.G. is supported in part by the American Cancer Society grant RSG-14-018-01-CCG.

References

Akhmanova A, Steinmetz MO (2010) Microtubule + TIPs at a glance. J Cell Sci 123:3415–3419. doi:10.1242/jcs.062414

Akiyoshi B, Sarangapani KK, Powers AF, Nelson CR, Reichow SL, Arellano-santoyo H, Gonen T, Ranish J a, Asbury CL, Biggins S (2010) Tension directly stabilizes reconstituted kinetochore- microtubule attachments. Nature 468:576–579. doi:10.1038/nature09594

Alexander SP, Rieder CL (1991) Chromosome motion during attachment to the vertebrate spindle: initial saltatory-like behavior of chromosomes and quantitative analysis of force production by nascent kinetochore fibers. J Cell Biol 113:805–815

Alushin G, Nogales E (2011) Visualizing kinetochore architecture. Curr Opin Struct Biol 21:661–669. doi:10.1016/j.sbi.2011.07.009

Alushin GM, Lander GC, Kellogg EH, Zhang R, Baker D, Nogales E (2014) High-resolution microtubule structures reveal the structural transitions in αβ-tubulin upon GTP hydrolysis. Cell 157:1117–1129. doi:10.1016/j.cell.2014.03.053

Alushin GM, Ramey VH, Pasqualato S, Ball DA, Grigorieff N, Musacchio A, Nogales E (2010) The Ndc80 kinetochore complex forms oligomeric arrays along microtubules. Nature 467:805–810. doi:10.1038/nature09423

Aravamudhan P, Felzer-Kim I, Gurunathan K, Joglekar AP (2014) Assembling the protein architecture of the budding yeast kinetochore-microtubule attachment using FRET. Curr Biol 24:1437–1446. doi:10.1016/j.cub.2014.05.014

Armond JW, Turner MS (2010) Force transduction by the microtubule-bound Dam1 ring. Biophys J 98:1598–1607. doi:10.1016/j.bpj.2010.01.004

Armond JW, Vladimirou E, Erent M, McAinsh AD, Burroughs NJ (2015) Probing microtubule polymerisation state at single kinetochores during metaphase chromosome motion. J Cell Sci 128:1991–2001. doi:10.1242/jcs.168682

Asbury CL, Gestaut DR, Powers AF, Franck AD, Davis TN (2006) The Dam1 kinetochore complex harnesses microtubule dynamics to produce force and movement. Proc Natl Acad Sci U S A 103:9873–9878. doi:10.1073/pnas.0602249103

Asbury CL, Tien JF, Davis TN (2011) Kinetochores' gripping feat: Conformational wave or biased diffusion? Trends Cell Biol 21:38–46. doi:10.1016/j.tcb.2010.09.003

Ayaz P, Munyoki S, Geyer EA, Piedra FA, Vu ES, Bromberg R, Otwinowski Z, Grishin NV, Brautigam CA, Rice LM (2014) A tethered delivery mechanism explains the catalytic action of a microtubule polymerase. Elife 3:e03069. doi:10.7554/eLife.03069

Barisic M, Aguiar P, Geley S, Maiato H (2014) Kinetochore motors drive congression of peripheral polar chromosomes by overcoming random arm-ejection forces. Nat Cell Biol 16:1249–1256. doi:10.1038/ncb3060

Bieling P, Kandels-Lewis S, Telley IA, van Dijk J, Janke C, Surrey T (2008) CLIP-170 tracks growing microtubule ends by dynamically recognizing composite EB1/tubulin-binding sites. J Cell Biol 183:1223–1233. doi:10.1083/jcb.200809190

Brouhard GJ, Stear JH, Noetzel TL, Al-Bassam J, Kinoshita K, Harrison SC, Howard J, Hyman AA (2008) XMAP215 Is a processive microtubule polymerase. Cell 132:79–88. doi:10.1016/j.cell.2007.11.043

Caplow M, Ruhlen RL, Shanks J (1994) The free energy for hydrolysis of a microtubule-bound nucleotide triphosphate is near zero: All of the free energy for hydrolysis is stored in the microtubule lattice. J Cell Biol 127:779–788. doi:10.1083/jcb.127.3.779

Cheerambathur DK, Desai A (2014) Linked in: Formation and regulation of microtubule attachments during chromosome segregation. Curr Opin Cell Biol 26:113–122. doi:10.1016/j.ceb.2013.12.005

Cheerambathur DK, Gassmann R, Cook B, Oegema K, Desai A (2013) Crosstalk between microtubule attachment complexes ensures accurate chromosome segregation. Science 342:1239–1242. doi:10.1126/science.1246232

Cheeseman IM, Desai A (2008) Molecular architecture of the kinetochore–microtubule interface. Nat Rev Mol Cell Biol 9:33–46. doi:10.1038/nrm2310
Chen Y, Rolls MM, Hancock WO (2014) An EB1-kinesin complex is sufficient to steer microtubule growth in vitro. Curr Biol 24:316–321. doi:10.1016/j.cub.2013.11.024
Cimini D, Cameron LA, Salmon ED, Hill C, Carolina N (2004) Anaphase spindle mechanics prevent mis-segregation of merotelically oriented chromosomes. Curr Biol 14:2149–2155. doi:10.1016/j.cub.2004.11.029
Civelekoglu-Scholey G, He B, Shen M, Wan X, Roscioli E, Bowden B, Cimini D (2013) Dynamic bonds and polar ejection force distribution explain kinetochore oscillations in PtK1 cells. J Cell Biol 201:577–593. doi:10.1083/jcb.201301022
Cooke CA, Schaar B, Yen TJ, Earnshaw WC (1997) Localization of CENP-E in the fibrous corona and outer plate of mammalian kinetochores from prometaphase through anaphase. Chromosoma 106:446–455. doi:10.1007/s004120050266
Cooper JR, Wordeman L (2009) The diffusive interaction of microtubule binding proteins. Curr Opin Cell Biol 21:68–73. doi:10.1016/j.ceb.2009.01.005
Coue M, Lombillo VA, McIntosh JR (1991) Microtubule depolymerization promotes particle and chromosome movement in vitro. J Cell Biol 112:1165–1175. doi:10.1083/jcb.112.6.1165
Desai A, Mitchison TJ (1997) Microtubule polymerization dynamics. Annu Rev Cell Dev Biol 13:83–117. doi:10.1146/annurev.cellbio.13.1.83
Dogterom M, Yurke B (1997) Measurement of the force-velocity relation for growing microtubules. Science 80(278):856–860. doi:10.1126/science.278.5339.856
Dong Y, Vanden Beldt KJ, Meng X, Khodjakov A, McEwen BF (2007) The outer plate in vertebrate kinetochores is a flexible network with multiple microtubule interactions. Nat Cell Biol 9:516–522. doi:10.1038/ncb1576
Doodhi H, Katrukha EAEA, Kapitein LCLC, Akhmanova A (2014) Mechanical and geometrical constraints control kinesin-based microtubule guidance. Curr Biol 24:322–328. doi:10.1016/j.cub.2014.01.005
Dumont S, Salmon ED, Mitchison TJ (2012) Deformations within moving kinetochores reveal different sites of active and passive force generation. Science 337:355–358. doi:10.1126/science.1221886
Efremov A, Grishchuk EL, McIntosh JR, Ataullakhanov FI (2007) In search of an optimal ring to couple microtubule depolymerization to processive chromosome motions. Proc Natl Acad Sci U S A 104:19017–19022. doi:10.1073/pnas.0709524104
Espeut J, Gaussen A, Bieling P, Morin V, Prieto S, Fesquet D, Surrey T, Abrieu A (2008) Phosphorylation relieves autoinhibition of the kinetochore motor Cenp-E. Mol Cell 29:637–643. doi:10.1016/j.molcel.2008.01.004
Ferraro-Gideon J, Sheykhani R, Zhu Q, Duquette ML, Berns MW, Forer A (2013) Measurements of forces produced by the mitotic spindle using optical tweezers. Mol Biol Cell 24:1375–1386. doi:10.1091/mbc.E12-12-0901
Gandhi SR, Gierliński M, Mino A, Tanaka K, Kitamura E, Clayton L, Tanaka TU (2011) Kinetochore-dependent microtubule rescue ensures their efficient and sustained interactions in early mitosis. Dev Cell 21:920–933. doi:10.1016/j.devcel.2011.09.006
Gestaut DR, Graczyk B, Cooper J, Widlund PO, Zelter A, Wordeman L, Asbury CL, Davis TN (2008) Phosphoregulation and depolymerization-driven movement of the Dam1 complex do not require ring formation. Nat Cell Biol 10:407–414. doi:10.1038/ncb1702
Gonen S, Akiyoshi B, Iadanza MG, Shi D, Duggan N, Biggins S, Gonen T (2012) The structure of purified kinetochores reveals multiple microtubule-attachment sites. Nat Struct Mol Biol 19:925–929. doi:10.1038/nsmb.2358
Grishchuk EL, Efremov AK, Volkov VA, Spiridonov IS, Gudimchuk N, Westermann S, Drubin D, Barnes G, McIntosh JR, Ataullakhanov FI (2008a) The Dam1 ring binds microtubules strongly enough to be a processive as well as energy-efficient coupler for chromosome motion. Proc Natl Acad Sci U S A 105:15423–15428. doi:10.1073/pnas.0807859105

Grishchuk EL, McIntosh JR (2006) Microtubule depolymerization can drive poleward chromosome motion in fission yeast. EMBO J 25:4888–4896. doi:10.1038/sj.emboj.7601353

Grishchuk EL, McIntosh JR, Molodtsov MI, Ataullakhanov FI (2012) Force generation by dynamic microtubule polymers. In: Goldman YE, Ostap EM (eds) Comprehensive biophysics. Elsevier, Netherlands, pp 93–117

Grishchuk EL, Molodtsov MI, Ataullakhanov FI, McIntosh JR (2005) Force production by disassembling microtubules. Nature 438:384–388. doi:10.1038/nature04132

Grishchuk EL, Spiridonov IS, Volkov V, Efremov A, Westermann S, Drubin D, Barnes G, Ataullakhanov FI, McIntosh JR (2008b) Different assemblies of the DAM1 complex follow shortening microtubules by distinct mechanisms. Proc Natl Acad Sci U S A 105:6918–6923. doi:10.1073/pnas.0801811105

Gudimchuk N, Vitre B, Kim Y, Kiyatkin A, Cleveland DW, Ataullakhanov FI, Grishchuk EL (2013) Kinetochore kinesin CENP-E is a processive bi-directional tracker of dynamic microtubule tips. Nat Cell Biol 15:1079–1088. doi:10.1038/ncb2831

Hayden JH, Bowser SS, Rieder CL (1990) Kinetochores capture astral microtubules during chromosome attachment to the mitotic spindle: direct visualization in live newt lung cells. J Cell Biol 111:1039–1045. doi:10.1083/jcb.111.3.1039

Hill TL (1985) Theoretical problems related to the attachment of microtubules to kinetochores. Cell Biol 82:4404–4408. doi:10.1073/pnas.82.13.4404

Hill TL, Kirschner MW (1982) Bioenergetics and kinetics of microtubule and actin filament assembly-disassembly. Int Rev Cytol 78:1–125. doi:10.1016/S0074-7696(08)60105-9

Howard J (2002) Mechanics of motor proteins. Physics of bio-molecules and cells. Physique des biomolécules et des cellules. Springer, Berlin, pp 69–94

Hunt AJ, McIntosh JR (1998) The dynamic behavior of individual microtubules associated with chromosomes in vitro. Mol Biol Cell 9:2857–2871. doi:10.1091/mbc.9.10.2857

Inoué S, Salmon ED (1995) Force generation by microtubule assembly/disassembly in mitosis and related movements. Mol Biol Cell 6:1619–1640. doi:10.1091/mbc.E12-02-0146

Joglekar AP, Bloom KS, Salmon ED (2010) Mechanisms of force generation by end-on kinetochore-microtubule attachments. Curr Opin Cell Biol 22:57–67. doi:10.1016/j.ceb.2009.12.010

Joglekar AP, Hunt AJ (2002) A simple, mechanistic model for directional instability during mitotic chromosome movements. Biophys J 83:42–58. doi:10.1016/S0006-3495(02)75148-5

Ke K, Cheng J, Hunt AJ (2009) The distribution of polar ejection forces determines the amplitude of chromosome directional instability. Curr Biol 19:807–815. doi:10.1016/j.cub.2009.04.036

Keener JP, Shtylla B (2014) A mathematical model of force generation by flexible kinetochore-microtubule attachments. Biophys J 106:998–1007. doi:10.1016/j.bpj.2014.01.013

Kerssemakers JWJ, Munteanu EL, Laan L, Noetzel TL, Janson ME, Dogterom M (2006) Assembly dynamics of microtubules at molecular resolution. Nature 442:709–712. doi:10.1038/nature04928

Khodjakov A, Rieder CL (1996) Kinetochores moving away from their associated pole do not exert a significant pushing force on the chromosome. J Cell Biol 135:315–327. doi:10.1083/jcb.135.2.315

Kim Y, Heuser JE, Waterman CM, Cleveland DW (2008) CENP-E combines a slow, processive motor and a flexible coiled coil to produce an essential motile kinetochore tether. J Cell Biol 181:411–419. doi:10.1083/jcb.200802189

Koshland DE, Mitchison TJ, Kirschner MW (1988) Polewards chromosome movement driven by microtubule depolymerization in vitro. Nature 331:499–504. doi:10.1038/331499a0

Lampert F, Hornung P, Westermann S (2010) The Dam1 complex confers microtubule plus end-tracking activity to the Ndc80 kinetochore complex. J Cell Biol 189:641–649. doi:10.1083/jcb.200912021

Lawrimore J, Bloom KS, Salmon ED (2011) Point centromeres contain more than a single centromere-specific Cse4 (CENP-A) nucleosome. J Cell Biol 195:573–582. doi:10.1083/jcb.201106036

Liao H, Li G, Yen TJ (1994) Mitotic regulation of microtubule cross-linking activity of CENP-E kinetochore protein. Science 80(265):394–398. doi:10.1126/science.8023161
Liu J, Onuchic JN (2006) A driving and coupling "Pac-Man" mechanism for chromosome poleward translocation in anaphase A. Proc Natl Acad Sci 103:18432–18437. doi:10.1073/pnas.0608962103
Lombillo VA, Nislow C, Yen TJ, Gelfand VI, McIntosh JR (1995) Antibodies to the kinesin motor domain and CENP-E inhibit microtubule depolymerization-dependent motion of chromosomes in vitro. J Cell Biol 128:107–115. doi:10.1083/jcb.128.1.107
Maddox P, Straight A, Coughlin P, Mitchison TJ, Salmon ED (2003) Direct observation of microtubule dynamics at kinetochores in Xenopus extract spindles: implications for spindle mechanics. J Cell Biol 162:377–382. doi:10.1083/jcb.200301088
Magidson V, O'Connell CB, Lončarek J, Paul R, Mogilner A, Khodjakov A (2011) The spatial arrangement of chromosomes during prometaphase facilitates spindle assembly. Cell 146:555–567. doi:10.1016/j.cell.2011.07.012
Sokolov Mai, Blumen A (2001) Directed particle diffusion under "burnt bridges" conditions. Phys Rev E—Stat Nonlinear, Soft Matter Phys 64:011102/1–011102/4. doi:10.1103/PhysRevE.64.011102
Maiato H, DeLuca J, Salmon ED, Earnshaw WC (2004) The dynamic kinetochore-microtubule interface. J Cell Sci 117:5461–5477. doi:10.1242/jcs.01536
Mandelkow E, Mandelkow E, Milligan RA (1991) Microtubule dynamics and microtubule caps: a time-resolved cryo- electron microscopy study. J Cell Biol 114:977–991. doi:10.1083/jcb.114.5.977
Maurer SP, Bieling P, Cope J, Hoenger A, Surrey T (2011) GTPgammaS microtubules mimic the growing microtubule end structure recognized by end-binding proteins (EBs). Proc Natl Acad Sci U S A 108:3988–3993. doi:10.1073/pnas.1014758108
Mayr MI, Storch M, Howard J, Mayer TU (2011) A non-motor microtubule binding site is essential for the high processivity and mitotic function of kinesin-8 Kif18A. PLoS ONE 6: e27471. doi:10.1371/journal.pone.0027471
McEwen BF, Hsieh C-EE, Mattheyses AL, Rieder CL (1998) A new look at kineochore structure in vertebrate somatic cells using high-pressure freezing an freeze substitution. Chromosoma 107:366–375. doi:10.1007/s004120050320
McIntosh JR, Grishchuk EL, Morphew MK, Efremov AK, Zhudenkov K, Volkov VA, Cheeseman IM, Desai A, Mastronarde DN, Ataullakhanov FI (2008) Fibrils connect microtubule tips with kinetochores: a mechanism to couple tubulin dynamics to chromosome Motion. Cell 135:322–333. doi:10.1016/j.cell.2008.08.038
McIntosh JR, Grishchuk EL, West RR (2002) Chromosome-microtubule interactions during mitosis. Annu Rev Cell Dev Biol 18:193–219. doi:10.1146/annurev.cellbio.18.032002.132412
McIntosh JR, O'Toole E, Zhudenkov K, Morphew M, Schwartz C, Ataullakhanov FI, Grishchuk EL (2013) Conserved and divergent features of kinetochores and spindle microtubule ends from five species. J Cell Biol 200:459–474. doi:10.1083/jcb.201209154
McIntosh JR, Volkov V, Ataullakhanov FI, Grishchuk EL (2010) Tubulin depolymerization may be an ancient biological motor. J Cell Sci 123:3425–3434. doi:10.1242/jcs.067611
Miller MP, Asbury CL, Biggins S (2016) A TOG protein confers tension sensitivity to kinetochore-microtubule attachments. Cell 165:1428–1439. doi:10.1016/j.cell.2016.04.030
Minoura I, Katayama E, Sekimoto K, Muto E (2010) One-dimensional brownian motion of charged nanoparticles along microtubules: a model system for weak binding interactions. Biophys J 98:1589–1597. doi:10.1016/j.bpj.2009.12.4323
Miranda JJL, De Wulf P, Sorger PK, Harrison SC (2005) The yeast DASH complex forms closed rings on microtubules. Nat Struct Mol Biol 12:138–143. doi:10.1038/nsmb896
Miranda JJL, King DS, Harrison SC (2007) Protein arms in the kinetochore-microtubule interface of the yeast DASH complex. Mol Biol Cell 18:2503–2510. doi:10.1091/mbc.E07-02-0135
Mitchison T, Evans L, Schulze E, Kirschner M (1986) Sites of microtubule assembly and disassembly in the mitotic spindle. Cell 45:515–527. doi:10.1016/0092-8674(86)90283-7

Mitchison TJ (1988) Microtubule dynamics and kinetochore function in mitosis. Annu Rev Cell Biol 4:527–549. doi:10.1146/annurev.cb.04.110188.002523

Mitchison TJ, Kirschner MW (1985) Properties of the kinetochore in vitro. II. Microtubule capture and ATP-dependent translocation. J Cell Biol 101:766–777. doi:10.1083/jcb.101.3.766

Mogilner A, Oster G (1996) Cell motility driven by actin polymerization. Biophys J 71:3030–3045. doi:10.1016/S0006-3495(96)79496-1

Mogilner A, Oster G (2003) Polymer motors: pushing out the front and pulling up the back. Curr Biol 13:R721–R733. doi:10.1016/j.cub.2003.08.050

Molodtsov MI, Grishchuk EL, Efremov AK, McIntosh JR, Ataullakhanov FI (2005) Force production by depolymerizing microtubules: a theoretical study. Proc Natl Acad Sci U S A 102:4353–4358. doi:10.1073/pnas.0501142102

Nagpal H, Fukagawa T (2016) Kinetochore assembly and function through the cell cycle. Chromosoma 125:645–659. doi:10.1007/s00412-016-0608-3

Nicklas RB (1965) Chromosome velocity during mitosis as a function of chromosome size and position. J Cell Biol 25:119–135. doi:10.1083/jcb.25.1.119

Nicklas RB (1983) Measurements of the force produced by the mitotic spindle in anaphase. J Cell Biol 97:542–548. doi:10.1083/jcb.97.2.542

Nicklas RB (1989) The motor for poleward chromosome movement in anaphase is in or near the kinetochore. J Cell Biol 109:2245–2255. doi:10.1083/jcb.109.5.2245

Nogales E (2001) Structural insights into microtubule function. Annu Rev Biochem 69:277–302. doi:10.1146/annurev.biochem.74.061903.155440

O'Connell CB, Khodjakov A, McEwen BF (2012) Kinetochore flexibility: creating a dynamic chromosome-spindle interface. Curr Opin Cell Biol 24:40–47. doi:10.1016/j.ceb.2011.12.008

Pearson K (1905) The problem of the random walk. Nature 72:294

Peskin CS, Odell GM, Oster GF (1993) Cellular motions and thermal fluctuations: the Brownian ratchet. Biophys J 65:316–324. doi:10.1016/S0006-3495(93)81035-X

Peskin CS, Oster GF (1995) Force production by depolymerizing microtubules: load-velocity curves and run-pause statistics. Biophys J 69:2268–2276. doi:10.1016/S0006-3495(95)80097-4

Pfarr CM, Coue M, Grissom PM, Hays TS, Porter ME, McIntosh JR (1990) Cytoplasmic dynein is localized to kinetochores during mitosis. Nature 345:263–265. doi:10.1038/345263a0

Powers AF, Franck AD, Gestaut DR, Cooper J, Gracyzk B, Wei RR, Wordeman L, Davis TN, Asbury CL (2009) The Ndc80 Kinetochore complex forms load-bearing attachments to dynamic microtubule tips via biased diffusion. Cell 136:865–875. doi:10.1016/j.cell.2008.12.045

Preciado López M, Huber F, Grigoriev I, Steinmetz MO, Akhmanova A, Koenderink GH, Dogterom M (2014) Actin-microtubule coordination at growing microtubule ends. Nat Commun 5:4778. doi:10.1038/ncomms5778

Putkey FR, Cramer T, Morphew MK, Silk AD, Johnson RS, McIntosh JR, Cleveland DW (2002) Unstable kinetochore-microtubule capture and chromosomal instability following deletion of CENP-E. Dev Cell 3:351–365. doi:10.1016/S1534-5807(02)00255-1

Rago F, Cheeseman IM (2013) The functions and consequences of force at kinetochores. J Cell Biol 200:557–565. doi:10.1083/jcb.201211113

Ramey VH, Wang H-W, Nakajima Y, Wong A, Liu J, Drubin D, Barnes G, Nogales E (2011) The Dam1 ring binds to the E-hook of tubulin and diffuses along the microtubule. Mol Biol Cell 22:457–466. doi:10.1091/mbc.E10-10-0841

Reimann P (2002) Brownian motors: noisy transport far from equilibrium. Phys Rep 361:57–265. doi:10.1016/S0370-1573(01)00081-3

Reithmann E, Reese L, Frey E (2015) Quantifying protein diffusion and capture on filaments. Biophys J 108:787–790. doi:10.1016/j.bpj.2014.12.053

Rieder CL (1982) The formation, structure, and composition of the mammalian kinetochore and kinetochore Fiber. Int Rev Cytol 79:1–58. doi:10.1016/S0074-7696(08)61672-1

Rieder CL, Salmon ED (1998) The vertebrate cell kinetochore and its roles during mitosis. Trends Cell Biol 8:310–318. doi:10.1016/S0962-8924(98)01299-9

Santaguida S, Musacchio A (2009) The life and miracles of kinetochores. EMBO J 28:2511–2531. doi:10.1038/emboj.2009.173

Schmidt JC, Arthanari H, Boeszoermenyi A, Dashkevich NM, Wilson-Kubalek EM, Monnier N, Markus M, Oberer M, Milligan RA, Bathe M, Wagner G, Grishchuk EL, Cheeseman IM (2012) The kinetochore-bound Ska1 complex tracks depolymerizing microtubules and binds to curved protofilaments. Dev Cell 23:968–980. doi:10.1016/j.devcel.2012.09.012

Shrestha RL, Draviam VM (2013) Lateral to end-on conversion of chromosome-microtubule attachment requires kinesins cenp-e and MCAK. Curr Biol 23:1514–1526. doi:10.1016/j.cub.2013.06.040

Shtylla B, Keener JP (2011) A mathematical model for force generation at the kinetochore-microtubule Interface. SIAM J Appl Math 71:1821–1848. doi:10.1137/100802645

Skibbens RV, Skeen VP, Salmon ED (1993) Directional instability of kinetochore motility during chromosome congression and segregation in mitotic newt lung cells: A push-pull mechanism. J Cell Biol 122:859–875. doi:10.1083/jcb.122.4.859

Skibbens RV, Salmon ED (1997) Micromanipulation of chromosomes in mitotic vertebrate tissue cells: tension controls the state of kinetochore movement. Exp Cell Res 235:314–324. doi:10.1006/excr.1997.3691

Steuer ER, Wordeman L, Schroer TA, Sheetz MP (1990) Localization of cytoplasmic dynein to mitotic spindles and kinetochores. Nature 345:266–268. doi:10.1038/345266a0

Su X, Qiu W, Gupta ML, Pereira-Leal JB, Reck-Peterson SL, Pellman D (2011) Mechanisms underlying the dual-mode regulation of microtubule dynamics by Kip3/Kinesin-8. Mol Cell 43:751–763. doi:10.1016/j.molcel.2011.06.027

Suzuki A, Badger BL, Salmon ED (2015) A quantitative description of Ndc80 complex linkage to human kinetochores. Nat Commun 6:8161. doi:10.1038/ncomms9161

Tanaka K, Kitamura E, Kitamura Y, Tanaka TU (2007) Molecular mechanisms of microtubule-dependent kinetochore transport toward spindle poles. J Cell Biol 178:269–281. doi:10.1083/jcb.200702141

Tanaka K, Mukae N, Dewar H, van Breugel M, James EK, Prescott AR, Antony C, Tanaka TU (2005) Molecular mechanisms of kinetochore capture by spindle microtubules. Nature 434:987–994. doi:10.1038/nature03483

Tanaka TU (2010) Kinetochore-microtubule interactions: steps towards bi-orientation. EMBO J 29:1–13. doi:10.1038/emboj.2010.294

Taylor EW (1965) Brownian and saltatory movements of cytoplasmic granules and the movement of anaphase chromosomes. In: Lee EH (ed) Proceedings of the fourth international congress on rheology. Interscience, New York, pp 175–191

Thomas GE, Bandopadhyay K, Sutradhar S, Renjith MR, Singh P, Gireesh KK, Simon S, Badarudeen B, Gupta H, Banerjee M, Paul R, Mitra J, Manna TK (2016) EB1 regulates attachment of Ska1 with microtubules by forming extended structures on the microtubule lattice. Nat Commun 7:11665. doi:10.1038/ncomms11665

Tirnauer JS, Canman JC, Salmon ED, Mitchison TJ (2002) EB1 targets to kinetochores with attached, polymerizing microtubules. Mol Biol Cell 13:4308–4316. doi:10.1091/mbc.E02

Tooley J, Stukenberg PT (2011) The Ndc80 complex: integrating the kinetochore's many movements. Chromosome Res 19:377–391. doi:10.1007/s10577-010-9180-5

Trushko A, Schäffer E, Howard J (2013) The growth speed of microtubules with XMAP215-coated beads coupled to their ends is increased by tensile force. Proc Natl Acad Sci U S A 110:14670–14675. doi:10.1073/pnas.1218053110

Umbreit NT, Gestaut DR, Tien JF, Vollmar BS, Gonen T, Asbury CL, Davis TN (2012) The Ndc80 kinetochore complex directly modulates microtubule dynamics. Proc Natl Acad Sci U S A 109:16113–16118. doi:10.1073/pnas.1209615109

Verma V, Mallik L, Hariadi RF, Sivaramakrishnan S, Skiniotis G, Joglekar AP (2015) Using protein dimers to maximize the protein hybridization efficiency with multisite DNA origami scaffolds. PLoS ONE 10(9):e0137125

Vichare S, Jain I, Inamdar MM, Padinhateeri R (2013) Forces due to curving protofilaments in microtubules. Phys Rev E 88:62708. doi:10.1103/PhysRevE.88.062708

Vitre B, Gudimchuk N, Borda R, Kim Y, Heuser JE, Cleveland DW, Grishchuk EL (2014) Kinetochore-microtubule attachment throughout mitosis potentiated by the elongated stalk of the kinetochore kinesin CENP-E. Mol Biol Cell 25:1–26. doi:10.1091/mbc.E14-01-0698

Volkov VA, Grissom PM, Arzhanik VK, Zaytsev AV, Renganathan K, McClure-Begley T, Old WM, Ahn N, Richard McIntosh J (2015) Centromere protein F includes two sites that couple efficiently to depolymerizing microtubules. J Cell Biol 209:813–828. doi:10.1083/jcb.201408083

Volkov VA, Zaytsev A V., Gudimchuk N, Grissom PM, Gintsburg AL, Ataullakhanov FI, McIntosh JR, Grishchuk EL (2013) Long tethers provide high-force coupling of the Dam1 ring to shortening microtubules. Proc Natl Acad Sci U S A 110:7708–7713. doi:10.1073/pnas.1305821110/-/DCSupplemental. www.pnas.org/cgi/doi/10.1073/pnas.1305821110

Wagoner JA, Dill KA (2016) Molecular motors: power strokes outperform brownian ratchets. J Phys Chem B 120:6327–6336. doi:10.1021/acs.jpcb.6b02776

Wan X, O'Quinn RP, Pierce HL, Joglekar AP, Gall WE, DeLuca JG, Carroll CW, Liu S-T, Yen TJ, McEwen BF, Stukenberg PT, Desai A, Salmon ED (2009) Protein architecture of the human kinetochore microtubule attachment site. Cell 137:672–684. doi:10.1016/j.cell.2009.03.035

Wang H-W, Ramey VH, Westermann S, Leschziner AE, Welburn JPI, Nakajima Y, Drubin DG, Barnes G, Nogales E (2007) Architecture of the Dam1 kinetochore ring complex and implications for microtubule-driven assembly and force-coupling mechanisms. Nat Struct Mol Biol 14:721–726. doi:10.1038/nsmb1274

Waters JC, Skibbens RV, Salmon ED (1996) Oscillating mitotic newt lung cell kinetochores are, on average, under tension and rarely push. J Cell Sci 109:2823–2831

Weaver LN, Ems-McClung SC, Stout JR, Leblanc C, Shaw SL, Gardner MK, Walczak CE (2011) Kif18A uses a microtubule binding site in the tail for plus-end localization and spindle length regulation. Curr Biol 21:1500–1506. doi:10.1016/j.cub.2011.08.005

Westermann S, Avila-Sakar A, Wang HW, Niederstrasser H, Wong J, Drubin DG, Nogales E, Barnes G (2005) Formation of a dynamic kinetochore-microtubule interface through assembly of the Dam1 ring complex. Mol Cell 17:277–290. doi:10.1016/j.molcel.2004.12.019

Westermann S, Wang H-W, Avila-Sakar A, Drubin DG, Nogales E, Barnes G (2006) The Dam1 kinetochore ring complex moves processively on depolymerizing microtubule ends. Nature 440:565–569. doi:10.1038/nature04409

Widlund PO, Stear JH, Pozniakovsky A, Zanic M, Reber S, Brouhard GJ, Hyman AA, Howard J (2011) XMAP215 polymerase activity is built by combining multiple tubulin-binding TOG domains and a basic lattice-binding region. Proc Natl Acad Sci U S A 108:2741–2746. doi:10.1073/pnas.1016498108

Wood KW, Sakowicz R, Goldstein LSB, Cleveland DW (1997) CENP-E is a plus end-directed kinetochore motor required for metaphase chromosome alignment. Cell 91:357–366. doi:10.1016/S0092-8674(00)80419-5

Yao X, Anderson KL, Cleveland DW (1997) The microtubule-dependent motor centromere-associated protein E (CENP-E) is an integral component of kinetochore corona fibers that link centromeres to spindle microtubules. J Cell Biol 139:435–447. doi:10.1083/jcb.139.2.435

Yardimci H, van Duffelen M, Mao Y, Rosenfeld SS, Selvin PR (2008) The mitotic kinesin CENP-E is a processive transport motor. Proc Natl Acad Sci U S A 105:6016–6021. doi:10.1073/pnas.0711314105

Ye AA, Cane S, Maresca TJ (2016) Chromosome biorientation produces hundreds of piconewtons at a metazoan kinetochore. Nat Commun 7:13221. doi:10.1038/ncomms13221

Zanic M, Stear JH, Hyman AA, Howard J (2009) EB1 recognizes the nucleotide state of tubulin in the microtubule lattice. PLoS ONE 4:e7585. doi:10.1371/journal.pone.0007585

Zakharov P, Gudimchuk N, Voevodin V, Tikhonravov A, Ataullakhanov FI, Grishchuk EL (2015) Molecular and mechanical causes of microtubule catastrophe and aging. Biophys J 109 (12):2574–2591. doi:10.1016/j.bpj.2015.10.048

Zaytsev AV, Ataullakhanov FI, Grishchuk EL (2013) Highly transient molecular interactions underlie the stability of kinetochore-microtubule attachment during cell division. Cell Mol Bioeng 6:393–405. doi:10.1007/s12195-013-0309-4

Zaytsev AV, Grishchuk EL (2015) Basic mechanism for biorientation of mitotic chromosomes is provided by the kinetochore geometry and indiscriminate turnover of kinetochore microtubules. Mol Biol Cell 26:3985–3998. doi:10.1091/mbc.E15-06-0384

Zaytsev AV, Mick JE, Maslennikov E, Nikashin B, DeLuca JG, Grishchuk EL (2015) Multisite phosphorylation of the NDC80 complex gradually tunes its microtubule-binding affinity. Mol Biol Cell 26:1829–1844. doi:10.1091/mbc.E14-11-1539

Molecular Mechanisms of Spindle Assembly Checkpoint Activation and Silencing

Kevin D. Corbett

Abstract In eukaryotic cell division, the Spindle Assembly Checkpoint (SAC) plays a key regulatory role by monitoring the status of chromosome-microtubule attachments and allowing chromosome segregation only after all chromosomes are properly attached to spindle microtubules. While the identities of SAC components have been known, in some cases, for over two decades, the molecular mechanisms of the SAC have remained mostly mysterious until very recently. In the past few years, advances in biochemical reconstitution, structural biology, and bioinformatics have fueled an explosion in the molecular understanding of the SAC. This chapter seeks to synthesize these recent advances and place them in a biological context, in order to explain the mechanisms of SAC activation and silencing at a molecular level.

1 Introduction

A critical decision point in the life of a eukaryotic cell is the mitotic metaphase-to-anaphase transition, when replicated chromosomes are segregated to opposite spindle poles prior to cell division. Before committing to anaphase, the cell must ensure that all chromosomes are attached to spindle microtubules, and that sister chromosomes (or homologs, in meiosis I) are bi-oriented; that is, attached to microtubules extending from opposite spindle poles. Failure to properly sense and respond to errors in microtubule attachment can lead to aneuploidy, a hallmark of cancer and (when it occurs in meiosis) a major cause of miscarriage and developmental disorders like Down Syndrome.

K.D. Corbett (✉)
Ludwig Institute for Cancer Research, San Diego Branch, La Jolla, CA, USA
e-mail: kcorbett@ucsd.edu

K.D. Corbett
Departments of Cellular & Molecular Medicine and Chemistry and Biochemistry, University of California, San Diego, La Jolla, CA 92093 USA

B.E. Black (ed.), *Centromeres and Kinetochores*, Progress in Molecular and Subcellular Biology 56, DOI 10.1007/978-3-319-58592-5_18

The metaphase-to-anaphase transition is controlled by the activity of a ubiquitin E3 ligase, the Anaphase-Promoting Complex/Cyclosome (APC/C) (Sudakin et al. 1995; King et al. 1995), which ubiquitinates and promotes the degradation of a number of substrates, most notably B-type cyclins and securin (Murray et al. 1989; Glotzer et al. 1991; Cohen-Fix et al. 1996; Morgan 1997; Shirayama et al. 1999). Securin is an inhibitor of a protease, separase, that when activated cleaves the Scc1 subunit of the cohesin complexes holding bi-oriented sister chromosomes together; this cleavage is the critical step initiating chromosome segregation in anaphase (Ciosk et al. 1998; Kamenz and Hauf 2016).

Prior to anaphase onset, the activity of the APC/C is inhibited by the spindle assembly checkpoint (SAC), which monitors the state of chromosome-microtubule attachment in the cell (reviewed in Musacchio and Salmon 2007; Lara-Gonzalez et al. 2012; Musacchio 2015; Zhang et al. 2016b; Etemad and Kops 2016). Microtubule attachment is mediated by kinetochores, complex protein assemblies with both DNA-binding and microtubule-binding subunits (reviewed in Pesenti et al. 2016; Nagpal and Fukagawa 2016). When kinetochores are not properly attached to microtubules, they mediate assembly of a soluble "wait anaphase" signal in the form of the four-protein Mitotic Checkpoint Complex (MCC), which directly binds and inhibits the APC/C. In this manner, a single unattached kinetochore is in most cases able to delay anaphase onset (Rieder et al. 1995).

Seminal work published in 1991 initiated study of the SAC by isolating the first mutants defective in this pathway, termed *mad* (mitotic arrest deficient) (Li and Murray 1991) and *bub* (budding uninhibited by benzimidazole) (Hoyt et al. 1991). Only now, however, are the detailed molecular mechanisms of SAC activation and silencing coming into sharp focus, thanks to a recent surge in structural and biochemical studies of the APC/C, its interactions with the MCC, and the mechanisms of MCC assembly and disassembly. This review covers several aspects of SAC function that have recently seen significant advances, beginning with the structure and function of the APC/C itself, and the mechanism of its inhibition by the MCC. I then move to the sites of MCC assembly—kinetochores—and outline the mechanisms of chromosome-microtubule attachment sensing and MCC assembly at unattached kinetochores. Finally, I address how the SAC is silenced after kinetochore-microtubule attachment, paying particular attention to a newly discovered pathway for direct MCC disassembly. Throughout, I attempt to place recent structural and biochemical work into the larger framework of SAC function that has been refined, through the work of many, over the 25 years since the discovery of this pathway.

2 The APC/C: Target of the Spindle Assembly Checkpoint

As the master regulator of anaphase onset, the mechanisms of the APC/C, particularly how it is regulated through the cell cycle and how it recognizes substrates, are of considerable interest. Because of its immense size and complexity, these questions were unanswerable until recent advances in cryo-electron microscopy (cryo-EM) began to provide high-resolution pictures of the APC/C in a variety of functional states (Chang et al. 2014, 2015; Brown et al. 2015, 2016; Zhang et al. 2016c; Yamaguchi et al. 2016; Alfieri et al. 2016). This structural work, coupled with in vitro and in vivo functional analysis, has significantly improved our understanding of APC/C substrate recognition, the role of "coactivator" proteins such as Cdc20 in that recognition, and how the APC/C is inhibited by the MCC. In particular, the previously enigmatic roles of Cdc20, a key APC/C coactivator that also acts as an inhibitor when incorporated into the MCC, have been significantly clarified.

2.1 Overall APC/C Architecture

The APC/C is a 19-subunit complex (20 when counting a bound coactivator; see below) with a total molecular weight of ~1.2 MDa (Fig. 1a) (Sudakin et al. 1995; King et al. 1995; Chang et al. 2014). It contains two E3 ubiquitin ligase subunits: Apc2 is related to the Cullin subunits of SCF ubiquitin ligases, while Apc11 contains a RING-type E3 ligase domain. These subunits bind several different E2 activating enzymes to mediate substrate ubiquitination, with different E2s responsible for ubiquitin chain initiation and elongation. The bulk of the APC/C forms two large structures, the so-called TPR lobe (or "arc lamp") named for the tetratricopeptide (TPR) repeats found in this lobe's subunits, and the platform. Together, the TPR lobe and platform define a large central cavity and serve to juxtapose functional modules responsible for substrate recognition with those responsible for ubiquitination (Fig. 1a). For a detailed discussion of APC/C architecture, the reader is referred to recent reviews on the subject (Primorac and Musacchio 2013; Chang and Barford 2014; Barford 2015).

2.2 APC/C Substrate Recognition Is Mediated by Coactivator Proteins

In order to recognize its substrates, the APC/C requires one of a family of "coactivator" proteins, which bind the APC/C in a cell cycle-regulated manner and dictate substrate specificity by binding directly to conserved "degron" motifs in those substrates. All APC/C coactivators are structurally related, with a central WD40 β-propeller domain responsible for degron recognition, and conserved motifs at the N- and C-terminus that mediate docking between the APC/C's TPR lobe and platform (Figs. 1a and 2b) (Zhang and Lees 2001; Schwab et al. 2001; Vodermaier et al. 2003; Thornton et al. 2006; Matyskiela and Morgan 2009; Izawa and Pines 2012; Chang et al. 2015; Zhang et al. 2016c). While detailed discussion is

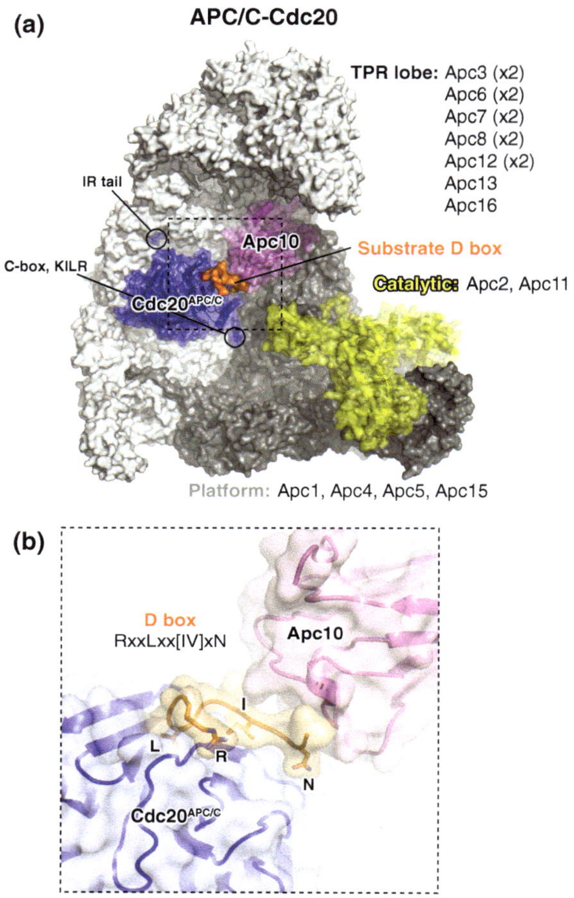

Fig. 1 Architecture of the APC/C and interactions with Cdc20, substrate degron motifs, and the MCC. **a** Structure of the human APC/C-bound to coactivator Cdc20 and high-affinity substrate Hsl1 (Zhang et al. 2016c) (PDB 5G04). The TPR lobe and platform of the APC/C are colored *light* and *dark gray*, respectively, with the catalytic module *yellow* and the substrate-recognition module *pink* (Apc10) and *blue* (Cdc20). Cdc20 binding to the APC/C is mediated by its N-terminal C box and KILR motifs, and the C-terminal IR tail (see Fig. 2b for Cdc20 domain structure). The D-box of Hsl1, *orange*, is sandwiched between Cdc20 and Apc10. **b** Close-up of Hsl1 D-box recognition by Cdc20 (*blue*) and Apc10 (*pink*). View is equivalent to (**a**)

outside the scope of this review, binding of the different coactivator proteins to the APC/C is regulated through phosphorylation of both the coactivators themselves (Zachariae et al. 1998; Jaspersen et al. 1999; Lukas et al. 1999; Kramer et al. 2000; Labit et al. 2012; Chang et al. 2015) and subunits of the APC/C (Lahav-Baratz et al. 1995; Shteinberg et al. 1999; Kramer et al. 2000; Golan et al. 2002; Kraft et al. 2003; Zhang et al. 2016c; Qiao et al. 2016). The end result of this regulation is that the bound coactivator, and therefore the APC/C's substrate specificity, depends on

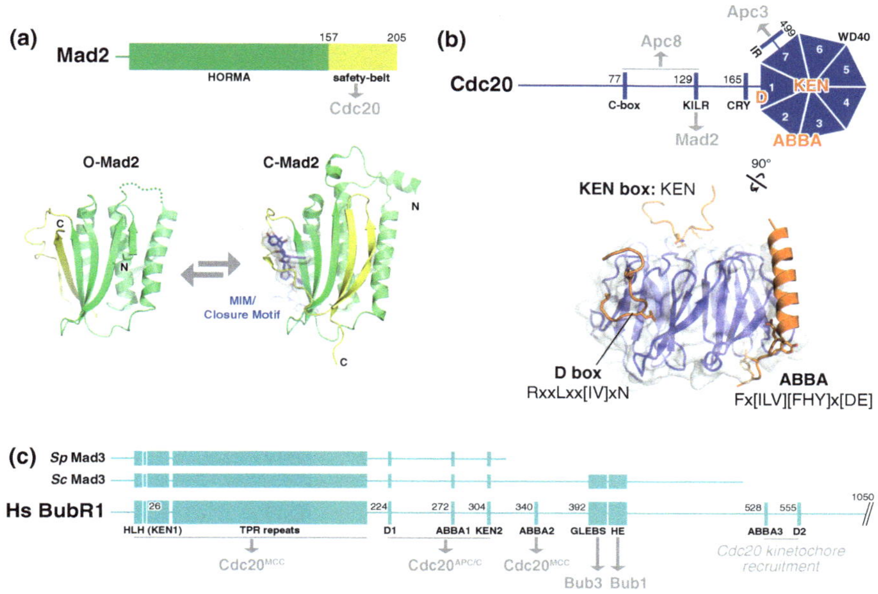

Fig. 2 Domain structure of core MCC subunits. **a** *Top* Domain structure of Mad2, with the HORMA domain core in green and the C-terminal safety belt *yellow*. *Bottom* Structures of Mad2 in the open (O-Mad2) and closed (C-Mad2) conformations. In O-Mad2, the safety belt occupies the binding site for Mad2-interacting motifs (MIMs). In C-Mad2, the safety belt wraps entirely around a bound MIM/closure motif (*blue sticks*), locking it in place. **b** *Top* Domain structure of Cdc20. The KILR motif (residues 129-132 in human Cdc20) can bind Mad2 as a MIM (in Cdc20^{MCC}) or contribute to APC/C binding (in Cdc20$^{APC/C}$). The CRY box (residues 165–167) of Cdc20^{MCC} is recognized by Cdc20$^{APC/C}$ in the APC/C-MCC complex (Alfieri et al. 2016) and contributes to Cdc20^{MCC} ubiquitination and degradation (Reis et al. 2006). *Bottom* Structure of *S. cerevisiae* Cdh1 (*blue*) bound to an inhibitor, Acm1 (*orange*), that contains all known degrons: D box, KEN box, and ABBA motif (He et al. 2013) (PDB 4BH6). Consensus residues in each degron motif are shown as sticks. **c** Domain structure and interactions of human BubR1 and its orthologs (called Mad3) in *S. pombe* and *S. cerevisiae*

cell-cycle stage: Cdh1 is bound during interphase, and Cdc20 is bound in mitosis. Here, I focus entirely on the APC/C-Cdc20 complex, which controls anaphase onset and constitutes the target of the SAC.

When bound to the APC/C as a coactivator, Cdc20 recognizes several different degron motifs via distinct surfaces on its central WD40 β-propeller domain (Figs. 1b, 2b and 3a) (reviewed in Davey and Morgan 2016). Recognition of one such motif, the destruction box (D-box) (Glotzer et al. 1991), is bipartite: this motif becomes sandwiched between Cdc20 and an adjacent APC/C subunit, Apc10 (Fig. 1b) (Buschhorn et al. 2011; da Fonseca et al. 2011; Chang et al. 2014). Recognition of the two other known degron motifs—the KEN box (named for its sequence: lysine-glutamate-asparagine) (Pfleger and Kirschner 2000) and ABBA motif (also termed A-motif, Phe-box, or IC20BD) (Burton et al. 2011; He et al.

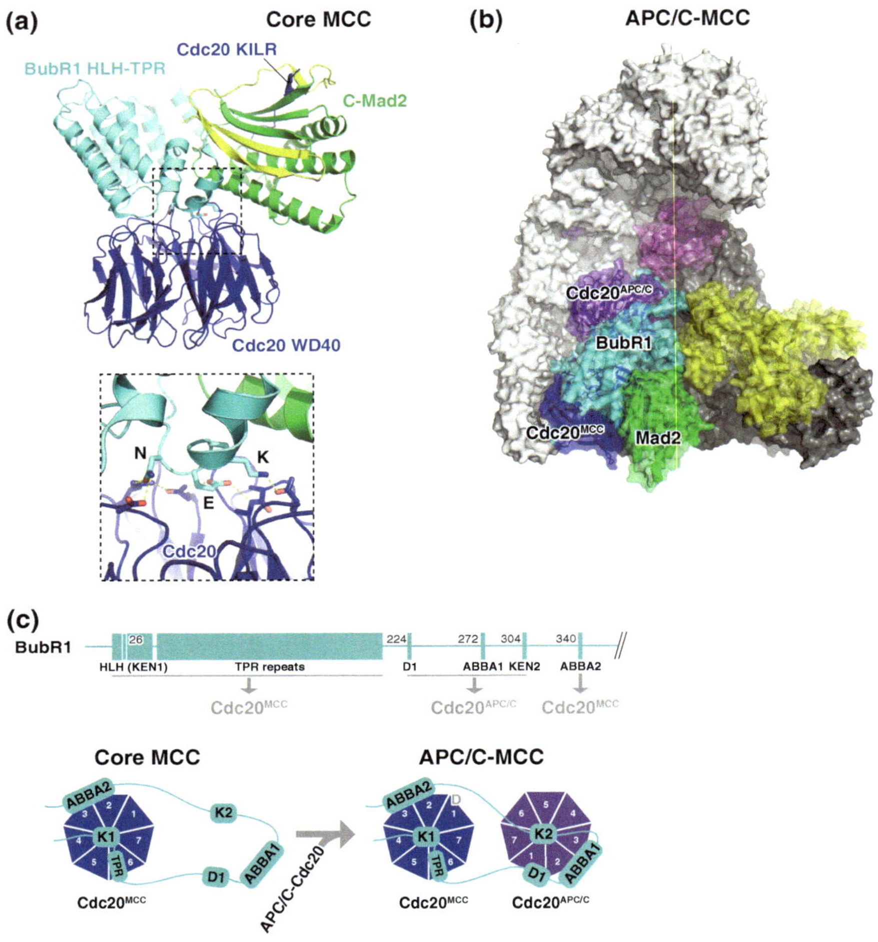

2013; Lischetti et al. 2014; Diaz-Martinez et al. 2015; Di Fiore et al. 2015)—is mediated solely by Cdc20 (Fig. 3a). Together, binding of one or more degrons by the APC/C-Cdc20 complex positions a substrate for ubiquitination by the catalytic module (Chang et al. 2014; Brown et al. 2015).

3 The Mitotic Checkpoint Complex Inhibits APC/C-Cdc20

The key element of SAC signaling is the four-protein MCC, which is generated at unattached kinetochores and directly binds and inhibits the APC/C-Cdc20 complex. The conserved "core" MCC comprises Mad2, Cdc20, and BubR1 (Mad3 in fungi),

◂**Fig. 3** MCC architecture and APC/C inhibition. **a** Structure of the *S. pombe* core MCC, containing the BubR1 (Mad3) N-terminal region (HLH and TPR; *cyan*), Cdc20 (*blue*), and Mad2 (*green*) (Chao et al. 2012) (PDB ID 4AEZ). *Inset* Close-up view of the BubR1 KEN1 motif interacting with Cdc20. **b** Structure of the human APC/C-Cdc20 bound to the MCC (Alfieri et al. 2016). Core MCC subunits are colored as in (**a**), and the APC/C is colored as in Fig. 1a except for $Cdc20^{APC/C}$ (*purple*). The C-terminal region of BubR1, as well as Bub3, are conformationally flexible and were not included in the model. MCC binding rotates $Cdc20^{APC/C}$ away from Apc10 and occupies the D-box binding site. BubR1 also occupies the space where E2 enzymes bind the catalytic module. **c** Interactions of BubR1 with $Cdc20^{MCC}$ and $Cdc20^{APC/C}$. In the core MCC, BubR1 interacts with $Cdc20^{MCC}$ mainly through its KEN1/TPR motif (Chao et al. 2012). In the APC-C-MCC structure, electron density in the ABBA-motif binding site of $Cdc20^{MCC}$ (Alfieri et al. 2016) was originally assigned to ABBA3, but recent data suggest that in fact ABBA2 occupies this site (Di Fiore et al. 2016). Core MCC assembly leaves the BubR1 D1, ABBA1, and KEN2 motifs available. Upon MCC binding to APC/C-Cdc20, these motifs bind $Cdc20^{APC/C}$ to mediate APC/C inhibition. While the structure of APC/C-MCC showed density in the $Cdc20^{MCC}$ D-box binding site (*gray* in schematic), this density cannot be confidently assigned to a specific region of BubR1 (Figure adapted from Chao et al. 2012 and Di Fiore et al. 2016)

with BubR1 forming a constitutive dimer with Bub3 in a subset of organisms including humans. Cdc20's role as an APC/C coactivator is outlined above; for many years, how Cdc20 also functions as an APC/C inhibitor was unknown. Recent structural work on both the isolated MCC and its complex with the APC/C have clarified this question, resulting in a simple, yet elegant, model for APC/C inhibition by the MCC and for the dual roles of Cdc20.

3.1 Mitotic Checkpoint Complex Architecture

Mad2 was the first protein demonstrated to bind and inhibit the APC/C (Li et al. 1997). Mad2 contains a HORMA domain (Aravind and Koonin 1998) that can adopt two different conformations: an inactive "open" conformation (O-Mad2), and a "closed" conformation (C-Mad2) that binds short peptide motifs called Mad2-interacting motifs (MIMs) or, more generally, closure motifs (Fig. 2a) (reviewed in Mapelli and Musacchio 2007; Luo and Yu 2008). These two conformations differ in the structure of the C-terminal region of the protein, termed the "safety belt": in C-Mad2, this segment wraps entirely around a bound closure motif to form a topologically linked complex (Luo et al. 2002). In O-Mad2, the safety belt is docked against the closure motif binding site (Luo et al. 2000), and the protein is therefore unable to bind a closure motif. The bulk of soluble Mad2 in the cell is in the O-Mad2 state (Luo et al. 2004); the rate-limiting step of MCC assembly is the recruitment of O-Mad2 to unattached kinetochores, where it is converted to C-Mad2 and associates with a closure motif in Cdc20, termed the KILR motif (Fig. 2b) (Hwang et al. 1998; Kim et al. 1998; Fang et al. 1998; Kallio et al. 1998; Luo et al. 2002).

BubR1 is the third member of the so-called "core" MCC (Hardwick et al. 2000; Tang et al. 2001; Sudakin et al. 2001; Fang 2002), and directly interacts with both

Mad2 and Cdc20, significantly stabilizing the overall complex (Figs. 2c and 3a) (Sczaniecka et al. 2008; Tipton et al. 2011; Chao et al. 2012; Faesen et al. 2017). BubR1, which arose along with its paralog Bub1 from a gene duplication event (Suijkerbuijk et al. 2012a; Vleugel et al. 2012; Di Fiore et al. 2016), has a complex domain structure featuring at least seven degron-like motifs: the N-terminal TPR-repeat domain contains a KEN box, and this domain is followed by a second KEN box, two D-boxes, and three ABBA motifs (Fig. 2c). In the core MCC, BubR1 binds Cdc20 through its N-terminal KEN box (KEN1) and the adjacent TPR-repeat domain (Sczaniecka et al. 2008; Chao et al. 2012), and also through one of its ABBA motifs (most likely ABBA2; Di Fiore et al. 2016). The TPR domain also binds MAD2, completing the cooperative assembly of the highly stable core MCC (Fig. 3a).

3.2 *APC/C-Cdc20 Binding and Inhibition by MCC*

The fully assembled MCC contains a copy of BubR1 with a series of degron motifs—D1, ABBA1, and KEN2—unoccupied (Fig. 3c). The presence of these degrons, and their importance for APC/C-Cdc20 inhibition by the MCC, led to a proposal that BubR1 could bind two copies of Cdc20, one as part of the MCC (termed $Cdc20^{MCC}$) and a second bound to the APC/C as a coactivator (termed $Cdc20^{APC/C}$) (Primorac and Musacchio 2013). An important biochemical and cryo-EM analysis of APC/C-Cdc20 and APC/C-MCC complexes purified from HeLa cells provided early evidence that this might be the case, as the stoichiometry of Cdc20 was doubled in APC/C-MCC versus APC/C-Cdc20 (Herzog et al. 2009). The relatively low resolution (by today's standards) of that study's EM analysis, however, prevented a clear visualization of the two copies of Cdc20 in APC/C-MCC. More recently, it was shown biochemically that the fully assembled MCC could bind a second copy of Cdc20 that was already bound to the APC/C, and that this binding was disrupted by mutating BubR1's D1 degron motif (Izawa and Pines 2014). More recent high-resolution structures of the APC/C-MCC complex have clearly shown the positions of two copies of Cdc20 in this complex, confirming the above findings (Fig. 3b) (Yamaguchi et al. 2016; Alfieri et al. 2016). These structures, plus detailed biochemical and genetic analysis with BubR1 mutants, also finally reveal the roles of BubR1's many degron-like motifs: in the APC/C-MCC complex, BubR1 winds between $Cdc20^{MCC}$ and $Cdc20^{APC/C}$, occupying all degron-binding sites of both copies (Fig. 3c) (Alfieri et al. 2016; Di Fiore et al. 2016). BubR1 binds $Cdc20^{MCC}$ through its KEN1 motif and TPR domain as described above, and also through its ABBA2 motif (Figs. 2c and 3c) (Diaz-Martinez et al. 2015; Alfieri et al. 2016; Di Fiore et al. 2016). Between these motifs, BubR1 wraps around the WD40 domain of $Cdc20^{APC/C}$, binding through its D1, ABBA1, and KEN2 motifs (Fig. 3c) and also causing a significant rotation of $Cdc20^{APC/C}$ that disrupts the bipartite D-box recognition site (Alfieri et al. 2016). APC/C-bound MCC also sterically occludes the binding of E2 enzymes to the APC/C catalytic module, further inhibiting activity (Yamaguchi et al. 2016; Alfieri et al. 2016). Thus, MCC targets already-assembled APC/C-Cdc20 for inhibition, binding through a series of degron motifs in BubR1. Because Cdc20$^{APC/C}$ remains bound to the APC/C in this complex, reactivation of the APC/C upon SAC silencing requires only removal or disassembly of the bound MCC (see Sect. 5 and Fig. 4).

4 Assembly of the Mitotic Checkpoint Complex at Unattached Kinetochores

The key molecular event monitored by the SAC is kinetochore-microtubule attachment. Kinetochores are complex multi-megadalton structures that assemble on each chromosome's centromere, where they both mediate chromosome-microtubule attachment and serve as signaling hubs for the checkpoints monitoring attachment status. The architecture and function of kinetochores are covered in recent excellent reviews (Pesenti et al. 2016; Nagpal and Fukagawa 2016); here I focus mainly on a conserved outer kinetochore complex, the KMN network, that serves as the major sensor of microtubule attachment and a scaffold for MCC assembly.

4.1 The KMN Network: A Scaffold for SAC Signaling and MCC Assembly

The KMN network is a highly conserved outer kinetochore complex that serves as both the main microtubule-binding component of the kinetochore, and a platform for MCC assembly when microtubules are not bound (Fig. 5a) (Cheeseman et al. 2006; Varma and Salmon 2012). The KMN network contains three subcomplexes with distinct roles: the Mis12 complex anchors the network to the inner kinetochore, the Ndc80 complex binds microtubules, and the Knl1 complex is responsible for recruiting SAC proteins.

Knl1 (Spc105 in *Saccharomyces cerevisiae*, Spc7 in *Schizosaccharomyces pombe*, Knl1/CASC5/Blinkin in humans) contains a large disordered N-terminal region with multiple conserved motifs. Nearest the N-terminus is a phosphatase-binding site, termed SILK/RVSF (Hendrickx et al. 2009; Liu et al. 2010). When kinetochores are not attached to microtubules, phosphatase binding is inhibited through phosphorylation of this site by the Aurora B kinase (Liu et al. 2010). Following the SILK/RVSF motif in Knl1 are multiple short motifs, termed MELT repeats (Desai et al. 2003; Nekrasov et al. 2003; Cheeseman et al. 2004; Vleugel et al. 2015b), that are phosphorylated by the Mps1 kinase when kinetochores are not attached to microtubules (London et al. 2012; Shepperd et al. 2012; Yamagishi et al. 2012). Phosphorylated MELT repeats (P-MELT) recruit the SAC

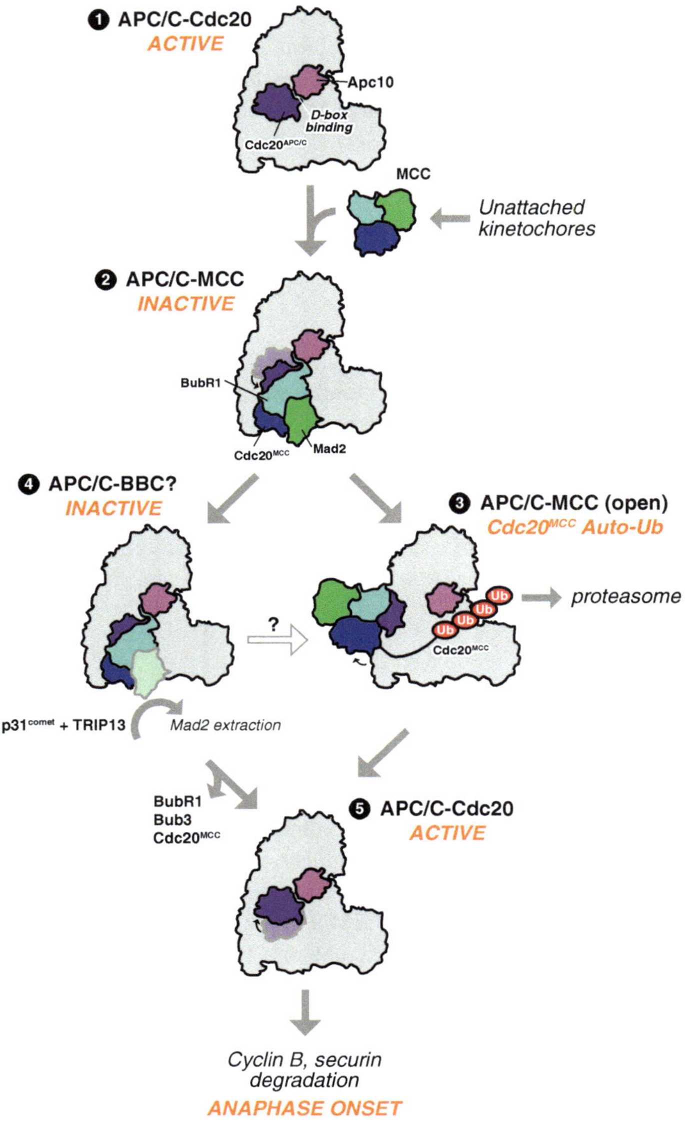

protein Bub3 along with its binding partners, Bub1 and BubR1 (Yamagishi et al. 2012; Primorac et al. 2013; Vleugel et al. 2013, 2015b; Krenn et al. 2014; Zhang et al. 2014; Overlack et al. 2015). As mentioned above, Bub1 and BubR1 are paralogs with similar overall structures, but each has evolved to fulfill distinct roles in the checkpoint: Bub1 serves as the major hub for MCC assembly by recruiting SAC proteins, and BubR1 is a subunit of the MCC (Bub1 and BubR1's evolution

◂**Fig. 4** Life cycle of the APC/C in mitosis. (*1*) After CDK phosphorylation (not shown) and binding of $Cdc20^{APC/C}$ (*purple*), the APC/C is active. (*2*) Unattached kinetochores trigger the assembly of the MCC (see Fig. 5), which binds and inhibits APC/C-Cdc20 by occupying all degron-recognition sites and rotating $Cdc20^{APC/C}$ away from Apc10 (Alfieri et al. 2016). (*3*) Upon SAC silencing, two pathways lead to APC/C reactivation. First, rotation of the bound MCC to the "open" position (stabilized by Apc15) allows auto-ubiquitination of the $Cdc20^{MCC}$ N-terminal tail, triggering proteasome-mediated destruction. (*4*) Second, $p31^{comet}$ and TRIP13 extract Mad2 from the MCC, potentially resulting in a partially bound state (BBC: BubR1-Bub3-Cdc20) in which the remaining MCC subunits are less-stably bound and prone to dissociation (potentially also involving $Cdc20^{MCC}$ ubiquitination; *outline arrow*). (*5*) After APC/C-Cdc20 reactivation, it ubiquitinates B-type cyclins and securin to promote anaphase onset and mitotic exit

from a bifunctional ancestor is discussed more fully in Suijkerbuijk et al. 2012a; Di Fiore et al. 2016). Both Bub1 and BubR1 bind Bub3 through their so-called GLEBS motifs (Taylor et al. 1998; Wang et al. 2001; Larsen et al. 2007), and the resulting complex is competent to bind Knl1 P-MELT repeats (Fig. 5b). Interestingly, Bub1:Bub3 binds much more strongly to P-MELT repeats than does BubR1:Bub3 (Primorac et al. 2013; Overlack et al. 2015), and the bulk of BubR1: Bub3 is recruited to kinetochores indirectly, through a pseudo-symmetric Bub1-BubR1 dimer interaction (Overlack et al. 2015; Zhang et al. 2015). Some BubR1:Bub3 is recruited directly to Knl1 P-MELT repeats, however, and preliminary evidence suggests that this pool may be the major source of BubR1:Bub3 that is incorporated into the MCC (see Sect. 4.3) (Zhang et al. 2016a). The requirement for BubR1 localization to kinetochores varies between organisms, however, as some BubR1 orthologs—such as *S. pombe* Mad3—lack both Bub3 and Bub1 binding motifs (Fig. 2c).

Once recruited to Knl1 P-MELT motifs, Bub1 and BubR1 recruit the remaining SAC components necessary for MCC assembly: Cdc20 and a complex of Mad1 bound to C-Mad2. Cdc20 is recruited by both Bub1 and BubR1, through homologous degron-like motifs C-terminal to these proteins' GLEBS motifs (BubR1 ABBA3 (Lischetti et al. 2014; Di Fiore et al. 2015), Bub1 KEN-ABBA (Vleugel et al. 2015a)). Mad1:Mad2 is also probably recruited by Bub1, with direct interactions between Mad1 and Bub1 having been reported in multiple organisms including fungi, nematodes, and humans (London and Biggins 2014; Moyle et al. 2014; Ji et al. 2017). Humans and other complex eukaryotes also possess a separate complex, known as RZZ (Rod-Zwilch-ZW10) that binds Bub1 and recruits Mad1: Mad2 (Wang et al. 2004; Kops et al. 2005; Buffin et al. 2005; Karess 2005; Barisic and Geley 2011; Zhang et al. 2015; Caldas et al. 2015; Silió et al. 2015). Regardless of its recruitment pathway, kinetochore-localized Mad1:Mad2 is necessary to recruit soluble O-Mad2 and mediate its conversion to C-Mad2, binding to the Cdc20 KILR motif, and assembly into the MCC (see Sect. 4.3).

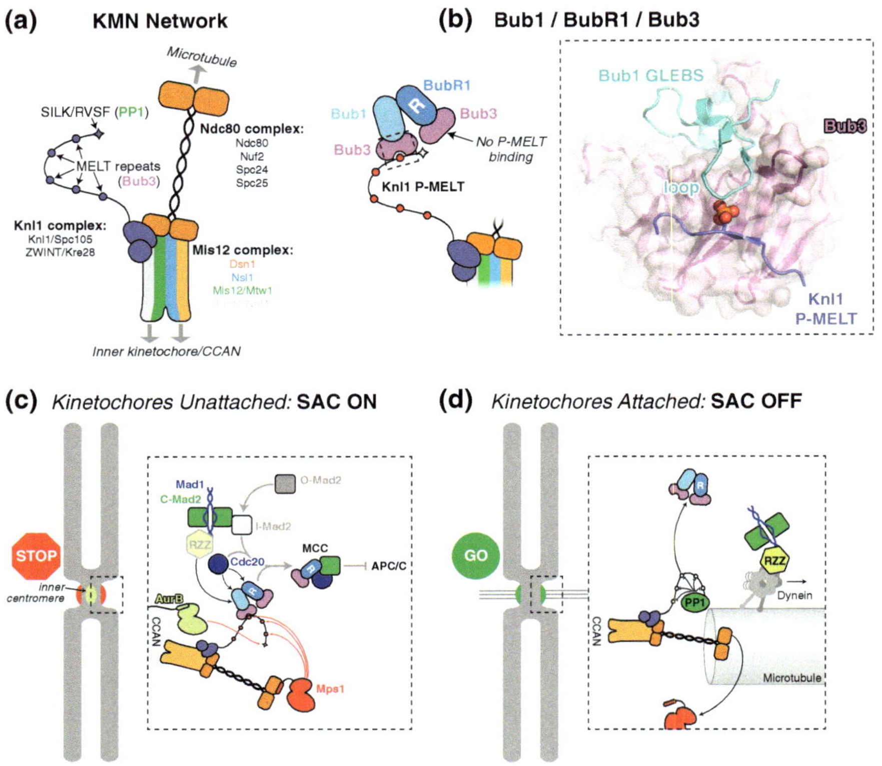

Fig. 5 Assembly of the MCC at unattached kinetochores. **a** Overall architecture of the outer kinetochore KMN network, consisting of the Knl1, Mis12, and Ndc80 subcomplexes. The Mis12 complex binds inner kinetochore protein complexes, also termed the "constitutive centromere-associated network" or CCAN. The Ndc80 complex contains the main microtubule-binding activity in the kinetochore, while Knl1 contains conserved SILK/RVSF and MELT motifs. **b** Interactions of Bub1, BubR1, and Bub3. Bub3 associates with P-MELT repeats when bound to the Bub1 GLEBS motif, but not the BubR1 GLEBS motif (detail view, right from PDB ID 4BL0 Primorac et al. 2013). BubR1 and Bub1 form a pseudo-symmetric dimer interaction involving their "helical extension" segments ("HE" in Fig. 2c) C-terminal to the GLEBS motif (Overlack et al. 2015). **c** MCC assembly at unattached kinetochores. Mps1 is recruited to Ndc80 in the absence of microtubules, and phosphorylates Knl1 MELT repeats to mediate recruitment of Bub1-Bub3 (*thin red arrows*). Aurora B (tethered at the inner centromere; *light yellow*) phosphorylates the Knl1 SILK/RVSF motif to inhibit PP1 binding (*thin red arrow*). Bub1 recruits Mad1:C-Mad2 (in some organisms, through the RZZ complex), and both Bub1 and BubR1 recruit Cdc20 (*thin black arrows*). Mad1-bound C-Mad2 converts soluble O-Mad2 to C-Mad2 concomitant with Cdc20 binding, followed by binding of BubR1 to complete MCC assembly (*thick gray arrows*). **d** After kinetochore-microtubule attachment, multiple mechanisms inactivate MCC assembly. In the absence of Mps1, PP1-mediated dephosphorylation of Knl1 MELT repeats causes loss of Bub1/Bub3, and in those organisms with RZZ, Mad1: C-Mad2 is actively "stripped" from kinetochores through coupling to dynein motors

4.2 The Mps1 Kinase Coordinates Attachment Sensing with MCC Assembly

As noted above, the kinase Mps1 phosphorylates Knl1 MELT repeats to initiate recruitment of SAC components to unattached kinetochores (the diverse roles of Mps1 are reviewed in Lan and Cleveland 2010; Liu and Winey 2012). Mps1 recruitment, therefore, must be responsive to kinetochore-microtubule attachment status. It is not surprising, therefore, that the key determinant of Mps1 recruitment is the Ndc80 complex, the major kinetochore complex responsible for microtubule binding (Martin-Lluesma et al. 2002; Nijenhuis et al. 2013; Zhu et al. 2013; Hiruma et al. 2015; Ji et al. 2015; Aravamudhan et al. 2015; Dou et al. 2015). Exactly how Mps1 kinase activity is coordinated with Ndc80-microtubule binding is not yet firmly established. One mechanistic model involves a direct competition between Mps1 and microtubules for Ndc80 binding. Supporting this idea, two groups recently showed that Mps1 binds directly to the Ndc80 CH domain, which is also responsible for microtubule binding (Wei et al. 2007; Ciferri et al. 2008; Hiruma et al. 2015; Ji et al. 2015). These studies showed that Mps1-Ndc80 binding is suppressed in vitro by microtubules, suggesting that Mps1 and microtubules compete directly for Ndc80 binding (Hiruma et al. 2015; Ji et al. 2015). Another possible mechanism is that Mps1 remains associated with Ndc80 even after microtubule attachment, but its ability to phosphorylate Knl1 is inhibited once attachment occurs (Aravamudhan et al. 2015). In either case, active Mps1 promotes MCC assembly at unattached kinetochores in several ways. First and most importantly, it phosphorylates the MELT repeats in the N-terminal region of Knl1 (London et al. 2012; Shepperd et al. 2012; Yamagishi et al. 2012), which in turn recruit Bub1:Bub3 as described above. Mps1 also phosphorylates Bub1 directly, and this phosphorylation was recently shown to be required for Bub1's ability to recruit Mad1:Mad2 to kinetochores (London and Biggins 2014; Ji et al. 2017). Finally, Mps1 phosphorylates Mad1 in its poorly characterized C-terminal RWD domain, promoting a direct Mad1-Cdc20 interaction that contributes to MCC assembly and SAC signaling (Hardwick et al. 1996; Faesen et al. 2017; Ji et al. 2017).

4.3 Assembling the MCC

Once all SAC components are recruited to unattached kinetochores, they participate in a complex structural dance, still incompletely understood, that ultimately results in fully assembled MCC. The first, and rate-limiting, step of MCC assembly is the association of Mad2 with Cdc20 (Simonetta et al. 2009; Faesen et al. 2017). This occurs when kinetochore-localized Mad1:Mad2 recruits soluble O-Mad2 to kinetochores through a pseudo-symmetric Mad2 homodimer interaction (Fig. 6) (Luo et al. 2004; Howell et al. 2004; Shah et al. 2004; de Antoni et al. 2005; Vink et al.

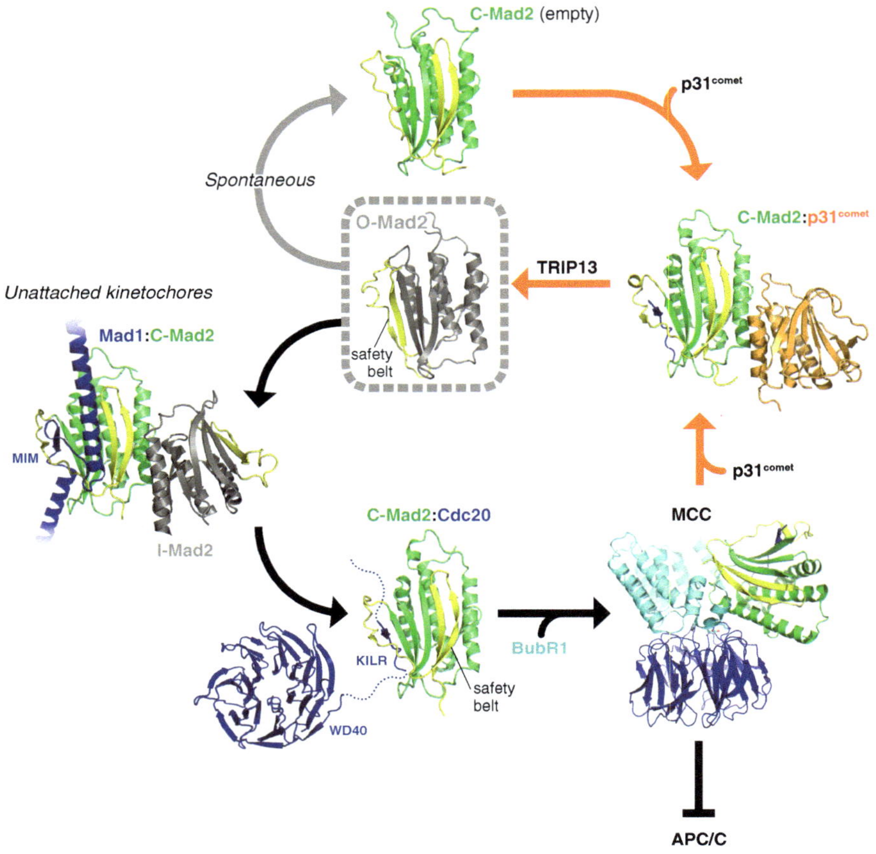

Fig. 6 The Mad2 conformational cycle. The majority of soluble cellular Mad2 is in the open (O-Mad2) conformation, with its C-terminal safety belt region (*yellow*) occluding the MIM/closure motif binding site (Luo et al. 2004) (structure from Luo et al. 2000); PDB 1DUJ). **SAC pathway** (*black arrows*): Upon SAC activation, unattached kinetochores recruit a complex of Mad1 bound to closed Mad2 (C-Mad2), which in turn dimerizes with soluble O-Mad2 to generate "intermediate" Mad2 (I-Mad2), primed for conversion to C-Mad2 and binding to Cdc20 [(Hara et al. 2015); structure shown is a composite of PDBs 1GO4 (Mad1:Mad2) (Sironi et al. 2002) and 2V64 (C-Mad2:I-Mad2) (Mapelli et al. 2007)]. After Mad2:Cdc20 binding, BubR1 (with associated Bub3, not shown) binds to complete the assembly of the MCC, which is competent for APC/C inhibition (Chao et al. 2012; PDB 4AEZ). **Recycling pathway** (*orange arrows*): p31^{comet} binds C-Mad2 in the MCC in a manner akin to Mad2 homodimerization (Yang et al. 2007; PDB 2QYF). C-Mad2:p31^{comet} is recognized and disassembled by TRIP13, reforming O-Mad2 (Ye et al. 2015). **Spontaneous pathway** (*gray arrow*): O-Mad2 spontaneously converts to ligand-free C-Mad2 with a lifetime (1/*k*) of ~10 h (Luo et al. 2004; PDB 1S2H). p31^{comet} and TRIP13 recycle this C-Mad2 to maintain a soluble pool of O-Mad2 for SAC activation (Ma and Poon 2016)

2006; Nezi et al. 2006). The resulting C-Mad2:O-Mad2 dimer has been visualized in two different x-ray crystal structures (Mapelli et al. 2007; Hara et al. 2015), and in both cases the O-Mad2 protomer adopts a subtly altered conformation compared to its structure in solution. This conformational shift is believed to promote dissociation of the C-terminal safety belt motif from its position occluding the closure motif binding site, resulting in a transient partially unfolded state (Mapelli and Musacchio 2007; Hara et al. 2015). Partially unfolded Mad2 is competent to associate with the Cdc20 KILR motif and refold into the closed state. The unique Mad1:Mad2-mediated conversion of soluble O-Mad2 to C-Mad2 has been termed the "Mad2 template model" (de Antoni et al. 2005; Musacchio and Salmon 2007; Mapelli and Musacchio 2007). After Mad2-Cdc20 binding, MCC assembly is completed when BubR1 binds both proteins as described above (Chao et al. 2012).

It has been understood for some time that in solution, O-Mad2 is less stable than C-Mad2 and will spontaneously convert to C-Mad2 with a half-time of several hours (Luo et al. 2004). Given that conversion of O-Mad2 to C-Mad2 is the rate-limiting step of MCC assembly, why does spontaneous conversion in solution not result in Cdc20 binding and MCC assembly? The key difference is likely the presence of Cdc20: binding of Mad2 to a closure motif can probably only occur in the transient partially unfolded state between O-Mad2 and C-Mad2, when the safety belt is disengaged from the HORMA domain core. Thus, the presence of Cdc20 at the time and place of Mad2 conformational conversion is likely key for complex formation. Further control over this assembly could be mediated by Mad1's functionally mysterious C-terminal RWD domain, which is phosphorylated by Mps1 at unattached kinetochores, interacts directly with Cdc20, and may be required for initial Cdc20-Mad2 association (Faesen et al. 2017; Ji et al. 2017). Away from kinetochores, spontaneous O-Mad2 to C-Mad2 conversion probably results in "empty" C-Mad2 that not only does not nucleate MCC assembly (Fig. 6), but is actively harmful in that it cannot be recruited to kinetochores when needed (as Mad1:Mad2 specifically recruits O-Mad2). For this reason, spontaneous O-Mad2 to C-Mad2 conversion must be continually counteracted in the cell to maintain a functional SAC (see Sect. 5.2 and Fig. 6) (Ma and Poon 2016).

5 Silencing the SAC

5.1 Kinetochore Transformations

After all kinetochores become attached to microtubules, the SAC must be silenced to allow anaphase onset. To accomplish SAC silencing, kinetochores undergo a number of structural and compositional changes. First, Ndc80 binding to microtubules suppresses Mps1 activity, either by removing it from kinetochores or spatially segregating it from its substrates. At the same time, the activity of the Aurora B kinase, which phosphorylates a number of outer kinetochore components when kinetochores are not attached, including the Knl1 SILK/RVSF motif, is suppressed (regulation of Aurora B is discussed in detail in Lampson and Cheeseman 2011; Carmena et al. 2012; van der Horst and Lens 2014; Krenn and Musacchio 2015). The loss of these two kinase activities alters the balance of kinase/phosphatase activity at the outer kinetochore, first enabling Protein Phosphatase 2A (PP2A), recruited by BubR1, to dephosphorylate the Knl1 SILK/RVSF motif (Espert et al. 2014; Nijenhuis et al. 2014). The SILK/RVSF motif then binds protein phosphatase 1 (PP1) (Liu et al. 2010; Rosenberg et al. 2011; Meadows et al. 2011; London et al. 2012), which in turn dephosphorylates the Knl1 MELT repeats, resulting in loss of Bub1:Bub3 and all associated SAC components. The delicate balance of kinase and phosphatase activities at kinetochores, and how this balance is affected by microtubule attachment and other events, is outside the scope of this review but is covered in detail elsewhere (Suijkerbuijk et al. 2012b; Foley and Kapoor 2013; Espert et al. 2014; Nijenhuis et al. 2014; Etemad and Kops 2016). Finally, in organisms that possess the RZZ complex, RZZ and an associated protein called Spindly mediate the active "stripping" of Mad1:Mad2 from kinetochores upon microtubule attachment by linking Mad1:Mad2 to the microtubule minus-end directed motor dynein (Starr et al. 1998; Howell et al. 2001; Gassmann et al. 2008, 2010; Yamamoto et al. 2008; Chan et al. 2009; Barisic et al. 2010). Thus, microtubule attachment sets in motion a series of events that result in the dissociation of all SAC components from kinetochores, thereby halting MCC assembly.

5.2 MCC Disassembly and Degradation

In addition to halting assembly of new MCC, SAC silencing requires that existing MCC, both soluble and APC/C-Cdc20 bound, be disassembled and/or degraded. Two separate pathways have been identified that contribute to MCC turnover, one involving ubiquitination and degradation of $Cdc20^{MCC}$, and the other involving direct disassembly of the MCC complex through the extraction of Mad2.

The first pathway for reactivation of inhibited APC/C-MCC complex involves the ubiquitination and subsequent degradation of $Cdc20^{MCC}$ (Pan and Chen 2004; King et al. 2007; Reddy et al. 2007; Ge et al. 2009; Foe et al. 2011). As noted above, the MCC not only occupies the degron-binding sites of the APC/C, it also sterically occludes binding of E2 enzymes to the APC/C catalytic module. Recent cryo-EM analysis of APC/C-MCC identified a minor conformational state (termed APC/C-MCC-open) in which the bound MCC is rotated away from the catalytic module, allowing binding of an E2 enzyme (Fig. 4) (Alfieri et al. 2016). A structure of the APC/C-MCC-open state with a bound E2, UbcH10, revealed how $Cdc20^{MCC}$ can be ubiquitinated while still bound to the APC/C (Alfieri et al. 2016). This work also revealed why the small APC/C subunit Apc15 is required for $Cdc20^{MCC}$ ubiquitination (Mansfeld et al. 2011; Foster and Morgan 2012; Uzunova et al.

2012): in the absence of Apc15, the APC/C-MCC-open state is not accessed, meaning that E2 binding and $Cdc20^{MCC}$ ubiquitination cannot occur (Alfieri et al. 2016). After ubiquitination of $Cdc20^{MCC}$, this protein is presumably targeted to the proteasome for degradation, resulting in the reactivation of APC/C-Cdc20.

A second pathway for MCC turnover involves the direct disassembly of MCC by two proteins, $p31^{comet}$ and TRIP13 (Pch2 in yeast). $p31^{comet}$ is a HORMA domain protein distantly related to Mad2, that was first identified as a Mad2-binding protein (Habu et al. 2002; Xia et al. 2004). TRIP13 is a AAA^{+} family ATPase, which was first identified as a regulator of the Mad2-related HORMAD proteins in meiotic prophase (San-Segundo and Roeder 1999; Borner et al. 2008; Wojtasz et al. 2009; Vader 2015). Recently, TRIP13 was found to cooperate with $p31^{comet}$ in MCC disassembly and SAC inactivation (Teichner et al. 2011; Tipton et al. 2012; Eytan et al. 2014; Wang et al. 2014; Miniowitz-Shemtov et al. 2015; Ma and Poon 2016). Together, $p31^{comet}$ and TRIP13 specifically recognize C-Mad2 and convert it to the unbound O-Mad2 conformation (Ye et al. 2015). This enzymatic activity has multiple important functions, depending on context. First, $p31^{comet}$ and TRIP13 can directly disassemble soluble MCC (Mansfeld et al. 2011; Eytan et al. 2014). $p31^{comet}$ can also bind to APC/C-MCC, and some evidence suggests that Mad2 can be extracted from within APC/C-bound MCC, albeit less efficiently than from soluble MCC (Mansfeld et al. 2011; Westhorpe et al. 2011). Mad2 extraction from either soluble MCC or APC/C-MCC is a possible source of the BBC complex (BubR1, Bub3, Cdc20), which has been found to bind and inhibit the APC/C under certain conditions (Nilsson et al. 2008; Kulukian et al. 2009; Westhorpe et al. 2011; Han et al. 2013). It is likely that after Mad2 extraction, the remaining subunits of the MCC are still able to inhibit APC/C-Cdc20 to some extent. Ultimately, however, Mad2 extraction by $p31^{comet}$ and TRIP13 would destabilize the MCC, promoting dissociation and APC/C-Cdc20 reactivation.

$p31^{comet}$ and TRIP13 are conserved in plants, animals, and insects, but a clear fungal homolog of $p31^{comet}$ is missing, raising doubts about the conservation of the $p31^{comet}$/TRIP13-mediated MCC disassembly pathway. Recently, however, a radically shortened $p31^{comet}$ relative, termed Tiny yeast comet 1 (Tyc1), has been identified in *S. cerevisiae* (Schuyler S.C., personal communication). The structural mechanisms of this protein, and how it relates to more canonical $p31^{comet}$ proteins, will be exciting to explore in the future.

Another question that remains largely unexplored is whether the two known pathways for MCC turnover, $Cdc20^{MCC}$ ubiquitination and $p31^{comet}$/TRIP13-mediated Mad2 extraction, are functionally linked. Addition of $p31^{comet}$ to cell extracts arrested in metaphase by nocodazole treatment has been shown to promote Cdc20 ubiquitination (Reddy et al. 2007), suggesting that Mad2 extraction might promote formation of the APC/C-MCC-open state and thereby promote $Cdc20^{MCC}$ ubiquitination. The two pathways are not perfectly intertwined, however, as RNAi depletion of $p31^{comet}$ and the priming E2 enzyme UbcH10 (necessary for Cdc20 ubiquitination) has an additive effect on SAC inactivation

(Reddy et al. 2007). Overall, the functional relationship between Cdc20^{MCC} ubiquitination and MCC disassembly by p31^{comet} and TRIP13 remains to be fully explored.

As noted above, O-Mad2 is unstable and spontaneously converts to C-Mad2 in vitro with a half-time of several hours (Luo et al. 2004). Based on the idea that this spontaneous conversion likely also occurs in the cell, we proposed that p31^{comet} and TRIP13 might be involved in "recycling" this C-Mad2 by converting it back to O-Mad2 (Fig. 6) (Ye et al. 2015). This was recently shown to be the case: knockout of TRIP13 in human cells causes a profound defect in SAC activation, rendering these cells unresponsive to microtubule poisons such as nocodazole (Ma and Poon 2016). Biochemical examination shows that, indeed, Mad2 overwhelmingly adopts the closed conformation in these cells, and there is no detectable Mad2 binding to other MCC subunits (Ma and Poon 2016). Addition of exogenous TRIP13 to extracts from TRIP13-knockout cells re-establishes the predominance of O-Mad2 in solution (Ma and Poon 2016). In overall agreement with these results, work in *Caenorhabditis elegans* has also shown that p31^{comet} and TRIP13 homologs (CMT-1 and PCH-2, respectively) are important for Mad2 recruitment to unattached kinetochores, and that loss of these factors causes defects in SAC activation (Nelson et al. 2015). Thus, p31^{comet} and TRIP13 contribute to both SAC activation and inactivation by catalyzing the closed-to-open conformational change in Mad2.

6 Conclusion

Recent years have seen tremendous advances in our understanding of the molecular structures and protein–protein interactions underlying the SAC: the structure and mechanisms of the APC/C, its mode of inhibition by the MCC, and the mechanisms of MCC assembly and disassembly. We know much less about what occurs at kinetochores, including how they promote MCC assembly, and how their structure and composition changes in response to microtubule attachment/detachment and other signals. Finally, our understanding of SAC dynamics, particularly how it is able to respond quickly to changes in kinetochore-microtubule attachment status, is in its infancy. Thus, while recent advances in molecular understanding of SAC mechanisms represent an important step forward, a true holistic understanding of this fascinatingly complex pathway still awaits.

Acknowledgements Thanks to Dhanya Cheerambathur, Pablo Lara-Gonzalez, and Arshad Desai for critical reading and input on the manuscript, and Andrea Musacchio and Scott Schuyler for sharing unpublished results. K.D.C. gratefully acknowledges support from the March of Dimes Foundation, National Institutes of Health (R01 GM104141), and the Ludwig Institute for Cancer Research.

References

Alfieri C, Chang L, Zhang Z et al (2016) Molecular basis of APC/C regulation by the spindle assembly checkpoint. Nature 536:431–436. doi:10.1038/nature19083

Aravamudhan P, Goldfarb AA, Joglekar AP (2015) The kinetochore encodes a mechanical switch to disrupt spindle assembly checkpoint signalling. Nat Cell Biol 17:868–879. doi:10.1038/ncb3179

Aravind L, Koonin EV (1998) The HORMA domain: a common structural denominator in mitotic checkpoints, chromosome synapsis and DNA repair. Trends Biochem Sci 23:284–286

Barford D (2015) Understanding the structural basis for controlling chromosome division. Philos Trans A Math Phys Eng Sci 373:20130392. doi:10.1098/rsta.2013.0392

Barisic M, Geley S (2011) Spindly switch controls anaphase: spindly and RZZ functions in chromosome attachment and mitotic checkpoint control. Cell Cycle 10:449–456. doi:10.4161/cc.10.3.14759

Barisic M, Sohm B, Mikolcevic P et al (2010) Spindly/CCDC99 is required for efficient chromosome congression and mitotic checkpoint regulation. Mol Biol Cell 21:1968–1981. doi:10.1091/mbc.E09-04-0356

Borner GV, Barot A, Kleckner N (2008) Yeast Pch2 promotes domainal axis organization, timely recombination progression, and arrest of defective recombinosomes during meiosis. Proc Natl Acad Sci U S A 105:3327–3332. doi:10.1073/pnas.0711864105

Brown NG, VanderLinden R, Watson ER et al (2015) RING E3 mechanism for ubiquitin ligation to a disordered substrate visualized for human anaphase-promoting complex. Proc Natl Acad Sci U S A 112:5272–5279. doi:10.1073/pnas.1504161112

Brown NG, VanderLinden R, Watson ER et al (2016) Dual RING E3 architectures regulate multiubiquitination and ubiquitin chain elongation by APC/C. Cell 165:1440–1453. doi:10.1016/j.cell.2016.05.037

Buffin E, Lefebvre C, Huang J et al (2005) Recruitment of Mad2 to the kinetochore requires the Rod/Zw10 complex. Curr Biol 15:856–861. doi:10.1016/j.cub.2005.03.052

Burton JL, Xiong Y, Solomon MJ (2011) Mechanisms of pseudosubstrate inhibition of the anaphase promoting complex by Acm1. EMBO J 30:1818–1829. doi:10.1038/emboj.2011.90

Buschhorn BA, Petzold G, Galova M et al (2011) Substrate binding on the APC/C occurs between the coactivator Cdh1 and the processivity factor Doc1. Nat Struct Mol Biol 18:6–13. doi:10.1038/nsmb.1979

Caldas GV, Lynch TR, Anderson R et al (2015) The RZZ complex requires the N-terminus of KNL1 to mediate optimal Mad1 kinetochore localization in human cells. Open Biol 5:150160. doi:10.1098/rsob.150160

Carmena M, Wheelock M, Funabiki H, Earnshaw WC (2012) The chromosomal passenger complex (CPC): from easy rider to the godfather of mitosis. Nat Rev Mol Cell Biol 13: 789–803. doi:10.1038/nrm3474

Chan YW, Fava LL, Uldschmid A et al (2009) Mitotic control of kinetochore-associated dynein and spindle orientation by human Spindly. J Cell Biol 185:859–874. doi:10.1083/jcb.200812167

Chang L, Barford D (2014) Insights into the anaphase-promoting complex: a molecular machine that regulates mitosis. Curr Opin Struct Biol 29:1–9. doi:10.1016/j.sbi.2014.08.003

Chang L, Zhang Z, Yang J et al (2014) Molecular architecture and mechanism of the anaphase-promoting complex. Nature 513:388–393. doi:10.1038/nature13543

Chang L, Zhang Z, Yang J et al (2015) Atomic structure of the APC/C and its mechanism of protein ubiquitination. Nature 522:450–454. doi:10.1038/nature14471

Chao WCH, Kulkarni K, Zhang Z et al (2012) Structure of the mitotic checkpoint complex. Nature 484:208–213. doi:10.1038/nature10896

Cheeseman IM, Niessen S, Anderson S et al (2004) A conserved protein network controls assembly of the outer kinetochore and its ability to sustain tension. Genes Dev 18:2255–2268. doi:10.1101/gad.1234104

Cheeseman IM, Chappie JS, Wilson-Kubalek EM, Desai A (2006) The conserved KMN network constitutes the core microtubule-binding site of the kinetochore. Cell 127:983–997. doi:10.1016/j.cell.2006.09.039

Ciferri C, Pasqualato S, Screpanti E et al (2008) Implications for kinetochore-microtubule attachment from the structure of an engineered Ndc80 complex. Cell 133:427–439. doi:10.1016/j.cell.2008.03.020

Ciosk R, Zachariae W, Michaelis C et al (1998) An ESP1/PDS1 complex regulates loss of sister chromatid cohesion at the metaphase to anaphase transition in yeast. Cell 93:1067–1076

Cohen-Fix O, Peters JM, Kirschner MW, Koshland D (1996) Anaphase initiation in Saccharomyces cerevisiae is controlled by the APC-dependent degradation of the anaphase inhibitor Pds1p. Genes Dev 10:3081–3093

da Fonseca PCA, Kong EH, Zhang Z et al (2011) Structures of APC/C(Cdh1) with substrates identify Cdh1 and Apc10 as the D-box co-receptor. Nature 470:274–278. doi:10.1038/nature09625

Davey NE, Morgan DO (2016) Building a regulatory network with short linear sequence motifs: lessons from the degrons of the anaphase-promoting complex. Mol Cell 64:12–23. doi:10.1016/j.molcel.2016.09.006

de Antoni A, Pearson CG, Cimini D et al (2005) The Mad1/Mad2 complex as a template for Mad2 activation in the spindle assembly checkpoint. Curr Biol 15:214–225. doi:10.1016/j.cub.2005.01.038

Desai A, Rybina S, Müller-Reichert T et al (2003) KNL-1 directs assembly of the microtubule-binding interface of the kinetochore in *C. elegans*. Genes Dev 17:2421–2435. doi:10.1101/gad.1126303

Di Fiore B, Davey NE, Hagting A et al (2015) The ABBA motif binds APC/C activators and is shared by APC/C substrates and regulators. Dev Cell 32:358–372. doi:10.1016/j.devcel.2015.01.003

Di Fiore B, Wurzenberger C, Davey NE, Pines J (2016) The mitotic checkpoint complex requires an evolutionary conserved cassette to bind and inhibit active APC/C. Mol Cell 64:1144–1153. doi:10.1016/j.molcel.2016.11.006

Diaz-Martinez LA, Tian W, Li B et al (2015) The Cdc20-binding Phe box of the spindle checkpoint protein BubR1 maintains the mitotic checkpoint complex during mitosis. J Biol Chem 290:2431–2443. doi:10.1074/jbc.M114.616490

Dou Z, Liu X, Wang W et al (2015) Dynamic localization of Mps1 kinase to kinetochores is essential for accurate spindle microtubule attachment. Proc Natl Acad Sci U S A 112:E4546–E4555. doi:10.1073/pnas.1508791112

Espert A, Uluocak P, Bastos RN et al (2014) PP2A-B56 opposes Mps1 phosphorylation of Knl1 and thereby promotes spindle assembly checkpoint silencing. J Cell Biol 206:833–842. doi:10.1083/jcb.201406109

Etemad B, Kops GJPL (2016) Attachment issues: kinetochore transformations and spindle checkpoint silencing. Curr Opin Cell Biol 39:101–108. doi:10.1016/j.ceb.2016.02.016

Eytan E, Wang K, Miniowitz-Shemtov S et al (2014) Disassembly of mitotic checkpoint complexes by the joint action of the AAA-ATPase TRIP13 and p31(comet). Proc Natl Acad Sci U S A 111:12019–12024. doi:10.1073/pnas.1412901111

Faesen AC, Thanasoula M, Maffini S et al (2017) In vitro reconstitution of spindle assembly checkpoint signaling identifies the determinants of catalytic assembly of the mitotic checkpoint complex. Nature 542:498–502. doi:10.1038/nature21384

Fang G (2002) Checkpoint protein BubR1 acts synergistically with Mad2 to inhibit anaphase-promoting complex. Mol Biol Cell 13:755–766. doi:10.1091/mbc.01-09-0437

Fang G, Yu H, Kirschner MW (1998) The checkpoint protein MAD2 and the mitotic regulator CDC20 form a ternary complex with the anaphase-promoting complex to control anaphase initiation. Genes Dev 12:1871–1883

Foe IT, Foster SA, Cheung SK et al (2011) Ubiquitination of Cdc20 by the APC occurs through an intramolecular mechanism. Curr Biol 21:1870–1877. doi:10.1016/j.cub.2011.09.051

Foley EA, Kapoor TM (2013) Microtubule attachment and spindle assembly checkpoint signalling at the kinetochore. Nat Rev Mol Cell Biol 14:25–37. doi:10.1038/nrm3494
Foster SA, Morgan DO (2012) The APC/C subunit Mnd2/Apc15 promotes Cdc20 autoubiquitination and spindle assembly checkpoint inactivation. Mol Cell 47:921–932. doi:10.1016/j.molcel.2012.07.031
Gassmann R, Essex A, Hu J-S et al (2008) A new mechanism controlling kinetochore-microtubule interactions revealed by comparison of two dynein-targeting components: SPDL-1 and the Rod/Zwilch/Zw10 complex. Genes Dev 22:2385–2399. doi:10.1101/gad.1687508
Gassmann R, Holland AJ, Varma D et al (2010) Removal of spindly from microtubule-attached kinetochores controls spindle checkpoint silencing in human cells. Genes Dev 24:957–971. doi:10.1101/gad.1886810
Ge S, Skaar JR, Pagano M (2009) APC/C- and Mad2-mediated degradation of Cdc20 during spindle checkpoint activation. Cell Cycle 8:167–171. doi:10.4161/cc.8.1.7606
Glotzer M, Murray AW, Kirschner MW (1991) Cyclin is degraded by the ubiquitin pathway. Nature 349:132–138. doi:10.1038/349132a0
Golan A, Yudkovsky Y, Hershko A (2002) The cyclin-ubiquitin ligase activity of cyclosome/APC is jointly activated by protein kinases Cdk1-cyclin B and Plk. J Biol Chem 277:15552–15557. doi:10.1074/jbc.M111476200
Habu T, Kim SH, Weinstein J, Matsumoto T (2002) Identification of a MAD2-binding protein, CMT2, and its role in mitosis. EMBO J 21:6419–6428
Han JS, Holland AJ, Fachinetti D et al (2013) Catalytic assembly of the mitotic checkpoint inhibitor BubR1-Cdc20 by a Mad2-induced functional switch in Cdc20. Mol Cell 51:92–104. doi:10.1016/j.molcel.2013.05.019
Hara M, Özkan E, Sun H et al (2015) Structure of an intermediate conformer of the spindle checkpoint protein Mad2. Proc Natl Acad Sci U S A 112:11252–11257. doi:10.1073/pnas.1512197112
Hardwick KG, Johnston RC, Smith DL, Murray AW (2000) MAD3 encodes a novel component of the spindle checkpoint which interacts with Bub3p, Cdc20p, and Mad2p. J Cell Biol 148:871–882. doi:10.1083/jcb.148.5.871
Hardwick KG, Weiss E, Luca FC et al (1996) Activation of the budding yeast spindle assembly checkpoint without mitotic spindle disruption. Science 273:953–956
He J, Chao WCH, Zhang Z et al (2013) Insights into degron recognition by APC/C coactivators from the structure of an Acm1-Cdh1 complex. Mol Cell 50:649–660. doi:10.1016/j.molcel.2013.04.024
Hendrickx A, Beullens M, Ceulemans H et al (2009) Docking motif-guided mapping of the interactome of protein phosphatase-1. Chem Biol 16:365–371. doi:10.1016/j.chembiol.2009.02.012
Herzog F, Primorac I, Dube P et al (2009) Structure of the anaphase-promoting complex/cyclosome interacting with a mitotic checkpoint complex. Science 323:1477–1481. doi:10.1126/science.1163300
Hiruma Y, Sacristan C, Pachis ST et al (2015) CELL DIVISION CYCLE. Competition between MPS1 and microtubules at kinetochores regulates spindle checkpoint signaling. Science 348:1264–1267. doi:10.1126/science.aaa4055
Howell BJ, McEwen BF, Canman JC et al (2001) Cytoplasmic dynein/dynactin drives kinetochore protein transport to the spindle poles and has a role in mitotic spindle checkpoint inactivation. J Cell Biol 155:1159–1172. doi:10.1083/jcb.200105093
Howell BJ, Moree B, Farrar EM et al (2004) Spindle checkpoint protein dynamics at kinetochores in living cells. Curr Biol 14:953–964. doi:10.1016/j.cub.2004.05.053
Hoyt MA, Totis L, Roberts BT (1991) *S. cerevisiae* genes required for cell cycle arrest in response to loss of microtubule function. Cell 66:507–517
Hwang LH, Lau LF, Smith DL et al (1998) Budding yeast Cdc20: a target of the spindle checkpoint. Science 279:1041–1044
Izawa D, Pines J (2012) Mad2 and the APC/C compete for the same site on Cdc20 to ensure proper chromosome segregation. J Cell Biol 199:27–37. doi:10.1083/jcb.201205170

Izawa D, Pines J (2014) The mitotic checkpoint complex binds a second CDC20 to inhibit active APC/C. Nature. doi:10.1038/nature13911
Jaspersen SL, Charles JF, Morgan DO (1999) Inhibitory phosphorylation of the APC regulator Hct1 is controlled by the kinase Cdc28 and the phosphatase Cdc14. Curr Biol 9:227–236
Ji Z, Gao H, Yu H (2015) Cell division cycle. Kinetochore attachment sensed by competitive Mps1 and microtubule binding to Ndc80C. Science 348:1260–1264. doi:10.1126/science.aaa4029
Ji Z, Gao H, Jia L et al (2017) A sequential multi-target Mps1 phosphorylation cascade promotes spindle checkpoint signaling. Elife 6:e22513. doi:10.7554/eLife.22513
Kallio M, Weinstein J, Daum JR et al (1998) Mammalian p55CDC mediates association of the spindle checkpoint protein Mad2 with the cyclosome/anaphase-promoting complex, and is involved in regulating anaphase onset and late mitotic events. J Cell Biol 141:1393–1406
Kamenz J, Hauf S (2016) Time to split up: dynamics of chromosome separation. Trends Cell Biol. doi:10.1016/j.tcb.2016.07.008
Karess R (2005) Rod-Zw10-Zwilch: a key player in the spindle checkpoint. Trends Cell Biol 15:386–392. doi:10.1016/j.tcb.2005.05.003
Kim SH, Lin DP, Matsumoto S et al (1998) Fission yeast Slp1: an effector of the Mad2-dependent spindle checkpoint. Science 279:1045–1047
King RW, Peters JM, Tugendreich S et al (1995) A 20S complex containing CDC27 and CDC16 catalyzes the mitosis-specific conjugation of ubiquitin to cyclin B. Cell 81:279–288
King EMJ, van der Sar SJA, Hardwick KG (2007) Mad3 KEN boxes mediate both Cdc20 and Mad3 turnover, and are critical for the spindle checkpoint. PLoS ONE 2:e342. doi:10.1371/journal.pone.0000342
Kops GJPL, Kim Y, Weaver BAA et al (2005) ZW10 links mitotic checkpoint signaling to the structural kinetochore. J Cell Biol 169:49–60. doi:10.1083/jcb.200411118
Kraft C, Herzog F, Gieffers C et al (2003) Mitotic regulation of the human anaphase-promoting complex by phosphorylation. EMBO J 22:6598–6609. doi:10.1093/emboj/cdg627
Kramer ER, Scheuringer N, Podtelejnikov AV et al (2000) Mitotic regulation of the APC activator proteins CDC20 and CDH1. Mol Biol Cell 11:1555–1569
Krenn V, Musacchio A (2015) The Aurora B kinase in chromosome bi-orientation and spindle checkpoint signaling. Front Oncol 5:225. doi:10.3389/fonc.2015.00225
Krenn V, Overlack K, Primorac I et al (2014) KI motifs of human Knl1 enhance assembly of comprehensive spindle checkpoint complexes around MELT repeats. Curr Biol 24:29–39. doi:10.1016/j.cub.2013.11.046
Kulukian A, Han JS, Cleveland DW (2009) Unattached kinetochores catalyze production of an anaphase inhibitor that requires a Mad2 template to prime Cdc20 for BubR1 binding. Dev Cell 16:105–117. doi:10.1016/j.devcel.2008.11.005
Labit H, Fujimitsu K, Bayin NS et al (2012) Dephosphorylation of Cdc20 is required for its C-box-dependent activation of the APC/C. EMBO J 31:3351–3362. doi:10.1038/emboj.2012.168
Lahav-Baratz S, Sudakin V, Ruderman JV, Hershko A (1995) Reversible phosphorylation controls the activity of cyclosome-associated cyclin-ubiquitin ligase. Proc Natl Acad Sci U S A 92:9303–9307
Lampson MA, Cheeseman IM (2011) Sensing centromere tension: Aurora B and the regulation of kinetochore function. Trends Cell Biol 21:133–140. doi:10.1016/j.tcb.2010.10.007
Lan W, Cleveland DW (2010) A chemical tool box defines mitotic and interphase roles for Mps1 kinase. J Cell Biol 190:21–24. doi:10.1083/jcb.201006080
Lara-Gonzalez P, Westhorpe FG, Taylor SS (2012) The Spindle Assembly Checkpoint. Curr Biol 22:R966–R980. doi:10.1016/j.cub.2012.10.006
Larsen NA, Al-Bassam J, Wei RR, Harrison SC (2007) Structural analysis of Bub3 interactions in the mitotic spindle checkpoint. Proc Natl Acad Sci U S A 104:1201–1206. doi:10.1073/pnas.0610358104
Li R, Murray AW (1991) Feedback control of mitosis in budding yeast. Cell 66:519–531
Li Y, Gorbea C, Mahaffey D et al (1997) MAD2 associates with the cyclosome/anaphase-promoting complex and inhibits its activity. Proc Natl Acad Sci U S A 94:12431–12436

Lischetti T, Zhang G, Sedgwick GG et al (2014) The internal Cdc20 binding site in BubR1 facilitates both spindle assembly checkpoint signalling and silencing. Nat Commun 5:5563. doi:10.1038/ncomms6563
Liu X, Winey M (2012) The MPS1 family of protein kinases. Annu Rev Biochem 81:561–585. doi:10.1146/annurev-biochem-061611-090435
Liu D, Vleugel M, Backer CB et al (2010) Regulated targeting of protein phosphatase 1 to the outer kinetochore by KNL1 opposes Aurora B kinase. J Cell Biol 188:809–820. doi:10.1083/jcb.201001006
London N, Biggins S (2014) Mad1 kinetochore recruitment by Mps1-mediated phosphorylation of Bub1 signals the spindle checkpoint. Genes Dev 28:140–152. doi:10.1101/gad.233700.113
London N, Ceto S, Ranish JA, Biggins S (2012) Phosphoregulation of Spc105 by Mps1 and PP1 regulates Bub1 localization to kinetochores. Curr Biol 22:900–906. doi:10.1016/j.cub.2012.03.052
Lukas C, Sørensen CS, Kramer E et al (1999) Accumulation of cyclin B1 requires E2F and cyclin-A-dependent rearrangement of the anaphase-promoting complex. Nature 401:815–818. doi:10.1038/44611
Luo X, Yu H (2008) Protein metamorphosis: the two-state behavior of Mad2. Structure 16:1616–1625. doi:10.1016/j.str.2008.10.002
Luo X, Fang G, Coldiron M et al (2000) Structure of the Mad2 spindle assembly checkpoint protein and its interaction with Cdc20. Nat Struct Biol 7:224–229. doi:10.1038/73338
Luo X, Tang Z, Rizo J, Yu H (2002) The Mad2 spindle checkpoint protein undergoes similar major conformational changes upon binding to either Mad1 or Cdc20. Mol Cell 9:59–71
Luo X, Tang Z, Xia G et al (2004) The Mad2 spindle checkpoint protein has two distinct natively folded states. Nat Struct Mol Biol 11:338–345. doi:10.1038/nsmb748
Ma HT, Poon RYC (2016) TRIP13 regulates both the activation and inactivation of the spindle-assembly checkpoint. Cell Reports 14:1086–1099. doi:10.1016/j.celrep.2016.01.001
Mansfeld J, Collin P, Collins MO et al (2011) APC15 drives the turnover of MCC-CDC20 to make the spindle assembly checkpoint responsive to kinetochore attachment. Nat Cell Biol 13:1234–1243. doi:10.1038/ncb2347
Mapelli M, Musacchio A (2007) MAD contortions: conformational dimerization boosts spindle checkpoint signaling. Curr Opin Struct Biol 17:716–725. doi:10.1016/j.sbi.2007.08.011
Mapelli M, Massimiliano L, Santaguida S, Musacchio A (2007) The Mad2 conformational dimer: structure and implications for the spindle assembly checkpoint. Cell 131:730–743. doi:10.1016/j.cell.2007.08.049
Martin-Lluesma S, Stucke VM, Nigg EA (2002) Role of Hec1 in spindle checkpoint signaling and kinetochore recruitment of Mad1/Mad2. Science 297:2267–2270. doi:10.1126/science.1075596
Matyskiela ME, Morgan DO (2009) Analysis of activator-binding sites on the APC/C supports a cooperative substrate-binding mechanism. Mol Cell 34:68–80. doi:10.1016/j.molcel.2009.02.027
Meadows JC, Shepperd LA, Vanoosthuyse V et al (2011) Spindle checkpoint silencing requires association of PP1 to both Spc7 and kinesin-8 motors. Dev Cell 20:739–750. doi:10.1016/j.devcel.2011.05.008
Miniowitz-Shemtov S, Eytan E, Kaisari S et al (2015) Mode of interaction of TRIP13 AAA-ATPase with the Mad2-binding protein p31comet and with mitotic checkpoint complexes. Proc Natl Acad Sci U S A 112:11536–11540. doi:10.1073/pnas.1515358112
Morgan DO (1997) Cyclin-dependent kinases: engines, clocks, and microprocessors. Annu Rev Cell Dev Biol 13:261–291. doi:10.1146/annurev.cellbio.13.1.261
Moyle MW, Kim T, Hattersley N et al (2014) A Bub1-Mad1 interaction targets the Mad1-Mad2 complex to unattached kinetochores to initiate the spindle checkpoint. J Cell Biol 204:647–657. doi:10.1083/jcb.201311015
Murray AW, Solomon MJ, Kirschner MW (1989) The role of cyclin synthesis and degradation in the control of maturation promoting factor activity. Nature 339:280–286. doi:10.1038/339280a0

Musacchio A (2015) The molecular biology of spindle assembly checkpoint signaling dynamics. Curr Biol 25:R1002–R1018. doi:10.1016/j.cub.2015.08.051
Musacchio A, Salmon ED (2007) The spindle-assembly checkpoint in space and time. Nat Rev Mol Cell Biol 8:379–393. doi:10.1038/nrm2163
Nagpal H, Fukagawa T (2016) Kinetochore assembly and function through the cell cycle. Chromosoma 125:645–659. doi:10.1007/s00412-016-0608-3
Nekrasov VS, Smith MA, Peak-Chew S, Kilmartin JV (2003) Interactions between centromere complexes in Saccharomyces cerevisiae. Mol Biol Cell 14:4931–4946. doi:10.1091/mbc.E03-06-0419
Nelson CR, Hwang T, Chen P-H, Bhalla N (2015) TRIP13/PCH-2 promotes Mad2 localization to unattached kinetochores in the spindle checkpoint response. J Cell Biol 211:503–516. doi:10.1083/jcb.201505114
Nezi L, Rancati G, de Antoni A et al (2006) Accumulation of Mad2-Cdc20 complex during spindle checkpoint activation requires binding of open and closed conformers of Mad2 in *Saccharomyces cerevisiae*. J Cell Biol 174:39–51. doi:10.1083/jcb.200602109
Nijenhuis W, von Castelmur E, Littler D et al (2013) A TPR domain-containing N-terminal module of MPS1 is required for its kinetochore localization by Aurora B. J Cell Biol 201: 217–231. doi:10.1083/jcb.201210033
Nijenhuis W, Vallardi G, Teixeira A et al (2014) Negative feedback at kinetochores underlies a responsive spindle checkpoint signal. Nat Cell Biol 16:1257–1264. doi:10.1038/ncb3065
Nilsson J, Yekezare M, Minshull J, Pines J (2008) The APC/C maintains the spindle assembly checkpoint by targeting Cdc20 for destruction. Nat Cell Biol 10:1411–1420. doi:10.1038/ncb1799
Overlack K, Primorac I, Vleugel M et al (2015) A molecular basis for the differential roles of Bub1 and BubR1 in the spindle assembly checkpoint. Elife 4:e05269. doi:10.7554/eLife.05269
Pan J, Chen R-H (2004) Spindle checkpoint regulates Cdc20p stability in *Saccharomyces cerevisiae*. Genes Dev 18:1439–1451. doi:10.1101/gad.1184204
Pesenti ME, Weir JR, Musacchio A (2016) Progress in the structural and functional characterization of kinetochores. Curr Opin Struct Biol 37:152–163. doi:10.1016/j.sbi.2016.03.003
Pfleger CM, Kirschner MW (2000) The KEN box: an APC recognition signal distinct from the D box targeted by Cdh1. Genes Dev 14:655–665
Primorac I, Musacchio A (2013) Panta rhei: the APC/C at steady state. J Cell Biol 201:177–189. doi:10.1083/jcb.201301130
Primorac I, Weir JR, Chiroli E et al (2013) Bub3 reads phosphorylated MELT repeats to promote spindle assembly checkpoint signaling. Elife 2:e01030. doi:10.7554/eLife.01030
Qiao R, Weissmann F, Yamaguchi M et al (2016) Mechanism of APC/CCDC20 activation by mitotic phosphorylation. Proc Natl Acad Sci U S A 113:E2570–E2578. doi:10.1073/pnas.1604929113
Reddy SK, Rape M, Margansky WA, Kirschner MW (2007) Ubiquitination by the anaphase-promoting complex drives spindle checkpoint inactivation. Nature 446:921–925. doi:10.1038/nature05734
Reis A, Levasseur M, Chang H-Y et al (2006) The CRY box: a second APCcdh1-dependent degron in mammalian cdc20. EMBO Rep 7:1040–1045. doi:10.1038/sj.embor.7400772
Rieder CL, Cole RW, Khodjakov A, Sluder G (1995) The checkpoint delaying anaphase in response to chromosome monoorientation is mediated by an inhibitory signal produced by unattached kinetochores. J Cell Biol 130:941–948
Rosenberg JS, Cross FR, Funabiki H (2011) KNL1/Spc105 recruits PP1 to silence the spindle assembly checkpoint. Curr Biol 21:942–947. doi:10.1016/j.cub.2011.04.011
San-Segundo PA, Roeder GS (1999) Pch2 links chromatin silencing to meiotic checkpoint control. Cell 97:313–324
Schwab M, Neutzner M, Möcker D, Seufert W (2001) Yeast Hct1 recognizes the mitotic cyclin Clb2 and other substrates of the ubiquitin ligase APC. EMBO J 20:5165–5175. doi:10.1093/emboj/20.18.5165

Sczaniecka M, Feoktistova A, May KM et al (2008) The spindle checkpoint functions of Mad3 and Mad2 depend on a Mad3 KEN box-mediated interaction with Cdc20-anaphase-promoting complex (APC/C). J Biol Chem 283:23039–23047. doi:10.1074/jbc.M803594200
Shah JV, Botvinick E, Bonday Z et al (2004) Dynamics of centromere and kinetochore proteins; implications for checkpoint signaling and silencing. Curr Biol 14:942–952. doi:10.1016/j.cub.2004.05.046
Shepperd LA, Meadows JC, Sochaj AM et al (2012) Phosphodependent recruitment of Bub1 and Bub3 to Spc7/KNL1 by Mph1 kinase maintains the spindle checkpoint. Curr Biol 22:891–899. doi:10.1016/j.cub.2012.03.051
Shirayama M, Toth A, Gálová M, Nasmyth K (1999) APC(Cdc20) promotes exit from mitosis by destroying the anaphase inhibitor Pds1 and cyclin Clb5. Nature 402:203–207. doi:10.1038/46080
Shteinberg M, Protopopov Y, Listovsky T et al (1999) Phosphorylation of the cyclosome is required for its stimulation by Fizzy/cdc20. Biochem Biophys Res Commun 260:193–198. doi:10.1006/bbrc.1999.0884
Silió V, McAinsh AD, Millar JB (2015) KNL1-Bubs and RZZ provide two separable pathways for checkpoint activation at human kinetochores. Dev Cell 35:600–613. doi:10.1016/j.devcel.2015.11.012
Simonetta M, Manzoni R, Mosca R et al (2009) The influence of catalysis on mad2 activation dynamics. PLoS Biol 7:e10. doi:10.1371/journal.pbio.1000010
Sironi L, Mapelli M, Knapp S et al (2002) Crystal structure of the tetrameric Mad1-Mad2 core complex: implications of a "safety belt" binding mechanism for the spindle checkpoint. EMBO J 21:2496–2506. doi:10.1093/emboj/21.10.2496
Starr DA, Williams BC, Hays TS, Goldberg ML (1998) ZW10 helps recruit dynactin and dynein to the kinetochore. J Cell Biol 142:763–774
Sudakin V, Chan GK, Yen TJ (2001) Checkpoint inhibition of the APC/C in HeLa cells is mediated by a complex of BUBR1, BUB3, CDC20, and MAD2. J Cell Biol 154:925–936. doi:10.1083/jcb.200102093
Sudakin V, Ganoth D, Dahan A et al (1995) The cyclosome, a large complex containing cyclin-selective ubiquitin ligase activity, targets cyclins for destruction at the end of mitosis. Mol Biol Cell 6:185–197
Suijkerbuijk SJE, van Dam TJP, Karagöz GE et al (2012a) The vertebrate mitotic checkpoint protein BUBR1 is an unusual pseudokinase. Dev Cell 22:1321–1329. doi:10.1016/j.devcel.2012.03.009
Suijkerbuijk SJE, Vleugel M, Teixeira A, Kops GJPL (2012b) Integration of kinase and phosphatase activities by BUBR1 ensures formation of stable kinetochore-microtubule attachments. Dev Cell 23:745–755. doi:10.1016/j.devcel.2012.09.005
Tang Z, Bharadwaj R, Li B, Yu H (2001) Mad2-Independent inhibition of APCCdc20 by the mitotic checkpoint protein BubR1. Dev Cell 1:227–237
Taylor SS, Ha E, McKeon F (1998) The human homologue of Bub3 is required for kinetochore localization of Bub1 and a Mad3/Bub1-related protein kinase. J Cell Biol 142:1–11
Teichner A, Eytan E, Sitry-Shevah D et al (2011) p31comet Promotes disassembly of the mitotic checkpoint complex in an ATP-dependent process. Proc Natl Acad Sci U S A 108:3187–3192. doi:10.1073/pnas.1100023108
Thornton BR, Ng TM, Matyskiela ME et al (2006) An architectural map of the anaphase-promoting complex. Genes Dev 20:449–460. doi:10.1101/gad.1396906
Tipton AR, Wang K, Link L et al (2011) BUBR1 and closed MAD2 (C-MAD2) interact directly to assemble a functional mitotic checkpoint complex. J Biol Chem 286:21173–21179. doi:10.1074/jbc.M111.238543
Tipton AR, Wang K, Oladimeji P et al (2012) Identification of novel mitosis regulators through data mining with human centromere/kinetochore proteins as group queries. BMC Cell Biol 13:15. doi:10.1186/1471-2121-13-15

Uzunova K, Dye BT, Schutz H et al (2012) APC15 mediates CDC20 autoubiquitylation by APC/C (MCC) and disassembly of the mitotic checkpoint complex. Nat Struct Mol Biol 19:1116–1123. doi:10.1038/nsmb.2412
Vader G (2015) Pch2(TRIP13): controlling cell division through regulation of HORMA domains. Chromosoma 124:333–339. doi:10.1007/s00412-015-0516-y
van der Horst A, Lens SMA (2014) Cell division: control of the chromosomal passenger complex in time and space. Chromosoma 123:25–42. doi:10.1007/s00412-013-0437-6
Varma D, Salmon ED (2012) The KMN protein network–chief conductors of the kinetochore orchestra. J Cell Sci 125:5927–5936. doi:10.1242/jcs.093724
Vink M, Simonetta M, Transidico P et al (2006) In vitro FRAP identifies the minimal requirements for Mad2 kinetochore dynamics. Curr Biol 16:755–766. doi:10.1016/j.cub.2006.03.057
Vleugel M, Hoogendoorn E, Snel B, Kops GJPL (2012) Evolution and function of the mitotic checkpoint. Dev Cell 23:239–250. doi:10.1016/j.devcel.2012.06.013
Vleugel M, Tromer E, Omerzu M et al (2013) Arrayed BUB recruitment modules in the kinetochore scaffold KNL1 promote accurate chromosome segregation. J Cell Biol 203: 943–955. doi:10.1083/jcb.201307016
Vleugel M, Hoek TA, Tromer E et al (2015a) Dissecting the roles of human BUB1 in the spindle assembly checkpoint. J Cell Sci 128:2975–2982. doi:10.1242/jcs.169821
Vleugel M, Omerzu M, Groenewold V et al (2015b) Sequential multisite phospho-regulation of KNL1-BUB3 interfaces at mitotic kinetochores. Mol Cell 57:824–835. doi:10.1016/j.molcel.2014.12.036
Vodermaier HC, Gieffers C, Maurer-Stroh S et al (2003) TPR subunits of the anaphase-promoting complex mediate binding to the activator protein CDH1. Curr Biol 13:1459–1468
Wang X, Babu JR, Harden JM et al (2001) The mitotic checkpoint protein hBUB3 and the mRNA export factor hRAE1 interact with GLE2p-binding sequence (GLEBS)-containing proteins. J Biol Chem 276:26559–26567. doi:10.1074/jbc.M101083200
Wang H, Hu X, Ding X et al (2004) Human Zwint-1 specifies localization of Zeste White 10 to kinetochores and is essential for mitotic checkpoint signaling. J Biol Chem 279:54590–54598. doi:10.1074/jbc.M407588200
Wang K, Sturt-Gillespie B, Hittle JC et al (2014) Thyroid hormone receptor interacting protein 13 (TRIP13) AAA-ATPase is a novel mitotic checkpoint-silencing protein. J Biol Chem 289:23928–23937. doi:10.1074/jbc.M114.585315
Wei RR, Al-Bassam J, Harrison SC (2007) The Ndc80/HEC1 complex is a contact point for kinetochore-microtubule attachment. Nat Struct Mol Biol 14:54–59. doi:10.1038/nsmb1186
Westhorpe FG, Tighe A, Lara-Gonzalez P, Taylor SS (2011) p31comet-mediated extraction of Mad2 from the MCC promotes efficient mitotic exit. J Cell Sci 124:3905–3916. doi:10.1242/jcs.093286
Wojtasz L, Daniel K, Roig I et al (2009) Mouse HORMAD1 and HORMAD2, two conserved meiotic chromosomal proteins, are depleted from synapsed chromosome axes with the help of TRIP13 AAA-ATPase. PLoS Genet 5:e1000702. doi:10.1371/journal.pgen.1000702
Xia G, Luo X, Habu T et al (2004) Conformation-specific binding of p31(comet) antagonizes the function of Mad2 in the spindle checkpoint. EMBO J 23:3133–3143. doi:10.1038/sj.emboj.7600322
Yamagishi Y, Yang C-H, Tanno Y, Watanabe Y (2012) MPS1/Mph1 phosphorylates the kinetochore protein KNL1/Spc7 to recruit SAC components. Nat Cell Biol 14:746–752. doi:10.1038/ncb2515
Yamaguchi M, VanderLinden R, Weissmann F et al (2016) Cryo-EM of mitotic checkpoint complex-bound APC/C reveals reciprocal and conformational regulation of ubiquitin ligation. Mol Cell 63:593–607. doi:10.1016/j.molcel.2016.07.003
Yamamoto TG, Watanabe S, Essex A, Kitagawa R (2008) SPDL-1 functions as a kinetochore receptor for MDF-1 in *Caenorhabditis elegans*. J Cell Biol 183:187–194. doi:10.1083/jcb.200805185
Yang M, Li B, Tomchick DR et al (2007) p31comet blocks Mad2 activation through structural mimicry. Cell 131:744–755. doi:10.1016/j.cell.2007.08.048

Ye Q, Rosenberg SC, Moeller A et al (2015) TRIP13 is a protein-remodeling AAA + ATPase that catalyzes MAD2 conformation switching. Elife 4:213. doi:10.7554/eLife.07367

Zachariae W, Schwab M, Nasmyth K, Seufert W (1998) Control of cyclin ubiquitination by CDK-regulated binding of Hct1 to the anaphase promoting complex. Science 282:1721–1724

Zhang Y, Lees E (2001) Identification of an overlapping binding domain on Cdc20 for Mad2 and anaphase-promoting complex: model for spindle checkpoint regulation. Mol Cell Biol 21:5190–5199. doi:10.1128/MCB.21.15.5190-5199.2001

Zhang G, Lischetti T, Nilsson J (2014) A minimal number of MELT repeats supports all the functions of KNL1 in chromosome segregation. J Cell Sci 127:871–884. doi:10.1242/jcs.139725

Zhang G, Lischetti T, Hayward DG, Nilsson J (2015) Distinct domains in Bub1 localize RZZ and BubR1 to kinetochores to regulate the checkpoint. Nat Commun 6:7162. doi:10.1038/ncomms8162

Zhang G, Mendez BL, Sedgwick GG, Nilsson J (2016a) Two functionally distinct kinetochore pools of BubR1 ensure accurate chromosome segregation. Nat Commun 7:12256. doi:10.1038/ncomms12256

Zhang H, Liu S-T, Department of Biological Sciences, University of Toledo, 2801 West Bancroft St., Toledo, OH 43606, USA (2016b) The mitotic checkpoint complex (MCC): looking back and forth after 15 years. AIMS Mol Sci 3:597–634. doi:10.3934/molsci.2016.4.597

Zhang S, Chang L, Alfieri C et al (2016c) Molecular mechanism of APC/C activation by mitotic phosphorylation. Nature 533:260–264. doi:10.1038/nature17973

Zhu T, Dou Z, Qin B et al (2013) Phosphorylation of microtubule-binding protein Hec1 by mitotic kinase Aurora B specifies spindle checkpoint kinase Mps1 signaling at the kinetochore. J Biol Chem 288:36149–36159. doi:10.1074/jbc.M113.507970

A Kinase-Phosphatase Network that Regulates Kinetochore-Microtubule Attachments and the SAC

Giulia Vallardi, Marilia Henriques Cordeiro and Adrian Thomas Saurin

Abstract The KMN network (for KNL1, MIS12 and NDC80 complexes) is a hub for signalling at the outer kinetochore. It integrates the activities of two kinases (MPS1 and Aurora B) and two phosphatases (PP1 and PP2A-B56) to regulate kinetochore-microtubule attachments and the spindle assembly checkpoint (SAC). We will first discuss each of these enzymes separately, to describe how they are regulated at kinetochores and why this is important for their primary function in controlling either microtubule attachments or the SAC. We will then discuss why inhibiting any one of them individually produces secondary effects on all the others. This cross-talk may help to explain why all enzymes have been linked to both processes, even though the direct evidence suggests they each control only one. This chapter therefore describes how a network of kinases and phosphatases work together to regulate two key mitotic processes.

1 Introduction

The kinetochore is a complex molecular machine consisting of over 100 different proteins (Nagpal and Fukagawa 2016; Pesenti et al. 2016). These proteins can be classified based on whether they perform a structural, functional, and/or regulatory role. This broadly divides the kinetochore into three core parts: (1) a constitutive inner network that is involved in tethering the outer kinetochore to chromosomes, (2) a constitutive outer network that reaches out to capture and hold on to microtubules, and (3) a dynamic regulatory set of regulatory components that transiently associate with the kinetochore at the appropriate times. These dynamic components are principally focussed on the KMN network, which is responsible

---

G. Vallardi · M.H. Cordeiro · A.T. Saurin (✉)
Division of Cancer Research, School of Medicine,
Jacqui Wood Cancer Centre, Ninewells Hospital
and Medical School, University of Dundee, Dundee DD1 9SY, UK
e-mail: a.saurin@dundee.ac.uk

B.E. Black (ed.), *Centromeres and Kinetochores*, Progress in Molecular and Subcellular Biology 56, DOI 10.1007/978-3-319-58592-5_19

for regulating the kinetochores two main functions: the physical attachment to microtubules and activation of the SAC.

The attachment to microtubules needs to be regulated to ensure that incorrect attachments, which fail to generate tension, can be removed in a process known as error-correction (Krenn and Musacchio 2015). The SAC, on the other hand, must be activated by unattached kinetochores to hold cells in mitosis and give time for tension-generating attachments to form (Joglekar 2016; Lischetti and Nilsson 2015; Musacchio 2015). This chapter focusses on the main kinetochore kinases and phosphatases that regulate these two processes: MPS1, Aurora B, PP1 and PP2A-B56.

At the beginning of mitosis, during prophase, kinase activities predominate at kinetochores: MPS1 phosphorylates proteins that initiate SAC signalling (symbolised by KNL1 in Fig. 1a), whereas Aurora B phosphorylates proteins to prevent their attachment to microtubules (symbolised by NDC80 in Fig. 1a). Upon nuclear envelope breakdown at the start of prometaphase, however, the phosphatases PP1 and PP2A-B56 are recruited to KNL1 where they begin to counteract the activity of the kinases. This reduces Aurora B activity to allow kinetochore-microtubule attachments to form and primes kinetochores to be ready to silence the SAC (Fig. 1b). When these microtubule attachments generate tension, kinase activities are reduced and phosphatase activities predominate, which stabilises these attachments and locally silences the SAC (Fig. 1c). When all kinetochores have achieved this stably attached state, the SAC is switched off globally, sister chromatids separate and the cell can exit mitosis.

As will become apparent, although these kinases and phosphatases perform very specific roles at kinetochores, there is also considerable interplay between them: the kinases regulate each other, the phosphatases regulate each other, and both kinases also regulate, directly or indirectly, both phosphatases (Fig. 1). Not surprisingly, this has contributed to a great deal of confusion in the field with regards to "who controls what". We shall attempt to clarify some of this confusion by first discussing the SAC and error-correction processes separately, from the point of view of their direct *bona fide* kinases and phosphatases. This will explain how these enzymes are regulated and why this is important to control either kinetochore-microtubule attachments or the SAC. We will then discuss the issue of cross-talk and explain why manipulating any one of these enzymes produces secondary effects on all the others. This demonstrates that all four enzymes form part of an interconnected network that regulates both mitotic processes.

2 MPS1: The SAC Kinase

Monopolar spindle 1 (MPS1) was originally identified as a gene that controls spindle pole body duplication in *Saccharomyces cerevisiae* (Winey et al. 1991), but later shown to be a dual-specificity protein kinase essential for the spindle checkpoint response (Lauze et al. 1995; Poch et al. 1994; Weiss and Winey 1996).

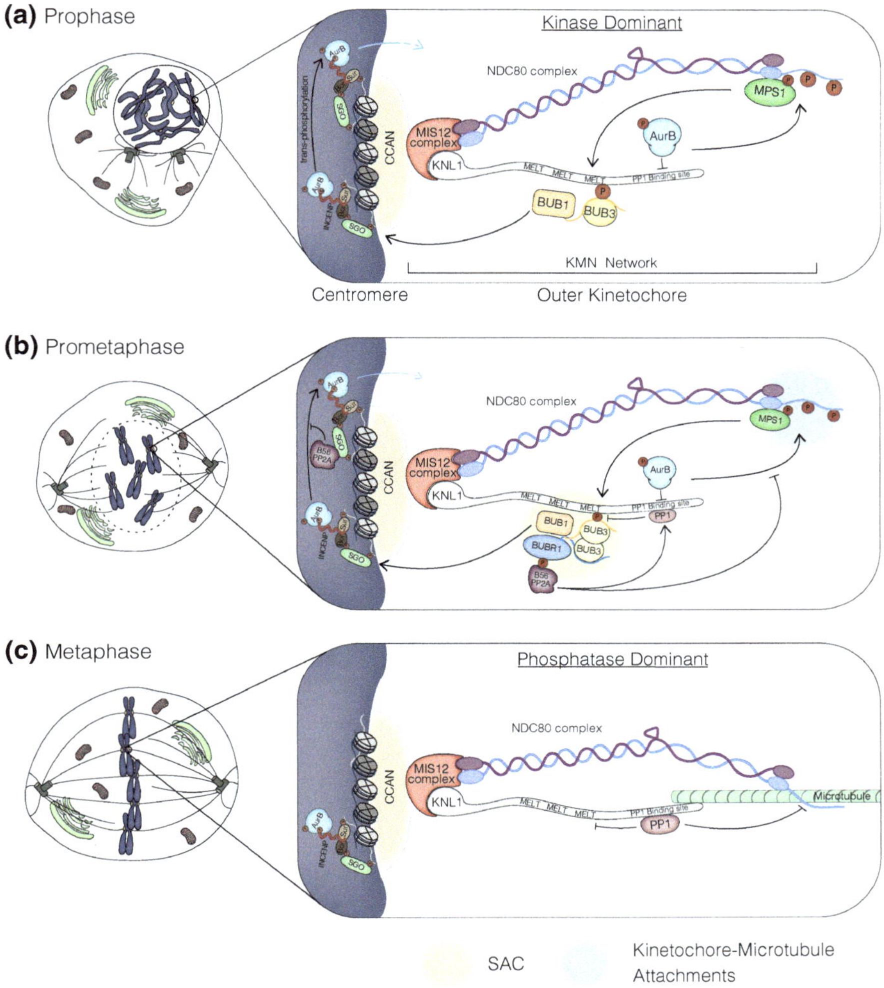

Fig. 1 Spatiotemporal control of signalling on the KMN network. Schematic to describe how a network of kinases and phosphatases regulate the SAC and kinetochore-microtubule attachments **a** In prophase, the kinase activities of MPS1 and Aurora B are unopposed and key signalling events are initiated **b** In prometaphase, on unattached kinetochores, PP2A-B56 is recruited via BUBR1, which begins to antagonise Aurora B to recruit PP1 to KNL1. PP2A-B56 also antagonises Aurora B to promote kinetochore-microtubule attachments, whereas PP1 antagonises MPS1 to limit KNL1-MELT phosphorylation and prime the SAC for rapid silencing **c** Following microtubule attachment at metaphase, PP1 reverses KNL1-MELT phosphorylation and antagonises Aurora B to stabilise kinetochore fibres

The first clues that MPS1 was an upstream component in the checkpoint signalling pathway came with the finding that overexpression of MPS1 alone is sufficient to cause a checkpoint-dependent arrest in *Schizosaccharomyces pombe* in the absence of spindle perturbations (Hardwick et al. 1996). MPS1 was subsequently shown to localise to unattached kinetochores (Abrieu et al. 2001; Castillo et al. 2002; Fisk and Winey 2001; Stucke et al. 2002) where it auto-phosphorylates on key residues in the activation loop and $P + 1$ loop (Kang et al. 2007; Mattison et al. 2007) in *trans* following dimerization (Hewitt et al. 2010). MPS1 kinetochore localisation is critical for SAC activity because N-terminal truncations abolish both kinetochore recruitment and the SAC, and artificially rescuing the localisation is sufficient to reactivate the SAC (Heinrich et al. 2012; Nijenhuis et al. 2013).

It is important to note that this kinetochore localisation is dynamic and MPS1 rapidly exchanges with the cytosol throughout mitosis (Howell et al. 2004; Jelluma et al. 2010). There is no known function of MPS1 in the cytosol, therefore it is tempting to speculate that the activity of MPS1 may be confined to the kinetochore. For example, auto-inhibitory mechanisms may be relieved upon kinetochore binding or active MPS1 may be susceptible to dephosphorylation by cytosolic phosphatases. The development of FRET-based reporters of MPS1 activity would help to shed light on exactly when and where MPS1 is active during mitosis.

When kinetochore-microtubule attachments form, the SAC needs to be silenced and a key event is the removal of MPS1 from kinetochores. Tethering MPS1 to kinetochores, by fusing to the outer kinetochores proteins MIS12 (Jelluma et al. 2010; Heinrich et al. 2012) or NDC80 (Ito et al. 2012), maintains MPS1 on attached kinetochores and prevents SAC silencing. Furthermore, re-recruitment of MPS1 to kinetochores that have already silenced the SAC at metaphase rapidly re-establishes the SAC and arrests mitotic exit (Kuijt et al. 2014; Ballister et al. 2014). This demonstrates that MPS1 recruitment is a key upstream event in SAC signalling, which explains why MPS1 activity is required for the recruitment of all known SAC components to the kinetochore (Heinrich et al. 2012; Santaguida et al. 2010; Kwiatkowski et al. 2010; Maciejowski et al. 2010; Hewitt et al. 2010; Sliedrecht et al. 2010). Furthermore, the fact that loss or gain of kinetochore MPS1 is sufficient to silence or activate the SAC, suggested that MPS1 could be the key "sensor" responsible for detecting kinetochore-microtubule attachment status and relaying this information to the spindle checkpoint machinery. The underlying basis for this sensing mechanisms was recently solved.

MPS1 binds directly to the NDC80 complex, the key microtubule attachment site at the outer kinetochore, via the Calponin Homology (CH) domains of NDC80 and NUF2 (Hiruma et al. 2015; Ji et al. 2015). One or both of these CH domains also bind to microtubules (Wilson-Kubalek et al. 2008; Alushin et al. 2010; Ciferri et al. 2008; Sundin et al. 2011), and the MPS1-NDC80 interaction is inhibited in the presence of microtubules in vitro (Hiruma et al. 2015; Ji et al. 2015). The prediction, therefore, is that competition between MPS1 and microtubules is likely to contribute to SAC silencing following stable kinetochore-microtubule attachment in vivo. It is important to note that although this mechanism was uncovered in humans, it may not have been conserved throughout evolution. MPS1 does not need to be removed from kinetochores to silence the SAC in budding yeast. Instead, a change in kinetochore structure following microtubule attachment is sufficient to spatially restrict access of MPS1 to its key kinetochore substrate, KNL1 (Aravamudhan et al. 2015; Joglekar 2016). In fact, others have argued that this may also be how MPS1 signalling is silenced in human cells (Musacchio 2015). The argument that it could at least contribute is valid and warrants further investigation.

An additional mechanism that could contribute to SAC silencing relates to inhibition of any one of the upstream inputs needed for MPS1-NDC80 interaction. Phosphorylation of MPS1 or NDC80 (by either Aurora B, MPS1 or CDK1) has been shown to enhance MPS1-NDC80 binding in vitro (Hiruma et al. 2015; Ji et al. 2015). Therefore, if phosphorylation is required for MPS1 kinetochore localisation in vivo, then a reduction in any one of these inputs following microtubule attachment/tension could contribute to MPS1 removal and SAC silencing. In this regard, Aurora B activity could well be the most relevant, because it is required for the localisation of MPS1 to kinetochores in the absence of microtubules (Jelluma et al. 2010; Santaguida et al. 2010; Saurin et al. 2011), and tension across kinetochores is known to reduce localised Aurora B activity (Liu et al. 2009). Furthermore, Aurora B-mediated phosphorylation of the NDC80 tail region, which is lost as microtubules make stable attachments with kinetochores, has been directly implicated in MPS1-NDC80 interaction in vitro (Ji et al. 2015) and MPS1 localisation in vivo (Zhu et al. 2013). Others have questioned the validity of the in vivo data (Etemad and Kops 2016), therefore, it will be important to clarify whether Aurora B controls MPS1 localisation directly in vivo and, if so, whether this occurs via NDC80 phosphorylation or alternative mechanisms.

The importance of MPS1 activity for its own localisation is still puzzling: MPS1 auto-phosphorylation enhances NDC80 interaction in vitro (Hiruma et al. 2015), and yet MPS1 inhibition increases kinetochore accumulation in cells (Jelluma et al. 2010). The reason for this difference is not clear, but the enhanced localisation in cells may result from feedback that is missing in vitro. A potential explanation could be provided by the recent discovery that ARHGEF17 controls MPS1 kinetochore localisation (Marquardt and Fisk 2016). MPS1 phosphorylates ARGEF17 to drive its own release from kinetochores, therefore MPS1 inhibition preserves kinetochore localisation in cells. It will be important in future to determine how ARGEF17 targets MPS1 to kinetochores (i.e. via NDC80 or not) and whether this is regulated directly by Aurora B.

The initiation of SAC downstream of MPS1 will be discussed in detail in an accompanying Chap. Molecular mechanisms of spindle assembly checkpoint activation and silencing. It is also the subject of some excellent recent reviews (Joglekar 2016; Lischetti and Nilsson 2015; Musacchio 2015). To understand the remainder of this chapter, however, it is important to state that a key event downstream of MPS1 is the phosphorylation of KNL1 on "MELT" repeats (Fig. 1b) (Yamagishi et al. 2012; London et al. 2012; Shepperd et al. 2012). This KNL1 phosphorylation initiates a cascade of events that culminate in the assembly of the mitotic checkpoint complex, which can diffuse throughout the cytoplasm to inhibit APC-CDC20 and prevent mitotic exit (Izawa and Pines 2015).

3 Aurora B: The Error-Correction Kinase

Aurora B is a member of the Aurora family of serine/threonine protein kinases, originally discovered in yeast (Chan and Botstein 1993) and later found to regulate chromosome segregation in several species including human (Glover et al. 1995; Gopalan et al. 1997; Kimura et al. 1997; Biggins et al. 1999; Hauf et al. 2003; Ditchfield et al. 2003). The main role of Aurora B in chromosome segregation is to regulate attachments between kinetochores and microtubules (Tanaka et al. 2002; Biggins et al. 1999; Hauf et al. 2003; Lampson et al. 2004); although it also has important roles in regulating sister chromatid cohesion (Losada et al. 2002; Nishiyama et al. 2013; Kim et al. 2013; Dai et al. 2006; Gimenez-Abian et al. 2004), assembling the outer kinetochore (Dimitrova et al. 2016; Petrovic et al. 2016; Kim and Yu 2015; Rago et al. 2015), and regulating the SAC (as discussed later) (Maldonado and Kapoor 2011; Santaguida et al. 2011; Saurin et al. 2011; Vader et al. 2007).

The principle behind its role in regulating attachments is that Aurora B can selectively destabilise microtubules from kinetochores that are incorrectly attached, whilst leaving correctly attached microtubules intact (Fig. 2). Therefore, although the chromosome attachment process is error-prone, the presence of Aurora B allows these errors to be corrected in iterative cycles of detachment and attachment until tension is achieved and kinetochore fibres remain stably bound. This immediately raises two important questions: how does Aurora B detach kinetochore-microtubules and how is that detachment process regulated by tension?

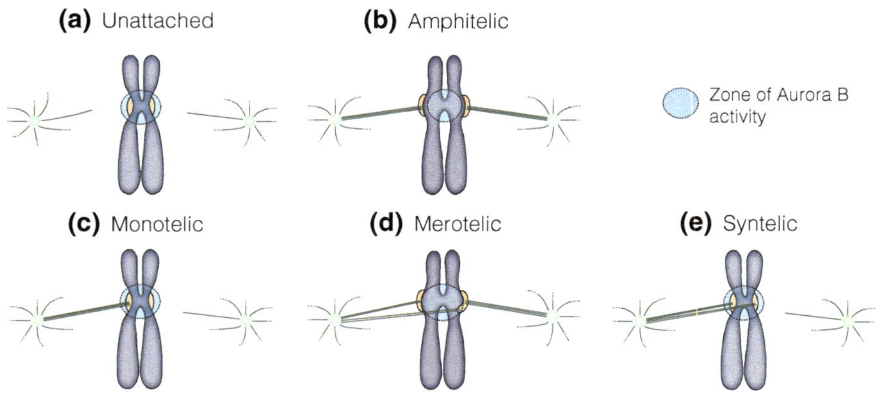

Fig. 2 The different types of kinetochore-microtubule attachments regulated during error-correction **a** A zone of Aurora B activity encompasses unattached kinetochores to destabilise any kinetochore-microtubule attachments that form **b** Following bipolar attachment, tension across the kinetochore stabilises microtubules by restricting Aurora B from phosphorylating key kinetochore substrates **c–e** The various types of microtubule attachments that do not generate sufficient tension and can therefore be destabilised by Aurora B

The mechanism by which Aurora B destabilises microtubule attachments is now well understood: it phosphorylates and inhibits key proteins at the outer kinetochore that are involved in stabilising those attachments. The NDC80 complex is one of the main outer kinetochore proteins that interacts with microtubules, via its CH domain and N-terminal tail, and Aurora B phosphorylates up to nine residues within this tail region to electrostatically interfere with microtubule binding (Miller et al. 2008; Guimaraes et al. 2008; Ciferri et al. 2008; DeLuca et al. 2006; Cheeseman et al. 2006; Alushin et al. 2010; Wei et al. 2007). In addition to simply binding microtubules, the kinetochore must also hold onto these microtubules as they depolymerise. This is believed to be dependent on two structurally unrelated, but functionally homologous, complexes in budding yeast (DAM1) or higher eukaryotes (SKA), both of which are phosphorylated by Aurora B to prevent microtubule interaction and/or kinetochore association (Asbury et al. 2006; Westermann et al. 2006; Welburn et al. 2009; Tanaka et al. 2007; Raaijmakers et al. 2009; Hanisch et al. 2006a; Gaitanos et al. 2009; Cheeseman et al. 2002; Chan et al. 2012; Schmidt et al. 2010). Finally, Aurora B also phosphorylates additional outer kinetochore components (Welburn et al. 2010; Hua et al. 2011), microtubule-binding proteins (Iimori et al. 2016), and microtubule-depolymerising kinesins (Zhang et al. 2007; Ohi et al. 2004; Knowlton et al. 2009; Lan et al. 2004; Andrews et al. 2004) to fine-tune the stability of microtubule attachments. Therefore, Aurora B phosphorylates multiple targets at the outer kinetochore to destabilise microtubule attachments, but why does it do this preferentially for kinetochore-microtubules that fail to generate tension? The short answer is that the activity of Aurora B is higher at these tensionless kinetochores (Liu et al. 2009). To explain the theories as to why, we must first explain how Aurora B activity and localisation is controlled.

Aurora B is part of the chromosome passenger complex (CPC), which also contains the Inner Centromere Protein (INCENP), Survivin and Borealin (Jeyaprakash et al. 2007). The Survivin and Borealin subunits are anchored to the N-terminus of INCENP, whereas the catalytic activity of the complex is provided by the Aurora B subunit bound to the C-terminus of INCENP (Krenn and Musacchio 2015). Activation of this catalytic subunit is controlled by multiple steps. First, a short region in the C-terminus of INCENP (termed the IN-box) binds near the active site to stimulate low levels of kinase activity. This allows subsequent auto-phosphorylation of the IN-box (on a TSS motif) and Aurora B itself (within the activation loop), which stimulates full activity of the complex (Sessa et al. 2005; Honda et al. 2003; Bishop and Schumacher 2002, Yasui et al. 2004). Auto-phosphorylation of at least the TSS motif is believed to occur in *trans* (Zaytsev et al. 2016; Sessa et al. 2005), which explains why artificial clustering of the CPC in vitro (Kelly et al. 2007) or in vivo (Wang et al. 2011a) activates Aurora B. This also explains why the endogenous pathways that promote CPC clustering in cells are so important for Aurora B activation.

At the start of mitosis the CPC initially clusters on chromatin before it concentrates at the centromere (Carmena et al. 2012b). This centromeric localisation pathway is controlled by multiple different kinase and phosphatase feedback loops that converge on key phospho-dependent interactions between histone tails and the Borealin or Survivin subunits of the CPC. In short, Histone H3-Thr3 (H3-T3) phosphorylation by Haspin creates a docking site for Survivin (Jeyaprakash et al. 2011; Kelly et al. 2010; Wang et al. 2010; Yamagishi et al. 2010; Niedzialkowska et al. 2012) and H2A-Thr120 (H2A-T120) phosphorylation by BUB1 allows the binding of Shugoshin proteins (SGO1 or SGO2), which also interact with CDK1-phosphorylated Borealin (or Survivin in fission yeast) (Kawashima et al. 2007, 2010; Yamagishi et al. 2010; Tsukahara et al. 2010). Aurora B activity reinforces its own localisation by activating Haspin (Ghenoiu et al. 2013; Wang et al. 2011b), promoting the kinetochore localisation of BUB1 (Vigneron et al. 2004), and inhibiting the chromatin localisation of the H3-T3 phosphatase RepoMan-PP1 (Qian et al. 2013). These pathways rapidly converge on the centromere because kinetochore-BUB1 phosphorylates H2A-T120 on histone tails in the centromeric region, and the H3-T3 kinase Haspin binds cohesin rings (Yamagishi et al. 2010), which are themselves concentrated at the centromere by a pathway involving BUB1/H2A-T120/SGO1 in human cells (Haarhuis et al. 2014). In addition, the inner centromeric recruitment of Haspin has recently been shown to be regulated by interaction with a SUMOylated form of Topoisomerase II (Edgerton et al. 2016; Yoshida et al. 2016).

The role of this centromeric CPC recruitment is to stimulate Aurora B activation by promoting *trans*-auto-phosphorylation, and to position this active pool close enough to the outer kinetochore to regulate microtubule attachment. The fact that Aurora B is "close" but not "at" the outer kinetochore is critical for the ability of the CPC to sense tension. The basic principle here is that the active pool of Aurora B is in range of its key substrates at kinetochores, but when tension is applied by microtubules, these substrates are pulled away from this zone of activity (Tanaka et al. 2002; Liu et al. 2009). Although the importance of the centromeric pool of Aurora B has recently been questioned (Campbell and Desai 2013), the principle of "spatial separation" is still a widely accepted model for tension-sensing. Exactly how this process works, however, is still very much a matter of debate (Krenn and Musacchio 2015; Lampson and Cheeseman 2011).

The debate principally concerns how Aurora B reaches its substrates from the centromere and why this is restricted by the tension exerted by microtubules. Aurora B could potentially reach from its anchor point at the centromere, assuming INCENP acts as a flexible linker to provide the required length. In this "leash" model, small changes in distance could physically pull the kinetochore out of reach of the Aurora B subunit (Santaguida and Musacchio 2009; Maresca and Salmon 2010). A simple prediction is that reducing the length of the leash should disturb error correction, however chromosome alignment is unaffected by deletion of the INCENP coiled-coil region (Vader et al. 2007). Alternatively, Aurora B could defuse away from the centromere to reach its outer kinetochore targets (Wang et al. 2011a). However, as pointed out previously by others (Krenn and Musacchio 2015), it is unlikely that simple diffusion alone could account for the very sharp cut-off in Aurora B activity that occurs within the 100 nm length scale of the kinetochore. At least part of the answer could be explained by the presence of phosphatases to sharpen this gradient near kinetochores.

Using a coupled in vitro kinase-phosphatase system, Aurora B was shown to exist in distinct high and low activity states, which may contribute to the required switch-like behaviour at kinetochores (Zaytsev et al. 2016). The role of phosphatase regulation in this regard is perhaps particularly important. Both PP1 and PP2A-B56 localise to the outer kinetochore and their regulated localisation and/or activity could contribute to the inactivation of kinetochore Aurora B or its substrates following microtubule attachment/tension (as discussed later). In fact, the possibility that tension may alter kinetochore phosphatase activity could potentially unite two camps that currently disagree over the tension-sensing mechanism (Krenn and Musacchio 2015; Campbell and Desai 2013). The debate principally concerns whether the important tension-dependent changes are intrinsic or extrinsic to the kinetochore. If intrinsic phosphatase activation worked together with an extrinsic increase in distance (to produce a sharp Aurora B gradient near kinetochores, for example), then both mechanisms could in fact contribute to tension-sensing. The important distance changes may depend on centromere-kinetochore distance, as originally proposed, or on the distance between the outer kinetochore and other pools of Aurora B that may reside in or around kinetochores. Furthermore, these ideas need not be mutually exclusive either, since these "other pools" could also indirectly depend on centromeric Aurora B for activity: the CPC could potentially use interactions with microtubules to position itself near to kinetochores (Banerjee et al. 2014; Krupina et al. 2016; Campbell and Desai 2013), or it may interact with components within the kinetochore itself. Although the existence of these kinetochore interaction interfaces is still speculative, it is important to note that several groups have reported a significant pool of active Aurora B that resides in or near the outer kinetochore (using the phospho-Aurora B-Thr232 activation loop antibody) (Posch et al. 2010; DeLuca et al. 2011; Caldas et al. 2013). How this active pool is regulated is unknown, but it may require the N-terminus of KNL1 (Caldas et al. 2013) and/or CPC dimerization (Bekier et al. 2015). It will be important in future to clarify whether this truly reflects an active pool of Aurora B (and not simply a cross-reacting Aurora B substrate) and, if it does, to determine whether this active pool is regulated by tension.

Finally, it is also important to note that microtubule attachment/tension is likely to feedback onto Aurora B and regulate its localisation to the centromere. Unaligned kinetochores have higher centromeric Aurora B levels than aligned kinetochores, which can be explained by the tension-sensitive removal of SGO1/2 due to the separation of kinetochore BUB1 away from its centromeric substrate Histone H2A (Salimian et al. 2011; Tanno et al. 2015; Nerusheva et al. 2014). Therefore, tension is likely to reduce the zone of Aurora B activity that originates from centromeres, in addition to separating the outer kinetochore away from this zone of activity (Fig. 2).

4 The Antagonising Phosphatases

Having discussed the main kinases that regulate the SAC and kinetochore-microtubule attachments, we will now discuss the phosphatases that antagonise these kinase inputs. The first evidence that a phosphatase was required for chromosome segregation came from genetic screens in fission yeast, which implicated protein phosphatase 1 (PP1) homologues in chromosome disjunction and mitotic exit (Ohkura et al. 1988, 1989; Booher and Beach 1989). Clues as to the molecular targets of PP1 came from subsequent work in budding yeast that showed how mutation of *glc7*, the PP1 homologue, can partially suppress mutation in *ipl1*, the Aurora B homologue (Francisco et al. 1994). This antagonism between Aurora B and PP1 was later demonstrated for the Aurora B substrate on chromatin, Histone-H3 (pSer10) (Murnion et al. 2001; Hsu et al. 2000). Although this explained why the chromatin-association of PP1 peaks during anaphase, when the bulk of Histone H3 is rapidly dephosphorylated, it did not explain why inhibiting PP1 caused a mitotic arrest much earlier in prometaphase. Furthermore, this mitotic arrest was associated with unstable kinetochore-microtubule attachments and persistent SAC activation, implying that the relevant PP1 substrates may reside at the kinetochore (Sassoon et al. 1999; Bloecher and Tatchell 1999). PP1 was subsequently confirmed as a kinetochore-localised phosphatase that peaks during metaphase, when kinetochore substrates that regulate either microtubule attachment or the SAC need to be dephosphorylated (Trinkle-Mulcahy et al. 2003, 2006; Alvarez-Tabares et al. 2007; Liu et al. 2010). Not surprisingly, PP1 was soon implicated in both of these processes.

The regulation of the SAC appears to be principally centred on the localised recruitment of PP1 to KNL1, which is a hub for SAC signalling at the kinetochore (Caldas and DeLuca 2014; Ghongane et al. 2014). PP1 associates with the N-terminus of KNL1 via short linear motifs (SILK and RVSF) (Liu et al. 2010; Rosenberg et al. 2011; Meadows et al. 2011; Espeut et al. 2012), which mediate the targeting of PP1 to many different regulatory subunits (Meiselbach et al. 2006; Hendrickx et al. 2009; Wakula et al. 2003; Egloff et al. 1997). This KNL1 interaction positions PP1 next to the MELT repeats, which are phosphorylated by MPS1 to establish SAC signalling at kinetochores (Yamagishi et al. 2012; London et al. 2012; Shepperd et al. 2012). This pool of PP1 is critical for dephosphorylating these MELT repeats and silencing the SAC in many different species, including humans (Pinsky et al. 2009; Rosenberg et al. 2011; Meadows et al. 2011; Espeut et al. 2012; Vanoosthuyse and Hardwick 2009; Nijenhuis et al. 2014). Although this KNL1-PP1 complex is the best validated PP1 complex at kinetochores, it is important to note that PP1 does bind to other kinetochore proteins as well (Hafner et al. 2014; Akiyoshi et al. 2009; Meadows et al. 2011; De Wever et al. 2014; Tang and Toda 2015; Kim et al. 2010; Sivakumar et al. 2016), some of which have also been implicated in SAC silencing (Meadows et al. 2011; Tang and Toda 2015; Sivakumar et al. 2016). It is also important to point out however, that in human cells at least, these other pools of PP1 cannot silence the SAC when MPS1 is inhibited in the absence of microtubules, because specific mutation of the RVSF motif in KNL1 abolishes SAC silencing under these conditions (Nijenhuis et al. 2014). However, the fact that some of these alternative pools of PP1 are delivered to kinetochores by microtubules suggests that they could aid in recruiting PP1 and silencing the SAC following microtubule attachment.

These data demonstrate that PP1 antagonises MPS1 to shut down the SAC, but does it also play a role in opposing Aurora B at the kinetochore? The early data in yeast certainly implies that it does, because here PP1 balances Aurora B activity to allow chromosome segregation (Francisco et al. 1994; Sassoon et al. 1999; Bloecher and Tatchell 1999; Pinsky et al. 2006). In human cells, however, although KNL1-PP1 does antagonise Aurora B at kinetochores, it only appears to do this following biorientation of those kinetochores (i.e. upon alignment to the metaphase plate) (Liu et al. 2010). Inhibiting KNL1-PP1 interaction destabilises established kinetochore fibres at metaphase, but it does not cause obvious defects in chromosome alignment. This implies that the main role of KNL1-PP1 is to keep Aurora B activity suppressed on kinetochores with bipolar (amphitelic) microtubule attachments. Aurora B activity also needs to be counteracted earlier to allow initial microtubule attachments to form. However, in human cells at least, this is the job of a secondary phosphatase—protein phosphatase 2A-B56 (PP2A-B56).

PP2A-B56 localises to unattached kinetochores via its regulatory B56 subunit, and siRNA-mediated depletion of all B56 isoforms causes a severe defect in kinetochore-microtubule attachment (Foley et al. 2011). Although PP2A-B56 has a well-established role in preserving cohesion at centromeres (Marston 2015), this attachment phenotype was unrelated to premature sister chromatid splitting, perhaps because siRNA-mediated depletion is unable to reduce total PP2A-B56 levels below the threshold required to maintain cohesion. Instead, the phenotype of PP2A-B56 depletion is associated with elevated kinetochore Aurora B activity, and inhibition of Aurora B is sufficient to rescue kinetochore-microtubule attachments (Foley et al. 2011). This was reminiscent of the phenotype seen following BUBR1 depletion (Lampson and Kapoor 2005), and subsequent work explained exactly why PP2A-B56 and BUBR1 are functionally linked.

PP2A-B56 localises to kinetochores by binding directly to BUBR1 on its kinetochore attachment regulatory domain (KARD) (Suijkerbuijk et al. 2012; Kruse et al. 2013; Xu et al. 2013). This interaction is strengthened by CDK1- and PLK1-mediated phosphorylation of the KARD, although recent structural and biochemical work suggests that CDK1 phosphorylation is the most functionally important (Hertz et al. 2016; Wang et al. 2016). Inhibiting kinetochore-PP2A-B56 directly, by mutating or deleting the KARD, elevates kinetochore Aurora B activity and disrupts kinetochore-microtubule attachments, which can be rescued by inhibiting Aurora B (Suijkerbuijk et al. 2012; Xu et al. 2013). The conclusion, therefore, is that PP2A-B56 is needed at kinetochores to suppress Aurora B activity and allow initial kinetochores-microtubules attachments to form. Once these attachments have formed correctly, PP2A-B56 is removed and PP1 is recruited, which could explain why PP1 is important to continue inhibiting Aurora B at this time.

These data therefore suggest that kinetochore PP1 and PP2A-B56 can be separated by time, if not by function. PP2A-B56 is needed to balance Aurora B activity on kinetochores prior to their alignment, whereas PP1 takes over to switch off the SAC and stabilise attachments after alignment. This handover is controlled by a switch in the phosphorylation state of the relevant recruitment motifs. The SILK and RVSF motifs in KNL1 are phosphorylated by Aurora B to inhibit PP1 binding (Liu et al. 2010), whereas, as discussed before, the KARD domain in BUBR1 is phosphorylated by CDK1/PLK1 to enhance PP2A-B56 interaction (Suijkerbuijk et al. 2012; Kruse et al. 2013; Xu et al. 2013; Hertz et al. 2016; Wang et al. 2016). Furthermore, BUBR1 itself is recruited to kinetochores by MPS1-dependent MELT phosphorylation (Yamagishi et al. 2012; London et al. 2012; Shepperd et al. 2012). Therefore, on unattached kinetochores, kinases help to elevate PP2A-B56 (MPS1, PLK1 and CDK1) and inhibit PP1 (Aurora B). Following stable microtubule attachment, however, many of these activities are lost, which removes kinetochore-PP2A and increases PP1. As discussed previously, this rise in kinetochore PP1 could also be aided by the localised delivery of extra pools of PP1 down microtubules (Hafner et al. 2014; Meadows et al. 2011; De Wever et al. 2014; Tang and Toda 2015; Kim et al. 2010; Sivakumar et al. 2016).

A final issue regarding phosphatase regulation, which is still relatively poorly understood, is whether catalytic activity can be regulated directly. In this regard, CDK1 activity may be particularly important. CDK1 phosphorylation has been shown to inhibit PP1 isoforms directly (Yamano et al. 1994; Kwon et al. 1997; Wu et al. 2009; Dohadwala et al. 1994; Grallert et al. 2015) and activate a kinetochore inhibitor of PP2A-B56 (Porter et al. 2013). CDK1 specifically localises to unattached kinetochores (Chen et al. 2008), therefore its removal upon microtubule attachment could contribute to the activation of kinetochore phosphatases at this time. It will be important in future to determine whether CDK1 activity is regulated at kinetochores and, if so, whether this impacts on localised phosphatase activation.

5 Signalling Cross-Talk at Kinetochores

5.1 *Feedback Between Kinases*

So far, we have presented two linear pathways to explain how the SAC is activated by MPS1 and how kinetochore-microtubule attachments are regulated by Aurora B. MPS1 phosphorylates KNL1 to recruit, directly or indirectly, all SAC components to the kinetochore. Aurora B phosphorylates outer kinetochore proteins to destabilise microtubule attachments. This explains why the SAC is particularly sensitive to MPS1 inhibition, whereas chromosomal alignment is particularly sensitive to Aurora B inhibition. This is an over-simplified picture, however, and there is considerable evidence of cross-talk between these two pathways. MPS1 knockdown or inhibition causes delays in chromosome alignment, which are associated with defective kinetochore-microtubule attachments (Jelluma et al. 2008;

Kwiatkowski et al. 2010; Maciejowski et al. 2010; Maure et al. 2007; Santaguida et al. 2010; Sliedrecht et al. 2010). Aurora B knockdown or inhibition causes defects in the SAC, even in the absence of microtubules, which suggests that Aurora B can activate the SAC directly (Maldonado and Kapoor 2011; Vader et al. 2007; Saurin et al. 2011; Santaguida et al. 2011). It is important to note that these phenotypes are considerably milder than those observed following inhibition of the *bona fide* upstream kinase. For example, Aurora B inhibition only sensitises the SAC (Saurin et al. 2011; Santaguida et al. 2011), whereas MPS1 inhibition overrides it completely (Hewitt et al. 2010; Kwiatkowski et al. 2010; Maciejowski et al. 2010; Santaguida et al. 2010; Sliedrecht et al. 2010). MPS1 inhibition only delays chromosome alignment (Hewitt et al. 2010), whereas Aurora B inhibition abolishes it completely (Ditchfield et al. 2003; Hauf et al. 2003). The obvious conclusion, therefore, is that MPS1 and Aurora B principally regulate the SAC and microtubule attachments, respectively, but they also have secondary effects that impinge on each other's pathways. What then, could be the molecular basis for this cross-talk?

As discussed previously, the localisation of Aurora B and MPS1 to centromeres and kinetochores is required for their initial activation during mitosis. Aurora B activity is needed to recruit MPS1 to kinetochores, and inhibiting Aurora B causes delays in activating MPS1 and establishing the SAC (Jelluma et al. 2010; Saurin et al. 2011; Santaguida et al. 2011). Similarly, MPS1 activity is needed to recruit Aurora B to centromeres, and MPS1 inhibition delays Aurora B activation and induces defects in kinetochore-microtubule attachment (van der Waal et al. 2012). Therefore, both kinases function in a positive feedback loop that contributes to their switch-like activation upon mitotic entry (Fig. 3a). This feedback also underlies much of the cross-talk between these two pathways, because uncoupling the feedback also uncouples the cross-talk. For example, recovering MPS1 at kinetochores (using a MIS12 fusion) bypasses any requirement for Aurora B activity in MPS1 activation and the SAC (Jelluma et al. 2010; Saurin et al. 2011). Conversely, recovering Aurora B at centromeres (using a CENPB-INCENP fusion) rescues most of the chromosome alignment defects seen following MPS1 inhibition (van der Waal et al. 2012).

These data suggest that the primary function of Aurora B in the SAC is to localise MPS1 to kinetochores, or alternatively, if Aurora B has additional functions, then these can be bypassed by artificial MPS1 recruitment (as discussed later). Furthermore, defective kinetochore-microtubule attachment upon MPS1 inhibition can be at least partially explained by the mislocalisation of Aurora B. There may well be additional effects of MPS1 inhibition, because restoring Aurora B to centromeres causes an incomplete, albeit substantial, rescue in chromosome alignment (van der Waal et al. 2012). This could reflect abnormal levels or turnover of Aurora B at CENP-B (i.e. Aurora B is rescued at centromeres but not to the exact wild type situation). Alternatively, MPS1 could have additional effects that are either independent of Aurora B activity (Maure et al. 2007; Espeut et al. 2008; Storchova et al. 2011) or Aurora B localisation (Bourhis et al. 2009; Jelluma et al. 2008). These additional effects may become more dominant when MPS1 is inhibited after mitotic entry, because this has minimal impact on Aurora B

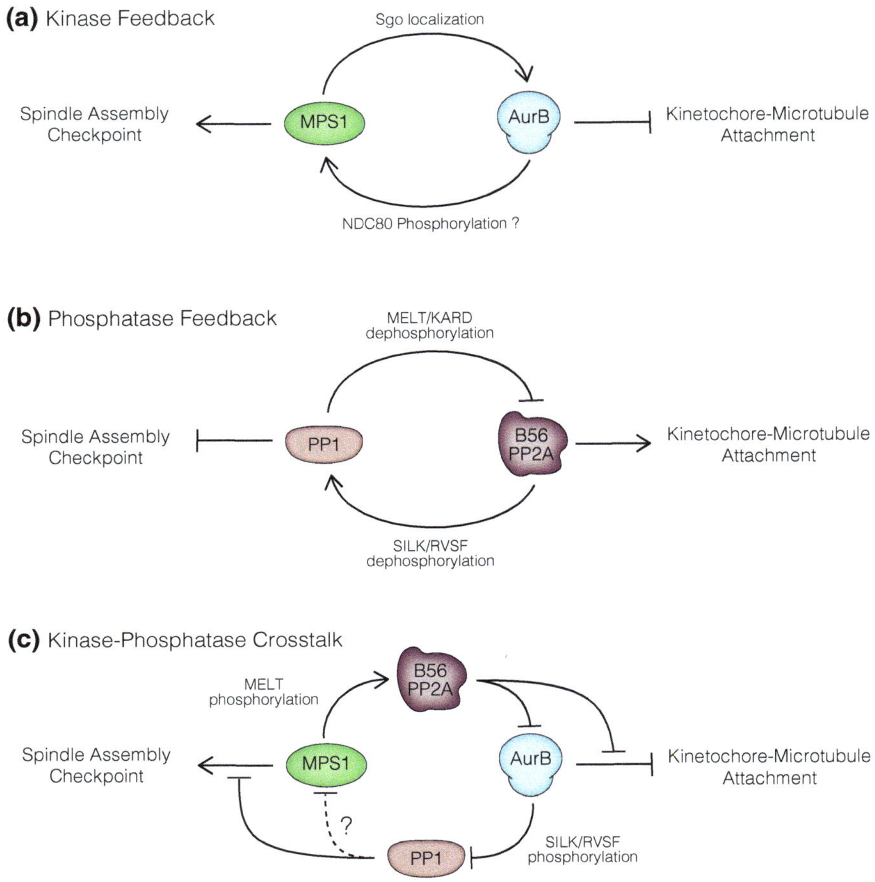

Fig. 3 Cross-talk between kinases and phosphatases at the kinetochore **a** A positive feedback loop that controls MPS1 and Aurora B localisation **b** A negative feedback loop that controls PP1 and PP2A-B56 localisation **c** Cross-talk between both kinases and both phosphatases at the kinetochore See text for full details of this cross-talk

localisation (van der Waal et al. 2012), but still causes significant defects in kinetochore-microtubule attachment (Sliedrecht et al. 2010; Hewitt et al. 2010).

5.2 Feedback Between Phosphatases

The kinetochore recruitment of PP1 and PP2A-B56 is regulated by multiple kinases: phosphorylation enhances kinetochore-PP2A-B56 and decreases kinetochore-PP1, therefore, by inference, dephosphorylation must decrease PP2A-B56 and increase PP1. This regulation is also critically dependent on cross-talk. PP1 dephosphorylates the KNL1-MELT and BUBR1-KARD motifs to decrease PP2A-B56, whereas PP2A-B56 dephosphorylates the KNL1-SILK and RVSF motifs to increase PP1 (Fig. 3b) (Nijenhuis et al. 2014). Therefore, both phosphatases are engaged in a negative feedback loop that regulates their own localisation to kinetochores. It is therefore important to be cautious when interpreting results following direct phosphatase inhibition, because inhibiting kinetochore PP2A-B56 also inhibits KNL1-PP1, whereas inhibiting KNL1-PP1 increases kinetochore PP2A-B56. Therefore, to untangle direct from indirect effects, it is important to compare the effects of inhibiting *both* phosphatases individually. This is best illustrated using the example of SAC regulation in human cells.

As discussed earlier, PP1 is well known to silence the SAC in many different species, from yeast to worms (Pinsky et al. 2009; Rosenberg et al. 2011; Meadows et al. 2011; Espeut et al. 2012; Vanoosthuyse and Hardwick 2009). Therefore, the discovery that PP2A-B56 is essential for SAC silencing in humans, on the fact of it, appears surprising (Espert et al. 2014; Nijenhuis et al. 2014). However, a close comparison of mutants that interfere with either kinetochore PP2A-B56 or KNL1-PP1 demonstrated that both phosphatase complexes are essential (Nijenhuis et al. 2014). In fact, the phenotypes are identical, except for one crucial difference: in both scenarios KNL1-PP1 is lost, but when targeting KNL1-PP1 directly, kinetochore PP2A-B56 is actually increased (due to the feedback, see Fig. 3b). Therefore, this PP2A-B56 pool cannot silence the SAC, which confirms that KNL1-PP1 is the direct SAC phosphatase in human cells, as it is in lower eukaryotes. PP2A-B56 is only essential to antagonise Aurora B and recruit PP1, because if Aurora B is inhibited then PP2A-B56 is no longer required (Nijenhuis et al. 2014).

In summary, the fact that these two phosphatases control each other's localisation to kinetochores makes it very difficult to tease apart their individual effects. It is particularly important to test carefully whether the effects of PP2A-B56 loss occurs directly, or indirectly, via PP1. Considering that PP1 and PP2A-B56 have both been shown to balance Aurora B and regulate kinetochore-microtubule attachments, it will be important to address the possible role of cross-talk here as well.

Finally, the phosphatase cross-talk discussed so far is based entirely on localisation, even though, as mentioned previously, both phosphatases can be inhibited directly by phosphorylation. It is possible therefore, that they could also reverse these phosphorylations and activate each other directly. In this regard, the recent discovery that $PP1^{Dis2}$ and $B56^{Par1}$ are reactivated sequentially upon mitotic exit in fission yeast is particularly interesting (Grallert et al. 2015). In this situation, $PP1^{Dis2}$ is able to auto-activate and then activate $B56^{Par1}$, both via dephosphorylation. It is hard to reconcile these findings with the fact that both of these enzymes need to be active much earlier in mitosis to function at kinetochores. One possible explanation is that they are globally inhibited by phosphorylation, but then locally reactivated exactly when and where they are needed. In this scenario, kinetochore clustering may elevate phosphatase activity enough to permit localised phosphatase reactivation when required. It will be important in the future to determine whether these phosphatases are indeed inhibited directly during mitosis, and if so, whether they are reactivated in specific locations via cross-talk.

5.3 Kinase-Phosphatase Cross-Talk

Having discussed the feedback between pairs of kinases or phosphatases, we would like to now finish by highlighting the cross-talk that exists between all four of these enzymes. In particular, we would like to discuss how kinases could mediated their effects via the phosphatases (Fig. 3c). This is important because it is very easy to adopt a kinase-centric view of signalling, even though phosphatase activity could be equally important. To highlight this point, let us now consider the role of Aurora B in SAC signalling.

As discussed previously, Aurora B activity is required to localise MPS1 to kinetochores and establish the SAC (Jelluma et al. 2010; Saurin et al. 2011; Santaguida et al. 2011). Although the molecular mechanism is still unclear, Aurora B is thought to control kinetochore MPS1 by phosphorylating either NDC80, MPS1, or some other protein required for MPS1-NDC80 interaction. Aurora B is well known to inhibit the phosphatase that antagonises MPS1 and the SAC (KNL1-PP1) (Liu et al. 2010), therefore the input of Aurora B could be wholly, or at least partially, explained by the inhibition of this phosphatase complex (Fig. 3c). KNL1-PP1 antagonises MPS1 at the level of MELT phosphorylation, and it may also inhibit MPS1 localisation and/or activity directly. Rescuing MPS1 localisation to kinetochores (with a MIS12-fusion), which bypasses any requirement for Aurora B in the SAC (Saurin et al. 2011), could also potentially rescue all these phosphatase-mediated effects (especially if kinetochore-MPS1 levels are artificially high). It will therefore be important in future to determine just how much of Aurora B's input into the SAC is mediated via KNL1-PP1.

It is important to also consider whether the effects of MPS1 inhibition could also be mediated, at least in part, via the phosphatases. MPS1 activity enhances the kinetochore localisation of PP2A-B56, via MELT phosphorylation, which can inhibit Aurora B substrates (Suijkerbuijk et al. 2012; Kruse et al. 2013; Xu et al. 2013) and Aurora B directly (Meppelink et al. 2015) (Fig. 3c). MPS1 also localises the acetyltransferase TIP60 to kinetochores, which is able to acetylate Aurora B and protect it from PP2A-mediated inactivation (Mo et al. 2016). It is therefore difficult to predict how MPS1 inhibition may impact on the PP2A-B56/Aurora B balance at kinetochores, but this is important to test because this balance is critical for regulating kinetochore-microtubule attachments.

5.4 Cross-Talk with Other Kinases

Finally, we have deliberately focussed this chapter on the principle kinases that regulate the SAC or microtubule attachment, simply to avoid confusion. Other kinases, however, are certainly important. PLK1 activity, for example, is needed to regulate the SAC (von Schubert et al. 2015; O'Connor et al. 2015) and kinetochore-microtubule attachments (Sumara et al. 2004; Hanisch et al. 2006b;

Peters et al. 2006; Lenart et al. 2007; Liu et al. 2012). It is also regulated by cross-talk with all four of the enzymes discussed above. PLK1 can phosphorylate MPS1 substrates directly (von Schubert et al. 2015; Espeut et al. 2015), MPS1 itself to inhibit kinetochore localisation (von Schubert et al. 2015), the KNL1-MELT and BUBR1-KARD motifs to recruit PP2A-B56 (Suijkerbuijk et al. 2012; Kruse et al. 2013; Xu et al. 2013; von Schubert et al. 2015), and the PP1 regulator SDS22 to inhibit the dephosphorylation and inactivation of Aurora B (Duan et al. 2016). Furthermore, PLK1 is itself activated and recruited to kinetochores by Aurora B (Shao et al. 2015; O'Connor et al. 2015; Carmena et al. 2012a) and inhibited and delocalised by PP2A-B56 (Foley et al. 2011). It will clearly be important in future to integrate PLK1 signalling into the complex network of other signals highlighted above.

6 Summary

We have learnt a great deal over the past few decades about how chromosome segregation is regulated. A key aspect of that concerns the kinetochore and, in particular, how this integrates the wide variety of signals needed to safeguard the microtubule attachment process. We now understand most of the enzymes that generate these signals, how these enzymes are regulated, and how they sense and transmit information to the chromosome segregation machinery. Perhaps all that remains is to learn how all this information is integrated together in a way that ensures chromosome segregation can proceed accurately and reliably. This "final frontier", however, may turn out to be the most challenging of all.

It is becoming increasingly clear that the more we learn about the connections between these signals the more confusing the situation becomes. The four main regulators highlighted here are embedded in a network that is, quite literally, entangled in cross-talk. It is hard for us to predict, even now, why all this cross-talk exists, so how will it feel when new connections are identified or, worse still, when new signals are integrated. The short section on PLK1 may provide a glimpse of the answer.

So, how will we rise to this challenge and what will it take to make the next big step? In the words of the late, great Richard Feynman: "What I cannot create, I do not understand." Or to put this another way, reductionist biology will only take us so far. Building the kinetochore may seem like a mammoth task, but there has been some remarkable recent progress in this area (Pesenti et al. 2016; Weir et al. 2016). There is also the alternative, but equally valid, approach of purifying kinetochores that could then be stripped of their regulatory components (Gonen et al. 2012; Akiyoshi et al. 2010). Reconstituting these in vitro systems with a defined set of enzymes would allow us to "create" the signalling network from scratch. If this could be allied with computational approaches to help us comprehend the systems-level behaviour in vitro, and then predict the behaviour in vivo, perhaps we will then be in a position to claim that we truly "understand".

References

Abrieu A, Magnaghi-Jaulin L, Kahana JA, Peter M, Castro A, Vigneron S, Lorca T, Cleveland DW, Labbe JC (2001) Mps1 is a kinetochore-associated kinase essential for the vertebrate mitotic checkpoint. Cell 106(1):83–93

Akiyoshi B, Nelson CR, Ranish JA, Biggins S (2009) Quantitative proteomic analysis of purified yeast kinetochores identifies a PP1 regulatory subunit. Genes Dev 23(24):2887–2899. doi:10.1101/gad.1865909

Akiyoshi B, Sarangapani KK, Powers AF, Nelson CR, Reichow SL, Arellano-Santoyo H, Gonen T, Ranish JA, Asbury CL, Biggins S (2010) Tension directly stabilizes reconstituted kinetochore-microtubule attachments. Nature 468(7323):576–579. doi:10.1038/nature09594

Alushin GM, Ramey VH, Pasqualato S, Ball DA, Grigorieff N, Musacchio A, Nogales E (2010) The Ndc80 kinetochore complex forms oligomeric arrays along microtubules. Nature 467 (7317):805–810. doi:10.1038/nature09423

Alvarez-Tabares I, Grallert A, Ortiz JM, Hagan IM (2007) Schizosaccharomyces pombe protein phosphatase 1 in mitosis, endocytosis and a partnership with Wsh3/Tea4 to control polarised growth. J Cell Sci 120(Pt 20):3589–3601. doi:10.1242/jcs.007567

Andrews PD, Ovechkina Y, Morrice N, Wagenbach M, Duncan K, Wordeman L, Swedlow JR (2004) Aurora B regulates MCAK at the mitotic centromere. Dev Cell 6(2):253–268

Aravamudhan P, Goldfarb AA, Joglekar AP (2015) The kinetochore encodes a mechanical switch to disrupt spindle assembly checkpoint signalling. Nat Cell Biol 17(7):868–879. doi:10.1038/ncb3179

Asbury CL, Gestaut DR, Powers AF, Franck AD, Davis TN (2006) The Dam1 kinetochore complex harnesses microtubule dynamics to produce force and movement. Proc Natl Acad Sci U S A 103(26):9873–9878. doi:10.1073/pnas.0602249103

Ballister ER, Riegman M, Lampson MA (2014) Recruitment of Mad1 to metaphase kinetochores is sufficient to reactivate the mitotic checkpoint. J Cell Biol 204(6):901–908. doi:10.1083/jcb.201311113

Banerjee B, Kestner CA, Stukenberg PT (2014) EB1 enables spindle microtubules to regulate centromeric recruitment of Aurora B. J Cell Biol 204(6):947–963. doi:10.1083/jcb.201307119

Bekier ME, Mazur T, Rashid MS, Taylor WR (2015) Borealin dimerization mediates optimal CPC checkpoint function by enhancing localization to centromeres and kinetochores. Nat Commun 6:6775. doi:10.1038/ncomms7775

Biggins S, Severin FF, Bhalla N, Sassoon I, Hyman AA, Murray AW (1999) The conserved protein kinase Ipl1 regulates microtubule binding to kinetochores in budding yeast. Genes Dev 13(5):532–544

Bishop JD, Schumacher JM (2002) Phosphorylation of the carboxyl terminus of inner centromere protein (INCENP) by the Aurora B Kinase stimulates Aurora B kinase activity. J Biol Chem 277(31):27577–27580. doi:10.1074/jbc.C200307200

Bloecher A, Tatchell K (1999) Defects in Saccharomyces cerevisiae protein phosphatase type I activate the spindle/kinetochore checkpoint. Genes Dev 13(5):517–522

Booher R, Beach D (1989) Involvement of a Type 1 protein phosphatase encoded by bwsl+ in fission yeast mitotic control. Cell 57:1009–1016

Bourhis E, Lingel A, Phung Q, Fairbrother WJ, Cochran AG (2009) Phosphorylation of a borealin dimerization domain is required for proper chromosome segregation. Biochemistry 48 (29):6783–6793. doi:10.1021/bi900530v

Caldas GV, DeLuca JG (2014) KNL1: bringing order to the kinetochore. Chromosoma 123 (3):169–181. doi:10.1007/s00412-013-0446-5

Caldas GV, DeLuca KF, DeLuca JG (2013) KNL1 facilitates phosphorylation of outer kinetochore proteins by promoting Aurora B kinase activity. J Cell Biol 203(6):957–969. doi:10.1083/jcb.201306054

Campbell CS, Desai A (2013) Tension sensing by Aurora B kinase is independent of surviving-based centromere localization. Nature 497(7447):118–121. doi:10.1038/nature12057

Carmena M, Pinson X, Platani M, Salloum Z, Xu Z, Clark A, Macisaac F, Ogawa H, Eggert U, Glover DM, Archambault V, Earnshaw WC (2012a) The chromosomal passenger complex activates Polo kinase at centromeres. PLoS Biol 10(1):e1001250. doi:10.1371/journal.pbio.1001250

Carmena M, Wheelock M, Funabiki H, Earnshaw WC (2012b) The chromosomal passenger complex (CPC): from easy rider to the godfather of mitosis. Nat Rev Mol Cell Biol 13 (12):789–803. doi:10.1038/nrm3474

Castillo AR, Meehl JB, Morgan G, Schutz-Geschwender A, Winey M (2002) The yeast protein kinase Mps1p is required for assembly of the integral spindle pole body component Spc42p. J Cell Biol 156(3):453–465. doi:10.1083/jcb.200111025

Chan CS, Botstein D (1993) Isolation and characterization of chromosome-gain and increase-in-ploidy mutants in yeast. Genetics 135(3):677–691

Chan YW, Jeyaprakash AA, Nigg EA, Santamaria A (2012) Aurora B controls kinetochore-microtubule attachments by inhibiting Ska complex-KMN network interaction. J Cell Biol 196(5):563–571. doi:10.1083/jcb.201109001

Cheeseman IM, Anderson S, Jwa M, Green EM, Kang J, Yates JR 3rd, Chan CS, Drubin DG, Barnes G (2002) Phospho-regulation of kinetochore-microtubule attachments by the Aurora kinase Ipl1p. Cell 111(2):163–172

Cheeseman IM, Chappie JS, Wilson-Kubalek EM, Desai A (2006) The conserved KMN network constitutes the core microtubule-binding site of the kinetochore. Cell 127(5):983–997. doi:10.1016/j.cell.2006.09.039

Chen Q, Zhang X, Jiang Q, Clarke PR, Zhang C (2008) Cyclin B1 is localized to unattached kinetochores and contributes to efficient microtubule attachment and proper chromosome alignment during mitosis. Cell Res 18(2):268–280. doi:10.1038/cr.2008.11

Ciferri C, Pasqualato S, Screpanti E, Varetti G, Santaguida S, Dos Reis G, Maiolica A, Polka J, De Luca JG, De Wulf P, Salek M, Rappsilber J, Moores CA, Salmon ED, Musacchio A (2008) Implications for kinetochore-microtubule attachment from the structure of an engineered Ndc80 complex. Cell 133(3):427–439. doi:10.1016/j.cell.2008.03.020

Dai J, Sullivan BA, Higgins JM (2006) Regulation of mitotic chromosome cohesion by Haspin and Aurora B. Dev Cell 11(5):741–750. doi:10.1016/j.devcel.2006.09.018

De Wever V, Nasa I, Chamousset D, Lloyd D, Nimick M, Xu H, Trinkle-Mulcahy L, Moorhead GB (2014) The human mitotic kinesin KIF18A binds protein phosphatase 1 (PP1) through a highly conserved docking motif. Biochem Biophys Res Commun 453(3):432–437. doi:10.1016/j.bbrc.2014.09.105

DeLuca JG, Gall WE, Ciferri C, Cimini D, Musacchio A, Salmon ED (2006) Kinetochore microtubule dynamics and attachment stability are regulated by Hec1. Cell 127(5):969–982. doi:10.1016/j.cell.2006.09.047

DeLuca KF, Lens SM, DeLuca JG (2011) Temporal changes in Hec1 phosphorylation control kinetochore-microtubule attachment stability during mitosis. J Cell Sci 124(Pt 4):622–634. doi:10.1242/jcs.072629

Dimitrova YN, Jenni S, Valverde R, Khin Y, Harrison SC (2016) Structure of the MIND complex defines a regulatory focus for yeast kinetochore assembly. Cell 167(4):1014–1027, e1012. doi:10.1016/j.cell.2016.10.011

Ditchfield C, Johnson VL, Tighe A, Ellston R, Haworth C, Johnson T, Mortlock A, Keen N, Taylor SS (2003) Aurora B couples chromosome alignment with anaphase by targeting BubR1, Mad2, and Cenp-E to kinetochores. J Cell Biol 161(2):267–280. doi:10.1083/jcb.200208091

Dohadwala M, Da Cruz EF, Da Cruz S, Hall FL, Williams RT, Carbonaro-Hall D, Nairn AC, Greengard P, Berndt N (1994) Phosphorylation and inactivation of protein phosphatase 1 by cyclin-dependent kinases. Proc Natl Acad Sci U S A 91:6408–6412

Duan H, Wang C, Wang M, Gao X, Yan M, Akram S, Peng W, Zou H, Wang D, Zhou J, Chu Y, Dou Z, Barrett G, Green HN, Wang F, Tian R, He P, Wang W, Liu X, Yao X (2016) Phosphorylation of PP1 regulator Sds22 by PLK1 ensures accurate chromosome segregation. J Biol Chem 291(40):21123–21136. doi:10.1074/jbc.M116.745372

Edgerton H, Johansson M, Keifenheim D, Mukherjee S, Chacon JM, Bachant J, Gardner MK, Clarke DJ (2016) A noncatalytic function of the topoisomerase II CTD in Aurora B recruitment to inner centromeres during mitosis. J Cell Biol 213(6):651–664. doi:10.1083/jcb.201511080

Egloff MP, Johnson DF, Moorhead G, Cohen PTW, Cohen P, Barford D (1997) Structural basis for the recognition of regulatory subunits by the catalytic subunit of protein phosphatase 1. EMBO J 16(8):1876–1887. doi:10.1093/emboj/16.8.1876

Espert A, Uluocak P, Bastos RN, Mangat D, Graab P, Gruneberg U (2014) PP2A-B56 opposes Mps1 phosphorylation of Knl1 and thereby promotes spindle assembly checkpoint silencing. J Cell Biol 206(7):833–842. doi:10.1083/jcb.201406109

Espeut J, Cheerambathur DK, Krenning L, Oegema K, Desai A (2012) Microtubule binding by KNL-1 contributes to spindle checkpoint silencing at the kinetochore. J Cell Biol 196(4):469–482. doi:10.1083/jcb.201111107

Espeut J, Gaussen A, Bieling P, Morin V, Prieto S, Fesquet D, Surrey T, Abrieu A (2008) Phosphorylation relieves autoinhibition of the kinetochore motor Cenp-E. Mol Cell 29(5):637–643. doi:10.1016/j.molcel.2008.01.004

Espeut J, Lara-Gonzalez P, Sassine M, Shiau AK, Desai A, Abrieu A (2015) Natural loss of Mps1 Kinase in nematodes uncovers a role for polo-like kinase 1 in spindle checkpoint initiation. Cell reports 12(1):58–65. doi:10.1016/j.celrep.2015.05.039

Etemad B, Kops GJ (2016) Attachment issues: kinetochore transformations and spindle checkpoint silencing. Curr Opin Cell Biol 39:101–108. doi:10.1016/j.ceb.2016.02.016

Fisk HA, Winey M (2001) The mouse Mps1p-like kinase regulates centrosome duplication. Cell 106(1):95–104

Foley EA, Maldonado M, Kapoor TM (2011) Formation of stable attachments between kinetochores and microtubules depends on the B56-PP2A phosphatase. Nat Cell Biol 13 (10):1265–1271. doi:10.1038/ncb2327

Francisco L, Wang W, Chan CSM (1994) Type 1 protein phosphatase acts in opposition to Ipll protein kinase in regulating yeast chromosome segregation. Mol Cell Biol 14(7):4731–4740

Gaitanos TN, Santamaria A, Jeyaprakash AA, Wang B, Conti E, Nigg EA (2009) Stable kinetochore-microtubule interactions depend on the Ska complex and its new component Ska3/C13Orf3. EMBO J 28(10):1442–1452. doi:10.1038/emboj.2009.96

Ghenoiu C, Wheelock MS, Funabiki H (2013) Autoinhibition and Polo-dependent multisite phosphorylation restrict activity of the histone H3 kinase Haspin to mitosis. Mol Cell 52 (5):734–745. doi:10.1016/j.molcel.2013.10.002

Ghongane P, Kapanidou M, Asghar A, Elowe S, Bolanos-Garcia VM (2014) The dynamic protein Knl1—a kinetochore rendezvous. J Cell Sci 127(Pt 16):3415–3423. doi:10.1242/jcs.149922

Gimenez-Abian JF, Sumara I, Hirota T, Hauf S, Gerlich D, de la Torre C, Ellenberg J, Peters JM (2004) Regulation of sister chromatid cohesion between chromosome arms. Curr Biol: CB 14 (13):1187–1193. doi:10.1016/j.cub.2004.06.052

Glover DM, Leibowitz MH, McLean DA, Parry H (1995) Mutations in aurora prevent centrosome separation leading to the formation of monopolar spindles. Cell 81(1):95–105

Gonen S, Akiyoshi B, Iadanza MG, Shi D, Duggan N, Biggins S, Gonen T (2012) The structure of purified kinetochores reveals multiple microtubule-attachment sites. Nat Struct Mol Biol 19 (9):925–929. doi:10.1038/nsmb.2358

Gopalan G, Chan CS, Donovan PJ (1997) A novel mammalian, mitotic spindle-associated kinase is related to yeast and fly chromosome segregation regulators. J Cell Biol 138(3):643–656

Grallert A, Boke E, Hagting A, Hodgson B, Connolly Y, Griffiths JR, Smith DL, Pines J, Hagan IM (2015) A PP1-PP2A phosphatase relay controls mitotic progression. Nature 517 (7532):94–98. doi:10.1038/nature14019

Guimaraes GJ, Dong Y, McEwen BF, Deluca JG (2008) Kinetochore-microtubule attachment relies on the disordered N-terminal tail domain of Hec1. Curr Biol: CB 18(22):1778–1784. doi:10.1016/j.cub.2008.08.012

Haarhuis JH, Elbatsh AM, Rowland BD (2014) Cohesin and its regulation: on the logic of X-shaped chromosomes. Dev Cell 31(1):7–18. doi:10.1016/j.devcel.2014.09.010

Hafner J, Mayr MI, Mockel MM, Mayer TU (2014) Pre-anaphase chromosome oscillations are regulated by the antagonistic activities of Cdk1 and PP1 on Kif18A. Nat Commun 5:4397. doi:10.1038/ncomms5397

Hanisch A, Sillje HH, Nigg EA (2006a) Timely anaphase onset requires a novel spindle and kinetochore complex comprising Ska1 and Ska2. EMBO J 25(23):5504–5515. doi:10.1038/sj.emboj.7601426

Hanisch A, Wehner A, Nigg EA, Sillje HH (2006b) Different Plk1 functions show distinct dependencies on Polo-Box domain-mediated targeting. Mol Biol Cell 17(1):448–459. doi:10.1091/mbc.E05-08-0801

Hardwick KG, Weiss E, Luca FC, Winey M, Murray AW (1996) Activation of the budding yeast spindle assembly checkpoint without mitotic spindle disruption. Science 273(5277):953–956

Hauf S, Cole RW, LaTerra S, Zimmer C, Schnapp G, Walter R, Heckel A, van Meel J, Rieder CL, Peters JM (2003) The small molecule Hesperadin reveals a role for Aurora B in correcting kinetochore-microtubule attachment and in maintaining the spindle assembly checkpoint. J Cell Biol 161(2):281–294. doi:10.1083/jcb.200208092

Heinrich S, Windecker H, Hustedt N, Hauf S (2012) Mph1 kinetochore localization is crucial and upstream in the hierarchy of spindle assembly checkpoint protein recruitment to kinetochores. J Cell Sci 125(Pt 20):4720–4727. doi:10.1242/jcs.110387

Hendrickx A, Beullens M, Ceulemans H, Den Abt T, Van Eynde A, Nicolaescu E, Lesage B, Bollen M (2009) Docking motif-guided mapping of the interactome of protein phosphatase-1. Chem Biol 16(4):365–371. doi:10.1016/j.chembiol.2009.02.012

Hertz EP, Kruse T, Davey NE, Lopez-Mendez B, Sigurethsson JO, Montoya G, Olsen JV, Nilsson J (2016) A conserved motif provides binding specificity to the PP2A-B56 phosphatase. Mol Cell 63(4):686–695. doi:10.1016/j.molcel.2016.06.024

Hewitt L, Tighe A, Santaguida S, White AM, Jones CD, Musacchio A, Green S, Taylor SS (2010) Sustained Mps1 activity is required in mitosis to recruit O-Mad2 to the Mad1-C-Mad2 core complex. J Cell Biol 190(1):25–34. doi:10.1083/jcb.201002133

Hiruma Y, Sacristan C, Pachis ST, Adamopoulos A, Kuijt T, Ubbink M, von Castelmur E, Perrakis A, Kops GJ (2015) Cell division cycle. Competition between MPS1 and microtubules at kinetochores regulates spindle checkpoint signaling. Science 348(6240):1264–1267. doi:10.1126/science.aaa4055

Honda R, Korner R, Nigg EA (2003) Exploring the functional interactions between Aurora B, INCENP, and survivin in mitosis. Mol Biol Cell 14(8):3325–3341. doi:10.1091/mbc.E02-11-0769

Howell BJ, Moree B, Farrar EM, Stewart S, Fang G, Salmon ED (2004) Spindle checkpoint protein dynamics at kinetochores in living cells. Curr Biol: CB 14(11):953–964. doi:10.1016/j.cub.2004.05.053

Hsu JY, Sun ZW, Li X, Reuben M, Tatchell K, Bishop DK, Grushcow JM, Brame CJ, Caldwell JA, Hunt DF, Lin R, Smith MM, Allis CD (2000) Mitotic phosphorylation of histone H3 is governed by Ipl1/aurora kinase and Glc7/PP1 phosphatase in budding yeast and nematodes. Cell 102(3):279–291

Hua S, Wang Z, Jiang K, Huang Y, Ward T, Zhao L, Dou Z, Yao X (2011) CENP-U cooperates with Hec1 to orchestrate kinetochore-microtubule attachment. J Biol Chem 286(2):1627–1638. doi:10.1074/jbc.M110.174946

Iimori M, Watanabe S, Kiyonari S, Matsuoka K, Sakasai R, Saeki H, Oki E, Kitao H, Maehara Y (2016) Phosphorylation of EB2 by Aurora B and CDK1 ensures mitotic progression and genome stability. Nat Commun 7:11117. doi:10.1038/ncomms11117

Ito D, Saito Y, Matsumoto T (2012) Centromere-tethered Mps1 pombe homolog (Mph1) kinase is a sufficient marker for recruitment of the spindle checkpoint protein Bub1, but not Mad1. Proc Natl Acad Sci U S A 109(1):209–214. doi:10.1073/pnas.1114647109

Izawa D, Pines J (2015) The mitotic checkpoint complex binds a second CDC20 to inhibit active APC/C. Nature 517(7536):631–634. doi:10.1038/nature13911

Jelluma N, Brenkman AB, van den Broek NJ, Cruijsen CW, van Osch MH, Lens SM, Medema RH, Kops GJ (2008) Mps1 phosphorylates Borealin to control Aurora B activity and chromosome alignment. Cell 132(2):233–246. doi:10.1016/j.cell.2007.11.046

Jelluma N, Dansen TB, Sliedrecht T, Kwiatkowski NP, Kops GJ (2010) Release of Mps1 from kinetochores is crucial for timely anaphase onset. J Cell Biol 191(2):281–290. doi:10.1083/jcb.201003038

Jeyaprakash AA, Basquin C, Jayachandran U, Conti E (2011) Structural basis for the recognition of phosphorylated histone h3 by the survivin subunit of the chromosomal passenger complex. Structure 19(11):1625–1634. doi:10.1016/j.str.2011.09.002

Jeyaprakash AA, Klein UR, Lindner D, Ebert J, Nigg EA, Conti E (2007) Structure of a Survivin-Borealin-INCENP core complex reveals how chromosomal passengers travel together. Cell 131(2):271–285. doi:10.1016/j.cell.2007.07.045

Ji Z, Gao H, Yu H (2015) CELL DIVISION CYCLE. Kinetochore attachment sensed by competitive Mps1 and microtubule binding to Ndc80C. Science 348(6240):1260–1264. doi:10.1126/science.aaa4029

Joglekar AP (2016) A cell biological perspective on past, present and future investigations of the spindle assembly checkpoint. Biology (Basel) 5(4). doi:10.3390/biology5040044

Kang J, Chen Y, Zhao Y, Yu H (2007) Autophosphorylation-dependent activation of human Mps1 is required for the spindle checkpoint. Proc Natl Acad Sci U S A 104(51):20232–20237. doi:10.1073/pnas.0710519105

Kawashima SA, Tsukahara T, Langegger M, Hauf S, Kitajima TS, Watanabe Y (2007) Shugoshin enables tension-generating attachment of kinetochores by loading Aurora to centromeres. Genes Dev 21(4):420–435. doi:10.1101/gad.1497307

Kawashima SA, Yamagishi Y, Honda T, Ishiguro K, Watanabe Y (2010) Phosphorylation of H2A by Bub1 prevents chromosomal instability through localizing shugoshin. Science 327 (5962):172–177. doi:10.1126/science.1180189

Kelly AE, Ghenoiu C, Xue JZ, Zierhut C, Kimura H, Funabiki H (2010) Survivin reads phosphorylated histone H3 threonine 3 to activate the mitotic kinase Aurora B. Science 330 (6001):235–239. doi:10.1126/science.1189505

Kelly AE, Sampath SC, Maniar TA, Woo EM, Chait BT, Funabiki H (2007) Chromosomal enrichment and activation of the aurora B pathway are coupled to spatially regulate spindle assembly. Dev Cell 12(1):31–43. doi:10.1016/j.devcel.2006.11.001

Kim HS, Kim SH, Park HY, Lee J, Yoon JH, Choi S, Ryu SH, Lee H, Cho HS, Lee CW (2013) Functional interplay between Aurora B kinase and Ssu72 phosphatase regulates sister chromatid cohesion. Nat Commun 4:2631. doi:10.1038/ncomms3631

Kim S, Yu H (2015) Multiple assembly mechanisms anchor the KMN spindle checkpoint platform at human mitotic kinetochores. J Cell Biol 208(2):181–196. doi:10.1083/jcb.201407074

Kim Y, Holland AJ, Lan W, Cleveland DW (2010) Aurora kinases and protein phosphatase 1 mediate chromosome congression through regulation of CENP-E. Cell 142(3):444–455. doi:10.1016/j.cell.2010.06.039

Kimura M, Kotani S, Hattori T, Sumi N, Yoshioka T, Todokoro K, Okano Y (1997) Cell cycle-dependent expression and spindle pole localization of a novel human protein kinase, Aik, related to Aurora of Drosophila and yeast Ipl1. J Biol Chem 272(21):13766–13771

Knowlton AL, Vorozhko VV, Lan W, Gorbsky GJ, Stukenberg PT (2009) ICIS and Aurora B coregulate the microtubule depolymerase Kif2a. Curr Biol: CB 19(9):758–763. doi:10.1016/j.cub.2009.03.018

Krenn V, Musacchio A (2015) The Aurora B Kinase in chromosome bi-orientation and spindle checkpoint signaling. Front Oncol 5:225. doi:10.3389/fonc.2015.00225

Krupina K, Kleiss C, Metzger T, Fournane S, Schmucker S, Hofmann K, Fischer B, Paul N, Porter IM, Raffelsberger W, Poch O, Swedlow JR, Brino L, Sumara I (2016) Ubiquitin receptor protein UBASH3B drives Aurora B recruitment to mitotic microtubules. Dev Cell 36(1):63–78. doi:10.1016/j.devcel.2015.12.017

Kruse T, Zhang G, Larsen MS, Lischetti T, Streicher W, Kragh Nielsen T, Bjorn SP, Nilsson J (2013) Direct binding between BubR1 and B56-PP2A phosphatase complexes regulate mitotic progression. J Cell Sci 126(Pt 5):1086–1092. doi:10.1242/jcs.122481

Kuijt TE, Omerzu M, Saurin AT, Kops GJ (2014) Conditional targeting of MAD1 to kinetochores is sufficient to reactivate the spindle assembly checkpoint in metaphase. Chromosoma 123 (5):471–480. doi:10.1007/s00412-014-0458-9

Kwiatkowski N, Jelluma N, Filippakopoulos P, Soundararajan M, Manak MS, Kwon M, Choi HG, Sim T, Deveraux QL, Rottmann S, Pellman D, Shah JV, Kops GJ, Knapp S, Gray NS (2010) Small-molecule kinase inhibitors provide insight into Mps1 cell cycle function. Nat Chem Biol 6(5):359–368. doi:10.1038/nchembio.345

Kwon YG, Lee SY, Choi Y, Greengard P, Nairn AC (1997) Cell cycle-dependent phosphorylation of mammalian protein phosphatase 1 by cdc2 kinase. Proc Natl Acad Sci U S A 94(6):2168–2173

Lampson MA, Cheeseman IM (2011) Sensing centromere tension: Aurora B and the regulation of kinetochore function. Trends Cell Biol 21(3):133–140. doi:10.1016/j.tcb.2010.10.007

Lampson MA, Kapoor TM (2005) The human mitotic checkpoint protein BubR1 regulates chromosome-spindle attachments. Nat Cell Biol 7(1):93–98. doi:10.1038/ncb1208

Lampson MA, Renduchitala K, Khodjakov A, Kapoor TM (2004) Correcting improper chromosome-spindle attachments during cell division. Nat Cell Biol 6(3):232–237. doi:10.1038/ncb1102

Lan W, Zhang X, Kline-Smith SL, Rosasco SE, Barrett-Wilt GA, Shabanowitz J, Hunt DF, Walczak CE, Stukenberg PT (2004) Aurora B phosphorylates centromeric MCAK and regulates its localization and microtubule depolymerization activity. Curr Biol: CB 14(4):273–286. doi:10.1016/j.cub.2004.01.055

Lauze E, Stoelcker B, Luca FC, Weiss E, Schutz AR, Winey M (1995) Yeast spindle pole body duplication gene MPS1 encodes an essential dual specificity protein kinase. EMBO J 14 (8):1655–1663

Lenart P, Petronczki M, Steegmaier M, Di Fiore B, Lipp JJ, Hoffmann M, Rettig WJ, Kraut N, Peters JM (2007) The small-molecule inhibitor BI 2536 reveals novel insights into mitotic roles of polo-like kinase 1. Curr Biol: CB 17(4):304–315. doi:10.1016/j.cub.2006.12.046

Lischetti T, Nilsson J (2015) Regulation of mitotic progression by the spindle assembly checkpoint. Mol Cell Oncol 2(1):e970484. doi:10.4161/23723548.2014.970484

Liu D, Davydenko O, Lampson MA (2012) Polo-like kinase-1 regulates kinetochore-microtubule dynamics and spindle checkpoint silencing. J Cell Biol 198(4):491–499. doi:10.1083/jcb.201205090

Liu D, Vader G, Vromans MJ, Lampson MA, Lens SM (2009) Sensing chromosome bi-orientation by spatial separation of aurora B kinase from kinetochore substrates. Science 323(5919):1350–1353. doi:10.1126/science.1167000

Liu D, Vleugel M, Backer CB, Hori T, Fukagawa T, Cheeseman IM, Lampson MA (2010) Regulated targeting of protein phosphatase 1 to the outer kinetochore by KNL1 opposes Aurora B kinase. J Cell Biol 188(6):809–820. doi:10.1083/jcb.201001006

London N, Ceto S, Ranish JA, Biggins S (2012) Phosphoregulation of Spc105 by Mps1 and PP1 regulates Bub1 localization to kinetochores. Curr Biol: CB 22(10):900–906. doi:10.1016/j.cub.2012.03.052

Losada A, Hirano M, Hirano T (2002) Cohesin release is required for sister chromatid resolution, but not for condensin-mediated compaction, at the onset of mitosis. Genes Dev 16(23):3004–3016. doi:10.1101/gad.249202

Maciejowski J, George KA, Terret ME, Zhang C, Shokat KM, Jallepalli PV (2010) Mps1 directs the assembly of Cdc20 inhibitory complexes during interphase and mitosis to control M phase

timing and spindle checkpoint signaling. J Cell Biol 190(1):89–100. doi:10.1083/jcb.201001050
Maldonado M, Kapoor TM (2011) Constitutive Mad1 targeting to kinetochores uncouples checkpoint signalling from chromosome biorientation. Nat Cell Biol 13(4):475–482. doi:10.1038/ncb2223
Maresca TJ, Salmon ED (2010) Welcome to a new kind of tension: translating kinetochore mechanics into a wait-anaphase signal. J Cell Sci 123(Pt 6):825–835. doi:10.1242/jcs.064790
Marquardt JR, Fisk HA (2016) ARHGEF17 sets the timer for retention of Mps1 at kinetochores. The Journal of cell biology 212(6):615–616. doi:10.1083/jcb.201602060
Marston AL (2015) Shugoshins: tension-sensitive pericentromeric adaptors safeguarding chromosome segregation. Mol Cell Biol 35(4):634–648. doi:10.1128/Mcb.01176-14
Mattison CP, Old WM, Steiner E, Huneycutt BJ, Resing KA, Ahn NG, Winey M (2007) Mps1 activation loop autophosphorylation enhances kinase activity. J Biol Chem 282(42):30553–30561. doi:10.1074/jbc.M707063200
Maure JF, Kitamura E, Tanaka TU (2007) Mps1 kinase promotes sister-kinetochore bi-orientation by a tension-dependent mechanism. Curr Biol: CB 17(24):2175–2182. doi:10.1016/j.cub.2007.11.032
Meadows JC, Shepperd LA, Vanoosthuyse V, Lancaster TC, Sochaj AM, Buttrick GJ, Hardwick KG, Millar JB (2011) Spindle checkpoint silencing requires association of PP1 to both Spc7 and kinesin-8 motors. Dev Cell 20(6):739–750. doi:10.1016/j.devcel.2011.05.008
Meiselbach H, Sticht H, Enz R (2006) Structural analysis of the protein phosphatase 1 docking motif: molecular description of binding specificities identifies interacting proteins. Chem Biol 13(1):49–59. doi:10.1016/j.chembiol.2005.10.009
Meppelink A, Kabeche L, Vromans MJ, Compton DA, Lens SM (2015) Shugoshin-1 balances Aurora B kinase activity via PP2A to promote chromosome bi-orientation. Cell reports 11(4):508–515. doi:10.1016/j.celrep.2015.03.052
Miller SA, Johnson ML, Stukenberg PT (2008) Kinetochore attachments require an interaction between unstructured tails on microtubules and Ndc80(Hec1). Curr Biol: CB 18(22):1785–1791. doi:10.1016/j.cub.2008.11.007
Mo F, Zhuang X, Liu X, Yao PY, Qin B, Su Z, Zang J, Wang Z, Zhang J, Dou Z, Tian C, Teng M, Niu L, Hill DL, Fang G, Ding X, Fu C, Yao X (2016) Acetylation of Aurora B by TIP60 ensures accurate chromosomal segregation. Nat Chem Biol 12(4):226–232. doi:10.1038/nchembio.2017
Murnion ME, Adams RR, Callister DM, Allis CD, Earnshaw WC, Swedlow JR (2001) Chromatin-associated protein phosphatase 1 regulates aurora-B and histone H3 phosphorylation. J Biol Chem 276(28):26656–26665. doi:10.1074/jbc.M102288200
Musacchio A (2015) The molecular biology of spindle assembly checkpoint signaling dynamics. Curr Biol: CB 25(20):R1002–R1018. doi:10.1016/j.cub.2015.08.051
Nagpal H, Fukagawa T (2016) Kinetochore assembly and function through the cell cycle. Chromosoma 125(4):645–659. doi:10.1007/s00412-016-0608-3
Nerusheva OO, Galander S, Fernius J, Kelly D, Marston AL (2014) Tension-dependent removal of pericentromeric shugoshin is an indicator of sister chromosome biorientation. Genes Dev 28(12):1291–1309. doi:10.1101/gad.240291.114
Niedzialkowska E, Wang F, Porebski PJ, Minor W, Higgins JM, Stukenberg PT (2012) Molecular basis for phosphospecific recognition of histone H3 tails by Survivin paralogues at inner centromeres. Mol Biol Cell 23(8):1457–1466. doi:10.1091/mbc.E11-11-0904
Nijenhuis W, Vallardi G, Teixeira A, Kops GJ, Saurin AT (2014) Negative feedback at kinetochores underlies a responsive spindle checkpoint signal. Nat Cell Biol 16(12):1257–1264. doi:10.1038/ncb3065
Nijenhuis W, von Castelmur E, Littler D, De Marco V, Tromer E, Vleugel M, van Osch MH, Snel B, Perrakis A, Kops GJ (2013) A TPR domain-containing N-terminal module of MPS1 is required for its kinetochore localization by Aurora B. J Cell Biol 201(2):217–231. doi:10.1083/jcb.201210033

Nishiyama T, Sykora MM, Huis in 't Veld PJ, Mechtler K, Peters JM (2013) Aurora B and Cdk1 mediate Wapl activation and release of acetylated cohesin from chromosomes by phosphorylating Sororin. Proc Natl Acad Sci U S A 110(33):13404–13409. doi:10.1073/pnas.1305020110

O'Connor A, Maffini S, Rainey MD, Kaczmarczyk A, Gaboriau D, Musacchio A, Santocanale C (2015) Requirement for PLK1 kinase activity in the maintenance of a robust spindle assembly checkpoint. Biol Open 5(1):11–19. doi:10.1242/bio.014969

Ohi R, Sapra T, Howard J, Mitchison TJ (2004) Differentiation of cytoplasmic and meiotic spindle assembly MCAK functions by Aurora B-dependent phosphorylation. Mol Biol Cell 15 (6):2895–2906. doi:10.1091/mbc.E04-02-0082

Ohkura H, Adachi Y, Kinoshita N, Niwa O, Toda T, Yanagida M (1988) Cold-sensitive and caffeine-supersensitive mutants of the Schizosaccharomyces pombe dis genes implicated in sister chromatid separation during mitosis. EMBO J 7(5):1465–1473

Ohkura H, Kinoshita N, Miyatani S, Toda T, Yanagida M (1989) The fission yeast disZ+ gene required for chromosome disjoining encodes one of two putative type 1 protein phosphatases. Cell 57:997–1007

Pesenti ME, Weir JR, Musacchio A (2016) Progress in the structural and functional characterization of kinetochores. Curr Opin Struct Biol 37:152–163. doi:10.1016/j.sbi.2016.03.003

Peters U, Cherian J, Kim JH, Kwok BH, Kapoor TM (2006) Probing cell-division phenotype space and Polo-like kinase function using small molecules. Nat Chem Biol 2(11):618–626. doi:10.1038/nchembio826

Petrovic A, Keller J, Liu Y, Overlack K, John J, Dimitrova YN, Jenni S, van Gerwen S, Stege P, Wohlgemuth S, Rombaut P, Herzog F, Harrison SC, Vetter IR, Musacchio A (2016) Structure of the MIS12 complex and molecular basis of its interaction with CENP-C at human kinetochores. Cell 167(4):1028–1040, e1015. doi:10.1016/j.cell.2016.10.005

Pinsky BA, Kung C, Shokat KM, Biggins S (2006) The Ipl1-Aurora protein kinase activates the spindle checkpoint by creating unattached kinetochores. Nat Cell Biol 8(1):78–83. doi:10.1038/ncb1341

Pinsky BA, Nelson CR, Biggins S (2009) Protein phosphatase 1 regulates exit from the spindle checkpoint in budding yeast. Curr Biol: CB 19(14):1182–1187. doi:10.1016/j.cub.2009.06.043

Poch O, Schwob E, de Fraipont F, Camasses A, Bordonne R, Martin RP (1994) RPK1, an essential yeast protein kinase involved in the regulation of the onset of mitosis, shows homology to mammalian dual-specificity kinases. Mol Gen Genet: MGG 243(6):641–653

Porter IM, Schleicher K, Porter M, Swedlow JR (2013) Bod1 regulates protein phosphatase 2A at mitotic kinetochores. Nat Commun 4:2677. doi:10.1038/ncomms3677

Posch M, Khoudoli GA, Swift S, King EM, Deluca JG, Swedlow JR (2010) Sds22 regulates aurora B activity and microtubule-kinetochore interactions at mitosis. J Cell Biol 191(1):61–74. doi:10.1083/jcb.200912046

Qian J, Beullens M, Lesage B, Bollen M (2013) Aurora B defines its own chromosomal targeting by opposing the recruitment of the phosphatase scaffold Repo-Man. Curr Biol: CB 23 (12):1136–1143. doi:10.1016/j.cub.2013.05.017

Raaijmakers JA, Tanenbaum ME, Maia AF, Medema RH (2009) RAMA1 is a novel kinetochore protein involved in kinetochore-microtubule attachment. J Cell Sci 122(Pt 14):2436–2445. doi:10.1242/jcs.051912

Rago F, Gascoigne KE, Cheeseman IM (2015) Distinct organization and regulation of the outer kinetochore KMN network downstream of CENP-C and CENP-T. Curr Biol: CB 25(5):671–677. doi:10.1016/j.cub.2015.01.059

Rosenberg JS, Cross FR, Funabiki H (2011) KNL1/Spc105 recruits PP1 to silence the spindle assembly checkpoint. Curr Biol: CB 21(11):942–947. doi:10.1016/j.cub.2011.04.011

Salimian KJ, Ballister ER, Smoak EM, Wood S, Panchenko T, Lampson MA, Black BE (2011) Feedback control in sensing chromosome biorientation by the Aurora B kinase. Curr Biol: CB 21(13):1158–1165. doi:10.1016/j.cub.2011.06.015

Santaguida S, Musacchio A (2009) The life and miracles of kinetochores. EMBO J 28(17):2511–2531. doi:10.1038/emboj.2009.173

Santaguida S, Tighe A, D'Alise AM, Taylor SS, Musacchio A (2010) Dissecting the role of MPS1 in chromosome biorientation and the spindle checkpoint through the small molecule inhibitor reversine. J Cell Biol 190(1):73–87. doi:10.1083/jcb.201001036

Santaguida S, Vernieri C, Villa F, Ciliberto A, Musacchio A (2011) Evidence that Aurora B is implicated in spindle checkpoint signalling independently of error correction. EMBO J 30 (8):1508–1519. doi:10.1038/emboj.2011.70

Sassoon I, Severin FF, Andrews PD, Taba MR, Kaplan KB, Ashford AJ, Stark MJ, Sorger PK, Hyman AA (1999) Regulation of Saccharomyces cerevisiae kinetochores by the type 1 phosphatase Glc7p. Genes Dev 13(5):545–555

Saurin AT, van der Waal MS, Medema RH, Lens SM, Kops GJ (2011) Aurora B potentiates Mps1 activation to ensure rapid checkpoint establishment at the onset of mitosis. Nat Commun 2:316. doi:10.1038/ncomms1319

Schmidt JC, Kiyomitsu T, Hori T, Backer CB, Fukagawa T, Cheeseman IM (2010) Aurora B kinase controls the targeting of the Astrin-SKAP complex to bioriented kinetochores. J Cell Biol 191(2):269–280. doi:10.1083/jcb.201006129

Sessa F, Mapelli M, Ciferri C, Tarricone C, Areces LB, Schneider TR, Stukenberg PT, Musacchio A (2005) Mechanism of Aurora B activation by INCENP and inhibition by hesperadin. Mol Cell 18(3):379–391. doi:10.1016/j.molcel.2005.03.031

Shao H, Huang Y, Zhang L, Yuan K, Chu Y, Dou Z, Jin C, Garcia-Barrio M, Liu X, Yao X (2015) Spatiotemporal dynamics of Aurora B-PLK1-MCAK signaling axis orchestrates kinetochore bi-orientation and faithful chromosome segregation. Sci Rep 5:12204. doi:10.1038/srep12204

Shepperd LA, Meadows JC, Sochaj AM, Lancaster TC, Zou J, Buttrick GJ, Rappsilber J, Hardwick KG, Millar JB (2012) Phosphodependent recruitment of Bub1 and Bub3 to Spc7/KNL1 by Mph1 kinase maintains the spindle checkpoint. Curr Biol: CB 22(10):891–899. doi:10.1016/j.cub.2012.03.051

Sivakumar S, Janczyk PL, Qu Q, Brautigam CA, Stukenberg PT, Yu H, Gorbsky GJ (2016) The human SKA complex drives the metaphase-anaphase cell cycle transition by recruiting protein phosphatase 1 to kinetochores. Elife 5. doi:10.7554/eLife.12902

Sliedrecht T, Zhang C, Shokat KM, Kops GJ (2010) Chemical genetic inhibition of Mps1 in stable human cell lines reveals novel aspects of Mps1 function in mitosis. PLoS ONE 5(4):e10251. doi:10.1371/journal.pone.0010251

Storchova Z, Becker JS, Talarek N, Kogelsberger S, Pellman D (2011) Bub1, Sgo1, and Mps1 mediate a distinct pathway for chromosome biorientation in budding yeast. Mol Biol Cell 22 (9):1473–1485. doi:10.1091/mbc.E10-08-0673

Stucke VM, Sillje HH, Arnaud L, Nigg EA (2002) Human Mps1 kinase is required for the spindle assembly checkpoint but not for centrosome duplication. EMBO J 21(7):1723–1732. doi:10.1093/emboj/21.7.1723

Suijkerbuijk SJ, Vleugel M, Teixeira A, Kops GJ (2012) Integration of kinase and phosphatase activities by BUBR1 ensures formation of stable kinetochore-microtubule attachments. Dev Cell 23(4):745–755. doi:10.1016/j.devcel.2012.09.005

Sumara I, Gimenez-Abian JF, Gerlich D, Hirota T, Kraft C, de la Torre C, Ellenberg J, Peters JM (2004) Roles of polo-like kinase 1 in the assembly of functional mitotic spindles. Curr Biol: CB 14(19):1712–1722. doi:10.1016/j.cub.2004.09.049

Sundin LJ, Guimaraes GJ, Deluca JG (2011) The NDC80 complex proteins Nuf2 and Hec1 make distinct contributions to kinetochore-microtubule attachment in mitosis. Mol Biol Cell 22 (6):759–768. doi:10.1091/mbc.E10-08-0671

Tanaka K, Kitamura E, Kitamura Y, Tanaka TU (2007) Molecular mechanisms of microtubule-dependent kinetochore transport toward spindle poles. J Cell Biol 178(2):269–281. doi:10.1083/jcb.200702141

Tanaka TU, Rachidi N, Janke C, Pereira G, Galova M, Schiebel E, Stark MJ, Nasmyth K (2002) Evidence that the Ipl1-Sli15 (Aurora kinase-INCENP) complex promotes chromosome bi-orientation by altering kinetochore-spindle pole connections. Cell 108(3):317–329

Tang NH, Toda T (2015) Alp7/TACC recruits kinesin-8-PP1 to the Ndc80 kinetochore protein for timely mitotic progression and chromosome movement. J Cell Sci 128(2):354–363. doi:10.1242/jcs.160036

Tanno Y, Susumu H, Kawamura M, Sugimura H, Honda T, Watanabe Y (2015) The inner centromere-shugoshin network prevents chromosomal instability. Science 349(6253):1237–1240. doi:10.1126/science.aaa2655

Trinkle-Mulcahy L, Andersen J, Lam YW, Moorhead G, Mann M, Lamond AI (2006) Repo-Man recruits PP1 gamma to chromatin and is essential for cell viability. J Cell Biol 172(5):679–692. doi:10.1083/jcb.200508154

Trinkle-Mulcahy L, Andrews PD, Wickramasighe S, Sleeman J, Prescott A, Lam YW, Lyon C, Swedlow JR, Lamond AI (2003) Time-lapse imaging reveals dynamic relocalization of PP1 throughout the mammalian cell cycle. Mol Bio Cell 14:107–117. doi:10.1091/mbc.E02-

Tsukahara T, Tanno Y, Watanabe Y (2010) Phosphorylation of the CPC by Cdk1 promotes chromosome bi-orientation. Nature 467(7316):719–723. doi:10.1038/nature09390

Vader G, Cruijsen CW, van Harn T, Vromans MJ, Medema RH, Lens SM (2007) The chromosomal passenger complex controls spindle checkpoint function independent from its role in correcting microtubule kinetochore interactions. Mol Biol Cell 18(11):4553–4564. doi:10.1091/mbc.E07-04-0328

van der Waal MS, Saurin AT, Vromans MJ, Vleugel M, Wurzenberger C, Gerlich DW, Medema RH, Kops GJ, Lens SM (2012) Mps1 promotes rapid centromere accumulation of Aurora B. EMBO Rep 13(9):847–854. doi:10.1038/embor.2012.93

Vanoosthuyse V, Hardwick KG (2009) A novel protein phosphatase 1-dependent spindle checkpoint silencing mechanism. Curr Biol: CB 19(14):1176–1181. doi:10.1016/j.cub.2009.05.060

Vigneron S, Prieto S, Bernis C, Labbe JC, Castro A, Lorca T (2004) Kinetochore localization of spindle checkpoint proteins: who controls whom? Mol Biol Cell 15(10):4584–4596. doi:10.1091/mbc.E04-01-0051

von Schubert C, Cubizolles F, Bracher JM, Sliedrecht T, Kops GJ, Nigg EA (2015) Plk1 and Mps1 cooperatively regulate the spindle assembly checkpoint in human cells. Cell Rep 12 (1):66–78. doi:10.1016/j.celrep.2015.06.007

Wakula P, Beullens M, Ceulemans H, Stalmans W, Bollen M (2003) Degeneracy and function of the ubiquitous RVXF motif that mediates binding to protein phosphatase-1. J Biol Chem 278 (21):18817–18823. doi:10.1074/jbc.M300175200

Wang E, Ballister ER, Lampson MA (2011a) Aurora B dynamics at centromeres create a diffusion-based phosphorylation gradient. J Cell Biol 194(4):539–549. doi:10.1083/jcb.201103044

Wang F, Dai J, Daum JR, Niedzialkowska E, Banerjee B, Stukenberg PT, Gorbsky GJ, Higgins JM (2010) Histone H3 Thr-3 phosphorylation by Haspin positions Aurora B at centromeres in mitosis. Science 330(6001):231–235. doi:10.1126/science.1189435

Wang F, Ulyanova NP, van der Waal MS, Patnaik D, Lens SM, Higgins JM (2011b) A positive feedback loop involving Haspin and Aurora B promotes CPC accumulation at centromeres in mitosis. Curr Biol: CB 21(12):1061–1069. doi:10.1016/j.cub.2011.05.016

Wang J, Wang Z, Yu T, Yang H, Virshup DM, Kops GJ, Lee SH, Zhou W, Li X, Xu W, Rao Z (2016) Crystal structure of a PP2A B56-BubR1 complex and its implications for PP2A substrate recruitment and localization. Protein Cell 7(7):516–526. doi:10.1007/s13238-016-0283-4

Wei RR, Al-Bassam J, Harrison SC (2007) The Ndc80/HEC1 complex is a contact point for kinetochore-microtubule attachment. Nat Struct Mol Biol 14(1):54–59. doi:10.1038/nsmb1186

Weir JR, Faesen AC, Klare K, Petrovic A, Basilico F, Fischbock J, Pentakota S, Keller J, Pesenti ME, Pan D, Vogt D, Wohlgemuth S, Herzog F, Musacchio A (2016) Insights from biochemical reconstitution into the architecture of human kinetochores. Nature 537(7619):249–253. doi:10.1038/nature19333

Weiss E, Winey M (1996) The Saccharomyces cerevisiae spindle pole body duplication gene MPS1 is part of a mitotic checkpoint. J Cell Biol 132(1–2):111–123

Welburn JP, Grishchuk EL, Backer CB, Wilson-Kubalek EM, Yates JR 3rd, Cheeseman IM (2009) The human kinetochore Ska1 complex facilitates microtubule depolymerization-coupled motility. Dev Cell 16(3):374–385. doi:10.1016/j.devcel.2009.01.011

Welburn JP, Vleugel M, Liu D, Yates JR 3rd, Lampson MA, Fukagawa T, Cheeseman IM (2010) Aurora B phosphorylates spatially distinct targets to differentially regulate the kinetochore-microtubule interface. Mol Cell 38(3):383–392. doi:10.1016/j.molcel.2010.02.034

Westermann S, Wang HW, Avila-Sakar A, Drubin DG, Nogales E, Barnes G (2006) The Dam1 kinetochore ring complex moves processively on depolymerizing microtubule ends. Nature 440(7083):565–569. doi:10.1038/nature04409

Wilson-Kubalek EM, Cheeseman IM, Yoshioka C, Desai A, Milligan RA (2008) Orientation and structure of the Ndc80 complex on the microtubule lattice. J Cell Biol 182(6):1055–1061. doi:10.1083/jcb.200804170

Winey M, Goetsch L, Baum P, Byers B (1991) MPS1 and MPS2: novel yeast genes defining distinct steps of spindle pole body duplication. J Cell Biol 114(4):745–754

Wu JQ, Guo JY, Tang W, Yang CS, Freel CD, Chen C, Nairn AC, Kornbluth S (2009) PP1-mediated dephosphorylation of phosphoproteins at mitotic exit is controlled by inhibitor-1 and PP1 phosphorylation. Nat Cell Biol 11(5):644–651. doi:10.1038/ncb1871

Xu P, Raetz EA, Kitagawa M, Virshup DM, Lee SH (2013) BUBR1 recruits PP2A via the B56 family of targeting subunits to promote chromosome congression. Biol Open 2(5):479–486. doi:10.1242/bio.20134051

Yamagishi Y, Honda T, Tanno Y, Watanabe Y (2010) Two histone marks establish the inner centromere and chromosome bi-orientation. Science 330(6001):239–243. doi:10.1126/science.1194498

Yamagishi Y, Yang CH, Tanno Y, Watanabe Y (2012) MPS1/Mph1 phosphorylates the kinetochore protein KNL1/Spc7 to recruit SAC components. Nat Cell Biol 14(7):746–752. doi:10.1038/ncb2515

Yamano H, Ishii K, Yanagida M (1994) Phosphorylation of dis2 protein phosphatase at the C-terminal cdc2 consensus and its potential role in cell cycle regulation. EMBO J 13 (22):5310–5318

Yasui Y, Urano T, Kawajiri A, Nagata K, Tatsuka M, Saya H, Furukawa K, Takahashi T, Izawa I, and Inagaki M (2004) Autophosphorylation of a newly identified site of Aurora-B is indispensable for cytokinesis. J Biol Chem 279(13):12997–13003. doi:10.1074/jbc.M311128200

Yoshida MM, Ting L, Gygi SP, Azuma Y (2016) SUMOylation of DNA topoisomerase IIalpha regulates histone H3 kinase Haspin and H3 phosphorylation in mitosis. J Cell Biol 213(6):665–678. doi:10.1083/jcb.201511079

Zaytsev AV, Segura-Pena D, Godzi M, Calderon A, Ballister ER, Stamatov R, Mayo AM, Peterson L, Black BE, Ataullakhanov FI, Lampson MA (1064) Grishchuk EL (2016) bistability of a coupled Aurora B kinase-phosphatase system in cell division. Elife 5:e10644. doi:10.7554/eLife.4

Zhang X, Lan W, Ems-McClung SC, Stukenberg PT, Walczak CE (2007) Aurora B phosphorylates multiple sites on mitotic centromere-associated kinesin to spatially and temporally regulate its function. Mol Biol Cell 18(9):3264–3276. doi:10.1091/mbc.E07-01-0086

Zhu T, Dou Z, Qin B, Jin C, Wang X, Xu L, Wang Z, Zhu L, Liu F, Gao X, Ke Y, Wang Z, Aikhionbare F, Fu C, Ding X, and Yao X (2013) Phosphorylation of microtubule-binding protein Hec1 by mitotic kinase aurora B specifies spindle checkpoint kinase Mps1 signaling at the kinetochore. J Biol Chem 288(50):36149–36159 doi:10.1074/jbc.M113.507970

Centromeric Cohesin: Molecular Glue and Much More

Mihailo Mirkovic and Raquel A. Oliveira

Abstract Sister chromatid cohesion, mediated by the cohesin complex, is a prerequisite for faithful chromosome segregation during mitosis. Premature release of sister chromatid cohesion leads to random segregation of the genetic material and consequent aneuploidy. Multiple regulatory mechanisms ensure proper timing for cohesion establishment, concomitant with DNA replication, and cohesion release during the subsequent mitosis. Here we summarize the most important phases of the cohesin cycle and the coordination of cohesion release with the progression through mitosis. We further discuss recent evidence that has revealed additional functions for centromeric localization of cohesin in the fidelity of mitosis in metazoans. Beyond its well-established role as "molecular glue", centromeric cohesin complexes are now emerging as a scaffold for multiple fundamental processes during mitosis, including the formation of correct chromosome and kinetochore architecture, force balance with the mitotic spindle, and the association with key molecules that regulate mitotic fidelity, particularly at the chromosomal inner centromere. Centromeric chromatin may be thus seen as a dynamic place where cohesin ensures mitotic fidelity by multiple means.

1 The Importance of Gluing DNA Molecules

Mitosis is one of the most dynamic periods in the life of the cell. In a short period of time, the cell condenses its DNA into discrete chromosomes, aligns them on the metaphase plane, and finally, destroys the forces that hold equal-DNA molecules together, creating two identical daughter nuclei in the process. The fidelity of this process relies on cells' ability to keep the two identical sister chromatids together from the moment of DNA replication until the later stages of mitosis, once (and only when) the conditions for their separation are met.

M. Mirkovic · R.A. Oliveira (✉)
Instituto Gulbenkian de Ciência, Rua da Quinta Grande,
2780-156 Oeiras, Portugal
e-mail: rcoliveira@igc.gulbenkian.pt

B.E. Black (ed.), *Centromeres and Kinetochores*, Progress in Molecular and Subcellular Biology 56, DOI 10.1007/978-3-319-58592-5_20

Sister chromatid cohesion provides cells with the ability to determine chromosome identity, as cohesed sister chromatids are identical and therefore need to be pulled to opposite poles. Moreover, sister chromatid cohesion provides the counterforce that resist the pulling force of the spindle, thus preventing premature sister chromatid separation (Oliveira et al. 2010; Tanaka et al. 2000), and random chromosome segregation. Cohesin is also essential for the correct geometry of the kinetochore region which promotes effective, stable capture of the kinetochores by the mitotic spindle, leading to the biorientation of chromosomes during metaphase (Ng et al. 2009; Sakuno et al. 2009; Stephens et al. 2013).

Therefore, to align chromosomes at the metaphase plane and segregate them symmetrically, chromosomal cohesive state must be maintained until anaphase at all cost. Premature separation of chromosomes renders the cell unable to align chromosomes correctly, causing random segregation of the genetic material and

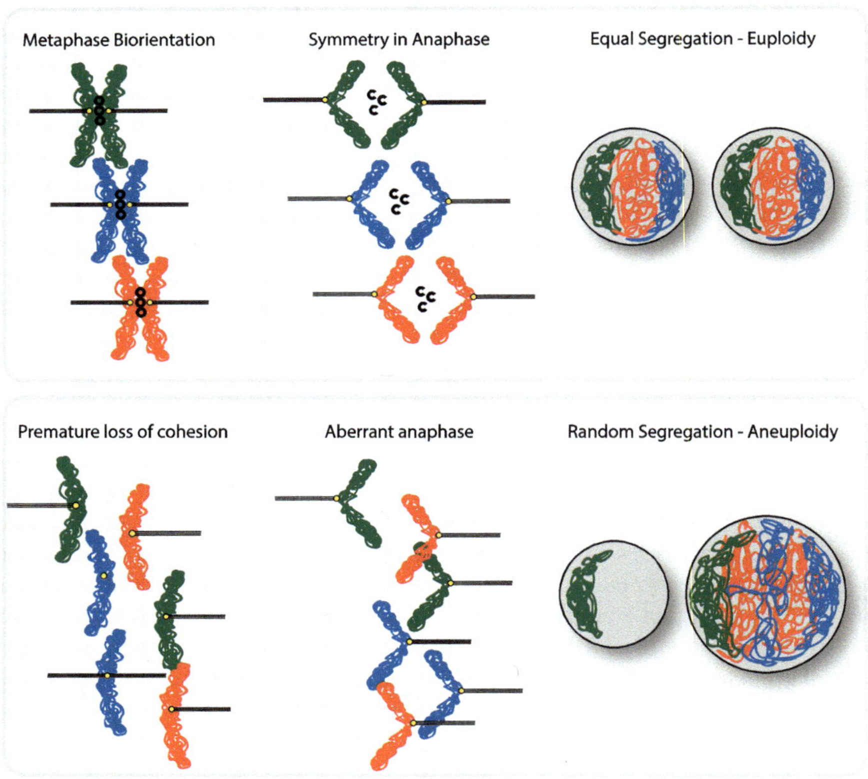

Fig. 1 Sister chromatid cohesion during mitosis. Cohesin is essential for biorientation of chromosomes on the metaphase plane and the symmetry of subsequent anaphase. Defects in sister chromatid cohesion result in premature separation of sister chromatids, resulting in random chromosome segregation and aneuploidy

consequent aneuploidy (Fig. 1), which is usually lethal and a common cause of human pathological conditions (Box 1).

Box 1—Sister chromatid cohesion defects and human disease
Proteins involved in keeping the two sister DNAs together have been linked to several human-health and reproduction conditions. Defects in cohesion and mechanisms regulating cohesin are common amongst **cancer** cells (De Koninck and Losada 2016; Losada 2014). Cancer cells display Chromosomal Instability (CIN) characterized by frequent gain or loss of chromosomes (Holland and Cleveland 2009). CIN enhances the speed at which the cancer cells can evolve, by gaining or losing whole chromosomes, making them highly adaptable to any possible treatment. Interestingly, recent studies have been able to reverse the CIN of multiple cancer-derived cells lines by reinstating the network associated with protection of cohesion (Tanno et al. 2015).

Age-related **female infertility** has also been proposed to relate with cohesion decay in aged oocytes, giving rise to genetic abnormalities such as Down's syndrome (Reviewed in Webster and Schuh 2016). "Cohesion fatigue", evidenced by decreased levels of cohesion is followed by segregation defects and decreased fertility in oocytes (Patel et al. 2015; Zielinska et al. 2015). It is currently thought that the meiotic cohesin variant is loaded into an oocyte only during the germ-line development (pre-meiotic S phase) without significant turnover (Burkhardt et al. 2016; Tachibana-Konwalski et al. 2013). This would mean that oocytes solely rely on cohesion established during their creation, and maintain it throughout the entire reproductive life cycle of the female, which lasts for decades in humans. Studies in human oocytes have revealed an increased distance between bivalents in meiosis of older females, leading to aberrant kinetochore attachments and segregation errors (Patel et al. 2015; Zielinska et al. 2015).

Other rare developmental disorders have also been linked to the cohesion process and are now known as "***Cohesinopathies***" (reviewed in references Dorsett 2007; Liu and Krantz 2008; Remeseiro et al. 2013). Most of these diseases are linked to the non-mitotic roles of the cohesion apparatus (e.g. regulation of transcription and genome architecture). However, a certain number of Cohesinopathies, such as the Roberts or Warsaw breakage syndromes exhibit cohesion defects between replicated chromatids during mitosis, resulting in aneuploidy and mitotic defects (Tomkins et al. 1979; van der Lelij et al. 2010).

In order to understand how defects in chromosome cohesion take place, it is fundamental to understand the molecular structure of the cohesin complex, as well as the principle mechanisms underlying its loading, establishment, and release during the cell cycle. Here we summarize our current knowledge on the regulation of sister chromatid cohesion. We further highlight the importance of such dynamic regulation for the efficiency of mitosis, in mechanisms that go far beyond cohesin's primary role in sister chromatid cohesion.

2 Cohesin: The Molecular Glue that Holds Chromosomes Together

The protein complex responsible for the pairing of replicated chromosomes is called cohesin (Guacci et al. 1997; Michaelis et al. 1997) (Fig. 2). Cohesin is a tripartite ring complex, which topologically entraps replicated DNA molecules keeping them together until the onset of anaphase (Haering et al. 2008; Ivanov and Nasmyth 2005). The core of this ring complex is composed out of three molecules: SMC 1 and SMC 3 (belonging to the Structural Maintenance of Chromosomes protein family) and the kleisin subunit Rad21/Scc1, which connects them (Nasmyth and Haering 2009; Peters et al. 2008) (Fig. 2). Additional proteins directly associate with the cohesin complex (Scc3/SA, Pds5, WAPL, Sororin) and are thought to have critical roles in cohesin dynamics, and consequently mitotic fidelity (summarized in Box 2).

Box 2—Cohesin and its regulators
See Fig. 3 and Table 1.

The most popular, and soundly tested cohesin model postulates that cohesin keeps sister chromatids together by entrapping sister DNA fibers within the same cohesin ring (Haering et al. 2008). EM-studies support that cohesin rings are about 40 nm in diameter (Haering et al. 2002) thus providing sufficient space for enclosing two 11 nm chromatin fibers. Other models have been proposed, such as the "handcuff" model, in which cohesion is mediated by two interlinked cohesin complexes, each entrapping its own DNA fiber (Diaz-Martinez et al. 2008; Guacci 2007). In either case, solid evidence supports that cohesin's interaction with DNA is of a topological nature (Haering et al. 2008; Ivanov and Nasmyth 2005),

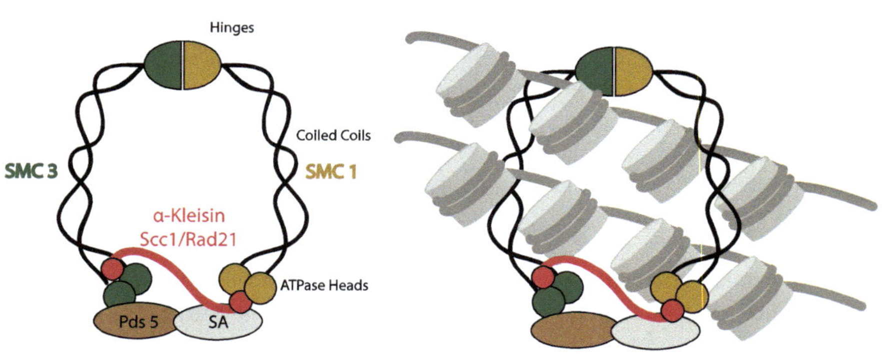

Fig. 2 The cohesin complex. Cohesin complex forms a ring-shaped molecule that topologically embraces sister DNA molecules inside its ring

emphasizing that regulation of cohesin binding and function relies on the opening and closing the interphases between the core components (discussed below).

Besides its role in sister chromatid cohesion, cohesin also regulates transcription, contributes to the DNA repair mechanisms, and participates in the organization of the genome in mitotic and post-mitotic tissues (Reviewed in Nasmyth and Haering 2009; Peters et al. 2008).

The distribution and presence of cohesin on chromatin during the cell cycle coincides with its multiple roles (Fig. 4). Cohesin is loaded onto chromatin during G1 phase in budding yeast (Guacci et al. 1997), and already in telophase in vertebrates (Losada et al. 1998). Fluorescence Recovery After Photobleaching (FRAP) studies have shown that during G1 phase cohesin is dynamically interacting with the DNA (Gerlich et al. 2006). Similar dynamics was observed in cells that are not undergoing mitotic divisions, for example endocycling *Drosophila melanogaster* salivary glands (Eichinger et al. 2013). This highly dynamic nature of cohesin–DNA interaction in non-dividing or non-replicated cells is believed to relate to cohesin's role in transcription regulation and interphase genome architecture.

Following the onset of S phase, a fraction of cohesin molecules establishes cohesion between newly replicated sister chromatids. Specific changes on the

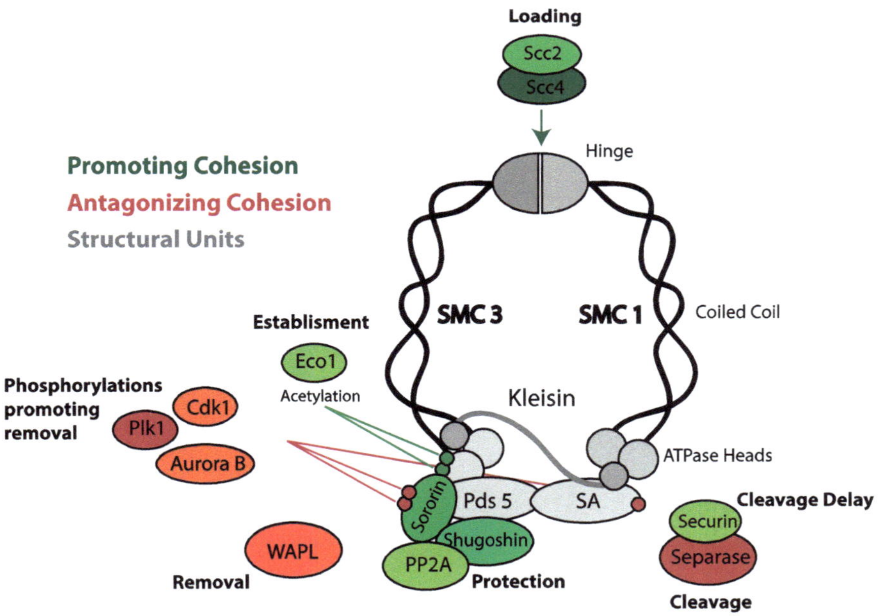

Fig. 3 Cohesin and associated molecules. The cohesin complex and the different associated molecules that modulate cohesin's function. Molecules are color-coded according to their influence on the stability of cohesin's association with chromatin (molecules that promote cohesion are in *green*; cohesion antagonists in red and proteins with dual effect in *orange*)

Table 1 Cohesin and its regulators

Name	Function	Mode of operation
SMC1	Structural/ATP binding and hydrolysis	Part of the ring; interacts with SMC3 through the hinge and the α-kleisin through the ATPase heads; binds and hydrolysis ATP
SMC3	Structural/ATP binding and hydrolysis	Part of the ring; interacts with SMC1 through the hinge and the α-kleisin through its coiled-coil; binds and hydrolysis ATP
α-kleisin/Rad21/Scc1	Structural/scaffold	Closes the ring by bridging the heads of SMC1/3; serves as a scaffold to other regulatory proteins
Pds5	Structural/scaffold	Connects the ring with various other molecules: (e.g. Sororin and WAPL)
SA/Scc3	Structural/scaffold	Essential for the ring structure; phosphorylated by Plk1 during prophase pathway
Scc2/NIPB/NIPBL	Loading	Required for cohesin loading; possibly opens SMC1/3 hinges
Scc4/Mau-2	Loading	Required for cohesin loading; possibly opens SMC1/3 hinges
Eco1/ESCO1	Establishment	Acetylates SMC3 heads, promoting sororin recruitment
WAPL	Removal	Interacts with Pds5 and disrupts the SMC3/kleisin interface, opening the ring
Sororin	Protection	Blocks WAPL interaction with Pds5
Cdk1/CycB	Removal/protection/cleavage inhibitor	Removes sororin; promotes shugoshin localization; inhibits separase
Aurora B	Removal/protection	Removes sororin; promotes shugoshin localization
Plk1	Removal/cleavage	Promotes cohesin release through SA phosphorylation; enhances Rad21/Scc1 cleavage by separase
Shugoshin	Protection	Recruits PP2A to the cohesin complex
PP2A	Protection	Dephosphorylates sororin
Securin	Cleavage inhibitor	Inhibits separase activation
Separase	Cleavage	Cleaves the Rad21/Scc1 subunit and opens the ring

cohesin complex (discussed in Sect. 3.2) ensure the post-replicative stabilization of cohesin–DNA interaction concomitantly or right after replication fork passage. This cohesive state is then maintained until the subsequent mitosis.

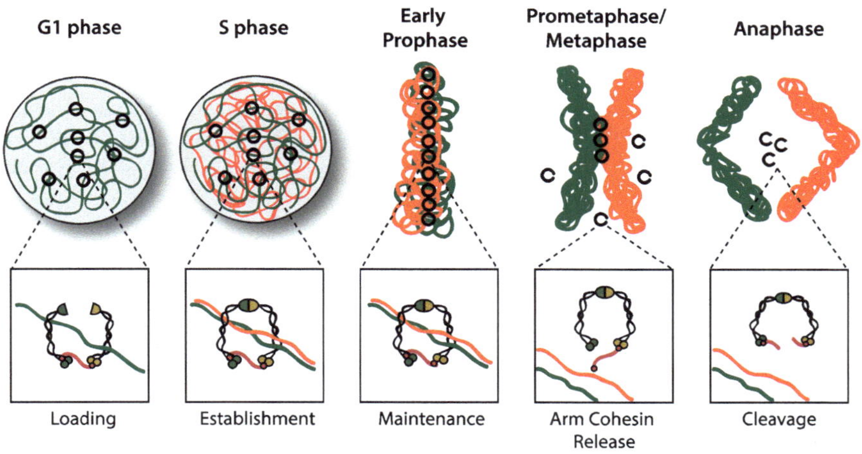

Fig. 4 Overview of the cohesin cycle. Cohesin is loaded in telophase or G1, and is dynamically associated with chromatin. Upon replication, cohesion is established, connecting two replicated strands. Non-centromeric cohesin is removed from chromosome arms during prophase in metazoans, resulting in X-shaped chromosomes in metaphase. Finally, cohesin is cleaved during anaphase, allowing for the separation of sister chromatids

In early mitosis, the majority of the cohesin complexes are released from chromosome arms. By the time cells reach metaphase, cohesion is solely maintained by a small pool of cohesin molecules retained at the centromeric and pericentromeric regions (Losada et al. 1998; Waizenegger et al. 2000; Warren et al. 2000), providing chromosome their characteristic "X-shape".

At the onset of anaphase, remaining centromeric cohesin is destroyed in a rapid and acute manner by a cysteine protease named separase, allowing the separation of sister chromatids by the spindle (Uhlmann et al. 1999). Separase cleaves the kleisin subunit Rad21/Scc1, releasing sister chromatids from topological entrapment. The destruction of cohesin during anaphase marks the point of no return for the mitotic cell: once cohesin is cleaved, separation of the chromatids is rapid and irreversible. Consequently, release of cohesin from mitotic chromosomes is a highly regulated affair.

The key surveillance mechanism governing cohesin release is the Spindle Assembly Checkpoint (SAC) (Reviewed in Musacchio and Salmon 2007). The SAC regulates cohesin cleavage by delaying the onset of anaphase until all the chromosomes are bioriented on the metaphase plane, with sister chromatids correctly oriented towards the opposite poles by the spindle (biorientation). SAC mediates this delay by directly inhibiting the Anaphase Promoting Complex/Cyclosome (APC/C), whose activity is needed for anaphase events. APC/C mediates cohesin cleavage through indirect activation of separase, the protease responsible for proteolytic opening of the cohesin ring.

Loss of cohesin or its regulators in virtually all organisms results in premature separation of sister chromatids (Guacci et al. 1997; Losada et al. 1998; Michaelis et al. 1997; Mirkovic et al. 2015; Sumara et al. 2000; Vagnarelli et al. 2004), arguing that cohesin is the most significant force that counteracts spindle forces. Nevertheless, it is conceivable that other forces may additionally play a role in chromosome cohesion. In particular, DNA–DNA intertwines (catenation) have long been argued to contribute to cohesion during mitosis (Reviewed in Diaz-Martinez et al. 2008; Guacci 2007; Liu et al. 2009b). Due to the helical nature of the DNA molecule, the replication fork passage creates tangles between replicated DNA molecules. These catenations need to be resolved before the onset of anaphase; otherwise, the entanglements will cause chromosome bridges and breakages in the DNA molecule. Topoisomerase II is the molecule responsible for decatenation of these linkages and inhibition of this enzyme leads to accumulation of catenations, which are sufficient to confer cohesion even in the absence of cohesin proteins (Toyoda and Yanagida 2006; Vagnarelli et al. 2004).

How much residual catenation contributes to cohesion during normal mitosis is a matter of debate. Although residual catenation has been observed even in anaphase segregating chromatids (Baumann et al. 2007), inhibition of topoisomerase II specifically during metaphase has only a small effect on the efficiency of chromosome segregation (Oliveira et al. 2010). This suggests that residual catenation may contribute to chromosome cohesion; yet, it is insufficient to resist the drastic spindle forces affecting chromosomes during mitosis. More importantly, unlike cohesin's destruction, which requires SAC silencing and APC/C activation, there is little to no evidence that removal of residual catenation is delayed by cell cycle progression checkpoints which control mitosis. SUMOylation of topoisomerase II has been proposed to restrict centromeric decatenation during mitosis (Bachant et al., 2002; Dawlaty et al. 2008; Ryu et al. 2010), but there is no evidence that this reaction is under surveillance of the SAC.

Thus, regulation of the cohesive state of chromosomes is mechanistically linked to the control of cohesin's association with chromatin throughout the cell cycle, which will be discussed below.

3 The Cohesin Cycle

3.1 *The Cohesin Cycle I: Chromatin Loading*

Cohesin loading onto chromatin is dependent on a two-protein complex known as Scc2/4, also known as NIPBL/MAU-2 (Ocampo-Hafalla and Uhlmann 2011) (Fig. 5). The Scc2/Scc4 loading complex is essential for sister chromatid cohesion during G1/S phase, but not during G2 (Ciosk et al. 2000; Uhlmann and Nasmyth 1998). This would entail that the Scc2/Scc4 has a primary function of loading cohesin onto the chromatin, but not in its stabilization or maintenance. Given the ring-like architecture of cohesin, its loading onto chromatin requires opening of the ring. Elegant experiments with fusion of interfaces between different cohesin components support that the entry gate for cohesin loading resides at the interface of the SMC1 and SMC3 hinge domains, in an ATP-dependent process

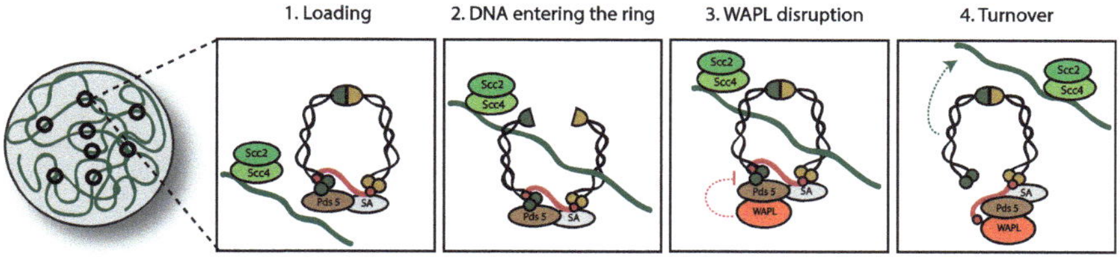

Fig. 5 Cohesin loading and turnover. Cohesin loading onto DNA depends on the Scc2/4 complex. DNA loading involves opening of the SMC1/3 interface, the hinge. Before replication, this interaction is dynamic, as loaded cohesin can be destabilized by WAPL, which opens the SMC3/Kleisin interface and releases cohesin from the chromatin

(Arumugam et al. 2003; Gruber et al. 2006; Weitzer et al. 2003). Nevertheless, the molecular mechanism by which Scc2/4 promote cohesin's loading remains unknown.

Sites of cohesin loading do not necessarily coincide with cohesin's accumulation sites. This is mostly due to the fact that once loaded, cohesin complexes can slide on the DNA molecule (Hu et al. 2011; Lengronne et al. 2004; Ocampo-Hafalla et al. 2016; Stigler et al. 2016). Additionally, before DNA replication, the cohesin molecules display a highly dynamic association with DNA (Gerlich et al. 2006). Dissociation of cohesin from un-replicated DNA molecules is mediated by Wings-apart like protein (WAPL) (Gandhi et al. 2006; Kueng et al. 2006; Verni et al. 2000). Upon binding to the cohesin complex, WAPL removes cohesin from chromatin by disrupting the interface between SMC3 and Rad21/Scc1 subunits (Buheitel and Stemmann 2013; Eichinger et al. 2013).

Cohesin loading is not a uniform event across the chromatin landscape and is found to be enriched at the centromeric/pericentromeric regions in most species studied so far (Blat and Kleckner 1999; Glynn et al. 2004; Hahn et al. 2013; Oliveira et al. 2014). Studies in budding yeast support that cohesin enrichment at the centromere is dependent on centromeric DNA sequences as well as proteins involved in kinetochore assembly (Megee and Koshland 1999; Tanaka et al. 1999; Weber et al. 2004). However, species with longer centromeric sequences, such as fission yeast, rely on heterochromatin rather than centromeric sequences for cohesin enrichment (Bernard et al. 2001; Nonaka et al. 2002). In accordance, recent studies in *D. melanogaster* showed that cohesin enrichment at ectopic regions of pericentromeric heterochromatin occurs in the absence of a proximal centromere, most likely due to preferential binding of the cohesin loading factor Scc2/Nipped B (Oliveira et al. 2014). The preferential activity of Nipped B at the centromeric region is thought to be due to the specific state of pericentromeric heterochromatin, mainly H4K20 and H3K9 methylations and the presence of HP1 protein, tough clear links have been controversial (Hahn et al. 2013; Koch et al. 2008; Oliveira et al. 2014).

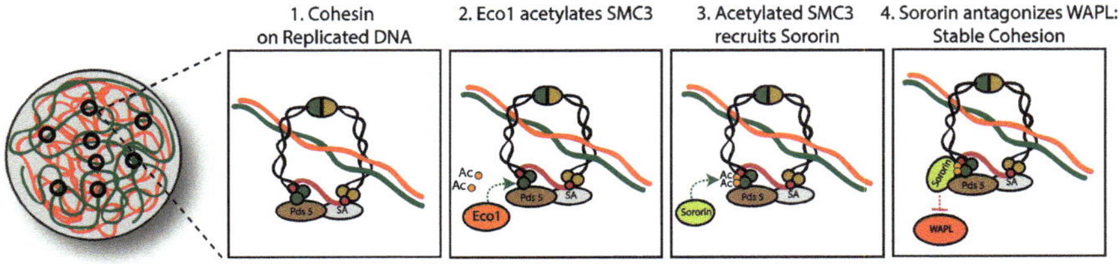

Fig. 6 Cohesion establishment during S phase. Upon DNA replication, a fraction of cohesin becomes stable on the chromatin. This happens due to SMC3 acetylation by Eco1 and recruitment of Sororin, protecting the cohesin complex from WAPL removal. This stable fraction of cohesin is considered "cohesive" cohesin, stably binding sister chromatids until the end of mitosis

3.2 *The Cohesin Cycle II: Cohesion Establishment*

Cohesin establishment occurs during replication, at the time the newly replicated DNA molecule is being formed. Disruption of cohesin loading during G1 results in sister chromatid defects, while disruption during G2 does not. This means that the "effective" cohesion is established during S phase, during DNA replication (Uhlmann and Nasmyth 1998). At the onset of replication, the dynamic properties of cohesin turnover change and a new pool of stable, "cohesive" cohesin can be identified by FRAP (Gerlich et al. 2006).

Stabilization of cohesin complexes upon replication depends on the Eco1 acetyl transferase (Skibbens et al. 1999; Toth et al. 1999) (Fig. 6). This enzyme acetylates cohesin associated with replicated DNA at specific lysine residues on SMC3 and failure to acetylate leads to cohesion defects and cell death. The mechanism by which SMC3 lysine acetylation prevents cohesin de-association once it is bound to chromatin is contentious (reviewed in Rudra and Skibbens 2013). Some studies propose models in which the acetylation locks the SMC3/kleisin interface, effectively closing the ring; however, these findings are inconsistent with the fact that SMC3 can be acetylated before replication (Rudra and Skibbens 2013). SMC3 acetylation during the S phase has also been shown to confer cohesin protection by aiding the recruitment of Sororin, which favors cohesion establishment by protecting acetylated cohesin complexes from WAPL-mediated removal (Nishiyama et al. 2010).

These stably associated cohesin molecules [$\sim$30% of total nuclear cohesin (Gerlich et al. 2006)] are responsible for sustaining cohesion from the time of DNA replication until the subsequent mitosis.

3.3 *The Cohesin Cycle III: Cohesin's Prophase Release and Retention at the Centromere*

Once the cell enters mitosis, profound changes in the distribution of cohesin begin to take place. Cohesin at the chromosome arms is removed while centromeric cohesion is retained (Losada et al. 1998; Waizenegger et al. 2000; Warren et al. 2000). The loss of

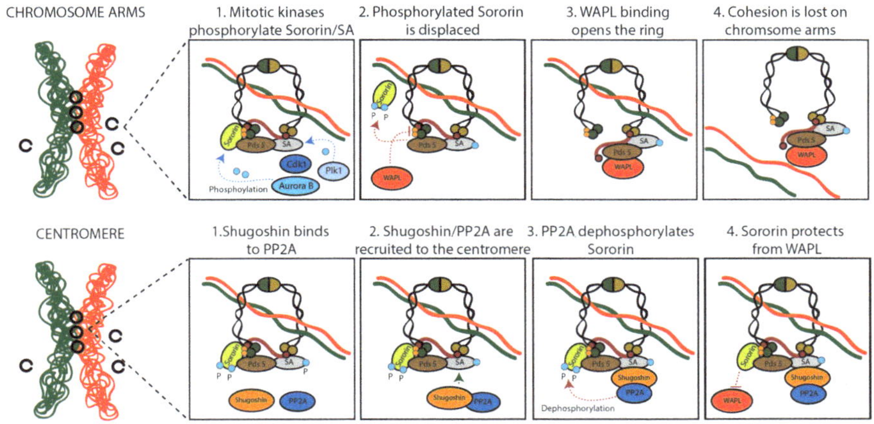

Fig. 7 Cohesin release during early mitosis and centromeric protection. In metazoans, cohesin is removed from the arms by the prophase pathway. Mitotic kinases phosphorylate sororin and SA. Phosphorylation induces sororin displacement, which allows WAPL to destabilize cohesin. Centromeric cohesin complex are protected from this removal process as Shugoshin/PP2A complex protects centromeric cohesion from WAPL-mediated removal

arm cohesion, coupled with centromeric retention gives the characteristic "X" shape to the metaphase chromosomes. The removal of cohesin from the arms in early mitosis is a consequence of the "prophase pathway" which mainly relies on action of WAPL protein (Gandhi et al. 2006; Kueng et al. 2006).

WAPL imposes opening of the cohesin ring by disrupting the interface between SMC3 and Rad21/Scc1 subunits (Buheitel and Stemmann 2013; Eichinger et al. 2013) (Fig. 7). Consequently, WAPL mutations or knockdown leads to the loss of the characteristic X shape of chromosomes, with cohesin remaining all over chromosome arms (Gandhi et al. 2006; Haarhuis et al. 2013; Kueng et al. 2006).

Several mitotic kinases contribute to the process of cohesin removal, by phosphorylating key proteins involved in the cohesin cycle. Aurora B and Cyclin-Dependent Kinase 1 (Cdk1) were shown to antagonize Sororin by phosphorylation, resulting in its dissociation from chromosome arms during prophase (Dreier et al. 2011; Nishiyama et al. 2013). WAPL and Sororin directly compete for the binding to the cohesin-associated protein Pds5 (Nishiyama et al. 2010). The removal of Sororin from chromosome arms during prophase favors WAPL binding, and consequently the removal of cohesin complexes from chromosome arms. In addition to antagonizing Sororin, Aurora B seems to participate in WAPL activation, thus directly promoting cohesin removal (Nishiyama et al. 2013). Polo-Like kinase (Plk) is another key mitotic kinase participating in the cohesin cycle. The phosphorylation activity of Plk1 is crucial for the release of cohesin during the prophase pathway by phosphorylation of SA (Hauf et al. 2005; Lenart et al. 2007; Sumara et al. 2002). The net result of these changes in the cohesin complex results in the removal of most of cohesin from chromosome arms but not from the centromeric region.

How are centromeric complexes protected from prophase pathway removal? A key molecule in the protection of centromeric cohesion is called Shugoshin, meaning "Guardian Spirit" in Japanese. Shugoshin confers protection of cohesin specifically at the centromere of both mitotic and meiotic cells (Kerrebrock et al. 1992; Kitajima et al. 2004; McGuinness et al. 2005).

Shugoshin is moved to the centromeric chromatin in complex with the PP2A phosphatase at the onset of mitosis (Kitajima et al. 2006; Liu et al. 2013b). Sugoshin/PP2A complex protects centromeric cohesin from WAPL-mediated removal by several means: It antagonizes the Aurora B-/Cdk1-mediated phosphorylation of Sororin and thereby favors Sororin interaction with Pds5, shifting the WAPL/Sororin competition for cohesin binding towards Sororin, preventing WAPL-mediated removal (Dreier et al. 2011; Liu et al. 2013b; Nishiyama et al. 2013). Aurora B and Cdk1 also phosphorylate and aid in the centromeric localization and activation of Shugoshin (Kitajima et al. 2006; Liu et al. 2013b; Tanno et al. 2010). This means that Cdk1 and Aurora B have conflicting roles in cohesin maintenance. They destabilize Sororin and thereby promote cohesin dissociation along chromosome arms, while at the same time localize and activate Shugoshin at the centromere, allowing for cohesin protection. Shugoshin also counteracts the effect of Plk-1 mediated phosphorylation of SA (Hauf et al. 2005; Kitajima et al. 2006; McGuinness et al. 2005). This was initially thought to rely on de-phosphorylation of this subunit by PP2A. However, recent evidence reveals that Shugoshin-bound centromeric cohesin complexes contain a phosphorylated form of SA (Liu et al. 2013b), suggesting alternative mechanisms may exist. In accordance, structural analyses propose that Shugoshin may additionally work by a direct competition with WAPL for the binding to cohesin (Hara et al. 2014).

This protection mechanism is of outmost importance as centromeric cohesin complexes are the only ones that suffice cohesion maintenance during prometaphase and metaphase, while chromosomes are under drastic pulling and pushing forces exerted by the mitotic spindle to accomplish chromosome alignment.

3.4 *The Cohesin Cycle IV: The Final Cut*

Mitosis is a process of trial and error, with a few decisive breakpoints. Mitotic events of chromosome attachment, substrate phosphorylation and biorientation are mostly redundantly regulated, and reversible. This allows for ample error correction in an otherwise error prone process. However, once the metaphase is formed, and chromosomes are bioriented, the cell reaches the point of no return: cohesin cleavage.

The cleavage of cohesin at the metaphase-to-anaphase transition is conducted by a large cysteine protease called separase, which cleaves the kleisin subunit, distancing the heads of SMC1 and SMC3 subunits (Lin et al. 2016; Uhlmann et al. 2000). This opens the cohesin ring, releasing sister DNA molecules from the proteinaceous cage (Fig. 8).

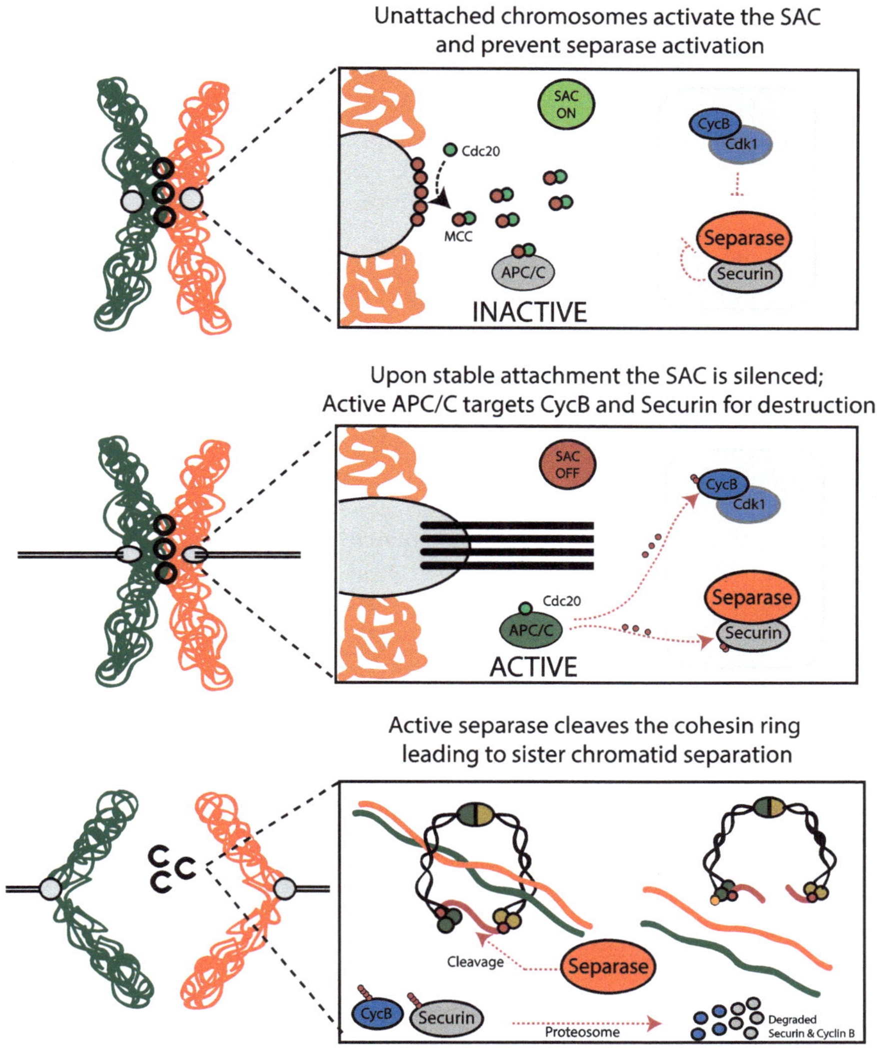

Fig. 8 Cohesin cleavage at the metaphase-to-anaphase transition. **a** In the presence of unattached kinetochores, the spindle assembly checkpoint is activated and generates the formation of the mitotic checkpoint complex (*MCC*) that prevents anaphase promoting complex/cyclosome activation. Separase is kept inactive by securin and Cdk1/CyclinB binding. **b** Upon bipolar attachment, the SAC signal is extinguished and the APC/C is activated. Active APC/C ubiquitinates securin and Cyclin B and targets them for degradation. **c** Active separase cleaves the Rad21/Scc1 subunit and causes ring opening. This opening allows the spindle to drag sister chromatids to opposite poles

Once the forces that hold chromosomes together are released, there is no going back: therefore, centromeric cohesin cleavage must occur only after multiple safeguard mechanisms have been satisfied. Separase activity is thus tightly regulated and inhibited through multiple mechanisms until the onset of anaphase.

First, separase is inhibited by the binding of Securin, whose degradation is a prerequisite for sister chromatid separation (Ciosk et al. 1998; Hirano et al. 1986; Zou et al. 1999). Securin inhibits separase by binding to its active site and abolishing its interaction with other substrates (Hornig et al. 2002; Lin et al. 2016). However, mutants for Securin in several organisms do not suffer from premature loss of cohesion, evidencing that other mechanisms of separase inhibition must be in place (Alexandru et al. 2001; Hellmuth et al. 2015) (see below). Furthermore, Securin has been proposed to work as a separase chaperone by binding to the nascent separase and aiding in its proper folding and activity (Jallepalli et al. 2001). Consequently, Securin was shown to be required for sister chromatid separation in fission yeast and *D. melanogaster* (Funabiki et al. 1996; Stratmann and Lehner 1996).

The second layer of separase inhibition is mediated by the Cdk1-Cyclin B complex. Cyclin B-Cdk1 phosphorylates separase and this phosphorylation promotes Cdk1-CycB-separase binding, preventing separase activation until the onset of anaphase (Gorr et al. 2005; Stemmann et al. 2001). The dual inhibition of separase by CycB-Cdk1/securin is lifted by the APC/C, an E3 ubiquitin ligase, which is the main effector of anaphase (reviewed in (Primorac and Musacchio 2013; Sullivan and Morgan 2007)). The APC/C ubiquitinates both securin and Cyclin B, which targets them for degradation by the proteasome and releases separase from its double leash. This, in turn, leads to cohesin cleavage and the onset of anaphase (Oliveira and Nasmyth 2010).

Given the importance of this transition, the APC/C itself is tightly regulated during mitosis by a surveillance mechanism known as the Spindle Assembly Checkpoint (SAC) (Musacchio and Salmon 2007; Sullivan and Morgan 2007) (Fig. 8a). The key effector of this mechanism is the Mitotic Checkpoint Complex (MCC). Unattached kinetochores catalyze the formation of this inhibitory complex, which sequesters Cdc20, a key activator required for APC/C activity (Musacchio and Salmon 2007; Sullivan and Morgan 2007). The MCC complex is composed of Mad2, BubR1, Bub3 and Cdc20 that form a complex that actively binds and inactivates the APC/C (Primorac and Musacchio 2013). As long as the SAC is active and the MCC is being produced at unattached kinetochores, the APC/C will not be activated by Cdc20, Cyclin B and Securin will remain intact, Separase inactive, and cohesin will not be cleaved.

This equilibrium changes once metaphase is achieved and chromosomes are bioriented (Fig. 8b). Stable chromosome attachments result in SAC satisfaction and the release of Cdc20 from the inhibitory MCC complex (Primorac and Musacchio 2013; Sullivan and Morgan 2007). Once this happens, APC/C binds Cdc20 becoming active to ubiquitinate Cyclin B and Securin. Ubiquitination promotes the proteasome-mediated degradation of these targets and consequently the release of Separase from its inhibition. Anaphase is imminent.

Since chromosome biorientation and microtubule attachment are highly dynamic processes, once all the chromosomes are bioriented, the decision to commit to anaphase must be rapid and the execution swift. Indeed, live imaging analysis revealed that separase-mediated cohesin cleavage happens within a few minutes during the metaphase-to-anaphase transition (Gerlich et al. 2006; Oliveira et al. 2014; Yaakov et al. 2012).

In order to achieve this sharp metaphase-to-anaphase transition and rapid cohesin cleavage, multiple positive feedback mechanisms are needed to create a molecular switch. First, Separase has autocatalytic activity, and once released from its Cyclin B-Cdk/Securin inhibition, it is able to cleave itself, and convert to an even more enzymatically potent form (Waizenegger et al. 2002). Furthermore, APC/C is constantly ubiquitinating the MCC and trying to pry away the Cdc20 subunit away from it, weakening the SAC signal in the process (He et al. 2011; Uzunova et al. 2012). In this way APC, accelerates its own release from SAC inhibition during anaphase.

In addition (or in parallel) to separase-mediated cleavage, the cohesin protection machinery is also released from centromeres at the metaphase-to-anaphase transition, which may accelerate cohesin release. Release of Shugoshin/PP2A from the centromeres may additionally promote the Plk1-mediated phosphorylation of Rad21/Scc1 (Plk1-mediated), which enhances its cleavage by the Separase (Alexandru et al. 2001; Hornig and Uhlmann 2004). Moreover, both Shugoshin and Sororin, two key molecules involved in cohesin protection, are directly targeted for degradation by the APC/C (Karamysheva et al. 2009; Rankin et al. 2005). Whether or not removal of the mechanisms involved in cohesin protection actively contribute to the sharp cohesion release at the metaphase-to-anaphase transition remains to be determined.

As discussed above, cohesin cleavage is only initiated once chromosome biorientation is achieved. Thus, given that chromosomes at this stage are being pulled by mitotic spindle, release of cohesin is on its own sufficient to trigger pole-ward chromosome movement (Oliveira et al. 2010; Uhlmann et al. 2000). This, however, is insufficient for efficient anaphase chromosome movement. Sister chromatid separation, when triggered alone, results in $\sim$ 1/3 slower movements, and concomitant re-activation of the SAC and error-correction mechanisms (Mirchenko and Uhlmann 2010; Oliveira et al. 2010). Uncoupling cohesin cleavage from Cyclin B destruction leads to similar failures in chromosome segregation (Parry et al. 2003; Vazquez-Novelle and Petronczki 2010; Vazquez-Novelle et al. 2014). Successful anaphase onset thus relies not only on a sharp anaphase transition but also on a synchrony between sister chromatid cohesion release and cell cycle progression. The fact that cohesin cleavage is regulated by the APC/C, which cleaves both securin (cohesin release) and Cyclin B (cohesin release + cell cycle transition) should in principle provide this synchrony. Additional feedbacks, however, further ensure that sister chromatid separation occurs in synchrony with inactivation of Cdk1 (reviewed in Kamenz and Hauf 2016).

4 Functional Implications for a Multiple-Step Cohesin Removal

Cohesin binding and release is a dynamic and multi-step process whose mechanisms are mostly conserved across species. Exception goes for the dual-step removal for cohesin during mitosis. In budding yeast, unlike in metazoans, arm cohesion is not removed at the onset of mitosis and the entire cohesin pool is removed at the metaphase-to-anaphase transition by Separase. The question does arise as to why do metazoans have a two-step removal of cohesin? Does accumulation and retention of cohesin specifically at the centromeric region play any specific function in metazoans? When considering the biological significance for cohesion removal during mitosis, one must have interphase functions of cohesin in mind. During prophase removal of cohesin, the Rad21/Scc1 subunit is not cleaved, but disengaged from SMC3 (see Sect. 3.3), leaving intact cohesin complexes in the cytoplasm. This cohesin is not reloaded during mitosis, possibly due to the dissociation of the Scc2/4 loading complex from chromosomes (Watrin et al. 2006; Woodman et al. 2014). However, these intact cohesin complexes can load freely during the impending telophase/G1 and preform roles in transcription regulation and interphase genome architecture early in the subsequent cell cycle. Thus, the prophase pathway may be seen as a recycling mechanism, protecting the majority of cohesin from cleavage during anaphase. It is nevertheless becoming more and more evident, however, that the concentration of cohesin specifically around the centromere fulfills important functions for the efficiency of mitosis, as outlined below.

4.1 *Sister Chromatid Resolution*

During replication, sister DNA molecules become heavily intertwined as a consequence of the unwinding of parental DNA strands and colliding replication forks. In order to segregate these tangled sister molecules into two daughter cells, their catenations must be resolved. Failure to resolve such DNA intertwines by topoisomerase II leads to breaks in the DNA molecules during anaphase, when chromosomes are pulled to the poles by the spindle.

Cohesin was shown to block the action of topoisomerase II (Farcas et al. 2011; Sen et al. 2016), possibly by keeping the two sisters in such close proximity that disfavors their efficient decatenation. Thus, cohesin removal from chromosome arms during prophase is believed to aid sister chromatid resolution along chromosome arms, providing topoisomerase II with enough space to resolve catenations (Fig. 9).

The degree to which sister chromatid resolution can occur in the presence of chromosome-bound cohesin has been hard to estimate. A recent study has elegantly shown that in the absence of WAPL, when cohesin is retained all over chromosome

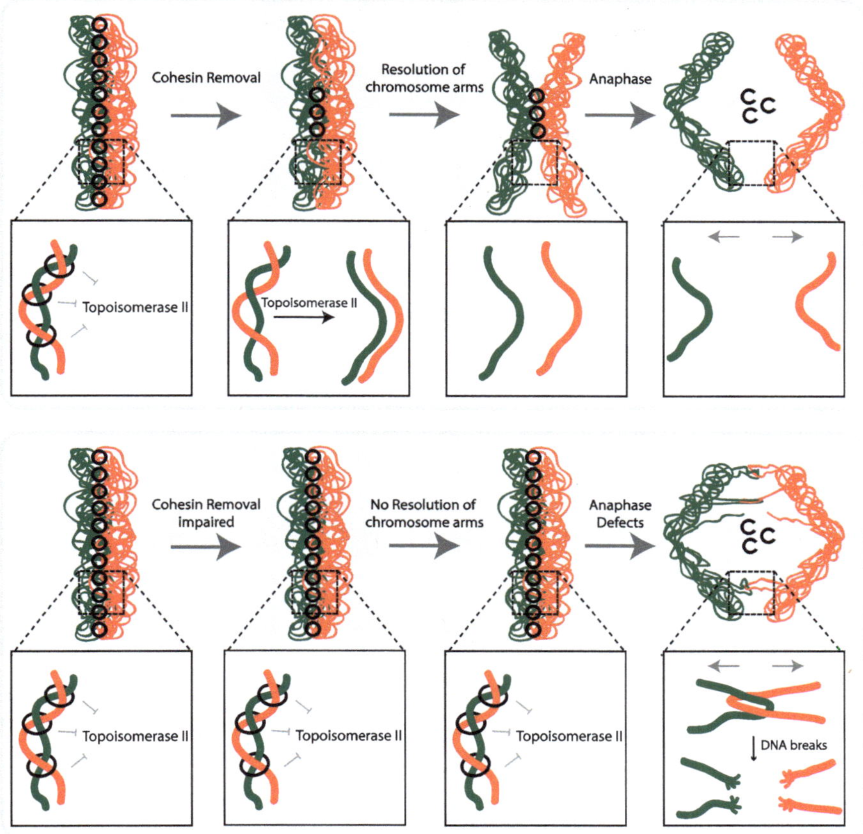

Fig. 9 Cohesin and sister chromatid resolution. Cohesin entrapment prevents efficient decatenation by topoisomerase II. Cohesin removal from chromosome arms ensures proper sister chromatid resolution. Abnormal retention of cohesin on the arms results in residual entanglements and consequently mitotic defects

arms, sister chromatids can be mostly resolved, at least at the limit of the cytological method applied to image differentially labeled sister chromatids (Nagasaka et al. 2016). Thus, although cohesin may impair efficient decatenation, the degree of chromosome intertwines even in the presence of cohesin must be residual.

These residual levels of chromosome intertwines are nevertheless sufficient to impair efficient chromosome segregation. When cohesin is not removed from chromosome arms in a timely manner, which happens if WAPL is downregulated and the prophase pathway inhibited, chromosomes lose their characteristic "X-shape" and cells undergo an erroneous anaphase, marked by detectable chromosome bridges during anaphase (Haarhuis et al. 2013; Tedeschi et al. 2013). Similar results were observed in cells expressing a modified version of Sororin that lacks its Cdk1-phosphorylation site. This version is not removed from chromosomes arms at the onset of mitosis leading to over-cohesion of metaphase chromosome arms and lagging chromosomes during anaphase (Nishiyama et al. 2013). Moreover, chromosome rearrangements that misplace pericentromeric heterochromatin away from the centromere were shown to abnormally accumulate non-centromeric cohesin (Oliveira et al. 2014). These chromosomes also exhibit chromatin stretching during anaphase, specifically at ectopic cohesin-retention sites. Thus, the spatial and temporal positioning of cohesin on the mitotic chromosome is crucial for timely chromosome resolution. Any disturbance, such as prolonged retention or enrichment of cohesin along chromosome arms leads to incomplete sister chromatid separation, followed by mitotic errors.

4.2 Inner-Centromere Defining Platform

Centromeric cohesin has recently emerged as a core component of the inner-centromere network and thereby influences the localization of important machinery that regulates mitotic fidelity (Fig. 10).

Kinetochore-microtubule attachments are regulated by the actions of Aurora B, a key mitotic kinase that destabilizes erroneous kinetochore-microtubule attachments. It is well established that Aurora B destabilizes attachments that are not under tension through the phosphorylation of key kinetochore substrates (Biggins and Murray 2001). This phosphorylation results in microtubule detachment and the

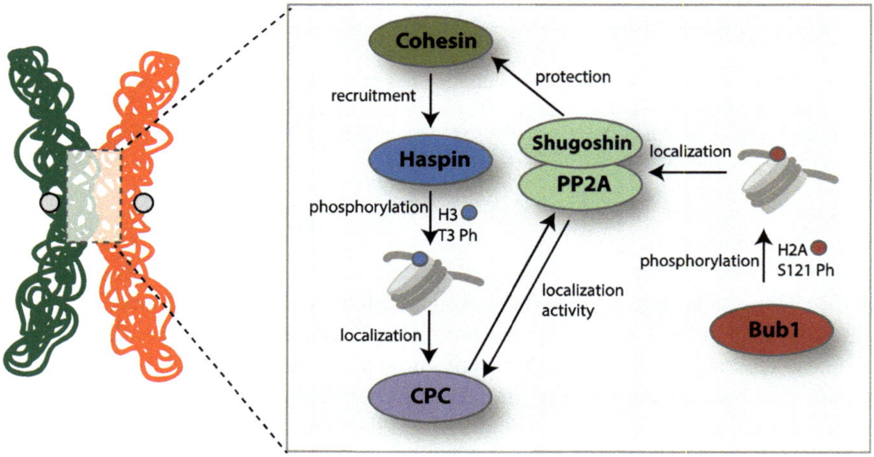

Fig. 10 The inner centromere network. Cohesin sets the blueprint for the inner centromere network, regulating chromosome architecture and microtubule attachment. Cohesin is needed for the recruitment of Haspin kinase, which triggers the cascade resulting in recruitment of CPC and Shugoshin to the pericentromeric region

creation of unattached kinetochores that can trigger SAC signaling. Aurora B, together with its regulatory partners INCENP, Borealin and Survivin, forms the Chromosome Passenger Complex (CPC). This complex decorates the entire chromosome length during early mitotic stages but dynamically shifts its localization towards prometaphase/metaphase, becoming highly enriched at the inner-centromeric region (Reviewed in (Carmena et al. 2012).

Cohesin's importance for CPC localization has been documented in several studies (Carretero et al. 2013; Haarhuis et al. 2013; Kenney and Heald 2006; Mirkovic et al. 2015; Sonoda et al. 2001; Vass et al. 2003) but only recently the mechanistic details for this interaction are being elucidated. CPC localization to the inner centromere was shown to depend on two histone marks: Histone H3 phosphorylation on Threonine 3 (H3pT3) and histone 2A-serine 121 (equivalent to H2A Threonine 120 in humans) phosphorylation (Yamagishi et al. 2010). The cohesin subunit PDS5A interacts with the Haspin Kinase, which is the kinase responsible for H3T3 phosphorylation (Yamagishi et al. 2010). Depletion of Pds5 or cohesin subunits result in delocalized Aurora B and possibly impaired error correction (Carretero et al. 2013; Mirkovic et al. 2015; Yamagishi et al. 2010). Interestingly enough, "too much" cohesin produces a similar phenotype, as WAPL depleted cells also exhibit delocalized Aurora B signals and defective error-correction capacity (Haarhuis et al. 2013).

In addition to CPC localization, cohesin also plays a role in the localization of another key inner centromere component: Shugoshin. Shugoshin interacts directly with cohesin and requires this interaction for its activity (Liu et al. 2013a, b). In this way, cohesin enhances its own centromeric protection but also contributes to other events that are governed by Sgo1 at the centromeres, namely biorientation of sister chromatids, localization of the CPC and SAC silencing (reviewed in Marston 2015).

Thus, while enhancing its own protection, cohesin plays a pivotal role in the establishment of the inner centromere network.

4.3 Force Balance

The binding and stability of microtubule attachments to the kinetochore is enhanced by the tension between the spindle and the kinetochore, both in vivo and in vitro (reviewed in Biggins 2015). Tension-dependent stabilization of kinetochore-microtubule interactions depends on an intrinsic stabilization ability of the mechanical force exerted by the microtubule pulling forces (Akiyoshi et al. 2010), as well as on biochemical changes that promote the stabilization of kinetochore-microtubule interactions. The latter are regulated by Aurora B kinase, responsible for the correction of erroneous microtubule-kinetochore interactions through the phosphorylation of key kinetochore substrates. Upon bipolar attachment, i.e. maximal tension, the increase in the distance between the inner-centromeric Aurora B and the kinetochore is believed to displace Aurora B away from its targets

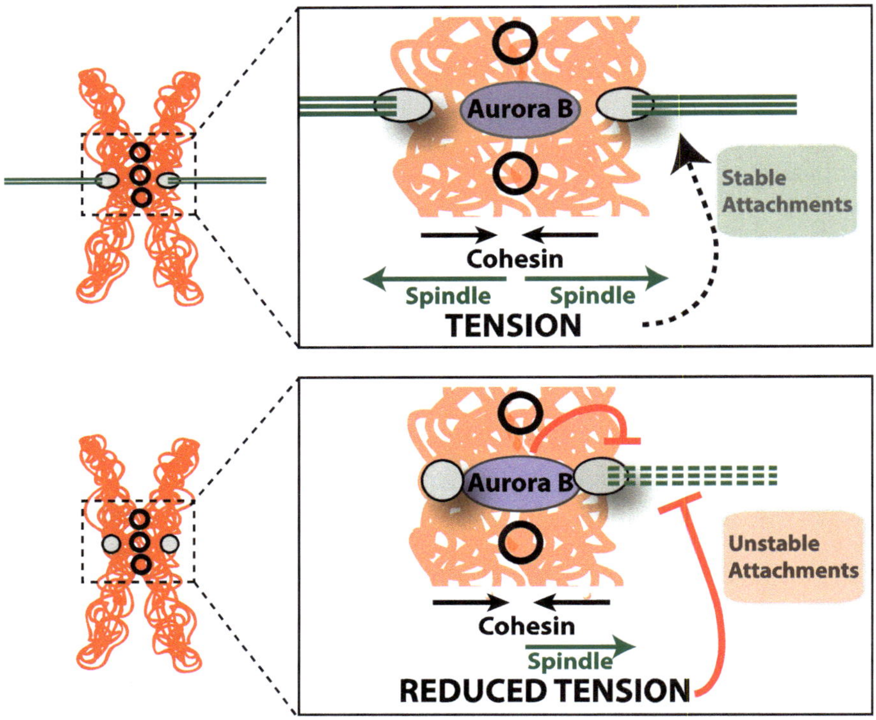

Fig. 11 Force balance. Cohesin is the major force resisting the mitotic spindle during metaphase. The antagonism between cohesin and the spindle results in sufficient tension that is required to stabilize the attachments of microtubules to the kinetochore. Erroneous attachments (e.g. mono-oriented chromosomes or chromosome with the two kinetochores bound to the same pole) are not under sufficient tension. This reduced tension leads to destabilization of these interactions by Aurora B kinase

thus reverting Aurora B-mediated destabilization of microtubule attachments (Liu et al. 2009a) (Fig. 11).

How chromosome tension is established, sensed and ultimately regulates kinetochore-microtubule interactions has been widely investigated. Bipolar attachment increases tension across the entire pericentromeric domain (inter-kinetochore tension), but also within each individual kinetochore, marked by the increase in the distance between the proteins of inner and outer kinetochore (reviewed in (Maresca and Salmon 2010). Both intra- and inter-kinetochore stretch require a counterforce to the spindle to generate stable microtubule attachment and tension. The cohesin ring presents the only force at the centromere that is able to resist the pulling forces of the spindle. Thus, centromeric cohesion contributes to the generation of tension needed for stable chromosome biorientation on the metaphase plane (Fig. 11). It provides the counterforce necessary to maintain a force-equilibrium between with the mitotic spindle, which can generate forces of up to hundreds of piconewtons (Nicklas et al. 1995; Ye et al. 2016). In agreement with cohesin's major role in the establishment of kinetochore tension, loss of cohesin prior to or during metaphase leads to extensive chromosome shuffling along the spindle, as attachments to isolated single sisters are highly unstable (Drpic et al. 2015; Mirkovic et al. 2015; Oliveira et al. 2010) (Fig. 11).

Whether or not cohesin could also contribute to tension sensing has also been speculated. Upon bipolar attachment, tension across sister chromatids will influence the entire pericentromeric domain and evidence suggests that this alone can lead to removal of centromeric cohesin complexes (Eckert et al. 2007; Ocampo-Hafalla et al. 2007). More distal pericentromeric domains would then provide the necessary antagonistic force to the spindle. This dynamic change on the cohesive forces could alone provide a cue to sense bipolar attachment. In agreement, cohesin-associated molecules, particularly Shugoshin, have been proposed to contribute to tension sensing and SAC silencing at the metaphase-to-anaphase transition (reviewed in Marston 2015).

However, inter-kinetochore stretch does not seem to be necessary for tension sensing as chromosomes in which two neighboring kinetochores were artificially tethered, preventing the inter-kinetochore stretch, still resulted in normal metaphase attachment. These experiments imply that mechanical tension exerted on the single kinetochore might be more important than the stretching between kinetochore pairs itself to stabilize chromosome attachments (Nannas and Murray 2014).

Regardless of the exact location that senses chromosome tension, the structure of the pericentromeric domain will likely play a major influence on the force provided by the chromosomes (Stephens et al. 2013). Does this force balance require a specific amount of cohesin at chromosomes and does centromeric accumulation play a role? It is conceivable that reaching the right spindle counterforce requires a fine-tuning of cohesin levels at chromosomes. This has been difficult to tackle experimentally as manipulating cohesin levels is not a trivial task. Metazoan chromosomes with artificial high levels of cohesin (e.g. WAPL knockdown) do display defects in chromosome attachment. Although these have been largely attributed to defects in the localization of the machinery that regulates microtubule-kinetochore attachments (see Sect. 4.2), it remains to be determined the consequences of too much cohesion on tension establishment and sensing, independently of Aurora B localization.

4.4 Anaphase Sharpness

Cohesin destruction marks the onset of anaphase, a point of no return for every dividing cell. As discussed above (see Sect. 3.4), several feedback loops operate at this stage to ensure efficient cohesin cleavage at this crucial transition. Restricting cohesin to centromeric region may be an additional mechanism to ensure fast anaphase onset and promote synchrony of anaphase movements, particularly in organisms containing variable chromosome sizes. Separase is functionally active along the entire chromosome, as evidenced by complete cohesin cleavage in WAPL mutants, in which cohesin is now all over chromosome arms, or in cells expressing Separase sensors targeted to the entire chromosomes (Haarhuis et al. 2013; Oliveira et al. 2014; Shindo et al. 2012; Yaakov et al. 2012).

Whether or not the efficiency of cohesin cleavage is the same all over the chromatin mass has been quite controversial. Direct measurements of Separase activity using engineered sensors at different chromosome loci in budding yeast, failed to detect any delay of cleaving telomeric versus centromeric sites (Yaakov et al. 2012). In contrast, other studies support that removal of cohesin at regions distal to the centromere is less efficient than at centromere-proximal ones (Oliveira et al. 2014; Renshaw et al. 2010). These studies thus suggest that although separase is capable of cleaving cohesin all over chromosome arms, coupling residual cohesion to the centromere may be an efficient way to accelerate cohesin degradation. This could be due to the pulling force of the spindle that could aid in cohesin release, or enhanced Separase activity at the centromeric region.

5 Concluding Remarks

In the cell biology field, centromeric cohesin is mostly viewed as an architectural molecule, a molecular glue linking sister chromatids and preventing random chromosome segregation. However, it is crucial to shift such a viewpoint in order to encompass all the diverse functions of cohesin during nuclear division. Restricting cohesion to the centromeric region during mitosis is of paramount importance for efficient chromosome resolution and segregation. Cohesin itself provides the main elastic force necessary to resist the metaphase spindle and establish biorientation of the chromosomes during metaphase. Cohesin is also crucial for the establishment of an inner-centromere network thus contributing to the localization and function of proteins involved in the regulation of chromosome attachments and spindle assembly checkpoint. As such, mitotic cohesin is way more than a pure "architectural" molecule and should be viewed as a dynamic scaffold for multiple mitotic processes, rather than a hinge keeping chromosomes together.

References

Akiyoshi B, Sarangapani KK, Powers AF, Nelson CR, Reichow SL, Arellano-Santoyo H, Gonen T, Ranish JA, Asbury CL, Biggins S (2010) Tension directly stabilizes reconstituted kinetochore-microtubule attachments. Nature 468:576–579

Alexandru G, Uhlmann F, Mechtler K, Poupart MA, Nasmyth K (2001) Phosphorylation of the cohesin subunit Scc1 by Polo/Cdc5 kinase regulates sister chromatid separation in yeast. Cell 105:459–472

Arumugam P, Gruber S, Tanaka K, Haering CH, Mechtler K, Nasmyth K (2003) ATP hydrolysis is required for cohesin's association with chromosomes. Curr Biol 13:1941–1953

Bachant J, Alcasabas A, Blat Y, Kleckner N, Elledge SJ (2002) The SUMO-1 isopeptidase Smt4 is linked to centromeric cohesion through SUMO-1 modification of DNA topoisomerase II. Mol Cell 9:1169–1182

Baumann C, Korner R, Hofmann K, Nigg EA (2007) PICH, a centromere-associated SNF2 family ATPase, is regulated by Plk1 and required for the spindle checkpoint. Cell 128:101–114

Bernard P, Maure JF, Partridge JF, Genier S, Javerzat JP, Allshire RC (2001) Requirement of heterochromatin for cohesion at centromeres. Science 294:2539–2542

Biggins S (2015) Under tension: kinetochores and basic research. Genetics 200:681–682

Biggins S, Murray AW (2001) The budding yeast protein kinase Ipl1/Aurora allows the absence of tension to activate the spindle checkpoint. Genes Dev 15:3118–3129

Blat Y, Kleckner N (1999) Cohesins bind to preferential sites along yeast chromosome III, with differential regulation along arms versus the centric region. Cell 98:249–259

Buheitel J, Stemmann O (2013) Prophase pathway-dependent removal of cohesin from human chromosomes requires opening of the Smc3-Scc1 gate. EMBO J 32:666–676

Burkhardt S, Borsos M, Szydlowska A, Godwin J, Williams SA, Cohen PE, Hirota T, Saitou M, Tachibana-Konwalski K (2016) Chromosome cohesion established by Rec8-cohesin in fetal oocytes is maintained without detectable turnover in oocytes arrested for months in mice. Curr Biol 26:678–685

Carmena M, Wheelock M, Funabiki H, Earnshaw WC (2012) The chromosomal passenger complex (CPC): from easy rider to the godfather of mitosis. Nat Rev Mol Cell Biol 13:789–803

Carretero M, Ruiz-Torres M, Rodriguez-Corsino M, Barthelemy I, Losada A (2013) Pds5B is required for cohesion establishment and Aurora B accumulation at centromeres. EMBO J 32:2938–2949

Ciosk R, Shirayama M, Shevchenko A, Tanaka T, Toth A, Shevchenko A et al (2000) Cohesin's binding to chromosomes depends on a separate complex consisting of Scc2 and Scc4 proteins. Mol Cell 5:243–254

Ciosk R, Zachariae W, Michaelis C, Shevchenko A, Mann M, Nasmyth K (1998) An ESP1/PDS1 complex regulates loss of sister chromatid cohesion at the metaphase to anaphase transition in yeast. Cell 93:1067–1076

Dawlaty MM, Malureanu L, Jeganathan KB, Kao E, Sustmann C, Tahk S, Shuai K, Grosschedl R, van Deursen JM (2008) Resolution of sister centromeres requires RanBP2-mediated SUMOylation of topoisomerase IIalpha. Cell 133:103–115

De Koninck M, Losada A (2016) Cohesin mutations in cancer. Cold Spring Harb Perspect Med 6: a026476

Diaz-Martinez LA, Gimenez-Abian JF, Clarke DJ (2008) Chromosome cohesion—rings, knots, orcs and fellowship. J Cell Sci 121:2107–2114

Dorsett D (2007) Roles of the sister chromatid cohesion apparatus in gene expression, development, and human syndromes. Chromosoma 116:1–13

Dreier MR, Bekier ME 2nd, Taylor WR (2011) Regulation of sororin by Cdk1-mediated phosphorylation. J Cell Sci 124:2976–2987

Drpic D, Pereira AJ, Barisic M, Maresca TJ, Maiato H (2015) Polar ejection forces promote the conversion from lateral to end-on kinetochore-microtubule attachments on mono-oriented chromosomes. Cell Rep 13:460–469

Eckert CA, Gravdahl DJ, Megee PC (2007) The enhancement of pericentromeric cohesin association by conserved kinetochore components promotes high-fidelity chromosome segregation and is sensitive to microtubule-based tension. Genes Dev 21:278–291

Eichinger CS, Kurze A, Oliveira RA, Nasmyth K (2013) Disengaging the Smc3/kleisin interface releases cohesin from Drosophila chromosomes during interphase and mitosis. EMBO J 32:656–665

Farcas AM, Uluocak P, Helmhart W, Nasmyth K (2011) Cohesin's concatenation of sister DNAs maintains their intertwining. Mol Cell 44:97–107

Funabiki H, Kumada K, Yanagida M (1996) Fission yeast Cut1 and Cut2 are essential for sister chromatid separation, concentrate along the metaphase spindle and form large complexes. EMBO J 15:6617–6628

Gandhi R, Gillespie PJ, Hirano T (2006) Human Wapl is a cohesin-binding protein that promotes sister-chromatid resolution in mitotic prophase. Curr Biol 16:2406–2417

Gerlich D, Koch B, Dupeux F, Peters JM, Ellenberg J (2006) Live-cell imaging reveals a stable cohesin-chromatin interaction after but not before DNA replication. Curr Biol 16:1571–1578

Glynn EF, Megee PC, Yu HG, Mistrot C, Unal E, Koshland DE, DeRisi JL, Gerton JL (2004) Genome-wide mapping of the cohesin complex in the yeast Saccharomyces cerevisiae. PLoS Biol 2:E259

Gorr IH, Boos D, Stemmann O (2005) Mutual inhibition of separase and Cdk1 by two-step complex formation. Mol Cell 19:135–141

Gruber S, Arumugam P, Katou Y, Kuglitsch D, Helmhart W, Shirahige K, Nasmyth K (2006) Evidence that loading of cohesin onto chromosomes involves opening of its SMC hinge. Cell 127:523–537

Guacci V (2007) Sister chromatid cohesion: the cohesin cleavage model does not ring true. Genes Cells 12:693–708

Guacci V, Koshland D, Strunnikov A (1997) A direct link between sister chromatid cohesion and chromosome condensation revealed through the analysis of MCD1 in *S. cerevisiae*. Cell 91:47–57

Haarhuis JH, Elbatsh AM, van den Broek B, Camps D, Erkan H, Jalink K, Medema RH, Rowland BD (2013) WAPL-mediated removal of cohesin protects against segregation errors and aneuploidy. Curr Biol 23:2071–2077

Haering CH, Farcas AM, Arumugam P, Metson J, Nasmyth K (2008) The cohesin ring concatenates sister DNA molecules. Nature 454:297–301

Haering CH, Lowe J, Hochwagen A, Nasmyth K (2002) Molecular architecture of SMC proteins and the yeast cohesin complex. Mol Cell 9:773–788

Hahn M, Dambacher S, Dulev S, Kuznetsova AY, Eck S, Worz S, Sadic D, Schulte M, Mallm JP, Maiser A et al (2013) Suv4-20h2 mediates chromatin compaction and is important for cohesin recruitment to heterochromatin. Genes Dev 27, 859–872

Hara K, Zheng G, Qu Q, Liu H, Ouyang Z, Chen Z, Tomchick DR, Yu H (2014). Structure of cohesin subcomplex pinpoints direct shugoshin-Wapl antagonism in centromeric cohesion. Nat Struct Mol Biol 21:864–870

Hauf S, Roitinger E, Koch B, Dittrich CM, Mechtler K, Peters JM (2005) Dissociation of cohesin from chromosome arms and loss of arm cohesion during early mitosis depends on phosphorylation of SA2. PLoS Biol 3:e69

He E, Kapuy O, Oliveira RA, Uhlmann F, Tyson JJ, Novak B (2011) System-level feedbacks make the anaphase switch irreversible. Proc Natl Acad Sci U S A 108:10016–10021

Hellmuth S, Pohlmann C, Brown A, Bottger F, Sprinzl M, Stemmann O (2015) Positive and negative regulation of vertebrate separase by Cdk1-cyclin B1 may explain why securin is dispensable. J Biol Chem 290:8002–8010

Hirano T, Funahashi S, Uemura T, Yanagida M (1986) Isolation and characterization of Schizosaccharomyces pombe cutmutants that block nuclear division but not cytokinesis. EMBO J 5:2973–2979

Holland AJ, Cleveland DW (2009) Boveri revisited: chromosomal instability, aneuploidy and tumorigenesis. Nat Rev Mol Cell Biol 10:478–487

Hornig NC, Knowles PP, McDonald NQ, Uhlmann F (2002) The dual mechanism of separase regulation by securin. Curr Biol 12:973–982

Hornig NC, Uhlmann F (2004) Preferential cleavage of chromatin-bound cohesin after targeted phosphorylation by Polo-like kinase. EMBO J 23:3144–3153

Hu B, Itoh T, Mishra A, Katoh Y, Chan KL, Upcher W, Godlee C, Roig MB, Shirahige K, Nasmyth K (2011) ATP hydrolysis is required for relocating cohesin from sites occupied by its Scc2/4 loading complex. Curr Biol 21:12–24

Ivanov D, Nasmyth K (2005) A topological interaction between cohesin rings and a circular minichromosome. Cell 122:849–860
Jallepalli PV, Waizenegger IC, Bunz F, Langer S, Speicher MR, Peters JM, Kinzler KW, Vogelstein B, Lengauer C (2001) Securin is required for chromosomal stability in human cells. Cell 105:445–457
Kamenz J, Hauf S (2016) Time to split up: Dynamics of chromosome separation. Trends Cell Biol 27:42
Karamysheva Z, Diaz-Martinez LA, Crow SE, Li B, Yu H (2009) Multiple anaphase-promoting complex/cyclosome degrons mediate the degradation of human Sgo1. J Biol Chem 284:1772–1780
Kenney RD, Heald R (2006) Essential roles for cohesin in kinetochore and spindle function in Xenopus egg extracts. J Cell Sci 119:5057–5066
Kerrebrock AW, Miyazaki WY, Birnby D, Orr-Weaver TL (1992) The Drosophila mei-S332 gene promotes sister-chromatid cohesion in meiosis following kinetochore differentiation. Genetics 130:827–841
Kitajima TS, Kawashima SA, Watanabe Y (2004) The conserved kinetochore protein shugoshin protects centromeric cohesion during meiosis. Nature 427:510–517
Kitajima TS, Sakuno T, Ishiguro K, Iemura S, Natsume T, Kawashima SA, Watanabe Y (2006) Shugoshin collaborates with protein phosphatase 2A to protect cohesin. Nature 441:46–52
Koch B, Kueng S, Ruckenbauer C, Wendt KS, Peters JM (2008) The Suv39 h-HP1 histone methylation pathway is dispensable for enrichment and protection of cohesin at centromeres in mammalian cells. Chromosoma 117:199–210
Kueng S, Hegemann B, Peters BH, Lipp JJ, Schleiffer A, Mechtler K, Peters JM (2006) Wapl controls the dynamic association of cohesin with chromatin. Cell 127:955–967
Lenart P, Petronczki M, Steegmaier M, Di Fiore B, Lipp JJ, Hoffmann M, Rettig WJ, Kraut N, Peters JM (2007) The small-molecule inhibitor BI 2536 reveals novel insights into mitotic roles of polo-like kinase 1. Curr Biol 17:304–315
Lengronne A, Katou Y, Mori S, Yokobayashi S, Kelly GP, Itoh T, Watanabe Y, Shirahige K, Uhlmann F (2004) Cohesin relocation from sites of chromosomal loading to places of convergent transcription. Nature 430:573–578
Lin Z, Luo X, Yu H (2016) Structural basis of cohesin cleavage by separase. Nature 532:131–134
Liu D, Vader G, Vromans MJ, Lampson MA, Lens SM (2009a) Sensing chromosome bi-orientation by spatial separation of aurora B kinase from kinetochore substrates. Science 323:1350–1353
Liu H, Jia L, Yu H (2013a) Phospho-H2A and cohesin specify distinct tension-regulated Sgo1 pools at kinetochores and inner centromeres. Curr Biol 23:1927–1933
Liu H, Rankin S, Yu H (2013b) Phosphorylation-enabled binding of SGO1-PP2A to cohesin protects sororin and centromeric cohesion during mitosis. Nat Cell Biol 15:40–49
Liu J, Krantz ID (2008) Cohesin and human disease. Annu Rev Genomics Hum Genet 9:303–320
Liu Z, Deibler RW, Chan HS, Zechiedrich L (2009b) The why and how of DNA unlinking. Nucleic Acids Res 37:661–671
Losada A (2014) Cohesin in cancer: chromosome segregation and beyond. Nat Rev Cancer 14:389–393
Losada A, Hirano M, Hirano T (1998) Identification of Xenopus SMC protein complexes required for sister chromatid cohesion. Genes Dev 12:1986–1997
Maresca TJ, Salmon ED (2010) Welcome to a new kind of tension: translating kinetochore mechanics into a wait-anaphase signal. J Cell Sci 123:825–835
Marston AL (2015) Shugoshins: tension-sensitive pericentromeric adaptors safeguarding chromosome segregation. Mol Cell Biol 35:634–648
McGuinness BE, Hirota T, Kudo NR, Peters JM, Nasmyth K (2005) Shugoshin prevents dissociation of cohesin from centromeres during mitosis in vertebrate cells. PLoS Biol 3:e86
Megee PC, Koshland D (1999) A functional assay for centromere-associated sister chromatid cohesion. Science 285:254–257

Michaelis C, Ciosk R, Nasmyth K (1997) Cohesins: chromosomal proteins that prevent premature separation of sister chromatids. Cell 91:35–45
Mirchenko L, Uhlmann F (2010) Sli15(INCENP) dephosphorylation prevents mitotic checkpoint reengagement due to loss of tension at anaphase onset. Curr Biol 20:1396–1401
Mirkovic M, Hutter LH, Novak B, Oliveira RA (2015) Premature sister chromatid separation is poorly detected by the spindle assembly checkpoint as a result of system-level feedback. Cell Rep 13:470–478
Musacchio A, Salmon ED (2007) The spindle-assembly checkpoint in space and time. Nat Rev Mol Cell Biol 8:379–393
Nagasaka K, Hossain MJ, Roberti MJ, Ellenberg J, Hirota T (2016) Sister chromatid resolution is an intrinsic part of chromosome organization in prophase. Nat Cell Biol 18:692–699
Nannas NJ, Murray AW (2014) Tethering sister centromeres to each other suggests the spindle checkpoint detects stretch within the kinetochore. PLoS Genet 10:e1004492
Nasmyth K, Haering CH (2009) Cohesin: Its roles and mechanisms. Annu Rev Genet 43:525–558
Ng TM, Waples WG, Lavoie BD, Biggins S (2009) Pericentromeric sister chromatid cohesion promotes kinetochore biorientation. Mol Biol Cell 20:3818–3827
Nicklas RB, Ward SC, Gorbsky GJ (1995) Kinetochore chemistry is sensitive to tension and may link mitotic forces to a cell cycle checkpoint. J Cell Biol 130:929–939
Nishiyama T, Ladurner R, Schmitz J, Kreidl E, Schleiffer A, Bhaskara V, Bando M, Shirahige K, Hyman AA, Mechtler K et al (2010) Sororin mediates sister chromatid cohesion by antagonizing Wapl. Cell 143:737–749
Nishiyama T, Sykora MM, Huis in 't Veld PJ, Mechtler K, Peters JM (2013) Aurora B and Cdk1 mediate Wapl activation and release of acetylated cohesin from chromosomes by phosphorylating Sororin. Proc Natl Acad Sci U S A 110:13404–13409
Nonaka N, Kitajima T, Yokobayashi S, Xiao G, Yamamoto M, Grewal SI, Watanabe Y (2002) Recruitment of cohesin to heterochromatic regions by Swi6/HP1 in fission yeast. Nat Cell Biol 4:89–93
Ocampo-Hafalla M, Munoz S, Samora CP, Uhlmann F (2016) Evidence for cohesin sliding along budding yeast chromosomes. Open Biol 6:150178
Ocampo-Hafalla MT, Katou Y, Shirahige K, Uhlmann F (2007) Displacement and re-accumulation of centromeric cohesin during transient pre-anaphase centromere splitting. Chromosoma 116:531–544
Ocampo-Hafalla MT, Uhlmann F (2011) Cohesin loading and sliding. J Cell Sci 124:685–691
Oliveira RA, Hamilton RS, Pauli A, Davis I, Nasmyth K (2010) Cohesin cleavage and Cdk inhibition trigger formation of daughter nuclei. Nat Cell Biol 12:185–192
Oliveira RA, Kotadia S, Tavares A, Mirkovic M, Bowlin K, Eichinger CS et al (2014) Centromere-independent accumulation of cohesin at ectopic heterochromatin sites induces chromosome stretching during anaphase. PLoS Biol 12:e1001962
Oliveira RA, Nasmyth K (2010) Getting through anaphase: splitting the sisters and beyond. Biochem Soc Trans 38:1639–1644
Parry DH, Hickson GR, O'Farrell PH (2003) Cyclin B destruction triggers changes in kinetochore behavior essential for successful anaphase. Curr Biol 13:647–653
Patel J, Tan SL, Hartshorne GM, McAinsh AD (2015) Unique geometry of sister kinetochores in human oocytes during meiosis I may explain maternal age-associated increases in chromosomal abnormalities. Biol Open 5:178–184
Peters JM, Tedeschi A, Schmitz J (2008) The cohesin complex and its roles in chromosome biology. Genes Dev 22:3089–3114
Primorac I, Musacchio A (2013) Panta rhei: the APC/C at steady state. J Cell Biol 201:177–189
Rankin S, Ayad NG, Kirschner MW (2005) Sororin, a substrate of the anaphase-promoting complex, is required for sister chromatid cohesion in vertebrates. Mol Cell 18:185–200
Remeseiro S, Cuadrado A, Losada A (2013) Cohesin in development and disease. Development 140:3715–3718

Renshaw MJ, Ward JJ, Kanemaki M, Natsume K, Nedelec FJ, Tanaka TU (2010) Condensins promote chromosome recoiling during early anaphase to complete sister chromatid separation. Dev Cell 19:232–244
Rudra S, Skibbens RV (2013) Cohesin codes—interpreting chromatin architecture and the many facets of cohesin function. J Cell Sci 126:31–41
Ryu H, Furuta M, Kirkpatrick D, Gygi SP, Azuma Y (2010) PIASy-dependent SUMOylation regulates DNA topoisomerase IIalpha activity. J Cell Biol 191:783–794
Sakuno T, Tada K, Watanabe Y (2009) Kinetochore geometry defined by cohesion within the centromere. Nature 458:852–858
Sen N, Leonard J, Torres R, Garcia-Luis J, Palou-Marin G, Aragon L (2016) Physical proximity of sister chromatids promotes top2-dependent intertwining. Mol Cell 64:134–147
Shindo N, Kumada K, Hirota T (2012) Separase sensor reveals dual roles for separase coordinating cohesin cleavage and cdk1 inhibition. Dev Cell 23:112–123
Skibbens RV, Corson LB, Koshland D, Hieter P (1999) Ctf7p is essential for sister chromatid cohesion and links mitotic chromosome structure to the DNA replication machinery. Genes Dev 13:307–319
Sonoda E, Matsusaka T, Morrison C, Vagnarelli P, Hoshi O, Ushiki T, Nojima K, Fukagawa T, Waizenegger IC, Peters JM et al (2001) Scc1/Rad21/Mcd1 is required for sister chromatid cohesion and kinetochore function in vertebrate cells. Dev Cell 1:759–770
Stemmann O, Zou H, Gerber SA, Gygi SP, Kirschner MW (2001) Dual inhibition of sister chromatid separation at metaphase. Cell 107:715–726
Stephens AD, Quammen CW, Chang B, Haase J, Taylor RM 2nd, Bloom K (2013) The spatial segregation of pericentric cohesin and condensin in the mitotic spindle. Mol Biol Cell 24:3909–3919
Stigler J, Camdere GO, Koshland DE, Greene EC (2016) Single-molecule imaging reveals a collapsed conformational state for dna-bound cohesin. Cell Rep 15:988–998
Stratmann R, Lehner CF (1996) Separation of sister chromatids in mitosis requires the Drosophila pimples product, a protein degraded after the metaphase/anaphase transition. Cell 84:25–35
Sullivan M, Morgan DO (2007) Finishing mitosis, one step at a time. Nat Rev Mol Cell Biol 8:894–903
Sumara I, Vorlaufer E, Gieffers C, Peters BH, Peters JM (2000) Characterization of vertebrate cohesin complexes and their regulation in prophase. J Cell Biol 151:749–762
Sumara I, Vorlaufer E, Stukenberg PT, Kelm O, Redemann N, Nigg EA, Peters JM (2002) The dissociation of cohesin from chromosomes in prophase is regulated by polo-like kinase. Mol Cell 9:515–525
Tachibana-Konwalski K, Godwin J, Borsos M, Rattani A, Adams DJ, Nasmyth K (2013) Spindle assembly checkpoint of oocytes depends on a kinetochore structure determined by cohesin in meiosis I. Curr Biol 23:2534–2539
Tanaka T, Cosma MP, Wirth K, Nasmyth K (1999) Identification of cohesin association sites at centromeres and along chromosome arms. Cell 98:847–858
Tanaka T, Fuchs J, Loidl J, Nasmyth K (2000) Cohesin ensures bipolar attachment of microtubules to sister centromeres and resists their precocious separation. Nat Cell Biol 2:492–499
Tanno Y, Kitajima TS, Honda T, Ando Y, Ishiguro K, Watanabe Y (2010) Phosphorylation of mammalian Sgo2 by Aurora B recruits PP2A and MCAK to centromeres. Genes Dev 24:2169–2179
Tanno Y, Susumu H, Kawamura M, Sugimura H, Honda T, Watanabe Y (2015) The inner centromere-shugoshin network prevents chromosomal instability. Science 349:1237–1240
Tedeschi A, Wutz G, Huet S, Jaritz M, Wuensche A, Schirghuber E, Davidson IF, Tang W, Cisneros DA, Bhaskara V et al (2013) Wapl is an essential regulator of chromatin structure and chromosome segregation. Nature 501:564–568
Tomkins D, Hunter A, Roberts M (1979) Cytogenetic findings in Roberts-SC phocomelia syndrome(s). Am J Med Genet 4:17–26

Toth A, Ciosk R, Uhlmann F, Galova M, Schleiffer A, Nasmyth K (1999) Yeast cohesin complex requires a conserved protein, Eco1p(Ctf7), to establish cohesion between sister chromatids during DNA replication. Genes Dev 13:320–333
Toyoda Y, Yanagida M (2006) Coordinated requirements of human topo II and cohesin for metaphase centromere alignment under Mad2-dependent spindle checkpoint surveillance. Mol Biol Cell 17:2287–2302
Uhlmann F, Lottspeich F, Nasmyth K (1999) Sister-chromatid separation at anaphase onset is promoted by cleavage of the cohesin subunit Scc1. Nature 400:37–42
Uhlmann F, Nasmyth K (1998) Cohesion between sister chromatids must be established during DNA replication. Curr Biol 8:1095–1101
Uhlmann F, Wernic D, Poupart MA, Koonin EV, Nasmyth K (2000) Cleavage of cohesin by the CD clan protease separin triggers anaphase in yeast. Cell 103:375–386
Uzunova K, Dye BT, Schutz H, Ladurner R, Petzold G, Toyoda Y, Jarvis MA, Brown NG, Poser I, Novatchkova M et al (2012) APC15 mediates CDC20 autoubiquitylation by APC/C (MCC) and disassembly of the mitotic checkpoint complex. Nat Struct Mol Biol 19:1116–1123
Vagnarelli P, Morrison C, Dodson H, Sonoda E, Takeda S, Earnshaw WC (2004) Analysis of Scc1-deficient cells defines a key metaphase role of vertebrate cohesin in linking sister kinetochores. EMBO Rep 5:167–171
van der Lelij P, Chrzanowska KH, Godthelp BC, Rooimans MA, Oostra AB, Stumm M, Zdzienicka MZ, Joenje H, de Winter JP (2010) Warsaw breakage syndrome, a cohesinopathy associated with mutations in the XPD helicase family member DDX11/ChlR1. Am J Hum Genet 86:262–266
Vass S, Cotterill S, Valdeolmillos AM, Barbero JL, Lin E, Warren WD, Heck MM (2003) Depletion of Drad21/Scc1 in Drosophila cells leads to instability of the cohesin complex and disruption of mitotic progression. Curr Biol 13:208–218
Vazquez-Novelle MD, Petronczki M (2010) Relocation of the chromosomal passenger complex prevents mitotic checkpoint engagement at anaphase. Curr Biol 20:1402–1407
Vazquez-Novelle MD, Sansregret L, Dick AE, Smith CA, McAinsh AD, Gerlich DW, Petronczki M (2014) Cdk1 inactivation terminates mitotic checkpoint surveillance and stabilizes kinetochore attachments in anaphase. Curr Biol 24:638–645
Verni F, Gandhi R, Goldberg ML, Gatti M (2000) Genetic and molecular analysis of wings apart-like (wapl), a gene controlling heterochromatin organization in Drosophila melanogaster. Genetics 154:1693–1710
Waizenegger I, Gimenez-Abian JF, Wernic D, Peters JM (2002) Regulation of human separase by securin binding and autocleavage. Curr Biol 12:1368–1378
Waizenegger IC, Hauf S, Meinke A, Peters JM (2000) Two distinct pathways remove mammalian cohesin from chromosome arms in prophase and from centromeres in anaphase. Cell 103: 399–410
Warren WD, Steffensen S, Lin E, Coelho P, Loupart M, Cobbe N, Lee JY, McKay MJ, Orr-Weaver T, Heck MM et al (2000) The Drosophila RAD21 cohesin persists at the centromere region in mitosis. Curr Biol 10:1463–1466
Watrin E, Schleiffer A, Tanaka K, Eisenhaber F, Nasmyth K, Peters JM (2006) Human Scc4 is required for cohesin binding to chromatin, sister-chromatid cohesion, and mitotic progression. Curr Biol 16:863–874
Weber SA, Gerton JL, Polancic JE, DeRisi JL, Koshland D, Megee PC (2004) The kinetochore is an enhancer of pericentric cohesin binding. PLoS Biol 2:E260
Webster A, Schuh M (2016) Mechanisms of aneuploidy in human eggs. Trends Cell Biol 21:55
Weitzer S, Lehane C, Uhlmann F (2003) A model for ATP hydrolysis-dependent binding of cohesin to DNA. Curr Biol 13:1930–1940
Woodman J, Fara T, Dzieciatkowska M, Trejo M, Luong N, Hansen KC, Megee PC (2014) Cell cycle-specific cleavage of Scc2 regulates its cohesin deposition activity. Proc Natl Acad Sci U S A 111:7060–7065
Yaakov G, Thorn K, Morgan DO (2012) Separase biosensor reveals that cohesin cleavage timing depends on phosphatase PP2A(Cdc55) regulation. Dev Cell 23:124–136

Yamagishi Y, Honda T, Tanno Y, Watanabe Y (2010) Two histone marks establish the inner centromere and chromosome bi-orientation. Science 330:239–243
Ye AA, Cane S, Maresca TJ (2016) Chromosome biorientation produces hundreds of piconewtons at a metazoan kinetochore. Nat Commun 7:13221
Zielinska AP, Holubcova Z, Blayney M, Elder K, Schuh M (2015) Sister kinetochore splitting and precocious disintegration of bivalents could explain the maternal age effect. Elife 4:e11389
Zou H, McGarry TJ, Bernal T, Kirschner MW (1999) Identification of a vertebrate sister-chromatid separation inhibitor involved in transformation and tumorigenesis. Science 285:418–422

Centromere Structure and Function

Kerry Bloom and Vincenzo Costanzo

Abstract The centromere is the genetic locus that specifies the site of kinetochore assembly, where the chromosome will attach to the kinetochore microtubule. The pericentromere is the physical region responsible for the geometry of bi-oriented sister kinetochores in metaphase. In budding yeast the 125 bp point centromere is sufficient to specify kinetochore assembly. The flanking region is enriched (3X) in cohesin and condensin relative to the remaining chromosome arms. The enrichment spans about 30–50 kb around each centromere. We refer to the flanking chromatin as the pericentromere in yeast. In mammals, a 5–10 Mb region dictates where the kinetochore is built. The kinetochore interacts with a very small fraction of DNA on the surface of the centromeric region. The remainder of the centromere lies between the sister kinetochores. This is typically called centromere chromatin. The chromatin sites that directly interface to microtubules cannot be identified due to the repeated sequence within the mammalian centromere. However in both yeast and mammals, the total amount of DNA between the sites of microtubule attachment in metaphase is highly conserved. In yeast the 16 chromosomes are clustered into a 250 nm diameter region, and 800 kb (16 $\times$ 50 kb) or $\sim$1 Mb of DNA lies between sister kinetochores. In mammals, 5–10 Mb lies between sister kinetochores. In both organisms the sister kinetochores are separated by about 1 μm. Thus, centromeres of different organisms differ in how they specify kinetochore assembly, but there may be important centromere chromatin functions that are conserved throughout phylogeny. Recently, centromeric chromatin has been reconstituted in vitro using alpha satellite DNA revealing unexpected features of centromeric DNA organization, replication, and response to stress. We will focus on the conserved features of centromere in this review.

K. Bloom (✉)
Department of Biology, University of North Carolina at Chapel Hill, 623 Fordham Hall CB#3280, Chapel Hill, NC 27599-3280, USA
e-mail: Kerry_bloom@unc.edu

V. Costanzo
DNA Metabolism Laboratory, IFOM, The FIRC Institute of Molecular Oncology, Vai Adamello 16, 21139 Milan, Italy
e-mail: Vincenzo.costanzo@ifom.eu

B.E. Black (ed.), *Centromeres and Kinetochores*, Progress in Molecular and Subcellular Biology 56, DOI 10.1007/978-3-319-58592-5_21

1 Characteristics of Centromere Chromatin—Primary Folding of a Nucleosome

1.1 Centromere-Specific Histone H3 Variant, Cse4 Yeast, CENP-A Mammals

Centromeres contain an atypical Histone H3, known as CENP-A, or Cse4 in budding yeast (Earnshaw et al. 2013). The functional role of CENP-A at the centromere is considerably less clear. A variety of studies reveal that CENP-A containing nucleoprotein complexes can adopt a number of conformations, including tetramers (Dalal et al. 2007; Shivaraju et al. 2012), hemisomes (Henikoff et al. 2014) or octamers (Camahort et al. 2009; Hasson et al. 2013); canonical left-handed octamers (Dechassa et al. 2011; Hasson et al. 2013), right-handed hemisomes or octamers (Furuyama and Henikoff 2009; Diaz-Ingelmo et al. 2015), and hexamers (Mizuguchi et al. 2007; Xiao et al. 2011). In budding yeast, with its point centromere and purportedly single Cse4-containing nucleosome (Lawrimore et al. 2011; Shivaraju et al. 2012; Aravamudhan et al. 2013; Haase et al. 2013; Wisniewski et al. 2014), there is controversy over the number of Cse4 molecules and the handedness of DNA wrap. At the heart of such controversies is the quest to determine critical features responsible for establishing a functional kinetochore. For Cse4 and CENP-A, the ability to swap centromere-specific domains with canonical Histone H3 reveals crucial molecular determinants unique to the centromere nucleosome. These include the Histone Fold Domain and the essential N-terminus (END) in Cse4 (Keith et al. 1999; Chen et al. 2000), the CENP-A targeting domain (CATD) within the HFD, essential N- and C-termini (Black et al. 2004; Logsdon et al. 2015), and insight into distinct requirements for establishment versus maintenance of the CENP-A nucleosome at the centromere (Logsdon et al. 2015). In addition to studies at the level of isolated nucleosomes, the centromere nucleosome must also be considered in the context of flanking chromatin. It is the intent of this review to focus on the higher order structures that lead to a bolus of CENP-A on the surface of the chromosome where they can interact and engage microtubules.

1.2 Centromere DNA Directs the Sequence Invariant Position of Cse4

The budding yeast centromere is unique in that the microtubule-DNA interphase is known at base pair resolution. There is a single position invariant Cse4 protein(s) at centromere DNA elements CDEI, II, and III. There are additional Cse4 molecules not confined to a specific sequence (between 3 and 5/chromosome), that can be observed as a cloud surrounding the centromere core (Haase et al. 2013). While evidence for and against the Cse4 cloud exists (Aravamudhan et al. 2013; Haase et al. 2013; Wisniewski et al. 2014), the position variable molecules can be eliminated in specific mutants (pat1, xrn1) as evidenced by the loss of the cloud and confirmed by a reduction in chromatin immunoprecipitation (Haase et al. 2013). We will focus on the sequence invariant molecules as the function of the position variable molecules is not known.

The centromere DNA elements are recognized by a unique DNA-binding complex, Centromere Binding Factor (CBF3, composed of Ndc10 (2), Cep3 (2), Ctf13, Skp1) (Espelin et al. 1997; Russell et al. 1999). Ndc10 is distantly related to the tyrosine DNA recombinases, such as CRE (Cho and Harrison 2012; Perriches and Singleton 2012). Ndc10 may be central to the controversy surrounding the Cse4 nucleosome, as well as the structure of the inner centromere. Ndc10 binds CDEII in the absence of other CBF3 components, as well as A + T rich domains elsewhere in the chromosome (Espelin et al. 2003). Thus Ndc10 is promiscuous in its DNA-binding properties, and is not restricted to kinetochores in vivo (Espelin et al. 2003). Within the kinetochore, Ndc10 localizes to the inner centromere proximal to the spindle axis, where it binds DNA as a dimer and has been proposed to promote DNA bending and/or looping (Pietrasanta et al. 1999). Ndc10 is essential for forming the yeast kinetochore and specifically in recruiting CENP-A (Cse4, Pearson et al. 2003).

The function of the variant centromere histone complex is not well understood. There are differing biological consequences imparted from the mechanics of DNA, whether it is wrapped in a canonical left- or non-canonical right-handed direction around the nucleosome. The determination of left- versus right-handed DNA wraps around the nucleosome can be assessed from indirect measurements of the number of supercoils in small circular plasmids. Circular centromere-containing plasmids contain fewer negative supercoils than acentric plasmids, indicative of a positive (right-handed wrap) supercoil around the centromere core (Bloom et al. 1983, 1984; Furuyama and Henikoff 2009; Diaz-Ingelmo et al. 2015). There are several alternative interpretations of the change in linking number observed in plasmids containing the centromere-specific Cse4 nucleosome, all of which are dependent on CDEIII and/or Ndc10 (Furuyama and Henikoff 2009; Gkikopoulos et al. 2011; Diaz-Ingelmo et al. 2015). One possibility is that that the functional centromere contains a left-handed wrap around Cse4, flanked by two DNAaseI hypersensitive nucleosome-depleted regions (~70 bp each) (Bloom and Carbon 1982). Loss of centromere function in ndc10 mutants leads to loss of the nuclease hypersensitivity (Saunders et al. 1988). If these sites become occupied by a nucleosome with canonical histones, plasmid DNA isolated from these mutants will appear to have additional negative superhelical turns (and hence plasmid from wild-type cells will a net change of +1 in linking number). Alternatively, if Ndc10 introduces a right-handed turn at the base of a DNA loop, this will also influence the net linking number in a positive fashion (Cho and Harrison 2012; Diaz-Ingelmo et al. 2015).

1.3 Primary Loop at the Cse4 Nucleosome

The looping function of Ndc10 and hypersensitivity of flanking sites (70 bp) to DNAase I provides important insight into the three-dimensional structure of the yeast centromere. One can build a unifying model that incorporates Ndc10 looping and dimerization function, the observed change in linking number and extent of micrococcal nuclease digestion (120 bp core (Cole et al. 2011; Krassovsky et al. 2012) and a larger 150–220 bp protected structure (Bloom and Carbon 1982; Funk et al. 1989; Gkikopoulos et al. 2011)). At the base of the CDEIII, Ndc10 binding results in a right-handed DNA wrap of the ends of a ~80 bp duplex that encircles the Cse4 nucleosome (Henikoff et al. 2014). This accounts for the core 80–120 bp seen in extensive MNase digests employed in ChIP seq studies (Cole et al. 2011; Krassovsky et al. 2012). DNA exiting the Ndc10 loop continues to wrap around the Cse4 nucleosome to what will be the surface of the chromosome (middle panel Fig. 1). This leaves about 70 bp of DNA surface exposed as evidenced by extreme DNAase I hypersensitivity. Considering the concentration of nucleosomes adjacent to CDEI-III, it is surprising that stochastic nucleosome dynamics [sliding as well as protein exchange (Verdaasdonk et al. 2012)] does not result in the nuclease hypersensitive sites becoming occupied by histones. These sites are kept nucleosome-free through the action of the ATP-dependent SWI/SNF chromatin remodeling complex, Snf2 (Gkikopoulos et al. 2011). Deletion of Snf2 results in a reduction in the nuclease hypersensitivity flanking CDEI-III, with no change in the flanking nucleosomes or other hypersensitive DNAase I sites.

In a three-dimensional model of the centromere, these nucleosome-free DNAase I sites of all 16 centromeres are clustered at the chromosome/microtubule interface. The 16 kinetochores are arrayed in a cylinder (~ 250 dia.) around the central spindle resulting in the confinement of DNA loops from all 16 chromosomes to the vicinity of the spindle. Clustering the DNAase I sites to the chromosome surface is a mechanism to distinguish the centromere from the chromosome arms. The active chromatin remodeling mechanism that prevents occlusion of these sites by histones ensures this unique attribute of centromeres. Exposure of such a large patch of naked DNA (~150 bp × 16–2500 bp) is likely to contribute to kinetochore microtubule capture and stabilization mechanisms.

The DNAase I hypersensitive sites are conserved in other fungi (*Kluyveromyces lactis*) harboring point centromeres (Heus et al. 1993a), even though the centromere sequences have diverged (Heus et al. 1993b). Centromeres from one organism do not confer segregation function in related species. In *K. lactis*, CDEII is about double the size of that in budding yeast S. cerevisiae. There could be two wraps of DNA around a Cse4 core, or there could be double the number of Cse4 molecules at the site of microtubule attachment. In either case, the model of extreme DNAase I sites on the chromosome surface may be applicable to other fungi as well as larger eukaryotes.

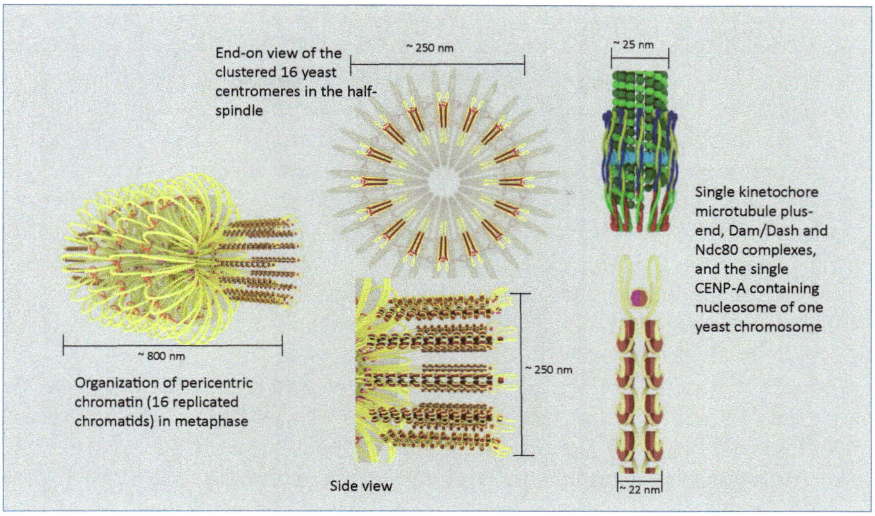

Fig. 1 A DNA basket on the surface of centromeres in metaphase. *Left* The pericentromere region of all 16 chromosomes in metaphase in budding yeast. The 125 bp CEN region (*pink* nucleosome at right-most edge of the nucleosome fiber, depicted as *yellow* DNA wrapped around *red* histones) lie at the apex of the pericentric chromatin loops (11 nm fiber, *yellow* strands). The centromere DNA containing loops extend perpendicular to the chromosome axis (Lawrimore et al. 2016). *Middle panels* Top: end-on view of the Cse4 containing nucleosomes, one from each centromere of the 16 sister chromatids are cylindrically arranged and lie on the surface of the chromosome. Bottom: side view of centromere DNA in metaphase. 80 bp of DNA is in direct contact with the Cse4-containing core (*pink*), flanking DNA follows a trajectory toward the kinetochore (*yellow* strands away from the pericentromere). DNA devoid of histones reflect the DNAase I hypersensitive regions (Bloom and Carbon 1982; Bloom et al. 1983) exiting and entering the Cse4 containing nucleosome that protrude from the surface of the chromosome to make a basket. *Far right* The Cse4 containing nucleosomes are proximal to the pericentric chromatin (*yellow* DNA strands, *bottom*), while the DNAase I hypersensitive sites protrude from the chromosome surface toward the kinetochore (*top*). Sister kinetochores lie ~800 nm away on the opposite surface of the sister strands

1.4 Pericentromeric Loop—Secondary Folding into Bottle Brush

Sister Cse4-containing nucleosomes are separated by 800 nm in metaphase. Each kinetochore lies at the apex of a DNA loop that extends on average about 400 nm from the sister chromatid axis in metaphase. The centromere loops have been found in 3C maps from yeast (Yeh et al. 2008; Duan et al. 2010), and Hi-C maps of lymphoblastoid cells in mammals (Dai et al. 2016). The size of the loops can be estimated from the region over which the structural maintenance of chromosome proteins (SMC's cohesin and condensin) are enriched. Loops from each chromosome are in turn looped, adopting loops within loop organization (Lawrimore et al. 2016) (Fig. 2). The emerging principle from the fractal organization of DNA loops is that stiffness is achieved within the pericentric chromatin. A high concentration of loops builds tension within the pericentromere and exerts a pushing force on centromere DNA at the apex of the loops (Cse4 and CDEI, II and III, including the hypersensitive sites). This pushing force is responsible for positioning the centromere to the chromosome surface. The concentration of loops is established through the energy-consuming process of loading and enriching cohesin in the pericentromere at the time of DNA replication. Once the pericentromere chromatin is enriched in SMC proteins, the fluctuations of the strands impose a tension force on the primary loop, where the centromere lies at the apex. This configuration, known as a bottle brush, is a common physical property of polymers that on solid substrates can generate nanonewtons of force, enough to break covalent bonds (Panyukov et al. 2009a, b).

The size and number of secondary loops has been estimated through the use of a chromosome dynamics simulator (Lawrimore et al. 2016). The bead-spring polymer model finds the thermodynamically favorable state of a string of beads with hinge forces connecting the beads tuned to give the strand the bending rigidity determined for DNA. In this model, cohesin is concentrated in the pericentromere and adopts a uniform and homogeneous distribution that surrounds the central spindle, but is radially displaced (Yeh et al. 2008; Stephens et al. 2011). There is no molecular

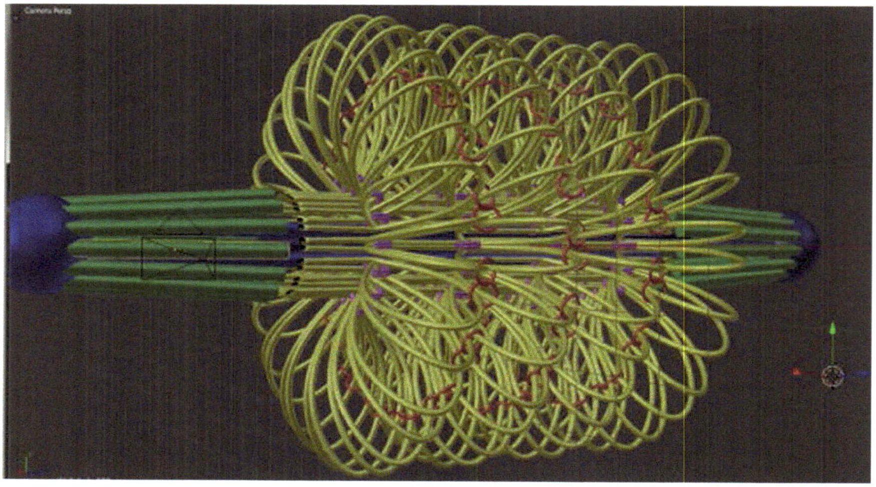

Fig. 2 Configuration of pericentric chromatin loops surrounding the spindle axis in the budding yeast. The *blue spheres* are spindle pole bodies, the *green* rods are kinetochore microtubules. The interpolar microtubules can be seen as blue rods extending about ¾ the length of the spindle from each pole. The *yellow* strands are pericentric chromatin. The primary loop (*horizontal*) is attached to a kinetochore microtubule. Chromosome arms (not shown) would extend north and south, from approximately the middle of the pericentromere. Condensin is at the base of the each of the radial subloops (*purple* staples). Cohesin are the *red* rings. The position of cohesin is the most thermodynamically favorable and matches the position observed experimentally with the size and number of loops modeled herein

mechanism that accounts for the position or appearance of a cohesin barrel. We have found that the size and number of DNA loops dictate the experimentally observed position. Loops that are approx. 10 kb give rise to a cohesin barrel that matches the dimension and homogeneity found in vivo (Lawrimore et al. 2016). The size of these loops is comparable to the 5 kb chromosome interaction domains found in yeast from Micro-C (Hsieh et al. 2015).

An alternative view of cohesin function at the centromere is provided in Hu et al. (2015). To reconcile the position of cohesin relative to the spindle axis with the canonical function of juxtaposing sister chromatids, Hu et al. propose that pericentric cohesin lies on average 5 kb from the centromere core and is restricted to sister chromatid interactions as observed in the arms. Stephens et al. (2011) found that a lacO array 6.8 kb from the centromere was separated greater than 65% of the time in wild-type cells, inconsistent with such a proposal. Furthermore, the radial distance of cohesin from the spindle, as well as the homogeneity of the barrel demands a physical explanation. Rather than holding sister chromatids in the pericentromere, intra-chromatid linkages as proposed in Lawrimore et al. (2016) account for the position and dynamics of pericentric cohesin. Behavior of the centromere linked lacO array as it extends and retracts thermodynamically, as well as in response to MT pulling and pushing forces also favors intra-chromosomal loops diagrammed in the model shown in Fig. 2. The intra-chromatid linkages contribute to the pushing mechanism predicted from the bottle brush, providing a novel view of centromere function in promoting sister separation and kinetochore tension at the metaphase plate (Lawrimore et al. 2015, 2016). Whether non-sister chromosome interactions are mediated by cohesin within the pericentromere remain controversial. Biophysical studies of cohesin diffusion on DNA indicates that the physiological pore size is much smaller than observed in electron micrographs of spread molecules (Stigler et al. 2016). Based upon the fluctuations of radially displaced LacO spots and correlated movement between pericentromeres of non-sister chromatids, models of cohesin rings that encompass a single chromatid and interact with cohesin rings from other chromosomes account for experimental observations (Stephens et al. 2011, 2013a; b). Interestingly, in experimental attempts to reduce the concentration of cohesin in the cell, the only region of the chromosome where cohesin was refractory to reduction was in the centromere (Heidinger-Pauli et al. 2010). Thus there are mechanisms to ensure the concentration remains in centromere, indicating this may be the critical site for the fidelity of chromosome segregation.

1.5 Loops on Mammalian DNA and Role of Satellite Repeats in Promoting DNA Looping

Recently, centromeric chromatin was reconstituted in cell-free extracts derived from *Xenopus laevis* eggs (Aze et al. 2016). For these experiments artificial chromosomes (BACs) containing large inserts (130–160 kb) of centromeric alpha satellite DNA sequences of different human centromeres were used. BACs were chosen for their large size, which allowed efficient formation of nuclear structures in egg extract (Aze et al. 2016). Electron microscopy (EM) based structural analysis of centromeric DNA isolated from egg extract highlighted the presence of numerous single-stranded DNA bubbles. These structures were due to intrinsic resistance of centromeric chromatin to psoralen-mediated cross-linking, a procedure required to prevent melting and branch migration of DNA preparations for EM analysis. Poor cross-linking was due to the presence of positively supercoiled DNA, which is known to prevent psoralen-mediated cross-linking (Bermudez et al. 2010). Formation of positively supercoiled DNA required Topoisomerase I activity. Chromatin composition analysis also revealed the enrichment of condensin, which can promote the formation of positively supercoiled DNA in closed plasmids in the presence of Topoisomerase I (Hirano 2012). These findings suggest that positive supercoils observed in centromeric DNA assembled in egg extract is an active phenomenon linked to the presence of Topoisomerase I and condensin.

When partially digested chromatin still retaining condensin was analyzed instead of naked DNA EM revealed the presence of long double stranded DNA loops embedded in a protein matrix (Fig. 3). The average size of the loops was around 1–1.5 kb but some of the loops reached 2.5–3 kb in size. Although the individual components of the protein matrix could not be resolved, electron dense particles corresponding to residual protein material left after partial proteolysis were spotted at the base of the loops, indicating a possible role in their formation or stabilization (Fig. 3). Consistent with this hypothesis removal of residual proteins by complete digestion dissolved the loop structure (Aze et al. 2016).

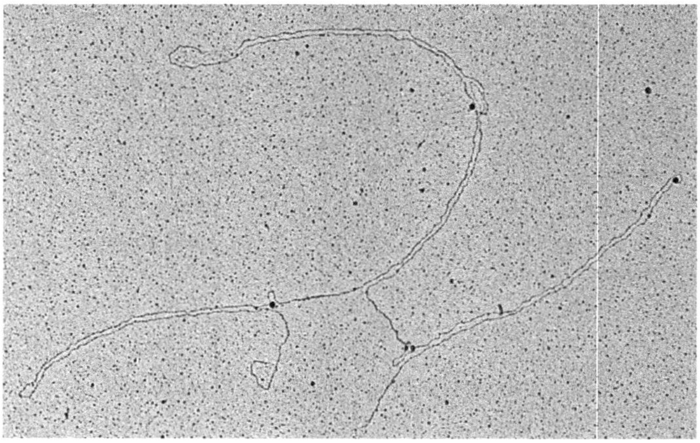

Fig. 3 Electron micrograph showing partially digested chromatin isolated from *Xenopus laevis* egg extract incubated with human alpha satellite DNA. Loops of double stranded DNA filaments running parallel to each other embedded in a protein matrix can be appreciated. Electron dense particles can be noticed at the base of some of the loops. Bar corresponds to 500 base pairs

Formation of loops observed on centromeric chromatin isolated from interphase egg extracts required DNA replication and active Topoisomerase I, although Topoisomerase II role could not be excluded. The presence of positively supercoiled DNA associated to the loops suggested that these structures formed behind replication forks as their presence ahead of them would heavily disturb fork progression (Branzei and Foiani 2010). Condensins play a major role in mitosis-dependent chromosome condensation in eukaryotic cells. A role for condensin during DNA replication has been documented in bacteria, where their activity is required for DNA loop formation. In this case condensins appear to resolve replicated replication origins by promoting the juxtaposition of DNA, drawing sister replication origins away from each other (Wang et al. 2015). To this end condensins encircle double-stranded DNA at their loading site and slide along it, tethering the two arms of double-stranded DNA together. If applied to mammalian cells this process could disentangle and individualize replicated sections of centromeric DNA promoting at same time DNA condensation, which could start at centromeres in interphase and then spread towards flanking regions of the chromosome during the mitotic phase. This process could help to disentangle highly repeated centromeric DNA sequences after replication and prevent their recombination.

The loops uncovered in this system could be related to the organization of centromeric chromatin in mitosis. The EM pictures evoke the bottlebrush structure proposed for the loop arrangement in yeast centromere. The smaller size of the loops identified in interphase by EM compared to the ones predicted by biophysical studies might reflect an early stage of centromere reorganization. The loops assembled in interphase could indeed be the basic unit of larger structures, which might grow further in mitosis. The requirement of Topoisomerase I and possibly Topoisomerase II activities for their formation and/or for their stabilization indicates the occurrence of rotational processes along DNA axes during loop formation and the possible presence of loop entanglements that stabilize these structures. Overall these features could be compatible with the formation of the centromeric spring.

CENP-A is a key factor able to trigger assembly of kinethocore proteins in vitro (Weir et al. 2016) and in Xenopus egg cytoplasm (Guse et al. 2011). However, the links between centromeric DNA and CENP-A as far as chromatin structure is concerned are largely unclear. Interestingly, CENP-A could be selectively loaded onto naked centromeric DNA containing alpha satellite in Xenopus egg extract (Aze et al. 2016). The ability of satellite DNA to induce loading of centromeric proteins was consistent with work performed on human artificial chromosomes (HACs), which are made with satellite DNA and are able attract centromere and kinetochore proteins when introduced into cells (Nakano et al. 2008; Bergmann et al. 2011; Kouprina et al. 2013). Replicative features such as invariant inter-origin distance between centromeric and non-centromeric DNA in Xenopus were also consistent with data obtained from studies on HAC replication in intact cells (Erliandri et al. 2014). These two systems could help to better define important aspects of centromeric chromatin assembly in the future.

It is generally assumed that CENP-A is loaded onto centromere chromatin between the end of mitosis and subsequent G1 phase of the cell cycle.

An epigenetic mechanism is involved in the deposition of new CENP-A onto chromatin regions with preexisting CENP-A. However, in vitro results suggested that de novo CENP-A loading can also take place in interphase and on naked centromeric DNA made of alpha satellite DNA. The reason why satellite DNA could trigger such a complex chain reaction of events is not known. Similar to the ability of CENP-A to induce kinetochore assembly (Guse et al. 2011) CENP-A could be an important trigger for centromeric loop formation. Centromeric chromatin reconstitution was achieved with DNA sequences containing the CENP-B box, a 17 bp element, which functions as a binding site for CENP-B protein (Fachinetti et al. 2015). CENP-B protein has been recently shown to be required for the stable loading of CENP-A on alpha satellite DNA (Fachinetti et al. 2015). Although satellite DNA has not been described yet in *X. laevis*, repetitive sequences containing a conserved CENP-B box able to bind *X. laevis* CENP-A have been isolated (Edwards and Murray 2005). Therefore, it is possible that the presence of CENP-B box facilitates recruitment of CENP-A onto human centromeric DNA in egg extract possibly through CENP-B. Consistent with this hypothesis proteomic analysis of centromeric chromatin assembled in egg extract revealed the selective loading of a number of centromeric proteins (Aze et al. 2016) among which a protein with weak homology to CENP-B. However, further studies are needed to confirm that this protein is the true CENP-B ortholog.

Other proteins found enriched in centromeric chromatin in many species such as condensins might be responsible for the centromere loop formation on satellite DNA. How condensins are specifically loaded on satellite DNA is not known. There could be a structural code in the repetitive DNA able to attract condensins and other centromeric proteins promoting the formation of complex structures. The existence of a structural code conserved from yeast to mammals has been proposed and it is related to an internal dyadic symmetry of individual tandem repeats and yeast centromeric sequences, which could give rise to mismatched hairpins (Koch 2000; Jonstrup et al. 2008). Some of these structures have been demonstrated in vitro but their significance in vivo has never been explored. Condensins might have affinity for hairpins and could recognize these structures. With the advent of techniques such as Crisper/Cas9 satellite DNA could be modified to test the in vivo relevance of this structural code. Condensins have also been implicated in the assembly of CENP-A chromatin in Xenopus and human cells (Samoshkin et al. 2009; Bernad et al. 2011) but their precise role in the process is still unclear. Recent work has shown that condensin subunit Cut3 in fission yeast mediates the organization of pericentromeric tandem repeats into a specific higher order structure, which helps to restrict CENP-A loading to centromeres (He et al. 2016). Interestingly, formation of neocentromeres triggered by CENP-A binding to extra-centromeric sites which do not contain centromeric repeats frequently occurs in genomic areas enriched for duplicated sequences (Marshall et al. 2008). Therefore, the tandem repeats or closely duplicated sequences themselves might act as signal to recruit condensins and CENP-A, giving rise to series of complex events required to build a mature centromere.

2 Mechanisms of Loop Formation

Several mechanisms of chromosome and centromere loop formation have been proposed in the literature. One is that natural fluctuations of the polymer chains lead to encounters that are stabilized by SMC proteins cohesin and condensin (Vasquez et al. 2016). Using bead-springs to model chromosome arms as polymer chains we have shown that domains of high interaction (intra-chromosome loops) arise naturally from polymer thermodynamics without the need for other mechanical or chemical potentials interacting via entropic potentials, such as the chromatin spring and excluded volume forces. These regions are not static and vary from cell to cell. In the centromere, such loops could be stabilized through the action of proteins such as topoisomerases, and SMCs that increase the lifetime of a given loop. A second model that has been proposed is the activity of loop extrusion enzymes (Alipour and Marko 2012). In this model, SMC proteins act as a machine that threads DNA into a loop as the enzymes translocate along the helix. There is considerable interest in this model from recent studies using 3C and Hi-C techniques (Fudenberg et al. 2016; Goloborodko et al. 2016a, b).

Chromatin remodeling proteins have also been shown to be loop extruding motors (De Cian et al. 2012; Sun et al. 2013), and may impart this function at the centromere. The centromere is enriched in several members of various chromatin remodeling protein families. Chromatin remodelers are able to mobilize nucleosomes, function to translocate linker DNA over the nucleosome, maintain nucleosome-free regions, such as around promoters, and loop DNA to tune topological domains. The yeast Isw1 complex (Imitation switch) is of particular interest as it has been shown in vitro to function as an inchworm that will form and propagate intra-stand DNA looping (Fig. 4) (De Cian et al. 2012). Isw1 is a member of the SWI/SNF ATPase complex and interacts with the centromere DNA-binding factor CBF1 (Moreau et al. 2003). Cbf1 is required to recruit Isw1 to sites of transcription (Moreau et al. 2003), and may recruit Isw1 to centromere as well. Other members of this family of ATPases, including Snf2 (Gkikopoulos et al. 2011) and PICH (Plk1-interacting checkpoint helicase) (Baumann et al. 2007) have been shown to have a direct role in centromere function. The actin containing chromatin remodeling ATPases have been implicated in chromatin structure of the pericentric domain (Chambers et al. 2012). These enzymes are not essential for centromere function, but mutations lead to decreased segregation fidelity. Their role in sculpting chromatin loops is consistent with the experimental evidence and the formation of loops may be a unifying function that enhances segregation fidelity. Loops also exhibit features observed in RNA processing and lariat intermediates. The base of secondary loops in the budding yeast centromere is proximal to the spindle, where condensin and dyskerin (CBF5) are concentrated (Snider et al. 2014). Dyskerin is a pseudouridine synthetase associated with the snoRNP complex (Zebarjadian et al. 1999). In other guises, these RNPs are able to make lariats (loops) in processing introns from primary transcripts. While the function on RNA and ssDNA is targeted to single-strand nucleic acid, perhaps the pericentromeric dyskerin-containing complexes have evolved a DNA based looping mechanism.

The Ndc10 protein of the CBF3 complex may also promote looping in the pericentromere. Ndc10 binds CDEIII where it recruits Cse4 (Pearson et al. 2003), but Ndc10 also localizes to pericentric chromatin along the spindle axis in live cells (Bouck and Bloom 2005). Based upon its homology to tyrosine site-specific recombinases, Ndc10 may be able to join distal sites to which it is bound (Jayaram et al. 2015). If chromatin-bound Ndc10 oligomerizes through its dimerization domain (Cho and Harrison 2012; Perriches and Singleton 2012), DNA loops will emerge.

3 Centromere Inactivation and de Novo Activation

If cohesin is not holding sister centromeres together, considering they are separated by 400–800 nm in metaphase, what other function might these molecules have. Using transcription as a mechanism to functionally inactivate the centromere we have found that cohesin contributes to the conformation of pericentric chromatin that is favorable for kinetochore assembly (Tsabar et al. 2016). It is unlikely that cohesin recruits kinetochore proteins as there are no direct interactions, and in vivo the pericentric cohesion barrel is well separated from the kinetochore/microtubule attachment complex (Yeh et al. 2008). It has been suggested that proteins such as Sgo1 contribute to the bias that favors sister centromeres to face opposite poles (Fernius and Hardwick 2007; Indjeian and Murray 2007). The barrel of pericentric cohesin could be the physical manifestation of such a mechanism. By assembling cohesin between sister centromeres, the centromeres will be inherently pushed apart, thereby favoring them to lie on the surface of the chromosome.

Chromatin remodeling complexes may also be critical in centromere architecture. Durand-Dubief (Durand-Dubief et al. 2012) found that like cohesin, the

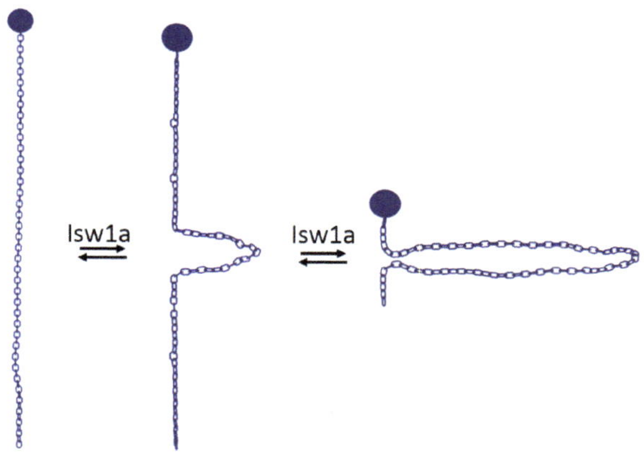

Fig. 4 Isw1a, a member of the SWI/SNF chromatin remodeling family functions as a loop extruding enzyme in vitro, adapted from (De Cian et al. 2012). Isw1 interacts with the yeast Cbf1 factor where it functions to maintain nucleosome-free regions at promoters. This and other ATPases at centromere could collectively promote pericentromeric looping

Swi/Snf-like chromatin remodeling complex, Fun30 contributes significantly to centromere function when centromeres are transcriptionally inactivated. In fun30 mutants, cells containing a chromosome with a transcriptionally inactivated centromere as the only centromere are completely inviable. Fun30 is therefore required to build a proper architecture that can compensate for centromere inactivation. Fun30 shares a phenotype with another centromere component Chl4. Chl4 is a non-essential kinetochore protein, associated with the inner kinetochore (Iml3 of the Ctf19 complex that interacts with Ctf19, Ctf3 and Mif2). Both Chl4 and Fun30, as well as Ctf19 are required for de novo centromere formation (Mythreye and Bloom 2003; Laha et al. 2011; Durand-Dubief et al. 2012).

4 Centromere DNA Replication and Response to Stress

Repetitive DNA sequences are generally unstable and prone to recombination (Branzei and Foiani 2010). The formation of positively supercoiled DNA loops might help to compact centromeric chromatin preventing inappropriate recombination. However, this structure could act as barrier to DNA replication machinery affecting the progression of replication fork. In this case the compact chromatin structure could act as double edge sword, protecting repetitive DNA from recombination on one hand and imposing replication roadblocks to replication fork progression on the other hand. In addition centromeric repetitive sequences can form secondary DNA structures such as hairpins with misaligned and mismatched bases during DNA replication, when the double-stranded DNA is unwound. Such hairpins have been described in vitro for individual satellite DNA repeats, which harbor internal dyadic symmetry conserved from yeast to primates, promoting in vitro self-annealing (Koch 2000; Jonstrup et al. 2008). Therefore chromatin and DNA conformation could be the source of replication stress ultimately provoking DNA breakage in the centromere. Inappropriate control of such stress together with the abnormal forces exerted in mitosis on centromere regions could lead to centromere DNA breakage. Consistent with this idea is that centromeres appear to be hotspots for chromosomal breakage and rearrangements in mammalian and yeast cells (Simi et al. 1998; McFarlane and Humphrey 2010). Induction of replication stress has also been linked to the formation of acentric broken chromosomes in human cells, in which the centromeric protein staining is completely lost (Burrell et al. 2013).

Surprisingly, reconstitution of centromeric chromatin revealed that centromeric DNA was efficiently replicated and replication efficiency was comparable to non-repetitive sequences (Aze et al. 2016). Therefore repetitive DNA did not cause apparent problems to the replication apparatus. These findings indicated the possible presence of specialized factors bound to centromeric chromatin that could help resolve replication roadblocks. Mass spectrometry analysis of the proteome associated with replicating centromeric DNA revealed the enrichment of several DNA repair and DNA structural proteins among which MSH2-6, the MRE11-RAD50 complex, HMGB1-3, XRCC1, XRCC5/DNA-PK, PARP1, ERCC6L/PICH helicase and MUS81 endonuclease (Aze et al. 2016). Some of these proteins were independently found on mouse centromeres (Saksouk et al. 2014). The accumulation of many of these repair factors was abolished by Geminin, an inhibitor of DNA replication fork assembly indicating that centromeric repair factors were loaded in a replication-dependent fashion. In contrast, other common replication players were underrepresented, such as the single-stranded DNA binding (ssDNA) complex RPA and ATR activator TopBP1. Accumulation of these proteins on DNA following induction of stalled replication forks was also diminished. Taken together these results indicated that centromeric DNA replication likely requires accessory DNA repair factors for accurate and efficient replication. This hypothesis was directly tested for the MSH2-6 complex, whose absence compromised centromeric replication. Furthermore, induction of replication stress by inhibition of DNA polymerases induced further recruitment of the MSH2-6 complex. In the same conditions accumulation of RPA, which usually follows formation of ssDNA induced by fork uncoupling triggered by polymerase arrest was not observed. These findings suggested that ssDNA arising at centromeric stalled forks does not get exposed to RPA and likely forms secondary structures containing mismatched bases, which attract MSH2-6.

Reduced RPA accumulation was also responsible for decreased levels of TopBP1 protein and suppression of ATR-dependent CHK1 phosphorylation following induction of stalled replication forks (Aze et al. 2016). Suppression of the ATR checkpoint was required for the efficient replication of repetitive centromeric DNA as forced activation of the checkpoint by interference with topoisomerase activity was indeed sufficient to selectively inhibit centromeric DNA replication. These results revealed two unexpected features of centromeric chromatin. One is that checkpoint suppression is an active phenomenon due to the topological arrangement of the centromeric chromatin in positively supercoiled loops, whose disruption restores the sensitivity to checkpoint activation. The second is that local suppression of ATR signaling facilitates replication of centromeric repetitive DNA, which would otherwise trigger continuous activation of ATR inhibiting replication origin firing.

The physiological roles of checkpoint suppression and sensitivity of centromeric DNA replication to its activation are unclear at the moment. One possibility is that ATR-dependent inhibition of centromeric DNA replication could play a role in preventing unscheduled chromosome segregation in response to stalled forks elsewhere on the chromosome. This could confer a selective advantage to chromosomes bearing DNA repeats at centromeres. It is worth noticing that similar to centromeric DNA other repetitive DNA regions such as the telomeres suppress checkpoint activation, form large loops and are organized in positively supercoiled domains (Benarroch-Popivker et al. 2016) indicating that there might be similarities between telomeres and centromeres in the way DNA is organized and respond to stress.

5 Centromere Breakage and Repair

Replication stress, topological constraints and pulling forces exerted on centromeres during chromosomes segregation might all cause centromere double strand break (DSBs). Recent evidence obtained in cancer cell lines documented increased incidence of DSBs at centromeres following replication stress induced by aphidicolin (Crosetto et al. 2013). The occurrence of DSBs at centromeres following replication stress could explain the centromeric enrichment of DNA repair such as DNA-PK, PARP1, MRE11 and MUS81 (Aze et al. 2016). These proteins are all involved at some level in DSB processing and repair, suggesting the occurrence of ongoing DSB repair at centromeres. Alternatively, DSB repair proteins might accumulate following induction of reversed forks (RFs), which form after the annealing of nascent DNA strands at stalled replication intermediates and mimic the occurrence of DSBs being double stranded (Errico et al. 2014). RFs might frequently occur on centromeric DNA due to their repetitive nature and might require nuclease such as MRE11 or MUS81 to be removed. DSBs might also occur during incomplete DNA decatenation, which requires Topoisomerase II as it is prevented by Topoisomerase II inhibitors (Liu et al. 2014). Interference with DNA decatenation induces formation at centromeres of ultrafine bridges (UFBs), mitotic DNA structures visible in the anaphase of mitosis due to the pulling of incompletely replicated or processed DNA intermediates during chromosome segregation (Liu et al. 2014), which could generate centromeric DSBs.

DSBs at centromeres could be rapidly rejoined through different DSB repair pathways. Centromeric DNA repair events might be facilitated by the presence of tandem repeats. Localization of homologous recombination (HR) and non-homologous end joining (NHEJ) repair proteins on DSBs centromeres has recently been confirmed in mammalian cells (Tsouroula et al. 2016). Interestingly, differently from other regions of the chromosome centromeric DSBs induced by Crispr/Cas9 were shown to bind both NHEJ and HR in all phases of the cell cycle, suggesting that both pathways are active at all times. Among the HR pathways single-strand annealing (SSA) and synthesis-dependent strand annealing (SDSA), might be particularly facilitated by repetitive sequences (Paques et al. 1998; Paques and Haber 1999). In SSA one end of the DSB with exposed ssDNA anneals to the other end and this event is facilitated by the presence of duplicated sequences. During SDSA the annealing phase is followed by replication-mediated extension of the invading strand, which uses the homologous sequence as template. Multiple direct repeats, as found in human centromeres, might also contribute to generate large ring structures via recombination between distant homologous direct repeats. Pairing between homologous sequences might stabilize such large rings. Abnormal metabolism and resolution of recombination products between these direct repeats would instead release the ring from the chromosome, forming extra-chromosomal circular molecules. Many extra-chromosomal circles of centromeric DNA have been observed to accumulate after replication stress in several species (Cohen et al. 2003; Cohen and Segal 2009).

Although studies of satellite-less neocentromeres have overshadowed the possible importance of repetitive DNA in established centromeres these observations indicate that the centromere might require repetitive DNA to ensure its optimal function, particularly under stress. It is likely that DNA repeats are actively maintained notwithstanding their problematic replication for a number of advantageous features such as promoting efficient repair of centromeres in case of breakage and functioning as checkpoint sensitive zone of the replication checkpoint. Unregulated recombination between repeated sequences could lead to the complete or partial loss of centromeric DNA following replication stress. Furthermore, abnormal resolution of SSA and SDSA repair events could promote contraction and expansion cycles of the intervening DNA repeats, respectively (Paques et al. 1998; Paques and Haber 1999). Loss of repetitive DNA could result in the shortening of centromeres in older cells, which have undergone several cell cycles. Centromere deterioration has been indeed observed in aging women although it is not clear if this depends on DNA replication (Nakagome et al. 1984).

6 Repetitive DNA and Loops Generate Centromere Force

The bottlebrush centromere provides a mechanistic understanding for the role of repeat DNA in the centromere and addresses several outstanding problems (Lawrimore et al. 2016). First, the organization of repeats into DNA loops via another SMC family member cohesin, has been well documented in the nucleolus (Harris et al. 2014). Second, the bottlebrush provides a physical basis for how a floppy DNA chain can be converted into a stiff (relative to an entropic chain) spring. A fluctuating chain in a thermal bath will find the most entropically favored state, that of a random coil, as a random coil chain is the most disordered. The addition of side chains relative to the primary chain (or primary axis), limits the ability of the primary chain to adopt a random coil, through limiting the number of states the primary chain can adopt. Additional side chains further restrict the motion of the primary axis until a point where they generate tension along the axis. In this fashion, enthalpic energy put into making chains, results in entropic forces of the brush where side chain fluctuation amplifies tension along the primary axis. Thus chromatin loops significantly change the state of the centromere from a floppy chromatin polymer into a stiff chromatin network.

The chromatin loops provide a mechanism for buffering changes in tension resulting from microtubule dynamics. Kinetochore microtubules are in a constant state of flux, and rare persistent growth and/or shortening events could result in large local changes in tension at individual kinetochores. DNA loops compensate for large changes in kinetochore microtubule length through their ability to convert between looped and stretched states (Stephens et al. 2013a). The looped state is radially displaced from the spindle axis. Increased tension at the kinetochore exerts a pulling force at the base of the loop, switching the loop to the stretched state and increasing the length of DNA along the spindle axis. The increase in axial DNA compensates for kinetochore microtubule shortening. Likewise persistent microtubule growth events will bias the DNA into the looped state, decreasing the length of axial DNA. Thus pericentric loops buffer the system such that tension can be maintained on a landscape of dynamic kinetochore microtubules.

The bottlebrush also helps us understand how the centromere retains morphology (and cohesin its uniform barrel structure) in a dynamic system where individual chromatin chains are fluctuating and microtubules are constantly prodding and probing the kinetochore resulting in its deformation surface of the centromere (Magidson et al. 2016). Chromosomes, like slip link gels are known to retain their elastic and tensile moduli properties over several orders of magnitude (Okumura and Ito 2001; Granick and Rubinstein 2004; Bloom and Joglekar 2010). It has been known for almost half a century that chromosomes expand and contract upon removal and return of mono- and divalent ions (Lezzi and Gilbert 1970), and depending on the treatment retain their original structure (e.g. such as their macroscopic banding pattern). The concentration of ring complexes such as cohesin and condensin in the pericentromere may be indicative of the slip-link property of centromeres. These complexes are able to compact chromatin, and in the case of cohesin, is able to compact DNA against a force of 0.45 pN (Sun et al. 2013).

7 Repeat Stability and Chromosome Loss

Loss of centromeric DNA repeats due to incorrect duplication or repair might ultimately cause impairment of microtubule attachment to chromosome. Replication stress has indeed been shown to be a primary cause of chromosome number abnormalities and instability (CIN) in cancer cells (Burrell et al. 2013). Complete loss of centromeric DNA following replication stress could indeed lead to segregation errors due to lack of functional kinetochores in case of acentric chromosomes, which are frequently encountered in CIN+ cells. This would suggest a pre-mitotic origin of some chromosome segregation errors (Burrell et al. 2013). This view was recently challenged by observations that the occurrence of acentric chromosomes is limited compared to whole lagging chromosomes caused by primary mitotic segregation errors in CIN+ cells (Bakhoum et al. 2014). However, complete loss of centromeric DNA causing formation of acentric chromosomes might not be the only pre-mitotic cause of chromosome segregation errors. A more limited loss of centromeric DNA similar to centromere degeneration observed in older cells, which is more difficult to detect by fluorescence-based techniques than complete loss, could lead to the formation of centromeres with suboptimal function. Some of these suboptimal centromeres might be responsible for microtubule-chromosome attachment problems leading to the formation of lagging chromosomes.

Among chromosome-microtubule issues there is merotelic attachment in which spindle fibers from both poles attach to a single kinetochore. This condition is particularly dangerous for chromosome stability as it often causes the occurrence of lagging chromosomes not sensed by the mitotic checkpoint apparatus. Lagging chromosomes can be inherited as extra-chromosomes in one of the daughter cells (Santaguida and Amon 2015). This extra chromosome could end up in micronuclei, in which replication and repair is inefficient due to limited availability of nuclear factors predisposing to catastrophic events such as chromosome pulverization (Crasta et al. 2012). In this case a small change in the centromeric DNA structure leading to weakening of chromosome attachment to spindle might lead to catastrophic consequences.

Merotelic attachment of mitotic origin uncorrected by the classic mitotic checkpoints could also generate forces sufficiently strong to physically shear the centromere leading to chromosome breakage (Guerrero et al. 2010). Loss of centromeric DNA and decreased loop formation could make the centromeric chromatin less resistant to such forces and more prone to break. Incomplete maturation and decreased levels of entanglements due to centromeric DNA abnormalities might also weaken the centromeric spring. Therefore, although it is not clear whether forces generated during mitosis are strong enough to break the normal chromatin or DNA structure present at centromeres (Ganem and Pellman 2012), centromere weakening might lower the tolerance to this type of mechanical stress.

Centromere DNA breakage might be an early event in cellular transformation. Inappropriate repair of centromere breakage in cells combined to inefficient DNA repair might predispose to breakage-fusion-bridge (BFB) cycles generating further chromosome instability (Martinez and van Wely 2011; Forsburg 2013). The BFB cycle has initially been described for telomeric end regions of chromosomes. However, although the telomere fusion hypothesis is believed to be the major engine of BFB it does not explain frequent loss and gains of whole chromosome arms in tumors (Beroukhim et al. 2010; Martinez and van Wely 2011; Forsburg 2013). Furthermore, cytogenetic and microarray analysis of breakage-fusion sites revealed frequent occurrence of tandem fusions of chromosome arms with parallel orientation containing interstitial centromere and telomere sequences (Martinez and van Wely 2011). Such events likely derive from a whole chromosome arm produced by centromere breakage that fused to an unprotected telomere of an intact chromosome, leading to the formation of a dicentric chromosome with two centromere-kinetochores structures. Dicentric chromosomes could undergo further breakage at random places when pulled by the spindle fibers bound to old and new centromeres. These events account for more than 50% of chromosome translocation in some tumors, whereas telomere–telomere fusions with antiparallel orientation produced by telomere erosion constitute only 1% of total translocations when analyzed with classical cytogenetic techniques (Martinez and van Wely 2011). These observations strongly suggest that centromere breakage is a major event in the BFB cycle.

To better understand these processes at molecular and structural level the genomic structure of centromeric DNA should be better characterized and annotated. The identification of unique non-repetitive sequences interspersed in centromeric DNA might be useful to design probes to monitor stability of centromeric DNA during cell cycle in unchallenged and stressful conditions. Such tools could be used to test the occurrence of contraction and expansion cycles at repetitive centromeric DNA during unperturbed and challenged DNA replication. These studies could help understanding why these repetitive DNA sequences are retained despite their problematic maintenance. Understanding how these processes occur at molecular level will therefore be essential to clarify the origin of genome instability predisposing to cancer.

References

Alipour E, Marko JF (2012) Self-organization of domain structures by DNA-loop-extruding enzymes. Nucleic Acids Res 40:11202–11212

Aravamudhan P, Felzer–Kim I, Joglekar AP (2013) The budding yeast point centromere associates with two Cse4 molecules during mitosis. Curr Biol 23:770–774

Aze A, Sannino V, Soffientini P, Bachi A, Costanzo V (2016) Centromeric DNA replication reconstitution reveals DNA loops and ATR checkpoint suppression. Nat Cell Biol 18:684–691

Bakhoum SF, Silkworth WT, Nardi IK, Nicholson JM, Compton DA, Cimini D (2014) The mitotic origin of chromosomal instability. Curr Biol 24:R148–R149

Baumann C, Korner R, Hofmann K, Nigg EA (2007) PICH, a centromere-associated SNF2 family ATPase, is regulated by Plk1 and required for the spindle checkpoint. Cell 128:101–114

Benarroch-Popivker D, Pisano S, Mendez-Bermudez A, Lototska L, Kaur P, Bauwens S, Djerbi N, Latrick CM, Fraisier V, Pei B, Gay A, Jaune E, Foucher K, Cherfils-Vicini J, Aeby E, Miron S, Londono-Vallejo A, Ye J, Le Du MH, Wang H, Gilson E, Giraud-Panis MJ (2016) TRF2-Mediated control of telomere dna topology as a mechanism for chromosome-end protection. Mol Cell 61:274–286

Bergmann JH, Rodriguez MG, Martins NM, Kimura H, Kelly DA, Masumoto H, Larionov V, Jansen LE, Earnshaw WC (2011) Epigenetic engineering shows H3K4me2 is required for HJURP targeting and CENP-A assembly on a synthetic human kinetochore. EMBO J 30: 328–340

Bermudez I, Garcia-Martinez J, Perez-Ortin JE, Roca J (2010) A method for genome-wide analysis of DNA helical tension by means of psoralen-DNA photobinding. Nucleic Acids Res 38:e182

Bernad R, Sanchez P, Rivera T, Rodriguez-Corsino M, Boyarchuk E, Vassias I, Ray-Gallet D, Arnaoutov A, Dasso M, Almouzni G, Losada A (2011) Xenopus HJURP and condensin II are required for CENP-A assembly. J Cell Biol 192:569–582

Beroukhim R, Mermel CH, Porter D, Wei G, Raychaudhuri S, Donovan J, Barretina J, Boehm JS, Dobson J, Urashima M, Mc Henry KT, Pinchback RM, Ligon AH, Cho YJ, Haery L, Greulich H, Reich M, Winckler W, Lawrence MS, Weir BA, Tanaka KE, Chiang DY, Bass AJ, Loo A, Hoffman C, Prensner J, Liefeld T, Gao Q, Yecies D, Signoretti S, Maher E, Kaye FJ, Sasaki H, Tepper JE, Fletcher JA, Tabernero J, Baselga J, Tsao MS, Demichelis F, Rubin MA, Janne PA, Daly MJ, Nucera C, Levine RL, Ebert BL, Gabriel S, Rustgi AK, Antonescu CR, Ladanyi M, Letai A, Garraway LA, Loda M, Beer DG, True LD, Okamoto A, Pomeroy SL, Singer S, Golub TR, Lander ES, Getz G, Sellers WR, Meyerson M (2010) The landscape of somatic copy-number alteration across human cancers. Nature 463:899–905

Black BE, Foltz DR, Chakravarthy S, Luger K, Woods VL Jr, Cleveland DW (2004) Structural determinants for generating centromeric chromatin. Nature 430:578–582

Bloom K, Joglekar A (2010) Towards building a chromosome segregation machine. Nature 463:446–456

Bloom KS, Carbon J (1982) Yeast centromere DNA is in a unique and highly ordered structure in chromosomes and small circular minichromosomes. Cell 29:305–317

Bloom KS, Fitzgerald-Hayes M, Carbon J (1983) Structural analysis and sequence organization of yeast centromeres. Cold Spring Harb Symp Quant Biol 47(Pt 2):1175–1185

Bloom K, Amaya E, Yeh E (1984) Centromeric DNA structure in yeast chromatin. In: Borisy GG, Cleveland DW, Murphy DB (Eds.), Molecular biology of the cytoskeleton. Cold Spring Harbor Laboratory Press, Cold Spring Harbor, pp 175–184

Bouck D, Bloom K (2005) The role of centromere-binding factor 3 (CBF3) in spindle stability, cytokinesis, and kinetochore attachment. Biochem Cell Biol 83:696–702

Branzei D, Foiani M (2010) Maintaining genome stability at the replication fork. Nat Rev Mol Cell Biol 11:208–219

Burrell RA, McClelland SE, Endesfelder D, Groth P, Weller MC, Shaikh N, Domingo E, Kanu N, Dewhurst SM, Gronroos E, Chew SK, Rowan AJ, Schenk A, Sheffer M, Howell M, Kschischo M, Behrens A, Helleday T, Bartek J, Tomlinson IP, Swanton C (2013) Replication stress links structural and numerical cancer chromosomal instability. Nature 494:492–496

Camahort R, Shivaraju M, Mattingly M, Li B, Nakanishi S, Zhu D, Shilatifard A, Workman JL, Gerton JL (2009) Cse4 is part of an octameric nucleosome in budding yeast. Mol Cell 35: 794–805

Chambers AL, Ormerod G, Durley SC, Sing TL, Brown GW, Kent NA, Downs JA (2012) The INO80 chromatin remodeling complex prevents polyploidy and maintains normal chromatin structure at centromeres. Genes Dev 26:2590–2603

Chen Y, Baker RE, Keith KC, Harris K, Stoler S, Fitzgerald-Hayes M (2000) The N terminus of the centromere H3-like protein Cse4p performs an essential function distinct from that of the histone fold domain. Mol Cell Biol 20:7037–7048

Cho US, Harrison SC (2012) Ndc10 is a platform for inner kinetochore assembly in budding yeast. Nat Struct Mol Biol 19:48–55

Cohen S, Segal D (2009) Extrachromosomal circular DNA in eukaryotes: possible involvement in the plasticity of tandem repeats. Cytogenet Genome Res 124:327–338

Cohen S, Yacobi K, Segal D (2003) Extrachromosomal circular DNA of tandemly repeated genomic sequences in Drosophila. Genome Res 13:1133–1145

Cole HA, Howard BH, Clark DJ (2011) The centromeric nucleosome of budding yeast is perfectly positioned and covers the entire centromere. Proc Natl Acad Sci U S A 108:12687–12692

Crasta K, Ganem NJ, Dagher R, Lantermann AB, Ivanova EV, Pan Y, Nezi L, Protopopov A, Chowdhury D, Pellman D (2012) DNA breaks and chromosome pulverization from errors in mitosis. Nature 482:53–58

Crosetto N, Mitra A, Silva MJ, Bienko M, Dojer N, Wang Q, Karaca E, Chiarle R, Skrzypczak M, Ginalski K, Pasero P, Rowicka M, Dikic I (2013) Nucleotide-resolution DNA double-strand break mapping by next-generation sequencing. Nat Methods 10:361–365

Dai C, Li W, Tjong H, Hao S, Zhou Y, Li Q, Chen L, Zhu B, Alber F, Zhou XJ (2016) Mining 3D genome structure populations identifies major factors governing the stability of regulatory communities. Nat Commun 7:11549

Dalal Y, Wang H, Lindsay S, Henikoff S (2007) Tetrameric structure of centromeric nucleosomes in interphase Drosophila cells. PLoS Biol 5:e218

De Cian A, Praly E, Ding F, Singh V, Lavelle C, Le Cam E, Croquette V, Pietrement O, Bensimon D (2012) ATP-independent cooperative binding of yeast Isw1a to bare and nucleosomal DNA. PLoS ONE 7:e31845

Dechassa ML, Wyns K, Li M, Hall MA, Wang MD, Luger K (2011) Structure and Scm3-mediated assembly of budding yeast centromeric nucleosomes. Nat Commun 2:313

Diaz-Ingelmo O, Martinez-Garcia B, Segura J, Valdes A, Roca J (2015) DNA topology and global architecture of point centromeres. Cell Rep 13:667–677

Duan Z, Andronescu M, Schutz K, McIlwain S, Kim YJ, Lee C, Shendure J, Fields S, Blau CA, Noble WS (2010) A three-dimensional model of the yeast genome. Nature 465:363–367

Durand-Dubief M, Will WR, Petrini E, Theodorou D, Harris RR, Crawford MR, Paszkiewicz K, Krueger F, Correra RM, Vetter AT, Miller JR, Kent NA, Varga-Weisz P (2012) SWI/SNF-like chromatin remodeling factor Fun30 supports point centromere function in *S. cerevisiae*. PLoS Genet 8:e1002974

Earnshaw WC, Allshire RC, Black BE, Bloom K, Brinkley BR, Brown W, Cheeseman IM, Choo KH, Copenhaver GP, Deluca JG, Desai A, Diekmann S, Erhardt S, Fitzgerald-Hayes M, Foltz D, Fukagawa T, Gassmann R, Gerlich DW, Glover DM, Gorbsky GJ, Harrison SC, Heun P, Hirota T, Jansen LE, Karpen G, Kops GJ, Lampson MA, Lens SM, Losada A, Luger K, Maiato H, Maddox PS, Margolis RL, Masumoto H, McAinsh AD, Mellone BG, Meraldi P, Musacchio A, Oegema K, O'Neill RJ, Salmon ED, Scott KC, Straight AF, Stukenberg PT, Sullivan BA, Sullivan KF, Sunkel CE, Swedlow JR, Walczak CE, Warburton PE, Westermann S, Willard HF, Wordeman L, Yanagida M, Yen TJ, Yoda K, Cleveland DW (2013) Esperanto for histones: CENP-A, not CenH3, is the centromeric histone H3 variant. Chromosome Res 21:101–106

Edwards NS, Murray AW (2005) Identification of xenopus CENP-A and an associated centromeric DNA repeat. Mol Biol Cell 16:1800–1810

Erliandri I, Fu H, Nakano M, Kim JH, Miga KH, Liskovykh M, Earnshaw WC, Masumoto H, Kouprina N, Aladjem MI, Larionov V (2014) Replication of alpha-satellite DNA arrays in endogenous human centromeric regions and in human artificial chromosome. Nucleic Acids Res 42:11502–11516

Errico A, Aze A, Costanzo V (2014) Mta2 promotes Tipin-dependent maintenance of replication fork integrity. Cell Cycle 13:2120–2128

Espelin CW, Kaplan KB, Sorger PK (1997) Probing the architecture of a simple kinetochore using DNA-protein crosslinking. J Cell Biol 139:1383–1396

Espelin CW, Simons KT, Harrison SC, Sorger PK (2003) Binding of the essential *Saccharomyces cerevisiae* kinetochore protein Ndc10p to CDEII. Mol Biol Cell 14:4557–4568

Fachinetti D, Han JS, McMahon MA, Ly P, Abdullah A, Wong AJ, Cleveland DW (2015) DNA sequence-specific binding of CENP-B enhances the fidelity of human centromere function. Dev Cell 33:314–327

Fernius J, Hardwick KG (2007) Bub1 kinase targets Sgo1 to ensure efficient chromosome biorientation in budding yeast mitosis. PLoS Genet 3:e213

Forsburg SL (2013) The CINs of the centromere. Biochem Soc Trans 41:1706–1711

Fudenberg G, Imakaev M, Lu C, Goloborodko A, Abdennur N, Mirny LA (2016) Formation of chromosomal domains by loop extrusion. Cell Rep 15:2038–2049

Funk M, Hegemann JH, Philippsen P (1989) Chromatin digestion with restriction endonucleases reveals 150–160 bp of protected DNA in the centromere of chromosome XIV in Saccharomyces cerevisiae. Mol Gen Genet 219:153–160

Furuyama T, Henikoff S (2009) Centromeric nucleosomes induce positive DNA supercoils. Cell 138:104–113

Ganem NJ, Pellman D (2012) Linking abnormal mitosis to the acquisition of DNA damage. J Cell Biol 199:871–881

Gkikopoulos T, Singh V, Tsui K, Awad S, Renshaw MJ, Scholfield P, Barton GJ, Nislow C, Tanaka TU, Owen-Hughes T (2011) The SWI/SNF complex acts to constrain distribution of the centromeric histone variant Cse4. EMBO J 30:1919–1927

Goloborodko A, Imakaev MV, Marko JF, Mirny L (2016a) Compaction and segregation of sister chromatids via active loop extrusion. Elife 5:e14864

Goloborodko A, Marko JF, Mirny LA (2016b) Chromosome compaction by active loop extrusion. Biophys J 110:2162–2168

Granick S, Rubinstein M (2004) Polymers: a multitude of macromolecules. Nat Mater 3:586–587

Guerrero AA, Gamero MC, Trachana V, Futterer A, Pacios-Bras C, Diaz-Concha NP, Cigudosa JC, Martinez AC, van Wely KH (2010) Centromere-localized breaks indicate the generation of DNA damage by the mitotic spindle. Proc Nat Acad Sci U S A 107:4159–4164

Guse A, Carroll CW, Moree B, Fuller CJ, Straight AF (2011) In vitro centromere and kinetochore assembly on defined chromatin templates. Nature 477:354–358

Haase J, Mishra PK, Stephens A, Haggerty R, Quammen C, Taylor RM 2nd, Yeh E, Basrai MA, Bloom K (2013) A 3D map of the yeast kinetochore reveals the presence of core and accessory centromere-specific histone. Curr Biol 23:1939–1944

Harris B, Bose T, Lee KK, Wang F, Lu S, Ross RT, Zhang Y, French SL, Beyer AL, Slaughter BD, Unruh JR, Gerton JL (2014) Cohesion promotes nucleolar structure and function. Mol Biol Cell 25:337–346

Hasson D, Panchenko T, Salimian KJ, Salman MU, Sekulic N, Alonso A, Warburton PE, Black BE (2013) The octamer is the major form of CENP-A nucleosomes at human centromeres. Nat Struct Mol Biol 20:687–695

He H, Zhang S, Wang D, Hochwagen A, Li F (2016) Condensin promotes position effects within tandem dna repeats via the rits complex. Cell reports 14:1018–1024

Heidinger-Pauli JM, Mert O, Davenport C, Guacci V, Koshland D (2010) Systematic reduction of cohesin differentially affects chromosome segregation, condensation, and DNA repair. Curr Biol 20:957–963

Henikoff S, Ramachandran S, Krassovsky K, Bryson TD, Codomo CA, Brogaard K, Widom J, Wang JP, Henikoff JG (2014) The budding yeast centromere DNA element II wraps a stable Cse4 hemisome in either orientation in vivo. Elife 3:e01861

Heus JJ, Bloom KS, Zonneveld BJ, Steensma HY, Van den Berg JA (1993a) Chromatin structures of *Kluyveromyces lactis* centromeres in *K. lactis* and *Saccharomyces cerevisiae*. Chromosoma 102:660–667

Heus JJ, Zonneveld BJ, de Steensma HY, van den Berg JA (1993b) The consensus sequence of *Kluyveromyces lactis* centromeres shows homology to functional centromeric DNA from *Saccharomyces cerevisiae*. Mol Gen Genet 236:355–362

Hirano T (2012) Condensins: universal organizers of chromosomes with diverse functions. Genes Dev 26:1659–1678

Hsieh TH, Weiner A, Lajoie B, Dekker J, Friedman N, Rando OJ (2015) Mapping nucleosome resolution chromosome folding in yeast by Micro-C. Cell 162:108–119

Hu B, Petela N, Kurze A, Chan KL, Chapard C, Nasmyth K (2015) Biological chromodynamics: a general method for measuring protein occupancy across the genome by calibrating ChIP-seq. Nucleic Acids Res 43:e132

Indjeian VB, Murray AW (2007) Budding yeast mitotic chromosomes have an intrinsic bias to biorient on the spindle. Curr Biol 17:1837–1846

Jayaram M, Ma CH, Kachroo AH, Rowley PA, Guga P, Fan HF, Voziyanov Y (2015) An overview of tyrosine site-specific recombination: from an Flp perspective. Microbiol Spectr 3

Jonstrup AT, Thomsen T, Wang Y, Knudsen BR, Koch J, Andersen AH (2008) Hairpin structures formed by alpha satellite DNA of human centromeres are cleaved by human topoisomerase IIalpha. Nucleic Acid Res 36:6165–6174

Keith KC, Baker RE, Chen Y, Harris K, Stoler S, Fitzgerald-Hayes M (1999) Analysis of primary structural determinants that distinguish the centromere-specific function of histone variant Cse4p from histone H3. Mol Cell Biol 19:6130–6139

Koch J (2000) Neocentromeres and alpha satellite: a proposed structural code for functional human centromere DNA. Hum Mol Genet 9:149–154

Kouprina N, Earnshaw WC, Masumoto H, Larionov V (2013) A new generation of human artificial chromosomes for functional genomics and gene therapy. Cell Mol Life Sci 70: 1135–1148

Krassovsky K, Henikoff JG, Henikoff S (2012) Tripartite organization of centromeric chromatin in budding yeast. Proc Natl Acad Sci U S A 109:243–248

Laha S, Das SP, Hajra S, Sanyal K, Sinha P (2011) Functional characterization of the Saccharomyces cerevisiae protein Chl1 reveals the role of sister chromatid cohesion in the maintenance of spindle length during S-phase arrest. BMC Genet 12:83

Lawrimore J, Bloom KS, Salmon ED (2011) Point centromeres contain more than a single centromere-specific Cse4 (CENP-A) nucleosome. J Cell Biol 195:573–582

Lawrimore J, Vasquez PA, Falvo MR, Taylor RM 2nd, Vicci L, Yeh E, Forest MG, Bloom K (2015) DNA loops generate intracentromere tension in mitosis. J Cell Biol 210:553–564

Lawrimore J, Aicher JK, Hahn P, Fulp A, Kompa B, Vicci L, Falvo M, Taylor RM 2nd, Bloom K (2016) ChromoShake: a chromosome dynamics simulator reveals that chromatin loops stiffen centromeric chromatin. Mol Biol Cell 27:153–166

Lezzi M, Gilbert LI (1970) Differential effects of K+ and Na+ on specific bands of isolated polytene chromosomes of *Chironomus tentans*. J Cell Sci 6:615–627

Liu Y, Nielsen CF, Yao Q, Hickson ID (2014) The origins and processing of ultra fine anaphase DNA bridges. Curr Opin Genet Dev 26:1–5

Logsdon GA, Barrey EJ, Bassett EA, DeNizio JE, Guo LY, Panchenko T, Dawicki-McKenna JM, Heun P, Black BE (2015) Both tails and the centromere targeting domain of CENP-A are required for centromere establishment. J Cell Biol 208:521–531

Magidson V, He J, Ault JG, O'Connell CB, Yang N, Tikhonenko I, McEwen BF, Sui H, Khodjakov A (2016) Unattached kinetochores rather than intrakinetochore tension arrest mitosis in taxol-treated cells. J Cell Biol 212:307–319

Marshall OJ, Chueh AC, Wong LH, Choo KH (2008) Neocentromeres: new insights into centromere structure, disease development, and karyotype evolution. Am J Hum Genet 82:261–282

Martinez AC, van Wely KH (2011) Centromere fission, not telomere erosion, triggers chromosomal instability in human carcinomas. Carcinogenesis 32:796–803

McFarlane RJ, Humphrey TC (2010) A role for recombination in centromere function. Trends Genet 26:209–213

Mizuguchi G, Xiao H, Wisniewski J, Smith MM, Wu C (2007) Nonhistone Scm3 and histones CenH3-H4 assemble the core of centromere-specific nucleosomes. Cell 129:1153–1164

Moreau JL, Lee M, Mahachi N, Vary J, Mellor J, Tsukiyama T, Goding CR (2003) Regulated displacement of TBP from the PHO8 promoter in vivo requires Cbf1 and the Isw1 chromatin remodeling complex. Mol Cell 11:1609–1620

Mythreye K, Bloom KS (2003) Differential kinetochore protein requirements for establishment versus propagation of centromere activity in *Saccharomyces cerevisiae*. J Cell Biol 160: 833–843

Nakagome Y, Abe T, Misawa S, Takeshita T, Iinuma K (1984) The "loss" of centromeres from chromosomes of aged women. Am J Hum Genet 36:398–404

Nakano M, Cardinale S, Noskov VN, Gassmann R, Vagnarelli P, Kandels-Lewis S, Larionov V, Earnshaw WC, Masumoto H (2008) Inactivation of a human kinetochore by specific targeting of chromatin modifiers. Dev Cell 14:507–522

Okumura Y, Ito K (2001) The polyrotaxane gel: a topological gel by figure-of-eight cross-links. Adv Mater 13:485–487

Panyukov S, Zhulina EB, Sheiko SS, Randall GC, Brock J, Rubinstein M (2009a) Tension amplification in molecular brushes in solutions and on substrates (dagger). J Phys Chem B 113:3750

Panyukov SV, Sheiko SS, Rubinstein M (2009b) Amplification of tension in branched macromolecules. Phys Rev Lett 102:148301

Paques F, Haber JE (1999) Multiple pathways of recombination induced by double-strand breaks in *Saccharomyces cerevisiae*. Microbiol and Mol Biol Rev 63:349–404

Paques F, Leung WY, Haber JE (1998) Expansions and contractions in a tandem repeat induced by double-strand break repair. Mol Cell Biol 18:2045–2054

Pearson CG, Maddox PS, Zarzar TR, Salmon ED, Bloom K (2003) Yeast kinetochores do not stabilize Stu2p-dependent spindle microtubule dynamics. Mol Biol Cell 14:4181–4195

Perriches T, Singleton MR (2012) Structure of yeast kinetochore Ndc10 DNA-binding domain reveals unexpected evolutionary relationship to tyrosine recombinases. J Biol Chem 287: 5173–5179

Pietrasanta LI, Thrower D, Hsieh W, Rao S, Stemmann O, Lechner J, Carbon J, Hansma H (1999) Probing the *Saccharomyces cerevisiae* centromeric DNA (CEN DNA)-binding factor 3 (CBF3) kinetochore complex by using atomic force microscopy. Proc Natl Acad Sci U S A 96: 3757–3762

Russell ID, Grancell AS, Sorger PK (1999) The unstable F-box protein p58-Ctf13 forms the structural core of the CBF3 kinetochore complex. J Cell Biol 145:933–950

Saksouk N, Barth TK, Ziegler-Birling C, Olova N, Nowak A, Rey E, Mateos-Langerak J, Urbach S, Reik W, Torres-Padilla ME, Imhof A, Dejardin J, Simboeck E (2014) Redundant

mechanisms to form silent chromatin at pericentromeric regions rely on BEND3 and DNA methylation. Mol Cell 56:580–594
Samoshkin A, Arnaoutov A, Jansen LE, Ouspenski I, Dye L, Karpova T, McNally J, Dasso M, Cleveland DW, Strunnikov A (2009) Human condensin function is essential for centromeric chromatin assembly and proper sister kinetochore orientation. PLoS ONE 4:e6831
Santaguida S, Amon A (2015) Short- and long-term effects of chromosome mis-segregation and aneuploidy. Nat Rev Mol Cell Biol 16:473–485
Saunders M, Fitzgerald-Hayes M, Bloom K (1988) Chromatin structure of altered yeast centromeres. Proc Natl Acad Sci U S A 85:175–179
Shivaraju M, Unruh JR, Slaughter BD, Mattingly M, Berman J, Gerton JL (2012) Cell-cycle-coupled structural oscillation of centromeric nucleosomes in yeast. Cell 150: 304–316
Simi S, Simili M, Bonatti S, Campagna M, Abbondandolo A (1998) Fragile sites at the centromere of Chinese hamster chromosomes: a possible mechanism of chromosome loss. Mutat Res 397:239–246
Snider CE, Stephens AD, Kirkland JG, Hamdani O, Kamakaka RT, Bloom K (2014) Dyskerin, tRNA genes, and condensin tether pericentric chromatin to the spindle axis in mitosis. J Cell Biol 207
Stephens AD, Haase J, Vicci L, Taylor RM 2nd, Bloom K (2011) Cohesin, condensin, and the intramolecular centromere loop together generate the mitotic chromatin spring. J Cell Biol 193:1167–1180
Stephens AD, Haggerty RA, Vasquez PA, Vicci L, Snider CE, Shi F, Quammen C, Mullins C, Haase J, Taylor RM 2nd, Verdaasdonk JS, Falvo MR, Jin Y, Forest MG, Bloom K (2013a) Pericentric chromatin loops function as a nonlinear spring in mitotic force balance. J Cell Biol 200:757–772
Stephens AD, Snider CE, Haase J, Haggerty RA, Vasquez PA, Forest MG, Bloom K (2013b) Individual pericentromeres display coordinated motion and stretching in the yeast spindle. J Cell Biol 203:407–416
Stigler J, Camdere GO, Koshland DE, Greene EC (2016) Single-molecule imaging reveals a collapsed conformational state for dna-bound cohesin. Cell Rep 15:988–998
Sun M, Nishino T, Marko JF (2013) The SMC1-SMC3 cohesin heterodimer structures DNA through supercoiling-dependent loop formation. Nucleic Acids Res 41:6149–6160
Tsabar M, Haase J, Harrison B, Snider CE, Eldridge B, Kaminsky L, Hine RM, Haber JE, Bloom K (2016) A cohesin-based partitioning mechanism revealed upon transcriptional inactivation of centromere. PLoS Genet 12:e1006021
Tsouroula K, Furst A, Rogier M, Heyer V, Maglott-Roth A, Ferrand A, Reina-San-Martin B, Soutoglou E (2016) Temporal and spatial uncoupling of dna double strand break repair pathways within mammalian heterochromatin. Mol Cell 63:293–305
Vasquez PA, Hult C, Adalsteinsson D, Lawrimore J, Forest MG, Bloom K (2016) Entropy gives rise to topologically associating domains. Nucleic Acids Res 44:5540–5549
Verdaasdonk JS, Gardner R, Stephens AD, Yeh E, Bloom K (2012) Tension-dependent nucleosome remodeling at the pericentromere in yeast. Mol Biol Cell 23:2560–2570
Wang X, Le TB, Lajoie BR, Dekker J, Laub MT, Rudner DZ (2015) Condensin promotes the juxtaposition of DNA flanking its loading site in Bacillus subtilis. Genes Dev 29:1661–1675
Weir JR, Faesen AC, Klare K, Petrovic A, Basilico F, Fischbock J, Pentakota S, Keller J, Pesenti ME, Pan D, Vogt D, Wohlgemuth S, Herzog F, Musacchio A (2016) Insights from biochemical reconstitution into the architecture of human kinetochores. Nature 537:249–253
Wisniewski J, Hajj B, Chen J, Mizuguchi G, Xiao H, Wei D, Dahan M, Wu C (2014) Imaging the fate of histone Cse4 reveals de novo replacement in S phase and subsequent stable residence at centromeres. Elife 3:e02203
Xiao H, Mizuguchi G, Wisniewski J, Huang Y, Wei D, Wu C (2011) Nonhistone Scm3 binds to AT-rich DNA to organize atypical centromeric nucleosome of budding yeast. Mol Cell 43:369–380

Yeh E, Haase J, Paliulis LV, Joglekar A, Bond L, Bouck D, Salmon ED, Bloom KS (2008) Pericentric chromatin is organized into an intramolecular loop in mitosis. Curr Biol 18:81–90
Zebarjadian Y, King T, Fournier MJ, Clarke L, Carbon J (1999) Point mutations in yeast CBF5 can abolish in vivo pseudouridylation of rRNA. Mol Cell Biol 19:7461–7472

The Role of Centromere Defects in Cancer

Thian Thian Beh and Paul Kalitsis

Abstract The accurate segregation of chromosomes to daughter cells is essential for healthy development to occur. Imbalances in chromosome number have long been associated with cancers amongst other medical disorders. Little is known whether abnormal chromosome numbers are an early contributor to the cancer progression pathway. Centromere DNA and protein defects are known to impact on the fidelity of chromosome segregation in cell and model systems. In this chapter we discuss recent developments in understanding the contribution of centromere abnormalities at the protein and DNA level and their role in cancer in human and mouse systems.

1 Cancer

Cancer, in the simplest textbook definition, is a disease where the balance between cellular proliferation and cell death has been distorted in favour of proliferation, causing uncontrollable cellular growth giving rise to tumours. However, the tumour mass itself is not homogeneous but instead is a complex tissue containing different interacting cell types—tumour-originating cells and also recruited normal cells, termed tumour-associated stromal cells, which have been found to play an active role in tumorigenesis (Hanahan and Weinberg 2011). Hence, the current endeavours in cancer research also encompass the role of the tumour microenvironment in tumour progression.

The appearance or phenotype of the different cancer cells has been used as the standard practice in cancer diagnosis and classification. These classifications could

T.T. Beh · P. Kalitsis (✉)
Murdoch Childrens Research Institute, Royal Children's Hospital,
Parkville, Melbourne 3052, Australia
e-mail: paul.kalitsis@mcri.edu.au

T.T. Beh · P. Kalitsis
Department of Paediatrics, University of Melbourne,
Parkville, Melbourne 3052, Australia

B.E. Black (ed.), *Centromeres and Kinetochores*, Progress in Molecular
and Subcellular Biology 56, DOI 10.1007/978-3-319-58592-5_22

include the cell-type origin of the cancer cells, the primary organ where the cancer cells were suspected to originate from, tumour staging and histopathological grading. In current practice, however, cytogenetic, genetic and protein marker information commonly further supplement the pathological analyses, enhancing the precision in diagnosis and subsequent treatment selection and prognosis of the disease (Song et al. 2015). Nevertheless, even after being defined by the conventional classification systems aforementioned, these different categories of cancer remain highly heterogeneous.

The complexity of cancer had therefore spurred the quest for biomarkers—a 'signature' or combination of genetic, epigenetic and protein level data, to better delineate the types of cancer. Such effort highlights the progress of the cancer research field towards precision medicine and targeted therapies. One of the major areas that have been proven to be useful for diagnosis of cancer is molecular cytogenetics. This area of research emerged owing to the discovery of the Philadelphia chromosome (translocation between chromosomes 9 and 22) in chronic myelogenous leukaemia (CML), the first evidence indicating that cancer is a genetic disease, and its ensuing development was due to the subsequent advancement in cytogenetic and then genomic technologies (Nowell 2007; Danielsen et al. 2016).

2 Aneuploidy and Cancer

2.1 What is Aneuploidy?

In humans, the normal number of chromosomes in somatic cells is 46 which was discovered in 1956 by Tjio and Levan (Tjio and Levan 1956). This discovery was made possible due to the use of the hypotonic shock method [first described in 1934 by Eleanor Slifer (Slifer 1934)] which improved the spread of nuclear content and the advent of cell culture technique combined with the use of colchicine treatment to release the chromosomes from spindle microtubules, increase the number of metaphase cells and to give the appearance of more condensed chromosomes (Gartler 2006).

Euploid refers to the exact multiple of the haploid chromosome complement and when the euploidy is more than two (diploidy), it is termed polyploidy. Aneuploidy ('not euploidy'), on the other hand, is the term subsequently coined to describe the karyotypic state deviating from the haploid multiple of chromosome number. It encompasses gain or loss of whole chromosome and chromosomal segments, termed whole chromosome aneuploidy and segmental or structural chromosome aneuploidy, respectively (Pavelka et al. 2010; Thompson and Compton 2011; Orr et al. 2015). Chromosome instability usually refers to the process of gains and loss of chromosomes leading to an aneuploidy state. The terms can often be interchanged.

2.2 Different Forms of Aneuploidy

Aneuploidy, especially in the short-term of its occurrence, generates large-scale gene copy number changes within the cells and had been demonstrated to cause transcriptomic and proteomic stresses which compromise the proliferative capability of these cells (Santaguida and Amon 2015; Dürrbaum and Storchová 2016). Hence, constitutional whole chromosome aneuploidies resulting from chromosome mis-segregation during germ cell formation often cause embryonic lethality and pregnancy loss at different stages, except for a small percentage of affected foetuses that developed to full term—the most common being trisomy 21 or better known as Down syndrome (Nagaoka et al. 2012; Santaguida and Amon 2015).

Somatic aneuploidies are also rare and have only been reported in a handful of congenital diseases such as mosaic variegated aneuploidy (MVA). MVA is diagnosed based on the cytogenetic observation where a fraction of cells are aneuploid and are commonly co-occurring with clinical features namely microcephaly, mental retardation and growth delay (Callier et al. 2005). BUB1B, was not only the first gene found to be associated with MVA but also the first mitotic spindle assembly checkpoint (SAC) gene where its allelic mutations in the germ-line were linked to a human disease (Hanks et al. 2004). Furthermore, BUB1B localises to kinetochores of chromosomes that have yet to attach to mitotic spindles in a bi-orientated manner.

2.3 How is Aneuploidy Linked to Cancer?

The potential causal link between aneuploidy and cancer advancement was proposed back in 1914 by Theodor Boveri (Boveri 1914). To date, the presence of aneuploid cells were found in approximately 90% of solid tumours and 50% of haematopoietic cancers (Mitelman et al. 2017). This high occurrence of aneuploidy especially in solid tumours combined with the rare incidence of non-malignancy-related aneuploidies led to the postulation that early carcinogenesis is sensitive to balancing the dosage effect from the copy number changes of oncogenes and tumour suppressors until tolerance is later acquired further advancing cancer development (Martínez-A and van Wely 2011; Santaguida and Amon 2015). Loss-of-function of some of the key tumour suppressors for instance Tp53 in mice (Baker et al. 2009) and deubiquitinating enzyme Ubp6 (mammalian homolog USP14) in yeast (Torres et al. 2010), and tetraploidisation of the cellular genome were proposed as some of the mechanisms that promote tolerance for aneuploidy (Gordon et al. 2012; Dewhurst et al. 2014).

One of the phenomena most highly associated with aneuploidy in the context of cancer is chromosomal instability (CIN). CIN is the rate of karyotypic change due to the gain or loss of whole (or part of) chromosomes during cell division which leads to the generation of cells with abnormal number (or organisation) of chromosomes (Holland and Cleveland 2009; Danielsen et al. 2016). CIN is believed to be the driver of cancer cell evolution and intratumor heterogeneity while both CIN and aneuploidy have been associated with poor prognostic outcome and resistance to therapy (Holland and Cleveland 2012).

3 Whole Chromosome Instability in Cancer

The centromere of mammalian chromosomes is made up of a tandemly repeated satellite DNA platform spanning up to several megabases unto which around 100 proteins assemble to form a mature mitotic structure known as the kinetochore (Fukagawa and Earnshaw 2014). A properly functioning centromere and its associated kinetochore are crucial in determining the fidelity of chromosome segregation during cell division. The centromere is the final locus of cohesion for sister chromatids before the SAC is satisfied and the cell signals to proceed into the irreversible transition from metaphase to anaphase. Hence, defects in the centromere- and kinetochore-associated component proteins have been proposed to contribute to aneuploidy (as an outcome of whole chromosome instability) (Orr et al. 2015). These include; (i) defective SAC components that fail in temporally halting the onset of anaphase, (ii) the failure of proteins involved in regulating the dynamics of kinetochore-microtubule interactions and (iii) the weakening of the chromosome cohesion complex causing premature separation of sister chromatids.

The first clues that mutations in kinetochore genes had a link with aneuploidy and cancer emerged from the association with functional mutations of the BUB1 gene in colon cancers (Cahill et al. 1998). SAC gene mutations displayed whole chromosome instability in contrast to previously identified colon cancer genes, which were characterised by microsatellite DNA instability linked to germ-line mutations of DNA mismatch repair genes.

More convincing evidence that kinetochore genes had a direct role in cancer predisposition arose from patients of MVA with germ-line BUB1B/BUBR1 biallelic missense mutations, mentioned in the previous section. Affected individuals also have a higher risk of developing cancer such as leukaemia, rhabdomyosarcoma and Wilms tumour (Hanks et al. 2004). Further karyotypic analyses performed on the cultured patient-derived fibroblast and lymphoblastoid cell lines showed ongoing segregation defect without preference for any chromosomes (Hanks and Rahman 2005). This suggests that a basal level of CIN exists in these MVA cells but the connection to cancer progression remained to be clarified.

Further evidence is beginning to accumulate for the association of SAC genes in cancer predisposition from patients with germ-line mutations. Haploinsufficient or heterozygous mutations of BUB1 and BUB3 have been identified in a small proportion of individuals with early-onset colon cancer (de Voer et al. 2013). Furthermore, cells from affected patients exhibit aneuploidy in multiple tissue types, showing that the mutations have a constitutive effect.

Since these initial findings of SAC gene mutations contributing to aneuploidy and cancer, many additional studies have reported the association of SAC and other kinetochore gene mutations and over-expression in a variety of cancer types. The challenge for researchers is to confirm the role of these mutations and mis-expression against the spectrum of many other gene mutations and structural variation observed in human cancer cells. High-throughput genomic technologies and mouse modelling is beginning to have an impact in this knowledge gap.

3.1 *Kinetochore Mutation Modelling in Mouse*

To understand the role of mutations and aberrant expression of kinetochore genes and cancer predisposition, researchers have tested this link using transgenic mouse models. Most reported studies have focussed on SAC genes (Table 1). Bub1b has been of special interest because of its link with premature ageing and aneuploidy in humans. Decreasing expression to half or less produces chromosome instability, tumour increase and decreased lifespan (Dai et al. 2004; Baker et al. 2004; Wijshake et al. 2012). By contrast, over-expression of Bub1b reduces aneuploidy rates in mutant backgrounds, protects against cancer and increases lifespan (Baker et al. 2013). It remains to be seen if sequence variants of BUB1B have any effect on longevity in humans.

Reducing the levels of other SAC proteins confirms the link between increased aneuploidy and cancer rates. Some cancer predisposition effects can be subtle and only surface when challenged with chemicals that damage DNA or microtubule spindles (Table 1). Another common phenotype is a reduced lifespan or premature aging which supports the hypothesis that elevated aneuploidy impacts on cell survival rates.

The kinetochore motor protein, Cenpe, follows similar trends in aneuploidy and cancer phenotypes when expression is reduced to half, however, higher rates of aneuploidy can have a protective effect in certain tissue types since cells die from too much genomic imbalance (Weaver et al. 2007; Silk et al. 2013). The Hec1 gene was originally named "highly expressed in cancer" and is a part of the Ndc80 structural sub-complex that links the kinetochore to spindle microtubules. This gene has also been subject to inducible tissue specific over-expression which leads to aneuploidy and tumour formation (Diaz-Rodríguez et al. 2008). In a similar conditional over-expression study, the chromosome passenger protein, Aurora kinase B (Aurkb) was shown to increase chromosome mis-segregation events and elevate tumour incidence by suppressing the cell cycle inhibitor p21(Cip1) (González-Loyola et al. 2015).

Many reports in human cancers have described increased expression levels in kinetochore genes (discussed in Sects. 3.1 and 5) but it is not clear whether the aberrant expression significantly contributes to cancer progression. Conditional over-expression of kinetochore genes in mouse models supports the hypothesis that

Table 1 Mouse kinetochore mutant models and cancer predisposition

Functional group	Gene	Expression	Aneuploidy	Cancer	References
KMN network	Ndc80/Hec1	Over	Yes	Yes	Diaz-Rodríguez et al. (2008)
Spindle assembly checkpoint complex	Bub1	Haplo	Yes	Yes—tumour suppressor dependent	Baker et al. (2009)
		Hypo	Yes	Yes	Schliekelman et al. (2009)
		Over	Yes	Yes	Ricke et al. (2011)
	Bub3	Haplo	Yes	No	Kalitsis et al. (2005)
		Haplo	Yes	Yes—carcinogen induced	Babu et al. (2003)
	Bubr1	Hypo	Yes	Yes—carcinogen induced	Baker et al. (2004)
		Haplo	Yes	Yes	Dai et al. (2004)
		Over	No	No	Baker et al. (2013)
		Haplo	Yes	Yes—carcinogen induced	Wijshake et al. (2012)
	Mad1	Haplo	Yes	Yes and microtubule poison induced	Iwanaga et al. (2007)
	Mad2	Haplo	Yes	Yes	Dobles et al. (2000), Michel et al. (2001)
		Over	Yes	Yes	Sotillo et al. (2007)
	Mps1	Deleted kinetochore localisation domain	Yes	Yes	Foijer et al. (2014)
Chromosome passenger complex	Aurkb	Over	Yes	Yes	González-Loyola et al. (2015)
Motor protein	Cenpe	Haplo	Yes	Yes–no	Weaver et al. (2007), Silk et al. (2013)

Expression categories; expression levels are higher than endogenous (over), half of endogenous (haplo) and less than half of endogenous (hypo)

mis-expression can be a driver of aneuploidy followed by tumour formation, with Bub1b over-expression being the exception (Table 1).

3.2 *Kinetochore Gene Over-expression in Human Cancers*

Whether kinetochore gene over-expression is involved in the early steps of cancer progression remains a hotly debated topic. One study supports an alternative hypothesis that mis-regulation of a key cell cycle transcription factor, FoxM1, which binds to most cell division promoters drive kinetochore gene mis-expression (Thiru et al. 2014). Upon analysing the publicly available expression compendia of human normal tissues and cancer samples, core kinetochore genes (which encode proteins functioning exclusively at the kinetochores) were found to be co-ordinately upregulated in cancers. Their upregulation coincided with the increased expression of other cell cycle and DNA replication genes suggesting a widespread activation of the cell division programme potentially poising the cells for division whenever the signal for cell cycle entry is received (Thiru et al. 2014). This concerted mis-expression of the core kinetochore genes may affect the stoichiometric balance of the protein complex and its probable connection with CIN and aneuploidy warrants further investigation. However, kinetochore gene mis-expression may still be a useful biomarker for cancer diagnosis (see Sect. 5).

4 Centromere-driven Structural Chromosome Instability in Cancer

In addition to aneuploidies, solid tumours exhibit complex karyotypes with high level of chromosomal aberrations (Thompson and Compton 2011). Complex rearrangements involving gains and losses of chromosomal segments generate structural rearrangements. These genome rearrangements contrast to aneuploidies and were thought to be caused by the defects in DNA replication, DNA decatenation, DNA repair, and also by telomere dysfunction and chromothripsis (Orr et al. 2015).

In malignancies, telomere crisis is where the telomere caps of the chromosomes are shortened to a critical length and produce chromosomes with sticky ends. These eroded chromosome ends can then fuse to other chromosomes giving rise to dicentric chromosomes (Fig. 1) (Mackinnon and Campbell 2011). Another route that produces dicentric chromosomes is a translocation event where a dicentric chromosome arises from a reciprocal exchange of the segments of two chromosomes carrying a centromere each and in parallel, an acentric chromosomal fragment is also produced (Mackinnon and Campbell 2011; Sarova et al. 2016).

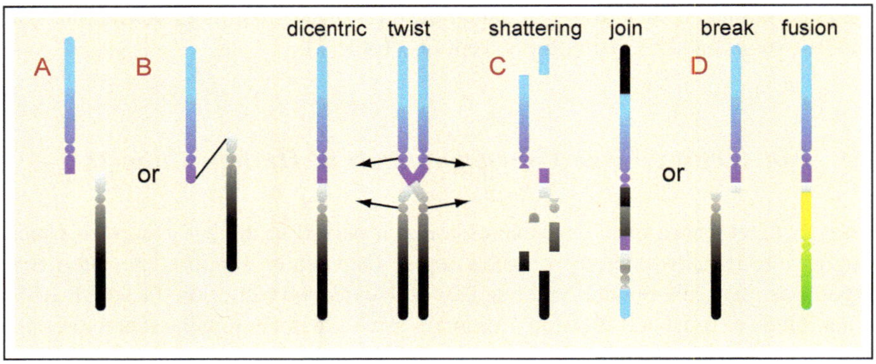

Fig. 1 Mechanism of dicentric-driven chromosome rearrangements. Initial chromosomal rearrangement is shown as two alternative forms. *A* Telomere attrition of two chromosome ends shown as a *straight line*, or *B* unbalanced translocation producing a functional dicentric chromosome (two active centromeres, *circles*). After DNA replication sister chromatids can twist between the two centromeres if they are sufficiently far apart. This event results in the centromeres of each chromatid binding to microtubules from opposite spindle poles. At the onset of chromosome segregation, each chromatid is pulled and *C* produces a chromosome bridge that is cut in G1 phase and prone to shattering, repair and joining, *D* alternatively, the dicentric chromosome stretches and breaks at a single site after cell division and can fuse to another chromosome (*yellow-green*). The generation of new dicentrics can perpetuate the breakage-fusion-bridge cycle

Dicentric chromosomes in cancers especially in (acute myeloid leukaemia) AML have been associated with complex karyotype and patients exhibiting complex karyotypes were reported to have bad clinical prognoses (Haferlach et al. 2012; Valcárcel et al. 2013; Sarova et al. 2016). Mechanistically, the presence of two functional centromeres on dicentric chromosomes poses the possibility of the two kinetochores on the same sister chromatid being attached to microtubules emanating from the opposite spindle poles (Fig. 1) (Beh and Kalitsis 2015). Hence, the probability of chromosome mis-segregation and chromosome breakage from anaphase to cytokinesis, is increased which might then cause further propagation of chromosomal rearrangements in subsequent cell cycles (Lo et al. 2002).

The classical process associated with the formation of dicentric chromosomes is the breakage-fusion-bridge (BFB) cycle, first described in maize by Barbara McClintock in 1939 (McClintock 1939). Dicentric chromosomes, either in linear or ring form, were thought to be mitotically unstable and were prone to undergo BFB. Hence, the occurrence of dicentrics combined with the increased tolerance for chromosome damage (as an outcome of BFB) in malignant cells had been postulated as the mechanism that generates constant genomic reorganisation and intra-tumour heterogeneity in cancer progression (Gisselsson et al. 2000).

Chromothripsis, a term first coined in 2011 to mean chromosome shattering, usually of one chromosome (Stephens et al. 2011). More recently, this mechanism has been linked to dicentric chromosomes (Fig. 1) reported in a study into paediatric acute lymphoblastic leukaemia (ALL) (Li et al. 2014). Li et al. (2014)

proposed that the combination of both mechanisms aforementioned, chromothripsis and BFB cycles, led to the generation of recurrent intrachromosomal amplification of chromosome 21 (iAMP21) which makes up 2% of the most common paediatric cancer, ALL. They categorised iAMP21 into two major forms; (i) sporadic cases that arose from telomere attrition or breakage of chromosome 21 followed by the fusion of sister chromatids generating a dicentric chromosome and BFB cycles which amplified the chromosome 21 regions before chromothripsis, and (ii) constitutional Robertsonian chromosome translocation (15;21) [rob(15;21)]. Robertsonian translocations involve inter-recombination events within the short arm region of acrocentric chromosomes. Human cells contain five acrocentric chromosomes, comprising; 13, 14, 15, 21 and 22. The translocation product can result from different combinations of the acrocentric chromosomes with the two originating centromeres located adjacent to one another. The occurrence of Robertsonian translocations is approximately 1 in 1000 live births (Jacobs et al. 1992), with the rob(15;21) translocation comprising only 1% of these. The presence of the rob(15;21) chromosome in an individual elevates the risk of ALL by a factor of 2700 (Li et al. 2014). Leukaemia cells from rob(15:21) patients show evidence of chromothripsis followed by duplication of the reassembled chromosome, similar to the model shown in Fig. 1. This was the first study to suggest that the dicentric chromosome is a trigger of chromothripsis and that the amplification of DNA segments are produced via the sequential effect of the two phenomena—BFB (with observable chromosome bridge intermediates) and chromothripsis. It still remains to be shown whether the two centromeres of the dicentric rob(15;21) chromosome remain active in all cells or whether one centromere is inactivated to stabilise the chromosome during cell division.

Another recent report describing the involvement of structural centromere aberrations in cancer development was in the appearance of giant rod or ring chromosomes also known as neochromosomes, commonly found in liposarcomas (Garsed et al. 2014). A single neochromosome was sorted from other chromosomes and sequenced. The sequencing revealed amplification of chromosomal segments containing oncogenes and evidence of multiple rounds of BFB. Interestingly, the appearance of a non-satellite DNA repeat centromere (neocentromere) was speculated to have arisen through the erosion of a canonical centromere (Kalitsis and Choo 2012).

4.1 Dicentric Chromosome Induction

Experimental systems employing mouse and human cell lines have been developed to test the link between the formation of a dicentric chromosome and the effects on genome instability and cancer predisposition. In the first study, a single dicentric chromosome was induced by the activation of an ectopic centromere on a single chromosome (Gascoigne and Cheeseman 2013). After dicentric induction, the chromosome failed to segregate accurately and showed complex genomic rearrangements. Furthermore, in vitro studies of the rearranged cell lines displayed hallmarks cellular transformation. In the second study, the human RPE-1 cell line was used to induce telomere fusions with the aid of an inducible dominant-negative allele of the telomere binding protein, TRF2 (Maciejowski et al. 2015). Dicentric chromosomes appeared via telomere fusions and were tracked using live-cell imaging. These chromosomes also showed signs of chromosome mis-segregation, DNA breakage and chromothripsis. Interestingly, the investigators show that the dicentric chromosome did not break in anaphase but actually later in G1 phase with the aid of 3′ repair exonuclease 1 (TREX1). More in vivo modelling experiments are needed to establish the link between dicentric chromosome formation and cancer predisposition.

5 Centromere Abnormalities as Biomarkers of Cancer

As aneuploidy has been associated with poor prognostic outcome for the different subtypes of various malignancies, it has been proposed that aneuploidy should be used as a biomarker to supplement the conventionally used clinicopathological information such as tumour stage, histological grade and other subtype-specific molecular markers (Danielsen et al. 2016).

Similarly, the presence of dicentric chromosomes could also be used as a supplementary biomarker, possibly as an indirect measure of ongoing aneuploidy and genome rearrangements, in assessing cancer. This had been demonstrated in a study where the presence of dicentric chromosomes was used alongside other cytogenetic and genomic information in the classification, risk stratification and subsequent treatment selection for the patients of childhood ALL (Moorman et al. 2014).

In a more recent study, the number of dicentric chromosomes present in an AML patient's cells served as a prognostic marker as exemplified by the significant difference in median survival between the group of patients with a single dicentric chromosome and the group with three dicentric chromosomes—5.8 months versus 1.8 months, respectively (Sarova et al. 2016). Further investigations will need to be carried out to determine the prognostic value of dicentric chromosomes in other cancer types besides ALL and AML aforementioned.

Gene expression of tumour samples has also been a valuable tool for the assessment and refinement of treatment of cancers. Most recently, a quantitative expression scoring system, Centromere and kinetochore gene Expression Score (CES) has been devised and scores were generated from 14 centromere and kinetochore protein genes. The data were used in a hypothesis-driven study to test the prognostic and predictive value of these mis-expressed genes in cancer patients (Zhang et al. 2016). All of the 14 genes were found to be upregulated, concurring with the finding of Thiru et al. (2014). High CES indicated undesirable patient outcomes including metastasis and poor overall survival, and was also shown to correlate with high genomic instability which sensitised these tumours to genotoxic agents. Hence, the CES system was demonstrated to have the ability to not only prognosticate cancer patients but also effectively predict clinical response to both adjuvant and radiotherapy (Zhang et al. 2016).

6 Conclusion

Evidence from patients and experimental model systems is beginning to accumulate in support for a role of centromere defects in cancer predisposition and progression. A group of kinetochore proteins involved in SAC signalling is especially noted for their involvement in increasing aneuploidy rates and risk of tumour progression. Other kinetochore genes have also been reported in cancers. The decreasing cost in genomic technologies will greatly assist in assessing whether other centromere and kinetochore genes contribute to the cancer burden in the aging human population. Furthermore, experimental model systems have and will continue to be employed in dissecting the role candidate genes and sequence variants. Centromere DNA abnormalities in the form of dicentric chromosomes are additionally being linked to further genomic instability which is a common feature of the cancer cell.

References

Babu JR, Jeganathan KB, Baker DJ et al (2003) Rae1 is an essential mitotic checkpoint regulator that cooperates with Bub3 to prevent chromosome missegregation. J Cell Biol 160:341–353. doi:10.1083/jcb.200211048

Baker DJ, Dawlaty MM, Wijshake T et al (2013) Increased expression of BubR1 protects against aneuploidy and cancer and extends healthy lifespan. Nat Cell Biol 15:96–102. doi:10.1038/ncb2643

Baker DJ, Jeganathan KB, Cameron JD et al (2004) BubR1 insufficiency causes early onset of aging-associated phenotypes and infertility in mice. Nat Genet 36:744–749. doi:10.1038/ng1382

Baker DJ, Jin F, Jeganathan KB, van Deursen JM (2009) Whole chromosome instability caused by Bub1 insufficiency drives tumorigenesis through tumor suppressor gene loss of heterozygosity. Cancer Cell 16:475–486. doi:10.1016/j.ccr.2009.10.023

Beh TT, Kalitsis P (2015) Centromeres in health, disease and evolution. In: Felekkis K, Voskarides K (eds) Genomic elements in health, disease and evolution. Springer, New York, pp 221–245

Boveri T (1914) Zur Frage der Entstehung maligner Tumoren. Gustav Fischer Verlag, Jena

Cahill DP, Lengauer C, Yu J et al (1998) Mutations of mitotic checkpoint genes in human cancers. Nature 392:300–303. doi:10.1038/32688

Callier P, Faivre L, Cusin V et al (2005) Microcephaly is not mandatory for the diagnosis of mosaic variegated aneuploidy syndrome. Am J Med Genet A 137:204–207. doi:10.1002/ajmg.a.30783

Dai W, Wang Q, Liu T et al (2004) Slippage of mitotic arrest and enhanced tumor development in mice with BubR1 haploinsufficiency. Cancer Res 64:440–445

Danielsen HE, Pradhan M, Novelli M (2016) Revisiting tumour aneuploidy—the place of ploidy assessment in the molecular era. Nat Rev Clin Oncol 13:291–304. doi:10.1038/nrclinonc.2015.208

de Voer RM, Geurts van Kessel A, Weren RDA et al (2013) Germline mutations in the spindle assembly checkpoint genes BUB1 and BUB3 are risk factors for colorectal cancer. Gastroenterology 145:544–547. doi:10.1053/j.gastro.2013.06.001
Dewhurst SM, McGranahan N, Burrell RA et al (2014) Tolerance of whole-genome doubling propagates chromosomal instability and accelerates cancer genome evolution. Cancer Discov 4:175–185. doi:10.1158/2159-8290.CD-13-0285
Diaz-Rodríguez E, Sotillo R, Schvartzman J-M, Benezra R (2008) Hec1 overexpression hyperactivates the mitotic checkpoint and induces tumor formation in vivo. Proc Natl Acad Sci U S A 105:16719–16724. doi:10.1073/pnas.0803504105
Dobles M, Liberal V, Scott ML et al (2000) Chromosome missegregation and apoptosis in mice lacking the mitotic checkpoint protein Mad2. Cell 101:635–645
Dürrbaum M, Storchová Z (2016) Effects of aneuploidy on gene expression: implications for cancer. FEBS J 283:791–802. doi:10.1111/febs.13591
Foijer F, Xie SZ, Simon JE et al (2014) Chromosome instability induced by Mps1 and p53 mutation generates aggressive lymphomas exhibiting aneuploidy-induced stress. Proc Natl Acad Sci U S A 111:13427–13432. doi:10.1073/pnas.1400892111
Fukagawa T, Earnshaw WC (2014) The centromere: chromatin foundation for the kinetochore machinery. Dev Cell 30:496–508. doi:10.1016/j.devcel.2014.08.016
Garsed DW, Marshall OJ, Corbin VDA et al (2014) The architecture and evolution of cancer neochromosomes. Cancer Cell 26:653–667. doi:10.1016/j.ccell.2014.09.010
Gartler SM (2006) The chromosome number in humans: a brief history. Nat Rev Genet 7:655–660. doi:10.1038/nrg1917
Gascoigne KE, Cheeseman IM (2013) Induced dicentric chromosome formation promotes genomic rearrangements and tumorigenesis. Chromosome Res 21:407–418. doi:10.1007/s10577-013-9368-6
Gisselsson D, Pettersson L, Höglund M et al (2000) Chromosomal breakage-fusion-bridge events cause genetic intratumor heterogeneity. Proc Natl Acad Sci U S A 97:5357–5362. doi:10.1073/pnas.090013497
González-Loyola A, Fernández-Miranda G, Trakala M et al (2015) Aurora B overexpression causes aneuploidy and p21Cip1 repression during tumor development. Mol Cell Biol 35:3566–3578. doi:10.1128/MCB.01286-14
Gordon DJ, Resio B, Pellman D (2012) Causes and consequences of aneuploidy in cancer. Nat Rev Genet 13:189–203. doi:10.1038/nrg3123
Haferlach C, Alpermann T, Schnittger S et al (2012) Prognostic value of monosomal karyotype in comparison to complex aberrant karyotype in acute myeloid leukemia: a study on 824 cases with aberrant karyotype. Blood 119:2122–2125. doi:10.1182/blood-2011-10-385781
Hanahan D, Weinberg RA (2011) Hallmarks of cancer: the next generation. Cell 144:646–674. doi:10.1016/j.cell.2011.02.013
Hanks S, Coleman K, Reid S et al (2004) Constitutional aneuploidy and cancer predisposition caused by biallelic mutations in BUB1B. Nat Genet 36:1159–1161. doi:10.1038/ng1449
Hanks S, Rahman N (2005) Aneuploidy-cancer predisposition syndromes: a new link between the mitotic spindle checkpoint and cancer. Cell Cycle 4:225–227
Holland AJ, Cleveland DW (2009) Boveri revisited: chromosomal instability, aneuploidy and tumorigenesis. Nat Rev Mol Cell Biol 10:478–487. doi:10.1038/nrm2718
Holland AJ, Cleveland DW (2012) Losing balance: The origin and impact of aneuploidy in cancer. EMBO Rep 13:501–514. doi:10.1038/embor.2012.55
Iwanaga Y, Chi Y-H, Miyazato A et al (2007) Heterozygous deletion of mitotic arrest-deficient protein 1 (MAD1) increases the incidence of tumors in mice. Cancer Res 67:160–166. doi:10.1158/0008-5472.CAN-06-3326
Jacobs PA, Browne C, Gregson N et al (1992) Estimates of the frequency of chromosome abnormalities detectable in unselected newborns using moderate levels of banding. J Med Genet 29:103–108
Kalitsis P, Choo KHA (2012) The evolutionary life cycle of the resilient centromere. Chromosoma 121:327–340. doi:10.1007/s00412-012-0369-6

Kalitsis P, Fowler KJ, Griffiths B et al (2005) Increased chromosome instability but not cancer predisposition in haploinsufficient Bub3 mice. Genes Chromosomes Cancer 44:29–36. doi:10.1002/gcc.20215

Li Y, Schwab C, Ryan SL et al (2014) Constitutional and somatic rearrangement of chromosome 21 in acute lymphoblastic leukaemia. Nature 508:98–102. doi:10.1038/nature13115

Lo AWI, Sabatier L, Fouladi B et al (2002) DNA amplification by breakage/fusion/bridge cycles initiated by spontaneous telomere loss in a human cancer cell line. Neoplasia 4:531–538. doi:10.1038/sj.neo.7900267

Maciejowski J, Li Y, Bosco N et al (2015) Chromothripsis and kataegis induced by telomere crisis. Cell 163:1641–1654. doi:10.1016/j.cell.2015.11.054

Mackinnon RN, Campbell LJ (2011) The role of dicentric chromosome formation and secondary centromere deletion in the evolution of myeloid malignancy. Genet Res Int 2011:643628. doi:10.4061/2011/643628

Martínez-A C, van Wely KHM (2011) Centromere fission, not telomere erosion, triggers chromosomal instability in human carcinomas. Carcinogenesis 32:796–803. doi:10.1093/carcin/bgr069

McClintock B (1939) The behavior in successive nuclear divisions of a chromosome broken at meiosis. Proc Natl Acad Sci U S A 25:405–416. doi:10.1073/pnas.25.8.405

Michel LS, Liberal V, Chatterjee A et al (2001) MAD2 haplo-insufficiency causes premature anaphase and chromosome instability in mammalian cells. Nature 409:355–359. doi:10.1038/35053094

Mitelman F, Johansson B, Mertens F (2017) Mitelman database of chromosome aberrations and gene fusions in cancer. https://cgap.nci.nih.gov/Chromosomes/Mitelman. Accessed 18 Jan 2017

Moorman AV, Enshaei A, Schwab C et al (2014) A novel integrated cytogenetic and genomic classification refines risk stratification in pediatric acute lymphoblastic leukemia. Blood 124:1434–1444. doi:10.1182/blood-2014-03-562918

Nagaoka SI, Hassold TJ, Hunt PA (2012) Human aneuploidy: mechanisms and new insights into an age-old problem. Nat Rev Genet 13:493–504. doi:10.1038/nrg3245

Nowell PC (2007) Discovery of the philadelphia chromosome: a personal perspective. J Clin Invest 117:2033–2035. doi:10.1172/JCI31771

Orr B, Godek KM, Compton D (2015) Aneuploidy. Curr Biol 25:R538–R542. doi:10.1016/j.cub.2015.05.010

Pavelka N, Rancati G, Li R (2010) Dr Jekyll and Mr Hyde: role of aneuploidy in cellular adaptation and cancer. Curr Opin Cell Biol 22:809–815. doi:10.1016/j.ceb.2010.06.003

Ricke RM, Jeganathan KB, van Deursen JM (2011) Bub1 overexpression induces aneuploidy and tumor formation through Aurora B kinase hyperactivation. J Cell Biol 193:1049–1064. doi:10.1083/jcb.201012035

Santaguida S, Amon A (2015) Short- and long-term effects of chromosome mis-segregation and aneuploidy. Nat Rev Mol Cell Biol 16:473–485. doi:10.1038/nrm4025

Sarova I, Brezinova J, Zemanova Z et al (2016) Molecular cytogenetic analysis of dicentric chromosomes in acute myeloid leukemia. Leuk Res 43:51–57. doi:10.1016/j.leukres.2016.01.001

Schliekelman M, Cowley DO, O'Quinn R et al (2009) Impaired Bub1 function in vivo compromises tension-dependent checkpoint function leading to aneuploidy and tumorigenesis. Cancer Res 69:45–54. doi:10.1158/0008-5472.CAN-07-6330

Silk AD, Zasadil LM, Holland AJ et al (2013) Chromosome missegregation rate predicts whether aneuploidy will promote or suppress tumors. Proc Natl Acad Sci U S A 110:E4134–E4141. doi:10.1073/pnas.1317042110

Slifer EH (1934) Insect development VI. The behavior of grasshopper embryos in anisotonic, balanced salt solutions. J Exp Zool 67:137–157. doi:10.1002/jez.1400670107

Song Q, Merajver SD, Li JZ (2015) Cancer classification in the genomic era: five contemporary problems. Hum Genomics 9:27. doi:10.1186/s40246-015-0049-8

Sotillo R, Hernando E, Díaz-Rodríguez E et al (2007) Mad2 overexpression promotes aneuploidy and tumorigenesis in mice. Cancer Cell 11:9–23. doi:10.1016/j.ccr.2006.10.019

Stephens PJ, Greenman CD, Fu B et al (2011) Massive genomic rearrangement acquired in a single catastrophic event during cancer development. Cell 144:27–40. doi:10.1016/j.cell.2010.11.055

Thiru P, Kern DM, McKinley KL et al (2014) Kinetochore genes are coordinately up-regulated in human tumors as part of a FoxM1-related cell division program. Mol Biol Cell 25:1983–1994. doi:10.1091/mbc.E14-03-0837

Thompson SL, Compton DA (2011) Chromosomes and cancer cells. Chromosome Res Int J Mol Supramol Evol Asp Chromosome Biol 19:433–444. doi:10.1007/s10577-010-9179-y

Tjio JH, Levan A (1956) The chromosome number of man. Hereditas 42:U1–U6

Torres EM, Dephoure N, Panneerselvam A et al (2010) Identification of aneuploidy-tolerating mutations. Cell 143:71–83. doi:10.1016/j.cell.2010.08.038

Valcárcel D, Ademà V, Solé F et al (2013) Complex, not monosomal, karyotype is the cytogenetic marker of poorest prognosis in patients with primary myelodysplastic syndrome. J Clin Oncol Off J Am Soc Clin Oncol 31:916–922. doi:10.1200/JCO.2012.41.6073

Weaver BAA, Silk AD, Montagna C et al (2007) Aneuploidy acts both oncogenically and as a tumor suppressor. Cancer Cell 11:25–36. doi:10.1016/j.ccr.2006.12.003

Wijshake T, Malureanu LA, Baker DJ et al (2012) Reduced life- and healthspan in mice carrying a mono-allelic BubR1 MVA mutation. PLoS Genet 8:e1003138. doi:10.1371/journal.pgen.1003138

Zhang W, Mao J-H, Zhu W et al (2016) Centromere and kinetochore gene misexpression predicts cancer patient survival and response to radiotherapy and chemotherapy. Nat Commun 7:12619. doi:10.1038/ncomms12619